Woodhead Publishing Series in Food Science, Technology and Nutrition:
Number 196

Tracing pathogens in the food chain

Edited by
Stanley Brul, Pina M. Fratamico and
Tom A. McMeekin

AF849199

WOODHEAD PUBLISHING

Oxford Cambridge Philadelphia New Delhi

Published by Woodhead Publishing Limited, Abington Hall, Granta Park, Great Abington. Cambridge CB21 6AH, UK
www.woodheadpublishing.com

Woodhead Publishing, 525 South 4th Street #241, Philadelphia, PA 19147, USA

Woodhead Publishing India Private Limited, G-2, Vardaan House, 7/28 Ansari Road, Daryaganj, New Delhi – 110002, India
www.woodheadpublishingindia.com

First published 2011, Woodhead Publishing Limited

British Library Cataloguing in Publication Data
A catalogue record for this book is available from the British Library.

ISBN 978-0-08-101479-0 (print)
ISBN 978-0-85709-050-8 (online)
ISSN 2042-8049 Woodhead Publishing Series in Food Science, Technology and Nutrition (print)
ISSN 2042-8057 Woodhead Publishing Series in Food Science, Technology and Nutrition (online)

The publisher's policy is to use permanent paper from mills that operate a sustainable forestry policy, and which has been manufactured from pulp which is processed using acid-free and elemental chlorine-free practices. Furthermore, the publisher ensures that the text paper and cover board used have met acceptable environmental accreditation standards.

Typeset by RefineCatch Limited, Bungay, Suffolk

Tracing pathogens in the food chain

Related titles:

Foodborne pathogens: hazards, risk analysis and control Second edition
(ISBN 978-1-84569-362-6)
Effective control of pathogens continues to be of great importance to the food industry. The first edition of *Foodborne pathogens* quickly established itself as an essential guide for all those involved in the management of microbiological hazards at any stage in the food production chain. This major new edition strengthens that reputation, with extensively revised and expanded coverage, including more than ten new chapters. Part I focuses on risk assessment and management in the food chain. Chapters in this section cover pathogen detection, microbial modeling, the risk assessment procedure, pathogen control in primary production, hygienic design and sanitation, among other topics. Parts II and III then review the management of key bacterial and non-bacterial foodborne pathogens.

Biofilms in the food and beverage industries
(ISBN 978-1-84569-477-7)
Biofilms in the food and beverage industries may form when bacteria attach to and colonise the surfaces of food handling and processing equipment and food products themselves. Human pathogens in biofilms can be harder to remove than free microorganisms and therefore may pose a more significant food safety risk. The opening chapters in this essential book consider fundamental aspects such as the ecology and characteristics of biofilms in food and beverage processing environments and methods for their detection. Part II then reviews biofilm formation by different microorganisms. Part III focuses on the significant issues of biofilm prevention and removal. Chapters on particular food industry sectors complete the collection.

Modelling microorganisms in food
(ISBN 978-1-84569-006-9)
While predictive microbiology has made a major contribution to food safety, there remain many uncertainties, e.g. growing evidence that traditional microbial inactivation models do not always fit the experimental data and an awareness that bacteria of one population do not behave homogeneously, that they may interact and behave differently in different food systems. These problems are all the more important because of the growing interest in minimal processing techniques that operate closer to death, survival and growth boundaries and thus require a greater precision from models. Edited by leading authorities, this collection reviews current developments in quantitative microbiology. Part I discusses best practice in constructing quantitative models and Part II looks at the specific areas in new approaches to modelling microbial behaviour.

Details of these books and a complete list of Woodhead's titles can be obtained by:

- visiting our web site at www.woodheadpublishing.com
- contacting Customer Services (e-mail: sales@woodheadpublishing.com; fax: +44 (0) 1223 893694; tel.: +44 (0) 1223 891358 ext. 130; address: Woodhead Publishing Limited, Abington Hall, Granta Park, Great Abington, Cambridge CB21 6AH, UK)

If you would like to receive information on forthcoming titles, please send your address details to: Francis Dodds (address, tel. and fax as above; e-mail: francis.dodds@woodheadpublishing.com). Please confirm which subject areas you are interested in.

Contents

Contributor contact details

(* = main contact)

Editors and Chapter 1

S. Brul
Molecular Biology & Microbial Food Safety
Swammerdam Institute for Life Sciences
University of Amsterdam
Nieuwe Achtergracht 166
1018 WV
Amsterdam
The Netherlands

E-mail: S.Brul@uva.nl

P. M. Fratamico
US Department of Agriculture
Agricultural Research Service
Eastern Regional Research Center
600 E Mermaid Lane
Wyndmoor, Pennsylvania 19038
USA

E-mail: pina.fratamico@ars.usda.gov

T. McMeekin
Tasmanian Institute of Agricultural Science/School of Agricultural Science
University of Tasmania
Private Bag 54
Hobart
Tasmania 7001
Australia

E-mail: Tom.McMeekin@utas.edu.au

Chapter 2

I. S. T. Fisher
HPA Centre for Infections
61 Colindale Avenue
Colindale
London
NW9 5EQ
UK

E-mail: ian.fisher@hpa.org.uk

Chapter 3

P. Gerner-Smidt
Enteric Diseases Laboratory Branch Centers for Disease Control and Prevention
Mail stop CO-3
1600 Clifton Road NE
Atlanta
Georgia 30333
USA

E-mail: plg5@cdc.gov

Chapter 4

C. Stein* and A. Ellis
Department of Food Safety and Zoonoses
World Health Organization
20 Avenue Appia
1211 Geneva 27
Switzerland

E-mail: steinc@who.int

T. Jones
Tennessee Department of Health
Nashville, Tennessee
USA

Chapter 5

T. Hald* and S. M. Pires
Division of Microbiology and Risk Assessment
National Food Institute
Technical University of Denmark
Mørkhøj Bygade 19
2860 Søborg
Denmark

E-mail: tiha@food.dtu.dk

Chapter 6

J. C. Buzby
US Department of Agriculture
Economic Research Service
1800 M Street, Room S2080
Washington, D.C. 20036
USA

E-mail: jbuzby@ers.usda.gov

Chapter 7

W. A. Gebreyes
Department of Veterinary Preventive Medicine
The Ohio State University
1920 Coffey Rd
Columbus
Ohio 43065
USA

E-mail: gebreyes@cvm.osu.edu

Chapter 8

K. L. F. Cooper
Enteric Disease Laboratory Branch
Centers for Disease Control and Prevention
1600 Clifton Road NE
Atlanta
Georgia 30333
USA

E-mail: KCooper@cdc.gov

Chapter 9

F. Pagotto* and A. Reid
Listeriosis Reference Service
Bureau of Microbial Hazards
Health Products and Food Branch
P/L 2204E, Room E412
Health Canada
Sir F.G. Banting Research Centre
251 Sir Frederick Banting Driveway
Ottawa
Ontario
K1A 0K9
Canada

E-mail: Franco.Pagotto@hc-sc.gc.ca

Chapter 10

E. K. Hyytia-Trees* and E. M. Ribot
Enteric Diseases Laboratory Branch
Centers for Disease Control and Prevention
Mail stop CO3
1600 Clifton Road
Atlanta
Georgia 30333
USA

E-mail: EHyytia-Trees@cdc.gov; ERibot@cdc.gov

Chapter 11

M. Wagner*
Institute for Milk Hygiene, Milk Technology and Food Science
Department for Farm Animals and Public Veterinary Health
Veterinärplatz 1
1210 Vienna
Austria

E-mail: martin.wagner@vetmeduni.ac.at

P. Rossmanith
Christian-Doppler Laboratory for Molecular Biological Food Analytics
Veterinärplatz 1
1210 Vienna
Austria

E-mail: Peter.Rossmanith@vetmeduni.ac.at

Chapter 12

J. van der Vossen, B. Keijser, F. Schuren, A. Nocker and R. Montijn
TNO Quality of Life
PO BOX 360
3700 AJ Zeist
The Netherlands

E-mail: jos.vandervossen@tno.nl; bart.keijser@tno.nl; roy.montijn@tno.nl

Chapter 13

R. A. Stabler, E. S. Nalerio, P. C. Strong and B. W. Wren*
Department of Infectious and Tropical Diseases
London School of Hygiene and Tropical Medicine
Keppel Street
London
WC1E 7HT
UK

E-mail: brendan.wren@lshtm.ac.uk

Chapter 14

J. P. Bowman
Food Safety Centre
Tasmanian Institute of Agricultural Research
University of Tasmania
Sandy Bay
Hobart
Tasmania 7005
Australia

E-mail: john.bowman@utas.edu.au

Chapter 15

T. M. Wassenaar
Molecular Microbiology and Genomics Consultants
Zotzenheim
Germany

E-mail: trudy@mmgc.eu

Chapter 16

A. C. M. van Zuijlen
Unilever R&D Vlaardingen
Biosciences Department
Olivier van Noortlaan 120
3133 AT Vlaardingen
The Netherlands

E-mail: andre-van.zuijlen@unilever.com

Chapter 17

J. Hoorfar*
National Food Institute
Technical University of Denmark
Mørkhøj Bygade 19
2860 Søborg
Denmark

E-mail: jhoo@food.dtu.dk

M. Wagner
Institute for Milk Hygiene
Technology and Food Science
Department of Farm Animal and Veterinary Public Health
University of Veterinary Medicine
Vienna
Austria

E-mail: martin.wagner@vu-wien.ac.at

K. Jordan
Teagasc
Moorepark Food Research Centre
Fermoy
Cork
Ireland

E-mail: kieran.jordan@teagasc.ie

G. C. Barker
Institute of Food Research
Norwich Research Park
Colney
Norwich
NR4 7UA
UK

E-mail: gary.barker@bbsrc.ac.uk

Chapter 18

P. Whyte*, S. Fanning, S. O'Brien and K. Solomon
Centre for Food Safety
School of Agriculture, Food Science and Veterinary Medicine
University College Dublin
Belfield
Dublin 4
Ireland

E-mail: paul.whyte@ucd.ie

P. Whyte, S. Fanning and L. O'Grady
Herd Health & Animal Husbandry Unit
School of Agriculture, Food Science and Veterinary Medicine
University College Dublin
Belfield
Dublin 4
Ireland

Chapter 19

B. T. Lunestad* and A. Levsen
National Institute of Nutrition and Seafood Research (NIFES)
P.O. Box 2029
Nordnes
5817 Bergen
Norway

E-mail: blu@nifes.no

J. T. Rosnes
NOFIMA Norconserv
Richard Johnsensgate 4
P.O. Box 327
4002 Stavanger
Norway

Chapter 20

K. L. Hiett
United States Department of Agriculture
Agricultural Research Service
Richard B. Russell Agricultural Research Center
Poultry Microbiological Safety Research Unit
950 College Station Road
Athens
Georgia 30605
USA

E-mail: Kelli.hiett@ars.usda.gov

Chapter 21

J. S. Van Kessel*, M. Santin-Duran and J. S. Karns
Environmental Microbial and Food Safety Laboratory
Animal and Natural Resources Institute
US Department of Agriculture
Agricultural Research Service
Beltsville
Maryland 20705
USA

E-mail: joann.vankessel@ars.usda.gov; monica.santin-duran@ars.usda.gov; jeffrey.karns@ars.usda.gov

Y. Schukken
Quality Milk Production Services
Department of Population Medicine and Diagnostic Sciences
College of Veterinary Medicine
Cornell University
Ithaca
New York 14850
USA

E-mail: yhs2@cornell.edu

Chapter 22

R. J. Lee* and R. E. Rangdale
Centre for Environment, Fisheries and Aquaculture Science
Barrack Road
The Nothe
Weymouth
Dorset
DT4 8UB
UK

E-mail: ron.lee@cefas.co.uk

Chapter 23

R. E. Mandrell
US Department of Agriculture
Agricultural Research Service
Western Regional Research Center
Produce Safety and Microbiology
Research Unit
800 Buchanan St
Albany
California 94710
USA

E-mail: robert.mandrell@ars.usda.gov

Woodhead Publishing Series in Food Science, Technology and Nutrition

1 **Chilled foods: a comprehensive guide** *Edited by C. Dennis and M. Stringer*
2 **Yoghurt: science and technology** *A. Y. Tamime and R. K. Robinson*
3 **Food processing technology: principles and practice** *P. J. Fellows*
4 **Bender's dictionary of nutrition and food technology Sixth edition** *D. A. Bender*
5 **Determination of veterinary residues in food** *Edited by N. T. Crosby*
6 **Food contaminants: sources and surveillance** *Edited by C. Creaser and R. Purchase*
7 **Nitrates and nitrites in food and water** *Edited by M. J. Hill*
8 **Pesticide chemistry and bioscience: the food-environment challenge** *Edited by G. T. Brooks and T. Roberts*
9 **Pesticides: developments, impacts and controls** *Edited by G. A. Best and A. D. Ruthven*
10 **Dietary fibre: chemical and biological aspects** *Edited by D. A. T. Southgate, K. W. Waldron, I. T. Johnson and G. R. Fenwick*
11 **Vitamins and minerals in health and nutrition** *M. Tolonen*
12 **Technology of biscuits, crackers and cookies Second edition** *D. Manley*
13 **Instrumentation and sensors for the food industry** *Edited by E. Kress-Rogers*
14 **Food and cancer prevention: chemical and biological aspects** *Edited by K. W. Waldron, I. T. Johnson and G. R. Fenwick*
15 **Food colloids: proteins, lipids and polysaccharides** *Edited by E. Dickinson and B. Bergenstahl*
16 **Food emulsions and foams** *Edited by E. Dickinson*
17 **Maillard reactions in chemistry, food and health** *Edited by T. P. Labuza, V. Monnier, J. Baynes and J. O'Brien*
18 **The Maillard reaction in foods and medicine** *Edited by J. O'Brien, H. E. Nursten, M. J. Crabbe and J. M. Ames*
19 **Encapsulation and controlled release** *Edited by D. R. Karsa and R. A. Stephenson*
20 **Flavours and fragrances** *Edited by A. D. Swift*

21 **Feta and related cheeses** *Edited by A. Y. Tamime and R. K. Robinson*
22 **Biochemistry of milk products** *Edited by A. T. Andrews and J. R. Varley*
23 **Physical properties of foods and food processing systems** *M. J. Lewis*
24 **Food irradiation: a reference guide** *V. M. Wilkinson and G. Gould*
25 **Kent's technology of cereals: an introduction for students of food science and agriculture Fourth edition** *N. L. Kent and A. D. Evers*
26 **Biosensors for food analysis** *Edited by A. O. Scott*
27 **Separation processes in the food and biotechnology industries: principles and applications** *Edited by A.S. Grandison and M. J. Lewis*
28 **Handbook of indices of food quality and authenticity** *R.S. Singhal, P. K. Kulkarni and D. V. Rege*
29 **Principles and practices for the safe processing of foods** *D. A. Shapton and N. F. Shapton*
30 **Biscuit, cookie and cracker manufacturing manuals Volume 1: ingredients** *D. Manley*
31 **Biscuit, cookie and cracker manufacturing manuals Volume 2: biscuit doughs** *D. Manley*
32 **Biscuit, cookie and cracker manufacturing manuals Volume 3: biscuit dough piece forming** *D. Manley*
33 **Biscuit, cookie and cracker manufacturing manuals Volume 4: baking and cooling of biscuits** *D. Manley*
34 **Biscuit, cookie and cracker manufacturing manuals Volume 5: secondary processing in biscuit manufacturing** *D. Manley*
35 **Biscuit, cookie and cracker manufacturing manuals Volume 6: biscuit packaging and storage** *D. Manley*
36 **Practical dehydration Second edition** *M. Greensmith*
37 **Lawrie's meat science Sixth edition** *R. A. Lawrie*
38 **Yoghurt: science and technology Second edition** *A. Y Tamime and R. K. Robinson*
39 **New ingredients in food processing: biochemistry and agriculture** *G. Linden and D. Lorient*
40 **Benders' dictionary of nutrition and food technology Seventh edition** *D A Bender and A. E. Bender*
41 **Technology of biscuits, crackers and cookies Third edition** *D. Manley*
42 **Food processing technology: principles and practice Second edition** *P. J. Fellows*
43 **Managing frozen foods** *Edited by C. J. Kennedy*
44 **Handbook of hydrocolloids** *Edited by G. O. Phillips and P. A. Williams*
45 **Food labelling** *Edited by J. R. Blanchfield*
46 **Cereal biotechnology** *Edited by P. C. Morris and J. H. Bryce*
47 **Food intolerance and the food industry** *Edited by T. Dean*
48 **The stability and shelf life of food** *Edited by D. Kilcast and P. Subramaniam*
49 **Functional foods: concept to product** *Edited by G. R. Gibson and C. M. Williams*
50 **Chilled foods: a comprehensive guide Second edition** *Edited by M. Stringer and C. Dennis*
51 **HACCP in the meat industry** *Edited by M. Brown*
52 **Biscuit, cracker and cookie recipes for the food industry** *D. Manley*
53 **Cereals processing technology** *Edited by G. Owens*
54 **Baking problems solved** *S. P. Cauvain and L. S. Young*
55 **Thermal technologies in food processing** *Edited by P. Richardson*
56 **Frying: improving quality** *Edited by J. B. Rossell*

57 **Food chemical safety Volume 1: contaminants** *Edited by D. Watson*
58 **Making the most of HACCP: learning from others' experience** *Edited by T. Mayes and S. Mortimore*
59 **Food process modelling** *Edited by L. M. M. Tijskens, M. L. A. T. M. Hertog and B. M. Nicolaï*
60 **EU food law: a practical guide** *Edited by K. Goodburn*
61 **Extrusion cooking: technologies and applications** *Edited by R. Guy*
62 **Auditing in the food industry: from safety and quality to environmental and other audits** *Edited by M. Dillon and C. Griffith*
63 **Handbook of herbs and spices Volume 1** *Edited by K. V. Peter*
64 **Food product development: maximising success** *M. Earle, R. Earle and A. Anderson*
65 **Instrumentation and sensors for the food industry Second edition** *Edited by E. Kress-Rogers and C. J. B. Brimelow*
66 **Food chemical safety Volume 2: additives** *Edited by D. Watson*
67 **Fruit and vegetable biotechnology** *Edited by V. Valpuesta*
68 **Foodborne pathogens: hazards, risk analysis and control** *Edited by C. de W. Blackburn and P. J. McClure*
69 **Meat refrigeration** *S. J. James and C. James*
70 **Lockhart and Wiseman's crop husbandry Eighth edition** *H. J. S. Finch, A. M. Samuel and G. P. F. Lane*
71 **Safety and quality issues in fish processing** *Edited by H. A. Bremner*
72 **Minimal processing technologies in the food industries** *Edited by T. Ohlsson and N. Bengtsson*
73 **Fruit and vegetable processing: improving quality** *Edited by W. Jongen*
74 **The nutrition handbook for food processors** *Edited by C. J. K. Henry and C. Chapman*
75 **Colour in food: improving quality** *Edited by D MacDougall*
76 **Meat processing: improving quality** *Edited by J. P. Kerry, J. F. Kerry and D. A. Ledward*
77 **Microbiological risk assessment in food processing** *Edited by M. Brown and M. Stringer*
78 **Performance functional foods** *Edited by D. Watson*
79 **Functional dairy products Volume 1** *Edited by T. Mattila-Sandholm and M. Saarela*
80 **Taints and off-flavours in foods** *Edited by B. Baigrie*
81 **Yeasts in food** *Edited by T. Boekhout and V. Robert*
82 **Phytochemical functional foods** *Edited by I. T. Johnson and G. Williamson*
83 **Novel food packaging techniques** *Edited by R. Ahvenainen*
84 **Detecting pathogens in food** *Edited by T. A. McMeekin*
85 **Natural antimicrobials for the minimal processing of foods** *Edited by S. Roller*
86 **Texture in food Volume 1: semi-solid foods** *Edited by B. M. McKenna*
87 **Dairy processing: improving quality** *Edited by G Smit*
88 **Hygiene in food processing: principles and practice** *Edited by H. L. M. Lelieveld, M. A. Mostert, B. White and J. Holah*
89 **Rapid and on-line instrumentation for food quality assurance** *Edited by I. Tothill*
90 **Sausage manufacture: principles and practice** *E. Essien*
91 **Environmentally-friendly food processing** *Edited by B. Mattsson and U. Sonesson*
92 **Bread making: improving quality** *Edited by S. P. Cauvain*

93 **Food preservation techniques** *Edited by P. Zeuthen and L. Bøgh-Sørensen*
94 **Food authenticity and traceability** *Edited by M. Lees*
95 **Analytical methods for food additives** *R. Wood, L. Foster, A. Damant and P. Key*
96 **Handbook of herbs and spices Volume 2** *Edited by K. V. Peter*
97 **Texture in food Volume 2: solid foods** *Edited by D. Kilcast*
98 **Proteins in food processing** *Edited by R. Yada*
99 **Detecting foreign bodies in food** *Edited by M. Edwards*
100 **Understanding and measuring the shelf-life of food** *Edited by R. Steele*
101 **Poultry meat processing and quality** *Edited by G. Mead*
102 **Functional foods, ageing and degenerative disease** *Edited by C. Remacle and B. Reusens*
103 **Mycotoxins in food: detection and control** *Edited by N. Magan and M. Olsen*
104 **Improving the thermal processing of foods** *Edited by P. Richardson*
105 **Pesticide, veterinary and other residues in food** *Edited by D. Watson*
106 **Starch in food: structure, functions and applications** *Edited by A-C Eliasson*
107 **Functional foods, cardiovascular disease and diabetes** *Edited by A. Arnoldi*
108 **Brewing: science and practice** *D. E. Briggs, P. A. Brookes, R. Stevens and C. A. Boulton*
109 **Using cereal science and technology for the benefit of consumers: proceedings of the 12th International ICC Cereal and Bread Congress, 24–26th May, 2004, Harrogate, UK** *Edited by S. P. Cauvain, L. S. Young and S. Salmon*
110 **Improving the safety of fresh meat** *Edited by J. Sofos*
111 **Understanding pathogen behaviour in food: virulence, stress response and resistance** *Edited by M. Griffiths*
112 **The microwave processing of foods** *Edited by H. Schubert and M. Regier*
113 **Food safety control in the poultry industry** *Edited by G. Mead*
114 **Improving the safety of fresh fruit and vegetables** *Edited by W. Jongen*
115 **Food, diet and obesity** *Edited by D. Mela*
116 **Handbook of hygiene control in the food industry** *Edited by H. L. M. Lelieveld, M. A. Mostert and J. Holah*
117 **Detecting allergens in food** *Edited by S. Koppelman and S. Hefle*
118 **Improving the fat content of foods** *Edited by C. Williams and J. Buttriss*
119 **Improving traceability in food processing and distribution** *Edited by I. Smith and A. Furness*
120 **Flavour in food** *Edited by A. Voilley and P. Etievant*
121 **The Chorleywood bread process** *S. P. Cauvain and L. S. Young*
122 **Food spoilage microorganisms** *Edited by C. de W. Blackburn*
123 **Emerging foodborne pathogens** *Edited by Y. Motarjemi and M. Adams*
124 **Bender's dictionary of nutrition and food technology Eighth edition** *D. A. Bender*
125 **Optimising sweet taste in foods** *Edited by W. J. Spillane*
126 **Brewing: new technologies** *Edited by C. Bamforth*
127 **Handbook of herbs and spices Volume 3** *Edited by K. V. Peter*
128 **Lawrie's meat science Seventh edition** *R. A. Lawrie in collaboration with D. A. Ledward*
129 **Modifying lipids for use in food** *Edited by F. Gunstone*
130 **Meat products handbook: practical science and technology** *G. Feiner*
131 **Food consumption and disease risk: consumer-pathogen interactions** *Edited by M. Potter*

132 **Acrylamide and other hazardous compounds in heat-treated foods** *Edited by K. Skog and J. Alexander*
133 **Managing allergens in food** *Edited by C. Mills, H. Wichers and K. Hoffman-Sommergruber*
134 **Microbiological analysis of red meat, poultry and eggs** *Edited by G. Mead*
135 **Maximising the value of marine by-products** *Edited by F. Shahidi*
136 **Chemical migration and food contact materials** *Edited by K. Barnes, R. Sinclair and D. Watson*
137 **Understanding consumers of food products** *Edited by L. Frewer and H. van Trijp*
138 **Reducing salt in foods: practical strategies** *Edited by D. Kilcast and F. Angus*
139 **Modelling microrganisms in food** *Edited by S. Brul, S. Van Gerwen and M. Zwietering*
140 **Tamime and Robinson's Yoghurt: science and technology Third edition** *A. Y. Tamime and R. K. Robinson*
141 **Handbook of waste management and co-product recovery in food processing: Volume 1** *Edited by K. W. Waldron*
142 **Improving the flavour of cheese** *Edited by B. Weimer*
143 **Novel food ingredients for weight control** *Edited by C. J. K. Henry*
144 **Consumer-led food product development** *Edited by H. MacFie*
145 **Functional dairy products Volume 2** *Edited by M. Saarela*
146 **Modifying flavour in food** *Edited by A. J. Taylor and J. Hort*
147 **Cheese problems solved** *Edited by P. L. H. McSweeney*
148 **Handbook of organic food safety and quality** *Edited by J. Cooper, C. Leifert and U. Niggli*
149 **Understanding and controlling the microstructure of complex foods** *Edited by D. J. McClements*
150 **Novel enzyme technology for food applications** *Edited by R. Rastall*
151 **Food preservation by pulsed electric fields: from research to application** *Edited by H. L. M. Lelieveld and S. W. H. de Haan*
152 **Technology of functional cereal products** *Edited by B. R. Hamaker*
153 **Case studies in food product development** *Edited by M. Earle and R. Earle*
154 **Delivery and controlled release of bioactives in foods and nutraceuticals** *Edited by N. Garti*
155 **Fruit and vegetable flavour: recent advances and future prospects** *Edited by B. Brückner and S. G. Wyllie*
156 **Food fortification and supplementation: technological, safety and regulatory aspects** *Edited by P. Berry Ottaway*
157 **Improving the health-promoting properties of fruit and vegetable products** *Edited by F. A. Tomás-Barberán and M. I. Gil*
158 **Improving seafood products for the consumer** *Edited by T. Børresen*
159 **In-pack processed foods: improving quality** *Edited by P. Richardson*
160 **Handbook of water and energy management in food processing** *Edited by J. Klemeš, R. Smith and J-K Kim*
161 **Environmentally compatible food packaging** *Edited by E. Chiellini*
162 **Improving farmed fish quality and safety** *Edited by Ø. Lie*
163 **Carbohydrate-active enzymes** *Edited by K-H Park*
164 **Chilled foods: a comprehensive guide Third edition** *Edited by M. Brown*
165 **Food for the ageing population** *Edited by M. M. Raats, C. P. G. M. de Groot and W. A Van Staveren*

166 **Improving the sensory and nutritional quality of fresh meat** *Edited by J. P. Kerry and D. A. Ledward*
167 **Shellfish safety and quality** *Edited by S. E. Shumway and G. E. Rodrick*
168 **Functional and speciality beverage technology** *Edited by P. Paquin*
169 **Functional foods: principles and technology** *M. Guo*
170 **Endocrine-disrupting chemicals in food** *Edited by I. Shaw*
171 **Meals in science and practice: interdisciplinary research and business applications** *Edited by H. L. Meiselman*
172 **Food constituents and oral health: current status and future prospects** *Edited by M. Wilson*
173 **Handbook of hydrocolloids Second edition** *Edited by G. O. Phillips and P. A. Williams*
174 **Food processing technology: principles and practice Third edition** *P. J. Fellows*
175 **Science and technology of enrobed and filled chocolate, confectionery and bakery products** *Edited by G. Talbot*
176 **Foodborne pathogens: hazards, risk analysis and control Second edition** *Edited by C. de W. Blackburn and P. J. McClure*
177 **Designing functional foods: measuring and controlling food structure breakdown and absorption** *Edited by D. J. McClements and E. A. Decker*
178 **New technologies in aquaculture: improving production efficiency, quality and environmental management** *Edited by G. Burnell and G. Allan*
179 **More baking problems solved** *S. P. Cauvain and L. S. Young*
180 **Soft drink and fruit juice problems solved** *P. Ashurst and R. Hargitt*
181 **Biofilms in the food and beverage industries** *Edited by P. M. Fratamico, B. A. Annous and N. W. Gunther*
182 **Dairy-derived ingredients: food and neutraceutical uses** *Edited by M. Corredig*
183 **Handbook of waste management and co-product recovery in food processing Volume 2** *Edited by K. W. Waldron*
184 **Innovations in food labelling** *Edited by J. Albert*
185 **Delivering performance in food supply chains** *Edited by C. Mena and G. Stevens*
186 **Chemical deterioration and physical instability of food and beverages** *Edited by L. Skibsted, J. Risbo and M. Andersen*
187 **Managing wine quality Volume 1: viticulture and wine quality** *Edited by A.G. Reynolds*
188 **Improving the safety and quality of milk Volume 1: milk production and processing** *Edited by M. Griffiths*
189 **Improving the safety and quality of milk Volume 2: improving quality in milk products** *Edited by M. Griffiths*
190 **Cereal grains: assessing and managing quality** *Edited by C. Wrigley and I. Batey*
191 **Sensory analysis for food and beverage control: a practical guide** *Edited by D. Kilcast*
192 **Managing wine quality Volume 2: oenology and wine quality** *Edited by A. G. Reynolds*
193 **Winemaking problems solved** *Edited by C. E. Butzke*
194 **Environmental assessment and management in the food industry** *Edited by U. Sonesson, J. Berlin and F. Ziegler*
195 **Consumer-driven innovation in food and personal care products** *Edited by S.R. Jaeger and H. MacFie*

196 **Tracing pathogens in the food chain** *Edited by S. Brul, P.M. Fratamico and T.A. McMeekin*
197 **Case studies in novel food processing technologies** *Edited by C. Doona, K Kustin and F. Feeherry*
198 **Freeze-drying of pharmaceutical and food products** *Tse-Chao Hua, Bao-Lin Liu and Hua Zhang*
199 **Oxidation in foods and beverages and antioxidant applications: Volume 1 Understanding mechanisms of oxidation and antioxidant activity** *Eric A. Decker, Ryan J. Elias and D. Julian McClements*
200 **Oxidation in foods and beverages and antioxidant applications: Volume 2 Management in different industry sectors** *Eric A. Decker, Ryan J. Elias and D. Julian McClements*
201 **Protective cultures, antimicrobial metabolites and bacteriophages for food and beverage biopreservation** *Christophe Lacroix*

Preface

Understanding the epidemiology of pathogens in production chains and processing environments is highly important for food safety control. Epidemiological surveillance is also an essential part of a food safety programme. The ever increasing possibilities of modern tools and techniques offer new options to trace unwanted organisms in the chain and take adequate timely measures. The original idea of combining experts from all fields of the food chain led to a combined editorship of the book that covers these aspects. The result is a comprehensive collection where epidemiology and surveillance set the scene for the application of the novel technologies for tracing pathogens in the food chain. We hope that this book will fill a critical void in the scientific literature and will prove of interest to food producers, governmental organizations, and the international scientific community. The editors would like to express here their gratitude to all who have contributed to the volume and to the excellent support at Woodhead Publishing. Acknowledgements also go to our families who had to endure our late-night readings or early weekend rises.

Stanley Brul, Pina M. Fratamico and Tom A. McMeekin

1

Introduction

S. Brul, University of Amsterdam, The Netherlands, P. Fratamico, ERRC, USA and T. McMeekin, University of Tasmania, Australia

Abstract: The presence of microorganisms in our day-to-day lives is briefly put into perspective with a strong focus on those bacteria that cause food safety concerns. From considerations on their initial presence, we move to their identification with state-of-the-art molecular tools and close the loop with attention to surveillance and microbial behaviour in specific chains. The contemporary concept in (predictive) food microbiology of the need to transform mechanistic data as much as possible to models at different organisational levels of biological structure and function is introduced.

Key words: foodborne pathogens, tracing and tracking microorganisms in the food chain, modelling microbial behaviour, systems analysis of microbial food preservation.

1.1 Microbes and the food chain

The world of microorganisms holds many promises and threats for the food manufacturer. On the one hand, the cells provide a rich source of functional molecules that can be introduced to the benefit of consumers into the diet through (food) fermentation (Nout, 2009; Kleerebezem *et al.*, 2010). On the other hand, the presence and proliferation of unwanted microorganisms can have consequences ranging from harmless but economically damaging food spoilage to dangerous food safety incidents (Havelaar *et al.*, 2010; Rajkovic *et al.*, in press). The aim of the food producer is to detect, identify, and enumerate the organisms of interest as soon as possible using the best methods that are currently available. Measures to deal with the unwanted presence of microbes can then rapidly be taken. For governmental organisations, this implies improved public health assurance (Havelaar *et al.*, 2010) while for the producer it means that standard operating procedures in food manufacturing, as well as adequate immediate action, including recalls if necessary, can be applied in an optimised manner (Jacxsens *et al.*, 2009). In all cases the consumer benefits from the improved response time and response quality. This book provides a topical overview of various important aspects involved, ranging from an assessment of the full food chain to genomics-based analysis of the isolated relevant microorganisms. Below we first give a bird's eye view of the various parts of the book and then provide a view on integration of the current topics in a systems approach using quantification tools at the various levels of complexity in the chain. The approach closely matches the views expressed in the European Technology Platform Food for Life (http://cordis.europa.eu/technology-platforms/pdf/foodforlife.pdf).

1.2 Where and in what 'state' noxious microbes are in our food chain

Here we highlight some of the main considerations of the various parts of the book. We will not dwell extensively on each individual chapter but rather highlight points to arouse interest in the reader who can then find the detail in the various chapters themselves.

1.2.1 Foodborne pathogen surveillance and outbreak investigation

This part deals with practical issues regarding outbreak detection and surveillance (see also the review by Pires *et al.*, 2009). Dr Fisher sets the scene by indicating the crucial role of an integrated approach to surveillance. He makes a plea for collaboration between services active on the human, veterinary and food aspects of outbreak surveillance and investigation. Dr Stein of the WHO provides an integrated view on outbreak detection and investigation. The data clearly shows the different challenges that the Western world faces compared to developing countries. Many of the issues can be addressed by available measures and appropriately applying well-known procedures. Also, sometimes mainly, economic considerations can be prime factors in determining the success of preventative, as well as curative strategies (Buzby and Roberts, 2009; Palou *et al.*, 2009). The chapter by Hald and Pires discusses strategies for addressing the attribution of foodborne infections along the chain. They elegantly show the various analyses, from epidemiological considerations to intervention studies and application of expert knowledge as primary input. Finally, a parameter already mentioned but still often imprudently easily tucked away is economic cost. The chapter by Buzby elaborates on such economic costs versus the global burden of foodborne disease. Managerial choices and scenarios are spelled out in cost–benefit scenarios.

1.2.2 Subtyping of foodborne pathogens

The characteristics of foodborne pathogens may well vary over time just as they do for any microbial population. Hence tools to monitor such change that are relevant at the physiological level are of major importance (discussed in Davidsen *et al.*, 2009; Hornstra *et al.*, 2009). The chapter by Gebreyes and Thakur sets the scene, and the use of phenotypic markers to assess behaviour at the cellular level is described. Cooper and coworkers then describe pulsed-field gel electrophoresis and related molecular methods for subtyping of bacteria. Pagotto adds a number of new 'pearls on the string' of subtyping tools. At all instances it is crucial that methods are validated and referenced. The chapter by Hyytia-Trees and Ribot does just that: referencing and standardization of (new) bacterial subtyping tools for use in the food chain.

1.2.3 Molecular methods, genomics and other emerging approaches in the surveillance and study of foodborne pathogens

As any other biology discipline, microbiology has recently made major leaps forward through the 'omics' revolution (Zhang *et al.*, in press). Transformation of the new data into knowledge as well as application in various applied settings is currently rapidly developing (Borneman *et al.*, 2007; Brul *et al.*, 2008). For applications in any field and thus also in food microbiology proper nucleic acid extraction protocols are essential. The chapter by Wagner and colleagues reiterates this point. In many cases of tracing pathogens in the food chains is all about finding the needle in the haystack, or, in other words, identifying the presence and pathogenic characteristics of low absolute numbers of microorganisms. The chapter by van der Vossen and coworkers next shows how various molecular typing methods can find practical application in surveying for *Campylobacter* occurrence in foods. Genome-wide typing methods are increasingly used as *Campylobacter* remains a major food safety concern. Costs for full genome sequencing are rapidly declining so that this is an increasingly viable option (Petrosino *et al.*, 2009).

The application of comparative genome-wide analysis is further discussed by Stabler *et al.* in a more generic sense. Non-nucleic acid methods that emerge for detection and characterisation are extensively discussed by Bowman from the University of Tasmania. Specific typing based on the presence of virulence genes is introduced by Wassenaar in Chapter 15. Toxigenic and spoilage bacterial spore-formers are a prime concern and economic problem to industry and society. Maybe even more than vegetative cells, spore-forming organisms often set the boundaries of what is possible in terms of food preservation treatments (van Zuijlen *et al.*, in press). While minimisation of heat application from sterilisation to near-pasteurisation values is highly desirable from a consumer's food quality perspective, food safety has to remain guaranteed throughout. The chapter by Van Zuijlen gives an industrial look at how ribotyping methods can be deployed to address recurring issues with bacterial spore-formers in the food manufacturing industry.

Finally, integration of microbiology data, complex systems analyses and stochastic models is discussed in the framework of 'Bio-tracing', an EU sponsored approach.

1.2.4 Tracing pathogens in particular food chains

The last part of the book is devoted to finding microbial pathogens in particular chains. An account is given of events in meat and game production by Whyte *et al*. This is a type of food chain that traditionally has significant issues with food safety. Fish production and the pathogen presence/behaviour in chicken is discussed by Lunestad *et al*. and Hiett *et al*., respectively. Dairy, shellfish and fresh produce are chains of generic interest that complete the overview given in the book. In all chains it is important to analyse the environmental conditions to and identify generically relevant parameters. These can be used to define better test conditions for more fundamental studies on the mechanisms involved in maintenance of microorganisms in a food chain environment. Such mechanistic basis will improve the robustness of predictive models of microbial behaviour in foods (McMeekin *et al*., 2007, 2010).

1.3 Towards integration

1.3.1 Modelling cellular behaviour

Many scientists in the field advocate an integrated 'systems' approach in tracing (and tracking) microorganisms in foods. Safety by design is seen as the way forward. Improved knowledge of the microorganisms under study is instrumental in providing novel options to test for their presence and behaviour. Systems biology is the analysis routine that is more and more applied in which the cycle between experiment and functional integrative genomics analysis of behaviour in a food-related environment is the key. In Fig. 1.1, this view is shown schematically for the analysis of the behaviour of bacterial spores. Clearly for this tool to be of use to food microbiology, models are needed that can operate at many different levels of complexity (Brul *et al*., 2008; McMeekin *et al*., 2007, 2010). Also, it is crucial that stochastic elements are introduced as it is more and more clear that heterogeneous behaviour, assessed on the basis of physiological end-points (growth, death, lag-time), of cells from genetically homogeneous populations is more the rule than the exception (Stringer *et al*., 2009).

Such heterogeneity is often ruled by fluctuations in gene expression (Pin *et al*., 2009). To assess the importance of various genes in regulatory pathways, sensitivity analysis regarding their control on signal output is key (Veening *et al*., 2008; Hornstra *et al*., 2009). To test for the expression levels of proteins, antibody-based staining is possible generally preceded by analysis of the behaviour of model organisms. The latter often allows for the full scope of genomics techniques to be used. In all this, data analysis is crucial. What will be needed increasingly are 'biological engineers' who are specialists with full appreciation of the biological complexity as well as a quantitative view on cellular biochemistry and molecular biology.

1.3.2 Modelling food microbiology in the framework of the food chain

When analysing the importance of molecular cell biology in food microbiology it is crucial to couple the stochastic data at the cellular level to the probability of cellular distribution over the chain (Havelaar *et al*., 2010). The data will have to

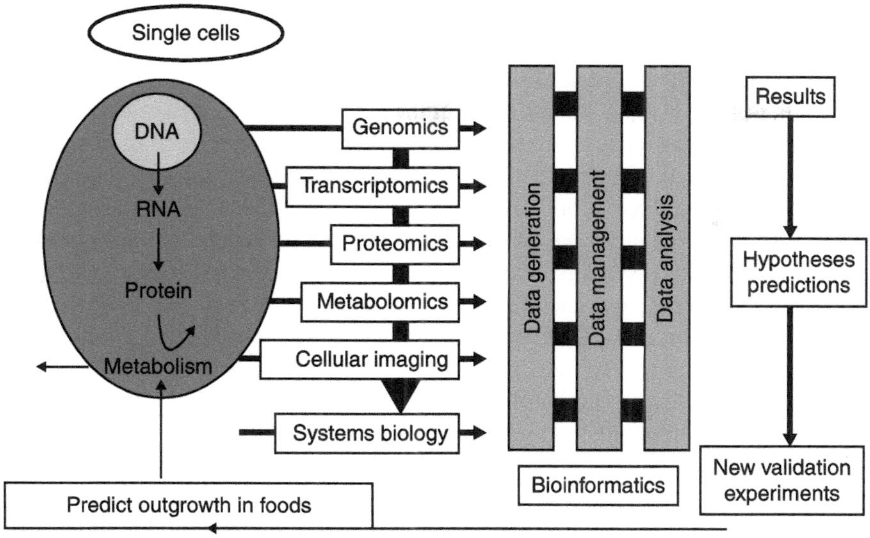

Fig. 1.1 A schematic representation of the sequence of events in systems biology as they pertain to food microbiology. Initially data is obtained at the various 'omics' levels. Such data needs to be complemented with proper image analysis or equivalents to assess the heterogeneity in expression of (various) signalling pathways in a genetically homogeneous population. The level of heterogeneity is increasingly seen as an important parameter determining the developmental state of (most) of the microorganisms in a genetically homogeneous population. Integration of the omics data as well as the single cell analysis data by 'biological engineers' (see main text) should provide models and lay the foundation for a systems biology approach to microbial food safety. Models are tested and improved through iterative cycling.

be input for techniques and approaches that are being developed for risk-benefit evaluation where consumer demands are well balanced against food safety risks. To profit from the improved biological understanding, it will be crucial to put the data on microbial cell physiology in relation to microbial exposure likelihood and disease risks at the consumer level. In doing so, the risk-benefit evaluation framework will have to be adapted to incorporate this new data efficiently. Here sustainable processing, preservation, packaging and logistic systems have to be developed. Balancing these often counteracting drivers of microbial food quality and safety requires developing models that can 'talk to each other'. That is meaning that improvements in the quantitative prediction of microbial physiology can be 'read' by models of cellular distribution and manufacturing, as well as by epidemiological risk assessment models (Havelaar *et al.*, 2010). Going one step further, biological models to establish the mechanistic basis of survival and virulence in the host (consumer) will be welcome to enhance the robustness of epidemiological models. Such models should also aim at being (more) proactive in identifying emerging pathogens and their characteristics.

To consider all these boundary conditions systematically is highly complex to set priorities in food microbiology research and continuous challenge of the criteria used will be important to reach a common practice.

In conclusion, in the overall context of the food chain, the value of tracing (and tracking) pathogens will only become proactively beneficial to food production if progress in molecular microbial physiology is translated by ‘biological engineers’ to outputs transferable to food engineers who can use these models to improve robustness and predictive powers of their food chain models. In these models either optimal safety settings or optimal quality settings may be taken as input. Thus, the engineering approach to the various areas of life science will be the key towards integration and effectively generating a ‘systems approach’ to microbial stability. This type of approach is indeed much advocated by the European Technology Platform Food for Life (Fig. 1.2 and http://cordis.europa.eu/technology-platforms/pdf/foodforlife.pdf).

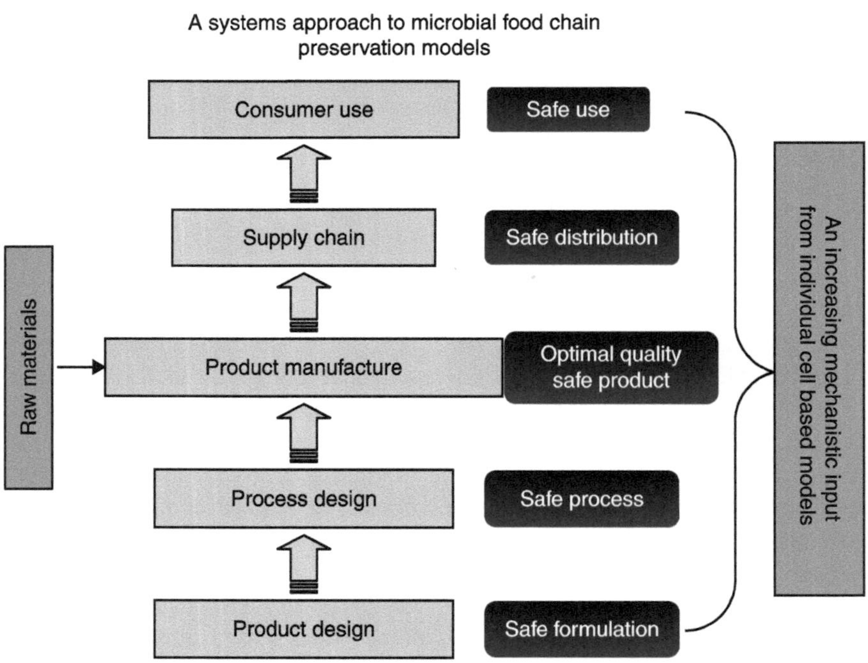

Fig. 1.2 The view on current and near future developments in modelling of microbial behaviour throughout the food chain. Clearly to develop models at the various levels of complexity of the chain will require input from agriculture to medicine and from consumer to producer (fork to farm). Such models will increasingly use individual cell based models as input. Modified from the European Technology Platform Food for Life (http://cordis.europa.eu/technology-platforms/pdf/foodforlife.pdf).

1.4 References

BORNEMAN A R, CHAMBERS P J and PRETORIUS I S (2007), 'Yeast systems biology: modelling the wine maker's art', *Trends in Biotechnology*, **25**, 349–55.

BRUL S, MENSONIDES F I C, HELLINGWERF K J and TEIXEIRA DE MATTOS M J (2008), 'Microbial systems biology: new frontiers open to predictive microbiology', *International Journal of Food Microbiology*, **128**, 16–21.

BUZBY J C and ROBERTS T (2009), 'The economics of enteric infections: human foodborne disease costs', *Gastroenterology*, **136**, 1851–62.

DAVIDSEN A O H, FRYE S A, BALASINGHAM S V, LAGESEN K, ROGNES T and TØNJUM T (2009), 'Genome dynamics in major bacterial pathogens', *FEMS Microbiology Reviews*, **33**, 453–70.

HAVELAAR A H, BRUL S, JONG A, DE JONGE R, DE ZWIETERING M H and TER KUILE B (2010), 'Future challenges to microbial food safety', *International Journal of Food Microbiology*, **139**(Suppl 1), S79–94.

HORNSTRA L, TERBEEK A, SMELT J P, KALLEMEIJN W and BRUL S (2009), 'On the origin of heterogeneity in (preservation) resistance of *Bacillus* spores: input for a "systems" analysis approach of bacterial spore outgrowth', *International Journal of Food Microbiology*, **134**, 9–15.

JACXSENS L, KUSSAGA J, LUNING P A, VAN DER SPIEGEL M, DE VLIEGHERE F and UYTTENDAELE M (2009), 'A microbial assessment scheme to measure microbial performance of food safety management systems', *International Journal of Food Microbiology*, **134**, 113–25.

KLEEREBEZEM M, HOLS P, BERNARD E, ROLAIN T, ZHOU M, SIEZEN R J and BRON P A (2010), 'The extracellular biology of lactobacilli', *FEMS Microbiology Review*, **34**, 199–230.

MCMEEKIN T A, HILL C, WAGNER M, DAHL A and ROSS T (2010), 'Ecophysiology of food-borne pathogens: essential knowledge to improve food safety', *International Journal of Food Microbiology*, **139**(Suppl 1), S64–78.

MCMEEKIN T A, MELLEFONT L A and ROSS T (2007), 'Predictive microbiology: past present and future. In Brul, S., van Gerwen, S. and Zwietering, M. (Eds): *Modelling Microorganisms in Food*, Woodhead, Cambridge, UK, pp. 7–21.

NOUT M J (2009), 'Rich nutrition from the poorest-cereal fermentations in Africa and Asia', *Food Microbiology*, **26**, 685–92.

PALOU A, PICO C and KEIJER J (2009), 'Integration of risk and benefit analysis – the window of benefit as a new tool?', *Critical Reviews in Food Science and Nutrition*, **49**, 670–80.

PETROSINO J F, HIGHLANDER S, LUNA R A, GIBBS R A and VERSALOVIC J (2009), 'Metagenomic pyrosequencing and microbial identification', *Clinical Chemistry*, **55**, 856–66.

PIN C, ROLFE M D, MUNÔZ-CUEVAS M, HINTON J C, PECK M W, WALTON N J and BARANYI J (2009), 'Network analysis of the transcriptional pattern of young and old cells of *Escherichia coli* during lag-phase', *BMC Systems Biology*, **3**, 108.

PIRES S M, EVERS E G, VAN PELT W, AYERS T, SCALLAN E, ANGULO F J, HAVELAAR A, HALD T and THE MED-VET-NET WORCKPACKAGE 28 WORKING GROUP (2009), 'Attributing the human disease burden of foodborne infections to specific sources', *Foodborne Pathogens and Disease*, **6**, 417–24.

RAJKOVIC A, SMIGIC N and DEVLIEGHERE F (in press) 'Contemporary strategies in combating microbial contamination in food chain', *International Journal of Food Microbiology*.

STRINGER S C, WEBB M D and PECK M W (2009), 'Contrasting effects of heat treatment and incubation temperature on germination and outgrowth of individual spores of non-proteolytic *Clostridium botulinum* bacteria. *Applied and Environmental Microbiology*, **75**, 2712–9.

VAN ZUIJLEN A, PERIAGO P M, AMÉZQUITA A, PALOP A, BRUL S and FERNÁNDEZ P S (in press) 'Characterization of *Bacillus sporothermodurans* IC4 spores: putative indicator microorganism for optimization of thermal processes in food sterilization', *Food Research International*.

VEENING J W, STEWART E J, BERNGRUBER T W, TADDEI F, KUIPERS O P and HAMOEN L W (2008), 'Bet-hedging and epigenetic inheritance in bacterial cell development', *Proceedings of the National Academy of Sciences of the United States of America*, **105**, 4393–8.

ZHANG W, LI F and NIE L (2010), 'Integrating multiple "omics" analysis for microbial biology: application and methodologies', *Microbiology*, **156**, 287–301.

Part I

Foodborne pathogen surveillance and outbreak investigation

2

Surveillance for foodborne pathogens in humans

I. S. T. Fisher, HPA Centre for Infections, UK

Abstract: Surveillance, in its short form, has been defined as 'information for action'. This chapter describes some of the surveillance methods that can be used and gives examples of their usage on a national and international basis. It describes some of the international outbreaks that have been identified with a range of pathogens and foodstuffs involved. It summarises the limitations of surveillance activities and some future trends. Most importantly, it emphasises the importance of collaboration between the human, veterinary and food aspects of foodborne surveillance to coordinate successful activities in this area.

Key words: human surveillance methods, international outbreaks, surveillance methods, International Health Regulations.

2.1 Introduction

Surveillance has been defined as the ongoing systematic collection, collation and analysis of data and the prompt dissemination of the resulting information to those who need to know so that an action can result. This can be more succinctly phrased as 'information for action' (Berkelman and Buehler, 1991). Information for action is most obvious in terms of foodborne disease surveillance in the identification of foodborne outbreaks and their causes and the implementation of public health measures to remove the vehicle of infection from the market place and hence stop further cases of illness. Although this is an immediate and effective method of stopping outbreaks, the longer term accumulation of evidence is often necessary for effective strategic intervention measures to be undertaken. An example of this is the knowledge that *Salmonella enterica* subsp. *enterica* serovar Enteritidis phage type (PT) 4 in the UK in the late 1980s and early 1990s was present in breeder and layer chicken flocks; however, it was not until almost ten years after it was recognised that intervention measures were implemented. The intervention measure initially introduced was the vaccination of breeder flocks in the UK to prevent onward transmission of *S.* Enteritidis PT4. While this showed a small but short-lived effect, it was not until the vaccination of layer flocks was brought in that a significant reduction in the level of human infections was achieved (Fig. 2.1). This has had an impact on eggs produced for domestic consumption, but it has not prevented problems with eggs imported into the UK (Little *et al.*, 2007). This demonstrates one of the difficulties with current food production and, more pertinently, consumption. The desire for seasonal foods all the year round has led to the increased shipping of products around the globe to fill this need. Hence, products can be harvested in countries where control measures can be less stringent and the potential for contamination much greater. Similarly, manufacturing practise has changed from products being made locally to being made at one site and shipped nationally, internationally or globally due to economies of scale; again, this allows the possibility of a contaminated product being distributed to consumers in many countries. These factors have led to many international outbreaks of foodborne pathogens occurring far from the source of the product. To demonstrate the diversity of the problem of international foodborne outbreaks, a selection of the pathogen/food combinations that have been identified are: *Salmonella*/ready to eat snacks (Killalea *et al.*, 1996), *Salmonella*/Halva (de Jong *et al.*, 2001), *Salmonella*/peanuts (Kirk *et al.*, 2004), *Salmonella*/chocolate (Werber *et al.*, 2005), *Salmonella*/fresh basil (Pezzoli *et al.*, 2008), *Escherichia coli*/spinach (Grant *et al.*, 2008; Wendel *et al.*, 2009), *Salmonella*/cooked meats (O'Flanagan *et al.*, 2008), norovirus/raspberries (Maunula *et al.*, 2009) and *Shigella*/baby corn (Lewis *et al.*, 2009). It has recently been estimated that approximately 30% of emerging pathogens identified over the past 60 years can be commonly transmitted by foods (Jones *et al.*, 2008). In a report published by

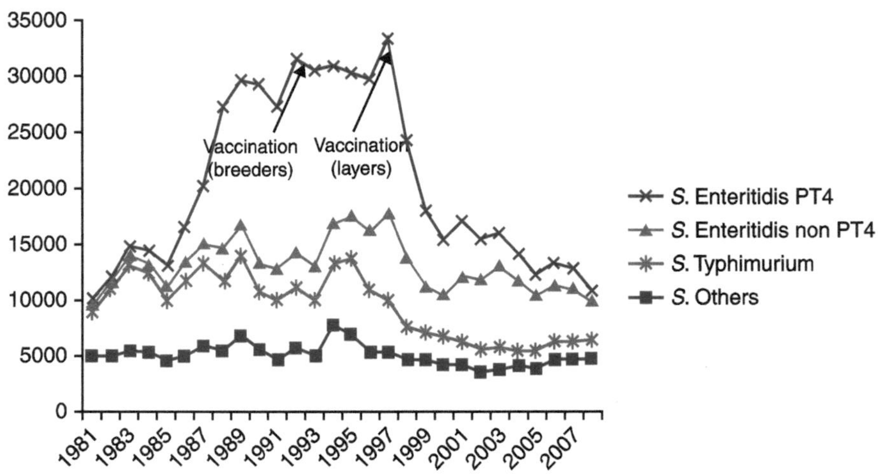

Fig. 2.1 Cases of human salmonellosis by year: England and Wales.

the World Health Organization (WHO) in 2008, the table of leading causes of death worldwide shows that diarrhoeal diseases (the majority of which are food- or waterborne) were the fifth most common cause in 2004 (WHO, 2008a). On a more positive note, the report also states that in children under the age of five, there is some evidence of a substantial decline.

Surveillance of foodborne pathogens is vital in that it provides the opportunity to assess trends in infections, identify new or emerging problems and assess the impact of interventions in a systematic way. This chapter introduces a summary of some of the common surveillance methodologies in use, gives some examples of their use in practise on both the national and international level, and some of the public health information systems and activities in place across the globe. It also looks at some developments for the future and areas whereby the surveillance of foodborne pathogens could be improved. It does not go into every single epidemiological method that can be employed but concentrates on those most applicable to the surveillance of foodborne pathogens.

2.2 Methods for the surveillance of foodborne pathogens

What is clear is that there is not, and cannot be, a single method in use for the surveillance of foodborne diseases. This is the case for other groups of infections, but is more critical for foodborne pathogens due to the very nature of the diverse sources and vehicles of infections involved. Surveillance methods in use range from notification-based, laboratory-based, to sentinel-based systems each with their advantages and disadvantages.

2.2.1 Statutory notification-based systems

These are usually legally binding systems whereby public health practitioners (often general practitioners) have an obligation to provide information to regulatory authorities for a range of notifiable diseases. These diseases are usually defined by national authorities, although some infections are also notifiable under the International Health Regulations (IHR) (see Section 2.3.6).

Advantages: They can provide an early alert of unusual events occurring within the community and will often include basic information that would not necessarily be available to other surveillance systems.

Disadvantages: Notifications are usually based on clinical diagnosis and as such lack laboratory confirmation of the pathogen. In the UK, notifications are made under the category of 'food poisoning' rather than being organism specific. This is defined as 'any disease of an infectious or toxic nature caused by or thought to be caused by the consumption of food or water' (Chief Medical Officer, 1992). There can be some delay in these notifications reaching the national level, which may make it difficult to identify diverse community outbreaks of foodborne infections.

2.2.2 Registration of deaths and hospital discharge data

In the majority of countries when a death occurs, a death certificate is completed. The certificate will include details of the cause of death and basic patient demographics. Similarly, hospitals will report the diagnoses and demographics of patients discharged from their care. When collated centrally, these can provide summaries of the magnitude of specific infections within countries.

Advantages: Registration of deaths and hospital discharge data do provide some data on the burden of illness of infections and can inform studies on the economic costs of these infections.

Disadvantages: These data are mainly acquired for more serious infections and do not represent the morbidity of pathogens in the general community. There is also a tendency to categorise infections under a 'miscellaneous' category rather than being specific, hence will not necessarily capture the full burden of death or hospitalisations for all infections.

2.2.3 Laboratory-based surveillance

Phenotypic methods

Reference laboratory confirmation is considered to be the gold standard of surveillance methods as it gives a precise identification of the pathogen involved. Samples are submitted to a laboratory for testing, and if they are sent to a national reference laboratory, they are then subjected to a range of different tests for characterisation. For bacteria, this would include phenotypic methods such as serotyping, phage typing, where relevant, and antimicrobial resistance testing.

Genotypic methods

The phenotypic characterisation of pathogens is enhanced by genotyping as well. This can be via a range of methods – pulsed-field gel electrophoresis (PFGE), multi-locus sequence typing, amplified fragment length polymorphism, plasmid profiling or multi-locus variable number of tandem repeat (VNTR) analysis – many of which help subdivide phenotypes and enable specific outbreak strains to be identified.

Advantages: Precise identification and characterisation of pathogens allows more accurate ascertainment of foodborne outbreaks of infection. Often outbreaks can only be discriminated from the general background noise by the small, but critical, differences from the routinely circulating pathogen causing many cases of sporadic infection. A small, isolated outbreak of *Salmonella* Typhimurium DT104 within a larger outbreak in the UK in 2000 was only differentiated from the epidemic strain by the identification of a 2.0 MDa plasmid uniquely associated with the smaller outbreak (Horby *et al*., 2003).

Disadvantages: There is inevitably a time delay in the onset of illness and the identification of the causative pathogen (due to time required to submit a specimen to a laboratory and subsequently to a reference laboratory for full characterisation). In addition, it is known that the numbers of cases that are identified in national statistics are only a small subset of people ill in the community. Figure 2.2 shows how the number of cases in the community can be lost between becoming ill and being included in national statistics. As illness resulting from foodborne pathogens tends to present with a mild, self-limiting disease, a large proportion of relatively healthy individuals will not even attend their local physician. Of those who do attend, not all will have a sample taken for testing. Routine diagnostics in the local laboratory may not be able to identify a pathogen, and even when a pathogen is identified, these samples may not be finally submitted to the National Surveillance Institute for inclusion in national statistics. Data from studies in England show that cases of salmonellosis are underreported by a factor of three, campylobacteriosis by a factor of eight and infection with norovirus by a factor of 1500 (Wheeler *et al.*, 1999). Similar studies in the Netherlands (de Wit *et al.*, 2001) and the USA (Mead *et al.*, 1999) have also shown high rates of underreporting, although the multipliers are different due to disparities in referral to primary health care systems and submission to reference laboratories for full characterisation. This loss of data means that the full extent of the burden of illness is often not fully quantified and that epidemiological studies may be hampered by not being able to investigate

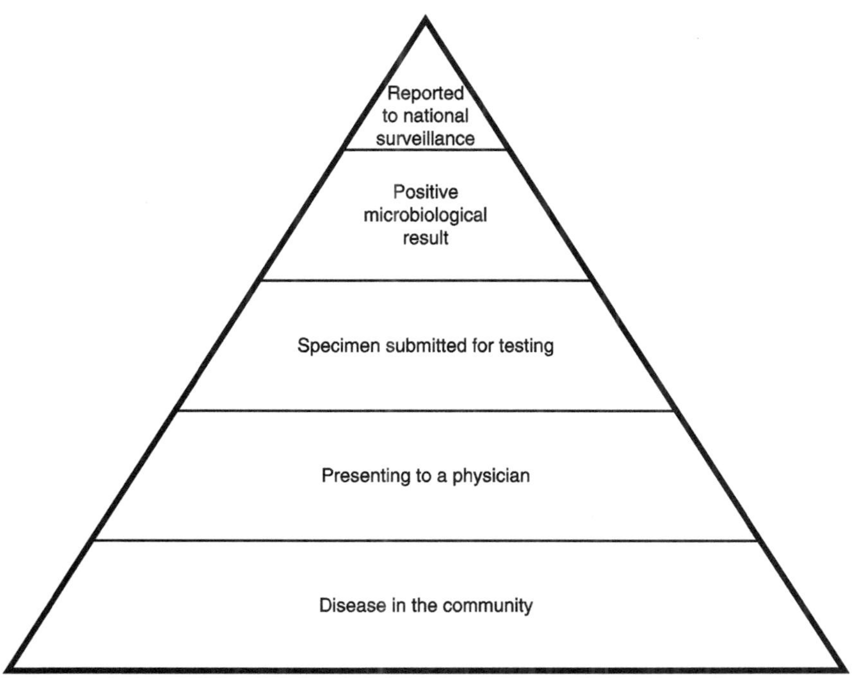

Fig. 2.2 Reporting pyramid with losses of data being reported to national surveillance.

every case. However, this does not prevent outbreaks being identified, successfully investigated and public health interventions implemented.

2.2.4 Sentinel surveillance methods

Sentinel surveillance is conducted on a subset of the national population into a range of infections, not just foodborne. Sentinel surveillance is often used to show the ebb and flow of infections that are usually diagnosed clinically, for example influenza. It can show the spread of infection across continents as epidemics evolve (Arkema *et al*., 2008). In terms of foodborne infections, it can be used as a measure for the incidence of illness in the population and can be tracked across many years. This allows baseline rates to be identified, targets set for reductions to be made and then tracked to determine the success (or failure) of intervention measures to achieve these. It allows for a more in-depth investigation into the risk factors for contracting infections from foodborne pathogens and into the modes of infection.

Advantages: Specific pathogens or syndromes (such as haemolytic uraemic syndrome, an indicator of *E. coli* O157 infection) can be targeted. Risk factors can be identified, subsequent control measures implemented and outcomes evaluated. Denominator data (the population within the sentinel surveillance area) can relatively easily be identified so that the rates of infection can be calculated and provide a definitive number to be measured and changes identified. Data from sentinel surveillance activities can also be used to extrapolate the national incidence of disease to provide the full measure of the burden of infections. Sentinel surveillance is also less resource intensive than investigating all sporadic cases of infection but can provide a large part of the data and information that such investigation would supply.

Disadvantages: Although it is less resource intensive, there is still an additional cost in conducting sentinel surveillance. While this can be done in relatively resource-rich countries, it is difficult to achieve in resource-poor countries that may have different public health priorities. Sentinel surveillance also has to strike a balance between being large enough to allow meaningful analysis and being representative of the population of a country as a whole to ensure that any extrapolations are as accurate as possible and are valuable in assessing the total burden of infection.

2.2.5 Geographical information systems

Geographical information systems (GIS) can be used to build a baseline of endemic levels of infections within a geographical area (usually national or sub-national) and identify when levels of infection breach those expected for that region and time of year. GIS can help illuminate the correlation between infections and environmental or other exposure factors that lead to that infection such as the proximity to sheep and cases of cryptosporidiosis. Mapping cases of infection is a powerful tool in analysing spatial and temporal trends and is of great relevance to outbreak investigation and control.

Advantages: These can provide evidence of levels of infection above that which is the expected amount and hence assist in identifying outbreaks or clusters of infection to inform investigations into sources of infection. GIS provides a visual reference to investigators and policy-makers that is easily understandable.

Disadvantages: GIS has a tendency to be limited to characterised phenotypes of infection and hence is dependent on laboratory-confirmed cases with the associated time delays. There is also a loss of data as only those cases for which a geographical reference can be identified can be included. This information can often be limited in the case of laboratory-confirmed diagnoses.

2.2.6 Enhanced surveillance for specific pathogens

Enhanced surveillance is valuable for infections with low numbers of cases, but where high public health impacts are involved (e.g. *Listeria monocytogenes*, shiga toxin-producing *E. coli*). Typically, once a case has been identified, either microbiologically or clinically presumptive, questionnaires are sent to patients or their carers to be completed. These ask about the consumption of various foods or possible environmental exposures that may have caused the infection. Results are collated into databases that can be analysed to provide information that would not be available in routine surveillance systems.

Advantages: Enhanced surveillance does provide more in-depth information than is readily accessible normally and hence can provide a greater insight into the exposures that have caused infection. This can be of particular value when different subtypes of bacteria inhabit different ecological niches and may require the provision of different public health messages. Such messages inform the population at risk of potential threats and enable informed action to be undertaken to lessen any possible infection.

Disadvantages: This is a resource-intensive method of gathering information. However, the information gathered and available for analysis can be useful in identifying specific risk factors that can help inform more precise interventions to prevent cases from occurring.

2.2.7 Outbreak investigations

Investigations into outbreaks of infections provide information that is invaluable in both the short term and long term. In the short term, the implementation of control measures to prevent the current outbreak is vital. It is important in outbreak investigations to try and ascertain information about the setting of outbreaks, the demographic details of those being taken ill (which can also provide valuable information on the vehicle of infection), what was the pathogen involved and what were the contributory factors that allowed the outbreak to occur. Evidence gathered from the investigation of outbreaks in the longer term help inform policy-makers of potential strategic interventions than can be implemented in the future.

Advantages: Full-scale epidemiological and microbiological investigations into all outbreaks may not be necessary on every occasion, but they can be beneficial in identifying new vehicles and modes of transmission that may not have been recognised before. The early part of the twenty-first century has seen a substantial increase in the number of outbreaks associated with salad products (Grant *et al.*, 2008; Greene *et al.*, 2008; Horby *et al.*, 2003; Wendel *et al.*, 2009). These are ready-to-eat products that are not intended for further heat treatment prior to consumption. If they do become contaminated during harvesting, then very few opportunities for a reduction in levels of pathogens on the food exist before being eaten. Outbreak investigations can also provide valuable information into microbial risk assessment, dose–response relationship and the ecology of foodborne pathogens.

Disadvantages: Outbreak investigations are resource intensive, and the benefits of investigating every outbreak may outweigh the cost of doing so. In addition, investigations can be hampered by being detected too late, allowing recall bias to influence the results of any study. There may not be any foodstuffs available for microbiological testing to be performed to confirm the pathogen in the food as well as in human cases involved in the outbreak, or an agent may not be identified at all. An example of the latter would be foodborne outbreaks associated with viruses that can be difficult to confirm, particularly in foods. Not all outbreak epidemiological studies will definitively identify a foodstuff associated with illness. This may be due to more than one food being contaminated, not asking the right questions in an investigation or an outbreak being too small to achieve a statistical significance. Nevertheless, such investigations are always useful as lessons can be learnt from even the most straightforward outbreaks.

2.2.8 New methods for surveillance of epidemic intelligence

New information systems based on the gathering of 'epidemic intelligence' (Paquet *et al.*, 2006) are becoming more common in collating information on events being identified by the media and on the Internet, thus bringing them to the attention of public health officials. These systems can be valuable in identifying events before they become formally known. Official systems currently in place include GPHIN (Public Health Agency of Canada, 2004), MEDYSIS and EPIS. GPHIN is the Global Public Health Information Network run by the Public Health Agency of Canada on behalf of the WHO. It trawls media sources across the globe and in multiple languages to identify events involving infectious diseases and circulates them to the public health community. MEDYSIS

(European Commission, n.d.) and EPIS, the Epidemiological Information System, are similar tools that have been developed by the European Commission and the European Centre for Disease Prevention and Control (ECDC), respectively. GPHIN, MEDISYS and EPIS are restricted access sites, although since December 2006, MEDISYS has also had a public domain version.

In addition to the above restricted or limited access sites, there are similar non-governmental electronic reporting systems such as ProMed-mail, which is the Program for Monitoring Emerging Diseases and is a program of the International Society for Infectious Diseases (ProMed-mail, n.d.) and the International Food Safety Network, which has been succeeded by Bites – Safe Food from Farm to Fork (Kansas State University, n.d.). Both of these systems either trawl public domain web pages or allow information on infectious disease to be posted on them. This information is then circulated (usually via email) to anyone who has subscribed to their distribution lists.

Advantages: The media (especially the Internet) is often the first to hear the instances of foodborne, or potentially foodborne, outbreaks of infection, and they often can be the first source of information received by public health agencies.

Disadvantages: The sheer volume of noise can make a full assessment of every report that may be an outbreak impossible to achieve. However, this useful source of information should not be ignored and can be used to complement information received from official sources.

2.3 National and international surveillance systems in use

Most countries have systems in place for collating data on notifiable diseases and registrations of deaths plus hospital discharge information along with phenotypic laboratory-based surveillance. These are successful to a greater or lesser extent depending on the public health infrastructure and commitment within each country. They are of value in highlighting unusual events that may be indicative of potential foodborne outbreaks of infection or in identifying deficiencies in food production practises that require measures to improve them. Much work remains to bring all countries up to a set standard. As all countries will benefit, this has to be a priority for public health in the future. The WHO is attempting to address these limitations by defining the set of 'core competencies' (see Section 2.3.7) that will set the baseline standards for the future.

2.3.1 Notifications, death registrations and hospital discharge data

Data on disease notifications, death registrations and hospital discharge are collated and published by national governments on a regular basis. They are not highly specific to particular foodborne pathogens but do provide assessments of the relative burden of these, particularly at the severe end of the spectrum.

2.3.2 Phenotypic and molecular typing networks – Global Salm-Surv, PulseNet US and PulseNet International

Phenotypic methods for typing foodborne pathogens have been an integral part of the identification of foodborne pathogens for many years, and as such have been invaluable in identifying causative agents and their sources. This has provided information to enable both short-term and long-term intervention measures to be put in place by policy-makers and industry. Global Salm-Surv is a WHO-led initiative that promotes capacity building in the detection, control and prevention of foodborne pathogens (Galanis *et al.*, 2006). Global Salm-Surv has been superseded by the Global Foodborne Infections Network. In the recent years, genotypic methods have come to the forefront. It has been stated that molecular typing has transformed public health by providing methods to further discriminate pathogens and identify clusters of cases that may not have been possible previously (Tauxe, 2006). This is certainly the case. The acknowledged method of choice in DNA-based methods has been that of PFGE to differentiate strains of bacterial pathogens. Standardised/harmonised PFGE methodology was initiated formally in the USA (Swaminathan *et al.*, 2001) and has subsequently been expanded to include other countries across the globe (Swaminathan *et al.*, 2006). PulseNet International brings together the six regional PulseNet networks (Asia-Pacific, Canada, Europe, Latin America and the Caribbean and the USA) into one microbiological network. PFGE is not the only method in use but is the most common. An advantage of using harmonised methodology such as the PulseNet system is that electronic images of strains can be compared directly without having to transfer strains to reference laboratories for characterisation. This negates the resulting additional cost involved in paying for the shipment of strains and inevitable time delay in getting them to their destination. An integral part of any such system has to be quality assurance to ensure that the right outputs are provided across laboratories. The key to creating successful laboratory networks that are able to identify and compare strains is by using validated methods for subtyping. PFGE, as described, has been the cornerstone of international public health microbiology, but new methods are being developed that can also be applied internationally. VNTR is rapidly becoming a method of choice for some *Salmonella* serovars, although currently only protocols for *S.* Typhimurium have been validated and published (Lindstedt *et al.*, 2004). Considerable work is currently being undertaken to develop VNTR techniques for other serovars (Larsson *et al.*, 2009) and other pathogens such as *E. coli*.

2.3.3 Sentinel surveillance

In the area of foodborne pathogens, the most obvious examples of sentinel surveillance are that of FoodNet in the USA and OzFoodNet in Australia (Kirk, McKay, *et al.*, 2008). The most recent FoodNet report gives preliminary details of the incidence of laboratory-confirmed bacterial and parasitic infections in ten participating states in the USA (CDC, 2009). It shows for a range of infections under surveillance the incidences of each of these in 2008 compared to the targets set for these pathogens for 2010. It also compares these data to those of previous years showing how they have changed. These data are invaluable in assessing trends and identifying which pathogens are not converging with their set targets. This provides information as to which areas in current food safety systems require particular attention or where public health messages need to be reinforced to ensure that targets can be achieved.

2.3.4 Geographical information systems

GIS have moved beyond the simple, but effective, mapping of cases of cholera in Broad Street, London, in 1854 by John Snow (Johnson, 2006), which marked the start of epidemiology as we know it today. They can be used to identify baseline levels of infection within countries or across borders and help identify and map outbreaks of infection in a more meaningful and understandable manner. This has been demonstrated by the development of the *Salmonella* Atlas as part of Workpackage 6 (use of GIS in epidemiological analyses of zoonotic diseases) of Med-Vet-Net (Med-Vet-Net *Salmonella* Atlas (2007), http://www.epigis.dk/). The outcomes show the evolution of various *Salmonella* serovars during the late 1990s and early part of the twenty-first century. Med-Vet-Net was a European Commission-funded Network of Excellence for zoonoses research (Med-Vet-Net, 2009). Resources for the development of this network were received under the Sixth Framework Programme of research within the 'Quality and Safety of Food' Priority Area (European Commission, n.d.). GIS are also valuable in mapping the evolution of infections such as the increase in the first few years of the twenty-first century of *Salmonella napoli* infection in Europe (Fisher *et al.*, 2009). In this case, GIS provided the evidence of an emerging problem so that investigations could be undertaken to control it.

2.3.5 Disease-specific networks in the EU

Disease surveillance networks have been an integral part of surveillance within the European Union (EU) for over 20 years for a range of infectious diseases. For foodborne pathogens, the relevant network was Enter-net (Fisher, 1999) – the surveillance network for the enteric pathogens, *Salmonella*, *E. coli* and *Campylobacter*. The network identified and investigated many outbreaks of foodborne infections both within Europe and across the globe. Many of these have been mentioned previously. Enter-n*et al*so created shared databases of laboratory-confirmed cases of infections that helped identify trends in pathogens internationally rather than just within individual countries. Production of annual reports as outputs from these databases showed changes in infections such as the increase of 60.5% in 21 countries in non-O157 *E. coli* infections during the first part of the new century (Anonymous, 2007). In the last quarter of 2007, this network was subsumed into the Food and Waterborne Disease (FWD) unit of the ECDC. The ECDC was set up in 2005 with the aim of providing a coordinated approach to EU-wide surveillance (ECDC, 2008). The FWD unit has continued the work of Enter-net in providing early alerts of unusual events in one country that may have an impact on others. When international outbreaks have been identified, the FWD unit has been involved in the coordination of any investigations undertaken.

2.3.6 International Health Regulations

The IHR to which all WHO Member States have signed up have been in place since 1969 and included a list of six diseases that should be notified to WHO (WHO, 2008a). Prior to this, they were known as the International Sanitary Regulations that were adopted in 1951. The 1969 IHR were amended in 1973 and 1981 to reduce the number of diseases to three (yellow fever, plague and cholera). The revised IHR have been extensively re-worked and were adopted on 23 May 2005 and entered into force on 15 June 2007 (WHO, 2008b). The purpose and scope are 'to prevent, protect against, control and provide a public health response to the international spread of disease in ways that are commensurate with and restricted to public health risks, and which avoid unnecessary interference with international traffic'. The main change is that the notification of diseases under IHR is no longer restricted to a limited number of specific infections. This allows for flexibility for the future as and when new infections or new transmission routes emerge. Annex. 2, which is the decision instrument for the notification of events to WHO, provides the algorithm that should be followed in order to identify what does need to be notified. This list does provide a list of infections that should be considered for notification, but foodborne outbreaks also fit these criteria. An analysis (Kirk, Musto, *et al.*, 2008) of foodborne outbreaks in Australia between 2001 and 2007 showed that seven outbreaks could have been reported under IHR.

In addition to defining how and what should be notified, Annex. 1 defines the core capacity requirements for surveillance and response from local/community level to the national level. These should help all countries achieve the capacity and capability to perform successful laboratory-based, or other, surveillance in the future. The regulations state that the core capacities should be in place by 2012.

2.3.7 WHO International Food Safety Authorities Network

In parallel to the development of the IHR, the WHO developed the International Food Safety Authorities Network (INFOSAN) information network in cooperation with the Food and Agriculture Organization of the United Nations (WHO, 2007). The network brings together the food standards agencies or Ministries of Health involved in food safety within each Member State. INFOSAN facilitates the exchange of information on food safety issues between members. Within INFOSAN lies the INFOSAN emergency system that informs members of food safety events of international concern. It also interacts with the Global Early Warning System for Major Animal Diseases.

2.3.8 Global burden of foodborne diseases

The WHO is also fostering collaborative efforts to improve the knowledge of the burden of foodborne diseases. One project in this field is the initiative to estimate the global burden of foodborne disease. This initiative is an integral part of the WHO global strategy for food safety (WHO, 2002). The initiative has the following four main objectives:

- To strengthen the capacity of countries in conducting burden of foodborne disease assessments and to increase the number of countries who have undertaken a burden of foodborne disease study
- To provide estimates on the global burden of foodborne diseases according to age, gender and regions for a defined list of causative agents of microbial, parasitic and chemical origin
- To increase awareness and commitment among Member States for the implementation of food safety standards
- To encourage countries to use burden of foodborne disease estimates for cost-effective analyses of prevention, intervention and control measures

These objectives should be able to provide accurate estimates as to the true burden of infection due to foodborne diseases within a country that will help in the cost–benefit analyses of any intended interventions. Public health is always faced with having to prioritise the limited resources available and studies to demonstrate the benefits by highlighting the areas in which interventions that can have an impact help in these negotiations.

2.4 Limitations to surveillance activities

There are limitations to all surveillance activities; these can range from insufficient resource to fund sentinel or enhanced surveillance, insufficient reference laboratory capacity and capability in identifying pathogens to political non-prioritorisation of public health activities. These have always been and are likely to remain problematic in the future. There is no single system that can answer all the public health questions that are raised and those that are used have their limitations. Laboratory-based systems are rarely universal in their coverage as has already been shown. Also, some infections may not be looked for in routine diagnostic microbiology. This is certainly the case when identifying non-O157 *E. coli* (Kraigher *et al.*, 2005). These serogroups do cause illness in humans and their under-reporting does not show the real problem with such infections. The true burden of foodborne infections is costly to calculate, but without this, the value of interventions is difficult to quantify. For foodborne pathogens, there certainly remains a need to collaborate fully and cooperate with other colleagues conducting surveillance activities within the rest of the food chain. However, public health can benefit from the implementation of some of the above-mentioned activities and a reduction in the morbidity and mortality associated with foodborne infections would be a bonus to all.

2.5 Future trends

Some of the trends foreseeable in the future might be near-patient testing whereby the identification of the pathogen can be made without sending a sample to a laboratory for characterisation. This may lead to a loss of specificity in diagnoses as full characterisation (often needed for outbreak identification and confirmation) may only be available at national reference laboratories. New technologies will be developed that should lead to better identification of infections. However, these need to be validated, particularly across countries, to ensure that international comparisons of pathogens can be made. The application of web-based reporting for cases and for outbreak investigations will become the norm. New surveillance tools based on Internet activity are becoming more common, a phenomenon that has been highlighted by the pandemic H1N1 activity in 2009 (Keller *et al.*, 2009; Valdivia and Monge-Corella, 2010; van Dijk *et al.*, 2009). There are even applications for mobile phones to track outbreaks (MIT media relations, 2009). Information on outbreaks can be fed into the HealthMap (http://www.healthmap.org/) system from anywhere in the world and helps report and track outbreaks of infectious diseases.

There should be an increased coordination of surveillance between the human, veterinary and food areas to improve coordination and cooperation between these fields. Initial discussions for a collaborative project in the UK and Republic of Ireland have already started to bring these disciplines together. Global databases have been developed that can provide a plethora of valuable data and information to all involved. These initiatives should be developed and improved upon to the benefit of all involved in the surveillance of foodborne infections.

2.6 Conclusions

National and international training is available for epidemiologists in many schemes (the US Epidemic Intelligence Service [http://www.cdc.gov/eis/index.html], the European Programme for Intervention Epidemiology Training [EPIET] [http://ecdc.europa.eu/en/epiet/Pages/HomeEpiet.aspx], the international Training Programs in Epidemiology and Public Health Interventions NETwork [http://tephinet.org/] and national Field Epidemiology Training Programs [http://www.ecdc.europa.eu/EN/ACTIVITIES/TRAINING/Pages/Training_FETP.aspx]), but it is only recently that a new training programme for public health microbiologists (European Public Health Microbiology Training Programme, ECDC) in tandem with epidemiologists in the EPIET programme has been created. In April 2009, a joint ECDC/WHO training course involving epidemiologists, microbiologists and veterinarians was held in Europe. This trend for multi-disciplinary collaboration must continue in the future as public health, particularly for foodborne pathogens, has to involve actors in all these fields. This will enable the achievement of effective and successful solutions to the burden of foodborne infections and the morbidity and mortality that ensues. It is encouraging to see that the WHO (South-East Asia Region), the Food and Agriculture Organization of the United Nations and the World Organisation for Animal Health have recently produced a guide to establishing collaboration between animal and human health sectors at the country level (WHO, 2008c).

There has also been a noticeable increase in the sharing of data and databases on infections across countries, although this does tend to be within single disciplines (Fisher and Threlfall, 2005). All the disease-specific networks in Europe had at their centre international databases. These activities have been taken forward by the ECDC in setting up the European Surveillance System database. This is a database for over 45 infections to which all EU Member States contribute. PulseNet Europe has created a database of PFGE images for *Salmonella*, *E. coli* and *Listeria* that is available to participants from the human, animal and food sectors on a secure website. This sharing of data will improve collaboration and communication across those involved in foodborne pathogen surveillance.

One issue that needs addressing is that of multiple systems that duplicate effort (e.g. GPHIN, MEDYSIS, EPIS). These need to be streamlined so that single sources of epidemic intelligence can be made available to public health professionals. There are public health professionals who have access to some (but not necessarily all) of the many and varied information systems, and a single gateway into all would improve the flow of communications and information significantly. This sounds easy to solve but is in fact problematic, as each of the different systems is 'owned' by different agencies and as such the solution may not be readily achievable.

Bacteria have demonstrated their ability to evolve and adapt, to move to new biological niches that are suited to their needs and allow them to infect people. In the future, they will continue do so. It is imperative that public health practitioners have the flexibility, capacity and capability to respond to these adaptations and develop suitable techniques and alliances to react accordingly. The global market provides opportunities for pathogens to affect people in many countries beyond their place of harvesting or production. Local and national investigators should be vigilant to the sources of outbreaks that have originated from outside their own geographical location or administrative boundary.

As evidenced by the multitude of disciplines involved in the production of this book, public health is not just about microbiology or epidemiology. It must involve collaborators across a range of specialities. In the field of foodborne disease surveillance, this should, as a minimum, include participants from human, veterinary and food agencies within and between countries. Considerable success has already been achieved in bringing together all the relevant disciplines involved in the surveillance of foodborne pathogens. This must not be lost and should be actively encouraged in the future. Foodborne pathogens do not respect national or indeed international boundaries, and similarly, those involved in their surveillance should not feel constrained to limit themselves to their own areas of activity.

2.7 References

ANONYMOUS (2007), *Enter-Net Annual Report: 2005 – Surveillance of Enteric Pathogens in Europe and beyond*, Enter-net surveillance hub, HPA, Centre for Infections, London.

ARKEMA J M, MEIJER A, MEERHOFF T J, VAN DER VELDEN J and PAGET W J (2008), 'Epidemiological and virological assessment of influenza activity in Europe, during the 2006–2007 winter', *Eurosurveillance*, **13**(34), 18958. Available from: http://www.eurosurveillance.org/ViewArticle.aspx?ArticleId=18958 [Accessed 27 May 2010].

BERKELMAN R L and BUEHLER J W (1991), 'Surveillance', in Holland, W.W., Detels, R. and Knox, G. (Eds): *Oxford Textbook of Public Health: Methods of Public Health*, Vol. 2, Oxford University Press, Oxford, pp. 161–176.

CDC (2009), 'Preliminary FoodNet data on the incidence of infection with pathogens transmitted commonly through food – 10 states, 2008', *MMWR*, **58**, 333–7.

CHIEF MEDICAL OFFICER (1992) *Definition of Food Poisoning (PL/CMO (92), 14)*, Department of Health, London.

DE JONG B, ANDERSSON Y, FISHER I S, O'GRADY K A and POWLING J (2001), 'International outbreak of *Salmonella* Typhimurium DT104 – update from Enter-net', *Eurosurveillance*, **5**(32), 1705. Available from: http://www.eurosurveillance.org/ViewArticle.aspx?ArticleId=1705 [Accessed 27 May 2010].

DE WIT M A, KOOPMANS M P, KORTBEEK L M, WANNET W J, VINJÉ J, VAN LEUSDEN F, ET AL. (2001), 'Sensor, a population-based cohort study on gastroenteritis in the Netherlands: incidence and etiology', *American Journal of Epidemiology*, **154**(7), 666–674.

ECDC (2008), *Surveillance of Communicable Diseases in the European Union: a Long-Term Strategy*. Available from: http://www.ecdc.europa.eu/en/aboutus/key%20documents/08-13_kd_surveillance_of_cd.pdf [Accessed 8 Dec 2009].

EUROPEAN COMMISSION (n.d.) *Medical Intelligence in Europe*. Available from: http://ec.europa.eu/health/ph_threats/com/preparedness/medical_intelligence_en.htm [Accessed 4 Jan 2010].

EUROPEAN COMMISSION (n.d.). *The Sixth Framework Programme of Community Activities in the Field of Research, Technological Development and Demonstration (RTD) for the Period 2002 to 2006*. Available from: http://ec.europa.eu/research/fp6/index_en.cfm [Accessed 14 Dec 2009].

FISHER I S T (ON BEHALF OF THE ENTER-NET PARTICIPANTS) (1999), 'The Enter-net international surveillance network – how it works', *Eurosurveillance*, **4**, 52–55.

FISHER I S T, JOURDAN-DA SILVA N, HÄCHLER H, WEILL F-X, SCHMID H, DANAN C, ET AL. (2009), 'Human infections due to *Salmonella* Napoli: a multi-country, emerging enigma recognized by the Enter-net international surveillance network', *Foodborne Pathogens and Disease*, **6**(5), 613–619.

FISHER I S T and THRELFALL E J (ON BEHALF OF THE ENTER-NET AND SALM-GENE PARTICIPANTS) (2005), 'The Enter-net and Salm-gene databases of foodborne bacterial pathogens that cause human infections in Europe and beyond: an international collaboration in surveillance and the development of intervention strategies', *Epidemiology and Infection*, **133**, 1–7.

GALANIS E, LO FO WONG D M A, PATRICK M, BINSZTEIN N, CIESLIK A, CHALERMCHAIKIT T, ET AL. (FOR WHO GLOBAL SALM-SURV) (2006), 'The use of web-based surveillance to describe the global distribution of *Salmonella* serotypes in humans and non-human sources, 2000–02', *Emerging Infectious Diseases*, **12**, 381–388.

GRANT J, WENDELBOE A M, WENDEL A, JEPSON B, TORRES P, SMELSER C, ET AL. (2008), 'Spinach-associated *Escherichia coli* O157:H7 outbreak, Utah and New Mexico, 2006', *Emerging Infectious* Diseases, **14**(10), 1633–1636.

GREENE S K, DALY E R, TALBOT E A, DEMMA L J, HOLZBAUER S, PATEL N J, ET AL. (2008), 'Recurrent multistate outbreak of *Salmonella* Newport associated with tomatoes from contaminated fields, 2005', *Epidemiology and Infection*, **136**(2), 157–165.

HORBY P W, O'BRIEN S J, ADAK G K, GRAHAM C, HAWKER J I, HUNTER P, ET AL. (2003), 'A national outbreak of multi-resistant *Salmonella enterica* serovar Typhimurium definitive type (DT) 104 associated with consumption of lettuce', *Epidemiology and Infection*, **130**(2), 169–178.

JOHNSON S (2006), *The Ghost Map: The Story of London's Most Terrifying Epidemic – and How it Changed Science, Cities and the Modern World*, Riverhead Books, New York, NY, p. 206.

JONES K E, PATEL N G, LEVY M A, STOREYGARD A, BALK D, GITTLEMAN J L, ET AL. (2008), 'Global trends in emerging infectious diseases', *Nature*, **451**, 990–993.

KANSAS STATE UNIVERSITY (n.d.) *Bites – Safe Food from Farm to Fork*. Available from: http://bites.ksu.edu/ [Accessed 3 Jan 2010].

KELLER M, BLENCH M, TOLENTINO H, FRIEFELD C C, MANDL K D, MAWUDEKU A, ET AL. (2009), *Use of Unstructured Event-Based Reports for Global Infectious Disease Surveillance*. Available from: http://www.cdc.gov/EID/content/15/5/689.htm [Accessed 23 Apr 2009].

KILLALEA D, WARD L R, ROBERTS D, DE LOUVOIS J, SUFI F, STUART J M, ET AL. (1996), 'International epidemiological and microbiological study of outbreak of *Salmonella* Agona infection from a ready to eat savoury snack – I: England and Wales and the United States', *BMJ*, **313**, 1105–1107.

KIRK M D, LITTLE C L, LEM M, FYFE M, GENOBILE D, TAN A, ET AL. (2004), 'An outbreak due to peanuts in their shell caused by *Salmonella enterica* serotypes Stanley and Newport – sharing molecular information to solve international outbreaks', *Epidemiology and Infection*, **132**, 571–577.

KIRK M D, MCKAY I, HALL G V, DALTON C B, STAFFORD R, UNICOMB L, ET AL. (2008), 'Foodborne disease in Australia: the OzFoodNet experience', *Clinical Infectious Diseases*, **47**(3), 392–400.

KIRK M D, MUSTO J, GREGORY J and FULLERTON K (2008) 'Obligations to report outbreaks of foodborne disease under the International Health Regulations (2005),', *Emerging Infectious Diseases*, **14**, 1440–1442.

KRAIGHER A, SEME K, KRT-LAH A and FISHER I (2005), 'Fatal case of HUS after VTEC *E. coli* O145 infection in Slovenia highlights importance of testing for this rare strain', *Eurosurveillance Weekly*, **10**, 150905.

LARSSON J T, TORPDAHL M, PETERSEN R F, SØRENSEN G, LINDSTEDT B A and NIELSEN E M (2009), *Development of a New Nomenclature for Salmonella Typhimurium Multilocus Variable Number of Tandem Repeats Analysis (MLVA). Eurosurveillance*, **14**(15), 19174. Available from: http://www.eurosurveillance.org/ViewArticle.aspx?ArticleId=19174 [Acccessed 27 May 2010].

LEWIS H C, ETHELBERG S, OLSEN K E, NIELSEN E M, LISBY M, MADSEN S B, ET AL. (2009), 'Outbreaks of Shigella sonnei infections in Denmark and Australia linked to consumption of imported raw baby corn', *Epidemiology and Infection*, **137**(3), 326–334.

LINDSTEDT B A, VARDUND T, AAS L and KAPPERUD G (2004), 'Multiple-locus variable-number tandem-repeats analysis of *Salmonella enterica* subsp. *enterica* serovar Typhimurium using PCR multiplexing and multicolor capillary electrophoresis', *Journal of Microbiological Methods*, **59**(2), 163–172.

LITTLE C L, SURMAN-LEE S, GREENWOOD M, BOLTON F J, ELSON R, MITCHELL R T, ET AL. (2007), 'Public health investigations of *Salmonella* Enteritidis in catering raw shell eggs, 2002–2004', *Letters in Applied Microbiology*, **44**(6), 595–601.

MAUNULA L, ROIVAINEN M, KERÄNEN M, MÄKELÄ S, SÖDERBERG K, SUMMA M, ET AL. (2009), 'Detection of human norovirus from frozen raspberries in a cluster of gastroenteritis outbreaks', *Eurosurveillance*, **14**(49), 19435. Available from: http://www.eurosurveillance.org/ViewArticle.aspx?ArticleId=19435 [Accessed 27 May 2010].

MEAD P, SLUTSKER L, DIETZ V, MCCAIG L, BRESEE J, SHAPIRO C, ET AL. (1999), 'Food-related illness and death in the United States', *Emerging Infectious Diseases*, **5**, 607–625.

MED-VET-NET (2009), *A European Network of Excellence Working for the Prevention and Control of Zoonoses and Foodborne Diseases*. Available from: http://www.medvetnet.org/cms/ [Accessed 4 Jan 2010].

MIT MEDIA RELATIONS (2009), *Outbreaks Near Me App Now Available for Android Mobile Phones*. Available from: http://web.mit.edu/press/2009/outbreaks-app.html [Accessed 2 Dec 2009].

O'FLANAGAN D, CORMICAN M, MCKEOWN P, NICOLAY N, COWDEN J, MASON B, ET AL. (2008), 'A multi-country outbreak of *Salmonella* Agona', *Eurosurveillance*, **13**(33), 18956. Available from: http://www.eurosurveillance.org/ViewArticle.aspx?ArticleId=18956 [Accessed 17 Nov 2009].

PAQUET C, COULOMBIER D, KAISER R and CIOTTI M (2006), 'Epidemic intelligence: a new framework for strengthening disease surveillance in Europe', *Eurosurveillance*, **11**(12), 665. Available from: http://www.eurosurveillance.org/ViewArticle.aspx?ArticleId=665 [Accessed 16 Nov 2009].

PEZZOLI L, ELSON R, LITTLE C L, YIP H, FISHER I, YISHAI R, ET AL. (2008), 'Packed with *Salmonella* – investigation of an international outbreak of *Salmonella* Senftenberg infection linked to contamination of pre-packed basil in 2007', *Foodborne Pathogena and Disease*, **5**(5), 661–668.

PROMED-MAIL (n.d.) *About PROMED-mail*. Available from: http://www.promedmail.org/pls/apex/f?p=2400:1950 [Accessed 31 Dec 2009].

PUBLIC HEALTH AGENCY OF CANADA (2004), *Global Public Health Intelligence Network (GPHIN)*. Available from: http://www.phac-aspc.gc.ca/media/nr-rp/2004/2004_gphin-rmispbk-eng.php#tphp [Accessed 4 Jan 2010].

SWAMINATHAN B, BARRETT T, HUNTER S, TAUXE R and FORCE C P T (2001), 'PulseNet, the molecular subtyping network for foodborne bacterial disease surveillance, United States', *Emerging Infectious Diseases*, **7**(3), 382–389.

SWAMINATHAN B, GERNER-SMIDT P, NG L-K, LUKINMAA S, KAM K-M, ROLANDO S, ET AL. (2006), 'Building PulseNet international: an interconnected system of laboratory networks to facilitate timely public health recognition and response to foodborne disease outbreaks and emerging foodborne diseases', *Foodborne Pathogens and Disease*, **3**, 36–50.

TAUXE R V (2006), 'Molecular subtyping and the transformation of public health', *Foodborne Pathogens and Disease*, **3**, 4–8.

VALDIVIA A and MONGE-CORELLA S (2010), 'Diseases tracked by using Google trends, Spain', *Emerging Infectious Diseases*, **16**(1). Available from: http://www.cdc.gov/EID/content/16/1/168.htm [Accessed 23 Dec 2009].

VAN DIJK A, ARAMINI A, EDGE G and MOORE K M (2009), 'Real-time surveillance for respiratory disease outbreaks, Ontario, Canada', *Emerging Infectious Diseases*, **15**(5), 799–801.

WENDEL A M, JOHNSON D H, SHARAPOV U, GRANT J, ARCHER J R, MONSON T, ET AL. (2009), 'Multistate outbreak of *Escherichia coli* O157:H7 infection associated with consumption of packaged spinach, August–September 2006: the Wisconsin investigation', *Clinical Infectious Dis*eases, **48**(8), 1079–1086.

WERBER D, DREESMAN J, FEIL F, VAN TREECK U, FELL G, ETHELBERG S, ET AL. (2005), 'International outbreak of *Salmonella* Oranienburg due to German chocolate', *BMC Infectious Diseases*, **5**, 7.

WHEELER J G, SETHI D, COWDEN J M, WALL P G, RODRIGUES L C, TOMPKINS D S, ET AL. (ON BEHALF OF THE INFECTIOUS INTESTINAL DISEASE STUDY EXECUTIVE) (1999), 'Study of infectious intestinal disease in England: rates in the community, presenting to general practice, and reported to national surveillance', *BMJ*, **318**, 1046–1050.

WHO (2002), *WHO Global Strategy for Food Safety: Safer Food for Better Health*. Available from: http://www.who.int/foodsafety/publications/general/global_strategy/en/ [Accessed 3 Jan 2010].

WHO (2007), *WHO INFOSAN Background and Information Notes*. Available from: http://www.who.int/foodsafety/fs_management/infosan_1007_en.pdf [Accessed 4 Dec 2009].
WHO (2008a), *The Global Burden of Disease: 2004 Update*. Available from: http://www.who.int/healthinfo/global_burden_disease/GBD_report_2004update_full.pdf [Accessed 9 Dec 2009].
WHO (2008b), *International Health Regulations*. Available from: http://www.who.int/ihr/9789241596664/en/index.html [Accessed 14 Dec 2009].
WHO (2008c), *Zoonotic Diseases: a Guide to Establishing Collaboration between Animal and Human Health Sectors at the Country Level*. Available from: http://www.searo.who.int/LinkFiles/Publication_Zoonotic.pdf [Accessed 14 Dec 2009].

3

Systems for real-time, linked foodborne pathogen surveillance

P. Gerner-Smidt, Centers for Disease Control and Prevention, USA

Abstract: This chapter describes the development of systems for real-time, linked foodborne pathogen surveillance using Salm-net/Enter-net and PulseNet as primary examples. Such systems should be seen in the context of the new International Health Regulations by the World Health Organization that were enacted in 2007 and set the rules for the reporting of public health events of international importance including foodborne infections. In order to fulfill the obligations established in these regulations, each country needs to have an efficient surveillance infrastructure in place for acute public health events. A core element in this respect is a system for real-time food pathogen surveillance from the farm to the fork. Most current systems focus on public health surveillance or surveillance of food and animals, although integrated systems have been implemented in some countries.

Key words: surveillance, integration, farm-to-fork, outbreaks, Salm-net, Enter-net, PulseNet, databases, networks.

3.1 Introduction

Foodborne infections are common in all parts of the world and although most infections are self-limiting bouts of gastroenteritis, they pose by their sheer numbers a significant burden to public health. Contamination of food may occur at any level in the food production 'farm-to-fork' continuum: at the point of production on the farm, during industrial processing, distribution, retail or in the home of the consumer. Additionally, foodborne infections do not respect any borders. Food is today often produced in one part of the world, exported to and consumed in a different country thousands of miles away. A foodborne infection may therefore have its ultimate origin in a different part of the world than the one where it was consumed.

Foodborne infections may be sporadic or occur in outbreaks. Outbreaks are acute events, and their rapid recognition and investigation is essential to identify their source, control them and prevent further disease. Whereas an outbreak investigation is a team work effort between microbiologists and epidemiologists, the most efficient method of outbreak detection is through active laboratory-based surveillance. Such surveillance systems may efficiently facilitate the recognition of both localized and diffuse outbreaks. Their efficiency depends on the speed of reporting, the completeness of the reporting, i.e. both the fraction of infections reported and the coverage of the food production chain, and the discriminatory power of the microbiological methods employed. The most efficient ones are the real-time laboratory-based surveillance systems that bridge public health and the full food production chain.

With the implementation of the World Health Organization's International Health Regulations (IHR, 2005) (World Health Organization, 2007a, 2007b), national foodborne infection events with international implications have become notifiable according to the decision tree presented in Fig. 3.1. In most instances foodborne infection events are notifiable if at least two of the following questions are answered affirmatively:

1 Is the public health impact of the event serious?
2 Is the event unusual or unexpected?
3 Is there a significant risk of international spread?
4 Is there significant risk of international travel or trade restrictions?

All countries are required to establish and maintain national surveillance systems for foodborne infections. The IHR are legally binding and may therefore provide the legal incentive to establish national and international systems for real-time, linked foodborne pathogen surveillance. This chapter will describe two such systems in detail, including an assessment of their strengths and weaknesses, and the author will give his view on what such a system ideally looks like and the prospects of getting there.

3.2 Models for real-time linked foodborne pathogen surveillance: the Salm-net/Enter-net model

Salm-net was one of the first international public health surveillance networks for foodborne infections. It was established in 1994 with 15 members from countries in the European Union (EU), Norway and Switzerland under the leadership of scientists from the Public Health Laboratory Service (PHLS, now Health Protection Agency, HPA), London, UK (Fisher, 1995). The members were epidemiologists and microbiologists from the national public health institute in each country responsible for surveillance of salmonellosis. The aim of Salm-Net was to prevent human *Salmonella* infections within the EU by strengthening international laboratory-based human *Salmonella* surveillance and creating an on-line European case-based database of compatible data available to all participants. Additionally, the phage typing schemes for the most common *Salmonella* serotypes, notably ser. Enteritidis and ser. Typhimurium, were

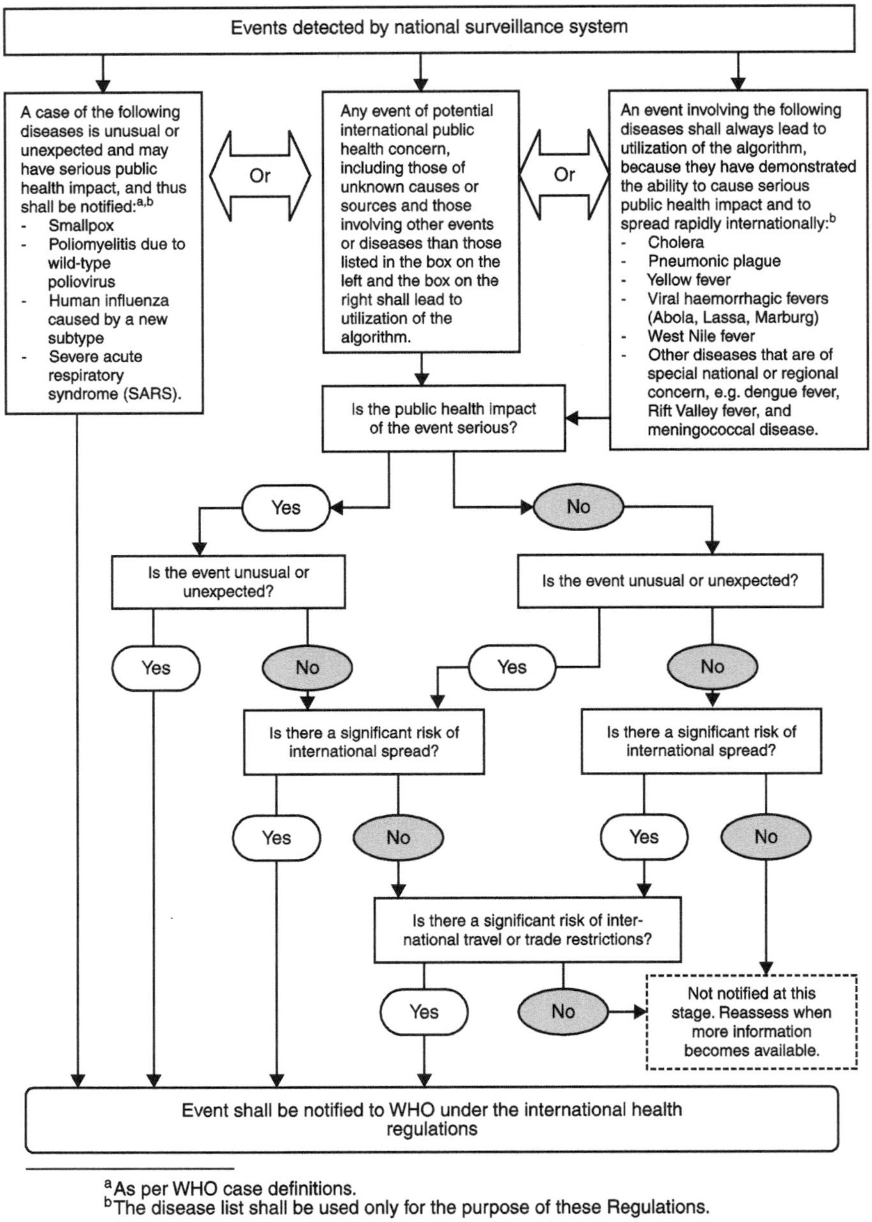

Fig. 3.1 Decision tree for reporting public health event of international importance to the World Health Organization according to the International Health Regulations (2005) (World Health Organization, 2007b) (reprinted with the permission of WHO).

harmonized and implemented in all participating countries. Finally, a moderated email rapid alert system was created where the participants could alert each other about national outbreaks with possible international expansion through the coordinating hub at PHLS.

The rapid alert system almost immediately showed its ability to recognize international outbreaks of foodborne infections; the first one was paradoxically caused by an agent not included in the project charter, *Shigella sonnei* (Fisher, 1995; Kapperud *et al.*, 1995). This outbreak occurred in Norway, Sweden and the UK and was traced to contaminated iceberg lettuce imported from Spain. During the first year of Salm-net an outbreak of *Salmonella* serotype Livingstone involving multiple countries associated with travel to Tunisia was retrospectively discovered through analysis of information submitted to the central hub in Colindale (Fisher and Enter-net participants, 1997). However, this was the only outbreak that was detected by analysis of data in the central database in the lifetime of the network.

In 1998, the scope of the network was expanded to include Shiga toxin-producing *Escherichia coli* (STEC, verocytotoxin producing *E. coli*, VTEC), especially the serotype O157:H7/H- (STEC O157) and the name of the network was changed to Enter-net (Fisher, 1999). At the same time, antimicrobial susceptibility data were added to the *Salmonella* reporting (Threlfall *et al.*, 1999) and an external quality assurance system for VTEC serotyping was established. By this time four non-European countries, Australia, Canada, Japan and South Africa, had joined the network. When the Enter-net contract was renewed with the European Commission in 2002, reporting of aggregated data on *Campylobacter* infections was added to the system (Fisher and Meakins, 2006). From 2003, the membership was expanded to include nine future member countries of the EU and Iceland (Fisher and Threlfall, 2005).

In 2007, the responsibility for the Enter-net surveillance was taken over by the Food and Waterborne Disease team at the newly established European Center for Disease Prevention and Control (ECDC) in Stockholm (http://ecdc.europa.eu/en/Activities/Surveillance/Enter-net/) and the data gradually integrated into the European Surveillance System database, TESSy, where the data will be available online to the participants.

3.2.1 Strengths of Salm-net/Enter-net

The major accomplishment of Salm-net and Enter-net was to bring microbiologists and epidemiologists together in a fruitful collaboration nationally and internationally. This is important because the surveillance of foodborne infections in some countries is the responsibility of the national reference laboratories and in others is coordinated by the state epidemiologists; it would likely not have been possible to introduce a comprehensive European surveillance of foodborne infections using either source alone. A key to the success of the network was the annual workshops where all the participants got to meet each other and discussed national and international technical and scientific experiences from the past year and strategies for the future of the network; this created the trust and the strong sense of common responsibility and ownership of the network between its members, which was a prerequisite for the numerous successful outbreak investigations performed by the network (Fisher, 1999; Fisher and Threlfall, 2005); some of these investigations also involved collaborations with non-EU countries (Killalea *et al.*, 1996; Lewis *et al.*, 2007; Mahon *et al.*, 1997; Pezzoli *et al.*, 2008; Threlfall *et al.*, 1998; VanBeneden *et al.*, 1999; Werber *et al.*, 2005) further illustrating the international impact of the network.

During different project phases the laboratory and epidemiological methods were harmonized and a minimum laboratory capacity was ensured in each country. Although molecular surveillance was never used for real-time surveillance as in PulseNet (see below) in the network, laboratory capacity for pulsed field gel-electrophoresis (PFGE) and other molecular subtyping methods using PulseNet-compatible protocols was developed in a project, Salm-gene, headed by the Enter-net coordinators (Fisher and Threlfall, 2005). Finally, Salm-net and Enter-net were instrumental in documenting changing trends in the occurrence of different *Salmonella* sero- and phage types in Europe (Fisher, 2004a, 2004b).

3.2.2 Weaknesses of Salm-net/Enter-net

The timeline from when a patient falls ill until his/her disease is reported to Salm-net/Enter-net may be divided into five phases with the typical duration of each phase as shown in Fig. 3.2(a). The phases are: (1) the patient needs to see a doctor and have a specimen collected, (2) the specimen needs to be cultured, (3) the culture needs to be forwarded to a public health laboratory, and (4) the isolate

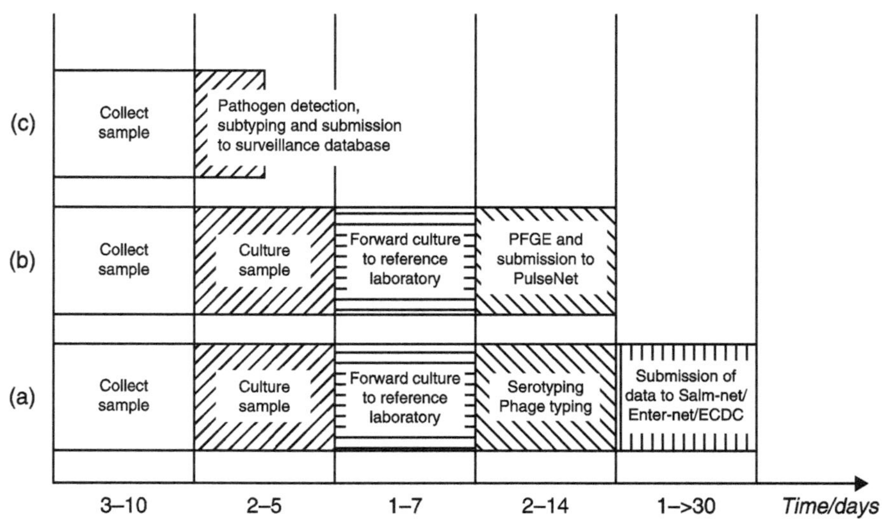

Fig. 3.2 Phases in the timeline from onset of illness until submission of data: (a) to Salm-net/Enter-net, (b) to PulseNet, and (c) to an optimal future real-time surveillance system.

needs to be subtyped, e.g. by serotyping or phage typing, and (5) finally, the results need to be reported to Salm-net/Enter-net. Although the aim of Salm-net/Enter-net was to achieve weekly updating of the network database, this goal was never achieved. At best, the participants uploaded data to the network on a monthly basis but not all participants were able to meet this goal. Additionally, despite good intentions by the network coordinators, the central database never became available online to the participants; although these are also the goals for the future surveillance under the auspices of ECDC, they have still not been achieved. This lack of timeliness of submission of data and lack of direct access to the central database by the participants limit its usefulness for outbreak investigations. Furthermore, the database only contains phenotypic information about the pathogens involved, i.e. sero-, phagetyping and antibiograms, which in general lack the high discriminatory power of molecular subtyping methods like PFGE that are often crucial for detection of clusters of infections and identification of case-patients in outbreaks. These shortfalls have been partially compensated for by the rapid alert system; however, even though the moderation of this email list ensures the uniformity and quality of the notifications, it also slows down the alerts. Since molecular subtyping is not an integral part of the European surveillance, it is not initiated in many countries before an alert has been received, additionally adding to the delay in the outbreak investigations.

Finally, a significant drawback of the European surveillance is that microbiological data from food and veterinary sources are not routinely integrated with the public health data, except at the national level in some countries; the prime example in this context being Denmark where data have been integrated in the Danish Zoonosis Centre since 1994 to ensure national compliance with the European 'Zoonosis Directive' (Anonymous, 2007a; European Commission, 1992). A growing number of countries in Europe and elsewhere have followed the Danish example of integration of public health, food and veterinary data on foodborne pathogens.

Presently, no real-time collection of data from the full farm-to-fork continuum takes place at the EU level. However, such data are gathered and published on an annual basis in the community summary report on trends and sources of zoonoses, zoonotic agents, antimicrobial resistance and foodborne outbreaks in the EU by the European Food Safety Authority (EFSA) and ECDC (Anonymous, 2007b).

3.3 Models for real-time linked foodborne pathogen surveillance: the PulseNet model

In the USA, a different model has been used to ensure timeliness of foodborne pathogen surveillance. This model is PulseNet, the national molecular subtyping network for foodborne bacterial disease surveillance, coordinated by CDC with the assistance of the Association of Public Health Laboratories (APHL) (Gerner-Smidt *et al.*, 2006; Swaminathan *et al.*, 2001). PulseNet was initiated in 1996 in response to an increasing number of outbreaks caused by STEC O157. Before

1996, STEC O157 outbreaks were detected locally by alert clinicians and/or through laboratory surveillance of foodborne pathogens identified to the species or serotype level. Once detected, outbreak-related isolates were forwarded to CDC for confirmation by more highly discriminatory methods. By the beginning of the 1990s, PFGE had become the new gold standard for highly discriminatory molecular subtyping of bacteria and was the method used by CDC. A drawback of the method was that only isolates subtyped on the same gel could reliably be compared. However, this changed with the informatics technology revolution and the introduction of software that enabled reliable comparison of PFGE patterns generated using the same subtyping procedure with appropriate reference profiles included on each gel. The software also enabled the creation of library databases containing not only PFGE patterns but also other strain-related information, demographic information and other phenotypic or molecular characteristics, e.g. source, serotype, phage type and antimicrobial susceptibility of the strain. However, isolates still had to be submitted to CDC for subtyping. With the introduction of PulseNet this changed. In this network, the subtyping procedure was decentralized out to public health laboratories in the states. The PFGE patterns produced at the local level were forwarded by email to the PulseNet database team at CDC to be analyzed, entered into a national database and then compared to each other and to patterns from other gels and laboratories. Crucial to the success of PulseNet was the use of the same highly standardized PFGE procedure. Many resources were allocated to develop and validate robust PFGE procedures and train the network participants in using them. One of the problems with PFGE at the time of introduction of PulseNet was the turn around time to perform the procedure; it was four or more working days for most published protocols, which was suboptimal for outbreak investigations where information needs to be available as soon as possible. When the first one-day PFGE protocol for subtyping of STEC O157 was introduced (Gautom, 1997), it was immediately adapted for PulseNet (Ribot *et al.*, 2006) and all PulseNet PFGE protocols developed since then have a one day turn around time.

The network started up with four participating state public laboratories in addition to the CDC laboratory but despite its small size, the utility of PulseNet in outbreak investigations became apparent almost immediately:

- PulseNet facilitated the detection of patient case-clusters that could represent common source outbreaks.
- PulseNet data were useful in the case definition to separate outbreak-related patients from sporadic cases infected with the same bacterial species or serotype.
- PulseNet provided microbiological evidence for the vehicle of an outbreak by confirming the presence of the outbreak subtype in the implicated food product.

Because of its national coverage, PulseNet could identify case-patients in different states belonging to the same outbreak. This was first shown in 1996 during an STEC O157 outbreak in the Northwestern states traced to unpasteurized apple juice (Cody *et al*., 1999). With centralized subtyping, this was impossible to do in real-time.

As the utility of the network became apparent, it grew rapidly both in number and types of participating laboratories, and the number of bacteria covered. Full national coverage was reached in 2001. Presently, the network has 75 participants from all state, and a number of city and county, public health laboratories as well as federal food regulatory agency laboratories from the United States Department of Agriculture (USDA) and the Food and Drug Administration (FDA).

PulseNet standardized PFGE protocols are available for *E. coli* O157, *Salmonella*, *Shigella*, *Listeria monocytogenes*, *Campylobacter jejuni/coli*, *Vibrio cholerae*, *Vibrio parahaemolyticus*, *Clostridium perfringens* (Cooper *et al*., 2006; Graves and Swaminathan, 2001; Maslanka *et al*., 1999; Parsons *et al*., 2007; Ribot *et al*., 2001, 2006) and the bioterrorism agent *Yersinia pestis* (http://www.cdc.gov/pulsenet/protocols.htm).

Today, PulseNet uses a customized version of the BioNumerics client/server software (Applied Maths, Sint-Martens-Latem, Belgium) to analyze the data generated by the participants. Each participant has their own client database where they enter and analyze their own data before uploading them through the internet to the national database on a server at CDC. As part of the analysis, the participants compare new profiles with existing ones in their database and in the national database; if a cluster of isolates that is clearly above the historical baseline is detected, the participant will notify the epidemiologists so that an outbreak investigation may be initiated; at the same time, the participant will notify the whole PulseNet community on the unmoderated PulseNet listserv. All clusters are confirmed by the database team at CDC, who also will check the national database for uploads of indistinguishable patterns from other states. In addition, the database team conducts their own national search every week, looking for clusters of patterns uploaded within the past 60 days (120 days for *Listeria*). All clusters identified are discussed with the epidemiologists at CDC, the PulseNet participants in the involved states are notified directly and the cluster is posted on the PulseNet listserv.

Because of its proven utility, PulseNet has become the *de facto* standard for molecular surveillance of foodborne infections. The model was used when the veterinary counterpart of PulseNet, the USDA VetNet, was established in 2003 (Jackson *et al*., 2007). The USDA VetNet database contains data on *Salmonella* and *Campylobacter* from diagnostic animal samples from veterinary sentinel laboratories and non-diagnostic samples originating from the animal arm of the National Antimicrobial Resistance Monitoring System (NARMS) (http://www.cdc.gov/NARMS/) and FSIS regulatory sampling of slaughter and processing plants. Since less than 10% of the data in the PulseNet databases originate from isolates from food, animals and the environment obtained during outbreak investigations and legally required surveillance of the food chain, the USDA VetNet complements the molecular surveillance of foodborne pathogens with more food and veterinary data. When a cluster of human isolates is detected in PulseNet, the cluster pattern is immediately compared with all non-human patterns in the PulseNet and the USDA VetNet databases in order to shed light on the source of cluster. Sometimes, linking human with non-human data provides microbiological evidence of the source of an outbreak even before it has been detected. This happened in the summer of 2002, when a sample of ground beef from a plant in Colorado contained an isolate of STEC O157 and a cluster of cases shortly afterwards was identified in patients in several states with a PFGE pattern that was indistinguishable from that of the meat isolate. A rapid epidemiological investigation confirmed the link; even though some meat had already been recalled, the recall was extended to include more batches based on the epidemiological evidence from the patients involved and the outbreak stopped (Gerner-Smidt *et al.*, 2005). The information in the PulseNet and USDA VetNet databases may also be used to study how and which food regulatory events, e.g. non-compliance with federal regulations, may lead to public health events, e.g. outbreaks, and to study the dynamics behind the emergence of new pathogenic strains. Such knowledge could be very useful to guide future food regulatory and public health prioritizations and interventions.

The international dimension of PulseNet became evident for the first time in 1998. That year, PulseNet linked shigellosis outbreaks in four states and two Canadian provinces to the same source, parsley, imported from the same producer in Mexico (CDC, 1999). Since then, sister PulseNet networks have been created in different regions of the world inspired by PulseNet USA (Swaminathan *et al.*, 2006). Currently PulseNet networks are operational in the Asia–Pacific region, Canada, Europe, Latin America and the Caribbean, and the Middle East. The networks collaborate under the umbrella of PulseNet International and investigate international clusters and outbreaks. They also collaborate in the development and validation of new methods for PulseNet (Cooper *et al.*, 2006; Kam *et al.*, 2008). The capacity and the surveillance infrastructure for foodborne infections are unevenly developed in different parts of the world and consequently the international PulseNet networks do not function the exact same way. In some regions, PulseNet is an adjunct to existing surveillance activities; in other regions the PulseNet network closely resembles the network in the USA. PulseNet Canada is an example of the latter. PulseNet USA and PulseNet Canada have direct access to and routinely query each other's database when a national cluster has been identified. This has proven very useful since the USA and Canada share the same food supply. In 2007, a large STEC O157 outbreak involving both countries was investigated (http://www.cdc.gov/ecoli/2007/october/100207.html; http://www.pulsenetinternational.org/pulsenet/casestudies.asp). The investigation led to the recall of 21.4 million pounds of frozen ground beef patties from a company in the USA that had imported the meat from a Canadian company.

3.3.1 Strengths of the PulseNet model

PulseNet has revolutionized the recognition of outbreaks of foodborne infections in the USA. Before PulseNet, laboratory surveillance was based on phenotypic information about the bacterial species and sometimes also the serotype gathered at the state level. These data were submitted at irregular intervals to CDC making it difficult and slow to recognize multi-state outbreaks. Local outbreaks were detected slowly and with a low sensitivity except for outbreaks associated with specific events, e.g. church suppers. With the introduction of routine high-discriminatory subtyping by PFGE at the state level, outbreaks that would not have been recognized by low-discriminatory subtyping methods are now identified; additionally, PulseNet facilitates the recognition of links between outbreaks occurring at the same time in different states; an example of this is a STEC O157 multistate outbreak that occurred in 2006 (CDC, 2006); it was detected simultaneously in Wisconsin and Oregon and was immediately linked by PulseNet to each other and to a third cluster in New Mexico. A joint epidemiological investigation rapidly identified fresh bagged baby spinach as the vehicle; PulseNet data facilitated the identification of cases in a total of 26 states. Before PulseNet, such clusters would have been investigated as separate outbreaks independently of each other and the link between them might never have been identified.

If the epidemiological follow-up of patients with foodborne infections is done concurrently with the real-time subtyping performed by PulseNet, common-source outbreaks involving very few cases may be detected and controlled. In Minnesota, patients diagnosed with a foodborne infection are interviewed using a standardized questionnaire as soon as they are reported to the state public health department and at the same time as the isolates are subtyped by PFGE. In June 2003, clinical isolates of STEC O157 with indistinguishable PFGE patterns were isolated from two patients in this state; from the questionnaires administered to the patients it was determined that both had consumed steaks purchased from a door-to-door vendor (Laine *et al.*, 2005) and an investigation was initiated. It turned out that the steaks had originated from the same vendor and the outbreak ceased following a recall the implicated meat. A total of only ten culture-confirmed cases were identified in five states during this outbreak.

The PulseNet database has also proven invaluable in linking seemingly sporadic case-patients infected by more common strains or case-patients separated widely in time or geographical location retrospectively. In 2007, CDC and state epidemiologists and other public health officials investigated a large multistate outbreak caused by *Salmonella* serotype Wandsworth associated with consumption of a puffed snack product (CDC, 2007). This outbreak was detected by prospective cluster searching by PulseNet. An outbreak of 18 cases infected with serotype Typhimurium within the large Wandsworth outbreak was initially overlooked because of its small size and the commonality of the pathogen, but was detected when samples of the implicated product were examined; the outbreak strain of serotype Wandsworth was found along with strains of six other serotypes including ser. Typhimurium. All strains were subtyped by PFGE and compared with the PulseNet database. Patients infected with strains displaying a pattern indistinguishable from any of the patterns of the isolates from the implicated product were interviewed and the smaller outbreak of serotype Typhimurium was identified.

PulseNet was also used to track a different outbreak of *Salmonella* serotype Typhimurium that was associated with handling of infected pet rodents that occurred over a 10-month period in 2003 and 2004 (CDC, 2005; Swanson *et al.*, 2007); 22 cases were identified when the PFGE pattern of a Typhimurium strain obtained from dead pet hamsters from a distributor in Minnesota in August 2004 was compared against the PulseNet database; upon interview all these patients had contacts to pet rodents. Whereas the first example illustrates the power of PulseNet to identify the extent of an outbreak to a fuller extent than is possible by prospective real-time surveillance alone, the second example illustrates how PulseNet was critical in identifying a link between what was thought only to be a veterinary problem to human disease.

3.3.2 Weaknesses of the PulseNet model

The timeline for reporting to PulseNet is shorter than in the Salm-net/Enter-net model. The length of sample collection, culture and culture-forwarding phases are the same for Salm-net/Enter-net and PulseNet. The subtyping phase in PulseNet consists of PFGE. Since the submission of the subtyping data occurs at the same time as the analysis is conducted in the local reference laboratory, the final submission phase does not exist (Fig. 3.2b). Nevertheless, the reporting delay that is inherent to all outbreak surveillance system also exists in PulseNet. It takes approximately two weeks from the time the first patient falls ill with an STEC O157 infection until a PulseNet cluster signal is generated (CDC, 2006) (Fig. 3.3). For *Salmonella* infections, the delay is even longer, approximately three

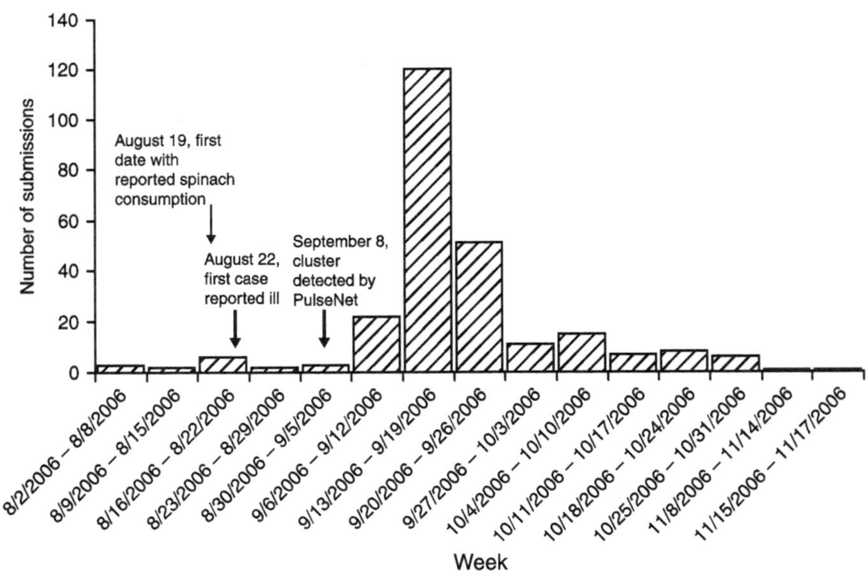

Fig. 3.3 Reporting delay of the PulseNet case cluster of *E. coli* O157:H7 representing an outbreak associated with consumption of fresh baby spinach (CDC, 2006). The histogram shows the number of submissions of the PFGE outbreak pattern to PulseNet by date. Note that patterns from isolates unrelated to the outbreak and indistinguishable from the outbreak pattern were submitted before the outbreak.

weeks (CDC, 2008). These delays are dependent on the resources available to the subtyping laboratories and include the time needed for forwarding of the isolates to the public health laboratories, subtyping and analysis. It is one of the goals for PulseNet laboratories to subtype all STEC O157 strains within four working days of receipt of the strains. No goal for *Salmonella* has been identified since not all laboratories have sufficient resources available to subtype all *Salmonella* isolates and salmonellosis in general is not as severe a disease as that caused by STEC O157. Many laboratories therefore batch the subtyping of salmonellas, which on average delays their submission to PulseNet by another week.

Even though results may be generated within 24 hours with PFGE, the preferred subtyping method in PulseNet, it is resource demanding and the results are not always easy to interpret. This is especially the case as the databases grow bigger. In 2008, PulseNet was forced to limit participants' access to its *Salmonella* database to the previous 2.5 years because access to the PulseNet server became severely impaired if more than a few participants accessed it at the same time leading to slow reaction times and frequent time-outs. PulseNet is therefore implementing new sequence-based methods like multi-locus variable number of tandem repeat analysis (MLVA) and single nucleotide polymorphism (SNP) analysis. The results of these methods are a sequence of numbers or characters, which require less computer resources for storage and analysis than PFGE. Additionally, more isolates may be subtyped in less time than by using PFGE, thereby potentially reduce the subtyping phase (Fig. 3.2) in PulseNet to one to two days. These new methods are described in more detail in Chapter 4.

The reason why PulseNet has not moved away from PFGE at the moment is that the new methods are not as universal as PFGE. The same PFGE protocol may be used to subtype a broad range of pathogens, e.g. all *Salmonella* serotypes, whereas sequence-based methods are much more specific and a large number of protocols will need to be developed and validated to cover the same range of pathogens as PFGE, e.g. the PulseNet MLVA protocol for STEC O157 (Hyytia-Trees *et al.*, 2006) will not work for any other STEC serotype.

Finally, PulseNet is a laboratory network and therefore only represents one side of the outbreak investigation. An outbreak investigation is the cross-disciplinary effort of laboratorians and epidemiologists. Today, PulseNet identifies more case-clusters than can be followed up epidemiologically by CDC and state epidemiologists with the available resources. Table 3.1 shows the number of case-clusters identified in the past five years. Approximately half of these are followed up actively; the other half are followed passively until the number of submissions of the cluster pattern to PulseNet drops without intervention. In approximately 10% of clusters an outbreak investigation leads to the identification of a common source. However, these numbers need to be interpreted with care. Since no subtyping method can efficiently separate all epidemiologically linked isolates from sporadic isolates, not all clusters represent common source outbreaks but are rather an accidental clustering of sporadic cases by a less than perfect subtyping method. Nevertheless, there are probably more common source outbreaks being detected than can be successfully investigated. A recent study (Hedberg *et al.*,

2007) looked at clusters identified retrospectively during a five-year period in the French molecular surveillance system for listeriosis to identify characteristics that could be used to predict the potential for identification of a common source. The results suggested that clusters with more than three cases identified within a six-week period and with six or more cases involved in total were more likely to have a common source identified. This approach could also be used to establish algorithms to be used for selecting PulseNet clusters of any pathogen for follow-up.

3.4 Future trends

The ideal system for real-time, linked foodborne pathogen surveillance is a network of integrated national, regional and global databases containing current information about human infections and pathogens present in food and animals. The information in the databases will be confidential and accessible only to the users to prevent misuse. The laboratory information will provide sufficient discrimination to allow for detection of clusters of human infections. The associated background demographic and epidemiological information will be sufficiently detailed to guide outbreak investigations, e.g. the non-human data provide accurate information down to the farm or production plant level and the epidemiological information from clinical isolates includes standardized questionnaires with exposure data adapted to each country/region. Access to the data is tailored to the need for each user, e.g. users from national public health institutes may have full access to public health data from their own region but more restricted access to non-human data and all data from different regions, and similarly a user from a national food regulatory agency may have full access to data from his/her own country but limited access to all other data. An international laboratory database coordinator may have access to all subtyping information in all databases but limited access to everything else. The access for a user who is a member of a cross-disciplinary outbreak investigation team may be expanded to include all data relevant to the outbreak investigation in question. In this ideal system, non-culture combined diagnostic and subtyping methods will be used in all diagnostic laboratories with real-time submission of data to the surveillance databases, thereby shortening the data submission delay by minimizing the culture phase and eliminating the culture-forwarding, subtyping and reporting phases in the public health laboratories (Fig. 3.2c). Algorithms to detect clusters of human infections and events in non-human data predictive of future public health adverse events will be run at regular intervals at different levels in the databases automatically alerting the relevant users when a signal is detected.

The current situation may seem far from this ideal. However, the surveillance infrastructure for a system like this is in place in many developed countries and regions: Enter-net has proven the power of international collaboration between laboratorians and epidemiologists for outbreak investigations and served as inspiration for this type of interaction internationally; PulseNet has proven the power of real-time highly discriminatory subtyping and analysis of foodborne pathogens for case-cluster detection and outbreak investigation. In Europe, data are gathered from the whole farm-to-fork continuum in the annual reports on trends and sources of zoonoses and foodborne outbreaks (Anonymous, 2007b) and in North America, the PulseNet databases in the USA and Canada are linked as are the PulseNet and the VetNet databases. The food, veterinary and public health authorities in the USA are working together in PulseNet, FoodNet, the Foodborne Disease Active Surveillance Network (Jones *et al.*, 2007) and NARMS (Gilbert *et al.*, 2007). Similar progress has been achieved in many countries. However, a number of countries still lack basic surveillance infrastructure; with the implementation of the IHR, countries will be forced to move forward in this area. WHO facilitates this process through their WHO Global Foodborne Infections network (formally WHO Global Salm Surv), which is dedicated to capacity building of surveillance (World Health Organization, 2007c). WHO Global Foodborne Infections also collaborates with PulseNet International in building laboratory capacity for high discrimination subtyping and countries with this capacity may share their data through participation in a regional PulseNet network (Swaminathan *et al.*, 2006). However, a number of technological hurdles need to be overcome before non-culture methods combining diagnostics and subtyping will be available. Harmonization and validation of such novel molecular detection and subtyping methods will be crucial to their successful implementation in the surveillance of foodborne infections at both the national and international levels. Finally, even though the tools and the infrastructure may be in place, political trust and goodwill need to be built before a true global system for real-time, linked foodborne pathogen surveillance can be established.

3.5 Sources of further information and advice

World Health Organization:
- International Health Regulations: http://www.who.int/csr/ihr/
- INFOSAN: http://www.who.int/foodsafety/fs_management/infosan/
- Global Salm-Surv: http://www.who.int/salmsurv/

European Centre for Disease Prevention and Control:
- Enter-net: http://ecdc.europa.eu/en/Activities/Surveillance/Enter-net/
- Program on food- and water-borne diseases and zoonoses: http://ecdc.europa.eu/en/Activities/Disease_Projects/_fwd/

PulseNet: http://www.pulsenetinternational.org
FoodNet: http://www.cdc.gov/foodnet
NARMS: http://www.fda.gov/cvm/narms_pg.html

3.6 Disclaimer

Use of trade names is for identification only and does not imply endorsement by the Centers for Disease Control and Prevention or by the US Department of Health and Human Services.

The findings and conclusions in this chapter have not been formally disseminated by the Centers for Disease Control and Prevention and should not be construed to represent any agency determination or policy.

3.7 References

ANONYMOUS (2007a), *Annual Report on Zoonoses in Denmark 2006, Ministry of Family and Consumer Affairs*, Copenhagen.

ANONYMOUS (2007b), 'The community summary report on trends and sources of zoonoses, zoonotic agents, antimicrobial resistance and foodborne outbreaks in the European Union in 2006', *The EFSA Journal*, 130; 3–352.

CDC (1999), 'Outbreaks of *Shigella sonnei* infection associated with eating fresh parsley – United States and Canada, July–August 1998', *MMWR Morbidity and Mortality Weekly Report*, **48**, 285–89.

CDC (2005), 'Outbreak of multidrug-resistant *Salmonella* Typhimurium associated with rodents purchased at retail pet stores – United States, December 2003–October 2004', *MMWR Morbidity and Mortality Weekly Report*, **54**, 429–33.

CDC (2006), 'Ongoing multistate outbreak of *Escherichia coli* serotype O157:H7 infections associated with consumption of fresh spinach – United States, September 2006', *MMWR Morbidity and Mortality Weekly Report*, **55**, 1045–6.

CDC (2007), Salmonella *Wandsworth Outbreak Investigation, June–July 2007*. Available from: http://www.cdc.gov/salmonella/wandsworth.htm [Accessed 8 December 2008].

CDC (2008), 'Outbreak of *Salmonella* serotype Saintpaul infections associated with multiple raw produce items – United States, 2008', *MMWR Morbidity and Mortality Weekly Report*, **57**(34), 929–34.

CODY S H, GLYNN M K, FARRAR J A, CAIRNS L, GRIFFIN P M, KOBAYASHI J, ET AL. (1999), 'An outbreak of *Escherichia coli* O157:H7 infection from unpasteurized commercial apple juice', *Annals of Internal Medicine*, **130**, 202–9.

COOPER K L, LUEY C K, BIRD M, TERAJIMA J, NAIR G B, KAM K M, ET AL. (2006), 'Development and validation of a PulseNet standardized pulsed-field gel electrophoresis protocol for subtyping of *Vibrio cholerae*', *Foodborne Pathogens and Disease*, **3**(1), 51–8.

EUROPEAN COMMISSION (1992), *Council Directive 92/117/EEC of 17 December 1992 Concerning Measures for Protection Against Specified Zoonoses and Specified Zoonotic Agents in Animals and Products of Animal Origin in Order to Prevent Outbreaks of Food-Borne Infections and Intoxications*, European Commission, Brussels.

FISHER I S (1995), 'Salm-Net: a network for human *Salmonella* surveillance in Europe', *Euro Surveillance*, 7–8: pii=194. Available from: http://www.eurosurveillance.org/ViewArticle.aspx?ArticleId=194.

FISHER I S (1999), 'The Enter-net international surveillance network – how it works', *Euro Surveillance*, **4**(5), 52–5.

FISHER I S (2004a), 'Dramatic shift in the epidemiology of *Salmonella enterica* serotype Enteritidis phage types in western Europe, 1998–2003 – results from the Enter-net international salmonella database', *Euro Surveillance*, **9**(11), 43–5.

FISHER I S (2004b), 'International trends in *Salmonella* serotypes 1998–2003 – a surveillance report from the Enter-net international surveillance network', *Euro Surveillance*, **9**(11), 45–7.

FISHER I S and ENTER-NET PARTICIPANTS (1997), 'An international outbreak of *Salmonella* Livingstone recognised by Enter/Salm-net', *Euro Surveillance*, **1**(34), pii=1006. Available from: http://www.eurosurveillance.org/ViewArticle.aspx?ArticleId=1006.

FISHER I S and MEAKINS S (2006), 'Surveillance of enteric pathogens in Europe and beyond: Enter-net annual report for 2004', *Euro Surveillance*, **11**(34), pii=3032. Available from: http://www.eurosurveillance.org/ViewArticle.aspx?ArticleId=3032.

FISHER I S and THRELFALL E J (2005), 'The Enter-net and Salm-gene databases of foodborne bacterial pathogens that cause human infections in Europe and beyond: an international collaboration in surveillance and the development of intervention strategies', *Epidemiology and Infection*, **133**(1), 1–7.

GAUTOM R K (1997), 'Rapid pulsed-field gel electrophoresis protocol for typing of *Escherichia coli* O157:H7 and other Gram-negative organisms in 1 day', *Journal of Clinical Microbiology*, **35**, 2977–80.

GERNER-SMIDT P, HISE K, KINCAID J, HUNTER S, ROLANDO S, HYYTIA-TREES E, ET AL. (2006), 'PulseNet USA: a five-year update', *Foodborne Pathogens and Disease*, **3**(1), 9–19.

GERNER-SMIDT P, KINCAID J, KUBOTA K, HISE K, HUNTER S B, FAIR M A, ET AL. (2005), 'Molecular surveillance of Shiga toxigenic *Escherichia coli* O157 by PulseNet USA', *Journal of Food Protection*, **68**, 1926–31.

GILBERT J M, WHITE D G and MCDERMOTT P F (2007), 'The US national antimicrobial resistance monitoring system', *Future Microbiology*, **2**(5), 493–500.

GRAVES L M and SWAMINATHAN B (2001), 'PulseNet standardized protocol for subtyping *Listeria monocytogenes* by macrorestriction and pulsed-field gel electrophoresis', *International Journal of Food Microbiology*, **65**(1–2), 55–62.

HEDBERG C, JACQUET C and GOULET V (2007), 'Surveillance of listeriosis in France, 2000–2004: evaluation of cluster investigation criteria', Paper presented at the *16th International Symposium on Problems of Listeriosis (ISOPOL XVI)*, 20–23 March, Savannah, GA, USA.

HYYTIA-TREES E, SMOLE S C, FIELDS P A, SWAMINATHAN B and RIBOT E (2006), 'Second-generation subtyping: a proposed PulseNet protocol for multiple-locus variable-number tandem repeat analysis of Shiga toxin-producing *Escherichia coli* O157 (STEC O157)', *Foodborne Pathogens and Disease*, **3**, 118–31.

JACKSON C R, FEDORKA-CRAY P J, WINELAND N, TANKSON J D, BARRETT J B, DOURIS A, ET AL. (2007), 'Introduction to United States Department of Agriculture VetNet: Status of *Salmonella* and *Campylobacter* databases from 2004 through 2005', *Foodborne Pathogens and Disease*, **4**(2), 241–8.

JONES T F, SCALLAN E and ANGULO F J (2007), 'FoodNet: overview of a decade of achievement', *Foodborne Pathogens and Disease*, **4**(1), 60–6.

KAM K M, LUEY C K, PARSONS M B, COOPER K L, NAIR G B, ALAM M, ET AL. (2008), 'Evaluation and validation of a PulseNet standardized pulsed-field gel electrophoresis protocol for subtyping *Vibrio parahaemolyticus*: an international multicenter collaborative study', *Journal of Clinical Microbiology*, **46**(8), 2766–73.

KAPPERUD G, RORVIK L M, HASSELTVEDT V, HOIBY E A, IVERSEN B G, STAVELAND K, ET AL. (1995), 'Outbreak of *Shigella sonnei* infection traced to imported iceberg lettuce'. *Journal of Clinical Microbiology*, **33**(3), 609–14.

KILLALEA D, WARD L R, ROBERTS D, DE LOUVOIS J, SUFI F, STUART J M, ET AL. (1996), 'International epidemiological and microbiological study of outbreak of *Salmonella agona* infection from a ready to eat savoury snack – I: England and Wales and the United States', *British Medical Journal*, **313**(7065), 1105–7.

LAINE E S, SCHEFTEL J M, BOXRUD D J, VOUGHT K J, DANILA R N, ELFERING K M, ET AL. (2005), 'Outbreak of *Escherichia coli* O157:H7 infections associated with nonintact blade-tenderized frozen steaks sold by door-to-door vendors', *Journal of Food Protection*, **68**(6), 1198–202.

LEWIS H C, KIRK M, ETHELBERG S, STAFFORD R, OLSEN K, NIELSEN E M, ET AL. (2007), 'Outbreaks of shigellosis in Denmark and Australia associated with imported baby corn, August 2007 – final summary', *Euro Surveillance*, **12**(40), pii=3279. Available from: http://www.eurosurveillance.org/ViewArticle.aspx?ArticleId=3279.

MAHON B E, PONKA A, HALL W N, KOMATSU K, DIETRICH S E, SIITONEN A, ET AL. (1997), 'An international outbreak of *Salmonella* infections caused by alfalfa sprouts grown from contaminated seeds', *Journal of Infectious Diseases*, **175**(4), 876–82.

MASLANKA S E, KERR J G, WILLIAMS G, BARBAREE J M, CARSON L A, MILLER J M, ET AL. (1999), 'Molecular subtyping of *Clostridium perfringens* by pulsed-field gel electrophoresis to facilitate food-borne-disease outbreak investigations', *Journal of Clinical Microbiology*, **37**(7), 2209–14.

PARSONS M B, COOPER K L F, KUBOTA K A, PUHR N, SIMININGTON S, CALIMLIM P S, ET AL. (2007), 'PulseNet USA standardized pulsed-field gel electrophoresis protocol for subtyping of *Vibrio parahaemolyticus*', *Foodborne Pathogens and Disease*, **4**(3), 285–92.

PEZZOLI L, ELSON R, LITTLE C L, YIP H, FISHER I, YISHAI R, ET AL. (2008), 'Packed with *Salmonella* – investigation of an international outbreak of *Salmonella senftenberg* infection linked to contamination of prepacked basil in 2007', *Foodborne Pathogens and Disease*, **5**(5), 661–8.

RIBOT E M, FAIR M A, GAUTOM R, CAMERON D N, HUNTER S B, SWAMINATHAN B, ET AL. (2006), 'Standardization of pulsed-field gel electrophoresis protocols for the subtyping of *Escherichia coli* O157:H7, *Salmonella*, and *Shigella* for PulseNet', *Foodborne Pathogens and Disease*, **3**(1), 59–67.

RIBOT E M, FITZGERALD C, KUBOTA K, SWAMINATHAN B and BARRETT T J (2001), 'Rapid pulsed-field gel electrophoresis protocol for subtyping of *Campylobacter jejuni*', *Journal of Clinical Microbiology*, **39**(5), 1889–94.

SWAMINATHAN B, BARRETT T J, HUNTER S B, TAUXE R V and THE CDC PULSENET TASK FORCE (2001), 'PulseNet: the molecular subtyping network for foodborne bacterial disease surveillance, United States', *Emerging Infectious Diseases*, **7**, 382–9.

SWAMINATHAN B, GERNER-SMIDT P, NG L K, LUKINMAA S, KAM K M, ROLANDO S, ET AL. (2006), 'Building PulseNet International: an interconnected system of laboratory networks to facilitate timely public health recognition and response to foodborne disease outbreaks and emerging foodborne diseases', *Foodborne Pathogens and Disease*, **3**(1), 36–50.

SWANSON S J, SNIDER C, BRADEN C R, BOXRUD D, WÜNSCHMANN A, RUDROFF J A, ET AL. (2007), 'Multidrug-resistant *Salmonella enterica* serotype Typhimurium associated with pet rodents', *New England Journal of Medicine*, **356**, 21–8.

THRELFALL E J, FISHER I S, WARD L R, TSCHAPE H and GERNER-SMIDT P (1999), 'Harmonization of antibiotic susceptibility testing for *Salmonella*: results of a study by 18 national reference laboratories within the European Union-funded Enter-net group', *Microbial Drug Resistance*, **5**(3), 195–200.

THRELFALL E J, WARD L R, HAMPTON M D, RIDLEY A M, ROWE B, ROBERTS D, ET AL. (1998), 'Molecular fingerprinting defines a strain of *Salmonella enterica* serotype Anatum responsible for an international outbreak associated with formula-dried milk', *Epidemiology and Infection*, **121**(2), 289–93.

VANBENEDEN C A, KEENE W E, STRANG R A, WERKER D H, KING A S, MAHON B, ET AL. (1999), 'Multinational outbreak of *Salmonella enterica* serotype Newport infections due to contaminated alfalfa sprouts', *Journal of the American Medical Association*, **281**(2), 158–62.

WERBER D, DREESMAN J, FEIL F, VAN TREECK U, FELL G, ETHELBERG S, ET AL. (2005), 'International outbreak of *Salmonella* Oranienburg due to German chocolate', *BMC Infectious Diseases*, **5**(1), 7.

WORLD HEALTH ORGANIZATION (2007a), *The Identification, Assessment and Management of Food Safety Events under the International Health Regulations (2005)*, World Health Organization, Geneva.

WORLD HEALTH ORGANIZATION (2007b), *International Health Regulations (2005)*, World Health Organization, Geneva.

WORLD HEALTH ORGANIZATION (2007c), *WHO Global Salm-Surv – A Surveillance Network for Foodborne Diseases*, World Health Organization, Geneva.

4

Detection, investigation and control of outbreaks of foodborne disease

C. Stein and A. Ellis, World Health Organization, Switzerland and T. Jones, Tennessee Department of Health, USA

Abstract: The investigation and control of foodborne disease outbreaks are multidisciplinary tasks requiring training and skills in clinical medicine, epidemiology, laboratory medicine, food microbiology and chemistry, food safety and food control, and risk communication and management. Many outbreaks of foodborne diseases are poorly investigated, if investigated at all, because these skills are often unavailable or because a field investigator is expected to master them all single-handedly without proper training. This chapter is written for all professionals involved in the investigation and control of foodborne disease outbreaks. It focuses largely on practical aspects of outbreak investigation and control, but in doing so provides generic guidance that can be adapted to local requirements. Detailed information about the steps of such efforts is given, including the planning and preparation, detection of outbreaks, investigation and control measures. Despite a clear focus on foodborne diseases, much of the material in this chapter is also applicable to the investigation of outbreaks of other communicable and non-communicable diseases.

Key words: outbreaks, foodborne illness, epidemiology, investigation.

4.1 Introduction

Foodborne diseases encompass a wide spectrum of illnesses caused by pathogens, parasites and chemical contaminants (WHO, WHO Initiative to Estimate the Global Burden of Foodborne Diseases). They lead to acute and chronic, and frequently fatal, conditions and are a global public health problem. The most commonly investigated outbreaks among foodborne illnesses, however, remain gastro-intestinal diseases. Acute diarrhoeal illness is very common worldwide. Diarrhoeal disease is estimated to account for 1.5 million childhood deaths annually, mostly in developing countries (WHO, 2008). The burden of diarrhoeal illness is also considerable in developed countries (Scallan *et al.*, 2005). Estimates of the burden of foodborne diseases are complicated by a number of factors: different definitions of acute diarrhoeal illness are used in various studies; most diarrhoeal illness is not reported to public health authorities; and few illnesses can be definitively linked to food. While not all gastroenteritis is foodborne, and not all foodborne diseases cause gastroenteritis, food does represent an important vehicle for pathogens of substantial public health significance. A global World Health Organization (WHO) initiative is under way to understand better the global public health burden of gastroenteritis and other foodborne diseases (WHO, WHO Initiative to Estimate the Global Burden of Foodborne Diseases).

There are many reasons why foodborne disease remains a global public health challenge. As some diseases are controlled, others emerge as new threats (Jones *et al.*, 2008). In many countries, the proportions of the population that are elderly, immunosuppressed or otherwise disproportionately susceptible to severe outcomes from foodborne diseases are growing. The globalization of the food supply has led to the rapid and widespread international distribution of foods. Travellers, refugees and immigrants may be exposed to unfamiliar foodborne hazards in new environments. Changes in microorganisms lead to the constant evolution of new pathogens, development of antibiotic resistance and changes in virulence in known pathogens. In many countries, as people increasingly consume food prepared outside the home, growing numbers are potentially exposed to the risks of poor hygiene in commercial food service settings (WHO, 1998). All of these emerging challenges require that public health workers continue to adapt to a changing environment with improved methods to investigate and combat these threats.

Foodborne disease outbreaks may often go unrecognized, unreported or uninvestigated. There are many resources available for the investigation of foodborne disease outbreaks, but few consider the specific needs of developing countries. WHO has produced investigation guidelines for the identification and investigation of foodborne disease outbreaks in a variety of settings (WHO, 2008). These 'Foodborne Disease Outbreaks – Guidelines for Investigation and Control' are WHO's recommendations on how best to investigate and control such events and they form the basis for this chapter. Numerous other resources are available for additional, more detailed information on surveillance, epidemiology, statistical analyses and the medical aspects of foodborne diseases. It is important to remember that no general guidelines will fit a specific situation perfectly, and the local environment will always necessitate modifying an investigation to account for the unique characteristics of every outbreak.

The investigation and control of foodborne disease outbreaks is a multidisciplinary task requiring skills in the areas of clinical medicine, epidemiology, laboratory medicine, food microbiology and chemistry, food safety and food control, and risk communication and management. This chapter describes the steps of an outbreak investigation in several sections, including:

- Section 1 (planning and preparation);
- Section 2 (outbreak detection);

- Section 3 (outbreak investigation);
- Section 4 (control measures).

For all technical background information, sample forms for data collection and analysis, questionnaires and other tools that may be useful during an investigation, the authors would like to refer the reader to the WHO 'Foodborne Disease Outbreaks – Guidelines for Investigation and Control' (WHO, 2008).

4.2 Planning and preparation

The responsibilities for the investigation and the management of outbreaks will vary between countries and depend on the nature and size of the outbreak, its importance with regard to the health of the public, its economic impact and other factors.

Typical steps in the investigation of a foodborne disease outbreak include:

- establish the existence of an outbreak;
- verify the diagnosis;
- define and count cases;
- determine the population at risk;
- describe the epidemiology;
- develop hypotheses;
- evaluate hypotheses;
- perform additional epidemiologic, environmental and laboratory studies, as necessary;
- implement control and prevention measures;
- communicate findings.

All responsible authorities should develop outbreak investigation and control plans to address:

- the arrangements for consulting and informing authorities at local, regional, national and international levels;
- the exact roles and responsibilities of organizations and individuals involved;
- the resources/facilities available to investigate outbreaks;
- when to convene an outbreak control team (OCT), its composition and its duties.

4.2.1 The outbreak control team

The criteria for convening a multidisciplinary OCT will vary according to the seriousness of the illness, its geographic spread, local circumstances and available resources. The role of the OCT is to coordinate all activities that are conducted to investigate and control an outbreak (Fig. 4.1). This may involve:

- deciding whether there is really an outbreak;
- deciding what type of investigations to be conducted;

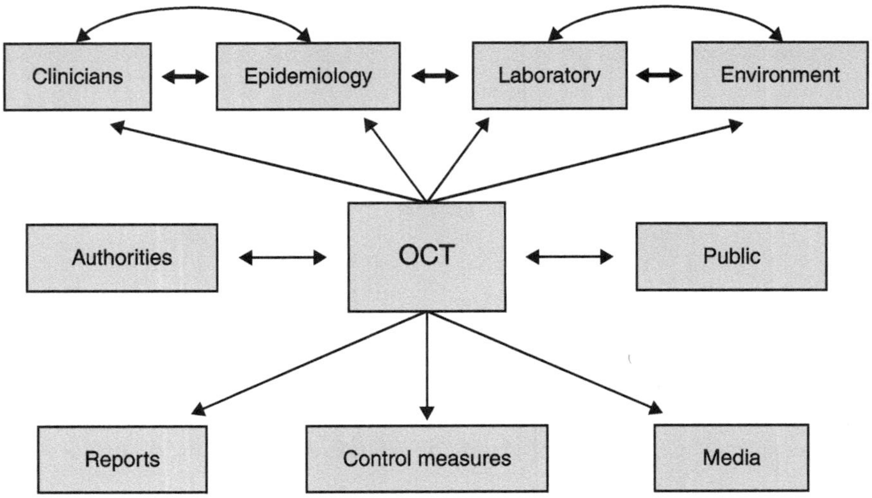

Fig. 4.1 Coordinating role of the OCT in an outbreak investigation. Reproduced with permission of the World Health Organization, 2008.

- case finding and interviews;
- planning the appropriate clinical and environmental sampling;
- ensuring that all collaborators use complementary methodology;
- environmental investigation of suspected food premises;
- agreeing and implementing control measures to prevent the further spread by means of exclusions, withdrawal of foods, closure of premises, etc.;
- working in concert with local medical providers to provide treatment and/or prophylaxis recommendations;
- organizing ongoing communications about the outbreak;
- making arrangements for media liaison;
- producing reports, including lessons learned, for health and other interested authorities;
- asking for external assistance, for example, secondment of a national investigation team.

Usually, the health authority in the area, which first identified and reported the outbreak, initiates proceedings to set up an OCT. In an outbreak that crosses administrative boundaries, the group itself should determine, at its first meeting, the extent of representation and who will act as a chairman. Once an OCT has been established, it should be in charge of all investigation and control activities.

OCT membership will vary according to circumstances but normally includes:

- a public health practitioner or epidemiologist under the authority of the public health officer in charge;
- a food safety control officer;
- a laboratorian (microbiologist, toxicologist or others as appropriate);
- secretarial and logistic support.

In addition, one or more of the following may be needed according to the presumed nature of the outbreak:

- food scientist (chemist, food microbiologist, technologist);
- clinician;
- veterinarian;
- toxicologist;
- virologist;
- other technical experts;
- press officer;
- representatives of local authorities (community leaders, etc.);
- hospital director, members of a hospital infection control group.

4.2.2 Record keeping

From the beginning of an outbreak, it is essential to ensure that all information received and all decisions taken by the OCT and others are recorded reliably and with the appropriate level of confidentiality. This means that:

- individual members of the OCT keep records of all activities performed in investigating the outbreak;
- minutes are kept and distributed;
- action notes are agreed upon and distributed immediately after OCT meetings;
- copies are kept of all communication with the public, including letters, fact sheets, public notices and media reports.

4.2.3 Communication

An important aspect of a successful outbreak management is effective communication. Throughout the course of an outbreak, it is important to share relevant information with:

- authorities and other professional groups;
- local health care providers (as appropriate);
- the media;
- the people directly affected;
- the general public.

Authorities and other professional groups

Authorities and other professional groups include local health authorities, food and water authorities, agricultural and veterinary authorities, educational organizations, and others. The objectives of informing these groups are to ensure accurate case finding and to facilitate the implementation of control measures, including recall of food products from the market. Where possible, efficient communication should be maintained with authorities and other professional groups through already established communication channels and by holding regular meetings.

Public

Public concern can become an important feature of an outbreak investigation. To achieve a proper balance between the scientific requirements of the investigation and responsiveness to public concern, public health authorities must deal actively with the need for public information. The objectives of informing the public in foodborne disease outbreaks are to provide:

- accurate information about the outbreak;
- information on implicated food products and how they should be handled;
- advice on personal hygiene measures to decrease the risk of person-to-person spread.

In some outbreaks, communication with the public will also help in identifying additional cases. Methods of communicating with the public will depend on local circumstances but may include regular press releases via newspapers, radio or television, public meetings, leaflets delivered to households and places of gathering, face-to-face advice in clinics, and messages on notice boards and to consumer groups. If a major outbreak is occurring or an outbreak has attracted intensive publicity, it may be necessary to establish a telephone helpline for the public. When establishing such helplines, it is important that staff has also been trained in gathering additional information (e.g. details about cases) from callers.

The information provided should always be objective and factual. Unconfirmed information usually should not be released. If a public health warning is required in the absence of confirmed results, the public should be informed why these measures have been taken and that they may have to be changed in the light of new knowledge.

Media

As the major interface between the general public and the health authorities, the media plays an important role in outbreak investigation and control. Developing good relationships with the media before an outbreak occurs may be very helpful in facilitating crisis-related communication. Accurate and comprehensive reporting of foodborne disease outbreaks by the media can:

- facilitate case finding through enhanced reporting of cases by the public and medical practitioners;
- inform the public about avoidance of risk factors for the particular illness in question, including how to manage the food risks, and appropriate preventive measures;
- maintain public and political support for disease investigation and control;
- minimize the release of conflicting information by different authorities that may undermine their credibility.

4.3 Outbreak detection

Outbreaks are frequently detected through public health surveillance systems. The primary goal of surveillance for foodborne disease outbreaks should be the prompt identification of any unusual clusters of disease potentially transmitted through food, which might require a public health investigation or response.

4.3.1 Definitions

Some key terms are defined here to ensure clarity.

- *Surveillance*: The systematic collection, analysis and interpretation of data essential to the planning, implementation and evaluation of public health practice, and the timely dissemination of this information for public health action.
- *Foodborne disease*: Any disease of an infectious or toxic nature caused by the consumption of food.
- *Foodborne disease outbreak*: Various definitions are in use. First, when the observed number of cases exceeds the expected number of cases, and second, the occurrence of two or more cases of a similar foodborne disease resulting from the ingestion of a common food.
- *Sporadic case*: A case that cannot be linked epidemiologically to other cases of the same illness.

Epidemiologists often use the terms 'cluster', 'outbreak' and 'epidemic' interchangeably. Typically, the term 'cluster' is used to describe a group of cases linked by time or place but without a common food or other source identified. In the foodborne disease context, 'outbreak' refers to two or more cases resulting from ingestion of a common food. The term 'epidemic' is often reserved for crises or situations involving larger numbers of people over a wide geographic area.

4.3.2 Data sources

Detecting outbreaks requires efficient mechanisms to capture and respond to a variety of data sources. In most countries, the main data sources for detecting foodborne disease outbreaks are the public, the media, reports of clinical cases from health care providers, surveillance data (laboratory reports, disease notifications) and food service facilities.

Surveillance activities are conducted at local, regional and national levels through a variety of systems, organizations and pathways (Borgdorff and Motarjemi, 1997). Among the many surveillance methods for foodborne disease, laboratory reporting and disease notification may contribute importantly to outbreak detection. Other types of surveillance that may be of value in detecting foodborne disease outbreaks are hospital-based surveillance, sentinel site surveillance and reports of death registration. Generally, these systems are not primary data sources for detecting outbreaks and their usefulness will depend on the quality of the system and the circumstances in which they are employed.

Laboratory-based surveillance

Laboratories receive and test clinical specimens from patients with suspected foodborne disease (e.g. fecal samples from patients with diarrhea). Often positive microbiological findings from these specimens are also sent by laboratories to the relevant public health authorities. In addition, some laboratories send patient material or isolates to a central reference laboratory for confirmation, typing or determination of resistance patterns. The collation of these reports and their systematic and timely analysis can provide useful information for detecting outbreaks, particularly when cases are geographically scattered or clinical symptoms are non-specific.

Detecting outbreaks is facilitated by early typing of isolates of foodborne pathogens. Routine typing may detect a surge of a particular subtype and link apparently unrelated infections. Interviewing these cases about their food consumption may then identify contaminated foods that may not have been recognized otherwise.

Traditional laboratory-based surveillance is 'passive' or dependent on laboratories to report cases to public health authorities. In some situations, such as when a potential problem is suspected, 'active surveillance' may be warranted for a period of time. In such a case, laboratories may be actively contacted by food safety or public health authorities prospectively on a regular basis to enquire about recent positive tests indicative of potential foodborne diseases.

Disease notification

In most countries, medical practitioners are required to notify public health authorities of all cases of specified diseases. Notification of cases is usually based on clinical judgement and may not require confirmation by other diagnostic means. Most statutory disease notification systems suffer from substantial under-reporting of diagnosed cases and long delays in notification. Also, many persons with foodborne disease do not seek medical advice or will not be diagnosed as suffering from a foodborne disease due to the lack of specificity of their symptoms. Thus, disease notification may be substantially more likely to occur for laboratory-confirmed illnesses. Medical practitioners who become aware of unusual clusters of diarrhoeal disease or other syndromes that may indicate foodborne disease should also be urged to report these promptly to public health authorities.

Other sources

Other sources may alert public health authorities to the occurrence of outbreaks. Often some creativity is needed to detect them as many of these sources were created for other purposes. Examples include reports of increased absenteeism from the workplace, schools or childcare facilities, pharmacy reports about increased drug sales, for example of anti-diarrheal medications, or consumer complaints to health departments or food regulators. Outbreaks may be anticipated after an increased risk of population exposure has been detected, for example contaminated drinking water or contamination of a commercially available food product.

4.3.3 Interpreting data sources

Outbreaks are often detected when ill persons share an easily recognized potential source of infection (such as in schools, hospitals, nursing facilities, correctional facilities, etc.). When such events are limited to small, well-defined populations, the number of ill persons can usually be quickly established. The main emphasis is on verifying that an outbreak has indeed occurred and controlling its spread. Detecting community outbreaks from surveillance data can be more difficult. Foremost, it requires timely collection and analysis of the data and interpretation of whether the number of observed cases exceeds expected numbers. This requires knowledge of the background rates or traditional disease patterns in a particular population at a particular time and place, including typical seasonal changes in disease occurrence. A small local outbreak may be missed by regional or national surveillance, or conversely, a widespread national outbreak may not be detectable by regional or local surveillance. A sudden increase in disease occurrence may clearly point towards an outbreak (Fig. 4.2) while small changes in a baseline can be difficult to interpret (Fig. 4.3). Even if the overall number of cases is not unusually high, a steep increase confined to a subgroup in the community or to a particular subtype of pathogen may be significant (Fig. 4.4).

There are causes other than outbreaks that may lead to increased numbers of observed or reported cases ('pseudo-outbreaks'). Examples include changes in local reporting procedures, changes in the case definition for reporting a specified disease, increased interest because of local or national awareness, changes in diagnostic procedures, or heightened concern among a specific population (e.g. 'psychogenic' outbreaks). In areas with sudden changes in population size such as resort areas, college towns and migrant farming areas, changes in the numerator (number of reported cases) may only reflect changes in the denominator (population size).

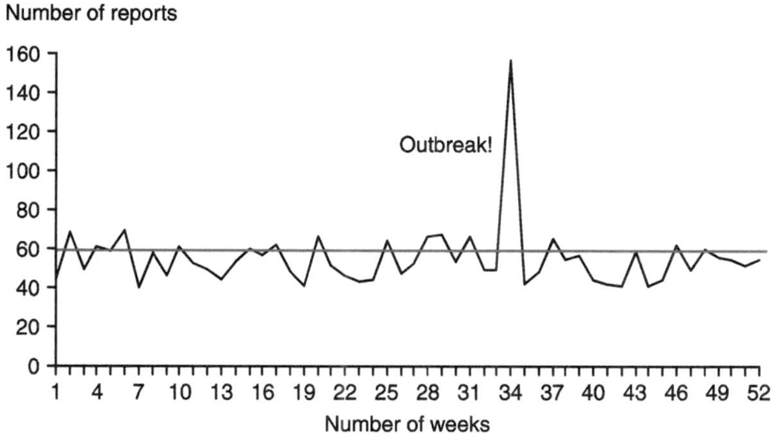

Fig. 4.2 Weekly number of reported cases indicating an outbreak in week 34. Reproduced with permission of the World Health Organization, 2008.

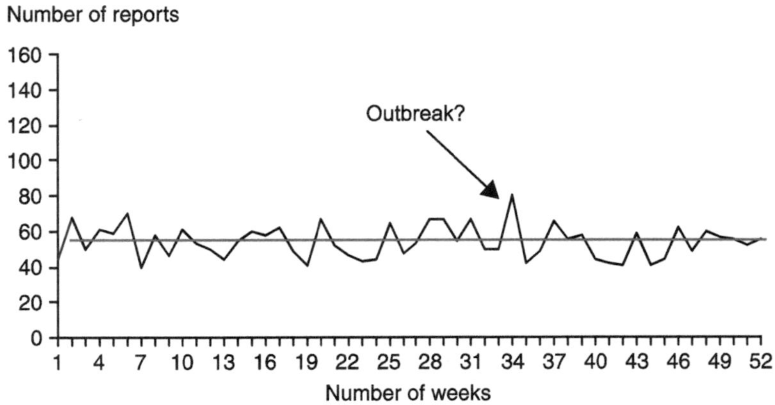

Fig. 4.3 Weekly number of reported cases where it is not clear whether or not the observed number of cases in week 34 has exceeded expected numbers. Reproduced with permission of the World Health Organization, 2008.

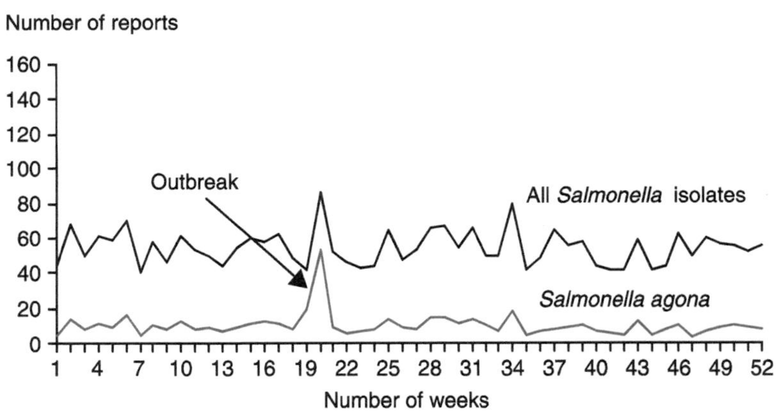

Fig. 4.4 Weekly number of *Salmonella* isolates. The outbreak of *S. agona* may have been missed without data on specific serotypes. Reproduced with permission of the World Health Organization, 2008.

4.4 Outbreak investigation

Foodborne disease outbreaks are investigated to prevent ongoing transmission of disease and prevent similar outbreaks in the future. Specific objectives include:

- to control ongoing outbreaks;
- to detect and remove implicated foods;
- to identify specific risk factors related to the host, the agent and the environment;

- to determine factors that contributed to contamination, growth, survival and dissemination of the suspected agent;
- to prevent future outbreaks and strengthen food safety policies;
- to provide epidemiological data for risk assessment of foodborne pathogens;
- to stimulate research for the prevention of similar outbreaks.

The scale of an outbreak may range from a local outbreak of a small number of linked cases with mild disease to a nation wide or international outbreak of severe disease with mobilization of public health resources from all levels. Irrespective of the scale of the outbreak, a full investigation of a foodborne disease outbreak will normally include:

- epidemiological investigations;
- environmental and food investigations;
- laboratory investigations.

4.4.1 Epidemiological investigations

Preliminary assessment of the situation

The investigation of a potential outbreak starts with the assessment of all available information to confirm or refute the existence of an outbreak and to establish a working case definition. This includes checking the validity of the information, identifying cases and obtaining information about them, and ensuring the collection of appropriate clinical specimens and food samples.

Once the validity of the reporting source has been verified, a group of the initial cases (e.g. five to ten persons) should be identified and interviewed as soon as possible. This critical step helps to better understand the clinical and epidemiological features of the affected group. Delays in performing these interviews can lead to recall bias or situations where patients will forget what they ate or did. The interviews should be open and comprehensive and include questions about:

- demographic details, including occupation;
- clinical details with date of onset, duration and severity of symptoms;
- visits to health care providers or hospitals;
- laboratory test results;
- contact with other ill persons;
- food consumption history;
- what the cases think caused their illness;
- if they know others with the same or similar illness;
- potential common exposures among those who have the same or a similar illness;
- date of exposure to suspected foods.

Clinical specimens (e.g. fecal samples, vomitus) from cases should be collected at the time of the initial contact. Laboratory confirmation of initial cases is essential to guide further investigation. Any available leftover foods that are suspected or were eaten during the potential incubation period should be sampled for laboratory examination. Information on the collection of clinical and food samples can be found in Section 4.8.

If the vehicle of infection is thought to be food, the premises where the suspect food has been produced, processed or handled should also be visited. It is important to visit these premises early as the amount of physical evidence that may have caused the outbreak will diminish with time. Appropriate food and environmental samples should be collected. It may also be appropriate to collect clinical specimens from food service workers at this time.

Develop preliminary hypotheses and plan further action

With the initial information from case interviews, the laboratory and the environmental inspection, it is often possible to describe the event in simple epidemiological terms and to form preliminary hypotheses about the cause of the outbreak. Apparent 'outliers' or unusual cases (e.g. the only case who resides in a different town, the oldest case, the youngest case) can often be good clues for generating hypotheses. General control measures and precautionary measures may be implemented at this stage. For example, suspect foods can be removed from sale or from the premises; ill food handlers should be excluded from work; and the public may be advised to avoid a certain food product or receive appropriate medical treatment. Obvious control measures must never be delayed at this early stage only because investigations are still under way. It is important to proceed with caution at this stage and acknowledge that initial hypotheses are still under investigation. Failure to do so can lead to the risk of implicating the wrong food and damage to the credibility of investigators and the food producer.

4.5 Descriptive epidemiological investigations

The descriptive epidemiology provides a picture of the outbreak by the three standard epidemiological parameters: *time, place* and *person*. This can guide immediate control measures, inform development of more specific hypotheses about the source and mode of transmission, suggest the need to collect further clinical, food or environmental samples, and guide development of further studies.

The steps of descriptive epidemiology include:

- establish a case definition;
- identify cases and obtain information from them;
- analyse the data by time, place and person characteristics;
- determine who is at risk of becoming ill;
- develop hypotheses about the exposure/vehicle that caused the disease;
- compare the developed hypotheses with the established facts;
- decide whether analytical studies are needed to test the hypotheses.

4.5.1 Establish a case definition

A case definition is a set of criteria for determining whether a person should be classified as being affected by the disease under investigation. As such, a case definition is an epidemiological tool to count cases; it is not used to guide clinical practice. Ideally, a case definition will include all cases (high sensitivity) but no person who does not have the illness (high specificity). A sensitive case definition will not only detect many cases but may also count as cases individuals who do not have the disease. A more specific case definition is more likely to include only persons who truly have the disease under investigation, but is more likely to miss some cases as well.

There are no rules about how sensitive or specific a case definition should be. At the early stage of an outbreak investigation, the aim is to detect as many cases as possible. This requires a sensitive case definition (e.g. a person with three or more loose stools in a 24-hour period). At a later stage, the clinical picture is often clearer and the diagnosis confirmed by laboratory means. This allows the use of a more specific case definition (e.g. laboratory-confirmed *Salmonella* infection), which may then be used to conduct further analytical studies. Criteria included in a case definition cannot be tested as a source of the outbreak or as risk factors in subsequent analyses.

A single case definition that suits all needs rarely exists. Thus, it is quite common to change case definitions during an investigation or to use different case definitions for different purposes. Many investigators use the following or similar case definitions in parallel:

- *Confirmed* cases that have a positive laboratory result (isolation of the causative agent or positive serological test). This case definition has a high specificity;
- *Probable* cases that have the typical clinical features of the illness but without laboratory confirmation;
- *Possible* cases that have fewer or atypical clinical features. This case definition has a high sensitivity.

Box 4.1 Example of case definitions in an investigation of an *E. coli* O157 outbreak

A case is defined as gastrointestinal illness in any resident of Area A within 5 days of attending the Area A Fair in June, 2003. Cases may be further categorized as:

Confirmed case:	Gastrointestinal illness in a person with microbiological confirmation of *E. coli* O157
Probable case:	Bloody diarrhoea or haemolytic uraemia syndrome in a person without microbiological confirmation
Possible case:	Non-bloody diarrhoea in a person without microbiological confirmation

4.5.2 Identify cases

The cases that prompt an outbreak investigation are often only a small fraction of the total number of people affected. To determine the full extent of the problem and the population at risk of illness, an active search should be conducted for additional cases.

Methods for finding additional cases will vary from outbreak to outbreak. Many foodborne disease outbreaks involve clearly identifiable groups, and case finding will be relatively self-evident (e.g. persons all attending the same wedding party). In other outbreaks, particularly those involving diseases with a long incubation period and/or with mild or asymptomatic illness, case finding may be quite difficult. Directly contacting physicians, hospitals, laboratories, schools or other populations at risk may help identify unreported cases.

In some outbreaks, public health officials decide to alert the public directly. For example, in outbreaks caused by a contaminated commercial food product, announcements in the media can alert the public to avoid the implicated product and to see a medical practitioner if they have symptoms compatible with the disease in question. When the outbreak involves a product that may have gone to other countries, notification of the event should be made to WHO. Information can then be shared with food safety authorities in other countries through the International Food Safety Authorities Network (INFOSAN), a joint initiative of Food and Agriculture Organization (FAO) and WHO that ensures the rapid sharing of information about food safety events that are international in scope. In addition, the event may be reportable under the International Health Regulations (WHO, 2005) and would be notified to WHO through the National IHR Focal Point.

If an outbreak affects a restricted population (e.g. students in a school or factory workers) and if a high proportion of cases is unlikely to be diagnosed, a survey of the entire population can be conducted. Questionnaires may be administered to determine the true incidence of clinical symptoms or to collect laboratory specimens to determine the number of asymptomatic cases.

Finally, cases can be found by reviewing laboratory surveillance data to find people with similar infections, assuming that the cause of the outbreak is known. Often a unique subtype, biochemical or molecular feature of the organism can help to identify cases that could be epi-linked. This can be particularly helpful in outbreaks caused by a widely distributed food product that crosses jurisdictional or even international boundaries.

4.5.3 Interview cases

Once cases are identified, information should be obtained about them in a systematic way by use of a standard questionnaire. This is in contrast to the preliminary phase of the investigation in which the interviews may be more wide-ranging and open-ended to allow for generation of hypotheses.

Questionnaires may be administered by an interviewer (face-to-face or by telephone) or by self-administration. Sometimes, patients themselves will not be interviewed but their parents, spouses or caretakers may provide data, and the sources of information should always be recorded on the questionnaire. Self-administered questionnaires may be distributed in person, by mail, email, fax or Internet.

Regardless of the disease under investigation, the following types of information should be collected about each case:

1 *Identifying information* – name, address, contact details (e.g. day time telephone number, work address) – to allow contacting patients for additional questions and notifying them of laboratory results and the outcome of the investigation. Names will help to check for duplicate records and addresses may allow mapping of cases. When recording, identifying information confidentiality issues must always be addressed in accordance with prevailing laws and regulations.
2 *Demographic information* – age, date of birth, sex, race and ethnicity, occupation, residence, etc. – to provide the 'person' characteristics of descriptive epidemiology that help define the population at risk of becoming ill.
3 *Clinical information* – to identify cases, verify that the case definition has been met, define the clinical syndrome or manifestations of disease, and help identify the following potential etiologies:
   (a) date and time of first signs and symptoms;
   (b) nature of initial and subsequent signs and symptoms;
   (c) severity and duration of symptoms;
   (d) medical visits and hospital admission;
   (e) treatment;
   (f) outcome of illness.
4 *Risk factor information* – to allow determining the source and the vehicle of the outbreak. This type of information will need to be tailored to the specific outbreak and the disease in question. Generally, the questionnaire will address food-related and personal risk factors:
   (a) Food-related risk factors:
      (i) detailed food history (see below);
      (ii) sources of domestic food and water supply;
      (iii) specific food handling practices, cooking preferences;
      (iv) eating out of the house.
   (b) Personal risk factors:
      (i) date and time of exposure to an implicated food or event (if this is known);
      (ii) contact with people with similar clinical signs and symptoms;
      (iii) information on recent (domestic and foreign) travel;
      (iv) recent group gatherings, visitors, social events;
      (v) recent farm visits;
      (vi) contact with animals;
      (vii) attending or working in a school, childcare facility, medical facility;

(viii) working as a food handler;
(ix) chronic illness, immunosuppression, pregnancy;
(x) recent changes in past medical history, regular medications;
(xi) recent immunizations, allergies.

Depending on the suspected pathogen and local patterns of food consumption and availability, enquiries should be conducted about any foods that could be a potential source of contamination in the outbreak. It is important to collect a thorough history of food consumption for the entire suspected incubation period (which is often three to five days prior to illness for many common foodborne pathogens). An accurate and thorough food history will often require direct questions about specific foods as well as open-ended questions. In addition, data should be collected on the amount/number of meals eaten, and the source and handling of suspected foods should be noted. Some examples of questionnaires are available from the WHO Guidelines (WHO, 2008).

If the pathogen is known, questions can focus on foods and other risk factors known to be associated with the particular pathogen. Knowing the incubation period of the pathogen can lead to the most likely period of exposure or identify an unusual event or a suspect meal. If certain foods are known to be associated with the pathogen, specific questions should be asked about them, though enquiries should not be limited to these foods.

If the pathogen is not known but the clinical details suggest a short incubation period, information should be gathered about all meals eaten within 72 hours before the onset of illness. As most persons cannot remember all foods eaten over a 72-hour period, it may be helpful to add a calendar to the questionnaire, the menu of a suspect meal, or a list of foods to aid in recall of relevant items.

In protracted outbreaks, when investigating illnesses with incubation periods longer than 72 hours (e.g. hepatitis A, typhoid fever, listeriosis) or when a person does not remember specific foods eaten, food preferences should be enquired about, that is foods usually eaten or routine dietary habits. Information should also be obtained about foods purchased within the range of the incubation period of the disease under suspicion.

4.5.4 Collate data

Once the first questionnaires have been completed, their information should be collated promptly to gain insight into the distribution of clinical symptoms and other factors among cases. The data can be summarized in a line listing, with each column representing a variable of interest and each row representing a case. New cases can be added conveniently to the list and updated as necessary (Table 4.1). A line list can be created directly by copying relevant information from the questionnaires or from a computerized database in which case data have been entered. Many types of computer software are available for this purpose, including

Table 4.1 Example of a line list to summarize case data

ID	Name	Age	Sex	Date and time of illness onset	Major signs and symptoms				Laboratory tests	
					D[a]	V[b]	F[c]	A[d]	Specimen	Results
1	MT	34	f	10/05, 2200	+	–	+	+	ND	
2	TG	45	f	11/05, 0800	+	–	dk	+	ND	
3	SH	23	m	11/05, 0500	+	–	+	+	faeces	*E. coli* 0157
4	RF	33	f	10/05, 1800	+B	+	+	+	faeces	Pending
5	SM	23	m	11/05, 1200	+	–	–	+	faeces	Pending
Etc.										

[a] diarrhoea.
[b] vomiting.
[c] fever.
[d] anorexia.

Note: B, bloody; dk, unknown/cannot remember; ND, not done. Reproduced with permission of the World Health Organization, 2008.

some that are available free of charge (such as EpiInfo, http://www.cdc.gov/epiinfo/ or EpiData, http://www.epidata.dk/).

While entering information, the consistency and quality of the data should be critically evaluated. If feasible, the respondents may be re-contacted to clarify illegible or ambiguous responses on the questionnaire.

4.5.5 Analyse data

Clinical details

The percentage of cases with a particular symptom or sign should be calculated and arranged in a table in a decreasing order. Organizing the information this way will help to determine whether the outbreak was caused by an intoxication, an enteric infection or a generalized illness. For example, if the predominant symptom is vomiting without fever and the incubation period is short (less than eight hours), intoxication by, for example, *Staphylococcus*, *Clostridium perfringens* or *Bacillus cereus* is likely, while fever in the absence of vomiting and an incubation period of more than 18 hours points towards an enteric infection such as *Salmonella, Shigella*, *Campylobacter* or *Yersinia* (see Section 4.6 for clinical features of foodborne pathogens).

Time

The time course of an outbreak is usually shown as a histogram, with the number of cases on the y-axis and the date of onset of illness on the x-axis. This graph, called an *epidemic curve*, may help in:

- confirming the existence of an epidemic;
- allowing forecasting of the further evolution of the epidemic;
- identifying the mode of transmission;

- determining the possible period of exposure and/or the incubation period of the disease under investigation;
- identifying outliers in terms of onset of illness that might provide important clues as to the source.

To draw an epidemic curve, the onset of illness must be known for each case. For diseases with long incubation periods, day of onset is sufficient. For diseases with a short incubation period – such as most foodborne pathogens – day and time of onset are more suitable.

The unit of time on the x-axis is usually based on the apparent incubation period of the disease and the length of time over which cases are distributed. As a rule of thumb, the x-axis unit should be no more than one-fourth of the incubation period of the disease under investigation (though this rule may not apply if the outbreak has occurred over a prolonged period of time). Thus, for an outbreak of salmonellosis with an average incubation period of 24 hours and cases confined to a few days, a 6-hour unit on the x-axis would be appropriate (Fig. 4.5).

If the disease and/or its incubation time are unknown, several epidemic curves with different units on the x-axis can be drawn to find one that portrays the data best. The pre-epidemic period on the graph should be shown to illustrate the background or 'expected' number of cases or the index case. If the outbreak is due to a known source (e.g. a particular food served at a common event such as a wedding), this information can also be labelled on the epidemic curve.

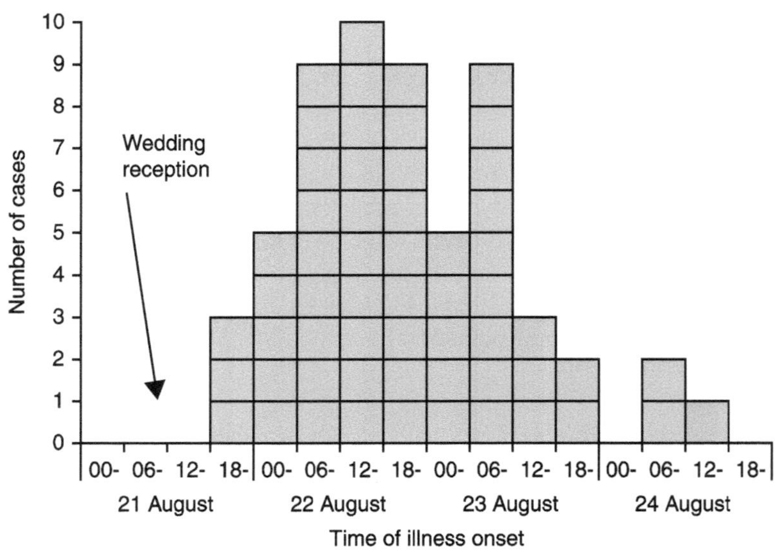

Fig. 4.5 Date and time of onset of illness among cases (n = 58), salmonellosis outbreak, wedding reception, Dublin, Ireland, 1996 (Source: Grein *et al.*, 1997).

The shape of an epidemic curve is determined by:

- the epidemic pattern (point source, common source or person-to-person spread);
- the period of time over which persons are exposed;
- the incubation period for the disease.

Place

Assessment of an outbreak by 'place' provides information on the geographic extent of the outbreak and may demonstrate clusters or patterns that provide important clues about its cause. Geographical information is best displayed by the use of maps. These can be produced by hand or using sophisticated geographic information systems.

Person

The purpose of describing an outbreak by 'person' characteristics is to identify features common to cases, as a clue to etiology or sources of infection. Age, sex, ethnicity, occupation or virtually any other characteristics can be used to describe the case population. If a single or specific characteristic emerges, this often points towards the population at risk and/or towards a specific exposure. For example, it may be apparent that only certain students in a school became ill, or only workers in a single factory or a group of people who attended a local restaurant were involved.

Determine who is at risk of becoming ill

A measure of disease frequency is important in characterizing an outbreak. The most common measure of disease frequency in epidemiology is the *rate*. Rates adjust for differences in population size and thus allow comparing the occurrence of disease among various subgroups (Table 4.2). Calculating rates of disease not only requires knowledge of the number of cases but also of the number of people in the population group(s) in which the disease can potentially occur in a given period of time (referred to as the *denominator*). This population group is called the *population at risk* and is usually defined on the basis of general demographic factors. For example, if the disease only affects children between 5 and 14 years of age, then the population at risk is the children in this age group living in the

Table 4.2 Cholera attack rate by age group, Mankhowkwe camp, Malawi, March–May 1988, showing that persons aged 15 years and older had the highest rates of disease

Age group (years)	Number of cases	Population	Attack rate (%)
< 5	131	5 303	2.5
5–14	261	12 351	2.1
≥ 15	392	12 091	3.2
Total	784	29 745	2.6

Note: Source: Moren, 1991.

area of the outbreak. Excluding population groups in which the disease does not occur helps the investigation to focus only on those affected, leading to clearer findings and more effective intervention and control activities. If only a certain ethnic group within a region is involved, for example, the investigation may focus on food items specific to that group.

A commonly used type of disease rate in outbreak investigations is the *attack rate*. An attack rate is a key factor in the formulation of hypotheses. It is calculated as the number of cases in the population at risk divided by the number of persons in the population at risk.

Sometimes, it is not possible to calculate *rates* because the population at risk is not known. In this situation, the distribution of cases themselves may help in formulating hypotheses.

Develop explanatory hypotheses

At this stage of the investigation, the data need to be summarized and hypotheses formulated to explain the outbreak. Hypotheses should address the source of the agent, the mode and the vehicle of transmission and the specific exposure that caused the disease.

If epidemiological, laboratory or food data strongly support a hypothesis, formal testing may be unnecessary. If such support is lacking or important questions remain unanswered, further studies may be needed. For example, often descriptive epidemiology will be able to explain the source of the outbreak and the general mode of transmission but will not reveal the specific exposure that caused the disease. Analytical epidemiological studies are then used to test the hypotheses.

4.6 Analytical epidemiological investigations

Analytical epidemiology studies frequently involve comparing the characteristics of a group of well persons to those of ill persons, in order to quantify the relationship between specific exposures and the disease under investigation. The two most commonly used types of analytical studies in outbreak investigations are *cohort studies* and *case–control studies*. When investigating outbreaks, a rapid result may be required to assist in control efforts and it may be advisable to conduct a limited analytical study initially. More thorough investigations can be conducted later, for example, to increase the knowledge basis associated with a particular food pathogen.

The value of a comparison group to identify specific exposures is illustrated by the following example: In a school outbreak of gastroenteritis, 30 cases are identified. Interviewing all 30 cases about their food consumption shows that all ate vanilla ice cream purchased from a street vendor one day before illness. Consumption of other foods was also enquired about, but no other food item was consumed by as many cases as vanilla ice cream.

Comparing the 30 cases to a group of 60 healthy students from the same school reveals that the entire group of healthy students also ate vanilla ice cream purchased from the same street vendor. By comparing other exposures, it is also revealed that most of the 30 cases had lunch in the school canteen the day before illness while most of the healthy students did not. This difference indicates that food from the school canteen is the more likely vehicle for the outbreak than vanilla ice cream. The finding that all cases had eaten vanilla ice cream merely reflects its popularity among the students.

4.6.1 Retrospective cohort studies

Retrospective cohort studies are feasible for outbreaks in small, well-defined populations in which all exposed and all non-exposed persons are identifiable. These studies compare the occurrence of disease among those who were exposed to a suspected risk factor with those who were not. For example, all persons attending a wedding reception (the 'cohort') may be interviewed to determine whether they became ill after the reception and to identify what foods and drinks they had consumed. After collecting information from each attendee, attack rates for illness are calculated for those who ate a particular food and for those who did not eat that food (Table 4.3).

In this example, out of a total of 68 persons who ate food 'A', 48 fell ill (attack rate 48/68 or 71%). The attack rate for those who did not eat food 'A' was two out of 102 or two per cent. Food 'A' is a likely risk factor for illness because:

- the attack rate is high among those exposed to food 'A' (71%);
- the attack rate is low among those not exposed to food 'A' (2%), so the difference (risk difference) between the two attack rates is high (69%);
- most cases (48/50 or 96%) were exposed to food 'A'.

Table 4.3 Cohort study

Exposure	Ill	Not ill	Total	Attack rate
Ate food 'A'	48	20	68	71%
Did not eat food 'A'	2	100	102	2%
Total	50	120	170	29%

Note: In this example, of a total of 68 persons who ate food 'A', 48 fell ill (attack rate 48/68 or 71%). The attack rate for those who did not eat food 'A' was 2/102 or 2%. Food 'A' is a likely risk factor for illness because:

- the attack rate is high among those exposed to food 'A' (71%);
- the attack rate is low among those not exposed to food 'A' (2%), so the difference (risk difference) between the two attack rates is high (69%);
- most cases (48/50 or 96%) were exposed to food 'A'.

In addition, a ratio of the two attack rates, known as the relative risk (RR), can be calculated in the following way:

$$\text{Relative risk (RR)} = \frac{\text{Attack rate for those who ate food 'A'}}{\text{Attack rate for those who did not eat food 'A'}} = \frac{71\%}{2\%} = 35.5$$

RR has no units and is a measure of the strength of association between the exposure and the disease. In the above example, the RR associated with eating food 'A' is 35.5. This means that persons who ate food 'A' were 35.5 times more likely to develop disease than those who did not.

4.6.2 Case–control study

In many circumstances, a clearly defined 'cohort' of all exposed and non-exposed persons cannot be identified or interviewed. In such situations, when cases have already been identified during a descriptive study and information gathered from them in a systematic way, a case–control study can be an efficient study design. In a case–control study, the distribution of exposures among cases and a group of healthy persons ('controls') are compared to each other (Table 4.4).

The questionnaire used for the controls is identical to that administered to the cases, except that questions about the details of clinical illness may not pertain to them. In this example, 96% of all cases had consumed food 'A' compared to only 17% of the controls. This suggests that consumption of food 'A' is associated with illness in one way or another. Unlike in a cohort study, attack rates (and therefore RR) cannot be calculated since the total number of persons at risk is not known. Instead, a different measure of association, called the *odds ratio* (OR), is used in case–control studies. The OR is calculated as the cross-product of a two-by-two table (the number of cases exposed times the number of controls not exposed, divided by the number of controls exposed times the number of cases not exposed; see Table 4.5).

For rare conditions (i.e. less than five per cent in the general population are affected), the OR is a good estimate of the RR. Thus, in this example, an exposure OR of 120 for food 'A' can be interpreted as 'the odds of having been exposed to the contaminated food in those who developed the disease "A" was 120 times that of people who did not eat food "A"'. This OR means that there is a very strong association between being a case and consumption of food 'A'. Like in a cohort study, statistical significance can be calculated to determine the probability that

Table 4.4 Case–control study

Exposure	Cases	Controls	Total
Ate food 'A'	48	20	68
Did not eat food 'A'	2	100	102
Total	50	120	170
Percent exposed	96%	17%	40%

Table 4.5 Example of a two-by-two table from a case–control study

	Cases	Controls	Total
Ate food 'A'	48	20	54
Did not eat food 'A'	2	100	21
Total	46	29	75

$$\text{Odds ratio} = \frac{[48 \times 100]}{[20 \times 2]} = 120$$

Chi square 92.6, p-value $< 6.10^{-22}$

such an OR could have occurred by chance alone. For the example above, this probability is extremely small ($1/6 \times 10^{22}$).

Choosing controls

An important decision in the design of a case–control study is defining who should be the controls. Conceptually, controls must not have the disease in question but should represent the population from which the cases come. In this way, controls provide the level of background exposure that one would expect to find among cases. If cases have a much higher exposure than controls, then exposure may be associated with disease.

In a community outbreak, a random sample of the healthy population may be the best control group. Sometimes, such community controls are identified by visits to randomly selected homes in the community of interest or telephone calls to randomly selected telephone numbers within the area. Other common control groups consist of:

- neighbours of cases;
- patients from the same physician practice or hospital who do not have the disease in question;
- family members or friends of cases;
- people who attended an implicated event but did not become ill;
- people who ate at an implicated food service facility during the time of exposure but did not become ill.

When designing a case–control study, the number of controls must be considered. While the number of cases is limited by the size of the outbreak, the number of potential controls will usually be more than needed. In general, the more the subjects who are included in a study, the easier it will be to find a statistical association between exposure and disease. In an outbreak of 50 or more cases, one control per case will usually suffice. In smaller outbreaks, one can use two, three or four controls per case. Increasing the number of controls beyond four per case, however, will rarely be worth the effort.

Box 4.2 Example of a cohort study

Table A is based on an outbreak of gastroenteritis following a church supper. Of the 80 persons attending the supper, 75 were interviewed. Forty-six met the case definition. Attack rates were calculated for those who did and did not eat each of the 14 food items.

Table A Attack rates by food items served at church supper, Oswego, New York, April 1940

	Number of persons who ate food item			Number of persons who did not eat food item		
	Ill	Total	Attack rate (%)	Ill	Total	Attack rate (%)
Baked ham	29	46	63	17	29	59
Spinach	26	43	60	20	32	62
Mashed potatoes	23	37	62	23	37	62
Cabbage salad	18	28	64	28	47	60
Jello	16	23	70	30	52	58
Rolls	21	37	57	25	38	66
Brown bread	18	27	67	28	48	58
Milk	2	4	50	44	71	62
Coffee	19	31	61	27	44	61
Water	13	24	54	33	51	65
Cakes	27	40	67	19	35	54
Vanilla ice cream	43	54	80	3	21	14
Choc. ice cream*	25	47	53	20	27	74
Fruit salad	4	6	67	42	69	61

* Excludes one person who was unsure of consumption

Looking at this table the most likely vehicle is vanilla ice cream. It has the highest attack rate (80%) for those who ate vanilla ice cream and the lowest for those who did not. Forty-three of the 47 cases can be "explained" by having eaten vanilla ice-cream. The attack rates for the other 13 food items do not display the same characteristics. Table B shows the same data for vanilla ice cream in the format of a two-by-two table which makes the calculation of attack rates, relative risks and statistical significance easier to visualize:

Table B Two-by-two table for consumption of vanilla ice cream (cohort study)

	Ill	Well	Total	Attack rate (%)
Ate vanilla ice cream	43	11	54	79.6
Did not eat vanilla ice cream	3	18	21	14.3
Total	46	29	75	61.3

RR = 79.6/14.3 = 5.6

The relative risk (RR) for eating vanilla ice cream is 79.6/14.3 or 5.6. This means that persons who ate vanilla ice cream were 5.6 times more likely to become ill than those who did not.

To determine the probability that the relative risk of 5.6 could have occurred by chance alone a statistical significance test can be calculated. This shows that the probability of obtaining a relative risk of 5.6 or even higher is 1/5 000 000 and therefore very unlikely to have occurred by chance alone.

Source: Goss, 1976

4.6.3 Dose response

A dose response is present if the risk of illness increases with increasing amount or duration of exposure. For example, if persons who ate two portions of a stew were more likely to become ill than people who ate only one portion, then this would suggest a 'dose response'. Finding a dose response supports the hypothesis that a particular exposure caused illness.

Looking for dose response is particularly important in outbreaks where cases and the comparison group (i.e. controls in case–control studies and unaffected persons in cohort studies) were exposed to the same risk factors. When the entire study population has been exposed to the same risk factors, demonstrating a dose response can be particularly helpful in assessing a situation. Additional information on these and other topics pertaining to epidemiologic and statistical aspects of investigating outbreaks is available free of charge on the Internet (WHO, 2002; http://Malaria.who.int/cmc_upload/0/000/015/866/basicepidemiology_tg-en.pdf, http://www.phppo.cdc.gov/phtn/catalog/pdf-file/Epi_Intro_1.pdf).

4.6.4 Addressing additional research issues

Outbreaks provide unique opportunities to address scientific questions above and beyond the immediate requirements of the investigations. While the rapid control of an outbreak must remain the primary objective for the investigator, additional research questions or collection of additional data related to the pathogen or the food under investigation may be addressed without jeopardizing this objective. Outbreak investigations can be an important opportunity to learn about a pathogen, the emergence of drug resistance and other important aspects of the epidemiology of foodborne disease.

Box 4.3 Example of a case-control study

Table A Odds ratios for exposure to foods served* in hospital "X", Dublin, Ireland, 1996

	Cases (n = 65)		Controls (n = 62)		
	Ate	**Did not eat**	**Ate**	**Did not eat**	**Odds ratio**
French onion soup	8	51	15	45	0.47
Baked ham	21	37	18	42	1.32
Parsley sauce	18	40	15	45	1.35
Cold salads	5	54	8	52	0.60
Creamed potatoes	23	35	23	35	1.00
Turnips and cabbage	30	29	21	38	1.87
Chicken curry rice	15	44	7	53	2.58
Sandwiches	6	53	3	56	2.11
Danish pastries	1	58	6	53	0.15
Chocolate mousse cake	42	16	5	53	27.83
Ice cream	10	48	16	43	0.56
Scones	1	58	4	56	0.24

* Persons who were uncertain about consumption of a particular food item are excluded

Table A is based on a salmonellosis outbreak in a hospital. Sixty-five patients and staff members met the case definition. Their exposures to specified foods were compared to those of 62 healthy patients and staff members. To determine the most likely vehicle of the outbreak, odds ratios were calculated for a total 56 food items served during breakfast, lunch and dinner over a three-day period (Table A shows only food items served during one lunch). The highest odds ratio was found for consumption of chocolate mousse cake.

Table B Two-by-two table for consumption of chocolate mousse cake (case–control study)

	Cases	**Controls**	**Total**
Ate chocolate mousse cake	42	5	47
Did not eat chocolate mousse cake	16	53	69
Total	58	58	115

$$\text{Odds ratio (OR)} = \frac{[42 \times 53]}{[5 \times 16]} = 27.8$$

The odds ratio for being exposed to chocolate mousse cake was 27.8. As salmonellosis is infrequent in the general population (and even in hospital) this odds ratio can be taken as a relative risk estimate, i.e. the risk of developing illness was much higher among persons who ate chocolate mousse cake than among those who did not.

Source: Grein, 1997

4.7 Environmental and food investigations

Food investigations (often also referred to as environmental or sanitary investigations) are conducted in coordination with epidemiologic and laboratory investigations to find out how and why an outbreak occurred, and most importantly, to institute corrective action to avoid similar occurrences in the future.

An environmental and food investigation performed in the context of a foodborne disease outbreak is distinctly different from a routine regulatory inspection to identify regulatory violations. Outbreak-related environmental investigations generally occur concurrently with laboratory and epidemiologic investigations and should be guided by data as it becomes available from other components of a multi-disciplinary investigation. During such investigations, efforts should be made to understand actual conditions at the time the suspected foods were prepared (i.e. prior to the outbreak), rather than simply observing current conditions. A thorough investigation should be done on each suspected food item that has been (or could be) implicated in the outbreak.

Examples of records that may be useful in an investigation include:

- menus, recipes, or product formulations;
- processing records;
- purchasing and inventory records;
- shipping records and other documentation of the source of an implicated product;
- hazard analysis and critical control points (HACCP) plans and records;
- corrective action records;
- flow diagrams;
- floor plans of the establishment;
- complaint records;
- cleaning records;
- food laboratory testing results;
- past inspection records;
- personnel records (including who was working when and absenteeism).

Because the amount of physical evidence may quickly diminish with time after an outbreak has been identified, associated food investigations should be carried out as soon as possible.

4.7.1 Investigation of food establishments

During a foodborne disease outbreak, investigation of a food establishment will often require:

- interviewing managers;
- interviewing any employees who may have played a role in the processing or preparation of suspected foods;
- reviewing of employee records (to determine whether some were out ill during the period of interest);

- a review of the overall operations and hygiene;
- a specific assessment of procedures that a suspect food underwent;
- food and environmental sampling;
- a review of food worker health and hygiene, including specimens for analysis;
- an assessment of the water system and supply;
- measuring temperatures, pH and water activity with appropriate equipment.

Investigations should be guided by what is already known about an outbreak from epidemiologic and laboratory investigations and the known reservoirs for the suspected agent. If a food has been incriminated epidemiologically, efforts should focus on how this particular food became contaminated. If laboratory investigations have identified a pathogen, efforts may focus on foods and conditions known to be associated with the particular pathogen. Food investigations without a clear focus can be expensive, time-consuming and of limited value.

4.7.2 Investigation of a suspect food

When investigating the role of a suspect food, its complete processing and preparation history should be reviewed, including sources and ingredients, persons who handled the items, procedures and equipment used, potential sources of contamination, and time and temperature conditions to which foods were exposed.

1 *Product description*: The suspect food should be fully described including information about:
   (a) all raw materials and ingredients used (menus, recipes, formulations);
   (b) sources of the ingredients;
   (c) physical and chemical characteristics (including pH, water activity [A_w]);
   (d) use of returned, reworked or leftover foods in processing;
   (e) intended use (e.g. home use, catering, for immediate consumption, for vulnerable groups).
2 *Observation of procedures from receipt to finish*: Observations must cover the entire range of procedures, focus on actual processes and work practices and include cleaning methods, schedules, personal hygiene of food handlers and other relevant information. The temperature history (temperature and duration) of the suspect food should be recorded as completely as possible, including while the food was stored, transported, prepared, cooked, heat-processed, held warm, chilled or re-heated. Observation of food handling practices may be valuable for small-scale operations and in the domestic setting as well as in commercial operations.
3 *Interviewing food handlers*: All food handlers who were directly involved in producing, preparing or handling suspect foods should be interviewed. Information should be obtained on the exact flow of the suspect food, its condition when received by the worker, the manner it was prepared or handled, and about unusual circumstances or practices during that time. Recent illnesses of food handlers (including before, during or after the date of the

outbreak exposure) and times of absence from work should also be noted. Specimens from ill food handlers should be obtained for microbial analysis. If any employee is found to be infected with the agent of concern, it is very important to differentiate whether he or she is infected because of having eaten the same food or he or she is a potential source of the problem. At every step of the process, data should be evaluated with respect to contamination, growth/proliferation and survival factors associated with the suspected pathogen(s).

4 *Making appropriate measurements*: Product temperatures during processing and storage and time sequences of operations should be measured and recorded as appropriate. This includes:
   (a) time and temperature conditions to which suspect foods were exposed;
   (b) water activity (A_w), moisture and pH of suspect foods;
   (c) size of containers used in procedures, depth of food in containers, etc.
   Again, attempting to understand actual conditions at the time foods implicated in the outbreak were prepared is paramount.
5 *Drawing a flow diagram of the operations*: All information and measurements should be entered in a flowchart to facilitate assessment of factors that may have contributed to the outbreak.
6 *Conducting an 'outbreak hazard analysis'*: Hazard analysis in an outbreak situation should address the following questions at each step of the processing of potentially implicated foods:
   (a) Could pathogens have been introduced at any stage?
   (b) Could pathogens already present have been able to grow at any stage?
   (c) Could pathogens have survived processes designed to kill them?

This should also include observation of the foodhandling environment, including assessing things such as the location and availability of sinks and appropriate hand-washing facilities and determining whether separate areas are maintained for the preparation of raw and ready-to-eat foods.

4.7.3 Food and environmental sampling

If laboratory facilities are available, appropriate food and environmental samples should be taken as early as possible as the amount of physical evidence will diminish with time. Always alert the laboratory before you leave to collect the samples. They can provide sampling materials regarding the type of specimens to be collected, their quantity, storage, packing and transport.

Food samples

Laboratory analysis of foods for microbial or chemical contamination is time- and resource-intensive and fraught with limitations regarding sampling and handling errors. Targeted sampling and laboratory analysis of foods should be directed by epidemiologic and environmental investigations. If an implicated food has not been identified at the time of sampling, a large number of specimens may be collected and stored for subsequent laboratory testing as additional information becomes available.

Examples of food samples that may be appropriate for collection and testing include:

- ingredients used to prepare incriminated foods;
- leftover foods from a suspect meal;
- foods from a menu that has been incriminated epidemiologically;
- foods known to be associated with the pathogen in question;
- foods in an environment that may have permitted the survival or growth of micro-organisms.

If a packaged food item is suspected of being involved in an outbreak, it is particularly important to collect unopened packages of that food from the same lot, if available. This can help determine whether the food was contaminated prior to receipt at the site of preparation. If there are no foods left from a suspect meal, samples of items that were prepared subsequently, but in a similar manner, may be collected instead, although findings from these tests must be interpreted carefully. If ingredients and raw items are still available, they should also be sampled. Storage areas should be checked for items that may have been overlooked. Even food retrieved from garbage containers may provide information useful in an investigation.

Environmental samples

The purpose of collecting environmental samples is to trace the sources and the extent of contamination that may have led to the outbreak. Samples may be taken from working surfaces, food contact surfaces of equipment, containers or other surfaces such as refrigerators, door handles, etc. Environmental samples may also include clinical specimens (such as fecal specimens, blood or nasal swabs) from food workers and water used for food processing.

Raw poultry, pork, beef and other meats are often contaminated with *Salmonella, Campylobacter jejuni, Yersinia enterocolitica, Clostridium perfringens, Staphylococcus aureus, Escherichia coli* O157 and other pathogens by the time they come into kitchens. If any of these agents are suspected in an outbreak, meat scraps, drippings on refrigerator floors and deposits on saws or other equipment can be helpful in tracing the source of contamination. Swabs can also be taken from tables, cutting boards, grinders, slicing machines and other utensils that had contact with the suspect food. However, as these pathogens are often present in such raw products, their detection does not automatically imply that they were the cause of the outbreak.

Food handlers

Food workers can be a source of foodborne contamination. Stool specimens or rectal swabs may be collected from food handlers for laboratory analysis to identify potential carriers or sources of contamination. Toxin-producing strains of *S. aureus* are carried in the nostrils, on the skin and occasionally in faeces of many healthy persons. If *S. aureus* intoxication is suspected, swabbing of the lower half-inch of the nostrils of food handlers can be performed. Swabs should also be taken from skin lesions (pimples, boils, infected cuts, burns, etc.) on unclothed areas of the body. Arrangements should be made for workers to be examined by a medical practitioner as appropriate. If hepatitis A virus (HAV) is suspected, blood from foodservice workers can be tested for IgM antibodies against HAV, which is an indication of acute infection (Heymann, 2004).

If ill food handlers are identified, an immediate decision will need to be made about excluding those people from work until symptoms have resolved or additional investigations have been done. Local jurisdictions may have different policies and rules regarding exclusion of foodservice workers, and criteria for allowing them to return to work, though guidelines exist (Heymann, 2004).

Food traceback

If a food investigation fails to identify a source of contamination at the place of preparation (e.g. infected worker or cross-contamination), contamination may have occurred before the food or ingredient arrived at the establishment. The simultaneous occurrence of multiple outbreaks due to the same pathogen at different sites is often evidence of primary contamination. Many raw foods may commonly be contaminated (primary contamination). Primary contamination may be more or less ubiquitous (e.g. *Bacillus cereus* in grain) or so common (e.g. *Salmonella* in poultry) that food safety measures will rely on subsequent procedures such as thorough cooking to ensure that food is fit for consumption. In these instances, the place of primary contamination may or may not be investigated depending on available resources, priorities and the epidemiological situation with regard to the outbreak.

Other situations in which tracing contamination to raw foods may be important and should be considered include the following:

- the pathogen is uncommon, newly emerging or re-emerging or causes serious disease (e.g. *E. coli* O157);
- it can be expected that foods will be eaten raw or lightly heated (e.g. shellfish, produce, shell eggs);
- little is known about a pathogen and there is a need to advance knowledge about its ecology;
- unlicensed or illegally sold foods were involved;
- there is a suspicion that foods were adulterated;
- the source of contamination is unusual;
- a new or unusual vehicle is involved.

In such situations, a ‘traceback’, or tracing of the implicated food backward through its distribution and production channels to its place of origin, is commonly performed. The purposes of ‘tracebacks’ of foods include:

- identifying the source and distribution of foods in order to alert the public and remove contaminated product from the marketplace;

- comparing distribution of illnesses and distribution of product in order to strengthen an epidemiologic association (sometimes referred to as an 'epi' traceback);
- determining the potential route or source of contamination by evaluating common distribution sites, processors or growers.

Food tracebacks are often resource-intensive investigations requiring the coordination of multiple investigators from different agencies and organizations, often spread across different jurisdictions. Such investigations frequently require the review of detailed data on dates, quantities, sources and conditions of foods received, collection of original shipping containers and labels or other documentation, and information on lot numbers, facilities involved, production dates and the like. Traceback investigations can result in irreparable damage to food firms. Therefore, it is critical that each piece of the investigation (epidemiologic, laboratory and environmental) is thorough, complete and accurate.

The investigation at a farm or dairy will follow the same principles as that for investigating a food establishment. However, depending on the type of food product or animal involved, specific knowledge and skills may be needed to carry out the actual investigations. Most commonly, veterinarians, agriculturists, microbiologists and water supply experts will conduct these investigations in collaboration with epidemiologists. Traceback investigations may lead to the identification of an ongoing public health threat and a need to take appropriate actions, including recall of foods, closing a facility, confiscating foods or warning consumers of a potential risk. Investigators should be prepared to coordinate activities closely with other appropriate agencies and organizations to ensure a prompt and effective response as necessary.

4.8 Laboratory investigations

Most outbreaks of foodborne disease are microbiological in origin and their investigation will usually require a microbiology laboratory. Although much less common than microbiological events, outbreaks due to chemically contaminated food also occur. Symptoms of both microbiological and chemical causes can be similar and may be difficult to distinguish even by laboratory tests. While the general principles of the investigation apply to both types of incidents, it is important to involve a chemical laboratory from the beginning if the possibility of a chemical cause seems likely.

The role of the *clinical laboratory* in foodborne disease outbreak investigations includes:

- ensuring that appropriate clinical specimens are collected;
- arranging appropriate laboratory investigations on clinical samples;
- working with other members of the investigation team to identify and characterize the pathogen involved in the outbreak.

The role of the *food laboratory* in foodborne disease outbreak investigations includes:

- advising on appropriate samples to be taken from food;
- performing appropriate laboratory investigations of the food to identify the suspect pathogens, toxins or chemicals;
- advising on further sampling when a specific agent is found in the food (e.g. guiding collection of clinical specimens from food handlers);
- working with the clinical laboratory to arrange for typing or additional characterization of organisms (e.g. serotyping, phage typing, molecular subtyping, antibiograms) as appropriate;
- supporting epidemiologic and environmental investigations in detecting the pathogen in the implicated food and understanding how the outbreak occurred.

4.8.1 Microbiological analyses

In any outbreak of a suspected foodborne disease, a microbiologist should be consulted as soon as possible. This person should be a member of the OCT.

Clinical samples

Diagnosis of most infectious diseases can only be confirmed if the etiological agent is isolated and identified from ill persons. This is particularly important when the clinical diagnosis is difficult to make because signs and symptoms are non-specific, as is the case with many foodborne diseases. Faecal samples are the most commonly collected specimens. Other specimens include vomitus, urine, blood and clinical specimens obtained from food handlers during the food investigations (e.g. swabs from rectum, nostrils, skin or nasopharynx).

The laboratory should be asked about appropriate methods for collection, preservation (including selection of the appropriate collection material) and shipment of specimens. An indication should be given of how many samples are likely to be sent for analysis and if the laboratory's resources are sufficient to deal with them.

Laboratory samples should be taken from ill persons as soon as possible. Whenever possible, samples should be taken from persons who have not undergone antibiotic treatment for their illness. In large outbreaks, specimens should be obtained from at least ten to 20 individuals (ideally 15–20% of all cases) who manifest illness typical of the outbreak and from some exposed, but not ill, persons. Once the diagnosis has been confirmed, there is usually no need to obtain additional samples if individuals manifest characteristic symptoms. In smaller outbreaks, specimens should be collected from as many cases as practicable.

All containers should be labelled with a waterproof marking pen before or immediately after collection with the patient's name, identification, date and time of collection and any other information required by the laboratory.

Molecular typing
Recent advances in laboratory methods have contributed substantially to improvements in the detection and investigation of foodborne disease outbreaks. Molecular microbiology technology has markedly changed the nature of many acute disease epidemiology investigations. Polymerase chain reaction (PCR) technology is increasingly being used to identify pathogens rapidly and, in many cases, allow determination of subtypes that previously required time-consuming and resource-intensive methods.

Pulsed-field gel electrophoresis (PFGE) can provide 'DNA fingerprints' of bacterial isolates from clinical and food specimens. If the PFGE patterns are the same, the investigators have additional evidence that the suspected food item is implicated in the event. PFGE can also help investigators include related cases and exclude concurrent cases that are epidemiologically unrelated to an outbreak. Such subtyping can be particularly useful when a pathogen implicated in an outbreak is very common and its presence in related specimens (e.g. cases, food and farm animals) may be purely coincidental. Further subdivision into types/subtypes may show them to be distinct and therefore unrelated, or indistinguishable, thus increasing the significance of their isolation.

Genetic sequencing technology has become more readily available and has been useful for assessing the relatedness of various pathogens involved in foodborne and waterborne outbreaks. For example, sequencing of HAVs collected during three large outbreaks associated with green onions demonstrated that similar virus strains caused all three outbreaks and that this strain was related to hepatitis A strains commonly isolated from patients living in the region where the green onions were grown. Sequencing of noroviruses is also becoming increasingly useful in identifying relatedness among potential outbreak-associated viruses.

Many subtyping and molecular microbiology tests are available only at specialized reference laboratories and may require coordination with the primary laboratory involved in an outbreak investigation.

Chemical investigations
In acute chemical exposures, most toxins or their metabolites are rapidly cleared from easily accessible specimens such as blood. Therefore, prompt collection and shipment of specimens is of critical importance.

When collecting samples for chemical analyses, it is important to closely collaborate with the analytical laboratory, make arrangements in advance for chemical samples to be analysed and seek advice about what specimens should be collected and how. The types of specimens to be collected will depend on the suspected chemicals. In an emergency where it is not possible to contact the laboratory, biological specimens (whole blood, serum, urine, vomitus) should be collected as soon as possible, sealed in a clean container and sent to the laboratory promptly. Substances from the ambient air, the collector's skin or clothes, or interfering substances in collection and storage supplies may be concentrated and measured along with the specimens, yielding inaccurate results.

Because care must be taken to avoid cross-contamination, contaminant-free materials (such as specialized collection containers) may be provided by the laboratory to ensure that extraneous contamination is kept at a minimum. Consultation with the testing laboratory is important in accurately interpreting results.

4.9 Control measures

The primary goal of outbreak investigations is to control ongoing public health threats and to prevent future outbreaks. Ideally, control measures are guided by the results of these investigations, but as this may delay the prevention of further cases, it is often unacceptable from a public health perspective. At the same time, specific interventions such as recalling a food product or closing a food premise can have serious economic and legal consequences and must be based on accurate information. Thus, implementing control measures is often a balancing act between the responsibility to prevent further cases and the need to protect the credibility of an institution.

4.9.1 Control of source

Once investigations have identified that a food or a food premise is associated with transmission of the suspected pathogen, measures should be taken to control the source. Steps may include:

- removing implicated foods from the market (food recall, food seizure);
- modifying a food production or preparation process;
- closing food premises or prohibiting the sale or use of foods.

4.9.2 Closing food premises

If on-site inspections reveal a situation that poses a continuing health risk to consumers, it may be advisable to close the premises until the problem has been solved. This may occur with the agreement of the business or be enforced by law (closing order). Once a premise has been closed, the responsible authorities should monitor the premises and ensure that they remain closed until reopening is authorized by appropriate authorities.

4.9.3 Removing implicated foods from the market

The objective of food recall and food seizure is to remove implicated foods as efficiently, rapidly and completely as possible from the market. A *food recall* is undertaken by any business that manufactures, wholesales, distributes or retails the suspect food, for example large corporations, partnerships or family-owned businesses. It may be initiated by the business itself or on request of an appropriate health authority. *Food seizure* is the process by which an appropriate authority removes a food product from the market if the business does not comply with a recall. In most cases, businesses will comply with a request for food recall to protect themselves from private lawsuits and damaged reputation where appropriate consumer protection legislation exists. Government regulatory agencies will often have an active role in removing implicated foods from distribution, which will vary by jurisdiction. In many situations, company recalls of products are voluntary at the suggestion of government authorities.

Once the appropriate authorities have decided to recall a food product, they should:

- communicate with and ensure the co-operation of the business(es) involved in the recall;
- directly advise local health authorities of the recall and any enforcement action required;
- ensure appropriate public notification;
- monitor the progress of the recall and its effectiveness;
- ensure that corrective actions are taken by the recalling business.

The recalling business is usually responsible for conducting the actual recall. The extent of recall will depend on the potential risk to the consumer. A business may conduct a recall to the retail level, or, if the public health is seriously jeopardized, to the individual consumer. Means of notifying will depend on the urgency of the situation and may include press releases, fax, telephone, radio, television or letters.

Efficient recall of a widely distributed product requires that a manufacturer can identify a product by production date or lot number and that distribution records for finished products are maintained for a period of time that exceeds the shelf life of the product.

Communication with the public

Although the business may have already issued a press release, the OCT or food safety committee itself may decide to notify the public. Preferably, this is done on the same day when the decision is taken to recall a food product. Information to the public should include:

- actions that consumers should take to prevent further exposure and illness;
- name and brand of the food product (including labelling) recalled;
- problem with product, reason for recall and how the problem was discovered;
- name and location of the producing establishment and point of contact;
- locations where product is likely to be found;
- numbers, amounts and distribution;
- means of notification, that is how the establishment is recalling the product;
- a description of common symptoms of illness associated with the suspected pathogen;
- appropriate food handling information for consumers;
- actions that consumers should take if illness occurs.

Post-recall reporting by the business

After the implementation of a recall, the business should provide the food safety committee or other appropriate authorities with an interim and a final report about the recall. The reports should contain the following information:

- copy of recall notice, letters to customers, retailers, etc.;
- circumstances leading to recall;
- action taken by the business;
- extent of distribution of relevant batch of food that was recalled;
- result of recall (percentage of stock recovered or accounted for);
- method of disposal or re-processing of recovered stock;
- difficulties experienced during recall;
- action proposed for the future to prevent a recurrence of the problem.

The interim and final reports give information about the effectiveness of the recall. If they are unsatisfactory, or evidence of corrective action is inadequate, further recall action may need to be considered.

4.9.4 Modifying a food production/preparation process

Once food investigations identify faults in production or preparation processes that may have contributed to the outbreak, corrective action must be taken immediately to avoid recurrences. Examples of corrective actions are modification of a recipe, change in a process, re-organization of working practices, change in storage temperatures or modification of instructions to consumers.

4.10 Control of transmission

4.10.1 Public advice

If a contaminated food product cannot be controlled at its source, steps need to be taken to eliminate or minimize the opportunities for further transmission of the pathogen. Depending on the situation, appropriate public advice may be issued during a period of hazard, for example:

- boiling of microbiologically contaminated water or avoidance if chemically contaminated;
- advice on proper preparation of foods (e.g.: WHO Five Keys to Safer Food);
- advice to dispose of foods;
- emphasizing personal hygiene measures.

4.10.2 Exclusion of infected persons from work and school

The risk of spreading infection by infected persons depends on their clinical picture and their standards of hygiene. Persons with diarrhoea present a far greater risk of spreading infection than asymptomatic persons with subclinical illness.

Decisions about exclusion from work must be made by health authorities depending on local laws and regulations. In general, the following persons with diarrhoea or vomiting should remain off work or school until no longer infectious:

- food workers whose duties involve touching unwrapped foods to be consumed raw or without further cooking or other forms of treatment;
- persons with direct contact to highly susceptible patients or persons in whom gastrointestinal infection would have particularly serious consequences (e.g. the young, the old, the immunocompromised);
- children aged less than five years;
- older children and adults with doubtful personal hygiene or with unsatisfactory toilet, hand-washing or hand-drying facilities at home, work or school.

Even if clinically well, no person should handle unpackaged food if having any of the following conditions:

- excretor of *Salmonella* Typhi or *Salmonella* Paratyphi;
- excretor of the aetiological agents of cholera, amoebic dysentery or bacillary dysentery;
- hepatitis A or hepatitis E and all other forms of acute hepatitis until diagnosed not to be hepatitis A or hepatitis E;
- *taenia solium* (pork tapeworm) infection;
- tuberculosis (in the infectious state).

Clinically healthy persons who are asymptomatic excreters of enteric pathogens and have good hygiene pose a minimal risk and do not need to be excluded from work or school. If an ill food handler was implicated in an outbreak, recommendations should be made for preventing such problems in the future. This may include suggestions for ensuring that mechanisms are in place for ongoing screening to prevent ill persons from working.

4.10.3 Advice on personal hygiene

Advice on personal hygiene should be issued to all persons with gastrointestinal disease and include the following:

- Avoid preparing food for other people until free from diarrhoea or vomiting.
- Thoroughly wash hands after defecation, urination and before meals. Thorough handwashing with soap in warm running water and drying is the most important factor in preventing the spread of enteric diseases.
- Use your own separate towels to dry hands. Institutions, particularly schools, should use liquid soaps and disposable towels or hand dryers.
- Clean toilet seats, flush handles, wash-hand basin taps and toilet door handles with disinfectant after use. If young children are infected, these cleaning procedures must be undertaken on their behalf. Similar arrangements may also be necessary in schools and residential institutions (if temporary exclusion is not possible).
- If employed in food preparation activities, scrub your nails with brush and soap.

4.10.4 Infection control precautions

Infection control precautions for hospitalized and institutionalized persons with infectious diarrhoea (particularly easily transmissible infections such as *S.* Typhi, *Shigella*, etc.) include:

- isolation of patients (e.g. in a private room with separate toilet if possible);
- barrier nursing precautions;
- strict control of the disposal or decontamination of contaminated clothing and bedding;
- strict observation of personal hygiene measures (see above).

4.10.5 Protecting risk groups

Certain groups are at a particularly high risk of severe illness and poor outcomes after exposure to a foodborne disease. Safe foodhandling practices, including strict adherence to thorough handwashing, should be particularly emphasized to such people. Specific advice for risk groups may be considered in some circumstances. Examples include:

- advising pregnant women against consumption of unpasteurized milk, unpasteurized cheeses and other foods potentially contaminated with *Listeria*;
- advising immunocompromised persons, such as those with HIV/AIDS, to avoid eating unpasteurized milk products, raw fish, etc.;
- advising persons with underlying liver disease to avoid consumption of raw oysters and other food that may transmit *Vibrio* bacteria;
- advising persons with underlying chronic hepatitis B or C or other liver disease to be vaccinated against hepatitis A if appropriate;
- advising day care centre personnel about receiving vaccination or immunoglobulin during a hepatitis A outbreak in the institution (although this may more likely protect against secondary spread than against foodborne transmission).

4.11 End of outbreak

4.11.1 Review of outbreak

The OCT should formally decide when an outbreak is over and issue a statement to this effect. A structured review should follow all outbreaks for which an OCT had been convened. It should include a formal debriefing meeting with all parties involved in the investigation. The aims of the debriefing meeting are:

- to ensure that control measures for the outbreak are effective;
- to identify long-term and structural control measures and plan their implementation;
- to assess whether further scientific studies should be conducted;
- to clarify resource needs, structural changes or training needs to optimize future outbreak response;

- to identify factors that compromised the investigations and seek solutions;
- to change current guidelines and develop new materials as required;
- to discuss legal issues that may have arisen;
- to arrange for completion of the final outbreak report.

4.11.2 Outbreak report

An interim report should be made available by the OCT within two to four weeks after the end of the investigations followed by a written final report. The final report should be comprehensive, protect confidentiality and be circulated to appropriate individuals and authorities. The report should follow the usual scientific format of an outbreak investigation report and include a statement about the effectiveness of the investigation, the control measures taken and recommendations for the future.

In addition, a summary report should be completed and forwarded to the appropriate authorities at the national level for collation, analysis and, when appropriate, reporting to the international level (e.g. SalmNet, WHO, etc.).

4.11.3 Future trends

Further studies may be conducted after completion of the initial investigations, particularly if new or unusual pathogens were involved or additional information for risk assessment of a particular pathogen is required. The need to catch up on routine work delayed due to the outbreak investigation often makes it difficult to conduct such follow-up studies. Still, it is important that these opportunities should be considered following each outbreak – either by OCT members themselves or by others who may be in a better position to do so. This may also include publication of the outbreak in an international journal to inform the scientific community at large.

Economic evaluations of outbreaks and associated control efforts can be important to assess the cost-effectiveness of outbreak investigations and food safety measures. Foodborne outbreaks will incur costs to the:

- health care providers (diagnostic and curative services);
- population (medication, time missed from school or work, reduced activity due to long-term sequelae, death);
- food industry (closure, adverse publicity, recall, litigation);
- agencies, laboratories and other persons and organizations involved in the investigation, response and control activities.

Costs associated with outbreaks can be enormous and quantifying them may help further motivate the commitment of the food industry and other agencies to food safety.

4.12 Acknowledgements

This chapter is based on our department's publication 'Foodborne Disease Outbreaks – Guidelines for Investigation and Control' to which many authors have originally contributed. While these guidelines are the property of the WHO, we would like to acknowledge these authors' contribution herewith.

4.13 References and useful reading

BORGDORFF M W and MOTARJEMI Y (1997), *Surveillance of Foodborne Diseases: What are the Options?* World Health Organization, Geneva, Switzerland.

CIFOR (2009), *Guidelines for Foodborne Disease Outbreak Response*. Council to improve foodborne disease outbreak response (CIFOR), Atlanta, USA.

Codex Alimentarius. World Health Organization, Food and Agriculture Organization of the United Nations. Available from: http://www.codexalimentarius.net/web/index_en.jsp# or http://www.ipfsaph.org/En/default.jsp.

DEPARTMENT OF HEALTH (1994), *Management of Outbreaks of Foodborne Illness*. HMSO, London. 1994.

FLINT J A, VAN DUYNHOVEN Y T, ANGULO F J, DELONG S M, BRAUN P, KIRK M, ET AL. (2005), 'Estimating the burden of acute gastroenteritis, foodborne disease, and pathogens commonly transmitted by food: and international review', *Clin Infect Dis*, **41**, 698–704.

GOSS M (1976), 'Oswego revisited', *Public Health Rep*, **91**, 168–70.

GREGG M (ED.) (2002), *Field Epidemiology*. 2nd ed. Oxford University Press Inc., Oxford, NYC, New York.

GREIN T, O'FLANAGAN D, MCCARTHY T and PRENDERGAST T (1997), 'An outbreak of *Salmonella* Enteritidis food poisoning in a psychiatric hospital in Dublin, Ireland', *Eurosurveillance*, **2**, 84–6.

GREIN T, O'FLANAGAN D, MCCARTHY T and BAUER D (1999), 'An outbreak of multidrug resistant *Salmonella* Typhimurium at a wedding reception', *Irish Medl J*, **92**, 238–41.

HEYMANN D L (ED.) (2004), *Control of Communicable Diseases Manual*. 18th ed. American Public Health Association, Washington, DC.

JONES K E, PATEL N G, LEVY M A, STOREYGARD A, BALK D, GITTLEMAN J L, ET AL. (2008), 'Global trends in emerging infectious diseases', *Nature*, **451**, 990–3.

MOREN A, STEFANAGGI S, ANTONA D, BITAR D, ETCHEGORRY M G, TCHATCHIOKA M, ET AL. (1991), 'Practical field epidemiology to investigate a cholera outbreak in a Mozambican refugee camp in Malawi, 1988', *J Trop Med Hygiene*, **94**, 1–7.

REINGOLD A (1998), 'Outbreak investigations – a perspective', *Emerg Infect Dis*, **4**, 21–7.

SCALLAN E, MAJOWICZ S, HALL G, BANERJEE A, BOWMAN C L, DALY L, ET AL. (2005), 'Prevalence of diarrhoea in the community in Australia, Canada, Ireland and the United States', *Int J Epid*, **34**, 454–60.

THAPAR N and SANDERSON I R (2004), 'Diarrhoea in children: an interface between developing and developed countries', *Lancet*, **363**, 641–53.

WHO (1997), *Introducing the Hazard Analysis and Critical Control Point System*. World Health Organization, Geneva (Document WHO/FSF/FO/97.2).

WHO (1998), *The Application of Risk Communication to Food Standards and Safety Matters, a Joint FAO/WHO Expert Consultation*. February 2–6, Rome, Italy. Available from: http://www.who.int/foodsafety/publications/micro/feb1998/en/index.html.

WHO (2005), *The International Health Regulations*. WHO, Geneva, Switzerland. Available from: http://whqlibdoc.who.int/publications/2008/9789241580410_eng.pdf
WHO (2008), *Foodborne Disease Outbreaks – Guidelines for Investigation and Control*, World Health Organization, Geneva.
WHO (2008), *Initiative to Estimate the Global Burden of Foodborne Diseases*. WHO, Geneva, Switzerland. Available from: http://www.who.int/foodsafety/foodborne_disease/ferg/en/index.html
WHO (2008), *The Global Burden of Disease, 2004 Update*, WHO, Geneva, Switzerland.

5

Attributing the burden of foodborne disease to specific sources of infection

T. Hald and S. M. Pires, Technical University of Denmark, Denmark

Abstract: This chapter describes the currently used methodologies for attributing the burden of human illness from diseases commonly transmitted through foods to specific sources. The chapter is introduced by putting the concept of attributing in context with disease surveillance, burden of disease studies and risk assessments, and emphasises the importance of attribution studies for risk management. This is followed by a discussion of the principles, strengths and limitations of different approaches for source attribution including references to practical examples. The chapter is concluded by a discussion of future perspective and recommendations.

Key words: human illness attribution, foodborne disease epidemiology, zoonoses, transmission routes, microbial risk assessment.

5.1 Introduction

Pathogens commonly transmitted to humans through foods are responsible for a high burden of human illness and death worldwide. The World Health Organization (WHO) estimates that 1.8 million children die each year from diarrhoea, and much of the childhood diarrhoea is caused by enteric pathogens that are commonly acquired from contaminated food or water. In developed countries, up to one-third of the population each year has an infection from a pathogen commonly transmitted through foods (WHO, 2005). Humans acquire these infections through a number of routes, including eating contaminated food, contact with live animals and contact with a contaminated environment. It is well recognised that foodborne transmission is responsible for a major proportion of these infections, and foodborne illnesses may involve many different food sources and commodities. Many countries have implemented intervention programmes during the last decades to prevent and control foodborne illnesses, particularly foodborne zoonoses (Wegener *et al.*, 2003; EFSA, 2009; Korsgaard *et al.*, 2009; Williman *et al.*, 2009). However, precise measurement of the public health impact of such interventions is difficult, in part because information on the attribution of the burden of foodborne illness to specific sources is often insufficient (Thorns, 2000).

For food safety policy and in order to prioritise effective food safety interventions, it is important to determine: (1) the total burden of human illness due to pathogens that are commonly (but not only) transmitted through food, (2) the fraction of the total burden that is attributable to foods, (3) the relative contribution of each food to that fraction, and (4) the most efficient points for intervention in the farm-to-consumption continuum.

The most widely used public health indicator to quantify the burden of illness in a given population is the reported incidence. Many countries have established some kind of surveillance system, often based on diagnostics submissions to microbiological laboratories and/or on outbreak reports. Generally, such systems capture only a small fraction of the true number of illnesses, and thus specific 'burden of illness' studies are required to estimate the true incidence. Examples of such studies include Mead *et al.* in the USA (1999), Hall *et al.* in Australia (2005), Kubota *et al.* in Japan (2008) and Wheeler *et al.* in the UK (1999). These studies account for underreporting of foodborne disease and make efforts to estimate the true number of illnesses in the population in a given time period. In recent years, attention towards sero-epidemiological studies has increased. In a recent Danish study, the detected level of antibodies against *Salmonella* in the population, obtained from population-based, age-stratified random serum samples, was combined with data on the decay rate of these antibodies to estimate the incidence of *Salmonella* infections (Simonsen *et al.*, 2009).

The burden of specific illnesses (e.g. salmonellosis and campylobacteriosis) is important for informing policy makers and provides the basis for estimating the 'burden of disease'. Burden of disease is a principle used by the Global Burden of Disease Study (WHO, 2008), and includes the quantification of morbidity, all disabling complications as well as mortality in a single summary measure: the disability adjusted life years (DALY). In other words, the DALY measure accounts for variations in both severities, durations and case-fatality ratios, and makes it possible to compare and rank different diseases (Havelaar *et al.*, 2007), which consequently provides a much more useful basis for policy choices on prevention and control than the burden of illness. Once decision makers have decided which diseases to focus on, more detailed insight on the main sources and reservoirs is needed.

Efforts to quantify the importance of specific sources (including foods) and animal reservoirs for human illness have been gathered under the term 'source attribution' or 'human illness attribution' (EFSA, 2008; Pires *et al.*, 2009). Knowledge of the most important (food) sources and animal reservoirs will ensure a more targeted control of the disease in question and support risk managers in their decision of allocating resources to achieve the highest possible public health benefit. Source attribution is therefore regarded as an important tool in the process of identifying and prioritising effective food safety interventions (Havelaar *et al.*, 2007).

A variety of general methods to attribute one or more foodborne diseases to specific sources has been developed, including microbiological approaches, epidemiological approaches, intervention studies and expert elicitation approaches. Each of these methods presents advantages and limitations, and the usefulness of each depends on the public health questions being addressed (Batz *et al.*, 2005). Several research groups are conducting attribution studies, but often different terminologies and food categorisation schemes are applied. Defining scientific concepts and harmonising terminology is essential for understanding and improving attribution methodologies and sharing of knowledge across the scientific community. This chapter describes a recently published proposal for harmonised nomenclature, and discusses the various approaches for human illness attribution. The chapter is to a wide extent based on the work by Pires *et al.* (2009).

5.2 Definitions

5.2.1 Human illness source attribution

Human illness *source attribution* is defined as the partitioning of the human disease burden of one or more foodborne infections to specific sources, where the term *source* includes animal reservoirs and vehicles, e.g. foods. A *reservoir* is restricted to an animal species or non-animal substance upon which the pathogen depends for its survival. Many infectious agents have more than one reservoir (Martin *et al.*, 1987). We define a *vehicle* as an inanimate communicator of a pathogen from its original reservoir until final exposure; vehicles of foodborne pathogens are traditionally food items, but other sources, e.g. contaminated drinking or recreational water, can be vehicles of the same pathogen.

Zoonotic pathogens can be transmitted from the animal reservoirs to humans through a variety of routes. Harmonisation of the categorisation of sources is important to compare and share the results of different source attribution studies, but standardised approaches are lacking. Categorisation systems should be hierarchical to accommodate different levels of detail required for different purposes. Ideally, the categorisation scheme should be in alignment with food consumption data and be internationally standardised. A standardised scheme for the categorisation of food sources has been proposed by Painter *et al.* (2009) and presented in Fig. 5.1.

5.2.2 Points of attribution

Human illness source attribution can take place at different points along the food chain (*points of attribution*), including at reservoir, processing and exposure. The point of reservoir is most closely represented by source attribution at the production stage. Because foodborne pathogens may enter the food distribution chain at different points and/or their load may change during production, the

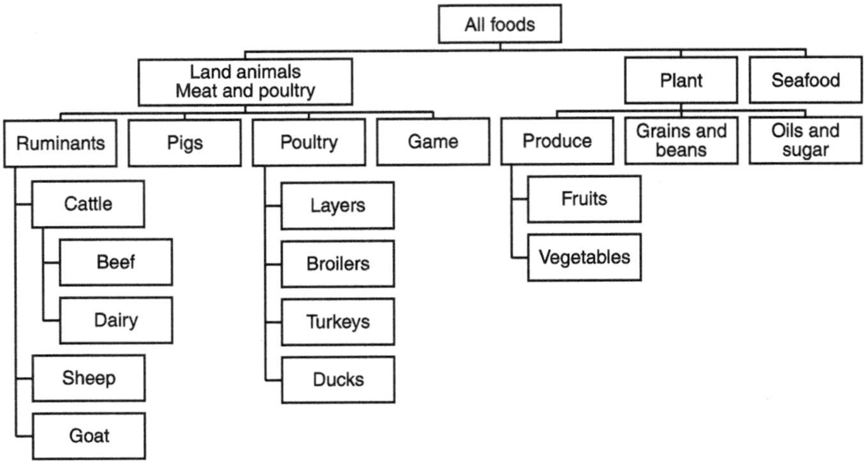

Fig. 5.1 Hierarchical scheme for categorising food items into categories within the main animal reservoirs (adapted from Painter *et al.*, 2009).

burden of disease attributed to specific sources may vary, depending on at which point along the food chain the approach focuses. For example, attribution of *Campylobacter* infections may partition more illness to chicken at the point of reservoir than it will partition to broiler meat at the point of consumption, since other foods, e.g. lettuce, may be cross-contaminated during preparation in the kitchen. Some of the methods to attribute foodborne diseases to specific sources work primarily at one point in the food chain (e.g. epidemiological approaches work primarily at the point of exposure), while other methods (e.g. expert elicitation approaches) can be more generally applied. The point of attribution consequently depends on the method chosen, which will depend on the availability of data and on the risk management question being addressed. Figure 5.2 presents the major transmission routes for foodborne infections, and indicates at which point in the transmission chain the different approaches attribute human illness.

5.3 Approaches for source attribution

Like epidemiological approaches in general, source attribution approaches describe an association between the outcome (i.e. the burden of a given disease) and the specific sources or exposures. Source attribution methods attempt to attribute the burden of disease at the population level, and do not describe causation of disease at the individual level.

Methods for source attribution of foodborne diseases include microbiological approaches, epidemiological approaches, intervention studies and expert

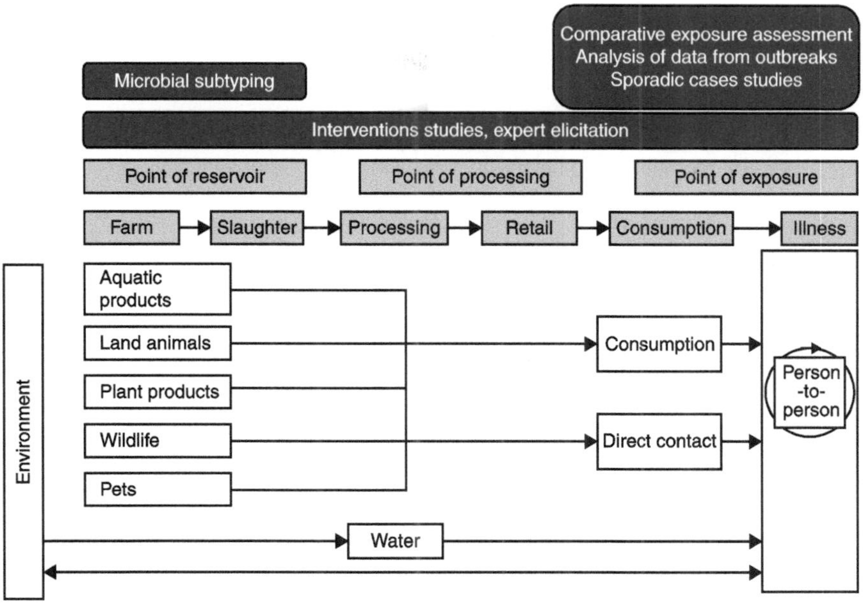

Fig. 5.2 Routes of transmission of zoonotic pathogens and points of human illness attribution (Pires *et al.*, 2009).

elicitations. The applicability of each method to address a given question will depend on a variety of factors, such as data requirements and availability, pathogen characteristics and the type of intervention aimed for. Each method presents strengths and limitations, and the utility of each will depend on the question being addressed. A summary of the main characteristics of all methods can be found in Table 5.1.

5.3.1 Microbial subtyping

The microbial subtyping approach involves characterisation of isolates of the pathogen by phenotypic and/or genotypic subtyping methods (e.g. serotyping, phage typing, antimicrobial susceptibility testing, pulsed-field gel electrophoresis and sequence-based subtyping). The principle is to compare the distribution of subtypes in potential sources (e.g. animals and food) with the subtype distribution in humans. The microbial subtyping approach is enabled by the identification of strong associations between some of the dominant subtypes and a specific reservoir or source, providing a heterogeneous distribution of subtypes among the sources. Subtypes exclusively, or almost exclusively, isolated from one source are regarded as indicators for the human health impact of that particular source, assuming that all human infections with these subtypes originate only from that source. Human infections caused by subtypes found in several

Table 5.1 Strengths and limitations of source attribution approaches

	Strengths	Limitations
Microbiological approaches		
Microbial subtyping	• Might be able to identify the most important reservoirs of the zoonotic agent, assisting prioritisation of where to focus control strategies at the animal level. • Attributes illness to the primary source, reducing uncertainty due to cross-contamination and the risk of attribution to an 'accidental' source. • Is able to follow trends over time. • Validation is possible using isolates from known sources.	• Limited to clonally disseminated pathogens that are heterogeneously distributed among the reservoirs. • No information provided on the different pathways through which the pathogen can be transmitted to humans. • Data intensive, requiring a collection of representative isolates from all (major) sources and therefore resource demanding. • Standardised subtyping methods are required (methods are changing over time).
Comparative exposure assessment	• Attributes illness to the point of exposure, taking into account the different transmission routes from the same reservoir. • Once a model is developed, new data can be easily included.	• Often limited by lack of sufficient data, which results in large uncertainties around the estimates. • Assumes linear relation between exposure and illness.
Epidemiological approaches		
Case-control studies (systematic review)	• Valuable tool to identify relevant risk factors for human infections, including sources of exposure, predisposing, behavioural or seasonal factors. • A systematic review of published case–control studies can provide an overview of the relevant exposures and risk factors for that infection. • Can identify a wide range of familiar and unfamiliar risk factors.	• Misclassification due to immunity may reduce attributable risk or even suggest protection. • Most studies only explain a small fraction of all cases. • Cases may reflect a mixture of possible sources of exposure, and it may be difficult to distinguish between these exposures. • Statistical power to determine the importance of common exposures often requires enrolment of many participants.

Table 5.1 Continued

	Strengths	Limitations
		• Misclassification of exposures due to lack of accuracy of recall may lead to an underestimation of the burden of illness attributed to specific exposures. • Need for standardisation of exposure measurement through questionnaires (e.g. *Campylobacter* is highly seasonal and in many studies cases and controls are asked questions in different time points).
Analysis of data from outbreaks	• Clear documentation that a specific pathogen was transmitted to humans via a specific food item can be available. • Data may capture the effect of contamination at multiple points from the farm to consumption chain. • A wide variety of food vehicles are represented, including less frequently identified food items. • Data from outbreak investigations may, in some countries or regions, be the most readily available source of information for source attribution.	• Outbreaks may not be detected, investigated or reported. • Quality of evidence varies and classification schemes for the data are not consistently used. • Large outbreaks, outbreaks associated with point sources, outbreaks that have short incubation periods, and outbreaks that cause serious illness, are more likely to be investigated. • Illnesses included in data from outbreak investigations may not be representative of all foodborne illnesses. • Certain food vehicles are more likely to be associated with reported outbreaks than others, which can lead to an overestimation of the proportion of human illnesses attributed to a specific food.
Intervention studies		
	• Allows for a direct measure of the impact of a given source on the number of human cases of infection, avoiding the account for the effect of external sources or risk factors.	• Interpretation of data from 'large-scale' interventions is difficult, since usually several interventions are implemented at the same time.

Continued

Table 5.1 Continued

	Strengths	Limitations
	• Large scale may reflect all complexities across the food chain.	• Complex and resource demanding studies. • Need for an extensive surveillance system in place. • Occurrence of natural experiments is unpredictable.
Expert elicitation		
	• Useful for providing interim results in countries where data are lacking. • Useful to evaluate the credibility of different attribution approaches.	• Conclusions are based on individual judgement, which may be misinformed or biased.

reservoirs are then distributed relative to the prevalence of the indicator types. This approach requires a collection of temporally and spatially related isolates from various sources and humans, and is consequently facilitated by an integrated foodborne disease surveillance programme focused on the collection of isolates from the major food animal reservoirs of foodborne diseases and from humans (Pires *et al.*, 2009).

The principle of comparing the distribution of subtypes found in animal and food sources with those found in humans to make inferences about the most important sources of human disease has been applied by several research groups (Rosef *et al.*, 1985; Van Pelt *et al.*, 1999; Sarwari *et al.*, 2001). One of the more advanced method has been developed to attribute human salmonellosis in Denmark (Hald *et al.*, 2004), which has been continuously improved to include data on antimicrobial susceptibility (Hald *et al.*, 2007) as well as data from multiple years (Pires and Hald, 2009). Using data from the integrated Danish *Salmonella* surveillance programme, a mathematical model was developed to quantify the contribution of each of the major food animal sources to human *Salmonella* infections. This model attributes domestically acquired laboratory-confirmed human *Salmonella* infections caused by different *Salmonella* subtypes (e.g. serotypes, phage types, antimicrobial resistant profiles) as a function of the prevalence of these subtypes in animal and food sources and the amount of each food source consumed, using a Bayesian framework with Markov Chain Monte Carlo simulation (Gilks *et al.*, 1996). This approach has proved to be a valuable tool in focusing food safety interventions to the appropriate animal reservoir in Denmark (Fig. 5.3; Wegener *et al.*, 2003, Korsgaard *et al.*, 2009), and the model has recently been adapted to attribute human salmonellosis in other countries (Pires *et al.*, 2008; Pires, 2009, manuscript VI; Mullner, 2009, chapter 6), as well

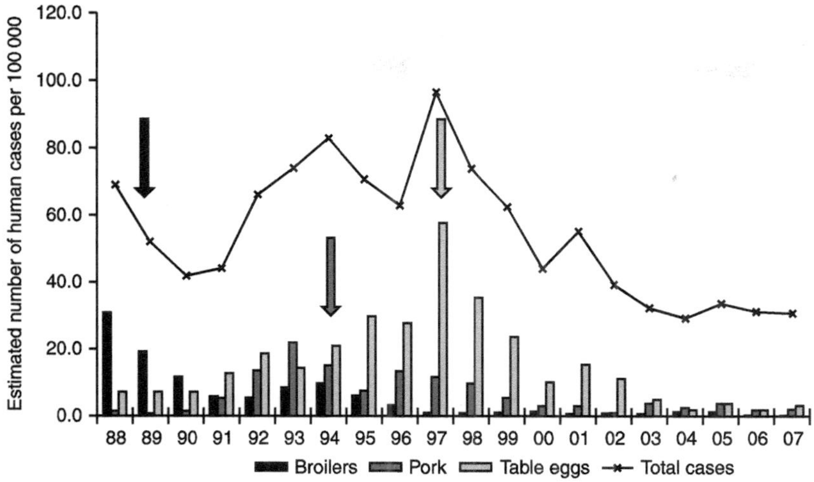

Fig. 5.3 Effects of *Salmonella* control programmes in Denmark as estimated by application of the microbiological subtyping approach on an annual basis. The arrows indicate the initiation of a new control programme in broiler chickens, in pigs and pork and laying hens, respectively. Remaining cases were attributable to beef, imported food products, infections acquired while travelling abroad and unknown sources. Source: Danish Zoonoses Centre, DTU National Food Institute

as human illness caused by other foodborne pathogens, e.g. *Listeria monocytogenes* (Little *et al.*, 2010).

Source attribution of human campylobacteriosis based on subtyping data has received increased attention in recent years, particularly after the application of Multi Locus Sequence Typing (MLST), which has shown that it is possible to identify some degree of host association between certain sequence types (ST) and a particular host reservoir despite the weakly clonal population structure of this pathogen (Dingle *et al.*, 2001; McCarthy *et al.*, 2007). In New Zealand, MLST data were recently applied to a modified Hald model to provide attribution estimates on the most important sources of human campylobacteriosis (Mullner *et al.*, 2009).

However, a whole new set of tools driven by the recent development of molecular typing techniques have emerged that all make inferences based on the population genetics of the pathogen. The basic assumption is that genetic relations between pathogen subtypes are indicators of host association or transmission pathways. These methods are also based on a comparison of subtypes from different sources and humans, but additionally take into account the genetic relatedness among the subtypes, i.e. how closely are they related and how they may be evolved from each other. Some of these methods directly provide attribution estimates, where a number or proportion of human cases are attributed to specific source. These include the Bayesian clustering algorithm STRUCTURE (Pritchard *et al.*, 2000) and the Asymmetric Island model (Wilson *et al.*, 2008). Other methods are based on clustering techniques that visualise the relatedness of bacterial subtypes using some graphical representation, for example the Minimum Spanning Trees (Feil *et al.*, 2004; Spratt *et al.*, 2004). Although such tools do not result in risk estimates, they still provide an increased insight into the population dynamics of a pathogen and can support the conclusion drawn from more mathematical models.

The mathematical modelling approaches applied for attributing human campylobacteriosis using MLST data include the STRUCTURE model and the Asymmetric Island model (Wilson *et al.*, 2008; Sheppard *et al.*, 2009; Mullner, 2009). In brief, the assumption is that the animal and environmental reservoirs of *Campylobacter* are separate populations within which the bacteria evolve through mutation and horizontal gene transfer (recombination) and between which genes may flow (migrate). Based on the estimated amount of mutation, recombination and migration, each human case is assigned probabilistically to the source populations. From these individual probabilities, the total amount of human disease attributable to each source is estimated. These techniques have so far primarily been used in attribution studies of human campylobacteriosis, but it is to be expected that they will also be applicable to other zoonotic diseases such as salmonellosis and listeriosis as sequence-based typing methods become more widely used.

The methods currently applied to subtype *Salmonella* isolates, mostly phenotypic methods, have limitations in their power to identify the origin of a given isolate, and consequently to assist source attribution studies. Molecular methods based on characterisation of the bacterial DNA have a considerably higher discriminatory power than the phenotypic methods and are increasingly being applied in outbreak investigation for pinpointing a particular source (Torpdahl *et al.*, 2007). The most recently developed methods typically target specific areas or genes of the genome and include the MLVA typing, where the numbers of repeat elements in specific loci are measured, and multi-locus sequence typing (MLST), where DNA sequences of specific genes are determined (Dingle *et al.*, 2001; Lindstedt *et al.*, 2004). Particularly for the sequence-based methods, a whole new research area has become available as the latest technology makes it possible to perform large scale sequencing at an affordable price. Still, the DNA-based methods value for source attribution of human salmonellosis has not been examined, and will undoubtedly challenge the optimal strategy: 'one typing method that fits all needs'. Very discriminatory methods are not necessarily the best solution for source attribution, where we are not looking for a single source for a particular outbreak, but rather want to relate groups of *Salmonella* strains with particular reservoirs/sources and then attribute human sporadic cases to these sources. Such a process must allow for some genetic diversity between strains from human and food sources even if they are epidemiologically related.

An important advantage of the microbial subtyping approach is that it allows for the identification of the most important reservoirs of the zoonotic agent, assisting risk managers to prioritise interventions and focus control strategies at the animal level. Additionally, attributing illness to the primary source reduces uncertainty due to cross-contamination and the risk of attribution to an 'accidental' source. Finally, when applied on a regular basis, the microbial subtyping approach allows for the analysis of the dynamic spread and trends of the most important sources of disease over time.

The microbial subtyping approach is limited to pathogens that are heterogeneously distributed among the reservoirs. This makes it appropriate for pathogens that are clonally distributed and present at least some host-associated subtypes. The approach is also quite data intensive, requiring a collection of isolates from all (major) sources that should to the extent possible represent what the human population is exposed to, i.e. the isolates from humans and sources should be related in time and space. This is in the authors' opinion a fundamental requirement, but it is often violated due to the lack of systematically collected surveillance data and/or the subsequent application of standardised subtyping methods.

5.3.2 Comparative exposure assessment approach

The comparative exposure assessment approach for source attribution makes use of stochastic modelling techniques similar to those used in quantitative microbial risk assessments. Nevertheless, the two methods differ in objectives and level of detail. A risk assessment typically aims at describing the complex dynamics of a pathogen in a single food commodity in the farm-to-consumption continuum, and predicting the public-health impact of interventions strategies. In contrast, the comparative exposure assessment approach aims at partitioning the observed (or predicted) human disease burden to all known transmission routes, including various foods, direct contact with live animals and environmental exposures. For this purpose, the various transmission routes are modelled in a more simplified fashion that represents only the main steps in the transmission pathway.

The principle of the comparative exposure assessment approach is to determine the relative importance of the known transmission routes by estimating the human exposure to that pathogen via each route. For each known transmission route, the following information is needed: prevalence and concentration (dose) of the pathogen in the source (e.g. in animal faeces or food products), the changes of the prevalence and concentration during the steps of the transmission chain that is modelled, and the frequency at which humans are exposed by the different routes (e.g. amount consumed and frequency of consumption). For food sources, the average exposure is estimated by multiplication of (averages of) the daily intake per person of the food product, the fraction of contaminated products at retail, the concentration of pathogens in contaminated products at retail and the fraction of pathogens that is eventually ingested by consumers after handling and preparation. For environmental (water) exposure, frequency, water intake and pathogen concentration are taken into account. For direct animal contact, calculations involve the frequency of human–animal contact, the (probability of) ingestion of faeces per contact, the fraction of contaminated animals and the pathogen concentration in faeces. All estimated exposures are then compared, and the human disease burden (e.g. the observed laboratory-confirmed infections or estimated total number of infections) caused by the specific pathogen is partitioned to each of the various transmission routes, proportionally to the size of the exposure dose. The estimates of exposure dose for each transmission route can, however, also be combined with a dose–response model to predict the number of infections from each route.

An example of the comparative exposure assessment approach for attribution is the Dutch exposure assessment for *Campylobacter* spp. (Evers *et al.*, 2008). The authors estimated the mean dose of *Campylobacter* ingested per person per day averaged over the entire Dutch population by different routes including consumption of food (animal or vegetable origin; raw or prepared), direct contact with animals (pets, farm animals and petting zoo animals) and water (swimming in or drinking water). Thirty-one routes related to these categories were investigated. Approximately two-thirds of the average exposure was related to direct contact with animals, whereas one-third was related to food. (Surface) water contributed only one per cent to the total exposure. Within the food routes, raw or partly cooked foods (chicken liver, milk, herring and vegetables) were the major sources of exposure, with chicken meat being the most important source of exposure from cooked meats.

Although not based on a formal mathematical model, Lake *et al.* (2006) have made a corresponding analysis for New Zealand by taking information from human disease surveillance and exposure information into account, and based on that drew conclusion regarding the most important food and non-food sources of human campylobacteriosis in New Zealand. McBride *et al.* (2005) used a model similar to that of Evers *et al.* (2008) to compare infection risks from four major pathways (food, recreational swimming, drinking water and occupational contact with livestock). In this study, the mean estimated daily exposures were combined with a dose-response model to estimate infection risks. As in the Dutch study, there were major uncertainties in the data underlying the model and the results are considered preliminary. Another comparative exposure assessment approach investigated the role of only foods, namely specific food dishes or ready-to-eat foods, for human cases of *Listeria monocytogenes* in the USA (FDA, 2003). This study identified deli meats as the major source of listeriosis in the US population.

The comparative exposure assessment approach attributes illness to the point of exposure, taking into account the different transmission routes from the same reservoir. This allows for the investigation of different pathways of exposure from the main reservoirs, as well as the estimation of the relative importance of each. However, the method requires many parallel exposure assessments to be made and is therefore data and labour intensive, even though the modelling is relatively simplistic. Modelling of the animal contact and environmental transmission routes requires information on contact frequencies and the probability of ingestion of the pathogen given contact. Such data are currently only available to a limited extent, or the parameters may be very difficult to measure in practice. Expert elicitation may in such cases be used to obtain parameter estimates as described by Havelaar *et al.* (2008). Currently many data gaps result in large uncertainty intervals and make uncertainty analysis an essential component of the results.

Another factor to consider is that there is no linear relation between exposure and illness, as the approach estimates mean exposures at the population level and not on an individual case basis. Therefore, source attribution estimates should be inferred with care and in light of factors that may influence the likelihood of estimated exposures to result in disease. For instance, immunity will probably play an important role in individuals frequently exposed through direct contact with animals (e.g. pets or occupational exposure to farm animals), since regular exposure to low doses of the same pathogen may induce an immune response, and consequently protect the individual from being infected after exposure. In contrast, sporadic exposures to specific animal species, although estimated to be low (e.g. exposure through petting zoo animals) may result in disease if individuals are exposed to pathogen subtypes that they have no acquired immunity against.

5.3.3 Epidemiological approaches

Epidemiological studies usually involve interviews of patients to elicit the patient's recall of foods consumed or other exposures before onset of illness. Methods used for human illness source attribution include studies of sporadic infections and analysis of data from outbreak investigations. Often such studies are built upon an existing public health surveillance infrastructure or involve additional interviews of patients and asymptomatic (control) individuals.

Generally, cases of foodborne illness are classified as either outbreak-related or sporadic cases. An outbreak is defined as (1) an incidence, observed under given circumstances, of two or more human cases of the same disease, or (2) a situation in which the observed number of cases exceeds the expected number and where the cases are linked (or are probably linked) to the same food source (Anonymous, 2003). Sporadic cases represent cases that have not been associated with known outbreaks (Engberg, 2006). An unknown proportion of cases classified as sporadic may be part of undetected outbreaks. Identification of possible sources of apparently sporadic infections and outbreaks may be undertaken using analytical epidemiological studies, which involve interviewing case-patients and asymptomatic (assumed to be non-infected) controls, or case–series studies, which involve interviewing only individuals who are ill.

Studies of sporadic infection

Several types of studies may be performed to identify possible sources of apparently sporadic human infections, including case–control, cohort and case–series studies. Case–control studies of sporadic infections are the most commonly applied approach for identifying possible exposures for sporadic foodborne disease. To allow for sufficient enrolment of patients, case–control studies of sporadic infections are often conducted over an extended period of time, and commonly use public health surveillance to ascertain culture-confirmed cases. Selected case-patients and a representative group of asymptomatic individuals (controls) are interviewed, and the relative role of exposures is estimated by comparing the frequency of exposures among cases and controls. When infections are associated with an exposure, the proportion of cases attributed to the exposure can be calculated and is defined epidemiologically as the 'population attributable fraction' (PAF) (Clayton and Hills, 1993). The population attributable fractions can be used to attribute the human disease burden to specific sources as done for example by Stafford *et al.* (2008). Based on the food-specific PAFs, the authors estimated for instance that 50 500 (95% credible interval 10 000–105 500) cases of *Campylobacter* infection in persons >5 years of age could be directly attributed each year to consumption of chicken in Australia.

Numerous case–control studies of sporadic infections of diseases commonly transmitted through food have been published, e.g. *Salmonella* Enteritidis infections in Denmark (Mølbak and Neimann, 2002), *Campylobacter* infections in Australia (Stafford *et al.*, 2008), Shiga toxin-producing *Escherichia coli* O157 infections in the USA (Voetsch *et al.*, 2007).

Case–control studies are a valuable tool to identify potential risk factors for human infections, including sources and predisposing, behavioural or seasonal factors (Engberg, 2006). Moreover, in addition to individual case–control studies, a systematic review of published case–control studies of sporadic infections of a given pathogen can provide an overview of the relevant exposures and risk factors for that infection, and a summary of the estimated population attributable fractions for each exposure. An overall population attributable fraction derived from a meta-analysis or weighted summary of several case–control studies of a certain pathogen can be combined with estimates of the burden of disease caused by that pathogen to estimate the burden of disease attributed to each exposure.

Limitations of case–control studies include that cases not associated with a recognised outbreak reflect a mixture of possible sources of exposure, and it may be difficult to distinguish between these exposures. Another limitation is misclassification due to immunity, which may reduce attributable risk or even suggest protection. Likewise, misclassification of exposures due to lack of accuracy of recall may lead to an underestimation of the burden of illness attributed to specific exposures. Most studies only explain a small fraction of all cases, and cases may reflect a mixture of possible sources of exposure, which can make it difficult to distinguish between these exposures. Lastly, statistical power to determine the importance of common exposures often requires enrolment of many participants.

Cohort studies are used less often for sporadic infections, since they usually require interviewing a high number of people, most of whom are not infected. Examples of cohort studies performed to determine the overall disease burden attributable to specific pathogens include the study of sporadic infections by de Wit *et al.* (2002) and the Infectious Intestinal Disease study in the United Kingdom (Wheeler *et al.*, 1999). Case–series studies of sporadic infections are commonly conducted, particularly for uncommon diseases that have a well-recognised source of infection, to which persons without the infection are infrequently exposed. Examples of case–series studies of sporadic infections and the common source for that disease include botulism associated with home-canned foods (Sobel *et al.*, 2004), *Vibrio vulnificus* and oysters (Shapiro *et al.*, 1998) and

Salmonella Typhi infections and foreign travel in the USA (Ackers *et al.*, 2000). When an exposure is uncommon in the general population, case–series studies of a sporadic infection can be used to determine the frequency of that exposure among the cases, and that frequency can be considered against the proportion of the burden of illness caused by that disease that is attributed to that specific exposure.

Analysis of data from outbreak investigations

An analysis of the information available from outbreak investigations can be utilised for source attribution. Many outbreak investigations are successful in identifying the specific source for the human infections. By conducting an analysis of data from these investigations, the most common food vehicles involved in outbreaks can be identified. A simple descriptive analysis or summary of outbreak data is useful for attributing illnesses to foods, but often the implicated food in an individual outbreak is a 'complex' food, containing several food items. Any of these foods could be the actual source of the infection. Several methods have been developed to use the information of simple foods involved in outbreaks to attribute human illness to sources (e.g. Adak *et al.*, 2005; Greig and Ravel, 2009). An alternative method for conducting an analysis of data from outbreak investigations was developed in the USA. In this method, food items are categorised into a hierarchical scheme according to their ingredients (Painter *et al.*, 2009). Foods that contain ingredients that are members of a single food category are considered 'simple foods', while foods that contain ingredients that are members of multiple food categories are considered 'complex foods'. As an example, steak is a simple food, whereas meat loaf is a complex food. Each implicated food is assigned to one or more mutually exclusive food categories according to its ingredients. For outbreaks that have implicated a simple food item, all illnesses are attributed to that single category. For outbreaks that have implicated a complex food item, illnesses are partitioned to each category in the complex food according to the proportion of illnesses attributed to each of those categories in outbreaks caused by simple foods. As a result, illnesses in an outbreak due to a complex food item are attributed to a category in the implicated complex food, only if that category has been implicated in at least one outbreak due to a simple food. The numbers of illnesses attributed to each category are then summed and used to determine the percentage of disease attributed to each category. To estimate the attribution of all foodborne illness in the population (and not just illnesses involved in reported outbreaks), the percentages of illness for each disease attributed to each category are weighted by the estimated annual burden of illness for each pathogen (Painter, 2006). This method has been adapted to attribute human salmonellosis and campylobacteriosis in Europe (Pires, 2009; Pires *et al.*, in press), and can be used to attribute foodborne illness in several countries or regions worldwide.

Outbreak investigations are often successful in identifying the food causing disease, and are thus able to provide clear documentation that a specific pathogen was transmitted to humans via a specific food item. A wide variety of food vehicles are represented, including less frequently identified food items. Additionally, depending on the level of traceback, the effect of contamination at multiple points on the farm-to-consumption chain may be investigated. A major advantage of the use of this approach is that these data are accessible in most countries, and often the only source of data available for source attribution in some countries or regions. Limitations of the use of outbreak data for attribution include that the quality of evidence varies between data sources and classification schemes for the data are not consistently used. Also, large outbreaks, outbreaks associated with point sources, outbreaks that have short incubation periods and outbreaks that cause serious illness are more likely to be investigated. Likewise, certain food vehicles are more likely to be associated with reported outbreaks than others, which can lead to an overestimation of the proportion of human illnesses attributed to a specific food. An important factor to consider is that illnesses included in data from outbreak investigations may not be representative of all foodborne illnesses. The fraction of the burden of foodborne disease that is associated with outbreaks varies between pathogens but is typically smaller than the correspondent to sporadic disease. Consequently, the extrapolation of source attribution estimates obtained through an analysis of data from outbreaks to the overall burden of disease should be made with care.

5.3.4 Intervention studies

Intervention studies can provide evidence of the burden of a foodborne disease attributable to specific sources through the measurement of the impact of interventions implemented in the food production chain in the number of human cases. Intervention studies can be designed as small or large scale studies to control a specific pathogen, and the measure of the impact of specific sources in human disease is facilitated when interventions are conducted in a randomised design. Examples of intervention studies include the reduction of the burden of human campylobacteriosis from poultry meat in Iceland (Stern *et al.*, 2003) and New Zealand (Williman *et al.*, 2009) and of human salmonellosis in Denmark (Wegener *et al.*, 2003).

In addition to implemented interventions, natural experiments or accidental interventions, where a change in exposure or consumption behaviour of the population resulted in a decrease in the number of reported cases of disease, have proven useful to assess the contribution of specific sources to the burden of disease. By analysing changes in consumption and numbers of reported cases of human illness, it is possible to estimate the number of human cases attributable to specific sources. Examples include the decline in the number of *Campylobacter* infections in the Netherlands in 2003, when the threat of avian influenza led to the depopulation of 30 million birds, and consequently a decrease in poultry consumption (Rosenquist *et al.*, 2004; Van Pelt *et al.*, 2009). Another example was the use of dioxin-contaminated chicken feed in Belgium, which resulted in the withdrawal of chicken meat and eggs from the market in June 1999 and subsequent decrease in reported human *Campylobacter* infections also occurred in the country. Public health surveillance data from preceding years (1994–1998) were used to predict the number of human *Campylobacter* infections expected in 1999. The number of reported cases in 1999 was substantially lower (40%) than the expected, suggesting that 40% of reported human *Campylobacter* infections in Belgium could be attributed to chicken (Velling and Van Lock, 2002).

Intervention studies have the advantage of allowing for a direct measure of the impact of a reservoir or source on the burden of human disease, excluding the interference of external sources or risk factors. Nonetheless, usually several interventions are implemented at the same time, which hampers the interpretation of data on the public health impact of a specific intervention and food/animal source.

5.3.5 Expert elicitations

Expert elicitations are often used to address data gaps during source attribution exercises, but a structured approach to use expert judgements can also be used to assist attribution. These structured approaches require more resources and technical expertise than conventional, unstructured evaluations and need a multidisciplinary approach, which may compromise their acceptance in practice. Expert estimates typically combine information from different sources, which can be considered both a strength and a weakness. To date, a number of studies have been performed using expert elicitation to attribute foodborne disease, e.g. in the USA (Mead *et al.*, 1999; Hoffmann *et al.*, 2007), in Australia (Hall *et al.*, 2005), and in the Netherlands (Havelaar *et al.*, 2008). Expert elicitation studies may be particularly useful to estimate the fraction of the burden of a disease that can be attributable to foods in general, followed by the application of another source attribution approach to estimate the relative importance of different food sources. This approach is often used to fill in data gaps or as an alternative source attribution method, for situations when none of the remaining available approaches can be applied. An important limitation is that conclusions are based on individual judgement, which may be misinformed or biased.

5.3.6 Combining different approaches

All described source attribution approaches have different data requirements and methodological characteristics, such as parameters included, estimation of uncertainty, prerequisites, assumptions and outcome. Additionally, the methods often focus on only one point of the transmission chain (e.g. point of reservoir or point of exposure), which means that choosing only one approach for source attribution may be inadequate to answer specific risk management questions. Such limitations may be overcome by the application of more than one source attribution method and subsequent comparison and combination of the results. Another possibility to improve estimates and the robustness of the results, and thus increase our confidence in the outcomes, is to integrate methods that attribute illness at the same or at different points of the farm-to-consumption continuum. Several synergies can be applied between microbiological approaches and between microbiological and epidemiological approaches.

Examples include:

1 A comparative exposure assessment framework on the basis of the estimates obtained by a microbial subtyping approach. In other words, the relative importance of different routes of transmission within reservoirs would be weighted by the proportion of disease attributed to that reservoir by microbial subtyping.
2 A comparative exposure assessment including routes from only one reservoir (estimated to be important by microbial subtyping), with the purpose of investigating which transmission route is contributing more to exposure to that reservoir. This would be very similar to a 'traditional' quantitative risk assessment.
3 A comparative exposure assessment incorporating subtyping data, distinguishing between pathogen subtypes and thus estimating the relative importance of routes for infection with different subtypes.
4 An example of integration of microbiological and epidemiological approaches could be the synergy between microbial subtyping and a systematic review of case–control studies. The latter provides a picture of the important risk factors, routes of transmission or sources, and these results can be used as a starting point for microbial subtyping exercises.
5 A microbial subtyping approach applied for specific subgroups of the population, estimated to be at higher risk by a systematic review of case–control studies.
6 Integration of case–control studies of sporadic disease and an analysis of data from outbreak investigations would be able to estimate the relative importance of sources of sporadic and outbreak-related cases, which for some pathogens may differ considerably.

5.4 Conclusions and recommendations

This chapter summarised the methods available for source attribution for human illness, and discussed strengths and weaknesses, as well as data requirements for each of the methods. Each method addresses different points in the food chain and different methods may therefore not result in the same conclusions. The choice of method depends on the specific question that needs answering and the data and resources available. Comparing and compiling results from more than one method and/or integrating several approaches analytically are recommended when possible in order to increase confidence in the results.

Source or reservoir attribution using microbial subtyping has gained increasing interest during recent years. The philosophy behind the approach is that control of the pathogen at the source of origin, i.e. the reservoir will prevent subsequent human exposure, regardless of the transmission route or vehicle. By collating results from surveillance programmes that are in place and comparing these with cases of human illness, the method provides added value to data that are already being collected. For *Salmonella*, serotyping and phage-typing have so far been the preferred typing methods for this purpose and the results have in some countries provided guidance to risk managers and policy makers on the implementation and evaluation of control strategies for major reservoirs. The recent development of new genotypic-based methods and the application of population-genetic analytical methods are expected to take source attribution by microbial subtyping to the next level. For *Campylobacter*, several papers on this have already been published and the approaches may also prove to be valuable for other pathogens such as VTEC and *Listeria monoctogenes*.

Although comparative exposure assessment is the only approach that in theory allows for the high level of detail needed for answering some risk management questions (e.g. estimating the proportion of cases attributable to minced meat or other meat categories), the method is seriously hampered by limited data availability making the results highly uncertain. Particularly, comparative exposure assessment between major categories (food, direct animal contact, environments, person-to-person) needs further development and more data to provide sufficient accurate results to be ready for decision support. Within the foodborne transmission route, comparative analysis of different transmission routes and sources is feasible if sufficient data is available.

Case–control studies of sporadic infections are a valuable tool to identify relevant risk factors to human foodborne infections, including sources of exposure and predisposing, behavioural or seasonal factors. By calculating the population-attributable fractions, the relative importance of the different risk factors can be estimated. A primary limitation of the method is the accuracy of the recall about exposures from interviewed participants, which can lead to either an over- or under-estimation of the contribution of specific sources. In addition, many participants will need to be enrolled in order to have sufficient statistical power to determine the importance of common exposures. Systematic reviews of case–control studies considering both periodical and geographical dimension may provide new insight into trends and dynamics of sources of human foodborne illnesses.

Outbreak investigations give public health officials important information about immediate control of individual events. In many countries outbreak investigations are undertaken and results summarising the result including the suspected or confirmed sources are available at a national or international level. Records over many years provide a relatively detailed dataset, making outbreak data attractive also for use in attribution models. The foods implicated in causing human disease can be assessed using aggregated data from many outbreak investigations and the most common food vehicles involved can be identified, with the caveat that the source of human infection is often not identified in a significant proportion of outbreak episodes. Although source attribution using outbreak data is a promising approach and often the only type of data available in a country, there are gaps in the datasets currently available. In addition, this approach is not appropriate for pathogens, which in general cause very few outbreaks, and/or if the most important sources differ significantly between sporadic and outbreak-related cases.

Structured expert elicitations have been used quite often for source attribution, and recently more explicit, when quantitative methods have been introduced. Experts are able to combine and weigh data from the different approaches for which currently no analytical methods exist. Protocols to quantify uncertainty and reduce bias in expert estimates have been developed in other areas of risk assessment but still need to be fully applied to source attribution. Although expert elicitation is commonly applied and often is the only method available, it is recognised that even when scientifically applied, expert elicitation is not a 'gold' standard for the assembly of scientific evidence. Formal expert elicitation should therefore to the extent possible be informed with all available results from other source attribution studies, outbreak data, prevalence surveys, etc.

Source attribution is increasingly used to partition human illness to the most important sources including food and as such support risk management strategies. This has identified a need for harmonising the terminology and definitions used as well as the categorisation of sources and food items taking into account the legal definition of water as food (EFSA, 2008). In a recent paper, we introduced nomenclature that should contribute to the harmonisation of concepts, definitions, and methods (Table 5.2) (Pires *et al.*, 2009). We encourage other scientists that conduct source attribution studies to clearly address and define as a minimum: (1) the sources considered, (2) the point(s) of attribution addressed, and (3) the attribution method(s) chosen, because this is considered crucial for knowledge sharing between research groups and for comparison of results among risk assessors and risk managers.

Data gathering for purposes of attribution should be question driven and the samples should be systematically collected and representative for what the human population is exposed to in terms of food and animal reservoirs. Attribution studies based on, for instance, clinical samples from diseased animals are therefore generally not a good indicator. In contrast, the Baseline Studies as carried out under the EU Zoonoses regulations are an important move in the right direction (Hald, 2008). Similarly, a common, e.g., multi-country approach to epidemiological studies is recommended.

Table 5.2 Proposal for harmonised terminology and definitions for attribution of human illness to specific sources

Concept	Definition
Human illness attribution	Partitioning of the human disease burden of one or more foodborne infections to specific sources.
Source	Origin of the pathogen causing infection, including animal reservoirs and vehicles, e.g. foods and water.
Points of attribution	Points in the food chain, where human illness attribution can take place, e.g. at production, distribution and consumption.

Source: Pires *et al.* (2009).

Source attribution studies are facilitated by the existence of an integrated surveillance and centralised collation of information. Surveillance of foodborne disease is a collective effort, depending on investigations performed by clinicians, veterinarians, microbiologists, epidemiologists, public health department officials and environmental health officers, culminating in data collation, data analysis and interpretation at the national level. Since the early to mid 1990s, Nordic countries (Denmark, Sweden, Norway and Finland) have implemented an integrated surveillance and established a network of institutions that collects and analyses zoonoses data. These Zoonosis Centres facilitate the analysis of all available information on foodborne zoonosis, and thus the identification and prioritisation of effective interventions and allocation of resources. In Denmark for instance, the planning and implementation of *Salmonella* control programmes has been done by close involvement of both microbiologists and epidemiologists, and the technical coordination of the programmes was made through committees with representatives from the industry, government bodies and the scientific community. In addition, there has been a very close collaboration between medical and veterinary epidemiologists in monitoring the effect of the programmes on the incidence of human infection, including sharing of relevant surveillance data.

In countries that have not established an integrated surveillance and/or a coordinating body that collects and analyses data, the different institutions responsible for monitoring and surveillance in the different public health components collect and compile the data. As a consequence, any epidemiological, risk assessment or source attribution study that requires data from different points of the farm-to-consumption continuum and human surveillance involves the solicitation of data to different parties, which often implies legal and time-consuming requirements. It is therefore recommended to establish a coordinated effort consisting of interdisciplinary collaboration between the public health and the veterinary and the food sector, between authorities, scientists and the industry, and between microbiologists and epidemiologists, as this is considered essential for integrated surveillance and thus for the successful control of foodborne diseases.

5.5 References

ACKERS M L, PUHR N D, TAUXE R V and MINTZ E D (2000), 'Laboratory-based surveillance of *Salmonella* serotype Typhi infections in the United States,' *Journal of the American Medical Association*, **283**, 2668–73.

ADAK G K, MEAKINS S M, YIP H, LOPMAN B A and O'BRIEN S J (2005), 'Disease risks from foods, England and Wales, 1996–2000', *Emerging Infectious Diseases*, **11**, 365–72.

ANONYMOUS (2003), 'Directive 2003/99/EC of The European Parliament and of the Council of 17 November 2003 on the monitoring of zoonoses and zoonotic agents, amending Council Decision 90/424/EEC and repealing Council Directive 92/117/EEC', *Official Journal of the European Union*, L **325**, 31–40.

BATZ M B, DOYLE M P, MORRIS J G, PAINTER J, SINGH R, TAUXE R V, TAYLOR M R, LO FO WONG D M A and THE FOOD ATTRIBUTION WORKING GROUP (2005), 'Attributing illness to food', *Emerging Infectious Diseases*, **11**, 993–9.

CLAYTON D and HILLS M (1993), *Statistical Models in Epidemiology*, Oxford University Press, Oxford/New York/Tokyo.

DE WIT M, KOOPMANS M P G, KORTBEEK L M, VAN LEEUWEN W J, BARTEDLS A I M, VAN DUYNHOVEN Y T H P (2002), 'Gastroenteritis in sentinel general practices, the Netherlands', *Emerging Infectious Diseases*, **1**, 82–91.

DINGLE K E, COLLES F M, WAREING D R, URE R, FOX A J, BOLTON F E, BOOTSMA H J, WILLEMS R J, URWIN R and MAIDEN M C (2001), 'Multilocus sequence typing system for *Campylobacter jejuni*', *Journal of Clinical Microbiology*, **39**(1), 14–23.

EFSA (2008), 'Scientific Opinion of the Panel on Biological Hazards on a request from EFSA on overview of methods for source attribution for human illness from food borne microbiological hazards', *The EFSA Journal*, **764**, 1–43.

EFSA (2009), 'The Community Summary Report on Trends and Sources of Zoonoses and Zoonotic Agents in the European Union in 2007', *The EFSA Journal*, **223**, 312.

ENGBERG J (2006), 'Contributions to the epidemiology of Campylobacter infections: A review of clinical and microbiological studies', *Danish Medical Bulletin*, **53**(4), 361–89.

EVERS E G, VAN DER FELS-KLERX H J, NAUTA M J, SCHIJVEN J F and HAVELAAR A H (2008), '*Campylobacter* source attribution by exposure assessment', *International Journal of Risk Assessment and Management*, **8**, 174–90.

FDA (2003), *Quantitative Assessment of Relative Risk to Public Health from Foodborne Listeria monocytogenes among Selected Categories of Ready-to-eat Foods*. Available from: http://www.fda.gov/Food/ScienceResearch/ResearchAreas/RiskAssessmentSafetyAssessment/ucm183966.htm [Accessed 30 May 2010].

FEIL E J, LI B C, AANENSEN D M, HANAGE W P and SPRATT B G (2004), 'eBURST: inferring patterns of evolutionary descent among clusters of related bacterial genotypes from multilocus sequence typing data', *Journal of Bacteriology*, **186**(5), 1518–30.

GILKS W R, RICHARDSON S and SPIEGELHALTER D J (1996), *Markov Chain Monte Carlo in Practice*, Chapman and Hall, London.

GREIG J D and RAVEL A (2009), 'Analysis of foodborne outbreak data reported internationally for source attribution', *International Journal of Food Microbiology*, **130**, 77–87.

HALD T (2008), 'EU-wide baseline studies: achievements and difficulties faced', *Trends in Food Science & Technology*, **19** (Suppl. 1), S40–S48.

HALD T, LO FO WONG D M and AARESTRUP F M (2007), 'The attribution of human infections with antimicrobial resistant *Salmonella* bacteria in Denmark to sources of animal origin', *Foodborne Pathogens and Disease*, **4**, 313–26.

HALD T, VOSE D, WEGENER H C and KOUPEEV T (2004), 'A Bayesian approach to quantify the contribution of animal-food sources to human salmonellosis', *Risk Analysis*, **24**, 251–65.

HALL G, KIRK M D, BECKER N, GREGORY J E, UNICOMB L, MILLARD G, STAFFORD R and LALOR K (2005), 'Estimating foodborne gastroenteritis, Australia', *Emerging Infectious Diseases*, **11**, 1257–64.

HAVELAAR A H, BRÄUNIG J, CHRISTIANSEN K, CORNU M, HALD T, MANGEN M J, MØLBAK K, PIELAAT A, SNARY E, VAN PELT W, VELTHUIS A and WAHLSTRÖM H (2007), 'Towards an integrated approach in supporting microbiological food safety decisions', *Zoonoses Public Health*, **54**, 103–17.

HAVELAAR A H, GALINDO A V, KUROWICKA D and COOKE R M (2008), 'Attribution of foodborne pathogens using structured expert elicitation', *Foodborne Pathogens and Disease*, **5**, 649–59.

HOFFMANN S, FISCHBECK P, KRUPNICK A and MCWILLIAMS M (2007), 'Using expert elicitation to link foodborne illnesses in the United States to foods', *Journal of Food Protection*, **70**, 1220–9.

KORSGAARD H, MADSEN M, FELD N C, MYGIND J and HALD T (2009), 'The effects, costs and benefits of *Salmonella* control in the Danish table-egg sector', *Epidemiology and Infection*, **137**(6), 828–36.

KUBOTA K, IWASAKI E, INAGAKI S, NOKUBO T, SAKURAI Y, KOMATSU M, TOYOFUKU H, KASUGA F, ANGULO F J and MORIKAWA K (2008), 'The human health burden of foodborne infections caused by *Campylobacter, Salmonella*, and *Vibrio parahaemolyticus* in Miyagi Prefecture, Japan', *Foodborne Pathogens and Disease*, **5**(5), 641–8.

LAKE R, CRESSEY P and GALLAGHER E (2006), 'Expert opinion as a method for estimating attribution of foodborne infection sources', *Abstracts of the Priority Setting of Foodborne and Zoonotic Pathogens*, Berlin, Germany, pp. 20–2.

LINDSTEDT B-A, VARDUND T, AAS L and KAPPERUD G (2004), 'Multiple-locus variable-number tandem-repeats analysis of *Salmonella enterica* subsp. *enterica* serovar Typhimurium using PCR multiplexing and multicolour capillary electrophoresis', *Journal of Microbiology Methods*, **59**, 163–72.

LITTLE C L, PIRES S M, GILLESPIE I A, GRANT K and NICHOLS G L (2010) 'Source attribution of *Listeria monocytogenes* in England and Wales: adaptation of the Hald *Salmonella* Source Attribution Model', *Foodborne Pathogens and Disease*, 7: 749–756.

MARTIN S W, MEEK A H and WILLEBERG P (1987), *Veterinary Epidemiology: Principles and Methods*, Iowa State University Press, Ames, IA.

MCBRIDE G, MELEASON M, SKELLY C, LAKE R, VAN DER LOGT P and COLLINS R (2005), *Preliminary Relative Risk Assessment for Campylobacter Exposure in New Zealand: 1. National Model for Four Potential Human Exposure Routes; 2. Farm Environmental Model.* NIWA Client Report: HAM2005-094. Available from: http://www.zoonosesresearch.org.nz/reports/PreliRelativeriskAssessment.pdf [Accessed 9 February 2010].

MCCARTHY N D, COLLES F M, DINGLE K E, BAGNALL M C, MANNING G, MAIDEN M C J and FALUSH D (2007), 'Host-associated Genetic Import in *Campylobacter jejuni*', *Emerging Infectious Diseases*, **13**(2), 267–72.

MEAD P S, SLUTSKER L, DIETZ V, MCCAIG L F, BRESEE J S, SHAPIRO C, GRIFFIN P M and TAUXE R V (1999), 'Food-related illness and death in the United States', *Emerging Infectious Diseases*, **5**(5), 607–25.

MØLBAK K and NEIMANN J (2002), 'Risk factors for sporadic infection with *Salmonella* Enteritidis, Denmark 1997–1999', *American Journal of Epidemiology*, **156**, 654–61.

MULLNER P (2009), *Estimating the Contribution of Different Sources to the Burden of Human Campylobacteriosis and Salmonellosis*, PhD Thesis, Massey University, Palmerston North, New Zealand.

MULLNER P, JONES G, NOBLE A, SPENCER S E, HATHAWAY S and FRENCH N P (2009), 'Source attribution of foodborne zoonoses in New Zealand: a modified Hald model', *Risk Analysis*, **29**(7), 970–84.

PAINTER J (2006), 'Estimating attribution of illnesses to food vehicle from reports of foodborne outbreak investigations', Paper presented at the *Society for Risk Analysis Annual Meeting*, Baltimore, December 3–6, 2006. Available from: http://birenheide.com/sra/2006AM/program/singlesession.php3?sessid—3-E [Accessed 9 February 2010].

PAINTER J A, AYERS T, WOODRUFF R, BLANTON E, PEREZ N, HOEKSTRA R M, GRIFFIN P M and BRADEN C (2009), 'Recipes for foodborne outbreaks: a scheme for categorizing and grouping implicated foods', *Foodborne Pathogens and Disease*, **6**(10), 1259–64.

PIRES S M (2009),. *Attributing Human Salmonellosis and Campylobacteriosis to Animal, Food and Environmental Sources*, PhD Thesis, ISBN 978-87-7611-311-7, Faculty of Life Sciences, University of Copenhagen, Denmark.

PIRES S M and HALD T (2009), 'Assessing the differences in public health impact of *Salmonella* subtypes using a Bayesian microbial subtyping approach for source attribution', *Foodborne Pathogens and Disease*, **7**(2), 143–51.

PIRES S M, EVERS E G, VAN PELT W, AYERS T, SCALLAN E, ANGULO F J, HAVELAAR A, HALD T and THE MED-VET-NET WORKPACKAGE 28 WORKING GROUP (2009), 'Attributing the human disease burden of foodborne infections to specific sources', *Foodborne Pathogens and Disease*, **6**(4), 417–24.

PIRES S M, NICHOLS G, WHALSTRÖM H, KAESBOHRER A, DAVID J, SPITZNAGEL H, VAN PELT W, BAUMANN A and HALD T (2008), '*Salmonella* source attribution in different European countries', *Proceeding in FoodMicro 2008*, Aberdeen, Scotland. Available from: http://www.foodmicro2008.org/fm08/resources/374/1143/pdf/FM2008_0211.pdf [Accessed 9 February 2010].

PIRES S M, VIGRE H, MAKELA P and HALD T (in press) 'Using outbreak data for source attribution of human salmonellosis and campylobacteriosis in Europe', *Foodborne Pathogens and Disease*.

PRITCHARD J, STEPHENS M and DONNELLY P (2000), 'Inference of Population Structure using Multilocus Genotype Data', *Genetics*, **155**, 945–59.

ROSEF O, KAPPERUD G, LAUWERS S and GONDROSEN B (1985), 'Serotyping of *Campylobacter* jejuni, *Campylobacter* coli and *Campylobacter* lardis from domestic and wild animals', *Applied and Environmental Microbiology*, **49**(6), 1507–10.

ROSENQUIST H, NIELSEN N L, SOMMER H M, NORRUNG B, VAN PELT W, WANNET W, VAN DE GIESSEN A W, MEVIUS D and VAN DUYNHOVEN Y (2004), 'Trends in gastroenteritis (GE) in the Netherlands, 1996–2003', *Infectieziekten Bulletin*, **15**, 335–41.

SARWARI A R, MAGDER L S, LEVINE P, MCNAMARA A M, KNOWER S, ARMSTRONG G L, ETZEL R, HOLLINGSWORTH J and MORRIS G JR. (2001), 'Serotype distribution of *Salmonella* isolates from food animals after slaughter differs from that of isolates found in humans', *Journal of Infectious Diseases*, **183**, 1295–9.

SHAPIRO R L, ALTEKRUSE S, HUTWAGNER L, BISHOP R, HAMMOND R, WILSON S, RAY B, THOMPSON S, TAUXE R V and GRIFFIN P M (1998), 'The role of Golf Coast oysters harvested in warmer months in *Vibrio vulnificus* infections in the United States, 1988–1996', *Journal of Infectious Diseases*, **178**, 752–9.

SHEPPARD S K, DALLAS J F, STRACHAN N J C, MACRAE M, MCCARTHY N D, WILSON D J, GORMLEY F J, FALUSH D, OGDEN I D, MAIDEN M C J and FORBES K J (2009), '*Campylobacter* genotyping to determine the source of human infection', *Clinical Infectious Diseases*, **48**(8), 1072–8.

SIMONSEN J, MØLBAK K, FALKENHORST G, KROGFELT K A, LINNEBERG A and TEUNIS P F M (2009), 'Estimation of incidences of infectious diseases based on antibody measurements', *Statistics in Medicine*, **28**(14), 1882–95.

SOBEL J, TUCKER N, SULKA A, MCLAUGHLIN J and MASLANKA S (2004), 'Foodborne botulism in the United States, 1990–2000', *Emerging Infectious Diseases*, **10**, 1606–11.

SPRATT B G, HANAGE W P, LI B, AANENSEN D M and FEIL E J (2004), 'Displaying the relatedness among isolates of bacterial species – the eBURST approach', *FEMS Microbiology Letters*, **241**(2), 129–34.

STAFFORD R J, SCHLUTER P J, WILSON A J, KIRK M D, HALL G and UNICOM L (2008), 'Population-attributable risk estimates for risk factors associated with *Campylobacter* infection, Australia', *Emerging Infectious Diseases*, **14**, 895–901.

STERN N J, HIETT K L, ALFREDSSON G A, KRISTINSSON K G, REIERSEN J, HARDARDOTTIR H, BRIEM H, GUNNARSSON E, GEORGSSON F, LOWMAN R, BERNDTSON E, LAMMERDING A M, PAOLI G M and MUSGROVE M T (2003), '*Campylobacter* spp. in Icelandic poultry operations and human disease', *Epidemiology and Infection*, **130**, 23–32.

THORNS C J (2000), 'Bacterial foodborne zoonoses', *Revue Scientifique et Technique*, **19**(1), 226–39.

TORPDAHL M, SØRENSEN G, LINDSTEDT B-A and NIELSEN E M (2007), 'Tandem repeat analysis for surveillance of human *Salmonella* Typhimurium infections', *Emerging Infectious Diseases*, **13**, 388–95.

VAN PELT W, VAN DE GIESSEN A W, VAN LEEUWEN W J, WANNET W, HENKEN A M and EVERS E G (1999), 'Oorsprong, omvang en kosten van humane salmonellose. Deel 1. Oorsprong van human salmonellose met betrekking tot varken, rund, kip, ei en overige bronnen', *Infectiezikten Bulletin*, pp. 240–3.

VAN PELT W, HAVELAAR A H, WESTRA P P and WAGENAAR J (2009), 'Strong regional reduction of campylobacteriosis during and after avian influenza poultry farm culling: a model for future intervention studies at primary production?', *Abstract book of the 5th Med Vet Net Annual Scientific Meeting*, El Escorial, Madrid, Spain.

VELLING A and VAN LOOCK F (2002), 'The dioxin crisis as an experiment to determine poultry-related *Campylobacte*r enteritis', *Emerging Infectious Diseases*, **8**, 19–22.

VOETSCH A C, KENNEDY M H, KEENE W E, SMITH K E, RABATSKY-HER T, ZANSKY S, ET AL. (2007), 'Risk factors for sporadic Shiga toxin-producing *Escherichia coli* O157 infections in FoodNet sites, 1999–2000', *Epidemiology and Infection*, **135**, 993–1000.

WEGENER H C, HALD T, LO FO WONG D M, MADSEN M, KORSGAARD H, BAGER F, GERNER-SMIDT P and MOLBAK K (2003), '*Salmonella* control programs in Denmark', *Emerging Infectious Diseases*, **9**, 774–80.

WHEELER J G, SETHI D, COWDEN J M, WALL P G, RODRIGUES L C, TOMPKINS D S, HUDSON M J and RODERICK P J (1999), 'Study of infectious intestinal disease in England: rates in the community, presenting to general practice, and reported to national surveillance', *British Medical Journal*, **318**, 1046–50.

WHO (2005), *The World Health Report 2005 – Making Every Mother and Child Count*, World Health Organization, Geneva. Available from: http://www.who.int/whr/2005/whr2005_en.pdf [Accessed 9 February 2010].

WHO (2008), *The Global Burden of Disease – 2004 Update*, World Health Organization, Geneva. Available from: http://www.who.int/healthinfo/global_burden_disease/2004_report_update/en/index.html [Accessed 9 February 2010].

WILLIMAN J, LIM E, PIRIE R, CRESSEY P and LAKE R (2009),. *Annual Report Concerning Foodborne Disease in New Zealand 2008*, Institute of Environmental Science & Research, Christchurch, New Zealand. Available from: http://www.nzfsa.govt.nz/science/research-projects/FW09062_FBI_report_2008_Final_June_2009_2.pdf [Accessed 5 January 2010].

WILSON D J, GABRIEL E, LEATHERBARROW A J, CHEESBROUGH J, GEE S, BOLTON E, FOX A, FEARNHEAD P, HART C A and DIGGLE P J (2008), 'Tracing the source of campylobacteriosis', *PLoS Genetics*, **4**(9), e1000203.

6

Determining the economic costs and global burden of foodborne disease

J. C. Buzby, Economic Research Service of the US Department of Agriculture, USA

Abstract: This chapter presents eight epidemiological and methodological challenges that hinder the estimation of the economic costs and burden of foodborne disease. The chapter begins with a brief overview of the different types of societal costs of foodborne disease, why they are important to estimate, the general principles of economic analysis for US food safety regulations and the World Health Organization's collaboration to estimate the global burden of foodborne disease. The chapter summarizes the main methods used to value the costs or burden of foodborne disease and provides examples of US and international estimates.

Key words: cost-of-illness, economic costs, foodborne disease, food safety, willingness-to-pay.

6.1 Introduction

This chapter discusses some of the published and ongoing analyses and estimates of the economic costs and global burden of foodborne disease. The World Health Organization (WHO) defines the *burden of disease* as 'the incidence and prevalence of morbidity, disability, and mortality associated with acute and chronic manifestations of diseases.' Here, we are concerned with diseases caused by pathogens commonly transmitted through food and focus on the economic costs of these diseases. The chapter begins with a brief overview of the different types of societal costs of foodborne disease, why they are important to estimate, the general principles of economic analysis for US food safety regulations and the WHO's collaboration to estimate the global burden of foodborne disease. The chapter summarizes the main methods used to value the costs or burden of foodborne disease and provides examples of US estimates, their use in benefit–cost analyses of US food safety interventions and international estimates. The chapter concludes with a discussion on future trends and sources of further information.

6.1.1 Overview of the types of costs of foodborne disease

The societal costs of foodborne disease are incurred by three main groups: individuals/households, the food industry, and the regulatory and public health sector (Table 6.1). Most analyses that estimate the costs of foodborne disease often include only a small subset of the more easily measurable costs, such as individual's/household's medical costs, costs of lost productivity and premature death, and exclude others due to lack of suitable data and analytical techniques.

6.1.2 Importance of determining the economic costs and burden of foodborne disease

Economic estimates of foodborne diseases can illustrate the economic costs and burden on a nation, other population subgroups or the world. These estimates can be used in three main ways to:

1 Evaluate the economic costs or burden of a single pathogen or a broader group of foodborne diseases.
2 Serve as input either in a cost-effectiveness analysis to obtain information needed in policy making or in a benefit–cost analysis of a proposed food safety rule. A *benefit–cost analysis* weighs the pros (i.e., benefits) and cons (i.e., costs) of a proposed regulation against a baseline (i.e., the best assessment of the current state without the regulation). In essence, the benefits are the costs of foodborne disease avoided or prevented because of a rule. Economic analyses help identify the cost-efficient strategies.
3 Aid comparisons and other analyses, such as to:
   (a) Compare estimated costs for different foodborne diseases. This information can help determine the order for targeting pathogen prevention and control efforts toward the most costly diseases. However, cost information on prevention and control strategies is also needed to determine whether and where there are *positive net benefits* from implementing a particular food safety rule (i.e., where benefits of the rule outweigh implementation and other costs).
   (b) Compare benefits and costs of alternative pathogen prevention and control rules, programs, and strategies to determine the most cost-effective interventions so that resources are appropriately allocated,
   (c) Evaluate and monitor foodborne disease over time, for example, to see which foodborne diseases are becoming more widespread and in which subpopulations and to see which diseases are becoming less common due to national, industry and other pathogen reduction efforts.

Table 6.1 Societal costs of foodborne illness

Costs to individuals/households[1]	Industry costs[2]	Regulatory and public health sector costs for foodborne pathogens
Human illness costs:	**Costs of animal production:**	**Disease surveillance costs to:**
Medical costs:	Morbidity and mortality of animals on farms	Monitor incidence/severity of foodborne disease
Physician visits	Reduced growth rate/feed efficiency and increased time to market	Monitor pathogen incidence in the food chain
Laboratory costs	Costs contaminated animal disposal (e.g., farm and slaughterhouse)	Develop integrated database from farm to table
Hospitalization or nursing home	Increased trimming or reworking at slaughter and processing plants	
Drugs and other medications	Illness among workers due to handling contaminated animals /products	**Research to:**
Ambulance or other travel costs	Increased meat product spoilage due to pathogen contamination	Identify new foodborne pathogens for acute and chronic illnesses
Income or productivity loss for:		Identify high-risk products
Ill person or person dying	**Control costs for pathogens at all links in the food chain:**	Identify high-risk production and consumption practices
Caregiver for ill person	New farm practices (age-segregated housing, sterilized feed, etc.)	Identify which consumers are at high-risk for which pathogens
Other illness costs:	Altered animal transport and marketing patterns (animal identification)	Develop cheaper and faster pathogen tests
Travel costs to visit ill person	New slaughterhouse procedures (hide wash, carcass sterilizing)	Risk assessment modeling for all links in the food chain
Vocational/physical rehabilitation	Altered product transport (increased use of time/temperature indicators)	
Child care costs	New wholesale/retail practices (pathogen tests, employee training)	**Outbreak costs:**
Special educational programs	Risk assessment modeling by industry for all links in the food chain	Outbreak investigation costs
Institutional care	Price incentives for pathogen-reduced product at each food chain link	Testing to contain an outbreak
Lost leisure time		
Home modifications		

Psychological costs: Pain, other psychological suffering Risk aversion **Averting behavior costs:** Extra cleaning/cooking time costs Extra cost of refrigerator, freezer, etc. Flavor change from traditional recipes Increased cost to buy safer foods **Altruism (willingness to pay for others to avoid illness)**	New processing procedures (pathogen tests, contract purchasing requirements) **Outbreak costs:** Herd slaughter/product recall Plant closings and cleanup Regulatory fines Product liability suits from consumers and other firms Reduced product demand because of outbreak: Generic animal product – all firms affected Reduction for specific firm at wholesale or retail level Increased advertising or consumer assurances following outbreak	Legal suits to enforce regulations that may have been violated[3] Cleanup costs **Other considerations:** Distributional effects in different regions, industries, etc. Equity considerations, such as special concern for children

[1] Willingness-to-pay estimate for reducing risks of foodborne disease is a comprehensive estimate of all these categories (assuming that the individuals have included employer-funded sick leave and medical programs in their estimates). The estimate is comprehensive and covers reduced risks for everyone – those who will become ill as well as those who will not.

[2] Some industry costs may fall with better pathogen control, such as reduced product spoilage, possible increases in product shelf-life, and extended shelf-life permitting shipment to more distant markets or lowering shipment costs to nearby markets.

[3] In adding up costs, care must be taken to assure that product liability costs to firms are not already counted in the estimated pain and suffering cost to individuals. However, the legal and court expenses incurred by all parties are societal costs.

Source: Adapted from Roberts, T. and E. Todd, 'Approaches to Estimating the Cost of Foodborne Disease,' WHO Consultation on the Economic Implications of Animal Production Food Safety, Washington DC, June 8–10, 1995.

6.1.3 WHO consultation to estimate the global burden of foodborne disease

In 2006, the WHO launched a initiative to estimate the global burden of foodborne disease to: (1) 'obtain reliable epidemiological estimates on current, projected and averted morbidity, disability, and mortality of foodborne diseases, and (2) provide countries with simple, user-friendly tools to conduct their own foodborne disease burden studies and examine the effectiveness of their prevention and intervention efforts' (WHO, 2006).

For this initiative, an international consultation of over 50 experts was convened in 2006, developed a set of recommendations, and provided a strategic framework to execute the recommendations. Importantly, the initiative established the Foodborne Disease Burden Epidemiology Reference Group (FERG) to implement the recommendations and estimate the burden (Stein *et al.*, 2007). These recommendations included elements of a standard protocol for conducting the studies in countries around the world to obtain estimates for all major causes of foodborne disease, including microbial, parasitic and chemical contamination of food. The effort will build upon existing protocols in countries and modify them according to regional or national capacities. The initiative encourages countries with limited data to estimate their foodborne disease burden because these estimates can catalyze valuable research and assist policy development (WHO, 2006).

One of the main recommendations is that the burden of disease studies be conducted in each country through a combined syndromic and etiologic agent-specific approach followed by an attribution of the proportion of *Disability Adjusted Life Years* (DALYs) that is estimated to be foodborne (WHO, 2006) [DALYs are discussed later in 6.3.2]. The *syndromic approach* may be a simple listing of the pathogens/toxins commonly transmitted through food by principal disease syndromes (e.g., a list of pathogens with reactive arthritis as a clinical syndrome, such as *Salmonella* sp., *Campylobacter* sp. and *Yersinia* sp.). The *etiologic approach* gathers data on pathogens causing laboratory-confirmed diarrheal diseases and separates these data by their most probable foodborne routes of transmission. Information generated from these steps will be extrapolated to all diarrheal diseases and adjusted for underreporting and population coverage.

6.1.4 General principles of economic analysis for US regulations

In the United States, Executive Order 12866 (and the updated version, Executive Order 13422) directs agencies to conduct an economic analysis of all proposed or existing regulations that meet certain conditions (i.e., including food safety regulations) (OMB, 1996, 2007). In particular, agencies are required to provide a quantitative and qualitative assessment of the anticipated benefits and costs of all Federal mandates resulting in annual expenditures of $100 million or more (in 1995 US dollars). This assessment must include the benefits and costs to State, local and tribal governments or private sector businesses. The purpose of these economic analyses is to inform decision makers about the consequences of alternative regulatory actions (Box 6.1).

Box 6.1 Main purposes of an economic analysis

According to Office of Management and Budget (OMB) Circular A-4 (2003), the economic analysis should provide sufficient information to decision makers so that they can determine whether:

- There is adequate information indicating the need for regulation and whether the government intervention is likely to do more good than harm;
- The potential benefits to society justify the potential costs. Given that not all benefits and costs can be described in monetary or even in quantitative terms, a statute may require another regulatory approach;
- The proposed action will maximize net benefits to society (including potential economic, environmental, public health and safety, and other advantages; distributional impacts; and equity), compared with other regulatory options;
- Where a statute requires a specific regulatory approach, the proposed action will be the most cost-effective, including reliance on performance objectives wherever feasible;
- Agency decisions are based on the best reasonably obtainable scientific, technical, economic, and other information.

There is no blueprint for conducting an economic analysis and different regulations may call for different analytical approaches and emphases (OMB, 2003). An economic analysis of an economically significant rule contains three elements:

1 A statement of the need for a proposed action. This addresses whether the problem presents a compelling public need.
2 An examination of alternative approaches and the agency's rationale for selecting the proposed action over other alternatives.
3 A *benefit–cost analysis*. Estimating the benefits and costs of regulations includes two conceptually consistent components:
   (a) A *risk assessment*, part of which characterizes the probabilities of occurrence of the different outcomes. The risk assessment used in a benefit–cost analysis is not a safety assessment; it does not tell us whether a food is 'safe', only what the probabilities are.
   (b) A *valuation of the levels and changes in risk* experienced by the populations that the regulation affects.

The Office of Management and Budget (OMB) reviews all mandated economic analyses written by federal agencies. Typically, if and when a rule is finalized, implementation is phased in over time. However, emergency measures may become effective immediately or shortly after the rule is finalized.

To the fullest extent possible, benefits and costs should be expressed in terms of the discounted currency (e.g., in constant dollars). Expressing both benefits and costs in dollars makes it easy to calculate *net benefits*. Negative net benefits means that the regulation may not be a good use of public expenditures, whereas positive net benefits means that it may be a good use. Additionally, expressing net benefits in dollars means that the cost of the proposed regulation can be compared with other options for public spending.

Monetization of both the major benefits and costs is preferred. However, a *cost-effectiveness analysis* (CEA) may be used when monetization of the benefits of reducing foodborne disease is not possible, such as when it is very difficult or controversial to estimate the benefits (e.g., benefits of reducing the costs to unpaid caregivers, value of a premature death of a fetus), or when there is a dispute of the existence of substantial benefits. In essence, CEA is an approach that compares two or more programs that address a particular problem by comparing costs with the 'number' of physical benefits, such as the number of adverse outcomes prevented per dollar spent on a regulation. OMB recommends that agencies do both a cost-effectiveness analyses and a benefit–cost analysis for all major rulemakings for which it is possible to develop measures of effectiveness for the expected health and safety outcomes (OMB, 2003). That is, a CEA provides complementary information to a benefit–cost analysis for decision makers to identify the most cost-effective means to attain a specific health target, like a certain number of illnesses averted or deaths avoided. CEA may be less useful when there are multiple health outcomes that have to be described and counted in multiple dimensions, such as the typical foodborne pathogen that can cause a mild illness, premature death and/or chronic sequelae.

Analysts never have complete information, and therefore, they must be attentive to the quality and reliability of data, models, assumptions, scientific inferences and other information. Sensitivity analyses may be conducted to cover a wider range of plausible model specifications and values of key parameters. A properly calibrated benefit–cost analysis provides information that helps risk managers make rational decisions about which risks are most amenable to reduction, given various financial and other constraints facing government, individuals and food companies. Additionally, a benefit–cost analysis can help risk managers modify policies so as to increase net benefits.

6.2 Challenges faced in estimating the impact of foodborne disease

Developing national or global estimates of the cost or burden of foodborne disease is challenging because data on foodborne disease and health may only be partially available, fragmented and may be concentrated on fatal health outcomes (WHO, September 2006). There are eight epidemiological and methodological challenges that hinder the estimation of the economic costs and burden of foodborne disease (Fig. 6.1).

6.2.1 Epidemiological challenges

Four epidemiological challenges confront attempts to estimate the economic costs and burden of foodborne disease.

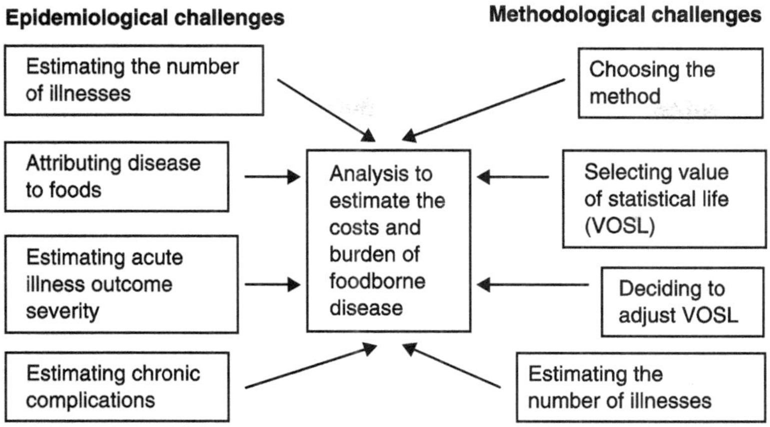

Fig. 6.1 Epidemiological and methodological challenges to estimate the costs and burden of foodborne disease.

Estimating the number of illnesses each year caused by a particular pathogen

It is difficult to estimate the annual number of foodborne illnesses caused by all of the different pathogens. In essence, for a person who becomes ill to be identified as a laboratory-confirmed case of foodborne disease in the United States and for the disease to be recorded in a national database, the ill individual must seek medical care, a specimen must be obtained, the laboratory must test for the particular foodborne disease that caused the illness, and the laboratory must confirm the causative agent and report the case to public health surveillance. However, the vast majority of people with a foodborne illness do not seek medical care, perhaps because their symptoms are limited to relatively mild gastrointestinal distress and/or because they consider their illness as a temporary inconvenience rather than an illness for medical health professionals to treat. Even if an ill person seeks medical care, relatively few provide a laboratory specimen for testing, and even if this lab testing is performed, it may not test for or identify the causative agent (CDC, 2008a).

The Foodborne Diseases Active Surveillance Network (FoodNet) consists of active surveillance and related epidemiologic studies for foodborne diseases in the United States. It is a collaborative project of the Centers for Disease Control and Prevention (CDC), ten Emerging Infections Program sites in different states, the US Department of Agriculture (USDA) and the Food and Drug Administration (FDA). In 2007, FoodNet covered 15% of the US population and found a total of 17 883 laboratory-confirmed cases of foodborne infection (CDC, 2008b).

Using FoodNet data and other sources of information, the total number of foodborne disease cases in the United States is estimated to be 76 million annually, according to a seminal article by Mead *et al.* (1999). Mead *et al.* also estimate that unknown agents accounted for roughly 81% of all foodborne illnesses in the

United States and 64% of the associated deaths. Frenzen (2003) estimates that gastroenteritis of unknown etiology is responsible for 4400 deaths in the United States each year. Deaths and chronic sequelae are much more costly than milder outcomes of all diseases. Illnesses due to unknown pathogens are excluded in many economic studies of foodborne disease, meaning that total costs will be underestimated.

Attributing foodborne disease to particular foods

There are limited data and other information on the share of foodborne illnesses attributed to specific foods. This information is critical in those economic analyses that estimate the number of foodborne disease cases that might be prevented by a proposed regulation. For example, the official regulatory impact analysis of the Pathogen Reduction/Hazard Analysis and Critical Control Point (HACCP) program for federally inspected meat and poultry slaughter and processing plants included assumptions about what portion of select foodborne illnesses (e.g., campylobacteriosis, salmonellosis and listeriosis) was due to meat and poultry (USDA, 1995, p. 6781). In general, the proportion attributed to foodborne transmission varies greatly from 1% for *Astrovirus* to 100% for *Clostridium perfringens* (Mead *et al.*, 1999).

Foodborne disease outbreak data often provide the best information linking specific pathogens to specific foods. In the United States, local and state health departments investigate and report roughly 400–500 foodborne outbreaks to CDC each year (CDC, 2008a). Outbreak information may help fill information and data gaps for those foodborne diseases that are not notifiable or under individual state surveillance. However, only 0.008% of all foodborne illness cases in the United States are identified in an outbreak (Mead *et al.*, 1999) and food vehicles implicated in outbreaks can differ considerably from those in sporadic cases (Denno *et al.*, 2009). Worldwide, food attribution data and approaches include case–control studies, analysis of outbreak data, expert judgment, microbial sub-typing and source tracking methods, among others (Batz *et al.*, 2005).

Estimating acute illness outcome severity

A thorough economic evaluation requires information on the severity of illness, the duration and outcomes, ranging from regaining full health to death, yet there are limited data on the distribution of outcome severity for most pathogens. FoodNet collects information on three illness severities (i.e., ill persons who visited a physician and tested positive for an infection from a foodborne pathogen, hospitalized patients, and those who died). However, FoodNet only conducts population-based surveillance for laboratory-confirmed cases of infection caused by nine foodborne pathogens (i.e., *Campylobacter*, *Cryptosporidium*, *Cyclospora*, *Listeria*, STEC O157 (formerly classified as *Escherichia coli* O157:H7), *Salmonella*, *Shigella*, *Vibrio* and *Yersinia* (Scallan, 2007; Jones *et al.*, 2007)) out of over 200 microbiological agents that cause foodborne disease (Mead *et al.*, 1999).

Estimating chronic complications
In addition to acute illness, many foodborne pathogens can cause one or more chronic sequelae such as arthritis, kidney failure, irritable bowel syndrome and Guillain–Barré syndrome (GBS). These sequelae can lead to costly lifetime health consequences and/or premature death. These chronic complications are usually not included in economic cost studies due to lack of data. Since 2000, FoodNet has conducted surveillance for post-diarrheal hemolytic uremic syndrome (HUS), which is a chronic complication of some acute infections from *E. coli* O157:H7. HUS is characterized by red blood cell destruction, kidney failure, and neurological complications such as seizures and strokes. When chronic sequelae are included, they can be associated with high average and total costs. For example, annual per patient direct costs for irritable bowel syndrome were estimated at $348–$8750 and indirect costs were in the range of $355–$3344 (Maxion-Bergemann *et al.*, 2006). Annual costs of *Campylobacter*-associated GBS linked to food were estimated at $136.0 million to $1.3 billion in 1995 (Buzby *et al.*, 1997a).

6.2.2 Methodological challenges

There are four methodological challenges in estimating the economic costs and burden of foodborne disease, which are described below.

Choosing the method
Deciding which method to use to estimate the impact of foodborne disease is the first methodological challenge. The choice is important, particularly if the resulting estimates will be incorporated in an economic analysis of proposed Federal food safety rule in that the resulting unit of measurement must be appropriate for the larger analysis and because the resulting estimates can influence real decisions on whether or not a policy should be implemented.

The goal of the analysis often determines the choice of the method among monetary (e.g., cost-of-illness (COI), willingness-to-pay (WTP)) and non-monetary methods (e.g., versions of the *health-adjusted life years* (HALYS) method) [The next two sections provide a general description of the main monetary and non-monetary methods]. If the goal is a benefit–cost analysis of a potential food safety regulation to see whether it is worthwhile to implement, then the WTP method should be used to estimate the benefits of the regulation, according to economic theory (Kuchler and Golan, 1999). Importantly, the OMB 'best practices' document states that the WTP is the preferred approach when valuing reductions in fatality risks that are expected by implementing a proposed regulation (OMB, 2003).

If the goal is to estimate foodborne disease costs and have a detailed breakdown of the cost components, then the COI method may be the preferred method. The COI method may also be used to value the benefits of anticipated reductions of foodborne disease following the implementation of a food safety regulation. As previously mentioned, monetization of both the major benefits and costs is preferred, yet sometimes monetization is not possible due to analytical limitations, ethical issues and other factors. This is when non-monetary methods like HALYs that look at health-related quality of life, such as in healthy-time equivalents, are used. HALYs are also sometimes used in a CEA.

Selecting the point or dollar estimate(s) of the value of statistical life

Selecting which point or dollar estimate(s) of the value of a statistical life to use is a particularly crucial methodological challenge. This is because the valuation of death can be the largest component of total estimated costs in a proposed regulation. The methodology for how economists value premature deaths has evolved over time, and as previously mentioned, WTP is currently considered as the best measure of benefits by economists. However, members of different parts of the US Federal Government use different WTP point or dollar estimates of the value of a statistical life (Golan *et al.*, 2005) and the choice matters in analyzing a proposed food safety rule. Some studies perform a sensitivity analysis using different value of statistical life scenarios (e.g., in Roberts *et al.*'s (1996) benefit–cost analysis of HACCP for meat and poultry slaughter and processing plants).

Deciding whether or not to adjust for pre-existing health status or age

Another methodological challenge is to decide whether the value of statistical life estimate(s) should be adjusted for pre-existing health status or age in a particular analysis. It is a matter for debate whether or not WTP for mortality risk reductions should be calculated on the expected remaining years of life lost due to the premature death or whether it should account for the pre-existing health status (Alberini *et al.*, 2004). For example, the elderly are more likely than younger individuals to have a co-morbidity that affects their ability to respond and therefore may be more susceptible to a foodborne illness. One example of a study that adjusted for age is Frenzen's 2005 analysis of *E. coli* O157, which grouped deaths into five-year intervals by the age of death. This study assumed that the costs of premature death ranged from $1.8 million for an adult over 85 years of age to $9.3 million for an infant. This study illustrates how estimated total costs could vary widely if the value of statistical life is adjusted for age and how total costs could vary as well, depending on the age breakdown of those who died prematurely because of the foodborne disease. Golan and Kuchler (1999) provide greater detail on the pros and cons of using single or multiple WTP values and discuss other ways in which WTP estimates can be adjusted (e.g., for income).

Estimating the anticipated impact of a proposed rule

Estimates of the anticipated numbers of foodborne disease cases and associated deaths that would be prevented if a proposed food safety rule was implemented are often an integral part of an economic analysis for that rule. Often, sensitivity analyses use a plausible range of model specifications to account for this uncertainty in a future outcome. Economists often rely on epidemiologists and risk assessors for this technical information.

6.3 Methods used to value the impact of foodborne disease

6.3.1 Monetary methods

Cost-of-illness method

Economists have developed both monetary and non-monetary methods for estimating the impact of foodborne disease and, in turn, for valuing the benefits of proposed food safety policies. One primary monetary method is the COI method, which is an accounting or tally of the dollars spent on *direct medical costs* (e.g., medical hospital inpatient expenses and physician outpatient care), *direct non-medical costs* (e.g., transportation to health care, relocation expenses) and *indirect costs* like the dollars of employment compensation that are foregone as a result of morbidity or mortality.

In general, a COI analysis for a particular foodborne disease traditionally starts with estimates of the annual number of cases and then divides this number into severity groups such as those who: (1) only had mild illnesses, (2) sought the advice of a physician, (3) were hospitalized, and (4) died prematurely because of their illness (for detailed food safety examples, see Buzby *et al.*, 1996). Those who develop chronic complications may be analyzed in a separate severity category. For example, Fig. 6.2 presents the schematic of a COI study on Shiga toxin-producing *E. coli* (STEC). The initial acute illness is subdivided by the severity of the outcome. The most severe complication in this example is HUS and some of these HUS cases develop end-stage renal disease (ESRD).

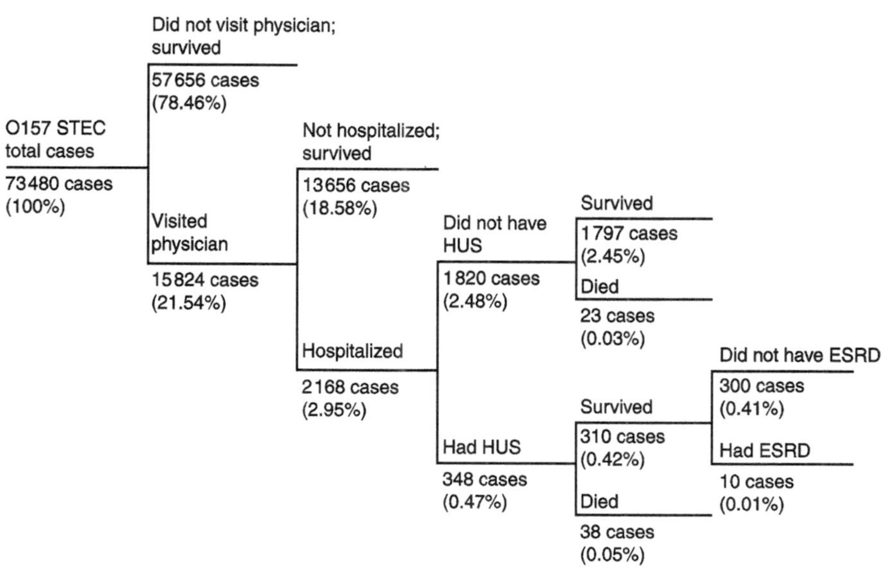

Fig. 6.2 ERS distribution of estimated annual US STEC O157 cases by disease outcome. Source: USDA/Economic Research Service.

Total costs for each severity group are then estimated. The bulk of a COI analysis usually entails calculating the corresponding medical costs, lost productivity costs, value of premature deaths and other illness-specific costs, such as special education and residential care, for a particular foodborne disease on an annual basis. In general, medical costs include physician and hospital services, supplies, medications and special procedures required for a specific foodborne illness. Hospitalization often accounts for a large share of medical costs.

Productivity loss typically measures the decline in production (output) because workers were ill and either missed work, performed poorly at work, regained only a portion of their pre-illness productivity, switched to less demanding and lower paying jobs, were unable ever to return to work or died prematurely. Productivity losses for those who died or were unable to return to work are usually calculated differently from those who missed some work but later resumed work. The *human capital approach* is often used for those cases in which work is interrupted temporarily. In essence, this approach uses the present value of future earnings that are foregone because of the illness. For example, this productivity loss may be calculated as the product of time lost from work multiplied by the corresponding wage rate such as the average daily wage. The total cost of lost productivity and premature deaths is the sum for all individuals affected, primarily the patients and, in the case of ill children, their parents or costs for paid caretakers.

Subsequent steps in the analysis identify what portion of these costs is attributable to foodborne sources and what portion of cases and deaths could potentially be averted by a proposed regulation. Assumptions must be made where data are not available, and these assumptions, as well as the rationale for the model, should be documented along with the results.

The advantages of the COI method are that it represents real costs to society and provides an easy-to-understand monetary measure of foodborne disease costs or the benefits of a program that reduces foodborne disease (Kuchler and Golan, 1999). Disadvantages include that it: (1) may not be a good measure of disease severity because estimates are influenced by income, education and other factors; (2) estimating direct medical expenses can be difficult because of the intricacies of disease coding and insurance arrangements; (3) estimating lost productivity costs may also be difficult because of the various forms of compensation available to employees and because large portions of the US population are not in the workforce, such as children, retirees or homemakers and home caregivers; and (4) it provides a partial estimate of economic costs as it excludes more difficult-to-measure costs to individual/households, industry and the regulatory/public health sector e.g., pain, suffering, lost leisure time, loss of business, liabilities from lawsuits (to the food industry) and the vast majority of chronic complications associated with foodborne disease (Kuchler and Golan, 1999). Therefore, COI estimates typically underestimate the actual benefits of a proposed food safety policy.

Willingness-to-pay approach

A second monetary method is the WTP approach, which 'measures the resources (dollars) individuals are willing and able to give up for a reduction in the probability of encountering a hazard that will compromise their health' (Kuchler and Golan, 1999). WTP estimates are often the result of labor market studies, which evaluate the small statistical risk of premature death and the increase in wages to compensate for taking this risk.

One advantage of the WTP approach is that it includes valuation of pain and suffering, lost leisure time, and other costs in a monetary measure. Also, the WTP approach is theoretically superior to other approaches because it reflects the observation that individual preferences for risk reduction are unique and that there is variation in the demand for risk reduction by individuals (Kuchler and Golan, 1999). As previously mentioned, the OMB best practices document (2003) states that the WTP approach is preferred when valuing reductions in fatality risks that are expected by implementing a proposed regulation.

The disadvantages of the WTP approach are that estimates are sensitive to the study populations, type of risk and level of risk, so these estimates may not be applicable if used in a different analysis. Also, it may not be practical to have a new study focus on the risk being evaluated and to develop new, targeted WTP estimates. The WTP also has some measurement difficulties, especially for non-fatal outcomes. As a result of these disadvantages, some analysts prefer simpler, non-monetary methods.

6.3.2 Non-monetary methods

Non-monetary methods deal with the health-related quality of life and provide estimates that are not in a monetary unit of measurement, such as in healthy-time equivalents. That is, these methods convert adverse health outcomes that compromise both lifespan and functional ability into a common unit of measurement (Golan *et al.*, 2005). HALY is an umbrella group of non-monetary methods that measure the years of full health lost because of living with morbidity. Two popular methods within this group are *quality-adjusted life years* (QALY) and *disability-adjusted life years* (DALY) (Gold *et al.*, 2002). QALYs assign values to health outcomes on a 0–1 scale, with 0 indicating death and 1 indicating perfect health. DALYs have the reverse 0–1 scale, meaning that food safety interventions would aim to maximize QALYs and minimize DALYs (Carabin *et al.*, 2005). Estimates from these methods are useful in cost-effectiveness analyses. Some analyses convert the estimates to dollars in a separate step by using a dollar value per HALY, QALY or DALY conversion factor. For example, the FDA has been monetizing QALYs in their benefit–cost analyses.

Disability-adjusted life years

The WHO initiative to estimate the global burden of foodborne disease ultimately aims to express the impact of each disease in DALYs because it is the WHO's preferred disease burden measure. According to WHO, 'the DALY measure combines the years of life lost due to premature death (YLL) and the years lived with disability (YLD) for varying degrees of severity, making time itself a common metric for death and disability. One DALY is a health measure, equating to one year of healthy life lost' (WHO, 2006).

The main advantage of the DALY approach is that it is an internally consistent common metric that can segregate co-morbidity (i.e., where several pathologies coexist and contribute and compete for the cause of death), thus separating epidemiology of the estimates from advocacy (WHO, 2006). It can develop and incorporate downstream effects on agricultural, social and trade costs that are traditionally missing from a cost or global burden analysis that targets one or more foodborne diseases for the human health implications. DALYs use the same value of a human life in rich and poor countries and also level the playing field between acute and chronic disease because it takes into account the duration of the syndrome and its severity (Carabin *et al.*, 2005).

The disadvantages of the DALY approach include that the approach still requires subjective value judgments on how to weight or discount for age of onset, disability weights and future losses (e.g., which discount rate to use) (WHO, 2006). Another disadvantage is that some social costs may be poorly represented by the DALY approach (e.g., reduced production and international trade because of a food safety issue).

6.4 Examples of the economic costs of foodborne disease and their use in cost–benefit analyses of food safety interventions

6.4.1 The United States

The USDA's Economic Research Service (ERS) conducted some of the earliest studies on the economic costs of foodborne illness in the United States (e.g., Roberts, 1989; Roberts and Frenkel, 1990; Roberts and Pinner, 1990). These initial estimates reflected the limited information then available about the incidence of foodborne illness and used the COI method to tally expenditures on medical care and lost productivity due to nonfatal illness and premature death. Over time, ERS updated and expanded these analyses using improved estimation methods and better data. Each series of ERS estimates incorporated better information on disease incidence, more detailed data on the health consequences of foodborne illness and advances in the economic methodologies for valuing health outcomes. See Golan *et al.* (2005) for greater detail on the ERS cost of foodborne illness literature.

ERS introduced their Foodborne Illness Cost Calculator in 2003. Currently, this calculator provides information on the assumptions behind foodborne illness cost estimates for *Salmonella* and STEC O157 and gives web users a chance to make their own assumptions and calculate their own cost estimates (ERS, 2008). ERS is currently expanding its cost calculator to incorporate additional foodborne diseases and their chronic complications. For example, Frenzen (2008) at ERS has updated earlier ERS estimates of the costs of *Campylobacter*-induced GBS (e.g., Buzby, Allos, *et al.*, 1997; Buzby, Roberts, *et al.*, 1997), but this analysis is not available at this time on the calculator. The calculator uses the COI method for nonfatal illnesses and WTP point or dollar estimates for the value assigned to the premature deaths.

In collaboration with the CDC's FoodNet, ERS prepared the *Salmonella* cost estimate for cases from all sources of transmission (Frenzen *et al.*, 1999; Breuer *et al.*, 2000). The study used new sources of data on medical costs and productivity losses, including a large commercial medical claims database and FoodNet surveillance data. Table 6.2 presents updated total case and total costs estimates for *Salmonella* from the ERS calculator and presents estimated foodborne cases and costs, assuming that 95% of salmonellosis cases are from foodborne transmission (i.e., from Mead *et al.* (1999) for nontyphoidal *Salmonella*). An implicit assumption is that salmonellosis cases from foodborne transmission incur the same average costs as cases transmitted via other routes of infection, such as from person-to-person transmission. In short, there are an estimated 1327328 cases in the United States, which cost $2417 million in 2007.

Also, in collaboration with FoodNet, ERS developed the total cost estimate for STEC O157 in 2005, using FoodNet surveillance data and a case–control study of STEC O157 patients (Table 6.3) (Frenzen, 2005, 2007). Assuming that 85% of STEC O157 cases are from foodborne transmission (i.e., from Mead *et al.*, 1999), then there are an estimated 62458 annual cases of STEC O157 in the United States at a cost of $390 million in 2007.

As Table 6.2 and Table 6.3 show, estimated costs per case vary by severity level (e.g., did not visit a physician vs. was hospitalized) and costs can be expensive, particularly for the more severe cases. The estimated average cost for STEC O157 is $6256 per case, far higher than the estimated average cost per salmonellosis case of $1821. However, the *Salmonella* estimate does not include costs for chronic sequelae.

Table 6.2 Estimated *Salmonella* costs in the United States, from all sources, 2007 dollars

Severity level	Total cases	Total costs	Average cost/case	Foodborne cases	Foodborne costs
	Number	*Million dollars*	*Dollars*	*Number*	*Million dollars*
Not hospitalized					
Did not visit physician; survived	1 224 547	61.7	50	1 163 320	58.6
Visited physician; survived	157 738	80.7	512	149 851	76.7
Hospitalized					
Survived	14 487	150.8	10 412	13 763	143.3
Died	415	2251.1	5 424 491	394	2 138.6
Total	1 397 187	2544.4	1821	1 327 328	2 417.2

Source: Total cases, total costs, and average costs estimates are from the Economic Research Service, http://www.ers.usda.gov/data/foodborneillness/salm_Intro.asp, accessed Oct. 8, 2008. Foodborne cases and costs assume 85% of total cases are foodborne (Mead *et al.*, 1999).

Table 6.3 Estimated STEC O157 costs in the United States, from all sources, 2007 dollars

Severity level	Total cases	Total costs	Average cost/case	Foodborne cases	Foodborne costs
	Number	*Million dollars*	*Dollars*	*Number*	*Million dollars*
Not hospitalized					
Did not visit physician; survived	57 656	1.6	29	49 008	1.4
Visited physician; survived	13 656	7.0	516	11 608	6.0
Hospitalized					
Did not have HUS; survived	1797	12.4	6922	1527	10.6
Had HUS but not ESRD; survived	300	11.6	38 695	255	9.9
Had HUS and ESRD; survived	10	58.7	5 868 761	9	49.9
Did not have HUS; died	23	103.7	4 506 837	20	88.1
Had HUS; died	38	264.6	6 963 826	32	224.9
Total	73 480	459.7	6256	62 458	390.8

Note: HUS = Hemolytic uremic syndrome, which is characterized by red blood-cell destruction, kidney failure, and neurological complications, such as seizures and strokes. ESRD = End-stage renal disease. Although these figures are for STEC O157 illnesses from all sources, average foodborne illness costs are usually considered to be the same as costs for those illnesses from non-foodborne sources.

Source: Total cases, total costs, and average costs estimates are from the Economic Research Service, http://www.ers.usda.gov/data/foodborneillness/ecoli_Intro.asp, accessed Oct. 8, 2008. Foodborne cases and costs assume 85% of total cases are foodborne (Mead *et al.*, 1999).

Table 6.4 provides a sample of the literature on the estimated costs and burden of foodborne disease in the United States. Some of the earliest studies focused on individual foodborne pathogens or extrapolated from these cost estimates to a wider group. For example, Roberts (1989) estimated annual costs for *Salmonella* and *Listeria* and extrapolated these costs to all other foodborne bacterial diseases.

In 1996, Buzby *et al.* published the most detailed and comprehensive report to that date documenting the analysis and cost estimates for six bacterial pathogens (*Campylobacter*, *C. perfringens*, *E. coli* O157:H7, *Listeria monocytogenes*, *Salmonella* and *Staphylococcus aureus*). Costs were then estimated by ERS for the foodborne parasite *Toxoplasma gondii* (Buzby and Roberts, 1996, 1997a). These estimates were used in the official regulatory impact analysis of the HACCP rule for federally inspected meat and poultry slaughter and processing plants (USDA, 1995, 1996).

Since then, there have been several detailed COI analyses for specific foodborne diseases. There have also been many cost and burden studies for

Table 6.4 Sample of estimated costs of foodborne diseases (FBD) in the United States

Author (year/method)	Foodborne disease cost study	Estimated costs
Todd (1989/COI)	Acute bacterial FBD	< $7 billion
Roberts (1989/COI)	All bacterial FBD	$4.8 billion
Roberts and Pinner (1990/COI)	*Listeria monocytogenes*	$480 million
Roberts and Frenkel (1990/COI)	Congenital toxoplasmosis	$0.4–$8.8 billion
Buzby *et al.* (1996/COI)	6 bacteria	$2.9–$6.7 billion
Barnard *et al.* (1995/COI)	FBD from meat consumption	$0.2–$5.5 billion
Buzby and Roberts (1997b/COI)	6 bacteria, 1 parasite	$6.5–$34.9 billion
Frenzen (1999/COI)	*Salmonella*	$464–$2399 million
Sandler *et al.* (2002/COI)	Select gastrointestinal FBD	$1,192 million
Frenzen *et al.* (2005/COI)	*E. coli* O157	$344 million
Roberts (2007/WTP)	All FBD	$1.4 trillion

Source: Compiled by author.

gastrointestinal disease, some of which specifically calculate costs for foodborne illness (e.g., Sandler *et al.*, 2002).

In the United States, COI estimates for foodborne disease have been used in food safety rulemaking by several agencies. For example, Buzby *et al.* (2006) provide a detailed summary of three food safety rules that incorporated the costs of foodborne illnesses in their economic analyses: (1) USDA's Food Safety and Inspection Service HACCP rule for meat and poultry in 1996, (2) the FDA *Salmonella* Enteritidis rule for shell eggs in 2000, and (3) FDA's proposed rule for ready-to-eat meat and poultry products (MPPs). Golan *et al.* (2005) also provide examples of US federal agencies' evaluations of the benefits of food safety programs using the value of reductions in foodborne health risks.

6.4.2 International

Several studies have estimated the economic costs and burden of foodborne disease outside of the United States (Table 6.5). The methodology and comprehensiveness of these studies have developed in parallel. That is, early international studies also focused on one or a few foodborne diseases and some served as a basis for extrapolations. For example, Todd (1989) extrapolated US estimates of foodborne illness cases and deaths from Bennett *et al.* (1987) to Canada with some adjustments. Another example is Razem and Katusin-Razem (1994). This study estimated the costs of notified food poisoning cases in Croatia by using salmonellosis cost data from four other countries (i.e., US, UK, Canada and Sweden) and by assuming that the cost per salmonellosis case would be representative of the average foodborne illness.

The cost of inpatient care in hospitals for acute infectious intestinal disease in England from 1991 to 1994 has been estimated at £24 million annually (Djuretic *et al.*, 1996). Roberts (1996) estimated the medical costs and value of lives lost

Table 6.5 Sample of estimated costs of foodborne diseases (FBD) outside the United States[1]

Author (year/method)	Foodborne disease cost study (country)	Estimated costs
Yule *et al.* (1988/COI)	Poultry-borne salmonellosis outbreak (Scotland)	£200 000–£900 000
Todd (1989/COI)	Acute bacterial FBD (Canada)	< $1.1 billion
Razem and Katusin-Razem (1994/COI)	All reported FBD (Croatia)	> $2 million
Djuretic *et al.* (1996/COI)	Acute infectious intestinal disease	£24 million
Roberts (1996)	5 FBD infections (England & Wales)	£300–£700 million
Roberts and Upton (2000/COI)	*E. coli* O157:H7 outbreak (UK)	£11 930 347
Havelaar *et al.* (2000/DALY)	*Campylobacter* sp. (Netherlands)	1400 DALYs
Scott *et al.* (2000/COI)	All FBD (New Zealand)	$55.1 million
Lindqvist *et al.* (2001/COI)	FBD (Sweden)	$123 million
Abe *et al.* (2002/COI)	*E. coli* outbreak (Japan)	¥82 686 000
Abelson *et al.* (2006/COI)	All FBD (Australia)	AU$1249.0 million
	Gastroenteritis	AU$1010.5 million
	Listeriosis	AU$85.5 million
	Toxoplasmosis	AU$40.2 million
	Hepatitis A	AU$36.7 million
	HUS	AU$25.6 million
	Irritable bowel syndrome	AU$19 million
	Guillain–Barré Syndrome	AU$3.2 million
	Reactive arthritis	AU$4.4 million
Kemmeren *et al.* (2006/DALY;COI)[2]	*Campylobacter* spp.	1,300 DALYs; €19.6 million
	Salmonella spp.	670 DALYs; €8.8 million
	Noroviris	55 DALYs; €25 million
	Rotovirus	370 DALYs; €21.7 million

[1] Some of the studies estimated costs of select chronic complications and most cost estimates are on an annual basis. Some estimates are for all routes of transmission and some estimates have been rounded.
[2] Undiscounted estimates in 2004 Euros shown here. Estimates are updated from earlier studies and are for all routes of transmission.
Source: Compiled by author.

from five infections in England and Wales at £300–£700 million annually. The foodborne percentage was not discussed in these studies but implied to be the bulk of the cases. Other studies focused on estimating the cost or burden of specific foodborne diseases or the costs of particular outbreaks.

Abelson *et al.* (2006) estimated that the annual cost of foodborne illness in Australia was AU$1249 million. This estimate includes AU$771.6 million in productivity and lifestyle costs to individuals and business, AU$231.5 million for premature mortality, AU$221.9 million for health care services, AU$14 million for food safety recalls and AU$10 million for government foodborne illness surveillance and investigation, and maintenance of food safety systems. Table 6.5 lists the estimated costs for health care, productivity, lifestyle and premature mortality costs for gastroenteritis, invasive listeriosis, toxoplasmosis, hepatitis A and select chronic sequelae (HUS, irritable bowel syndrome, GBS and reactive arthritis). The study used the COI method and a human capital valuation of a life lost at $2.5 million when the data were not age specific, otherwise the study converted this amount to an annual figure for use for each year of life lost (i.e., AU$108 000).

Kemmeren *et al.* (2006) used the DALY and COI approaches to estimate the burden of seven foodborne pathogens in the Netherlands. Of these seven, annual costs were estimated for four pathogens (i.e., *Campylobacter* spp., *Salmonella* spp., noroviris and rotovirus). The two viruses had the highest costs, largely due to the productivity losses while on sick leave. Estimates were updated from earlier studies and are for all routes of transmission.

Table 6.5 is not all inclusive – studies have also been conducted to estimate foodborne disease costs in Finland, Germany, Denmark and other countries. In general, obtaining and understanding studies published in different languages complicates a comprehensive analysis, particularly if they are unpublished or working papers. Inter-country comparisons of foodborne illness cases, deaths and associated costs are hindered by differences in surveillance systems and methodology. Snowdon *et al.* (2002) speculate that the cost of foodborne illness in developing countries may be greater than that in industrialized countries because of high infant mortality, malnutrition, chronic diarrhea, lost work and childcare, and that the economic consequences of foodborne disease are substantial in both industrialized and developing nations.

The current status of the WHO consultation to estimate the global burden of foodborne disease is that there are preliminary findings from studies conducted in some countries, such as Jordan and Vietnam. For example, the Jordan burden of illness study estimated that the annual incidence of foodborne disease in Jordan was 123 per 100 000 for *Salmonella*, 130 per 100 000 for *Brucella* and 306 per 100 000 for *Shigella* (WHO, 2006). At the time of writing this chapter, published DALYs from this study were not available.

6.5 Future trends

All aspects of the economic analyses for food safety regulations could benefit from better data. Importantly, CDC is currently updating Mead *et al.*'s (1999) seminal report on food-related illness and death in the United States. As previously mentioned, unknown pathogens represent one substantial information gap as they are estimated to account for roughly 81% of foodborne illnesses and hospitalizations and roughly 64% of deaths (Mead *et al.*, 1999). With more and improved information and data on foodborne disease, COI studies and other economic analyses could more accurately portray actual foodborne disease costs. For example, studies could incorporate more chronic sequelae from foodborne illnesses such as reactive arthritis and liver disease.

The methodology used to estimate the economic costs or burden of foodborne disease is evolving alongside the food safety regulations themselves. In particular, the use of point or dollar estimates of the value of premature deaths used in economic analyses that support food safety regulations has changed over the past few decades. In the late 1980s and earlier in this decade, most federal agencies used a value of approximately $5 million per statistical life based on midrange estimates of compensating wages (Golan and Kuchler, 1999). More recently, agencies and researchers have increasingly adjusted value statistical life estimates for age and other measures. Additionally, there is a trend toward using non-monetary measures to estimate the economic costs and burden of foodborne disease.

The total estimated cost and economic burden of all foodborne diseases in the United States and the world is not known. Current estimates of foodborne disease in the US tend to undervalue true societal costs for several reasons. First, most COI estimates cover costs for one or more foodborne diseases, whereas over 200 microbiological agents are known to be transmitted through food and there are numerous foodborne disease agents that are not microbiological (e.g., mad cow disease). Costs for *all* foodborne diseases – viral, bacterial, parasitic, fungal and all the non-microbiological agents – need to be included for a complete assessment of food safety-related costs. Additionally, the cause of most cases of gastrointestinal distress is never identified. Hence, the *actual* spectrum of the causes of foodborne disease and the extent of acute and chronic manifestations are unknown. As a result, the overall economic analysis will be similarly restricted.

Second, economic estimates underestimate the true impact of foodborne disease because the cost estimates tend to focus primarily on medical costs and lost productivity while excluding more difficult-to-measure costs to individuals/households, the food industry and the regulatory/public health sector. For example, the pain, suffering and lost leisure time of the victim and the victim's family are usually not included in cost estimates. Neither the loss of business nor liabilities from lawsuits (to the food industry) are included. Likewise, the value of preventive action is not included nor are the resources spent by federal, state and local governments to investigate the source and epidemiology of foodborne disease.

Third, costs for the vast majority of chronic complications associated with foodborne disease are not estimated. Because these complications can be life-long and may cause premature death, including these in an economic analysis can greatly increase estimated costs.

Despite these data limitations, economic estimates of the costs and burden of foodborne disease can enhance our understanding of food safety issues, such as which pathogens are causing the greatest economic damage. Additionally, some of these estimates can be used within cost–benefit analyses and cost-effectiveness analyses to help countries set priorities by determining the most efficient food safety prevention and control regulations, given limited federal resources.

6.6 Sources of further information and advice

RTI International's website (2008) provides extensive information on cost of illness studies, including a primer by Segal (2006) on the methods used when conducting cost of illness studies plus reviews of current studies and summaries for selected conditions. The website also includes information on the National Institutes of Health costs of illness studies. Although the majority of this information is not specifically on analyses of foodborne disease costs, the site provides background material on the COI method.

The Institute of Medicine's 'Valuing Health for Regulatory Cost-Effectiveness Analysis' (2006) is a good source for information on how to estimate health-related consequences of regulations that reduce health, safety and environmental risks and how to conduct and use CEA. The book is also a good reference for benefit–cost analysis, HALYs and QALYs.

The WHO web site (http://www.who.int/foodborne_disease/burden/en/) provides information on the initiative to estimate the global burden of foodborne disease. The site includes biographies of the members of the FERG and related meetings and reports. WHO's website also provides introductory information on DALYs and disability weights, discounting, and age weighting.

6.7 Disclaimer

The views expressed here are those of the author and may not be attributed to the Economic Research Service or the US Department of Agriculture.

6.8 References

ABE K, YAMAMOTO S and SHINAGAWA K (2002), 'Economic impact of an *Escherichia coli* O157:H7 in Japan', *J Food Prot*, **65**(1), 66–72.

ABELSON P, FORBES, M P and HALL G (2006, Mar) *The Annual Cost of Foodborne Illness and Australia*, Commonwealth of Australia, Australian Government Department of Health and Ageing, Canberra.

ALBERINI A, CROPPER M, KRUPNICK A and SIMON N B (2004), 'Does the value of a statistical life vary with age and health status? Evidence from the US and Canada', *J Environ Econ Manage*, **48**, 769–92.

BARNARD N D, NICHOLSON A and HOWARD J L (1995, Nov) 'The medical costs attributable to meat consumption', *Prev Med*, **24**(6), 656–7.

BATZ M B, DOYLE M P, MORRIS G, PAINTER J, SINGH R, TAUXE R V, ET AL. (2005, July), 'Attributing illness to food', *Emerg Inf Dis*, **11**(7), 993–9.

BENNETT J V, HOLMBERG S D, ROGERS M F and SOLOMON S L (1987), 'Infectious and parasitic diseases', in Amler, R.W. and Dull, H.B. (Eds): *Closing The Gap: The Burden of Unnecessary Illness*, Oxford University Press, New York.

BREUER T, BUZBY J, RIGGS T L and ROBERTS T (2000, Mar) 'Medical costs, productivity losses and premature deaths associated with foodborne salmonellosis', *J Assoc Food Drug Off*, **64**(1), 6–11.

BUZBY J C, ROBERTS T, LIN C T J and MACDONALD J M (1996), *Bacterial foodborne disease: medical costs and productivity losses*. Economic Research Service, U.S. Dept. of

Agriculture, AER-741. Available from: http://www.ers.usda.gov/Publications/AER741/ [Accessed 28 Nov 2008].

BUZBY J, SPINELLI F and NARDINELLI C (2006), 'Evaluating U.S. food safety regulations which use benefit and cost information', in Roberts, J. (Ed.): *The Economics of Infectious Disease*, Oxford University Press, Oxford, pp. 299–325.

BUZBY J C and ROBERTS T (1996, Sep–Dec), 'ERS updates U.S. foodborne disease costs for seven pathogens', Economic Research Service, U.S. Dept. of Agriculture, *Food Rev*, **19**(3), 20–5.

BUZBY J C and ROBERTS T (1997a, Sep–Dec) 'Guillain–Barré syndrome increases foodborne disease costs', Economic Research Service, U.S. Dept. of Agriculture, *Food Rev*, **20**(3), 36–42.

BUZBY J C and ROBERTS T (1997b), 'Economic costs and trade impacts of microbial foodborne illness,' *World Health Stat Quarterly*, **50**(1/2), 57–66.

BUZBY J C, ALLOS B M and ROBERTS T (1997), 'The economic burden of *Campylobacter*-associated GBS', *J Inf Dis*, **176**(Suppl 2), S192–7.

BUZBY J C, ROBERTS T and ALLOS B M (1997, July) *Estimated annual costs of Campylobacter-associated Guillain–Barré syndrome*. Economic Research Service, U.S. Dept. of Agriculture, AER-756. Available from: http://www.ers.usda.gov/publications/Aer756/ [Accessed 28 Nov 2008].

CARABIN H, BUDKE C M, COWAN L D, WILLINGHAM A L and TORGERSON P R (2005, July) 'Methods for assessing the burden of parasitic zoonoses: echinococcosis and cysticercosis', *Trends Parasit*, **21**(7), 327–33.

CDC (CENTERS FOR DISEASE CONTROL AND PREVENTION) (2008a), *Foodborne illness*. CDC, Atlanta, Georgia. Available from: http://www.cdc.gov/ncidod/dbmd/diseaseinfo/foodborneinfections_g.htm.

CDC (2008b), 'Preliminary FoodNet data on the incidence of infection with pathogens transmitted commonly through food: 10 States, 2007', *MMWR Weekly*, **57**(14), 366–70.

DENNO D M, KEENE W E, HUTTER C M, KOEPSELL J K, PATNODE M, FLODIN-HURSH D, ET AL. (2009), 'A tri-county comprehensive assessment of risk factors for sporadic bacterial reportable enteric infections in children', *J Inf Dis*, **199**(4), 467–76.

DJURETIC T, RYAN M J and WALL P G (1996, Apr), 'The cost of inpatient care for acute infectious intestinal disease in England from 1991 to 1994', *CDR Rev*, **6**, R78–80.

ERS (ECONOMIC RESEARCH SERVICE) (2008), U.S. Dept. of Agriculture. *Foodborne illness cost calculator*. Available from: http://www.ers.usda.gov/Data/FoodborneIllness/ [Accessed 18 Aug 2008].

FRENZEN P D (2003, Feb) 'Mortality due to gastronenteritis of unknown etiology in the United States', *J Infect Dis*, **187**, 441–52.

FRENZEN P D (2005), 'Economic cost of illness due to *Escherichia coli* O157 infections in the United States', *J Food Prod*, **68**(12), 2623–30.

FRENZEN P D (2007, Sep) *An online cost calculator for estimating the economic cost of illness due to Shiga toxin-producing E. coli (STEC) O157 infections*. Economic Research Service, U.S. Dept. of Agriculture, EIB-28. Available from: http://www.ers.usda.gov/Publications/EIB28/ [Accessed 26 November 2008].

FRENZEN P D (2008), 'Economic cost of Guillain–Barré Syndrome in the United States', *Neurology*, **71**, 21–7.

FRENZEN P D, RIGGS T L, BUZBY J C, BREUER T, ROBERTS R, VOETSCH D, ET AL. and THE FOODNET WORKING GROUP (1999, May–Aug) '*Salmonella* cost estimate update using FoodNet Data', Economic Research Service, U.S. Dept. of Agriculture. *Food Rev*, **22**(2), 10–15.

GOLAN E, BUZBY J, CRUTCHFIELD S, FRENZEN P, KUCHLER F, RALSTON K, ET AL. (2005), 'The value to consumers of reducing foodborne risks', in Hoffmann, S. and Taylor, M. (Eds): *Toward Safer Food: Perspectives of Risk and Priority Setting*, Resources for the Future, Washington, DC, pp. 129–58.

GOLAN E and KUCHLER F (1999), 'Willingness to pay for food safety: costs and benefits of accurate measures', *Am J Agr Econ*, **81**(5), 1185–91.

GOLD M R, STEVENSON D and FRYBACK D G (2002), 'HALYs and QALYs and DALYs, oh my: similarities and differences in summary measures of population health', *Annu Rev Pub Health*, **23**, 115–34.

HAVELAAR A H, DE WIT M A S, VAN KONINGSVELD R and VAN KEMPEN E (2000), 'Health burden in the Netherlands due to infection with thermophilic *Campylobacter* spp.', *Epidemiol Infect*, **125**, 505–22.

INSTITUTE OF MEDICINE (2006), *Valuing Health for Regulatory Cost-Effectiveness Analysis*, The National Academies Press, Washington, DC.

JONES T F, MCMILLIAN M B, SCALLION E, FRENZEN P D, CRONQUIST A B, THOMAS S, ET AL. (2007), 'A population-based estimate of the substantial burden of diarrhoeal disease in the United States; FoodNet, 1996–2003', *Epidemiol Infect*, **135**, 293–301.

KEMMEREN J M, MANGEN M J J, VAN DUYNHOVEN Y T H P and HAVELAAR A H (2006), *Priority Setting of Foodborne Pathogens: Disease Burden and Costs of Selected Enteric Pathogens*, National Institute for Public Health and the Environment, RIVM Report 330080001.

KUCHLER F and GOLAN E (1999), *Assigning Values to Life: Comparing Methods for Valuing Health Risks*, Economic Research Service, U.S. Department of Agriculture, AER-784. Available from: http://www.ers.usda.gov/Publications/AER784/ [Accessed 28 November 2008].

LINDQVIST R, ANDERSSON Y, LINDBÄCK J, WEGSCHEIDER M, ERIKSSON Y, TIDESTRÖM L, ET AL. (2001, June) 'A one-year study of foodborne illnesses in the municipality of Uppsala, Sweden', *Emerg Infect Dis*, **7**(3 Suppl), 588–92.

MAXION-BERGEMANN S, THIELECKE F, ABEL F and BERGEMANN R (2006), 'Costs of irritable bowel syndrome in the UK and US', *Pharmacoecon*, **24**(1), 21–37.

MEAD P S, SLUTSKER L, DIETZ V, MCCAIG L F, BRESEE J S, SHAPIRO C, ET AL. (1999, Sep–Oct) 'Food-related illness and death in the United States', *Emerg Infect Dis*, **5**(5), 607–25.

OMB (1996, Jan 11) *Economic analysis of federal regulations under executive order 12866.* Available from: http://www.whitehouse.gov/omb/inforeg/riaguide.html [Accessed 6 November 2001].

OMB (OFFICE OF MANAGEMENT AND BUDGET) (2003, Sep 17) *Circular A-4, regulatory analysis*. Available from: http://www.whitehouse.gov/omb/circulars/a004/a-4.html [Accessed 28 November 2008].

OMB (OFFICE OF MANAGEMENT AND BUDGET) (2007, Jan 18) *Executive order: further amendment to executive order 12866 on regulatory planning and review.* Available from: http://www.whitehouse.gov/news/releases/2007/01/20070118.html [Accessed 15 October 2008].

RAZEM D and KATUSIN-RAZEM B (1994, Aug) 'The incidence and costs of foodborne diseases in Croatia', *J Food Prot*, **57**(8), 746–52.

ROBERTS J A (1996), *Economic Evaluation of Surveillance*, Department of Public Health, and Policy, London.

ROBERTS T (1989, May) 'Human illness costs of foodborne bacteria', *Am J Agr Econ*, **71**(2), 468–74.

ROBERTS T (2007), 'WTP estimates of the societal cost of US foodborne illness', *Am J Agr Econ*, **5**, 1183–8.

ROBERTS T and PINNER R (1990), 'Economic impact of disease caused by *Listeria monocytogenes*', in Miller, A.J., Smith, J.L. and Somkuti, G.A. (Eds): *Foodborne Listeriosis*, Elsevier Science Publishing Co., Inc., Amsterdam, The Netherlands, pp. 137–49.

ROBERTS T and FRENKEL J K (1990), 'Estimating income losses and other preventable costs caused by congenital toxoplasmosis in people in the United States', *J Am Vet Med Assoc*, **196**(2), 249–56.

ROBERTS J A and UPTON P A (2000), *E. coli O157: An Economic Assessment of an Outbreak*, Lothian Health Board, Edinburgh.

ROBERTS T, BUZBY J and OLLINGER M (1996, Dec) 'Using benefit and cost information to evaluate a food safety regulation: HACCP for meat and poultry', *Am J Agr Econ*, **78**, 1297–301.

ROBERTS T and TODD E (1995, June 8–10) 'Approaches to estimating the cost of foodborne disease', in *WHO Consultation on the Economic Implications of Animal Production Food Safety*, Washington, DC.

RTI INTERNATIONAL (2008), *Cost-of-illness studies*. Washington, DC. Available from: http://www.rti.org/page.cfm?nav=443&objectid=CA1E1F48-8B6C-4F07-849D6A4C12CBF3C3

SANDLER R S, EVERAHART J E, DONOWITZ M, ADAMS E, CRONIN K, GOODMAN C, ET AL. (2002, May) 'The burden of selected digestive diseases in the United States', *Gastroenterology*, **122**(5), 1500–11.

SCALLAN E (2007, Mar 1) 'Activities, achievements, and lessons learned during the first 10 years of foodborne diseases active surveillance network: 1996–2005', *Clin Infect Dis*, **44**, 718–25.

SCOTT W G, SCOTT H M, LAKE R J and BAKER, M G (2000, July) 'Economic cost to New Zealand of foodborne infectious disease', *New Zealand Med J*, **113**(1113), 281–4.

SEGAL J E (2006, Jan) *Costs-of-illness studies – a primer*. RTI International, RTI-UNC Center of Excellence in Health Promotion Economics. Available from: www.rti.org/pubs/COI_primer.pdf [Accessed 25 August 2008].

SNOWDON J A, BUZBY J C and ROBERTS T (2002), 'Epidemiology, cost, and risk of foodborne disease', in Cliver, D. and Riemann H. (Eds): *Foodborne Diseases*, 2nd ed., Elsevier Press, Ltd., New York, NY, pp. 31–51.

STEIN C, KUCHENMÜLLER T, HENDRICKX S, PRÜSS-ŰSTÜN A, WOLFSON L, ENGELS D, ET AL. (2007), 'The global burden of disease assessments – WHO is responsible? *Negl Trop Dis*, **1**(3), 1–8.

TODD E C D (1989, Aug) 'Preliminary estimates of costs of foodborne disease in Canada and costs to reduce salmonellosis', *J Food Prot*, **52**(8), 586–94.

USDA (1995, Feb 3) 'Pathogen reduction: hazard analysis and critical control point (HACCP) systems: proposed rule', *Fed Reg*, **9**(306), 6774–889.

USDA (1996, July 25) 'Pathogen reduction; hazard analysis and critical control point (HACCP) systems, final rule', *Fed Reg*, **61**(144), 38856–906.

WHO (WORLD HEALTH ORGANIZATION) (2006), *WHO Consultation to Develop a Strategy to Estimate the Global Burden of Foodborne Diseases: Taking Stock and Charting the Way Forward*, WHO, Geneva, September 25–27.

YULE B F, MACLEOD A F, SHARP J C, ET AL. (1988, Feb) 'Costing of a hospital-based outbreak of poultry-borne Salmonellosis', *Epid Infect*, **100**(1), 35–42.

Part II

Subtyping of foodborne pathogens

7

Phenotypic subtyping of foodborne pathogens

W. A. Gebreyes, The Ohio State University, USA and S. Thakur, North Carolina State University, USA

Abstract: This chapter discusses the various phenotyping methods that are currently used to characterize foodborne pathogens. The methods that are predominantly used and described in this chapter include serotyping, phage typing, biotyping, antibiotyping and multi-locus enzyme electrophoresis. In this chapter, the principles of each of these methods, technical aspects, applications, interpretation of findings, strengths and weaknesses are described. Further reading on less common phenotypic methods is included as a concluding remark.

Key words: foodborne pathogens, phenotyping, serotyping, antimicrobial resistance patterns, phage typing, biotyping, multi-locus enzyme electrophoresis.

7.1 Overview of phenotypic subtyping

Various phenotypic schemes have been used for subtyping of foodborne pathogens for more than half a century. These approaches have been used for subtyping of pathogens based on antigenic properties, metabolic activities, colony morphology, resistance to antimicrobial agents and other chemicals and resistance to bacteriophages (Bettleheim and Taylor, 1969; Hinton and Allen, 1982). Even though the development of genotypic approaches minimized the usefulness of phenotyping schemes, phenotypic approaches remain an important component of typing foodborne pathogens due to their unique functional (phenotypic) significance. Most phenotypic methods, such as serotyping and antibiotyping are currently commonly used to subtype pathogens as a means of detecting outbreaks and also as a tool to supplement genotyping. In subtyping, a single approach does not often result in the best discrimination and may not provide the most comprehensive information regarding the characteristics of organisms. Similar to other subtyping approaches, selection of phenotypic subtyping methods is conducted based on specific criteria, including discriminatory ability, reproducibility, typability, cost and other epidemiologically important parameters. This chapter mainly focuses on the predominant phenotyping methods often used to subtype foodborne pathogens: serogrouping/serotyping, biotyping, phage typing, antibiotyping and multi-locus enzyme electrophoresis.

7.2 Serogrouping and serotyping

Serotyping is phenotypic characterization of a pathogen based on the immunologic reactivity and the antigenic determinants present on the cell surface. These surface antigens include the O-specific polysaccharide of the lipopolysaccharide (O antigen), the capsular polysaccharides (K antigen) and the protein H flagellar antigens (H antigen) to which specific antibodies are generated in animals. Identification of variations in the molecular epitopes comprising these antigens within the different species of a pathogen through use of specific antibodies forms the basis of serotyping. Serotypic identification is based on agglutination reactions in which antibodies and antigen bind and form clumps on a glass slide or react in wells of a microtiter plate. Characterization of pathogens by serotyping has been employed for many decades and is extensively used in categorizing important bacterial pathogens including *Salmonella* (Gebreyes, Davies *et al.*, 2004) and *Escherichia coli* (Durso *et al.*, 2007). This phenotyping method is widely used in foodborne pathogens, most commonly in *Salmonella* and *E. coli*. Over 2500 different serotypes of *Salmonella* and 700 serotypes of *E. coli* have been identified so far. With this vast array of serotypes that have been identified from humans, animals and the environment, serotyping has been used to assist in categorizing the pathogenic strains from the non-pathogenic strains of bacterial pathogens and associate host predilection. As such, serotyping tremendously aids in outbreak investigations and in understanding the epidemiology of bacterial pathogens in the host and their environment. Serotyping schemes are also available for other important bacterial pathogens including *Listeria, Campylobacter, Streptococcus, Kleibsella, Pneumococcus, Vibrio* and *Haemophilus* to name a few (Penner, 1988; Abdelnour *et al.*, 2009; Turton *et al.*, 2007; Gil *et al.*, 2007; Lin *et al.*, 2006; Inzana and Mathison, 1987). However, for this chapter, we will focus on *Salmonella* as the candidate bacterial pathogen for describing the method and the usefulness of serotyping in greater detail.

Serotyping scheme for the genus *Salmonella* is based on the Kauffman-White typing scheme established in 1929 and maintained by the Institute Pasteur that updates it every year (Popoff *et al.*, 2001). Taxonomically, the genus *Salmonella* comprises two species, *Salmonella enterica* and *Salmonella bongori. S. enterica* is further subdivided into six subspecies including I (enterica), II (salmonae), IIIa (arizonae), IIIb (diarizonae), IV (houtenae) and VI (indica), whereas *S. bongori* contains formerly subspecies V. More than 99% of the human isolates fall in subspecies I. *Salmonella* serotyping is challenging and laborious and, as such, only a few dedicated laboratories perform the process on a regular basis. More than 2523 serotypes of *Salmonella* are present with newer serotypes identified regularly. Differentiation of the *Salmonella* serotypes is performed by specific agglutination of antibodies with the polysaccharides (O antigen) and the flagellar proteins (H antigen) (Fig. 7.1). The O antigen is an immunogenic carbohydrate composed of different sugars, and the combinations and structures of these different sugars govern the type of antigen present in a particular serotype (Fitzgerald *et al.*, 2003). There are currently 46 O antigens that are coded by the *rfb* gene clusters and an additional 11 O antigens that are not coded by these genes but are either encoded on a plasmid or by bacteriophages (Fitzgerald *et al.*, 2003). The 119 H antigen results due to variation in the gene that results in the composition of the exposed flagella. The flagellum in *Salmonella* is encoded by two different genes (*fliC* and *flhB*) that work in coordination with each other for flagella development. That is, when one gene is turned on, the other gene is turned off and not expressed. There are some strains of *Salmonella* that carry only one of the genes and are known as monophasic strains. Grouping of isolates based on only O-antigen epitopes is also commonly practiced. This variation of serotyping is known as serogrouping and mainly involves characterizing isolates by polyvalent and group specific antisera against the O antigens.

The first step in sample preparation for serotyping involves preparation of a fresh bacterial culture that has been incubated and grown overnight on a plate or a slant. A single colony is picked and determination of the heat stable somatic O antigens is performed by agglutination tests using polyvalent antisera. A positive reaction is, for example, clumping with a clear background with latex agglutination reactions. Appropriate quality control strains are also used routinely. O-antigen testing is done first followed by identification of the H antigen. Reactions are carried out in pre-marked glass slides, and a reaction of 3+ (75% of cells agglutinate) or greater is considered satisfactory. It is important to remember that *Salmonella* strains can also

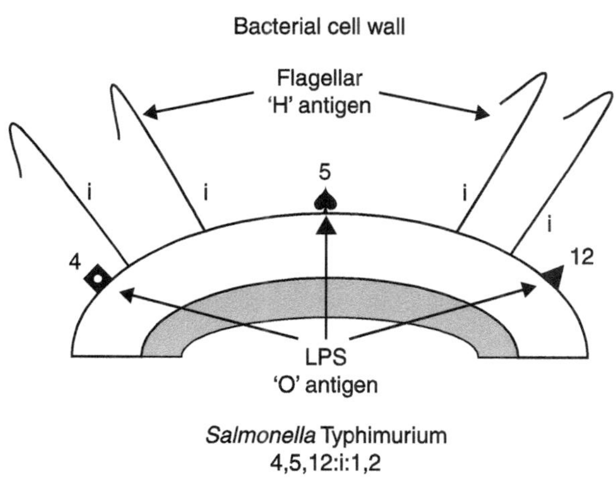

Fig. 7.1 Schematic demonstration of somatic (O) and flagellar (H) antigens of *Salmonella* serovar Typhimurium showing various epitopes.

be monophasic in which only one gene that encodes for the flagella is functional. In very rare cases, one may also encounter both the phases at the same time.

Serotyping has been of tremendous aid in epidemiological studies (Gebreyes and Altier, 2002; Gebreyes, Davies *et al.*, 2004), surveillance (NARMS, 2009) and in investigating foodborne outbreaks (Vugia *et al.*, 2009). *Salmonella* is responsible for a very high number of bacterial foodborne infections causing an estimated 7444 cases of illness in the USA in 2009 (Vugia *et al.*, 2009). Serotyping of these strains is very important for health care agencies in epidemiological investigations to trace outbreaks and mount appropriate response to control outbreaks. Serotyping data is also useful in identifying the potential sources of infection and in trace back investigations. For *Salmonella*, serotyping data is the first piece of information sought by agencies to determine their response. Specific *Salmonella* serotypes have a higher predilection to cause disease outbreaks in both humans and food animals. These include serotypes Typhimurium, Newport, Heidelberg, Enteritidis, Derby and Kentucky (Gebreyes and Thakur, 2005). The non-typhoidal *Salmonella* Typhimurium phage type DT 104 is an antimicrobial resistant clone of *Salmonella* that has been responsible for many outbreaks worldwide. It has a wide host range and is able to survive well in humans, animals and the environment. Another important serotype is S. Newport, which has been isolated from the US food supply and from animals and humans and has acquired extended spectrum β-lactamase enzymes, thereby rendering it resistant to multiple antimicrobials (Varma *et al.*, 2006).

Serotyping has inherent limitations, which can range from improper interpretation of results to cross reactivity. Furthermore, the entire process is laborious and time consuming. In addition, there is serological cross reactivity that has been shown between different serotypes including serotype 4 and 7 in *Actinobacillus pleuropneumoniae* (Mittal and Bourdan, 1991), Enterobacteriaceae (Hofstra and Dankert, 1979) and *Streptococcus* (Väkeväinen *et al.*, 2001). However, serotyping still remains a valuable tool for epidemiological and trace back studies due to its robustness and value as a phenotypic tool. The future of serotyping lies in the development and improved antisera and new molecular tools to enhance the sensitivity, specificity and ease of data interpretation. A recent development in genotyping is replacing the use of phenotypic serotyping via oligonucleotide-based microarrays that target the specific antigens of different bacterial pathogens. A method of serotyping *Salmonella, Streptococcus* and *E. coli* has been developed (Yoshida *et al.*, 2007; Wen *et al.*, 2006; Karin *et al.*, 2007). Initial work has been focused on targeting the most common *Salmonella* serotypes that are encountered. The microarray technology is bound to ultimately place all the oligonucleotide on the array chip in the near future. This method is described in greater detail in other sections.

7.3 Biotyping

Biotyping is a method of subtyping organisms using their biochemical utilization profiles. There is a wide range of approaches in biotyping. This subtyping approach has been used very commonly in *E. coli* as early as 1975 (Van Der Waaij *et al.*, 1975). Typically, the subtypes are determined based on enzyme profiles, carbohydrate assimilation patterns and in some cases resistance to acids such as boric acid for yeasts (Pizzo *et al.*, 2005). In other cases, subtyping methods that are based on specific gene loci are also referred to as biotyping. However, the presentation in this chapter is limited to the phenotypic approaches as the genotypic approaches are covered in other chapters.

Biotyping methods are conducted using various approaches. The most common approach is determining fermentation patterns for carbohydrates tested on acid-base indicators such as phenol red (Okerman and Devriese, 1985). In Gram-negative Enterobacteriaceae such as *E. coli*, fermentation to the following 14 carbohydrates is commonly used: adonitol, L-arabinose, dulcitol, n-inositol, lactose, maltose, D-raffinose, L-rhamnose, salicin, D-sorbitol, sorbose, sucrose, D-trehalose and D-xylose. Other approaches of biotyping include testing for the production of enzymes such as β-hemolysin and kinases (Devriese and Oeding, 1976) and determining the types of growth in specific agar. Fermentation of carbohydrates is detected based on color change of the colonies when acids are formed as a result of the fermentation process. The change in color could vary depending on the organism and the type of carbohydrate. For instance, *E. coli* forms yellow colonies when adonitol, dulcitol, raffinose, rhamnose or sorbose is fermented. On the other hand, colonies change to black when there is an aesculin hydrolysis (Hinton and Allen, 1982). Other enzyme tests such as the decarboxylase test are also determined based on color changes. Decarboxylase activity results in a change in the color of the medium to purple indicating alkalinity. A relatively more recent technology development has resulted in the use of an automated plate reader for determining these color changes, for example, using the Titertek plate reader developed by ICN Pharmaceuticals Ltd. (Crichton and Taylor, 1995).

Biotyping results are often interpreted descriptively using the patterns of fermentation for carbohydarates or the colony characteristics following growth on specific agars. Often the fermentation patterns are analyzed in terms of association with other characteristics such as the presence of specific virulence factors and also with clinical manifestations. This analysis enables detection of highly virulent strains and the association of specific patterns to clinically important phenotypes. This approach has been used primarily for *E. coli* strain characterization. In addition, this method has also been used for characterization of *Yersinia enterocolitica, Staphylococcus aureus* and other bacterial and fungal agents (Devriese and Oeding, 1976; Polonelli *et al.*, 1989; Pham *et al.*, 1991).

Biotyping approaches have been commonly used in epidemiological investigations and characterization of important foodborne pathogens, including *Yersinia enterocolitica. Y. enterocolitica* is a heterogeneous species, which can be divided into six biotypes (1A, 1B, 2, 3, 4 and 5) on the basis of variation in biochemical reactions (Wauters *et al.*, 1987). Subdivision into these six biotypes is based on the biochemical reactions to the following: pyrazinamidase activity, esculin hydrolysis, tween-esterase activity, indole production, and xylose and

Table 7.1 *Yersinia enterocolitica* biotyping. BT = biotype. All biochemical tests done at 25°C

Biochemical reaction	*Y. enterocolitica*					
	BT 1A	BT 1B	BT 2	BT 3	BT 4	BT 5
Pyrazinamidase	+	–	–	–	–	–
Esculin	+	–	–	–	–	–
Tween	+	+	–	–	–	–
Indole	+	+	+	–	–	–
Xylose	+	+	+	+	–	–
Trehalose	+	+	+	+	+	–

Modified from Wauters *et al.*, 1987 [permission pending].

trehalose acidification (Table 7.1). Biotypes 1B and 2–5 include strains that are associated with disease in humans and animals. Biotype 1A consists of nonpathogenic strains. Biotype 1A can, however, cause opportunistic infections in immunocompromised individuals, has been reported in patients with gastroenteritis (Burnens *et al.*, 1996; Bottone, 1997; Grant *et al.*, 1998; Tennant *et al.*, 2003), and is also widely distributed in water, feces and food (Bottone, 1999). *Y. enterocolitica* biogroup 4 and serotype O:3 is one of the most commonly isolated serotypes from human clinical yersiniosis cases (Bottone, 1999; Fredriksson-Ahomaa *et al.*, 2001). Previous work done in the USA and elsewhere indicated that serotype O:3 is commonly recovered from swine and swine products (Korte *et al.*, 2004; Bhaduri and Wesley, 2006). Bhaduri and Wesley reported serogroup O:3 as the most dominant virulence serogroup presently associated with swine in the USA (2006). *Y. enterocolitica* infection in infants and children through household preparation of chitterlings has been reported (Lee *et al.*, 1990; Jones *et al.*, 2003). However, McNally and colleagues indicated that in Great Britain, during 1999–2000, the most common biotype from pig samples was 1a (53.4%), and only 5% were biotype 4 (O:3) (McNally *et al.*, 2004).

Biotyping is currently infrequently used mainly due to the fact that numerous highly discriminatory genotypic approaches have been adopted in the last two decades. The main advantage of this approach is the fact that biotyping requires a minimal amount of equipment or technical sophistication. In contrast, the major disadvantage is that this approach is very laborious. This method has a relatively less discriminatory power as compared to other phenotypic and genotypic approaches. A comparison between serotyping and biotyping in various Enterobacteriaceae was conducted by Van Der Waiij *et al.* (1975). The results showed that while serotyping was found to be more discriminatory, biotyping was found to have several advantages. In addition to the high typability of the method, the test was relatively less laborious, and the binary coding system of the results was easily interpretable. Investigation of 70 *Y. enterocolitica* isolates from different sources (humans, swine, sheep and cattle) in the United Kingdom showed that biotyping was as useful as Amplified Fragment Length Polymorphism (AFLP) genotyping (Fearnley *et al.*, 2005).

Overall, while biotyping is among the early methods that served as a predecessor to the current subtyping schemes, this approach is less and less used. In this era of genomics and technical sophistication, the use of this approach will remain minimal for subtyping and tracking of foodborne pathogens.

7.4 Phage typing

Bacteriophages are viruses that have the ability to invade and lyse specific bacterial cells. They were first discovered in 1896 by Ernest Hankin and have been used for a variety of purposes ranging from typing to control of bacterial pathogens in food (Hagens and Loessner, 2007). The specificity of the phage is mediated by the proteins associated with the tails that bind to target surface molecules on the bacterial cell. The variability in the susceptibility of bacterial strains to different phages determines their phage type. This unique property of phages has been used as a tool to differentiate between different isolates of bacteria belonging to the same species. The phage typing tool has been valuable for outbreak investigations and is a useful method for epidemiological studies because of the ease of use and simplicity and the ability to employ phage typing in combination with other differentiation methods (Laconcha *et al.*, 1998). The high specificity of phages for specific bacterial pathogens also makes them ideal candidates to detect target bacteria and lyse them without disturbing the natural micro flora, thereby making them an important tool for enhancing food safety. Bacteriophages usually belong to the order Caudovirales, which are double-stranded DNA viruses. Phages can either be temperate (lysogenic), which do not lyse but integrate into the genome of the host or lytic which kill their host and are therefore more appropriate for typing and control purposes. For typing, the bacterial strain to be tested is spread on a plate, which is then inoculated with a different phage. The plate is incubated at the designated temperature and time duration. The development of plaques (clear zone) indicates lytic phage activity and this plaque distribution is recorded and compared to a chart to determine the phage type.

Phage typing has been used extensively for discriminating within non-typhoidal *Salmonella* serotypes and is primarily used in serotype Typhimurium and Enteritidis (Gebreyes, *et al.*, 2004). Similar typing schemes have also been developed and used for *E. coli*, *Campylobacter*, *Listeria* and *Streptococcus* (Kim *et al.*, 2009; Hart and Smith, 2009; Bull *et al.*, 2006; Domelier *et al.*, 2009). Phage typing has been highly successful for understanding the epidemiology of *Salmonella* serotypes in swine, poultry and humans (Gebreyes, *et al.*, 2004; Kalender *et al.*, 2009). We have detected penta-resistant *S.* Typhimurium phage type DT 104 in swine and humans with the resistance determinants encoded on the chromosome (Gebreyes and Altier, 2002). Another phage type of interest commonly detected in swine is *S.* Typhimurium phage type DT 193 that also exhibits multidrug resistance with the genes encoded on a plasmid instead of the chromosome. These results indicated that phage typing is a valuable tool in a situation where serotyping alone was unable to discriminate between the *S.* Typhimurium isolates. Similarly, phage typing has been used to study the phenotypic diversity of *Campylobacter* isolates from humans and poultry (Wareing *et al.*, 2002; Bull *et al.*, 2006) and *E. coli* O157:H7 in cattle, beef and chicken (Sánchez *et al.*, 2009; Chinen *et al.*, 2009).

There are challenges to phage typing, which are primarily attributed to the presence of untypable strains due to lower sensitivity to the phages. Another issue is the development of bacteriophage-insensitive mutants, which do not become infected with the phage. Even though the method is straightforward, it is difficult to maintain the live cultures of phages under laboratory conditions. As such, there are only a few diagnostic laboratories that routinely perform phage typing at a cost. Phages have the unique ability to discriminate between live and dead bacterial cells and will always have great acceptance among epidemiologists and in trace back investigations. Their recent use as an alternate to antimicrobials for killing bacterial pathogens has gathered attention and may play an important role in public health and food safety.

7.5 Antibiotyping (antibiogram)

A routinely used phenotypic subtyping method is antibiotyping. Antibiotyping (antibiogram) is a method of subtyping microorganisms, particularly bacterial pathogens using their pattern of resistance to various classes of antimicrobials. This approach is an extension to the clinical or research use of antimicrobial resistance testing of isolates. Antimicrobial resistance testing is performed using various types of methods. While some of the methods enable distinguishing whether an isolate is resistant or susceptible to antimicrobials in a binomial way (such as disc diffusion), other methods give detailed information on the levels of inhibitory concentration often referred to as minimum inhibitory concentration (MIC). The Kirby–Bauer agar disc diffusion approach is one of the oldest and most flexible testing method. This test enables qualitative grouping of isolates into three (resistant, intermediate or susceptible) categories. The MIC method based on agar dilution and broth dilution gives quantitative information by determining the concentration at which each antimicrobial inhibits growth of the isolates. The agar disc diffusion (also known as Kirby–Bauer) method is a low cost and flexible technique for conducting these tests as compared to MIC-based methods. The qualitative data generated from this test are also often sufficient for typing purposes. For instance in the USA the National Antimicrobial Resistance Monitoring System (NARMS) uses a set of antimicrobial panels for foodborne pathogens (NARMS, 2009).

The application of antibiotyping for epidemiology of foodborne pathogens has become a more common approach paralleling the emergence of multi drug-resistant strains. In the last decade, numerous epidemiological studies used antibiotyping as primary approach for phenotyping, often supplementing a genotyping approach (Frost *et al.*, 1982; Threlfall *et al.*, 1985; Hampton *et al.*, 1995; Low *et al.*, 1997; Besser *et al.*, 2000; Horby *et al.*, 2003; Gebreyes, *et al.*, 2004; Gebreyes and Thakur, 2005; Molla *et al.*, 2006; Zewde *et al.*, 2009). In a longitudinal study that involved characterization of Typhimurium and Copenhagen, phage types were further subtyped based on their antibiotypes, which showed a better discriminatory ability than phage typing (Table 7.2).

The major advantage of antibiotyping is its dual purpose use for determining the occurrence of antimicrobial resistance while at the same time allowing typing of strains. This approach has been found to have a fairly good discriminatory power. A Simpson's index of diversity (DI) comparison conducted by Gebreyes *et al.* (2006) in *Salmonella* showed that antibiotyping has a moderate DI value of 0.579, lower than most genotypic methods; however, it was similar to that of other phenotypic approaches. One major disadvantage of this approach is the variability in findings and the subjective interpretation of results. One inherent problem that affects the interpretation of MIC-based antibiotyping approaches is the variability in dilution ranges and MIC values. The qualitative Kirby–Bauer approach does not have this issue, however, when the zone of inhibition is not so defined or in the intermediate category, interpretation may be subjective.

Table 7.2 Phage types, serotypes and antibiotyping (R-type) of *Salmonella* isolated from pigs

Phage type (369)	Serotype (n)	Antibiotypes (R-type)
DT104 (125)	Copenhagen (121)	AxACSSuT(102), SSu (10), AxACeCSSuT (2), Other (7)
	Typhimurium (4)	AxACSSuT (4)
DT21 (96)	Copenhagen (5)	AxACKSSuT (1), AxACeKSSuT (1), ACeCKSSuT (1), ACeKSSuT (1), AKSSuT (1)
	Typhimurium (91)	AKSSuT (66), AxAKSSuT (12), None (3), Other (10)
DT193 (91)	Copenhagen (67)	AKSSuT (50), AxAKSSuT (3), AxACeCKSSuT (3), Other (11)
	Typhimurium (24)	AxASSuT (8), ASSuT (6), AKSSuT (4), Other (6)
DT208 (18)	Copenhagen (5)	Te (4), AKSSuT (1)
	Typhimurium (13)	AKSSuT (4), AxACKSSuT (4), Other (5)
DT12 (16)	Copenhagen (16)	AKSSuT (13), Other (3)
U302 (14)	Copenhagen (12)	AxACSSuT (11), AxASSuT (1)
	Typhimurium (2)	AKSSuT (2)
DT120 (1)	Copenhagen (1)	AxACSSuT (1)
DT169 (1)	Copenhagen (1)	AxACSSuT (1)
Untypable (5)	Copenhagen (5)	AKSSuT (3), AxACSSuT (2)
RDNCa (2)	Copenhagen (1)	AKSSuT (1)
	Typhimurium (1)	AKSSuT (1)

Adopted from Gebreyes *et al.*, 2004 with permission.

7.6 Multi-locus enzyme electrophoresis (MLEE)

MLEE has been used for more than two decades as a subtyping approach, as well as for population genetics studies in eukaryotic and prokaryotic organisms (Selander *et al.*, 1986). As the name implies, the main principle of this approach is studying polymorphic variation of enzymes using gel electrophoresis. The rate of migration of a protein during electrophoresis is charge dependent based on its amino acid sequence. Therefore, mobility variations among enzymes (electromorphs or allozymes) can be directly associated with alleles at the corresponding structural gene locus. Previous studies reported that 80–90% of amino acid substitutions can be detected by electrophoresis (Ramshaw *et al.*, 1979; Shumaker *et al.*, 1982).

MLEE has been used to study the epidemiology of various bacterial organisms, including foodborne (Bishai and Sears, 1993) and nosocomial pathogens (Catalano, 1994). The investigation of Enteritidis using MLEE was also used to provide evidence of inter-serovar relatedness. In a study reported by Stanley and Baquar (1994), MLEE defined *S.* Enteritidis as a polyphyletic serovar closely related to *S.* Dublin, *S.* Gallinarum and *S.* Pullorum. In an investigation on dissemination of vancomycin-resistant enterococci (VRE) from food animals to humans, MLEE analyses showed that VRE were disseminated via meat products and were also found in faecal samples of non-hospitalized humans (Witte, 1997).

MLEE has a unique strength as it deals with functional units (enzymes) that can directly affect the biology of the organisms. A recent review on parasitic systematics reported that MLEE is an underutilized phenotypic approach, and its benefit is masked by the recent emergence of various genotypic methods. It was recommended that MLEE be used to complement genotypic approaches to fully understand the dynamics of parasites (Andrews and Chilton, 1999). Two major weaknesses of MLEE include a post-translational modification error and also a synonymous substitution in the respective enzyme gene loci. Post-translational modification error could result in changes in amino acids within the functional proteins while indeed isolates are clonal genotypically as well as functionally (translation). This may result in categorizing two clonally related strains into different groups. In the latter case, while strains of bacteria may have changes in the genetic makeup, these mutations may be a synonymous substitution that will not result in changes of the amino acid. Since MLEE is based on the functional amino acid/protein unit characterization, it would not be able to detect such base pair change unless the mutations result in different amino acid substitution. Currently, MLEE is less commonly used since genotypic approaches such as multi-locus sequence typing (MLST) and also other highly discriminatory but less laborious methods that give unambiguous results have been introduced. Regardless, MLEE will most likely remain a useful tool due to its unique functional strength until genomic and proteomic tools become more routinely used. In a very recent report, Tibayrenc (2009) reported that while refined data can be obtained from various newly developed genotypic methods, MLEE has a unique strength and can still be used when inexpensive and effective approaches are needed.

7.7 Hemagglutination

Hemagglutination (HA) is a method often used for quantification of viruses or bacteria. This method is widely used beyond foodborne pathogens for investigation of various pathogenic organisms. The main principle of HA is based on the characteristics of pathogens' surface envelope proteins that are able to agglutinate to human or animal red blood cells and bind to its N-acetylneuraminic acid. This process is very commonly used in characterization of influenza viruses. The hemagglutinin of influenza virus binds to erythrocytes causing the formation of lattice. On the other hand detection and quantification of host (humans or animals) infected with these types of organisms can be done by using hemagglutinin inhibition (HI) assay to detect the occurrence of a pathogen in a host using antibodies against the specific hemagglutinins. One of the most common typing systems used in influenza viruses is the hemagglutinin (H) and neuraminidase (N) types. Often subtypes of this virus are designated based on their H and N types. Immunoassays based on monoclonal antibodies have been developed since 1986 that are commonly used for subtyping these pathogens. In this assay, HI is complemented with neuraminidase inhibition (NI) to type the strains of influenza (Walls *et al.*, 1986).

7.8 Conclusions

In general, phenotyping approaches have been used for about a century. Selection of subtyping methods should be based on specific criteria whose priority may depend on the purpose. Some of the key criteria include their discriminatory power, reproducibility, typability, cost, labor and turnaround time. While the methodological approaches of phenotyping are quite diverse and thus could have varying degrees of discriminatory power as well as extent of laboriousness, the majority of the methods remain useful primarily since they deal with functional attributes of the pathogens. In addition, some of the phenotypic systems are inherently being used in conjunction with genotypic approaches, particularly those used for detection systems, such as Enzyme-Linked Immuno Sorbent Assay (ELISA). For instance, PCR-ELISA approaches are commonly used for detection of various pathogens as well as characterization of immunologic units (Eyigor *et al.*, 2007). Beyond detection, various other serologic-based typing methods are also used in various pathogens. Some of the methods include immunodiffusion, serum neutralization antigenic correlation and others. As such methods are rarely used in foodborne pathogen typing, less emphasis is given in this chapter. Additionally, newly developed methods such as Matrix-Assisted Laser Desorption/Ionization Time-Of-Flight (MALDI TOF) mass spectrometry and others are emerging. However, these methods are often used for detection purposes rather than subtyping or as part of genotyping such as for detection and subtyping based on single nucleotide polymorphism. A recent study differentiating clinical isolates of *Campylobacter*, *Helicobacter* and *Arcobacter* (Alispahic *et al.*, 2009) and a review of this method for clinical application (Tost and Gut, 2005) are recommended for further reading to understand the principles and applications of this approach. Despite the increasing number of genotypic approaches introduced for use in foodborne pathogens, phenotypic approaches remain an important complementation to the new sophisticated genotyping systems.

7.9 References

ABDELNOUR A, SOLEY C, GUEVARA S, DAGAN N P and ARGUEDAS A (2009), '*Streptococcus pneumoniae* Serotype 3 among Costa Rican children with otitis media: clinical, epidemiological characteristics and antimicrobial resistance patterns', *BMC Pediatrics*, **9**, 52.

ALISPAHIC M, HUMMEL K, JANDRESKI-CVETKOVIC D, NÖBAUER K, RAZZAZI-FAZELI E, HESS M, ET AL. (2009), 'Species-specific identification and differentiation of *Arcobacter, Helicobacter* and *Campylobacter* by full-spectral MALDI-TOF MS analysis', *Journal of Medical Microbiology*, **59**(3), 295–301.

ANDREWS R H and CHILTON N B (1999), 'Multilocus enzyme electrophoresis: a valuable technique for providing answers to problems in parasite systematics', *International Journal for Parasitology*, **29**, 213–53.

BESSER T E, GOLDOFT M, PRITCHETT L C, KHAKHRIA R, HANCOCK D D, RICE D H, ET AL. (2000), 'Multiresistant *Salmonella* Typhimurium DT104 infections of humans and domestic animals in the Pacific Northwest of the United States', *Epidemiology and Infection*, **124**, 193–200.

BETTLEHEIM K A and TAYLOR J (1969), 'A study of *Escherichia coli* isolated from chronic urinary infection', *Journal of Medical Microbiology*, **2**, 225–36.

BHADURI S and WESLEY I (2006), 'Isolation and characterization of *Yersinia enterocolitica* from swine feces recovered during the National Animal Health Monitoring System's Swine 2000 study', *Journal of Food Protection* **69**, 2107–12.

BISHAI W R and SEARS C L (1993), 'Food poisoning syndromes', *Gastroenterology Clinics of North America*, **22**, 579–608.

BOTTONE E (1997), '*Yersinia enterocolitica*: the charisma continues', *Clinical Microbiology Reviews*, **10**, 257–76.

BOTTONE E J (1999), '*Yersinia enterocolitica*: overview and epidemiologic correlates', *Microbes and Infection*, **1**, 323–33.

BULL S A, ALLEN V M, DOMINGUE G, JØRGENSEN F, FROST J A, URE R, ET AL. (2006), 'Sources of *Campylobacter* spp. colonizing housed broiler flocks during rearing', *Applied and Environmental Microbiology*, **72**, 645–52.

BURNENS A, FREY A and NICOLET J (1996), Association between clinical presentation, biogroups and virulence attributes of *Yersinia enterocolitica* strains in human diarrhoeal disease', *Epidemiology and Infection*, **116**, 27–34.

CATALANO M (1994), 'Bacterial genotyping in nosocomial infections', *Medicina (B Aires)*, **54**, 596–603.

CHINEN I, EPSZTEYN S, MELAMED C L, AGUERRE L, MARTÍNEZ ESPINOSA E, MOTTER M M, ET AL. (2009), 'Shiga toxin-producing *Escherichia coli* O157 in beef and chicken burgers, and chicken carcasses in Buenos Aires, Argentina', *International Journal for Food Microbiology*, **132**, 167–71.

CRICHTON P B and TAYLOR A (1995), 'Biotyping of *Escherichia coli* in microwell plates', *British Journal of Biomedical Sciences*, **52**, 173–7.

DEVRIESE L A and OEDING P (1976), 'Characteristics of *Staphylococcus aureus* strains isolated from different animal species', *Research in Veterinary Science*, **21**, 284–91.

DOMELIER A S, VAN DER MEE-MARQUET N, SIZARET P Y, HÉRY-ARNAUD G, LARTIGUE M F, MEREGHETTI L, ET AL. (2009), 'Molecular characterization and lytic activities of *Streptococcus agalactiae* bacteriophages and determination of lysogenic-strain features', *Journal of Bacteriology*, **191**, 4776–85.

DURSO L M, BONO J L and KEEN J E (2007), 'Molecular serotyping of *Escherichia coli* O26:H11', *Applied and Environmental Microbiology*, **71**, 4941–4.

EYIGOR A, GONCAGUL G and CARLI K T (2007), 'A PCR-ELISA for the detection of *Salmonella* from chicken intestine', *Journal of Biological and Environmental Sciences*, **1**, 45–9.

FEARNLEY C, ON S L, KOKOTOVIC B, MANNING G, CHEASTY T and NEWELL D G (2005), 'Application of fluorescent amplified fragment length polymorphism for comparison of human and animal isolates of *Yersinia enterocolitica*', *Applied and Environmental Microbiology*, **71**(9), 4960–5.

FITZGERALD C, SHERWOOD R, GHEESLING L L, BRENNER F W and FIELDS P I (2003), 'Molecular analysis of the rfb O Antigen Gene Cluster of *Salmonella* enterica Serogroup O:6,14 and development of a serogroup-specific PCR assay', *Applied and Environmental Microbiology*, **69**, 6099–105.

FREDRIKSSON-AHOMAA M, HALLANVUO S, KORTE T, SIITONEN A and KORKEALA H (2001), 'Correspondence of genotypes of sporadic *Yersinia enterocolitica* bioserotype 4/O:3 strains from human and porcine sources', *Epidemiology and Infection*, **127**, 37–47.

FROST J A, ROWE B, WARD L R and THRELFALL E J (1982), 'Characterization of resistance plasmids and carried phages in an epidemic clone of multi-resistant *Salmonella* Typhimurium in India', *Journal of Hygiene*, **88**, 193–204.

GEBREYES W A and ALTIER C (2002), 'Molecular characterization of multidrug-resistant *Salmonella* enterica subsp. enterica serovar Typhimurium isolates from swine', *Journal of Clinical Microbiology*, **40**, 2813–22.

GEBREYES W A, ALTIER C and THAKUR S (2006), 'Molecular epidemiology and diversity of *Salmonella* serovar Typhimurium in pigs using phenotypic and genotypic approaches', *Epidemiology and Infection*, **134**, 187–98.

GEBREYES W A, DAVIES P R, TURKSON P K, MORROW W E, FUNK J A, ALTIER C, ET AL. (2004), 'Characterization of antimicrobial-resistant phenotypes and genotypes among *Salmonella enterica* recovered from pigs on farms, from transport trucks, and from pigs after slaughter', *Journal of Food Protection*, **67**, 698–705.

GEBREYES W A and THAKUR S (2005), 'Multidrug-resistant *Salmonella enterica* serovar Muenchen from pigs and humans and potential interserovar transfer of antimicrobial resistance', *Antimicrobial Agents and Chemotherapy*, **49**, 503–11.

GEBREYES W A, THAKUR S, DAVIES P R, FUNK J A and ALTIER C (2004), 'Trends in antimicrobial resistance, phage types and integrons among *Salmonella* serotypes from pigs, 1997–2000', *Journal of Antimicrobial Chemotherapy*, **53**, 997–1003.

GIL A I, MIRANDAB H, LANATAA C F, PRADAC A, HALLD E R, BARRENOA C M, ET AL. (2007), 'O3:K6 Serotype of *Vibrio parahaemolyticus* identical to the global pandemic clone associated with diarrhea in Peru', *International Journal of Infectious Diseases*, **11**, 324–8.

GRANT T, BENNETT-WOOD V and ROBINS-BROWNE R M (1998), 'Identification of virulence associated characteristics in clinical isolates of *Yersinia enterocolitica* lacking classical virulence markers', *Infection and Immunity*, **66**, 1113–20.

HAGENS S and LOESSNER M J (2007), 'Application of bacteriophages for detection and control of foodborne pathogens', *Applied Microbiology and Biotechnology*, **76**(3), 513–19.

HAMPTON M D, THRELFALL E J, FROST J A, WARD L R and ROWE B (1995), '*Salmonella* Typhimurium DT 193: differentiation of an epidemic phage type by antibiogram, plasmid profile, plasmid fingerprint and *Salmonella* plasmid virulence (*spv*) gene probe', *Journal of Applied Bacteriology*, **78**, 402–8

HART J and SMITH G (2009), 'Verocytotoxin-producing *Escherichia coli* O157 outbreak in Wrexham, North Wales, July 2009', *Euro Surveillance*, **14**(32), pii=19300.

HINTON M and ALLEN V (1982), 'The biotyping of *Escherichia coli* isolated from healthy farm animals', *Journal of Hygiene*, **88**, 543–55.

HOFSTRA H and DANKERT J (1979), 'Antigenic cross-reactivity of major outer membrane proteins in enterobacteriaceae species', *Journal of General Microbiology*, **111**(2), 293–302.

HORBY P W, O'BRIEN S J, ADAK G K, GRAHAM C, HAWKER J I, HUNTER P, ET AL.; PHLS OUTBREAK INVESTIGATION TEAM (2003), 'A national outbreak of multi-resistant *Salmonella enterica* serovar Typhimurium definitive phage type (DT) 104 associated with consumption of lettuce', *Epidemiology and Infection*, **130**, 169–78.

INZANA T J and MATHISON B (1987), 'Serotype specificity and immunogenicity of the capsular polymer of *Haemophilus pleuropneumoniae* serotype 5', *Infection and Immunity*, **55**, 1580–7.

JONES T F, BUCKINGHAM S C, BOPP C A, RIBOT E and SCHAFFNER W (2003), 'From pig to pacifier: chitterling-associated yersiniosis outbreak among black infants', *Emerging Infectious Diseases*, **9**, 1007–9.

KALENDER H, SEN S, HASMAN H, HENDRIKSEN R S and AARESTRUP F M (2009), 'Antimicrobial susceptibilities, phage types, and molecular characterization of *Salmonella enterica* serovar Enteritidis from chickens and chicken meat in Turkey', *Foodborne Pathogens and Disease*, **6**, 265–71.

KARIN B, BOZENA K, PETER K, PETER S, RALF E and HERBERT H (2007), 'Fast DNA serotyping of *Escherichia coli* by use of an oligonucleotide microarray', *Journal of Clinical Microbiology*, **45**, 370–9.

KIM S, KIM S H, PARK J H, LEE K S, PARK M S and LEE B K (2009), 'Clustering analysis of *Salmonella enterica* serovar Typhi isolates in Korea by PFGE, ribotying, and phage typing', *Foodborne Pathogens and Disease*, **6**, 733–8.

KORTE T, FREDRIKSSON-AHOMAA M, NISKANEN T and KORKEALA H (2004), 'Low prevalence of yadA-positive *Yersinia enterocolitica* in sows', *Foodborne Pathogens and Disease*, **1**, 45–52.

LACONCHA I, PEZ-MOLINA N L, REMENTERIA A, AUDICANA A, PERALES I and GARAIZAR J (1998), 'Phage typing combined with pulsed-field gel electrophoresis and random amplified polymorphic DNA increases discrimination in the epidemiological analysis of *Salmonella* Enteritidis strains', *International Journal of Food Microbiology*, **40**, 27–34.

LEE L, GERBER A, LONSWAY D, SMITH J, CARTER G, PUHR N, ET AL. (1990), '*Yersinia enterocolitica* O:3 infections in infants and children, associated with the household preparation of chitterlings. *New England Journal of Medicine,* **322**, 984–7.

LIN J, KALTOFT M S, BRANDAO A P, ECHANIZ-AVILES G, BRANDILEONE M C C, HOLLINGSHEAD S K, ET AL. (2006), 'Validation of a Multiplex Pneumococcal Serotyping Assay with Clinical Samples', *Journal of Clinical Microbiology*, **44**, 383–8.

LOW J C, ANGUS M, HOPKINS G, MUNRO D and RANKIN S C (1997), 'Antimicrobial resistance of *Salmonella enterica* Typhimurium DT104 isolates and investigation of strains with transferable apramycin resistance', *Epidemiology and Infection*, **118**, 97–103.

MCNALLY A, CHEASTY T, FEARNLEY C, DALZIEL R W, PAIBA G A, MANNING G, ET AL. (2004), 'Comparison of the biotypes of *Yersinia enterocolitica* isolated from pigs, cattle and sheep at slaughter and from humans with yersiniosis in Great Britain during 1999–2000', *Letters in Applied Microbiology*, **39**, 103–8.

MITTAL K R and BOURDON S (1991), 'Cross-reactivity and antigenic heterogeneity among *Actinobacillus pleuropneumoniae* strains of serotypes 4 and 7, *Journal of Clinical Microbiology*, **29**, 1344–7.

MOLLA B, BERHANU A, MUCKLE A, COLE L, WILKIE E, KLEER J, ET AL. (2006), 'Multidrug resistance and distribution of *Salmonella* serovars in slaughtered pigs', *Journal of Veterinary Medicine. B, Infectious Diseases and Veterinary Public Health*, **53**, 28–33.

NARMS (NATIONAL ANTIMICROBIAL RESISTANCE MONITORING SYSTEM) (2009), *Antimicrobial Tested, Concentration Ranges and Break Points*. Available from: http://ars.usda.gov/Main/docs.htm?docid=6750&page=3 [Accessed 25 September 2009].

OKERMAN L and DEVRIESE L (1985), 'Biotypes of enteropathogenic *Escherichia coli* strains from rabbits', *Journal of Clinical Microbiology*, **22**, 955–8.

PENNER J L (1988), 'The genus *Campylobacter*: a decade of progress', *Clinical Microbiology Reviews*, **1**, 157–72.

PHAM J N, BELL S M and KANZARONE Y M (1991), 'Biotype and antibiotic sensitivity of 100 clinical isolates of *Yersinia enterocolitica*', *Journal of Antimicrobial Chemotherapy*, **28**, 13–8.

PIZZO G, GIAMMANCO G M, PECORELLA S, CAMPISI G, MAMMINA C and D'ANGELO M (2005), 'Biotypes and randomly amplified polymorphic DNA (RAPD) profiles of subgingival *Candida albicans* isolates in HIV infection', *New Microbiologica*, **28**, 75–82.

POLONELLI L, CONTI S, MAGLIANI W and MORACE G (1989), 'Biotyping of pathogenic fungi by the killer system and with monoclonal antibodies', *Mycopathologia*, **107**, 17–23.

POPOFF M Y, BOCKEMÜHL J, BRENNER F W and GHEESLING L L (2001), 'Supplement 2000 (no. 44) to the Kauffmann-White scheme', *Research in Microbiology*, **152**(10), 907–9.

RAMSHAW J A, COYNE J A and LEWONTIN R C (1979), 'The sensitivity of gel electrophoresis as a detector of genetic variation', *Genetics*, **93**, 1019–37.

SÁNCHEZ S, MARTÍNEZ R, GARCÍA A, BLANCO J, ECHEITA A, HERMOSO DE MENDOZA J, ET AL. (2009), 'Shiga toxin-producing *Escherichia coli* O157:H7 from extensive cattle of the fighting bulls breed', *Research in Veterinary Science*, **88**(2), 208–10.

SELANDER R K, CAUGANT D A, OCHMAN H, MUSSER J M, GILMOUR M N and WHITTAM T S (1986), 'Methods of multilocus enzyme electrophoresis for bacterial population genetics and systematics', *Applied and Environmental Microbiology*, **51**, 873–84.

SHUMAKER K M, ALLARD R W and KAHLER A L (1982), 'Cryptic variability at enzyme loci in three plant species: *Avena barbata, Hordeum vulgaris*, and *Zea myes*', *Journal of Heredity*, **73**, 86–90.

STANLEY J and BAQUAR N (1994), 'Phylogenetics of *Salmonella* Enteritidis', *International Journal of Food Microbiology*, **21**, 79–87.

TENNANT S, GRANT T and ROBINS-BROWNE R (2003), 'Pathogenicity of *Yersinia enterocolitica* biotype 1A', *FEMS Immunology and Medical Microbiology*, **38**, 127–37.

THRELFALL E J, ROWE B, FERGUSON J L and WARD L R (1985), 'Increasing incidence of resistance to gentamicin and related aminoglycosides in *Salmonella* Typhimurium phage type 204c in England, Wales and Scotland', *Veterinary Record*, **117**, 355–7.

TIBAYRENC M (2009), 'Multilocus enzyme electrophoresis for parasites and other pathogens', *Methods in Molecular Biology*, **551**, 13–25.

TOST J and GUT I G (2005), 'Genotyping single nucleotide polymorphisms by MALDI mass spectrometry in clinical applications', *Clinical Biochemistry*, **38**, 335–50.

TURTON J F, ENGLENDER H, GABRIEL S N, TURTON S E, KAUFMANN M E and PITT T L (2007), 'Genetically similar isolates of *Klebsiella pneumoniae* serotype K1 causing liver abscesses in three continents', *Journal of Medical Microbiology*, **56**, 593–7.

VÄKEVÄINEN M, EKLUND C, ESKOLA J and KÄYHTY H (2001), 'Cross-reactivity of antibodies to type 6B and 6A polysaccharides of *Streptococcus pneumoniae*, evoked by pneumococcal conjugate vaccines, in infants', *Journal of Infectious Diseases*, **184**(6), 789–93.

VAN DER WAAIJ D, SPELTIE T M, GUINEE P A M and AGTERBERG C (1975), 'Serotyping and biotyping of 160 *Escherichia coli* strains: Comparative study', *Journal of Clinical Microbiology*, **1**, 237–8.

VARMA J K, MARCUS R, STENZEL S A, HANNA S S, GETTNER S, ANDERSON B J, ET AL. (2006), 'Highly resistant *Salmonella* Newport-MDRAmpC transmitted through the domestic US food supply: a FoodNet case-control study of sporadic *Salmonella* Newport infections, 2002–2003', *Journal of Infectious Diseases*, **194**, 222–30.

VUGIA D, CRONQUIST A, CARTTER M, TOBIN-D'ANGELO M, BLYTHE D, SMITH K, ET AL. (2009), 'Preliminary FoodNet data on the incidence of infection with pathogens transmitted commonly through food – 10 states, 2008', *Morbidity & Mortality Weekly Report*, **58**, 333–7.

WALLS H H, HARMON H W, SLAGLE J J, STOCKSDALE C and KENDALE A P (1986), 'Characterization and evaluation of monoclonal antibodies developed for typing influenza A and influenza B viruses', *Journal of Clinical Microbiology*, **23**, 240–5.

WAREING D R, BOLTON F J, FOX A J, WRIGHT P A and GREENWAY D L (2002), 'Phenotypic diversity of *Campylobacter* isolates from sporadic cases of human enteritis in the UK', *Journal of Applied Microbiology*, **92**, 502–9.

WAUTERS G, KANDOLO K and JANSSENS M (1987), 'Revised biogrouping scheme of *Yersinia enterocolitica*', *Contributions to Microbiology*, **9**, 14–21.

WEN L, WANG Q, LI Y, KONG F, GILBERT G L, CAO B, ET AL. (2006), 'Use of a serotype-specific DNA microarray for identification of group B *Streptococcus* (*Streptococcus agalactiae*)', *Journal of Clinical Microbiology*, **44**, 1447–52.

WITTE W (1997), 'Impact of antibiotic use in animal feeding on resistance of bacterial pathogens in humans', *Ciba Foundation Symposium*, **207**, 61–71.

YOSHIDA C, FRANKLIN K, KONCZY P, MCQUISTON J R, FIELDS P I, NASH J H, ET AL. (2007), 'Methodologies towards the development of an oligonucleotide microarray for determination of *Salmonella* serotypes', *Journal of Microbiological Methods*, 70; 261–71.

ZEWDE B M, ROBBINS R, ABLEY M J, HOUSE B, MORROW W E and GEBREYES W A (2009), 'Comparison of Swiffer wipes and conventional drag swab methods for the recovery of *Salmonella* in swine production systems', *Journal of Food Protection*, **72**, 142–6.

8

Pulsed-field gel electrophoresis and other commonly used molecular methods for subtyping of foodborne bacteria

K. L. F. Cooper, Centers for Disease Control and Prevention, USA

Abstract: The application of molecular methods to bacterial subtyping has greatly enhanced the sensitivity with which foodborne disease clusters are detected. This chapter will discuss the basic principles, inherent strengths and weaknesses, and general applicability of each of some of the most commonly used non-gene-specific molecular methods for the subtyping of foodborne bacterial pathogens.

Key words: molecular subtyping, pulsed-field gel electrophoresis (PFGE), polymerase chain reaction (PCR)-based techniques, ribotyping, foodborne disease surveillance.

8.1 Introduction

The application of subtyping tools for the characterization and subtyping of bacteria has greatly enhanced our ability to establish links between phenotypes and genotypes, and transformed the way surveillance of illness caused by pathogenic bacteria is conducted. Molecular subtyping is the examination of specific characteristics of a group of strains or isolates that provide us with the means to catalogue and discriminate between them beyond the species or serotype level using molecular (i.e. nucleic acid based) methods. Subtyping tools are the cornerstone of molecular epidemiology studies, e.g. in outbreak investigations, where molecular data are used to assess genetic relatedness between a given number of strains or isolates as a means to determine which ones are likely to be part of an outbreak. In the context of 'common source' outbreaks, strain relatedness is determined by comparing genotypic data and identifying the strains that produce subtypes that are 'indistinguishable' from one another. Strains that yield subtypes that are different from the 'outbreak' are, typically, not considered to be part of the outbreak. Such classifications help investigators narrow down the focus of epidemiologic investigations and, as a result, saving valuable time and resources.

Bacterial subtyping was first introduced to assess strain diversity by employing methods that focus on phenotypic characteristics such as serotyping, antimicrobial susceptibility testing, toxin typing, phage typing, and multilocus enzyme electrophoresis (Arbeit, 1995). Additional information on this subject can be found in the previous chapter of this book. While some of these methods are still regularly used, many of them have been plagued with limitations such as insufficient discrimination, poor reproducibility, use of highly specialized reagents, and lack of versatility and applicability to a wide range of bacterial pathogens. Much progress has been made in recent years in the development of more robust and sophisticated DNA-based subtyping techniques. The application of these techniques has forever changed the field of strain subtyping by providing us with tools that help us achieve greater levels of strain differentiation, enhanced reproducibility, and are more amenable to standardization as compared to their phenotypic counterparts (Olive and Bean, 1999; Wiedmann, 2002).

Due to the utility of these methods in bacterial foodborne disease surveillance, the number of new molecular approaches created over the last decade has rapidly expanded (Hahm *et al.*, 2003; Killgore *et al.*, 2008; Lim *et al.*, 2005; Wiedmann, 2002) and promises to continue doing so in the future. This chapter presents the basic principles behind some of the most commonly used non-sequencing-based methods applied in recent years, and highlights their inherent strengths and weaknesses, primarily within the context of bacterial subtyping of foodborne pathogens.

8.2 Technical overview

While the repertoire of molecular techniques available for subtyping bacteria continues to expand at an accelerated pace, some of the most widely used genotypic methods have their foundation on two general approaches: restriction enzyme analysis (REA) based methods and polymerase chain reaction (PCR) based techniques. Methods using REA include ribotyping and pulsed-field gel electrophoresis (PFGE), as well as a variety of micro-restriction approaches. Some of the most commonly used PCR-based methods include PCR-ribotyping, random amplified polymorphic DNA analysis (RAPD), arbitrarily primed PCR (AP-PCR), repetitive element PCR (rep-PCR), and amplified fragment length polymorphism (AFLP), among others. A diagram of the procedural steps involved in the performance of these methods is included in Figure 8.1.

8.2.1 Restriction enzyme analysis (REA) based techniques

REA approaches rely on the use of restriction endonucleases that recognize and cleave specific sequences in the DNA to generate fragments that may be polymorphic in size when comparing the resulting profile from one strain to

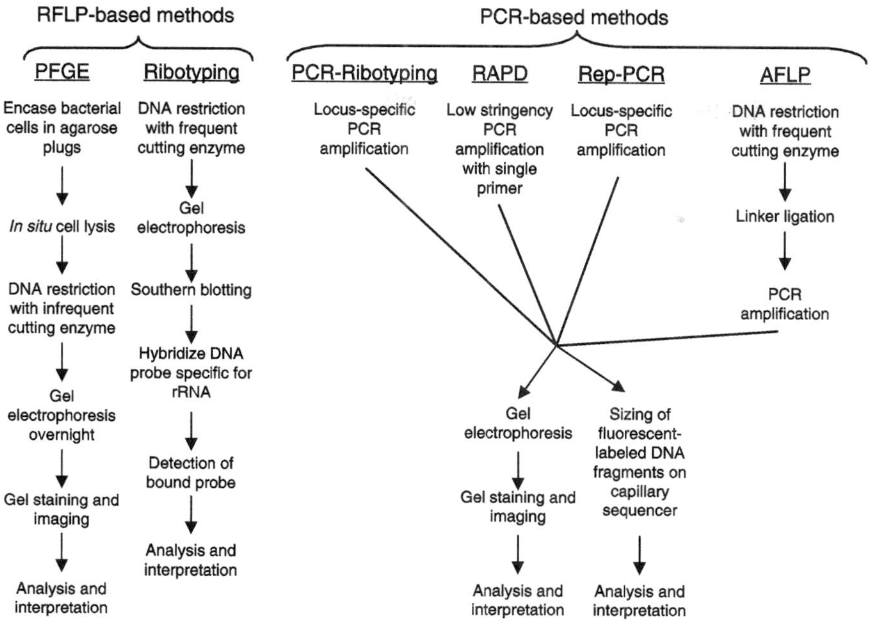

Fig. 8.1 Diagram of the procedural steps for the various molecular methods discussed in this chapter.

another. Restriction fragment length polymorphisms can arise from restriction site mutations, recombination, and insertions or deletions that occur within the genome at or between restriction sites. Variations in the size of the resulting DNA fragments can be resolved in a size-dependent fashion via gel electrophoresis. RFLP fingerprints can be used to make assumptions about genetic relatedness between two or more strains; the larger the number of restriction fragments differences (polymorphisms) the farther apart, genetically, the strains are believed to be (Tenover *et al.*, 1995). In early RFLP methods, genomic DNA was typically treated with high frequency restriction endonucleases which resulted in DNA fingerprints containing 3–500 restriction fragments ranging in size from <1 to 30 kilobases (kb). The resulting DNA fragments were often difficult to resolve and too complex to analyze accurately and consistently. These limitations led researchers to focus on the development of alternative approaches that would maintain a high level of discriminatory power and at the same time simplify the analysis and interpretation of the fingerprints. The two methods that emerged from these efforts have been commonly accepted by the scientific community: PFGE and ribotyping.

Pulsed-field gel electrophoresis (PFGE)

In PFGE, the DNA fingerprint is simplified by reducing the number of estriction fragments through the use of infrequently cutting restriction enzymes (macro-restriction). PFGE generates fingerprints by restricting intact genomic

DNA released from cells that have been embedded in a solid agarose matrix. The agarose matrix (plug) acts as a stabilizing medium that prevents the DNA from breaking or shearing as the cells are subjected to *in situ* chemical lysis. The immobilization of cells in a matrix prior to lysis is necessary to produce clean fingerprints comprised of fragments that are typically larger than 20 Kb; a process that cannot be achieved with liquid-based DNA isolation procedures which cause random shearing of the DNA due to exposure to chemical and mechanical forces. The intact genomic DNA is then digested with an infrequent cutting restriction enzyme that ideally produces between 8 and 25 DNA fragments ranging from 20 kb to >1 Mb in size. Fragments in this size range cannot be resolved with conventional electrophoresis which is typically limited to the resolution of fragments smaller than 20–30 kb. This limitation is overcome by subjecting the agarose plug containing macro-restriction DNA mixture to electrophoresis in a 'pulsed' electric field (Schwartz and Cantor, 1984). Since 1984, several different PFGE formats have been developed (Carle and Olson, 1984; Carle *et al.*, 1986; Chu *et al.*, 1986), however common to them all is the application of an electric field that alternates in direction (pulses) in a continuous manner throughout the course of the electrophoresis run. The principle behind this approach is that smaller DNA fragments will reorient to a directional change in the electrical field faster than larger fragments and therefore will move more rapidly through the agarose matrix resulting in resolution of size-dependent DNA banding patterns. Some platforms have the ability to change the time interval the electrical current is applied in one direction before it switches to another direction. The switch times are typically short at the beginning, and are increased (ramped) over the course of a run. Once the run is complete, gels are stained with a fluorescent dye allowing bands to be visualized and stored with a digital imaging system. The digitized banding patterns can be analyzed and compared using a variety of commercially available software systems (Gerner-Smidt *et al.*, 1998).

Because different genera of bacteria vary in the guanine and cytosine content of their DNA, the optimal restriction enzyme for PFGE may vary from organism to organism. For example, *Xba*I and *Bln*I are commonly used for PFGE analysis of *E. coli, Salmonella*, and *Shigella* (Ribot *et al.*, 2006), while *Asc*I and *Apa*I are used for *Listeria monocytogenes* (Graves *et al.*, 2005a). The selection of a restriction enzyme(s) for a particular organism is based on the generation of an appropriate number of restriction fragments. In general, these patterns should be sufficiently complex to provide meaningful subtyping information, while at the same time be composed of a manageable number of fragments (e.g., more than 10 but less than 25–30) that display a good size distribution such that the resulting restriction fragment patterns can be accurately and reproducibly analyzed.

Ribotyping

In ribotyping, the RFLP fingerprint is simplified by only visualizing the fragments that carry the rRNA encoding genes (rDNA). In this approach, whole genomic DNA is restricted into approximately 300–500 DNA fragments ranging from 1 to 30 kb in size. Once electrophoresed, the resulting DNA fragments are transferred from the agarose gel to either a nitrocellulose or nylon membrane by Southern blotting (Southern, 1975). The membrane-bound nucleic acid is then hybridized to probes containing rDNA sequences. In this way, only 5–15 fragments from each isolate will be visualized. This resulting DNA fingerprint is referred to as the ribotype. rRNA is present in all organisms and the sequence is highly conserved, therefore the same type of probes can be used for most organisms. A commercially available automated ribotyping system has been developed by Qualicon-DuPont (Wilmington, DE), which has been evaluated for the subtyping of several foodborne pathogens (Clermont *et al.*, 2001; De Cesare *et al.*, 2001).

8.2.2 Polymerase chain reaction (PCR)-based techniques

PCR has served as the technical basis for multiple subtyping methods by allowing the examination of sequence variation with in genomic DNA that may result in detectable differences in amplicon length and/or number that is often indicative of strain variation. Below we describe four of the most common PCR-based subtyping techniques: (1) PCR-ribotyping; (2) PCR-amplified restriction length polymorphism AFLP; (3) rep-PCR; and (4) RAPD or AP-PCR. While PCR is the central technical platform, each of these techniques varies on the amplification approach and/or the genomic region they interrogate. Following amplification, the resulting banding patterns are separated on an agarose or small polyacrylamide gel, and visualized by staining with ethidium bromide. Alternatively, more sophisticated approaches have been described that utilize fluorescently labeled primers for PCR amplification (Indra *et al.*, 2008; Le *et al.*, 1998; Lindstedt *et al.*, 2000a, 2000b) for the visualization of the amplicons by capillary electrophoresis within an automated DNA sequencing instrument (Figure 8.1).

PCR-ribotyping

In PCR-ribotyping, the intergenic region between the genes encoding 16S and 23S rRNA is amplified. The basis behind this approach is that the genome of most organisms contains more than one rDNA operon. The size of the intergenic region and flanking the 16S and the 23S rDNA vary both within the same strain and between different strains (Bouchet *et al.*, 2008). The resulting amplicons are separated by agarose gel electrophoresis and visualized following ethidium bromide staining. Various bacterial species differ in the number of rDNA operons they possess. PCR-ribotyping is particularly suited for subtyping organisms that contain many rDNA operons, due to the potential complexity of the resulting patterns. For example, *Clostridium difficile* strains generally produce 7–15 bands (Bidet *et al.*, 2000). Due to this relatively large number of fragments, ribotyping has been found to be particularly useful and widely used in subtyping clostridia (Cohen, 2007).

Amplified fragment length polymorphism (AFLP)

AFLP is a fingerprinting technique based on the selective amplification of a subset of DNA fragments generated by restriction enzyme digestion. The initial step in AFLP is restriction endonuclease digestion of total genomic DNA. This can be performed with a single restriction enzyme but is most commonly done with the combination of two different enzymes thereby generating large numbers of restriction fragments. Oligonucleotide adaptors are then ligated to the sticky ends of the resulting restriction fragments. The resulting fragments are used as the template for a subsequent PCR amplification reaction. The PCR amplification is performed with primers that are complementary to the adaptor(s) and contain one to two selective bases at their 3′ end. The addition of the selective bases that extend into the restriction fragments ensures that only restriction fragments that contain target sequences that are complimentary to the selective bases are amplified (Vos *et al.*, 1995). The resulting AFLP fragments may be separated and visualized following agarose or polyacrylamide gel electrophoresis. Alternatively, fluorescently labeled primers can be used to generate amplicons that can then be separated by capillary electrophoresis using automated DNA sequencers (Lindstedt *et al.*, 2000a, 2000b).

Repetitive element PCR (rep-PCR)

Rep-PCR targets are naturally-occurring interspersed repetitive DNA elements found in the genome of virtually every bacterium studied thus far. These repetitive elements are classified into several different families based on their coding capabilities, size, distribution (scattered or localized), and the frequency of occurrence in the genome. Several different families of repetitive sequences have been pursued as targets in various subtyping schemes (Lupski and Weinstock, 1992; Rasschaert *et al.*, 2005; Versalovic *et al.*, 1991). A commonly used target is the repetitive extragenic palindromic elements which are 33–40 bp in unit length, and are typically present in approximately 500–1,000 copies per bacterial genome. These repetitive elements are commonly found outside structural genes within the *Escherichia coli* and *Salmonella* serotype Typhimurium bacterial genomes (Stern *et al.*, 1984). Another commonly used target is the 124–127 bp enterobacterial repetitive intergenic consensus elements which are prevalent in a large number of gram-negative species and are present in approximately 30–150 copies per genome (Sharples and Lloyd, 1990). Finally, the BOX sequence is a 154 bp repetitive element that is composed of three subunit sequences (boxA, boxB, and boxC) that can be present in various combinations within the BOX sequence (Martin *et al.*, 1992). The BOX element was initially identified in *Streptococcus pneumoniae*, but has also been found in and used for subtyping foodborne bacterial pathogens (Hahm *et al.*, 2003).

Rep-PCR relies on the PCR amplification of repetitive sequences that are dispersed in multiple locations throughout the bacterial genome and can be found in either strand (orientation) of the DNA. PCR amplification is performed with a single PCR primer containing a portion of the repeat sequence that is designed to amplify outward from the repeat sequence. When inverted repeats of the repeat element are in sufficiently close proximity, the genomic region located between the repeat elements is amplified. The resulting amplicons can be resolved by electrophoresis to generate a fingerprint pattern for that strain. The number of these repetitive sequence elements scattered throughout the genome differs between various organisms, therefore the complexity and utility of the fingerprint patterns produced may also vary. Ultimately, each repeat region must be evaluated to determine it's usefulness for subtyping of a particular organism (Rasschaert *et al.*, 2005; Mohapatra and Mazumder, 2007).

Rep-PCR is the basis for the DiversiLab System (bioMérieux, Marcy l'Etoile, France), a commercially available system that utilizes a semi-automated platform for data generation and analysis. The use of a highly standardized system, such as the DiversiLab System, is advantageous since it reduces the variability of results that is often seen with 'home-made' systems by providing quality-controlled reagents in a kit format, microfluidics-based DNA amplicon fractionation and detection, and internet-based computer-assisted analysis, reporting, and data storage (Healy *et al.*, 2005).

Arbitraily primed (AP)-PCR/Random amplified polymorphic DNA analysis (RAPD)

AP-PCR or RAPD is a fingerprinting technique that is based on the use of short (approximately 10 bp) primer target sequences chosen arbitrarily (Welsh and McClelland, 1990). Under low-stringency conditions, these primers are capable of binding to many sequences with a variety of DNA mismatches. PCR amplicons are generated when primers bind to genomic regions close enough and in the correct orientation to allow for efficient amplification of the region between the primer binding sites. Positional variation in primer binding sites among strains can result in the generation of a strain-specific DNA fingerprint pattern that can be visualized by agarose gel electrophoresis (Hopkins and Hilton, 2001). Multiple random sequences have been described and evaluated for use in randomly amplified schemes for subtyping various organisms (Hopkins and Hilton, 2001; Stephan, 1996).

8.3 Comparison of molecular methods

The value of any bacterial typing method is measured by how well the method performs in regard to the intended objective. This is assessed by a variety of criteria, which includes the ability to provide relevant epidemiological or evolutionary data and more practical considerations such as the application and performance of the method. Some of the technical factors that must be carefully evaluated when choosing a subtyping method include typeability, reproducibility, discriminatory power, and potential for standardization. From the practical side, there are a number of convenience factors to consider such as cost-effectiveness, ease of use, time for assay completion (speed), and sample throughput (Struelens, 1996). Currently no subtyping method is optimal in all of these factors, therefore a deep understanding of how each step in the process can impact on the usefulness of each method is instrumental in helping determine which technique is best suited to achieve the intended objective.

8.3.1 Typeability

Typeability refers to the ability of a method to produce an unambiguous result, or subtype, from each isolate tested. The typeability of a method can be impacted by strain specific characteristics that interfere with the performance of the method. For example, strains that have the ability to modify their own DNA can negatively impact the typeability of REA methods if the modifications result in masking the restriction sites needed to generate the fingerprint or profile. Methylation of nucleotides within a restriction site is the most commonly encountered modification that prevents restriction enzymes from cleaving DNA at that site thus eliminating the possibility of generating a RFLP pattern for that strain. In some cases, this problem can be overcome by using enzymes known as isoschizomers. Isoschizomers recognize the same restriction site (sequence) but are unaffected by the position at which the methyl group is added. However, isoschizomers are not available for all enzymes and may also be blocked by DNA modifications.

Similarly, a rare problem encountered with some strains is a generalized degradation of the DNA that occurs during the electrophoresis step of PFGE protocols that utilize Tris-based buffers (Koort *et al.*, 2002). While the complete mechanism by which this occurs is still not completely understood, it appears that the DNA within these strains is susceptible to shearing during the electrophoresis run in a phenomenon believed to be mediated by the formation of Tris radicals and superoxide molecules generated during electrophoresis (Ray *et al.*, 1992). For many years, these strains were considered to be untypeable strains by PFGE, however Romling *et al.* (Romling and Tummler, 2000) recognized that the addition of 50 mM Thiourea to the electrophoresis running buffer quenches these reactive molecules and results in the generation of a pattern for almost all species and strains tested. Strains exhibiting DNA degradation sensitivity have been described in a wide variety of bacterial species, however multiple studies have shown that all of the tested strains are resolved with the addition of thiourea thereby significantly increased the overall typeability of PFGE (Romling and Tummler, 2000; Sawabe *et al.*, 2007; Silbert *et al.*, 2003).

The typeability of PCR-based methods is dependent on the proper selection and design of primers for the amplification reactions. Rep-PCR and PCR-ribotyping target specific genomic regions, therefore knowledge of the target genomic sequence is required for appropriate primer design. Fortunately, the target sequences are highly conserved between species enabling the design of 'universal' primers that can be used for the examination of different organisms with a single amplification scheme. AFLP and RAPD approaches are more universal in design since they target random genomic sequences rather than specific target sequences. However, in AFLP care needs to be taken when selecting the appropriate primers for the different restriction enzyme/adapters and primer combinations to ensure sufficiently complex DNA banding patterns are obtained to provide bacterial subtyping data for a particular organism. Similarly, since multiple RAPD primers have been described, the most appropriate primers must be determined empirically as to which generate the best DNA pattern for differentiation of a specific organism. This is related to the overall discriminatory power of the method that is described in more detail later in this chapter.

In general, most molecular methods have a very a high level of typeability. The advantage of the methods described in this chapter is that they retain this high level of typeability when the same basic method is used for subtyping a diverse range of pathogens, thereby making these methods more broadly applicable than phenotypic subtyping methods.

8.3.2 Reproducibility and stability

A reproducible technique is capable of producing the same result upon repeated testing of the same bacterial isolate. Lack of reproducibility generally results from technical variation associated with the performance of the method, such as inconsistent electrophoresis run lengths which can affect fragment resolution. Additionally, biological variation in the marker analyzed during in vivo or in vitro passage of the organism during the course of a study or outbreak investigation can reduce the reproducibility of a method when, in fact, the differences being observed are associated with the genetic stability of the strain rather than a problem intrinsic to the method. A high level of reproducibility is particularly important when comparing results generated in different experiments or laboratories. The construction of reliable databases and accurate interpretation of the fingerprint data can be achieved only if the method used is highly reproducible.

The reproducibility of RAPD is hindered by the use of low-stringency conditions for the amplification process. Likewise, the high tolerance for mismatches in the Rep-PCR annealing step for most organisms also negatively impacts the reproducibility of this method. Variations in primer concentration, DNA template concentration and quality, *Taq* polymerase concentration and type, $MgCl_2$ concentration, number of amplification cycles, and brand and type of thermocycler can greatly impact the reproducibility of RADP and Rep-PCR. While standardization of these techniques has improved the inter- and intra-laboratory reproducibility that was lacking in earlier years (Tyler *et al.*, 1997), it is very difficult to control all the factors that lead to artificial differences in the profiles (particularly in regard to reagent and equipment variability). However, significant effort has been placed in the development of kits and all inclusive platforms that may assist in increasing reproducibility. For example, the DiversiLab System mentioned above claims to have solved the problems relating to the poor reproducibility issues of Rep-PCR (Healy *et al.*, 2005).

PFGE, traditional ribotyping, and PCR-ribotyping exhibit a high level of reproducibility when used under standardized conditions. The use of such standardized protocols, as well as the availability of increasingly sophisticated image acquisition and analysis software, has greatly enhanced the reproducibility of these methods. For example, PFGE patterns can be quite complex in regard to the number and distribution of the bands within the pattern. As a result, data analysis is prone to error due to the subjective nature of band assignment during analysis. One of the most crucial components impacting PFGE analysis is the use of common electrophoresis parameters, such as a standard gel volume and concentration of agarose, the ionic strength and volume of running buffer used, the electrophoresis buffer temperature during the run, the voltage, and the switching times of electrophoresis (Murchan *et al.*, 2003). In addition to improving inter- and intra-laboratory reproducibility, standardization of the method is needed to ensure that all parameters integrated into the performance of the protocol facilitate the analysis of the results generated. The standardization process used by PulseNet by which the different factors that can influence the analysis are tested is discussed in detail in Chapter 10.

As mentioned previously, some of the methods described in this chapter have traditionally relied on the separation of banding patterns in an agarose or small polyacrylamide gels, and visualization of the patterns following ethidium bromide staining. However, in some PCR-based methods (Indra *et al.*, 2008; Lindstedt *et al.*, 2000a, 2000b) fluorescently labeled amplicons are separated by capillary electrophoresis using automated DNA sequencers in a rapid and high throughput manner. The use of an internal size standard allows for the sizing of PCR fragments with a high degree of accuracy and enables the direct comparison of fingerprint patterns with previously stored fingerprints. Despite the advantages of this detection method, the cost of equipment and necessary reagents are prohibitively expensive for most laboratories.

The genomic instability of an organism should not be confused with lack of reproducibility of a method. Bacterial strains have the potential to change with the passage of time regardless of whether they are linked to a common source outbreak or used as a reference strain in a research study. It has been shown that genetic changes leading to differences in the PFGE patterns of *E. coli* O157 strains are primarily caused by insertions, deletions and rearrangements (Kudva *et al.*, 2002). In some instances, extreme genomic instability may be observed. For instance, if a temperate phage is inserted unstably in the genome this may cause unpredictable multi-band changes in the PFGE pattern of a single isolate upon repeat testing (Bielaszewska *et al.*, 2006). More commonly, the excision or insertion of genetic elements, such as transposons in the bacterial chromosome, causes minor changes in the appearance of the restriction profile. Even the acquisition or loss of a non-integrative extra-chromosomal DNA molecule, e.g. a plasmid, could introduce a difference that could cause a strain to be 'mis-classified' or excluded from a cluster. One must recognize that the rate with which this type of variability occurs varies significantly from organism to organism making interpretation of data, even with strains of the same serotype, potentially challenging. While extensive studies aimed at the evaluation of subtyping methods for this kind of instability are extremely useful in developing an understanding of the rate and process of genetic instability for a particular method and organism, these types of studies are extremely resource demanding and are therefore rarely performed. Ultimately, the potential for variability associated with these genetic changes should be acknowledged and considered when evaluating subtyping data.

8.3.3 Discriminatory power

The discriminatory power of a typing technique is an estimate of the method's ability to differentiate between unrelated strains. Methods that possess a high discriminatory power are able to clearly differentiate between epidemiologically unrelated strains and, at the same time, group all strains with 'like' characteristics. The discriminatory power of a technique can be quantified using the Simpson's Index of Discrimination, which is an assessment of the probability that two unrelated strains will be characterized as different subtypes (Hunter and Gaston, 1988). However, caution must be exercised when interpreting results that are obtained with this index since the values used for calculation, such as the population size and number of types, can vary dramatically from study to study, and are therefore generally not comparable between different studies. Furthermore, the resulting index values do not always represent a fair or realistic estimate of the discriminatory power of a given subtyping method. In cases where subtypes are overrepresented in the population, the Simpson's Index may underestimate the discrimination of the method. Therefore, when evaluating the discriminatory power of a subtyping method it is also important to carefully examine the distribution of the subtypes that are generated for a particular pathogen. For example, five PFGE patterns account for over 78% of the 31 000 patterns of *Salmonella* serotype Enteritidis in the PulseNet USA database (Figure 8.2). In this instance, the Simpson's Index of Diversity is relatively low. However, it is important to realize that a cluster of case-patients infected with one of the rare

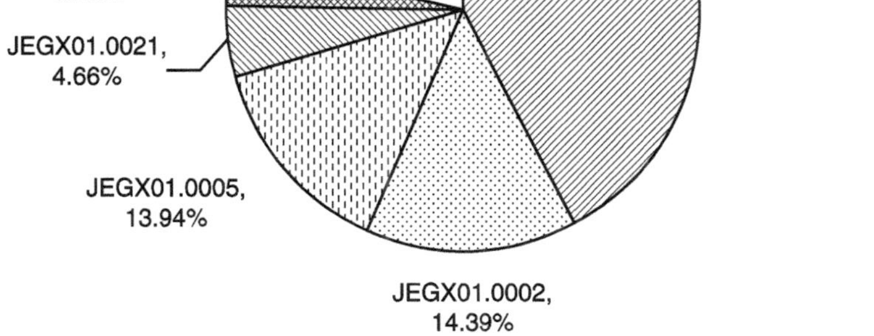

Fig. 8.2 Representation of the percentage distribution of *Salmonella enterica* serotype Enteritidis isolates with various PFGE patterns. Percentages were determined as the total number of isolates containing the top five patterns (JEGX01.0004, JEGX.0002, JEGX.0005, JEGX01.0021, and JEGX01.0034) or all other patterns within a database of 31 500 *Salmonella enterica* serotype Enteritidis PFGE patterns.

subtypes likely represents an outbreak thereby providing epidemiological important information.

Molecular subtyping methods can vary greatly in their discriminatory power primarily because the evolutionary clock of the genetic region targeted by the method may differ significantly from one method to the next. Identifying a method with a 'clock' speed that is appropriate for the intended objective requires extensive testing of isolates for which a link or correlation to another trait (characteristic) has been clearly established. For example, within evolutionary studies it is often appropriate to focus on methods that interrogate conserved genomic areas, whereas in epidemiology it is generally more desirable to target more rapidly evolving regions or targets (Struelens *et al.*, 1998; van Belkum *et al.*, 2007).

While discriminatory power is a useful factor to assess the performance of subtyping techniques, the ultimate goal for laboratory-based surveillance systems is to identify clusters of strains with 'like' characteristics or linking a subtype to phenotypic trait. In the case of outbreak investigations, the objective is to employ methodologies that produce epidemiologically relevant information and help the investigator focus on the investigation by identifying the strains that are likely to be part of the outbreak from those that are not. The bacterial genome is amazingly fluid and differences can be found even within highly related isolates. Therefore, molecular data are only part of the picture and must be correlated with available epidemiological information (Barrett *et al.*, 2006).

Many studies have been performed to compare the discrimination achieved by different subtyping methods on a set of isolates (Hahm *et al.*, 2003; Jonas *et al.*, 2003; Vogel *et al.*, 2004). Unrelated isolates are often grouped differently by different methods, although epidemiologically related isolates are typically grouped together regardless of the method used (Hahm *et al.*, 2003). This is a function of the region or regions that the subtyping method interrogated and demonstrates the variable information that may be obtained about the inter-relationship of isolates depending on the chosen method. Thus, subtyping methods that build on different principles often supplement each other, and it is often possible to increase the discriminatory power through the integration of multiple methods in a subtyping scheme.

The various subtyping methods discussed in this chapter differ in regard to discriminatory ability among foodborne pathogens (Lim *et al.*, 2005; Lukinmaa *et al.*, 2004; Vogel *et al.*, 2004). In regard to PCR-based methods, an obvious approach to enhance the discriminatory power associated with these techniques is to increase the number of meaningful targets used in the assay. However, this requirement will increase the cost and decrease the speed and throughput of a test. Several studies suggest that AFLP has the highest discriminatory power of the PCR-based methods discussed here and, in some cases, provides similar or slightly better discriminatory power than PFGE (Lindstedt *et al.*, 2000a; Nair *et al.*, 2000). However, PFGE has achieved 'gold standard' status for most foodborne bacteria due to a high level of reproducibility and an excellent discriminatory power as well as correlation of PFGE results with epidemiological data. The use of additional enzymes can be helpful to provide additional discrimination with PFGE and has become a common practice within foodborne outbreak investigations. It is important to note that the majority of the studies comparing PFGE to other subtyping methods are based on single-enzyme PFGE data, and therefore often underestimate the power of discrimination of this method. Ribotyping displays only a moderate discriminatory power for most bacterial species primarily due to the limited region of the genome being queried.

8.3.4 Convenience parameters

In determining the cost-effectiveness of a method, one must weigh the tangible and intangible factors that could impact its implementation in individual laboratories or a network and the value contributions it will make to our scientific knowledge. Some of the tangible factors that directly impact cost-effectiveness of a method are; cost of equipment, reagents and supplies, data analysis and management systems, as well as personnel time. Typically, the cost associated with a method is offset by less tangible factors, in particular if the method is used to assist in epidemiologic investigations where the expected outcomes measure in the ability to detect, control, and prevent the spread of illness in a community. In this context, the cost-effectiveness of a method must be evaluated based on the epidemiological relevance of the data that it produces. This is an important factor since as the cost and complexity associated with the performance of a method increase, the application of the method for routine subtyping may be beyond the capabilities of some laboratories. While the methods discussed in this chapter do not differ dramatically in regard to the cost of reagents and materials, the most noticeable difference is the labor and time required for performance of these methods. Due to the limited number of procedural steps, RAPD and Rep-PCR are the most straightforward and rapid of the PCR-based methods. Conversely, AFLP is more labor-intensive and time-consuming due to the additional restriction and ligation steps. Similarly, a recognized limitation of PFGE is the labor and time required for analysis, which is greater than most PCR-based methods despite recent protocol modifications that have reduced the majority of PFGE protocols to a 24-hour procedure. PFGE is also impacted by a relatively low throughput as compared to PCR-based methods, which consequently increases the time and labor required for testing large numbers of isolates. Despite the recognized limitations, PFGE analysis is still considered to be one of the most cost-effective methods for subtyping most organisms due to the high reproducibility and discriminatory power associated with this method.

8.3.5 PFGE as the 'gold standard'

While other methodologies may present a more rapid and technically simpler subtyping process, PFGE has remained the 'gold standard' for subtyping many bacterial species due to its universal applicability, obtainable inter-laboratory reproducibility and superior correlation with epidemiological information. The latter point is truly what has raised PFGE to 'gold standard' status as a subtyping tool for foodborne bacterial pathogens. For over a decade now PFGE has been used to accurately and consistently discriminate between outbreak and non-outbreak isolates obtained from a number of outbreak investigations of foodborne bacterial pathogens. The first documented example of the utility of PFGE was demonstrated by the ability of the method to group outbreak related isolates within a retrospective study conducted on isolates from an *Escherichia coli* O157:H7 outbreak associated with contaminated hamburgers served in a fast-food restaurant chain in 1993 (Barrett *et al.*, 1994). The useful information provided by the subtyping data cemented the utility of the PFGE as a molecular subtyping tool in outbreak investigations and was the catalyst for the development of the national network for foodborne disease surveillance known as PulseNet USA (Gerner-Smidt *et al.*, 2006; Swaminathan *et al.*, 2001).

PulseNet USA is the molecular subtyping network for foodborne disease surveillance with participation of more than 75 state, city and county public health laboratories, as well as agricultural and federal food regulatory agency laboratories. Laboratory-based surveillance with PulseNet helps detect clusters of foodborne infections, and assists epidemiologists to narrow down the number of cases included in the investigations by separating cases included in the cluster from sporadic ones. Once a cluster is detected through PulseNet, an epidemiologic investigation is initiated to determine whether an epidemiological link can be identified between the cases. If as association is found, the cluster is classified as an outbreak. PFGE can also provide confirmation of epidemiologically implicated sources by matching patterns from patient isolates to those obtained from the contaminated source or vehicle. The utility of the network has been demonstrated repeatedly in numerous foodborne outbreak investigations (CDC, 2006; Gerner-Smidt *et al.*, 2006; Graves *et al.*, 2005b). By conducting laboratory-based surveillance in real-time at the local and national level, PulseNet has been able to uncover many multi-locality outbreaks by linking cases that would have been considered to be 'sporadic' years earlier. The success of the PulseNet Program has led to its continued expansion both nationally and internationally. PulseNet is dealt with in more detail in Chapter 3 of this book.

8.4 Library subtyping

To perform active large-scale laboratory-based surveillance, reference libraries (databases) must be established in a manner that allows the data manager to accurately analyze data generated in different laboratories and at different times. The idea is for individual laboratories to generate and analyze data locally and then submit patterns to a central database where patterns can be compared at will. This decentralized approach provides each laboratory with the ability to monitor changes in bacterial populations and to identify clusters that may represent common source outbreaks in real-time. The key ingredient in this formula is the implementation of subtyping schemes that are robust and highly reproducible. The implementation of standardized protocols with strictly controlled reagents and performance parameters has greatly enhanced the reproducibility and comparability of data generated with a variety of subtyping methods when used in different laboratories (Struelens *et al.*, 1998). Additionally, the development of library reference systems has been aided by the availability of instrumentation and software for the digital capture and analysis of gel images. These software programs have the ability to normalize the individual banding patterns obtained over multiple gel images through the use of standard reference patterns present within each of the gels. The normalized DNA fingerprints can be stored and used for the comparison of large numbers of strain profiles over time, a task that would be cumbersome, if not completely impossible, with visual banding pattern comparison. The use of band-based clustering algorithms allows for the construction of dendograms that visually display pattern similarities and differences between groups of strains (Figure 8.3) (Rementeria *et al.*, 2001).

8.4.1 Standardization of subtyping methods

The ability to generate intra- and inter-laboratory reproducible data depends on the understanding of and ability to control each step in the method or protocol that could introduce unwanted variability and lead to inaccurate analysis or interpretation of results. In order to be considered 'standardized', a protocol must undergo intense scrutiny to evaluate the performance of each step of the protocol. The standardization process is typically performed in multiple phases which include development, expanded internal testing (validation), and external testing. This process ensures that standardized protocols retain a high level of robustness and reproducibility over multiple phases of testing. The standardization process is discussed in more detail in Chapter 10 of this book.

Even with a standardized protocol certain aspects are difficult, if not impossible, to control. One of these is differences that arise from variation in the quality or performance of some of the reagents or equipment available in different laboratories. Some techniques are more susceptible to these variations and therefore the data they produce are not suitable to populate reference library systems. For example, issues associated with the generation of reproducible data with RAPD have been observed due the low-stringency conditions in the amplification process utilized (Tyler *et al.*, 1997). Despite efforts to attempt to alleviate these issues and develop optimized protocol (Hopkins and Hilton, 2001), RAPD is not widely viewed as an appropriate method for development of a library subtyping system due to the potential variation that could be introduced as a result in differences in the performance of the equipment, particularly thermocyclers, and reagents between different laboratories. Conversely, methods for which a high level of standardization can be achieved are more amenable for integration into reference libraries. Library typing systems have been described for ribotyping, PFGE, AFLP, rep-PCR and PCR-ribotyping. Some of these methods have seen limited use, while others (i.e. PFGE within PulseNet) have been widely applied for use in library subtying systems.

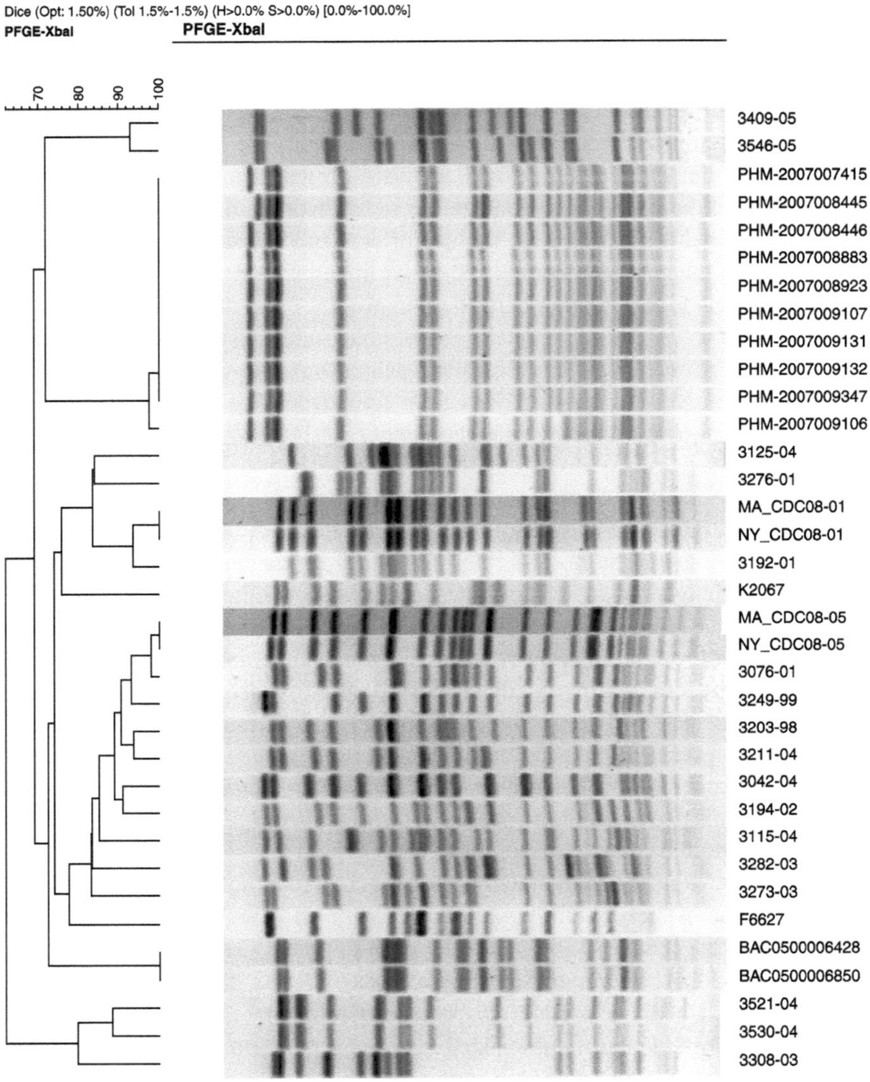

Fig. 8.3 Dendrogram of pulsed-field gel electrophoresis analysis of *Escherichia coli* isolates from the PulseNet national database.

8.4.2 Quality assurance/quality control

Quality assurance/quality control (QA/QC) is also an essential component of any library-based subtyping system to ensure that the data submitted to the reference library are of the highest quality possible. The PulseNet QA/QC program consists of two main elements: a certification process and a proficiency testing challenge. In the certification process, laboratories must demonstrate the ability to generate high-quality data with a set of strains sent to them by the CDC PulseNet laboratory before they are allowed to upload data directly to the national database (Gerner-Smidt *et al.*, 2006). Once certified, participating laboratories are required to participate in an annual proficiency-testing challenge to ensure that they continue to produce high-quality data. Finally, the use of common standards and controls by all participants is essential to demonstrate that protocols are followed accurately and to allow for normalization and comparison of data generated in sequential runs or in multiple laboratories.

8.5 Data interpretation for foodborne disease surveillance and outbreak investigation

Correct interpretation of subtyping data may be a complex task, but it is vital to the outcome of any epidemiological investigation. Without appropriate interpretation guidelines, different investigators may reach quite different conclusions about the relatedness of a set of isolates when analyzing the same data set. Bacterial subtyping methods sometimes detect small genetic differences (e.g., 2–3 different bands in PFGE) during the course of an outbreak that may not be epidemiologically significant (Tenover *et al.*, 1995). Conversely, the detection of an identical subtype does not necessarily imply an epidemiological relationship or a link between these two isolates. The development of interpretation guidelines is dependent on an understanding of the organism (clonal vs. highly mutable), the subtyping method (discriminatory power and reproducibility), and the intended use of the subtyping data. This understanding is only achieved through the analysis of subtyping data from large numbers of investigations and the correlation of these results with available epidemiological data. The following paragraphs will address some considerations in the evaluation of interpretation criteria as well as present some commonly cited interpretation guidelines and their appropriate application to evaluation of subtyping data. Overall, such guidelines should be viewed as recommendations, rather than rules, that can be revised as additional information and understanding becomes available.

It is important to note that methodological artifacts may lead to variations in the results that do not represent true differences. For instance, due to the comparative nature of gel-based subtyping methods, differences may exist in resolution (i.e. one versus two bands that may be resolved on separate runs or with different equipment). Closely matching patterns should be carefully examined to determine whether differences in resolution are apparent and, if so, isolates should be re-run on the same gel for confirmation. Similarly, experimental variation associated with DNA amplification or endonuclease restriction can also result in pattern differences. As mentioned previously, pattern variation may also be a consequence of genetic variation in the marker analyzed by the subtyping method during the course of an investigation (Barrett *et al.*, 2006). Awareness of methodological artifacts and other factors that influence the reproducibility of a method is important for accurate data analysis and interpretation. While the use of standardized protocols and reagents has assisted in reducing variation, it is still observed and may confound data analysis.

Due to the widespread use of PFGE, several interpretation guidelines have been described which differ based on the organisms and applications for which they were developed. Tenover et. al. (1995) established interpretation criteria in which patterns that differ from the outbreak pattern by two or three bands are considered to be subtypes of the outbreak pattern and profiles that differ by up to six bands are considered possibly related. This recommendation was based on the fact that a single genetic event (rearrangement, point mutation in a restriction site, an insertion, or a deletion) could potentially lead to a three-band difference depending on the location of the mutation relative to the restriction site. These interpretation guidelines were primarily developed for monitoring hospital-acquired infections of nosocomial pathogens. Hospital or community outbreaks are frequently prolonged, with strains being passed from person to person. Changes in PFGE patterns are not uncommon in settings with significant person-to-person transmission events. For over a decade, these guidelines have been extensively applied and have been instrumental in the interpretation of PFGE data from nosocomial pathogens (Barrett *et al.*, 2006; Muldrew *et al.*, 2008).

In contrast, Barrett et. al. (2006) described PFGE interpretation guidelines for foodborne pathogens which represents a different scenario than previously described. Foodborne disease outbreaks often result from a single contamination event where all of the patients are exposed to the same contaminated food over a relatively short period of time. The source of infection is usually identified through epidemiological studies (case-control or cohort) in which recent food exposure histories are compared between patients and controls. The results of these interview studies are dependent on the correct classification of patients as being outbreak related or not (i.e. a high specificity in the case definition). The erroneous inclusion of even a limited number of sporadic cases in the outbreak group may completely obscure an association with the implicated food product in an interview study. Additionally, the majority of enteric pathogens responsible for foodborne diseases are clonal. Due to these characteristics, variability among isolates from patients is expected to be limited. Therefore, it is recommended that, in the absence of supporting epidemiological information, only indistinguishable patterns are considered to be part of a foodborne disease cluster. Even if such a strict cluster definition is used, it should be remembered that the detection of an identical subtype does not necessarily imply a causal relationship or a link between two isolates. Clearly, molecular subtyping information must be analyzed in conjunction with epidemiologic data to determine causal relationships between isolates before public health action is taken.

One of the most challenging issues in the evaluation of a cluster happens when it is caused by a common subtyping pattern (Borucki *et al.*, 2004; Hedberg and Besser, 2006). When common patterns are encountered in an outbreak, it is often impossible to differentiate between outbreak-related isolates and unrelated sporadic infections. This needs to be kept in mind when such clusters are investigated. Common patterns are a considerable concern within highly clonal pathogens such as the *Salmonella* serovar Typhimurium DT104 complex (Liebana *et al.*, 2002), *Salmonella* serovar Enteritidis (Ridley *et al.*, 1998), and *E. coli*

O157:H7 (Hyytia-Trees *et al.*, 2007). The issue of common patterns highlights the need to analyze molecular subtyping information in conjunction with epidemiologic data. As mentioned before, additional discrimination may often be achieved by adding supplementary subtyping methods as part of the investigation.

8.6 Future trends

The ideal subtyping method is universally applicable, rapid, reproducible, inexpensive, easy to perform, and produces objective data that differentiates all epidemiologically unrelated strains from each other and clusters together all isolates associated with a common source. Among the currently used bacterial subtyping methods, none meets this ideal. While the utility of PFGE for foodborne pathogen surveillance has been demonstrated by over a decade of use and several hundreds of thousands of patterns stored within the PulseNet national database, the extensive experience with this technique and scope of these databases has also highlighted some of the inherent weaknesses associated with this technique. PFGE is a somewhat cumbersome, resource intensive, and the data it generates are not always easy to interpret.

As discussed earlier, the observation of common subtyping patterns can be problematic within outbreak investigations also with PFGE. The issue of common patterns is a function of the discriminatory power (or lack thereof) of the specific method and underscores the need to pursue the development of alternative subtyping methods. Current techniques could be complemented or replaced by new ones if they are shown to enhance the sensitivity of bacterial subtyping, particularly in regard to these clonal pathogens.

It is safe to say that bacterial subtyping is a technology-driven field where researchers often try to adapt the latest instrumentation or analysis software to the field on interest. As new or improved molecular biology techniques become available, it is inevitable that they will be evaluated for their potential to subtype bacterial populations. The convergence of ever-increasing amounts of comparative genomic data with these technological advances has resulted in a wide variety of potential bacterial subtyping platforms some of which are described in later chapters in this book. The majority of the methods being pursued are moving away from the general genomic comparison that is the hallmark of the methods discussed in this chapter, to specific interrogation of the DNA sequence (Hyytia-Trees *et al.*, 2007). Many of these methods are highly attractive due the potential for increased automation, throughput, and speed but they still are too organism specific to be applied on a large scale, e.g. in PulseNet. The ultimate utility of these techniques is still being assessed, however their success will depend largely on their ability to fit within existing networks and meet practical considerations such as equipment availability, reagent costs, and the compatibility and reliability of new data relative to existing methods in particular PFGE.

Active laboratory foodborne surveillance is only as effective as the coverage that the system possesses. The expansion of PulseNet nationally (all 50 states) and internationally has improved the ability of the program to detect foodborne outbreaks in the USA and abroad (Swaminathan *et al.*, 2006). Likewise, enhanced collaborations with all players in the food safety field will allow clinical data to be compared to that from food and veterinary sources. Such comparisons are essential for an integrated surveillance system for outbreak detection and response. Additionally, such broad collaborations are critical for applied food safety research and future risk assessments.

8.7 Disclaimer

Use of trade names is for identification only and does not imply endorsement by the Centers for Disease Control and Prevention or by the US Department of Health and Human Services.

The findings and conclusions in this chapter have not been formally disseminated by the Centers for Disease Control and Prevention and should not be construed to represent any agency determination or policy.

8.8 References

ARBEIT, R D (1995), 'Laboratory procedures for the epidemiologic analysis of microorganisms', in Murray, P.R., Baron, E.J., Pfaller, M.A., Tenover, F.C. and Yolken, R.H. (Eds): *Manual of Clinical Microbiology*, ASM Press, Washington DC, pp. 116–37.

BARRETT T J, GERNER-SMIDT P and SWAMINATHAN B (2006), 'Interpretation of pulsed-field gel electrophoresis patterns in foodborne disease investigations and surveillance', *Foodborne Pathogens and Disease*, **3**, 20–31.

BARRETT T J, LIOR H, GREEN J H, KHAKHRIA R, WELLS J G, BELL B P, ET AL. (1994), 'Laboratory investigation of a multistate foodborne outbreak of *Escherichia coli* O157:H7 by using pulsed-field gel electrophoresis and phage typing', *Journal of Clinical Microbiology*, **32**(12), 3013–17.

BIDET P, LALANDE V, SALAUZE B, BURGHOFFER B, AVESANI V, DELMEE M, ET AL. (2000), 'Comparison of PCR-Ribotyping, Arbitrarily Primed PCR, and Pulsed-Field Gel Electrophoresis for Typing *Clostridium difficile*', *Journal of Clinical Microbiology*, **38**, 2484–7.

BIELASZEWSKA M, PRAGER R, ZHANG W, FRIEDRICH A W, MELLMANN A, TSCHAPE H, ET AL. (2006), 'Chromosomal dynamism in progeny of outbreak-related sorbitol-fermenting enterohemorrhagic *Escherichia coli* O157:NM', *Applied Environmental Microbiology*, **72**, 1900–9.

BORUCKI M K, REYNOLDS J, GAY C C, MCELWAIN K L, KIM S H, KNOWLES D P, ET AL. (2004), 'Dairy farm reservoir of *Listeria monocytogenes* sporadic and epidemic strains', *Journal of Food Protection*, **67**, 2496–9.

BOUCHET V, HUOT H and GOLDSTEIN R (2008), 'Molecular Genetic Basis of Ribotyping', **21**, 262–73.

CARLE G F, FRANK M and OLSON M V (1986), 'Electrophoretic separations of large DNA molecules by periodic inversion of the electric field', *Science*, **232**, 65–8.

CARLE G F and OLSON M V (1984), 'Separation of chromosomal DNA molecules from yeast by orthogonal-field-alternation gel electrophoresis', *Nucleic Acids Research*, **12**, 5647–64.

CDC (2006), 'Ongoing multistate outbreak of *Escherichia coli* serotype O157:H7 infections associated with consumption of fresh spinach – United States, September 2006. *MMWR Morb Mortal Wkly Rep*, **55**, 1045–6.

CHU G, VOLLRATH D and DAVIS R W (1986), 'Separation of large DNA molecules by contour-clamped homogeneous electric fields', *Analytical Biochemistry*, **234**, 1582–5.

CLERMONT O, CORDEVANT C, BONACORSI S, MARECAT A, LANGE M and BINGEN E (2001), 'Automated ribotyping provides rapid phylogenetic subgroup affiliation of clinical extraintestinal pathogenic *Escherichia coli* strains', *Journal of Clinical Microbiology*, **39**, 4549–53.

COHEN S (2007), '9th annual symposium on advances in separation science and mass spectrometry', *Expert Review of Proteomics*, **4**, 443–4.

DE CESARE A, MANFREDA G, DAMBAUGH T R, GUERZONI M E and FRANCHINI A (2001), 'Automated ribotyping and random amplified polymorphic DNA analysis for molecular typing of *Salmonella enteritidis* and *Salmonella typhimurium* strains isolated in Italy', *Journal of Applied Microbiology*, **91**, 780–5.

GERNER-SMIDT P, GRAVES L M, HUNTER S and SWAMINATHAN B (1998), 'Computerized analysis of restriction fragment length polymorphism patterns: Comparative evaluation of two commercial software packages', *Journal of Clinical Microbiology*, **36**, 1318–23.

GERNER-SMIDT P, HISE K, KINCAID J, HUNTER S, ROLANDO S, HYYTIÃ-TREES E, ET AL. (2006), 'PulseNet USA: A five-year update', *Foodborne Pathogens and Disease*, **3**, 9–19.

GRAVES L M, HUNTER S B, ONG A R, SCHOONMAKER-BOPP, D., HISE K, KORNSTEIN, L., ET AL. (2005B), 'Microbiological aspects of the investigation that traced the 1998 outbreak of listeriosis in the United States to contaminated hot dogs and establishment of molecular subtyping-based surveillance for Listeria monocytogenes in the PulseNet network', *Journal of Clinical Microbiology*, **43**, 2350–5.

HAHM B-K., MALDONADO Y, SCHREIBER E, BHUNIA A K and NAKATSU C H (2003), 'Subtyping of foodborne and environmental isolates of *Escherichia coli* by multiplex-PCR, rep-PCR, PFGE, ribotyping and AFLP', *Journal of Microbiological Methods*, **53**, 387–99.

HEALY M, HUONG J, BITTNER T, LISING M, FRYE S, RAZA S, ET AL. (2005), 'Microbial DNA typing by automated repetitive-sequence-based PCR', *Journal of Food Protection*, **43**, 199–207.

HEDBERG C W and BESSER J M (2006), 'Commentary: Cluster evaluation, PulseNet, and public health practice', *Foodborne Pathogens and Disease*, **3**, 32–35.

HOPKINS K L and HILTON A C (2001), 'Optimization of random amplification of polymorphic DNA analysis for molecular subtyping of *Escherichia coli* O157', *Letters in Applied Microbiology*, **32**, 126–30.

HUNTER P R and GASTON M A (1988), 'Numerical index of the discriminatory ability of typing systems: an application of Simpson's index of diversity', *Journal of Clinical Microbiology*, **26**, 2465–6.

HYYTIA-TREES, E.K., COOPER K, RIBOT E M and GERNER-SMIDT, P. (2007), 'Recent developments and future prospects in subtyping of foodborne bacterial pathogens', *Future Microbiology*, **2**, 175–85.

INDRA A, HUHULESCU S, SCHNEEWEIS M, HASENBERGER P, KERNBICHLER S, FIEDLER A, ET AL. (2008), 'Characterization of *Clostridium difficile* isolates using capillary gel electrophoresis-based PCR ribotyping', *Journal of Medical Microbiology*, **57**, 1377–82.

JONAS D, SPITZMÜLLER, B, WEIST K, RÜDEN, H. and DASCHNER F D (2003), 'Comparison of PCR-based methods for typing *Escherichia coli*', *Clinical Microbiology and Infection*, **9**, 823–31.

KILLGORE G, THOMPSON A, JOHNSON S, BRAZIER J, KUIJPER E, PEPIN J, ET AL. (2008), 'Comparison of seven techniques for typing international epidemic strains of *Clostridium difficile*: Restriction endonuclease analysis, pulsed-field gel electrophoresis, PCR-ribotyping, multilocus sequence typing, multilocus variable-number tandem-repeat analysis, amplified fragment length polymorphism, and surface layer protein a gene sequence typing', *Journal of Clinical Microbiology*, **46**, 431–7.

KOORT J M K, LUKINMAA S, RANTALA M, UNKILA E and SIITONEN A (2002), 'Technical improvement to prevent DNA degradation of enteric pathogens in pulsed-field gel electrophoresis', *Journal of Clinical Microbiology*, **40**, 3497–8.

KUDVA I T, EVANS P S, PERNA N T, BARRETT T J, AUSUBEL F M, BLATTNER F R, ET AL. (2002), 'Strains of *Escherichia coli* O157:H7 Differ Primarily by Insertions or Deletions, Not Single-Nucleotide Polymorphisms', *Journal of Bacteriology*, **184**, 1873–9.

LE H, FUNG D C Y, YU B and TRENT R J (1998), 'Capillary electrophoresis: New technology for DNA diagnostics', *Pathology*, **30**, 304–8.

LIEBANA E, GARCIA-MIGURA, L., CLOUTING C, CLIFTON-HADLEY, F.A., LINDSAY E, THRELFALL E J, ET AL. (2002), 'Multiple genetic typing of *Salmonella enterica* serotype typhimurium isolates of different phage types (DT104, U302, DT204b, and DT49) from animals and humans in England, Wales, and Northern Ireland', *Journal of Clinical Microbiology*, **40**, 4450–6.

LIM H, LEE K H, HONG C-H., BAHK G-J. and CHOI W S (2005), 'Comparison of four molecular typing methods for the differentiation of *Salmonella* spp', *International Journal of Food Microbiology*, **105**, 411–8.

LINDSTEDT B-A., HEIR E, VARDUND T and KAPPERUD, G. (2000a), 'Fluorescent amplified-fragment length polymorphism genotyping of *Salmonella enterica* subsp. enterica serovars and comparison with Pulsed-Field Gel electrophoresis typing', *Journal of Clinical Microbiology*, **38**, 1623–1627.

LINDSTEDT B-A., HEIR E, VARDUND T, MELBY K K and KAPPERUD, G. (2000b), 'Comparative fingerprinting analysis of *Campylobacter jejuni* subsp. jejuni strains by amplified-fragment length polymorphism genotyping', *Journal of Clinical Microbiology*, **38**, 3379–87.

LUKINMAA S, NAKARI U, EKLUND M and SIITONEN A (2004), Application of molecular genetic methods in diagnostics and epidemiology of foodborne bacterial pathogens', *Pathologica, Microbiologica Et Immunologica Scandinavica*, **112**, 908–29.

LUPSKI J R and WEINSTOCK G M (1992), 'Short, interspersed repetitive DNA sequences in prokaryotic genomes', *Journal of Bacteriology*, **174**, 4525–9.

MARTIN B, HUMBERT O, CAMARA M, GUENZI E, WALKER J, MITCHELL T, ET AL. (1992), 'A highly conserved repeated DNA element located in the chromosome of *Streptococcus pneumoniae*', *Nucleic Acids Research*, **20**, 3479–83.

MOHAPATRA B R and MAZUMDER K B (2007), 'Comparison of five rep-PCR genomic fingerprinting methods for differentiation of fecal *Escherichia coli* from humans, poultry and wild birds', *FEMS Microbiology Letters*, **277**, 98–106.

MULDREW K L, TANG Y-W., LI H and STRATTON C W (2008), 'Clonal dissemination of *Staphylococcus epidermidis* in an oncology ward', *Journal of Clinical Microbiology*, **46**, 3391–6.

MURCHAN S, KAUFMANN M E, DEPLANO A, DE RYCK R, STRUELENS M, ZINN C E, ET AL. (2003), 'Harmonization of pulsed-field gel electrophoresis protocols for epidemiological typing of strains of methicillin-resistant *Staphylococcus aureus*: A single approach developed by consensus in 10 European laboratories and its application for tracing the spread of related strains', *Journal of Clinical Microbiology*, **41**, 1574–85.

NAIR S, SCHREIBER E, THONG K-L., PANG T and ALTWEGG M (2000), 'Genotypic characterization of *Salmonella typhi* by amplified fragment length polymorphism fingerprinting provides increased discrimination as compared to pulsed-field gel electrophoresis and ribotyping', *Journal of Microbiological Methods*, **41**, 35–43.

OLIVE D M and BEAN P (1999), 'Principles and applications of methods for DNA-based typing of microbial organisms', *Journal of Clinical Microbiology*, **37**, 1661–9.

RASSCHAERT G, HOUF K, IMBERECHTS H, GRIJSPEERDT K, DE ZUTTER L and HEYNDRICKX M (2005), Comparison of five repetitive-sequence-based PCR typing methods for molecular discrimination of *Salmonella enterica* isolates', *Journal of Clinical Microbiology*, **43**, 3615–23.

RAY T, WEADEN J and DYSON P (1992), 'Tris-dependent site-specific cleavage of *Streptomyces lividans* DNA', *FEMS Microbiology Letters*, **96**, 247–52.

REMENTERIA A, GALLEGO L, QUINDÓS G and GARAIZAR J (2001), 'Comparative evaluation of three commercial software packages for analysis of DNA polymorphism patterns', *Clinical Microbiology & Infection*, **7**, 331–336.

RIBOT E M, FAIR M A, GAUTOM R, CAMERON D N, HUNTER S B, SWAMINATHAN B, ET AL. (2006), 'Standardization of pulsed-field gel electrophoresis protocols for the subtyping of *Escherichia coli* O157:H7, *Salmonella*, and *Shigella* for PulseNet', *Foodborne Pathogens and Disease*, **3**, 59–67.

RIDLEY A M, THRELFALL E J and ROWE B (1998), 'Genotypic characterization of *Salmonella enteritidis* phage types by plasmid analysis, ribotyping, and pulsed-field gel electrophoresis', *Journal of Clinical Microbiology*, **36**, 2314–21.

ROMLING U and TUMMLER B (2000), 'Achieving 100% typeability of *Pseudomonas aeruginosa* by pulsed-field gel electrophoresis', **38**, 464–5.

SAWABE E, KATO H, OSAWA K, CHIDA T, TOJO N, ARAKAWA Y, ET AL. (2007), 'Molecular analysis of *Clostridium difficile* at a university teaching hospital in Japan: a shift in the predominant type over a five-year period', *European Journal of Clinical Microbiology and Infectious Diseases*, **26**, 695–703.

SCHWARTZ D C and CANTOR C R (1984), 'Separation of yeast chromosome-sized DNAs by pulsed-field gradient gel electrophoresis', *Cell*, **37**, 67–75.

SHARPLES G J and LLOYD R G (1990), 'A novel repeated DNA sequence located in the intergenic regions of bacterial chromosomes', *Nucleic Acids Research*, **18**, 6503–8.

SILBERT S, BOYKEN L, HOLLIS R J and PFALLER M A (2003), 'Improving typeability of multiple bacterial species using pulsed-field gel electrophoresis and thiourea', *Diagnostic Microbiology and Infectious Disease*, **47**, 619–21.

SOUTHERN E M (1975), 'Detection of specific sequences among DNA fragments separated by gel electrophoresis', *Journal of Molecular Biology*, **98**, 503–17.

STEPHAN R (1996), 'Randomly amplified polymorphic DNA (RAPD) assay for genomic fingerprinting of *Bacillus cereus* isolates', *International Journal of Food Microbiology*, **31**, 311–16.

STERN M J, AMES G F-L, SMITH N H, CLARE ROBINSON E and HIGGINS C F (1984), 'Repetitive extragenic palindromic sequences: A major component of the bacterial genome', *Cell*, **37**, 1015–26.

STRUELENS M, DE GHELDRE, Y. and DEPALANO A (1998), 'Comparative and library epidemiological typing systems: outbreak investigations versus surveillance systems', *Infection Control and Hospital Epidemiology*, **19**, 565–9.

STRUELENS M J (1996), 'Consensus guidelines for appropriate use and evaluation of microbial epidemiologic typing systems', *Clinical Microbiology and Infection*, **2**(1), 2–11.

SWAMINATHAN B, BARRETT T J, HUNTER S B and TAUXE R V (2001), 'PulseNet: The molecular subtyping network for foodborne bacterial disease surveillance, United States', *Emerging Infectious Diseases*, **7**, 383–9.

SWAMINATHAN B, GERNER-SMIDT, P., NG L-K., LUKINMAA S, KAM K-M., ROLANDO S, GUTIÉRREZ, E.P.R. and BINSZTEIN N (2006), 'Building PulseNet international: An interconnected system of laboratory networks to facilitate timely public health recognition and response to foodborne disease outbreaks and emerging foodborne diseases', *Foodborne Pathogens and Disease*, **3**, 36.

TENOVER F C, ARBEIT R D, GOERING R V, MICKELSEN P A, MURRAY B E, PERSING D H and SWAMINATHAN B (1995), 'Interpreting chromosomal DNA restriction patterns produced by pulsed-field gel electrophoresis: criteria for bacterial strain typing', *Journal of Clinical Microbiology*, **33**, 2233–9.

TYLER K D, WANG G, TYLER S D and JOHNSON W M (1997), 'Factors affecting reliability and reproducibility of amplification-based DNA fingerprinting of representative bacterial pathogens', *Journal of Clinical Microbiology*, **35**, 339–46.

VAN BELKUM A, TASSIOS P, DIJKSHOORN L, HAEGGMAN S, COOKSON B, FRY N, ET AL. (2007), 'Guidelines for the validation and application of typing methods for use in bacterial epidemiology', *Clinical Microbiology and Infection*, **13**, 1–46.

VERSALOVIC J, KOEUTH T and LUPSKI R (1991), 'Distribution of repetitive DNA sequences in eubacteria and application to finerpriting of bacterial genomes', *Nucleic Acids Research*, **19**, 6823–31.

VOGEL B F, FUSSING V, OJENIYI B, GRAM L and AHRENS P (2004), 'High-resolution genotyping of listeria monocytogenes by fluorescent amplified fragment length polymorphism analysis compared to pulsed-field gel electrophoresis, random amplified polymorphic DNA analysis, ribotyping, and PCR-restriction fragment length polymorphism analysis', *Journal of Food Protection*, **67**, 1656–65.

VOS P, HOGERS R, BLEEKER M, REIJANS M, LEE T V D, HORNES M, ET AL. (1995), 'AFLP: A new technique for DNA fingerprinting', *Nucleic Acids Research*, **23**, 4407–14.

WELSH J and MCCLELLAND M (1990), 'Fingerprinting genomes using PCR with arbitrary primers', *Nucleic Acids Research*, **18**, 7213–18.

WIEDMANN M (2002), 'Subtyping of bacterial foodborne pathogens', *Nutrition Reviews*, **60**, 201–8.

9

Emerging methods for foodborne bacterial subtyping

F. Pagotto and A. Reid, Health Canada, Canada

Abstract: In this chapter, emerging methods for the generation of bacterial fingerprints are discussed. Molecular typing is often carried out to define clonal lineages, compare species from differing geographical regions and/or time of isolation, or to estimate variability within a specific bacterial population. In the context of food microbiology and epidemiological investigations during an outbreak, a major goal of molecular typing is to provide supportive laboratory evidence showing that isolates recovered are genetically related and thus represent the same (i.e. outbreak) strain. Considerations and challenges are discussed, with the major nucleic acid- and protein-based typing schemes/methods for the characterization of bacterial foodborne pathogens being considered as possible replacements to currently used methods, of which pulsed-field gel electrophoresis is considered the 'gold standard'. Advantages and disadvantages of individual techniques are highlighted.

Key words: molecular characterization, typing, VNTR, MLVA, AFLP, MLST, microarray, MALDI–TOF, mass spectrometry, antibodies, comparative genomics, SNPs, surface plasmon resonance.

9.1 Introduction

There are many reasons for performing characterization or fingerprinting of etiological agents of human disease. Typeability, the ability to apply a typing scheme to distinguish strains or isolates of a given bacterial species, and discriminatory power, the ability to distinguish unrelated pathogens, are two major considerations when developing a molecular characterization scheme. These two characteristics, once satisfied for a particular method and foodborne bacterial pathogen, allow an investigator to provide laboratory evidence to support epidemiological investigations to confirm or refute that a given organism is, in fact, responsible for a particular public health threat (e.g. that an outbreak is occurring). In this chapter, emerging methods for the characterization of bacteria are described, with a focus on the typing of foodborne pathogens. While typing or fingerprinting can rely on phenotypic or genotypic traits, characterization at the molecular level is often required for characterization beyond the species and/or subspecies level.

In the context of, *Listeria* species, for example, phenotypic characterization such as serotyping is often carried out as the first characterization step, followed by genotyping using pulsed-field gel electrophoresis (PFGE), the method currently considered as the 'gold standard' for the characterization of this and other bacterial foodborne pathogens. How well a technique performs really depends on its intended outcome. For instance, the ability to differentiate strains that are considered unrelated by epidemiological data is of essence when conducting outbreak and/or sporadic case investigations. While most outbreaks of listeriosis are due to the same three serovars, serology can be a useful first step in assessing the involvement of a strain in an outbreak. Serology is quicker than PFGE, although not as discriminatory. The ideal method for this application (PFGE, or emerging technique) should be able to link epidemiologically related food (or environmental) and clinical isolates, while distinguishing them from unrelated strains. As a result, it becomes easy to understand why the discriminatory power must be at a suitable level. If a typing scheme is too discriminatory, the investigation will not be able to draw a link between the source of the isolates and the clinical isolates themselves. Defining the level of discrimination then becomes the empirical exercise.

The one underlying assumption in foodborne pathogen-related investigations is that the pathogen will not significantly mutate and will therefore be recognized by the phenotypic and/or genotypic typing scheme being used. For example, consider a bacterial pathogen as it survives its journey from a farm to a factory deli slicer to the retail store to the consumer (where it causes illness) to the agar plate on which it is isolated in the hospital; any mutation resulting in an altered fingerprint would impair association or traceback (i.e. source attribution) of this organism. Characterization schemes have to be able to address this 'evolution' in order to be successful.

Other aspects that are important when considering a typing scheme include the method's ability to estimate population variation, to define clonal lineages and to discriminate isolates across geographical regions and/or over time (i.e. endemic strains). How feasible is it to discriminate sporadic cases versus outbreaks of foodborne disease? Ideally, the method will be able to distinguish unrelated strains, identify and group epidemiologically related strains, be inexpensive, rapid, easy-to-use, have an objective interpretation, be amenable to automation, and possibly be portable. The ability to combine two or more of these characteristics makes routine surveillance a powerful tool in defining the impact of foodborne pathogens on society. Related topics such as standardization, quality assurance, the use of databases, etc., are covered in detail in other chapters, and as such will not be discussed here.

Historically, phenotype-based methods for typing of bacterial foodborne pathogens have been used to characterize observable (and stable) characteristics of an organism. Often, the genetic components were unknown. In general, there is a movement away from phenotype-based methods, in favour of nucleic acid-based methods, as genotypic approaches offer improved stability and discriminatory power. However, some phenotype-based methods, such as serotype and phage-type analyses, continue to play an important role in the early stages of characterization for many bacterial pathogens. Because these early methods have been in use for a long time, large pools of data are available, allowing for easy comparisons between labs/organizations and over time. Nucleic acid-based methods are more commonly used to detect differences in the genomes of strains, such as the number of repetitive elements in a region or even single nucleotide polymorphisms (SNPs). Schemes aimed at differences at the genome level may provide higher resolution within a bacterial species, where phenotyping is limited by growth conditions, availability of antibodies, phages, etc.

Protein-based technologies for bacterial subtyping have the ability to detect differences between the proteomes of two bacteria, which may not necessarily be reflected by changes in nucleic acid sequences. For instance, post-translational modification of proteins and/or expression levels of some proteins may differ between bacteria, despite the genes encoding these proteins being identical. Conversely, these methods would not be capable of detecting silent mutations in DNA sequences. Once a protein marker has been identified as being specific for a given organism or a group of organisms, the DNA sequence encoding this protein may be useful for DNA-based approaches [e.g. target for polymerase chain reaction (PCR) amplification or DNA microarray]. Identification of a protein biomarker could also be followed by the expression and purification of the protein target, for use in immunological detection methodologies such as protein microarrays. Despite these advantages, the use of protein-based technologies for bacterial subtyping is relatively new, and the instances where these methods have been used to accurately type bacteria beyond the genus and species levels are few.

9.2 Nucleic acid-based technologies

9.2.1 Variable number of tandem repeats (VNTR)/multiple-locus VNTR analysis (MLVA)

One of the more discriminatory PCR-based molecular typing methods is the variable number of tandem repeats (VNTR) assay (Fig. 9.1). This technique has a strong basis in human genetics, forensics and DNA fingerprinting. As a result of increased output of fully sequenced bacterial genomes, bacteria have been shown to contain tandem sequences of repetitive DNA elements. These repeats have been observed to vary within a bacterial species in terms of the number of times a particular repeat sequence is present (referred to as the copy number), making them useful molecular fingerprints. Such tandem repeats appear to be ample in prokaryotic genomes and many have a polymorphism in terms of the number of repeats present across a particular species. The mutation rates of

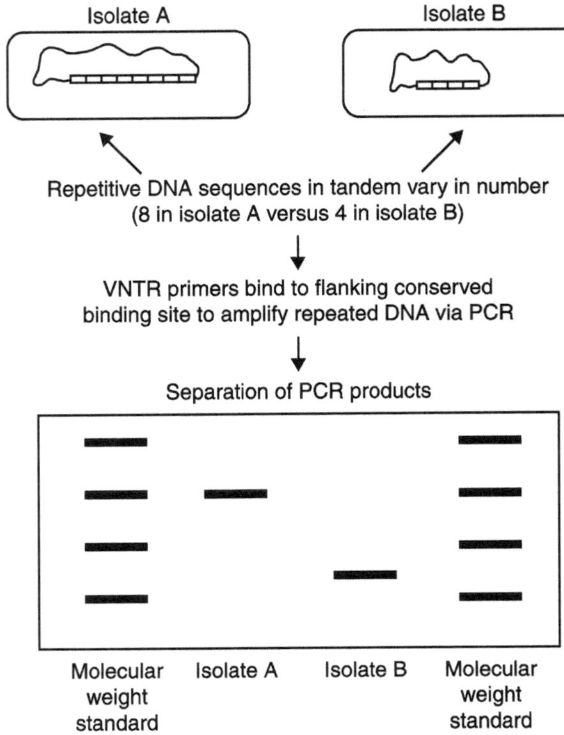

Fig. 9.1 Variable number of tandem repeats (VNTR) analyses. A single primer pair is used to amplify repetitive DNA sequences found in tandem but which vary in number. Amplification products are then analysed by agarose gel electrophoresis (lower panel). Alternatively, primers can be labelled and products analysed using capillary electrophoresis.

these VNTR loci range from 10^{-4} to 10^{-6}, depending on the organism (Vogler *et al.*, 2006, 2007).

Generally speaking, these repeated sequences are classified based on their size. If the repeats range from one to 13 base pairs (bp), they are referred to as microsatellites, as they contain short sequence repeats; those with 10–100 bp are called minisatellites. The variability in terms of repeat numbers found in a species helps fulfil the definition of them being classified as the VNTR loci (Nakamura *et al.*, 1987). It appears that the major mechanisms for variability in VNTR loci involve DNA polymerase slippage and/or recombination events (van Belkum *et al.*, 1998).

One of the first bacterial organisms analysed by this technique was *Mycobacterium tuberculosis*, in which the repeats were termed MIRUs, or mycobacterial interspersed repeat units (Magdalena *et al.*, 1998). Since then, many software programs have been developed to help identify these tandem repeats. One commonly used program is the Tandem Repeat Finder (Benson, 1999), although there are many other search tools available (Grissa, Bouchon, *et al.*, 2008; Grissa, Vergnaud, *et al.*, 2008). While the VNTR-related software(s) help identify repeats in a genome, they are not able to predict whether a given tandem repeat will have sufficient polymorphism(s) to be of value for foodborne (and other) investigations.

The VNTR technique uses a non-ambiguous format, based on the number of repetitive sequences present, determined by DNA sequencing of amplified products or by separation of these products by agarose gel electrophoresis. While the VNTR method was developed to address a single repetitive element, this technique has been modified to include more than one genome target per assay. Multiple-locus VNTR analysis, also known as MLVA, is considered to be the next generation typing method, possibly replacing PFGE as the 'gold standard' for molecular characterization of bacterial pathogens. With MLVA/VNTR, a common approach is to use primer(s) tagged with a fluorescent molecule, which allows separation and analysis of products by capillary electrophoresis (i.e. DNA sequencer). The use of a DNA sequencer is of great value with the MLVA approach, as different genomic regions can be identified by different fluorescent tags, and thus analysed simultaneously.

MLVA has been used as an alternative method to PFGE for discriminating epidemiologically unrelated isolates of highly related organisms such as *Escherichia coli* O157:H7 (Noller *et al.*, 2003). This makes the method attractive for use in foodborne outbreak situations, in which microbial source tracking and attribution are highly important in identification of the source (Chiou, 2010).

A number of reviews have described VNTR/MLVA methodologies and their applications (Grissa, Bouchon, *et al.*, 2008; Hyytia-Trees *et al.*, 2007; Ramazanzadeh and McNerney, 2007). Interestingly, a similar approach is lacking for parasitic and viral foodborne pathogens, as well as for the bacterial enteric pathogen *Campylobacter* (van Belkum, 2007), in which multi-locus sequence typing (MLST) appears to be more suitable (Hyytia-Trees *et al.*, 2007). A VNTR/MLVA method requires that genomes should be analysed for repeated motifs. It is possible that in these latter organisms, insufficient information is available and that, more likely, there are repeated sequences, but these are not useful or discriminatory enough for epidemiological investigations. While a powerful tool for molecular characterization and subtyping below the species level, it appears that some aspects of the mechanism(s) of variability in the context of a biological role or evolutionary aspect may not be completely addressed by the VNTR method (Mazars *et al.*, 2001).

9.2.2 Amplified fragment length polymorphism (AFLP)

Amplified fragment length polymorphism (AFLP) (Fig. 9.2, reproduced from Pagotto *et al.*, 2005) is a specific marker-based identification assay that allows for the detection of polymorphisms in genomes of interest (Mueller and Wolfenbarger, 1999; Vos and Kuiper, 1997). AFLP combines aspects of restriction fragment length polymorphism (RFLP) and randomly amplified polymorphic DNA (RAPD) methodologies (Bensch and Akesson, 2005). In a typical assay, the isolated bacterial genome is subject to digestion using two restriction endonucleases

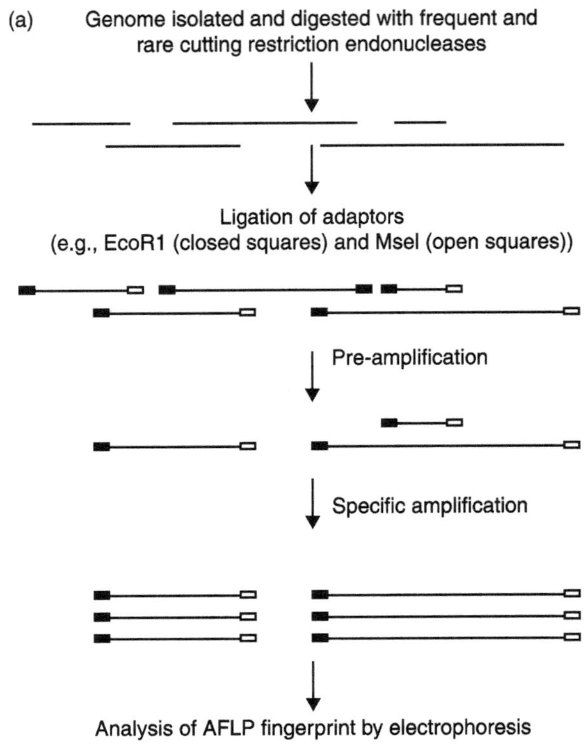

Fig. 9.2 Amplified fragment length polymorphism (AFLP). (a) General procedure illustrating the overall method. *Opposite* (b) Diagrammatic example showing pre-amplification and subsequent amplification using primers 1 and 2 (one of many possible primers). Previous genomic information is not required with the AFLP technique. Current AFLP methods incorporate the labelling of either primer 1 or 2 with a fluorescent dye molecule.

(Fig. 9.2a). This is followed by ligation of double-stranded adapters to the ends and a subsequent selective amplification of a subset of the adapted fragments. These can be visualized on denaturing polyacrylamide gels with the aid of fluorescence techniques (or autoradiography). The adapters used in AFLP are complementary to two restriction sites targeted in the initial chromosome digest (Fig. 9.2b). This adapted DNA serves as a template for primers complementary to the adaptor sequences (Fig. 9.2b, middle). However, these have one additional base extending into the 'unknown' region of DNA, in order to reduce the number of fragments to be amplified. Specific amplification occurs with the use of primers 1 and 2 (Fig. 9.2b, bottom), where preamplified products are targeted, as they have the correct nucleotides for primer annealing. Products may be then visualized and analysed using appropriate software programs and algorithms.

While AFLP technology is based on selective amplification of a subset of the genomic restriction fragments, it remains a random amplification technique.

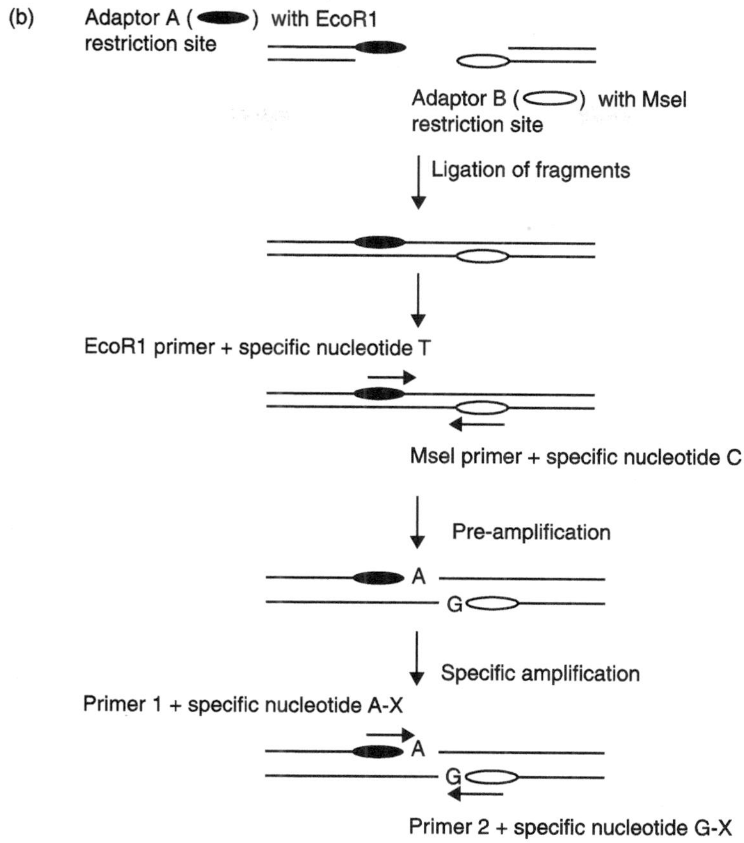

Fig. 9.2 Continued.

However, a major advantage over other non-specific amplification methods such as RAPD is that AFLP is able to use stringent PCR amplification conditions. Unlike RAPD primers, AFLP primers are usually 17–21 nucleotides in length and anneal perfectly to their complementary sequence target sites, which improves the reproducibility and resolution of the method. AFLP is most useful for detecting polymorphisms within a total genome.

An advanced application of AFLP is cDNA-AFLP, which allows the identification and quantification of expressed mRNAs (Breyne *et al.*, 2003). This AFLP-based transcript profiling targets genome-wide expression analyses in a species where little information about the genome may be available. Differences in the size of products (or even in the amount of product) provide useful information on the expression levels of various genes. While currently used to study gene expression, it may have a potential role in molecular characterization if isolated pathogens would be, for example, grown up under similar conditions prior to having the technique applied. In this context, for example, virulence gene expression typing might be possible. This has been shown in a study by Levterova *et al.* (2010) in which genetic markers involved with antimycotic drug resistance in medically relevant fungi were correlated with changing levels of RNA under various typing conditions. This approach may be further developed to be used as an alternative to microarray technology (Reijans *et al.*, 2003) or protein-based gel electrophoretic (i.e. proteomic) assays (Hooshdaran *et al.*, 2004).

9.2.3 Multi-locus sequence typing (MLST)

The multi-locus sequence typing (MLST) approach serves as the basis for many nucleotide-based methods currently used to generate molecular profiles of bacterial pathogens (Maiden *et al.*, 1998). Originally based on multi-locus enzyme electrophoresis (Selander *et al.*, 1986), where organisms were characterized based on the electrophoretic mobilities of approximately 20 of their housekeeping enzymes, MLST categorizes allelic types from the nucleotide sequences of these housekeeping genes rather than by the electrophoretic mobilities of their respective gene products. Maiden *et al.* (1998) initially described how bacterial isolates could be typed directly based on the sequence of internal fragments (approximately 450–500 bp) of approximately seven housekeeping genes. The sequence differences found between all the isolates are each assigned a distinct allele number (Fig. 9.3). Originally, the method was designed to unambiguously characterize isolates by simultaneously considering the allele designations for the seven housekeeping genes (Enright and Spratt, 1999). The obvious advantage of MLST is that it is based on DNA sequence data, providing unambiguous and highly discriminatory information. In addition, MLST allows for the ease of exchange of data amongst laboratories and/or countries, making global epidemiology possible (Enright and Spratt, 1999; Maiden *et al.*, 1998).

MLST has since evolved to include genetic (i.e. sequence) information from virulence genes to help provide additional and appropriate levels of discrimination. Surveying the genetic variation in virulence genes may also allow for a better understanding of the pathogenic potential of the organism under study.

Sequence types may then be determined from the concatenated code of allelic assignments for each of the genes involved. Concatenation of genes is often used to infer phylogenies (Gadagkar *et al.*, 2005) and optimizes relational analyses due to the fact that more information is being considered. Furthermore, recombination events tend to be addressed and their impact on classification minimized when compared to the use of single genes for analyses (Pascual *et al.*, 2010). As such, various combinations of housekeeping, virulence and stress genes are now being explored for their potential use in characterization of bacteria at or beyond the species level (Pascual *et al.*, 2010). In the example shown in Fig. 9.3, sequences from genes A, B, C and D would be concatenated and the resulting dendrogram be generated using phylogenetic analyses such as the neighbour-joining clustering method (Costas *et al.*, 1990). Internet-based MLST databases are available at http://www.mlst.net as well as http://pubmlst.org.

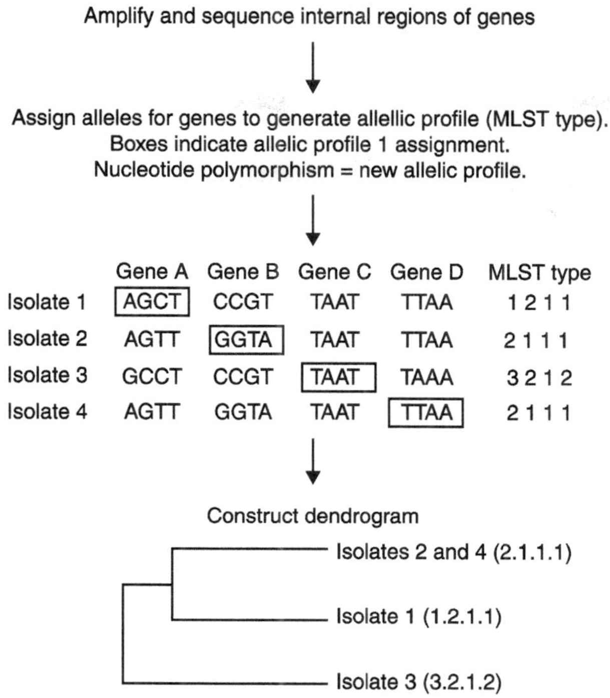

Fig. 9.3 Multi-locus sequencing typing (MLST). Genes of interest (e.g. housekeeping, or other) are amplified and sequenced. Each gene is then assigned a numerical value, corresponding to an allelic profile that is compared to other genes from other species being examined. Strains that have differences in their gene sequences are given different allelic numbers. The allelic profiles, combinations of several genes, are then used to generate a sequence type. The sequence types (allelic profiles) are used to generate comparative dendrograms (lower panel). For each gene, a different isolate is designated an allelic profile of 1.

9.2.4 Genome sequencing and comparative genomics

The free-living human bacterial pathogen, *Haemophilus influenzae*, was the first genome to be sequenced in its entirety (Fleischmann *et al.*, 1995). Sequencing technologies have since evolved, and the cost for sequencing a typical bacterial genome (not including the cost for bioinformatic analyses) is now well within reach for many researchers (~$5000 per genome). At the time this chapter was drafted, there were over 1000 complete prokaryotic genomes available (http://www.ncbi.nlm.nih.gov/genomes/lproks.cgi), a ten-fold increase over the last seven years (Nelson, 2003).

Genomics uses a holistic approach to study the mechanisms that determine biological processes such as development, disease and responses to environmental stimuli. It allows for the study of the structure and expression of the entire genetic complement of an organism. The use of genomics and comparative genomics (comparisons and subsequent analyses of the genomes of different species, although in the context of this chapter, it could also apply to multiple serovars or subtypes within a genus) permits investigations into multiple strains within a species. Information on repetitive elements, bacteriophage(s) and their insertion site(s), SNPs, insertion and deletions of genetic material, recombination events and other alterations of the chromosomal content become available.

Clearly, once the genome information is obtained using traditional sequencing approaches or alternative technologies such as pyrosequencing and resequencing microarray technologies (Clarke, 2005; Zhang *et al.*, 2006), there is a strong need for bioinformatics. A major tool is the basic local alignment search tool (BLAST) algorithm, useful for comparing nucleotide (and amino acid) sequences of the genome in question against a standardized database (Altschul *et al.*, 1997).

While large-scale comparative genomics is not commonplace in the world of bacterial subtyping (in terms of epidemiological investigations), its usefulness was shown recently in a publication where genomes of strains of *Listeria monocytogenes* implicated in a large, multi-province outbreak of listeriosis in Canada were sequenced during the outbreak (Gilmour *et al.*, 2010). This work demonstrated that, while PFGE revealed two closely related yet distinct pulsotypes that may have excluded some strains from being implicated in the same outbreak, comparative genomics was able to demonstrate that all PFGE types were from a common source.

9.2.5 Single nucleotide polymorphism (SNP) analysis

The human genome project was influential in making single nucleotide polymorphism, or SNP analysis, a popular method of choice for the study of molecular genetics (Weiner and Hudson, 2002). With the increasing number of bacterial genomes available sequences, it has been recognized that the SNP technique would be equally valuable for both characterization and differentiation of strains, especially for those organisms that are considered monomorphic (i.e. having low levels of sequence diversity) (Achtman, 2008; Foley *et al.*, 2009; Weiner and Hudson, 2002; Zhang *et al.*, 2006). The classic foodborne pathogens considered monomorphic include *E. coli* O157:H7, *Salmonella enterica* serovar Typhi and *Shigella sonnei* (Achtman, 2008; Kidgell *et al.*, 2002; Pupo *et al.*, 2000; Zhang *et al.*, 2006).

SNP analysis is able to map loci in bacterial genomes where evolution occurs. This evolution presents itself via horizontal gene transfer events, addition and/or deletion of sequences, recombination events and single nucleotide mutations (Achtman, 2008; Foley *et al.*, 2007, 2009). The SNPs have been detected in many different ways, using techniques based on sequencing, mass spectrometry (MS), amplification (PCR)-based and through the use of DNA microarrays (Cebula *et al.*, 2005; Foley *et al.*, 2009; Hyytia-Trees *et al.*, 2007; Lechner *et al.*, 2002; Mortimer *et al.*, 2004; Roos *et al.*, 2006; Tyagi *et al.*, 1998; Weiner and Hudson, 2002; Zhang *et al.*, 2006).

An interesting application of the SNP technique has been demonstrated by Ward and coworkers, who focussed on lineage I strains of *L. monocytogenes* (Ducey *et al.*, 2007; Ward *et al.*, 2008). Lineage I strains are overrepresented in human listeriosis cases and appear to be responsible for the majority of outbreaks and sporadic cases (Gray *et al.*, 2004; Jeffers *et al.*, 2001). A multilocus genotyping (MLGT) assay of SNP sites using flow cytometry for lineage I strains was developed that uses 60 allele-specific probes. The MLGT technique was also compared against PFGE and an MLST assay based on housekeeping and virulence genes. The MLGT approach was shown to be superior to the MLST developed by Revazishvili *et al.* (2004). However, PFGE still showed greater strain discriminatory power (Ward *et al.*, 2008). This new approach, however, shows promise and may be further developed to help overcome the limitations of PFGE (Doumith *et al.*, 2006; Ducey *et al.*, 2007; Gerner-Smidt *et al.*, 2006; Pagotto *et al.*, 2005). Currently, most SNP assays appear to focus on Gram-negative bacterial pathogens (reviewed in Foley *et al.*, 2007, 2009).

A primer extension-based SNP analysis, known as the single nucleotide primer extension (SNuPE), can be used for the detection of a nucleotide located at a given polymorphic location (Kuppuswamy *et al.*, 1991; Nikolausz *et al.*, 2009). Applications of this method have been described for phylotyping of *L. monocytogenes* and *E. coli* (Ducey *et al.*, 2007; Hommais *et al.*, 2005; Rudi *et al.*, 2003), and are reviewed by Nikolausz *et al.* (2009).

9.2.6 DNA microarrays

Early DNA microarrays were based on miniaturization of the classic nucleic acid hybridization technique, more specifically, the reverse dot-blot assay (Maskos and Southern, 1993a, 1993b; Saiki *et al.*, 1989). In typical microarray experiments, probes (often referred to as features) are deposited onto solid support matrices (Corneau *et al.*, 2001; Pagotto *et al.*, 2005). These probes can consist of PCR amplicons, full-length cDNAs or short, 25–50 mer oligonucleotides (Ammar *et al.*, 2009). Genomes are then fragmented, labelled and hybridized to the DNA chip. While most microarrays use chemically modified microscope slides as the support for generating high-density probe arrays, alternative supports such as the Luminex platforms are available, where beads carry the features of interest on the outside and a unique dye on the inside, allowing up to 100 assays (i.e. features to be tested) within a single sample (Garaizar *et al.*, 2006). While there are limitations in terms of features vis-à-vis the traditional DNA chips, the Luminex technology can provide reduced time to results with more reproducible results due to the reaction kinetics taking place in the liquid solution.

Genomotyping using DNA microarrays is a common approach in comparative genomics (Borucki *et al.*, 2003; Call *et al.*, 2003; Clewley, 2002; Huyghe *et al.*, 2009; Pettigrew, 2004; Zhang *et al.*, 2004). Early investigations have focused on detection and molecular characterization, sequencing, mutation/polymorphism, and gene discovery or gene expression (Borucki *et al.*, 2004), though other applications are possible (Bekal *et al.*, 2003; Borucki *et al.*, 2005; Call *et al.*, 2001, 2003; Hadd *et al.*, 2005; Pagotto *et al.*, 2008).

A common DNA assay is for the study of gene expression where one sample is subjected to a 'treatment' and compared to an untreated control. Figure 9.4(a)

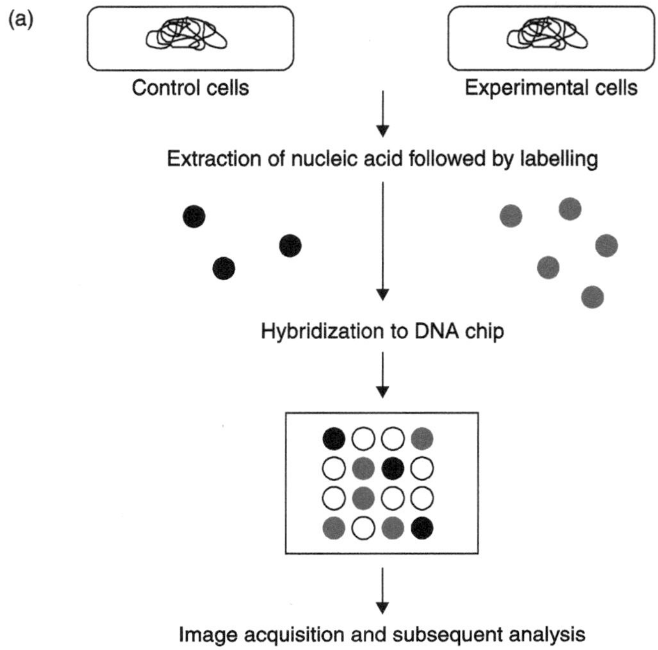

Fig. 9.4 DNA microarray method. (a) The same bacterial cells grown under different conditions have their RNA extracted and labelled with two different dyes (grey and black). The labelled cDNAs are then hybridized to a DNA chip that contains features specific for different genes (lower panel). Bioinformatics-based analyses aid in revealing genes expressed in the control (e.g. black) versus the experimental condition (grey). *Opposite* (b) In another experiment, the DNA could be extracted from two similar bacterial species, labelled in a similar manner, and then hybridized to a DNA chip to reveal differences in genome content. In this example, five genes were present in the clinical strain but absent or significantly different in the environmental strain. Three genes present in the environmental strain were not seen in the clinical strain.

illustrates a generic approach, where the same array is used to analyse two strains that have had their RNAs labelled differentially. As shown in Fig. 9.4(a), experimental cells have five genes that are turned on when compared to the control cells (grey spots); the control cells have three genes that are turned off under the experimental conditions (black spots); and there are genes that do not have a significant change in expression in either conditions and so would have a hybrid colour between the two labelling dyes (shown in the figure as empty spots).

An alternate approach to the use of DNA microarrays would be through the use of a comparative genomics approach. Nucleic acid (i.e. DNA) from two strains being investigated would be labelled with different dyes, hybridized to a DNA chip, and then scored on the presence or absence of the features that were positive (Clewley, 2002). Figure 9.4(b) is a representative schematic of this approach. It is possible to identify genes that are present in clinical strains that have 'evolved' to become more virulent in humans, for example. An attractive feature of this approach

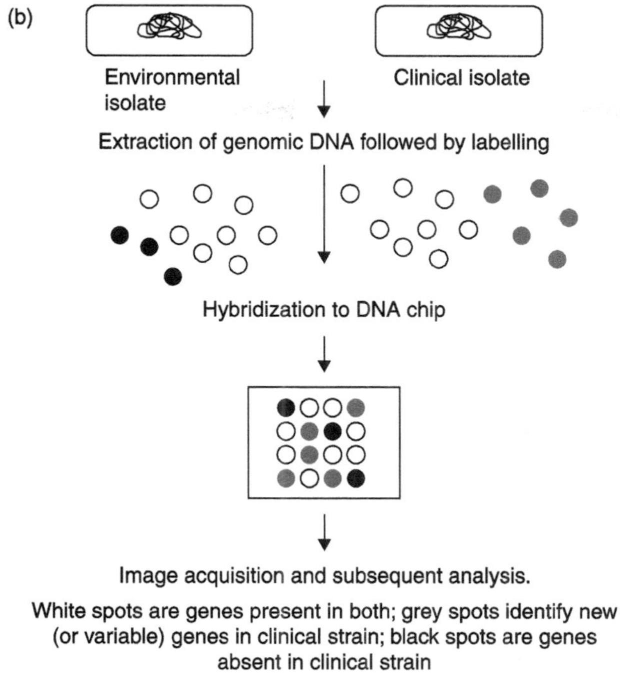

Fig. 9.4 Continued.

to this is that the genomes of the strains being examined do not need to be known. Mixed-genome microarrays have been used to elucidate differences that exist in *L. monocytogenes* (Borucki *et al.*, 2004; Call *et al.*, 2003).

While bioinformatics-based, downstream analyses allow for the querying of thousands of these unique probes in a single experiment, making the use of such genome-wide studies powerful, caution is warranted. It has been noted that errors appear to be compounded exponentially when using a large number of features (i.e. markers) for the analyses of bacterial genomes (Call *et al.*, 2008). An alternative to this might be the VNTR approach, where effects of compounding error were shown to have a lesser impact (Call *et al.*, 2008).

9.3 Protein-based technologies

9.3.1 Gel-based and gel-free proteomics

Either gel-based [1D or 2D sodium dodecyl sulphate-polyacrylamide gel electrophoresis (SDS-PAGE)] or newer gel-free (i.e. MS-based) approaches can be used to generate a picture of the proteome of a given bacterium. Such proteomes can serve as fingerprints for bacterial identification/characterization or can be used to identify proteins whose expression or expression levels are unique to a given subtype (e.g. a protein whose expression is unique to a given *Salmonella* serovar). Biomarkers identified in this manner can then be specifically targeted for typing of bacterial isolates. A subset of the bacterial proteome can also be targeted (e.g. soluble cytoplasmic proteins, outer membrane proteins, etc.). Much like genome-wide approaches for bacterial typing, proteomic approaches to bacterial typing allow the simultaneous monitoring of a number of markers, providing a higher degree of discriminatory power.

Single-dimension SDS-PAGE approaches have been used for bacterial typing since 1990 (Costas *et al.*, 1990). Briefly, cell lysates (or subcellular fractions, e.g. periplasmic proteins) are separated by discontinuous SDS-PAGE based on their molecular mass, the proteins are stained for visualization (e.g. silver staining), and the image is digitized (e.g. using a densitometer) and compared to the profile generated from a reference isolate (or used as a query to probe a database) (Fig. 9.5a). Normalization of the gel images is required to correct for aberrant migration, and the relatedness of isolates is determined by statistical analysis. This approach is still used to distinguish bacteria at the species level (Baele *et al.*, 2008; Benito *et al.*, 2008; Devriese *et al.*, 2002; Liu *et al.*, 2006) but does not appear to have sufficient discriminatory power for further subtyping of isolates.

Two-dimensional electrophoresis separates proteins based on both their charge (first dimension) and molecular mass (second dimension), allowing greater coverage of the proteome and distinction of protein isoforms (Fig. 9.5b). Labelling of cell lysates or subcellular fractions with distinct fluorescent dyes, for example, allows two or more samples to be run on the same gel, increasing the ease of sample comparison (2D-difference gel electrophoresis or 2D-DIGE). *Haemophilus influenzae* isolates were shown to harbour many differences in their 2D gel electrophoresis profiles, many due to different charge states of proteins (Cash, 2000). A 2D gel approach was also used to identify serovar-specific protein isoforms or proteins with distinct expression levels between serovars of *S. enterica* subsp. *enterica* (Encheva *et al.*, 2007) and was shown to provide additional discrimination of *E. coli* O157:H7 isolates relative to PFGE analysis (Yokoyama *et al.*, 2001). Despite improved protein resolution in 2D gels, a large proportion of the bacterial proteome remains unresolved by this approach, for instance hydrophobic integral membrane proteins, low-copy number protein and proteins at the extremes of the pI and/or molecular mass scale. Furthermore, the execution of 2D gel electrophoresis and downstream data analysis remains laborious, despite the availability of pre-cast gels, labelling technologies for 2D-DIGE and improved scanners/data analysis software (Cash, 2009). Methods for sample preparation and gel preparation/running need to be standardized, to allow comparisons to be made between laboratories and with proteomic databases. For a recent review describing the use of 1D and 2D gel-based proteomics for bacterial typing, consult the study by Cash (2009).

More recently, gel-free proteomic approaches have emerged [see Gevaert *et al.* (2007) for a review]. These typically involve enzymatic digestion (e.g. using trypsin) of the protein complement of an isolate, followed by chromatographic separation [liquid chromatography (LC)] and peptide identification by MS. In order to decrease sample complexity and improve proteome coverage, additional

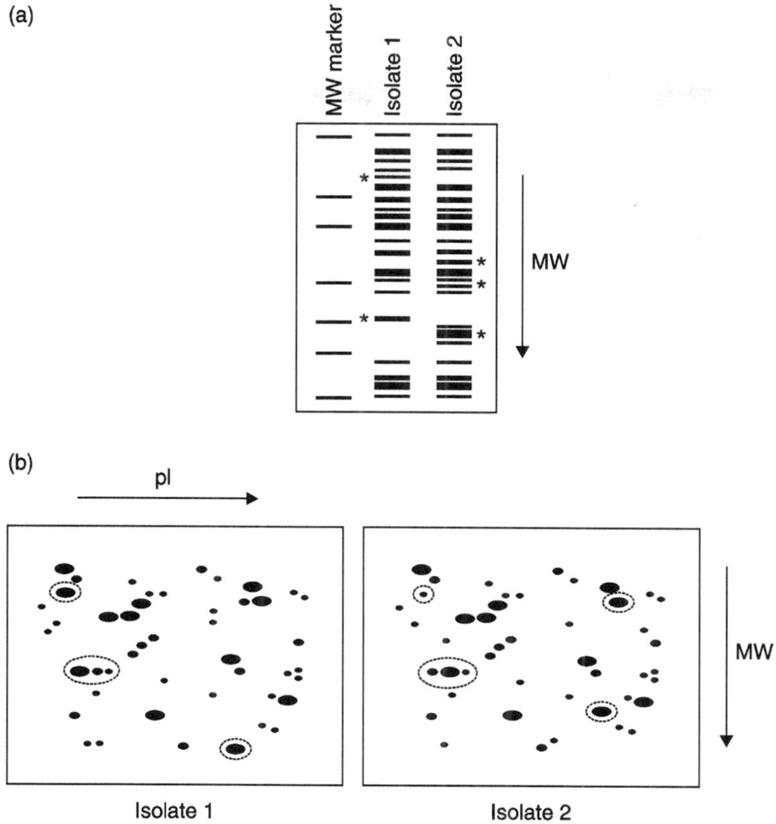

Fig. 9.5 One- and two-dimensional protein electrophoresis. (a) Schematic diagram illustrating a 1D SDS-PAGE gel, where proteins from a cell lysate or subcellular fraction are separated based on molecular mass. In this case, isolates 1 and 2 are readily distinguished from each other based on the presence or absence of specific protein bands (indicated by asterisks). (b) Schematic diagram representing a 2D SDS-PAGE analysis, where proteins are separated based on charge (pI) and molecular mass. The two isolates pictured here can be distinguished from each other based on the presence/absence of protein spots as well as differences in the intensity of protein spots (circled). MW = molecular weight; pI = isoelectric point.

steps are introduced prior to LC-MS analysis, such as tagging and isolating a subset of peptides [e.g. isotope-coded affinity tag (ICAT) tagging of Cys-containing peptides]. In order to ensure optimal proteome coverage, several upstream strategies should be pursued for each sample. While it is clear that gel-free approaches show improved protein coverage relative to gel-based technologies (Schmidt *et al.*, 2004; Wolff *et al.*, 2007), these methods can suffer from bias (e.g. ICAT-LC/MS shows bias for high molecular mass proteins) (Schmidt *et al.*, 2004). Thus, gel-based and gel-free technologies may provide complementary data and enable the generation of a more complete proteome map (see, e.g. Wolff *et al.*, 2007), which could improve the discriminatory capacity of proteomics for bacterial subtyping.

9.3.2 Matrix-assisted laser desorption-ionization time-of-flight (MALDI–TOF) mass spectrometry (MS)

Since 1996 (Claydon *et al.*, 1996; Holland *et al.*, 1996), whole-cell matrix-assisted laser desorption-ionization time-of-flight (MALDI–TOF) MS has been widely used to identify and characterize bacteria. Briefly, the method consists of spotting bacterial cells or cell extracts on a plate, extracting peptides/proteins *in situ* (if whole cells are spotted), adding suitable matrix and subjecting the spot to MALDI–TOF MS (Fig. 9.6). Spectra obtained reveal high- and low-molecular weight intact protein ions [typically 4000–13 000 mass units, see Dieckmann *et al.* (2008) and references therein], largely corresponding to high-abundance, basic, cytosolic proteins such as ribosomal proteins and nucleic acid-binding proteins, which form a bacterial 'fingerprint'. These fingerprints can be compared to reference libraries of spectra to identify organisms, or peak lists can be compared to available protein (or translated genomic) databases to identify likely matches. The latter can be complicated by post-translational modifications, which are often not accounted for in genomic databases. Alternatively, *in situ* enzymatic digests or peptide sequencing can be used to identify peaks of interest. In some cases, attempts are made to identify marker ions specific for a given genus or species of bacteria. These marker ions can allow differentiation between highly related organisms, whose overall fingerprints may be largely indistinguishable.

Intact-cell (IC) MALDI–TOF MS has been successfully used to discriminate foodborne and clinically relevant bacteria at the genus and species levels [e.g. *Neisseria* (Ilina *et al.*, 2009), *Escherichia* (Mazzeo *et al.*, 2006), *Campylobacter* (Kolinska *et al.*, 2008; Mandrell *et al.*, 2005; Winkler *et al.*, 1999), *Salmonella* (Mazzeo *et al.*, 2006), *Yersinia* (Mazzeo *et al.*, 2006), *Helicobacter* (Winkler *et al.*, 1999) and *Listeria* (Mazzeo *et al.*, 2006)]. While intraspecies-level discrimination has been possible for some organisms [e.g. biomarkers specific for subspecies (Dieckmann *et al.*, 2008) or serovars (Mazzeo *et al.*, 2006) of *Salmonella*, for lineages of *L. monocytogenes* (Barbuddhe *et al.*, 2008), O157 versus non-O157 *E. coli* (Mazzeo *et al.*, 2006), intraspecies mass variation of a *Campylobacter jejuni*-specific biomarker (Mandrell *et al.*, 2005)], subspecies-level differentiation remains a challenge, in part because MS fingerprints within a species are very alike. To achieve subspecies-level discrimination of *Salmonella*, for instance, protocols had to be optimized to generate spectra with a much higher number of reproducible protein peaks (> 300) than generally acquired (usually ~30). Because of this limitation, the utility of MALDI–TOF MS for high-level discrimination, such as that required for source tracking during an outbreak, remains to be seen.

A good biomarker ion should be consistently present at intensities well above background. The stability of candidate biomarker ions needs to be assessed, under various growth conditions, as media composition can affect peak presence and

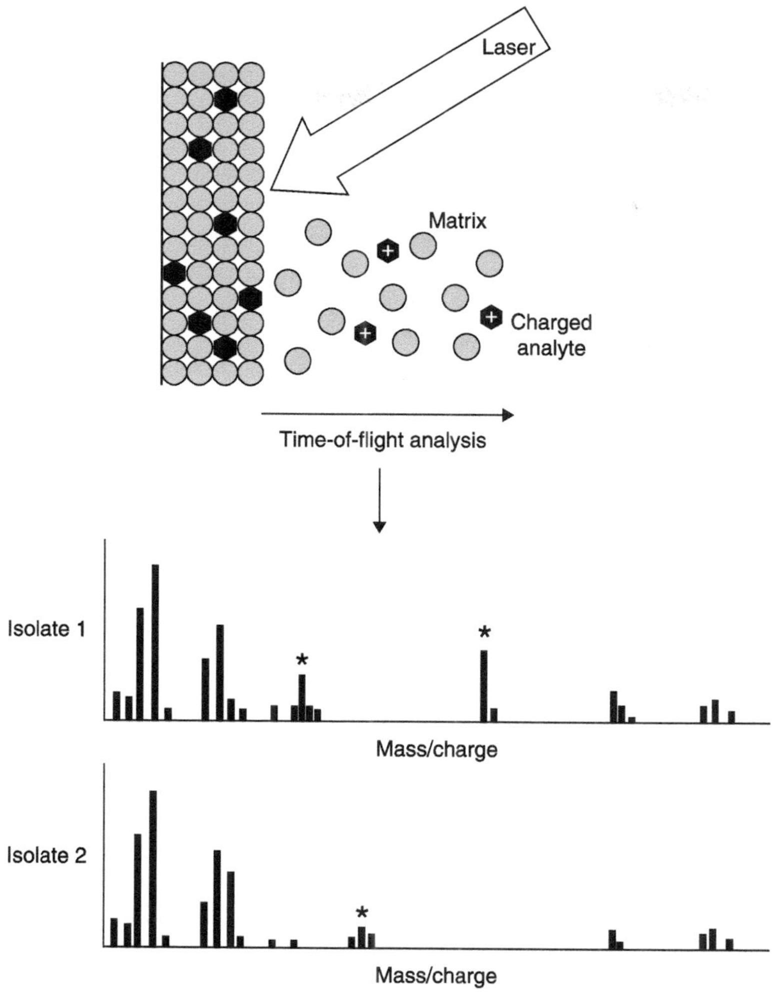

Fig. 9.6 Matrix-assisted laser desorption-ionization time-of-flight mass spectrometry (MALDI–TOF MS). The sample to be analysed is mixed with matrix on a plate and ionized using a laser. The resulting charged analytes are resolved based on their mass-to-charge ratios, to yield spectra or bacterial 'fingerprints'. In this case, isolates 1 and 2 can be distinguished based on the presence/absence of select peaks (indicated by asterisks).

intensity (Mazzeo *et al.*, 2006; Russell, 2009) and variability between replicates (Saenz *et al.*, 1999; Walker *et al.*, 2002). The issue of intra- and inter-laboratory variability also needs to be addressed. The inclusion of standardized growth and sample preparation methodologies may go a long way towards ensuring reproducible spectra across different laboratories, as these parameters have been shown to influence the quality and reproducibility of spectra (Gantt *et al.*, 1999; Wang *et al.*, 1998; Williams *et al.*, 2003). In addition, strong bioinformatics tools such as complete libraries of reference spectra, algorithms for protein database searches, software to affect spectra comparisons and analyses are a necessity to improve automation and limit bias and error that could be caused by spectrum interpretation. While some tools are already available (Bright *et al.*, 2002; Jarman *et al.*, 2000), further advances in this field are likely to vastly improve the discriminatory capability of this technique.

Ideally, high molecular weight protein ions should be well represented in the spectrum, as the likelihood of amino acid differences between strains increases with increasing protein size. Optimization of sample preparation (e.g. media used for bacterial growth, cell concentration and protein extraction protocols), matrix and solvent composition becomes crucial when seeking to obtain reproducible spectra capable of subspecies-level discrimination.

Advantages of this method include the speed (results in minutes) and accuracy of data acquisition, the small amount of material required (loopful of a bacterial colony), its tolerance to contaminants (e.g. media components) and the possibility of analysing mixed cultures and of automating data acquisition and analysis (e.g. use of software such as SARAMIS to identify biomarker peaks). The availability of databases, such as the Rapid Microorganism Identification Database [http://www.rmidb.org; (Pineda *et al.*, 2003)], will greatly aid in the routine identification of microbes from mass spectra. Clear disadvantages include the initial cost of the instrument (though this is comparable to the cost of a DNA sequencer) and the currently limited successes of this method for subspecies-level differentiation.

Alternative approaches and technologies include the use of IC MALDI–TOF MS, where cells are added directly to a MALDI plate, without prior or subsequent extraction steps. IC-MS analysis yields spectra containing peaks corresponding to surface components of the bacterial cell, and thus is likely to yield different candidate biomarker ions than the conventional approach (Bright *et al.*, 2002; Walker *et al.*, 2002). In addition, these surface-derived markers may correlate with phenotypic characteristics such as serotype and phage type (Edwards-Jones *et al.*, 2000). The ability to generate MALDI spectra from single cells could also revolutionize bacterial identification by MS, for instance by allowing the identification of multiple components of a mixed culture, possibly precluding the need to start with an isolated bacterial colony. This technology has already been applied to the detection of *Bacillus subtilis* spores (Stowers *et al.*, 2000; van Wuijckhuijse *et al.*, 2005), and cells of *E. coli* (van Wuijckhuijse *et al.*, 2005) and *Erwinia herbicola* (Kleefsman *et al.*, 2007), though sensitivity at higher mass values remains a hurdle for the application of this technology to bacterial typing (Russell, 2009).

9.3.3 Electrospray ionization (ESI) mass spectrometry (MS)

Electrospray ionization-MS (ESI–MS), with or without prior analyte separation by LC (Fig. 9.7), has also been applied to the identification and characterization of bacteria. ESI–MS generates species with multiple charges, which increases mass accuracy and allows better resolution of high mass proteins. ESI–MS data are generally more reproducible and quantitative data (peak intensity) are more

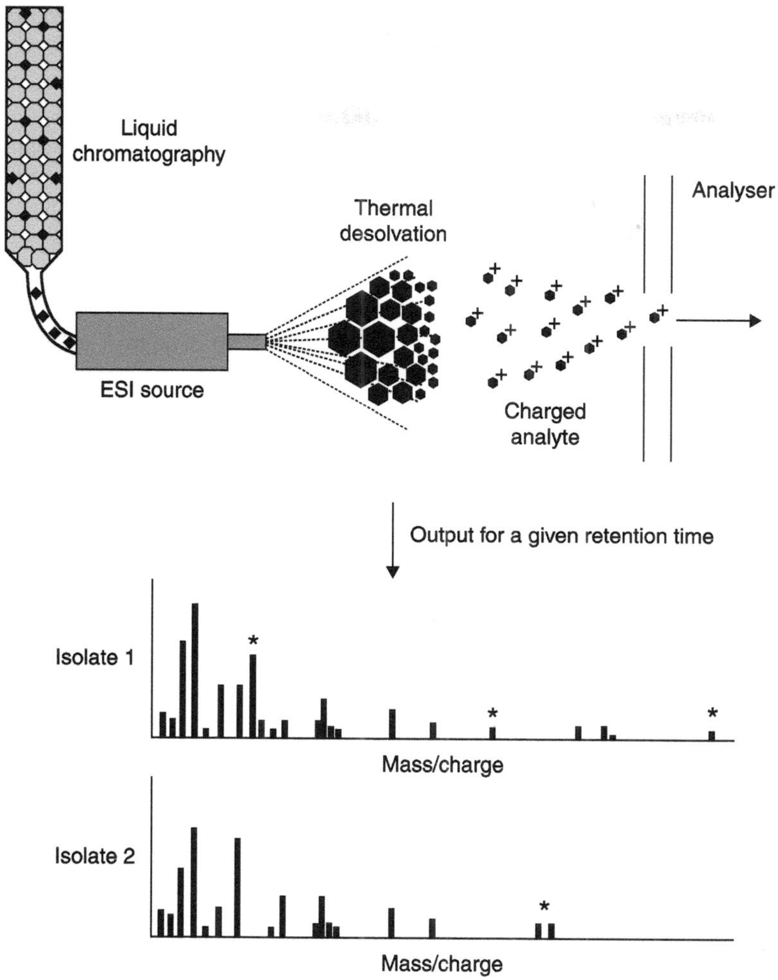

Fig. 9.7 Liquid chromatography–electrospray ionization mass spectrometry (LC–ESI–MS). The sample to be analysed is fractionated by liquid chromatography, and the resulting fractions are dispersed by the electrospray source into a fine aerosol. Resulting charged analytes are resolved based on their mass-to-charge ratios, yielding spectra or bacterial 'fingerprints'. In this case, isolates 1 and 2 can be readily distinguished based on the presence/absence of specific peaks (indicated by asterisks).

reliable, not subject to uneven protein distribution seen in MALDI spots or subject to differences in the number and position of MALDI laser hits. The coupling of protein/peptide separation by LC with ESI–MS analysis has one important advantage over the routine MALDI–TOF MS approach. For instance, proteins with identical masses may exhibit different retention times in the LC phase, allowing them to be distinguished. This enhanced separation results in a greater number of protein peaks, which increases the likelihood of detecting suitable biomarkers. Disadvantages of ESI–MS include low tolerance to contaminants in the sample and increased complexity of the resulting spectra due to the presence of multiple charge states (though spectrum deconvolution yields data of comparable simplicity to MALDI–TOF data). With LC–ESI–MS, sample processing and data analysis are more time-consuming (hours as opposed to minutes), and consideration needs to be given to shifting retention times of biomarker peaks due, for instance, to column wear or subtle differences in solvent composition, which could impact reproducibility and inter-laboratory variability. A known amount of a standard of known retention time could be included to correct for such shifts and to serve as a standard for determination of relative peak intensity (Everley *et al.*, 2008).

Much like MALDI–TOF MS analysis of whole bacteria, whole-cell ESI–MS has been applied to bacterial characterization. This procedure eliminates the time constraints related to extensive sample preparation and chromatography prior to MS analysis, using washed bacterial cells resuspended in a suitable solvent system as sample. Whole-cell ESI–MS spectra could be used to distinguish between *E. coli* and *Bacillus cereus* cells, and showed further discrimination between isolates of a single species (Goodacre *et al.*, 1999). This approach was also able to distinguish *Bacillus* species, as well as to provide further discrimination between strains of *B. subtilis* (Vaidyanathan *et al.*, 2001). The latter study revealed that spectra are influenced by changes in the cone potential in the electrospray ion source, and that the application of different voltages could enhance the discriminatory ability of this method. Finally, prior knowledge of differences between *Clostridium botulinum* flagellin proteins fuelled the application of a top-down ESI–MS approach to identify marker ions with potential for identifying and characterizing strains of *C. botulinum* (Twine *et al.*, 2008). Purified flagellin protein was analysed by ESI–MS, and a *C. botulinum* flagellin-specific marker ion was identified, as well as a second marker ion that was specific to certain strains of *C. botulinum* (Twine *et al.*, 2008).

The strength of a coupled LC–ESI–MS approach was illustrated in a study by Everley *et al.* (2008), where LC–ESI–QTOF–MS was used to identify biomarkers for *E. coli* and *Shigella* species. Cell lysates (from plate-grown cells) were separated by reversed-phase chromatography prior to ionization and detection by ESI–QTOF–MS (Everley *et al.*, 2008). This approach allowed discrimination between non-pathogenic *E. coli*, O157 EHEC and non-O157 EHEC, as well as between *Shigella flexneri* and *S. sonnei* (Everley *et al.*, 2008).

9.3.4 Serotyping

Components at the bacterial surface are in direct interaction with the environment, and as such are especially subject to selective pressure, which often leads to antigenic variability. For instance, structures such as lipopolysaccharides (LPS) and capsular polysaccharides (CPS) can be made up of a variety of sugars and linkages, and be decorated with non-stoichiometric modifications. In *E. coli* alone, greater that 170 distinct O and 80 distinct K antigens are recognized

(Orskov *et al.*, 1977). Furthermore, the expression of structures such as flagella and CPS can be regulated by environmental cues and/or phase variability, which can lead to alterations in surface composition and result in immune avoidance in the host or enhanced survival in harsh environments.

Serotyping schemes based on LPS O antigens (O typing), CPS antigens (K typing) and flagellar antigens (H typing) have long been used to distinguish bacteria belonging to a given species but harbouring variations in surface components. Serotyping has been applied to many enteric pathogens, including but not limited to *E. coli, S. enterica* and *L. monocytogenes*. Assays typically involve mixing bacterial cells with antiserum (on glass slides or in test tubes) and observing for agglutination caused by antibody-mediated cross-links formed between bacterial cells. For H typing, it can be necessary to passage the bacterial strain several times to obtain a flagellated isolate. For *S. enterica* serovars, which express one of the two distinct flagellar antigens, complete serotyping requires selection for flagellar phase variants. While these assays are relatively simple to perform, their interpretation can be subjective, and it can be cost-prohibitory to maintain a complete panel of antisera for serotyping and time-consuming to obtain complete antigenic formulae, particularly in instances where multiple passages of the bacterial strain are required. Furthermore, in instances such as foodborne disease outbreaks, it is often necessary to discriminate bacterial strains beyond the level of serotype.

9.3.5 Protein microarrays

Similar to DNA-based microarrays, protein-based microarrays can be used to detect and characterize bacteria. Rather than oligonucleotide probes, purified proteins (typically antibodies) are spotted onto a solid support, where they act to trap molecules (e.g. proteins, carbohydrates, haptens) from a sample. Trapped molecules are typically detected and quantified using fluorescence, from either a labelled captured antigen or a secondary detection molecule (Fig. 9.8).

This approach has been used to detect and characterize bacteria from a variety of sample matrices, including food [see Wingren and Borrebaeck (2009) for a recent review]. Antibody arrays could allow for the detection of many bacterial species and/or serotypes simultaneously, by including species- and/or serotype-specific antibodies on the array [e.g. detection of *E. coli* O157:H7 and *Salmonella typhimurium* from culture-enriched ground beef filtrate (Gehring *et al.*, 2008)]. One application of this technology is in developing alternatives to traditional bacterial serotyping, which can be time-consuming, labour-intensive and costly. Antibodies against somatic (O) and/or flagellar (H) antigens are used to generate arrays that detect whole bacterial cells and allow determination of serotype. This approach has been used to develop an O serotyping array for 17 O-types of *E. coli* (Anjum *et al.*, 2006) and an O and H serotyping array suitable for the characterization of 20 common *Salmonella* serovars (Cai *et al.*, 2005). Considerations include the type (monoclonal, polyclonal, fragments), availability, purity and concentration of the antibodies printed on the array, as well as their

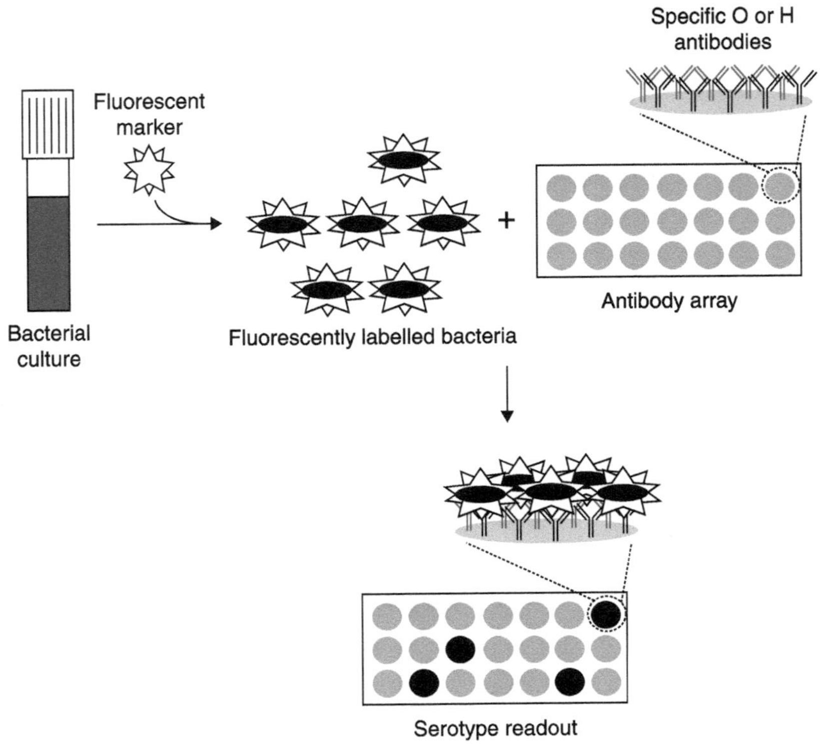

Fig. 9.8 Antibody microarray. Bacteria within a sample are labelled with a cell-permeable fluorescent marker and applied to a microarray slide on which antibodies against specific O- and/or H-antigens have been spotted. Interactions between the labelled bacteria and specific antibodies are visualized by fluorescence and yield a given pattern (which translates, in this example, into the organism's serotype).

performance in an array-based assay [see Wingren and Borrebaeck (2009)]. These should ensure high sensitivity but minimal cross-reactivity. Stability and shelf-life of the arrays must also be ascertained. The use of a miniaturized system for the arrays, such as the ArrayTube system (CLONDIAG) (Anjum *et al.*, 2006), or a 96-well format would allow for simultaneous serotyping of many isolates. Another advantage of this method is that it does not require prior knowledge of DNA sequence for every potential isolate and requires only small amounts of the sample. In addition to the use of an antibody array for the typing of isolated, intact bacteria, such arrays have also been used for the detection of bacterial toxins from complex samples, including food (Ligler *et al.*, 2003; Shriver-Lake *et al.*, 2003). Analysis of these arrays has to date been largely adapted from DNA microarrays. There is currently no standard method for antibody array normalization, though procedures based on the amount of antibody spotted or a reference protein added at a known concentration to the mix prior to labelling, among others, have been reported (Wingren and Borrebaeck, 2009).

9.3.6 Surface plasmon resonance (SPR) biosensors

The use of surface plasmon resonance (SPR) biosensors allows detection of analytes (e.g. bacteria, toxins) from a sample, without the use of labelled antibodies for detection. Briefly, recognition molecules (e.g. antibodies against *E. coli* O157:H7) are immobilized onto a metal sensor surface (e.g. gold-coated chip), the sample is flowed across the chip, and the binding of analyte (e.g. *E. coli* O157:H7) to the recognition molecule causes a change in the refractive index near the chip surface, which is detected by an optical reader (detects resulting changes in the characteristics of a plasmon-coupled light wave) (Fig. 9.9). The magnitude

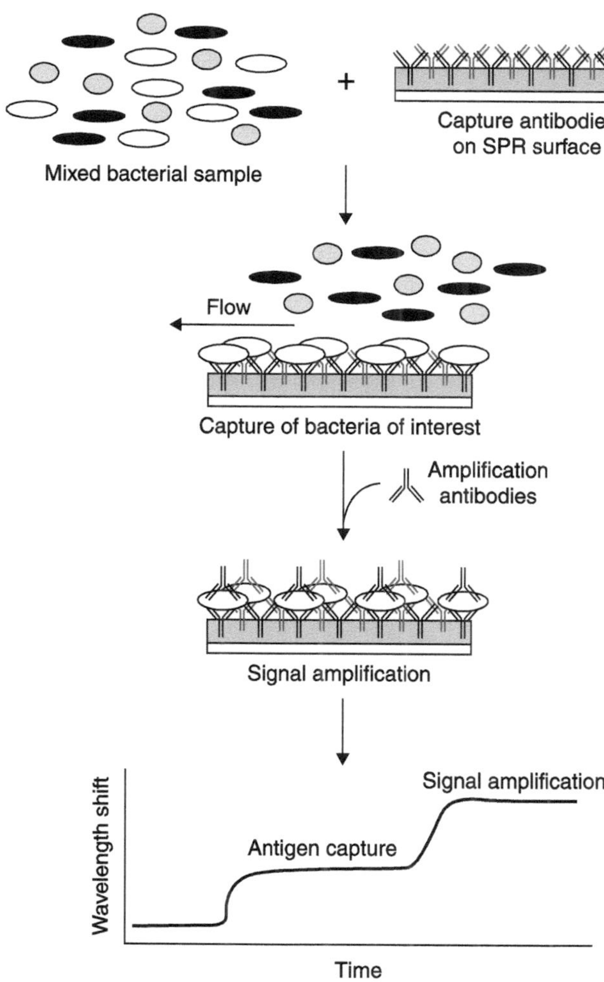

Fig. 9.9 Surface plasmon resonance (SPR) biosensor. Capture antibodies are immobilized on an SPR chip, and bacteria of interest interact with the capture antibodies as the sample is applied across the chip surface. Secondary antibodies specific to the target bacteria are applied to the chip in order to amplify the signal and increase specificity. The interaction is measured by changes in the refractive index near the surface of the chip caused by analyte binding.

of the refractive index change is proportional to the concentration of bound analyte [for a recent review describing SPR technology, see (Piliarik, Parova, *et al.*, 2009; Piliarik, Vaisocherova, *et al.*, 2009)]. This approach has been used to detect a given pathogen (*E. coli* O157:H7, *S.* Typhimurium, *Legionella pneumophila* or *Yersinia enterocolitica*) from a pooled sample containing all four pathogens (Oh *et al.*, 2005). Alternatively, a sandwich approach can be used to improve the specificity and the limit of detection. In this approach, the sensor surface is coated with a primary recognition molecule (e.g. antibody against *E. coli* O157:H7), the analyte is captured from the sample, and a secondary recognition molecule (e.g. different antibody against *E. coli* O157:H7) is applied to the chip and binds to the bound analyte. This approach compensates for non-specific binding of bacteria to sensor surfaces and has been used successfully for the detection of *E. coli* O157:H7, *S. typhimurium, L. monocytogenes, C. jejuni* (Taylor *et al.*, 2006) and various *Salmonella* serovars (Bokken *et al.*, 2003).

For analytes too small to generate a substantial change in refractive index near the metal's surface, a competitive assay can be more useful. In this assay, the sample (e.g. food rinse containing bacterial toxin) and a known amount of a recognition element (e.g. antibody against the toxin) are allowed to interact in solution. The mixture is then applied to the chip, where analyte (e.g. toxin) is immobilized onto the metal surface. Only those recognition molecules that are not bound to analyte from the sample are available to interact with the sensor, providing an indirect measurement of the amount of analyte present in the sample.

Other biosensors rely on the use of different detection technologies to monitor analyte capture (e.g. alternate optical, evanescent-wave, electrochemical or piezoelectric-based sensors). While many of the current biosensors rely on antibody-based capture, emerging recognition elements such as protein nucleic acids (PNAs) are likely to find their way into these devices (Patel, 2006).

While this technology has been used to detect specific bacteria (e.g. *E. coli* O157:H7) from a mixture, its usefulness is limited by the fact that only a single organism can be detected at a time. This technology could however be coupled with a multichannel device, with each channel containing a distinct capture antibody, thus allowing simultaneous detection of multiple bacteria. The main limitation of this methodology is the availability of capture molecules (e.g. antibodies) specific to a given species, subspecies, serovar, etc. of bacteria. For this reason, it remains difficult to imagine how this approach could allow for a level of discrimination beyond that of the serotype.

9.4 Other emerging technologies

9.4.1 Carbohydrate- and lectin-based microarrays

Carbohydrate-based arrays have recently emerged as a tool for bacterial detection. Many bacteria rely on binding to carbohydrate molecules within the host as a means of initial attachment, subsequent colonization and/or invasion. Differences in carbohydrate binding affinities may be responsible for different host and/or tissue specificities, and it may be possible to discriminate between bacterial pathogens based on their affinities for a range of carbohydrates.

Much like with DNA or protein arrays, carbohydrate arrays are solid supports onto which glycans are covalently affixed via linkers in discrete locations. Bacteria within a sample are labelled (typically with a cell-permeable fluorescent dye) and applied to the array, unbound cells are washed away, and bound cells are detected (typically via fluorescence). This type of array allowed differences to be observed in mannose-binding affinity between a mutant strain of *E. coli* and its parent (Disney and Seeberger, 2004).

While not yet developed to such a level, an extensive carbohydrate microarray could allow the generation of a glycan binding fingerprint for a given bacterial isolate, which could then be compared to a library of glycan binding data for known pathogens or to other isolates in order to determine relatedness. It remains to be seen whether sufficient differences exist in carbohydrate affinities within a given bacterial species, subspecies or serovar to make this a viable technology for bacterial typing and/or subtyping.

While carbohydrate microarrays yield a fingerprint based on bacterial binding to carbohydrates, such as those found on host cells, lectin-based microarrays yield a fingerprint based on the carbohydrates expressed on the bacterial cell surface. As with glycan arrays, lectins (carbohydrate-binding proteins) are immobilized onto a solid support, and labelled bacteria are applied to the array and subsequently detected. This approach enabled the discrimination of two derivatives of *E. coli* K-12 (JM101 and HB191) based on their lectin fingerprint (Hsu and Mahal, 2006; Hsu *et al.*, 2006). The main limitation of this method is the availability of lectins for inclusion in the array, as lectins that recognize unique bacterial sugars are lacking.

9.5 Conclusions and future trends

While other chapters in this book have undoubtedly covered many of the challenges and considerations related to various aspects of tracing pathogens in the food chain, those most relevant to the information provided in this chapter are presented below.

Sample preparation is often ignored in many aspects of method development. In this chapter, it is assumed that a purified, isolated bacterial colony serves as the starting material. Most often, sample preparation will require steps to address intrinsic characteristics of the organism(s) under study, such as conditions required for efficient cell lysis, different properties seen in Gram-positive versus Gram-negative bacteria, initial setup costs, extraction of genomic DNA, conditions for extraction of proteins, etc. However, sample preparation becomes a real problem for the characterization of 'troublesome pathogens'. These include viruses and parasites that cannot be cultured, as well as viable but non-culturable bacterial organisms (VBNC). Here, the challenges lie in the development of a typing scheme that can be done on a small, potentially representative component of the organism. This will most often involve amplification of specifically targeted nucleic acids.

One successful approach that addresses some of these challenges was described by Pagotto *et al.* (2008), where a simultaneous detection and genotyping platform was designed to overcome the fact that noroviruses cannot be cultured *in vitro*. Briefly, a fragment consisting of 917 bp was amplified from different norovirus positive samples using reverse-transcriptase PCR (RT-PCR) that encompassed major regions used for genotyping and detection. This fragment was hybridized to a DNA microarray containing features designed to differentiate human norovirus genogroups GI and GII.

Many available technologies, whether nucleic acid- or protein-based, require prior knowledge of the specific agents so that specific targets (i.e. genetic or antibody probes) can be used. As such, it is expected that variations on a theme and/or combinations will be one way to overcome the remaining challenges. For example, antibody-derived structures such as scFvs, Fabs, bi- and multi-functional derivatives, aptamers and peptide nucleic acids may help achieve greater discrimination of 'difficult' pathogens. The technologies employed in other fields, such as chemistry, biochemistry and physics, may be important sources to draw from. One example is the use of Fourier transform infrared spectroscopy (FTIR) in discriminating strains of *C. botulinum* (Kirkwood *et al.*, 2006; Leclair *et al.*, 2006).

9.6 References

ACHTMAN M (2008), 'Evolution, population structure, and phylogeography of genetically monomorphic bacterial pathogens', *Annu Rev Microbiol*, **62**, 53–70.

ALTSCHUL S F, MADDEN T L, SCHAFFER A A, ZHANG J, ZHANG Z, MILLER W, ET AL. (1997), 'Gapped BLAST and PSI-BLAST: a new generation of protein database search programs', *Nucleic Acids Res*, **25**, 3389–402.

AMMAR R, SMITH A M, HEISLER L E, GIAEVER G and NISLOW C (2009), 'A comparative analysis of DNA barcode microarray feature size', *BMC Genomics*, **10**, 471.

ANJUM M F, TUCKER J D, SPRIGINGS K A, WOODWARD M J and EHRICHT R (2006), 'Use of miniaturized protein arrays for *Escherichia coli* O serotyping', *Clin Vaccine Immunol*, **13**, 561–7.

BAELE M, DECOSTERE A, VANDAMME P, CEELEN L, HELLEMANS A, MAST J, ET AL. (2008), 'Isolation and characterization of *Helicobacter suis* sp. *nov.* from pig stomachs', *Int J Syst Evol Microbiol*, **58**, 1350–8.

BARBUDDHE S B, MAIER T, SCHWARZ G, KOSTRZEWA M, HOF H, DOMANN E, ET AL. (2008), 'Rapid identification and typing of *Listeria* species by matrix-assisted laser desorption ionization-time of flight mass spectrometry', *Appl Environ Microbiol*, **74**, 5402–7.

BEKAL S, BROUSSEAU R, MASSON L, PREFONTAINE G, FAIRBROTHER J and HAREL J (2003), 'Rapid identification of *Escherichia coli* pathotypes by virulence gene detection with DNA microarrays', *J Clin Microbiol*, **41**, 2113–25.

BENITO M J, SERRADILLA M J, MARTIN A, ARANDA E, HERNANDEZ A and CORDOBA M G (2008), 'Differentiation of Staphylococci from Iberian dry fermented sausages by protein fingerprinting', *Food Microbiol*, **25**, 676–82.

BENSCH S and AKESSON M (2005), 'Ten years of AFLP in ecology and evolution: why so few animals?', *Mol Ecol*, **14**, 2899–914.

BENSON G (1999), 'Tandem repeats finder: a program to analyze DNA sequences', *Nucleic Acids Res*, **27**, 573–80.

BOKKEN G C, CORBEE R J, VAN KNAPEN F and BERGWERFF A A (2003), 'Immunochemical detection of *Salmonella* group B, D and E using an optical surface plasmon resonance biosensor', *FEMS Microbiol Lett*, **222**, 75–82.

BORUCKI M K, GAY C C, REYNOLDS J, MCELWAIN K L, KIM S H, CALL D R, ET AL. (2005), 'Genetic diversity of *Listeria monocytogenes* strains from a high-prevalence dairy farm', *Appl Environ Microbiol*, **71**, 5893–9.

BORUCKI M K, KIM S H, CALL D R, SMOLE S C and PAGOTTO F (2004), 'Selective discrimination of *Listeria monocytogenes* epidemic strains by a mixed-genome DNA microarray compared to discrimination by pulsed-field gel electrophoresis, ribotyping, and multilocus sequence typing', *J Clin Microbiol*, **42**, 5270–6.

BORUCKI M K, KRUG M J, MURAOKA W T and CALL D R (2003), 'Discrimination among *Listeria monocytogenes* isolates using a mixed genome DNA microarray', *Vet Microbiol*, **92**, 351–62.

BREYNE P, DREESEN R, CANNOOT B, ROMBAUT D, VANDEPOELE K, ROMBAUTS S, ET AL. (2003), 'Quantitative cDNA-AFLP analysis for genome-wide expression studies', *Mol Genet Genomics*, **269**, 173–9.

BRIGHT J J, CLAYDON M A, SOUFIAN M and GORDON D B (2002), 'Rapid typing of bacteria using matrix-assisted laser desorption ionisation time-of-flight mass spectrometry and pattern recognition software', *J Microbiol Met*, **48**, 127–38.

CAI H Y, LU L, MUCKLE C A, PRESCOTT J F and CHEN S (2005), 'Development of a novel protein microarray method for serotyping *Salmonella enterica* strains', *J Clin Microbiol*, **43**, 3427–30.

CALL D R, BORUCKI M K and BESSER T E (2003), 'Mixed-genome microarrays reveal multiple serotype and lineage-specific differences among strains of *Listeria monocytogenes*', *J Clin Microbiol*, **41**, 632–9.

CALL D R, BROCKMAN F J and CHANDLER D P (2001), 'Detecting and genotyping *Escherichia coli* O157:H7 using multiplexed PCR and nucleic acid microarrays', *Int J Food Microbiol*, **67**, 71–80.

CALL D R, ORFE L, DAVIS M A, LAFRENTZ S and KANG M S (2008), 'Impact of compounding error on strategies for subtyping pathogenic bacteria', *Foodborne Pathog Dis*, **5**, 505–16.

CASH P (2000), 'Proteomics in medical microbiology', *Electrophoresis*, **21**, 1187–201.

CASH P (2009), 'Proteomics in the study of the molecular taxonomy and epidemiology of bacterial pathogens', *Electrophoresis*, **30**(Suppl 1), S133–41.

CEBULA T A, JACKSON S A, BROWN E W, GOSWAMI B and LECLERC J E (2005), 'Chips and SNPs, bugs and thugs: a molecular sleuthing perspective', *J Food Prot*, **68**, 1271–84.

CHIOU C S (2010), 'Multilocus variable-number tandem repeat analysis as a molecular tool for subtyping and phylogenetic analysis of bacterial pathogens', *Expert Rev Mol Diagn*, **10**, 5–7.

CLARKE S C (2005), 'Pyrosequencing: nucleotide sequencing technology with bacterial genotyping applications', *Expert Rev Mol Diagn*, **5**, 947–53.

CLAYDON M A, DAVEY S N, EDWARDS-JONES V and GORDON D B (1996), 'The rapid identification of intact microorganisms using mass spectrometry', *Nat Biotechnol*, **14**, 1584–6.

CLEWLEY J P (2002), 'Genomotyping: comparative bacterial genomics using arrays', *Commun Dis Public Health*, **5**, 258–9.

CORNEAU N, PAGOTTO F and FARBER J M (2001), 'DNA microarrays: a high-throughput technology', *Canadian Meat Science Association News*, July 3–7.

COSTAS M, HOLMES B and SLOSS L L (1990), 'Comparison of SDS-PAGE protein patterns with other typing methods for investigating the epidemiology of "*Klebsiella aerogenes*" ', *Epidemiol Infect*, **104**, 455–65.

DEVRIESE L A, VANCANNEYT M, DESCHEEMAEKER P, BAELE M, VAN LANDUYT H W, GORDTS B, ET AL. (2002), 'Differentiation and identification of *Enterococcus durans, E. hirae* and *E. villorum*', *J Applied Microbiol*, **92**, 821–7.

DIECKMANN R, HELMUTH R, ERHARD M and MALORNY B (2008), 'Rapid classification and identification of salmonellae at the species and subspecies levels by whole-cell matrix-assisted laser desorption ionization-time of flight mass spectrometry', *Appl Environ Microbiol*, **74**, 7767–78.

DISNEY M D and SEEBERGER P H (2004), 'The use of carbohydrate microarrays to study carbohydrate-cell interactions and to detect pathogens', *Chem Biol*, **11**, 1701–7.

DOUMITH M, JACQUET C, GOULET V, OGGIONI C, VAN LOOCK F, BUCHRIESER C, ET AL. (2006), 'Use of DNA arrays for the analysis of outbreak-related strains of *Listeria monocytogenes*', *Int J Med Microbiol*, **296**, 559–62.

DUCEY T F, PAGE B, USGAARD T, BORUCKI M K, PUPEDIS K and WARD T J (2007), 'A single-nucleotide-polymorphism-based multilocus genotyping assay for subtyping lineage I isolates of *Listeria monocytogenes*', *Appl Environ Microbiol*, **73**, 133–47.

EDWARDS-JONES V, CLAYDON M A, EVASON D J, WALKER J, FOX A J and GORDON D B (2000), 'Rapid discrimination between methicillin-sensitive and methicillin-resistant *Staphylococcus aureus* by intact cell mass spectrometry', *J Med Microbiol*, **49**, 295–300.

ENCHEVA V, WAIT R, BEGUM S, GHARBIA S E and SHAH H N (2007), 'Protein expression diversity amongst serovars of *Salmonella enterica*', *Microbiology*, **153**, 4183–93.

ENRIGHT M C and SPRATT B G (1999), 'Multilocus sequence typing', *Trends Microbiol*, **7**, 482–7.

EVERLEY R A, MOTT T M, WYATT S A, TONEY D M and CROLEY T R (2008), 'Liquid chromatography/mass spectrometry characterization of *Escherichia coli* and *Shigella* species', *J Amer Soc Mass Spec*, **19**, 1621–8.

FLEISCHMANN R D, ADAMS M D, WHITE O, CLAYTON R A, KIRKNESS E F, KERLAVAGE A R, ET AL. (1995), 'Whole-genome random sequencing and assembly of *Haemophilus influenzae* Rd', *Science*, **269**, 496–512.

FOLEY S L, LYNNE A M and NAYAK R (2009), 'Molecular typing methodologies for microbial source tracking and epidemiological investigations of Gram-negative bacterial foodborne pathogens', *Infect Genet Evol*, **9**, 430–40.

FOLEY S L, ZHAO S and WALKER R D (2007), 'Comparison of molecular typing methods for the differentiation of *Salmonella* foodborne pathogens', *Foodborne Path Dis*, **4**, 253–76.

GADAGKAR S R, ROSENBERG M S and KUMAR S (2005), 'Inferring species phylogenies from multiple genes: concatenated sequence tree versus consensus gene tree', *J Exp Zool B Mol Dev Evol*, **304**, 64–74.

GANTT S L, VALENTINE N B, SAENZ A J, KINGSLEY M T and WAHL K L (1999), 'Use of an internal control for matrix-assisted laser desorption/ionization time-of-flight mass spectrometry analysis of bacteria', *J Am Soc Mass Spec*, **10**, 1131–7.

GARAIZAR J, REMENTERIA A and PORWOLLIK S (2006), 'DNA microarray technology: a new tool for the epidemiological typing of bacterial pathogens?', *FEMS Immunol Med Microbiol*, **47**, 178–89.

GEHRING A G, ALBIN D M, REED S A, TU S I and BREWSTER J D (2008), 'An antibody microarray, in multiwell plate format, for multiplex screening of foodborne pathogenic bacteria and biomolecules', *Anal Bioanal Chem*, **391**, 497–506.

GERNER-SMIDT P, HISE K, KINCAID J, HUNTER S, ROLANDO S, HYYTIA-TREES E, ET AL. (2006), 'PulseNet USA: a five-year update', *Foodborne Pathog Dis*, **3**, 9–19.

GEVAERT K, VAN DAMME P, GHESQUIERE B, IMPENS F, MARTENS L, HELSENS K, ET AL. (2007), 'A la carte proteomics with an emphasis on gel-free techniques', *Proteomics*, **7**, 2698–718.

GILMOUR M W, GRAHAM M, VAN DOMSELAAR G, TYLER S, KENT H, TROUT-YAKEL K M, ET AL. (2010), 'High-throughput genome sequencing of two *Listeria monocytogenes* clinical isolates during a large foodborne outbreak', *BMC Genomics*, **11**, 120.

GOODACRE R, HEALD J K and KELL D B (1999), 'Characterisation of intact microorganisms using electrospray ionisation mass spectrometry', *FEMS Microbiol Lett*, **176**, 17–24.

GRAY M J, ZADOKS R N, FORTES E D, DOGAN B, CAI S, CHEN Y, ET AL. (2004), '*Listeria monocytogenes* isolates from foods and humans form distinct but overlapping populations', *Appl Environ Microbiol*, **70**, 5833–41.

GRISSA I, BOUCHON P, POURCEL C and VERGNAUD G (2008), 'On-line resources for bacterial micro-evolution studies using MLVA or CRISPR typing', *Biochimie*, **90**, 660–8.

GRISSA I, VERGNAUD G and POURCEL C (2008), 'CRISPRcompar: a website to compare clustered regularly interspaced short palindromic repeats', *Nucl Acids Res*, **36**, W145–8.

HADD A G, BROWN J T, ANDRUSS B F, YE F and WALKERPEACH C R (2005), 'Adoption of array technologies into the clinical laboratory', *Expert Rev Mol Diagn*, **5**, 409–20.

HOLLAND R D, WILKES J G, RAFII F, SUTHERLAND J B, PERSONS C C, VOORHEES K J, ET AL. (1996), 'Rapid identification of intact whole bacteria based on spectral patterns using matrix-assisted laser desorption/ionization with time-of-flight mass spectrometry', *Rap Commun Mass Spec*, **10**, 1227–32.

HOMMAIS F, PEREIRA S, ACQUAVIVA C, ESCOBAR-PARAMO P and DENAMUR E (2005), 'Single-nucleotide polymorphism phylotyping of *Escherichia coli*', *Appl Environ Microbiol*, **71**, 4784–92.

HOOSHDARAN M Z, BARKER K S, HILLIARD G M, KUSCH H, MORSCHHAUSER J and ROGERS P D (2004), 'Proteomic analysis of azole resistance in *Candida albicans* clinical isolates', *Antimicrob Agents Chemother*, **48**, 2733–5.

HSU K L and MAHAL L K (2006), 'A lectin microarray approach for the rapid analysis of bacterial glycans', *Nat Prot*, **1**, 543–9.

HSU K L, PILOBELLO K T and MAHAL L K (2006), 'Analyzing the dynamic bacterial glycome with a lectin microarray approach', *Nat Chem Biol*, **2**, 153–7.

HUYGHE A, FRANCOIS P and SCHRENZEL J (2009), 'Characterization of microbial pathogens by DNA microarrays', *Infect Genet Evol*, **9**, 987–95.

HYYTIA-TREES E K, COOPER K, RIBOT E M and GERNER-SMIDT P (2007), 'Recent developments and future prospects in subtyping of foodborne bacterial pathogens', *Future Microbiol*, **2**, 175–85.

ILINA E N, BOROVSKAYA A D, MALAKHOVA M M, VERESHCHAGIN V A, KUBANOVA A A, KRUGLOV A N, ET AL. (2009), 'Direct bacterial profiling by matrix-assisted laser desorption-ionization time-of-flight mass spectrometry for identification of pathogenic *Neisseria*', *J Mol Diag*, **11**, 75–86.

JARMAN K H, CEBULA S T, SAENZ A J, PETERSEN C E, VALENTINE N B, KINGSLEY M T, ET AL. (2000), 'An algorithm for automated bacterial identification using matrix-assisted laser desorption/ionization mass spectrometry', *Anal Chem*, **72**, 1217–23.

JEFFERS G T, BRUCE J L, MCDONOUGH P L, SCARLETT J, BOOR K J and WIEDMANN M (2001), 'Comparative genetic characterization of *Listeria monocytogenes* isolates from human and animal listeriosis cases', *Microbiology*, **147**, 1095–104.

KIDGELL C, REICHARD U, WAIN J, LINZ B, TORPDAHL M, DOUGAN G, ET AL. (2002), '*Salmonella typhi*, the causative agent of typhoid fever, is approximately 50,000 years old', *Infect Genet Evol*, **2**, 39–45.

KIRKWOOD J, GHETLER A, SEDMAN J, LECLAIR D, PAGOTTO F, AUSTIN J W, ET AL. (2006), 'Differentiation of group I and group II strains of *Clostridium botulinum* by focal plane array Fourier transform infrared spectroscopy', *J Food Prot*, **69**, 2377–83.

KLEEFSMAN I, STOWERS M, VERHEIJEN P, WUIJCKHUIJSE A, KIENTZ C and MARIJNISSEN J (2007), 'Bioaerosol analysis by single particle mass spectrometry', *Part Part Sys Char*, **24**, 85–90.

KOLINSKA R, DREVINEK M, JAKUBU V and ZEMLICKOVA H (2008), 'Species identification of *Campylobacter jejuni* ssp. *jejuni* and *C. coli* by matrix-assisted laser desorption/ionization time-of-flight mass spectrometry and PCR', *Folia Microbiol*, **53**, 403–9.

KUPPUSWAMY M N, HOFFMANN J W, KASPER C K, SPITZER S G, GROCE S L and BAJAJ S P (1991), 'Single nucleotide primer extension to detect genetic diseases: experimental application to hemophilia B (factor IX) and cystic fibrosis genes', *Proc Natl Acad Sci U S A*, **88**, 1143–7.

LECHNER D, LATHROP G M and GUT I G (2002), 'Large-scale genotyping by mass spectrometry: experience, advances and obstacles', *Curr Opin Chem Biol*, **6**, 31–8.

LECLAIR D, PAGOTTO F, FARBER J M, CADIEUX B and AUSTIN J W (2006), 'Comparison of DNA fingerprinting methods for use in investigation of type E botulism outbreaks in the Canadian Arctic', *J Clin Microbiol*, **44**, 1635–44.

LEVTEROVA V, PANAIOTOV S, BRANKOVA N and TANKOVA K (2010), 'Typing of genetic markers involved in stress response by fluorescent cDNA-amplified fragment length polymorphism technique', *Mol Biotechnol*, **45**(1), 34–8.

LIGLER F S, TAITT C R, SHRIVER-LAKE L C, SAPSFORD K E, SHUBIN Y and GOLDEN J P (2003), 'Array biosensor for detection of toxins', *Anal Bioanal Chem*, **377**, 469–77.

LIU B, LI H, WU S, ZHANG X and XIE L (2006), 'A simple and rapid method for the differentiation and identification of thermophilic bacteria', *Can J Microbiol*, **52**, 753–8.

MAGDALENA J, VACHEE A, SUPPLY P and LOCHT C (1998), 'Identification of a new DNA region specific for members of *Mycobacterium tuberculosis* complex', *J Clin Microbiol*, **36**, 937–43.

MAIDEN M C, BYGRAVES J A, FEIL E, MORELLI G, RUSSELL J E, URWIN R, ET AL. (1998), 'Multilocus sequence typing: a portable approach to the identification of clones within populations of pathogenic microorganisms', *Proc Natl Acad Sci U S A*, **95**, 3140–5.

MANDRELL R E, HARDEN L A, BATES A, MILLER W G, HADDON W F and FAGERQUIST C K (2005), 'Speciation of *Campylobacter coli, C. jejuni, C. helveticus, C. lari, C. sputorum*, and *C. upsaliensis* by matrix-assisted laser desorption ionization-time of flight mass spectrometry', *Appl Environ Microbiol*, **71**, 6292–307.

MASKOS U and SOUTHERN E M (1993a), 'a novel method for the analysis of multiple sequence variants by hybridisation to oligonucleotides', *Nucl Acids Res*, **21**, 2267–8.

MASKOS U and SOUTHERN E M (1993b), 'a study of oligonucleotide reassociation using large arrays of oligonucleotides synthesised on a glass support', *Nucl Acids Res*, **21**, 4663–9.

MAZARS E, LESJEAN S, BANULS A L, GILBERT M, VINCENT V, GICQUEL B, ET AL. (2001), 'High-resolution minisatellite-based typing as a portable approach to global analysis of *Mycobacterium tuberculosis* molecular epidemiology', *Proc Natl Acad Sci U S A*, **98**, 1901–6.

MAZZEO M F, SORRENTINO A, GAITA M, CACACE G, DI STASIO M, FACCHIANO A, ET AL. (2006), 'Matrix-assisted laser desorption ionization-time of flight mass spectrometry for the discrimination of foodborne microorganisms', *Appl Environ Microbiol*, **72**, 1180–9.

MORTIMER C K, PETERS T M, GHARBIA S E, LOGAN J M and ARNOLD C (2004), 'Towards the development of a DNA-sequence based approach to serotyping of *Salmonella enterica*', *BMC Microbiol*, **4**, 31.

MUELLER U G and WOLFENBARGER L L (1999), 'AFLP genotyping and fingerprinting', *Trends Ecol Evol*, **14**, 389–94.

NAKAMURA Y, JULIER C, WOLFF R, HOLM T, O'CONNELL P, LEPPERT M, ET AL. (1987), 'Characterization of a human "midisatellite" sequence', *Nucleic Acids Res*, **15**, 2537–47.

NELSON K E (2003), 'The future of microbial genomics', *Environ Microbiol*, **5**, 1223–5.

NIKOLAUSZ M, CHATZINOTAS A, TANCSICS A, IMFELD G and KASTNER M (2009), 'The single-nucleotide primer extension (SNuPE) method for the multiplex detection of various DNA sequences: from detection of point mutations to microbial ecology', *Biochem Soc Trans*, **37**, 454–9.

NOLLER A C, MCELLISTREM M C, PACHECO A G, BOXRUD D J and HARRISON L H (2003), 'Multilocus variable-number tandem repeat analysis distinguishes outbreak and sporadic *Escherichia coli* O157:H7 isolates', *J Clin Microbiol*, **41**, 5389–97.

OH B K, LEE W, CHUN B S, BAE Y M, LEE W H and CHOI J W (2005), 'The fabrication of protein chip based on surface plasmon resonance for detection of pathogens', *Biosens Bioelectron*, **20**, 1847–50.

ORSKOV I, ORSKOV F, JANN B and JANN K (1977), 'Serology, chemistry, and genetics of O and K antigens of *Escherichia coli*', *Bacteriol Rev*, **41**, 667–710.

PAGOTTO F, CORNEAU N, MATTISON K and BIDAWID S (2008), 'Development of a DNA microarray for the simultaneous detection and genotyping of noroviruses', *J Food Prot*, **71**, 1434–41.

PAGOTTO F, CORNEAU N, SCHERF C, LEOPOLD P and FARBER J M (2005), 'Molecular typing and differentiation of foodborne bacterial pathogens', in Fratamico P M, Bhunia, A.K. and Smith, J.L. (Eds): *Foodborne Pathogens: Microbiology and Molecular Biology*, Caister Academic Press, Norfolk, pp. 51–75.

PASCUAL J, MACIAN M C, ARAHAL D R, GARAY E and PUJALTE M J (2010), 'Multilocus sequence analysis of the central clade of the genus Vibrio by using the 16S rRNA, recA, pyrH, rpoD, gyrB, rctB and toxR genes', *Int J Syst Evol Microbiol*, **60**, 154–65.

PATEL P D (2006), 'Overview of affinity biosensors in food analysis', *J AOAC Int*, **89**, 805–18.

PETTIGREW M M (2004), 'An array of diverse microbial genomes', *Trends Biotechnol*, **22**, 491–3.

PILIARIK M, PAROVA L and HOMOLA J (2009), 'High-throughput SPR sensor for food safety', *Biosens Bioelectron*, **24**, 1399–404.

PILIARIK M, VAISOCHEROVA H and HOMOLA J (2009), 'Surface plasmon resonance biosensing', *Met Mol Biol*, **503**, 65–88.

PINEDA F J, ANTOINE M D, DEMIREV P A, FELDMAN A B, JACKMAN J, LONGENECKER M, ET AL. (2003), 'Microorganism identification by matrix-assisted laser/desorption ionization mass spectrometry and model-derived ribosomal protein biomarkers', *Anal Chem*, **75**, 3817–22.

PUPO G M, LAN R and REEVES P R (2000), 'Multiple independent origins of *Shigella* clones of *Escherichia coli* and convergent evolution of many of their characteristics', *Proc Natl Acad Sci U S A*, **97**, 10567–72.

RAMAZANZADEH R and MCNERNEY R (2007), 'Variable number of tandem repeats (VNTR) and its application in bacterial epidemiology', *Pak J Biol Sci*, **10**, 2612–21.

REIJANS M, LASCARIS R, GROENEGER A O, WITTENBERG A, WESSELINK E, VAN OEVEREN J, ET AL. (2003), 'Quantitative comparison of cDNA-AFLP, microarrays, and GeneChip expression data in *Saccharomyces cerevisiae*', *Genomics*, **82**, 606–18.

REVAZISHVILI T, KOTETISHVILI M, STINE O C, KREGER A S, MORRIS JR J G and SULAKVELIDZE A (2004), 'Comparative analysis of multilocus sequence typing and pulsed-field gel electrophoresis for characterizing *Listeria monocytogenes* strains isolated from environmental and clinical sources', *J Clin Microbiol*, **42**, 276–85.

ROOS A, DIELTJES P, VOSSEN R H, DAHA M R and DE KNIJFF P (2006), 'Detection of three single nucleotide polymorphisms in the gene encoding mannose-binding lectin in a single pyrosequencing reaction', *J Immunol Methods*, **309**, 108–14.

RUDI K, KATLA T and NATERSTAD K (2003), 'Multi locus fingerprinting of *Listeria monocytogenes* by sequence-specific labeling of DNA probes combined with array hybridization', *FEMS Microbiol Lett*, **220**, 9–14.

RUSSELL S C (2009), 'Microorganism characterization by single particle mass spectrometry', *Mass Spec Rev*, **28**, 376–87.

SAENZ A J, PETERSEN C E, VALENTINE N B, GANTT S L, JARMAN K H, KINGSLEY M T, ET AL. (1999), 'Reproducibility of matrix-assisted laser desorption/ionization time-of-flight mass spectrometry for replicate bacterial culture analysis', *Rap Com Mass Spec*, **13**, 1580–5.

SAIKI R K, WALSH P S, LEVENSON C H and ERLICH H A (1989), 'Genetic analysis of amplified DNA with immobilized sequence-specific oligonucleotide probes', *Proc Natl Acad Sci U S A*, **86**, 6230–4.

SCHMIDT F, DONAHOE S, HAGENS K, MATTOW J, SCHAIBLE U E, KAUFMANN S H, ET AL. (2004), 'Complementary analysis of the *Mycobacterium tuberculosis* proteome by two-dimensional electrophoresis and isotope-coded affinity tag technology', *Mol Cell Prot*, **3**, 24–42.

SELANDER R K, CAUGANT D A, OCHMAN H, MUSSER J M, GILMOUR M N and WHITTAM T S (1986), 'Methods of multilocus enzyme electrophoresis for bacterial population genetics and systematics', *Appl Environ Microbiol*, **51**, 873–84.

SHRIVER-LAKE L C, SHUBIN Y S and LIGLER F S (2003), 'Detection of staphylococcal enterotoxin B in spiked food samples', *J Food Prot*, **66**, 1851–6.

STOWERS M A, VAN WUIJCKHUIJSE A L, MARIJNISSEN J C, SCARLETT B, VAN BAAR B L and KIENTZ C E (2000), 'Application of matrix-assisted laser desorption/ionization to on-line aerosol time-of-flight mass spectrometry', *Rapid Com Mass Spec*, **14**, 829–33.

TAYLOR A D, LADD J, YU Q, CHEN S, HOMOLA J and JIANG S (2006), 'Quantitative and simultaneous detection of four foodborne bacterial pathogens with a multi-channel SPR sensor', *Biosens Bioelectron*, **22**, 752–8.

TWINE S M, PAUL C J, VINOGRADOV E, MCNALLY D J, BRISSON J R, MULLEN J A, ET AL. (2008), 'Flagellar glycosylation in *Clostridium botulinum*', *FEBS J*, **275**, 4428–44.

TYAGI S, BRATU D P and KRAMER F R (1998), 'Multicolor molecular beacons for allele discrimination', *Nat Biotechnol*, **16**, 49–53.

VAIDYANATHAN S, ROWLAND J J, KELL D B and GOODACRE R (2001), 'Discrimination of aerobic endospore-forming bacteria via electrospray-ionization mass spectrometry of whole cell suspensions', *Anal Chem*, **73**, 4134–44.

VAN BELKUM A (2007), 'Tracing isolates of bacterial species by multilocus variable number of tandem repeat analysis (MLVA)', *FEMS Immunol Med Microbiol*, **49**, 22–7.

VAN BELKUM A, SCHERER S, VAN ALPHEN L and VERBRUGH H (1998), 'Short-sequence DNA repeats in prokaryotic genomes', *Microbiol Mol Biol Rev*, **62**, 275–93.

VAN WUIJCKHUIJSE A L, STOWERS M A, KLEEFSMAN W A, VAN BAAR B L M, KIENTZ C E and MARIJNISSEN J C M (2005), 'Matrix-assisted laser desorption/ionisation aerosol time-of-flight mass spectrometry for the analysis of bioaerosols: development of a fast detector for airborne biological pathogens', *J Aer Sci*, **36**, 677–87.

VOGLER A J, KEYS C, NEMOTO Y, COLMAN R E, JAY Z and KEIM P (2006), 'Effect of repeat copy number on variable-number tandem repeat mutations in *Escherichia coli* O157:H7', *J Bacteriol*, **188**, 4253–63.

VOGLER A J, KEYS C E, ALLENDER C, BAILEY I, GIRARD J, PEARSON T, ET AL. (2007), 'Mutations, mutation rates, and evolution at the hypervariable VNTR loci of *Yersinia pestis*', *Mutat Res*, **616**, 145–58.

VOS P and KUIPER M (1997), 'AFLP analyses', in Caetano-Anollés, G. and Gresshoff, P.M. (Eds): *DNA Markers: Protocols Applications, and Overviews*, Wiley, Hoboken, NJ, pp. 115–31.

WALKER J, FOX A J, EDWARDS-JONES V and GORDON D B (2002), 'Intact cell mass spectrometry (ICMS) used to type methicillin-resistant *Staphylococcus aureus*: media effects and inter-laboratory reproducibility', *J Microbiol Met*, **48**, 117–26.

WANG Z, RUSSON L, LI L, ROSER D C and LONG S R (1998), 'Investigation of spectral reproducibility in direct analysis of bacteria proteins by matrix-assisted laser desorption/ionization time of flight mass spectrometry', *Rapid Com Mass Spec*, **12**, 456–64.

WARD T J, DUCEY T F, USGAARD T, DUNN K A and BIELAWSKI J P (2008), 'Multilocus genotyping assays for single nucleotide polymorphism-based subtyping of *Listeria monocytogenes* isolates', *Appl Environ Microbiol*, **74**, 7629–42.

WEINER M P and HUDSON T J (2002), 'Introduction to SNPs: discovery of markers for disease', *Biotechniques*, **10**(Suppl 4–7), 12–13.

WILLIAMS T L, ANDRZEJEWSKI D, LAY J O and MUSSER S M (2003), 'Experimental factors affecting the quality and reproducibility of MALDI TOF mass spectra obtained from whole bacteria cells', *J Am Soc Mass Spec*, **14**, 342–51.

WINGREN C and BORREBAECK C A (2009), 'Antibody-based microarrays', *Met Mol Biol*, **509**, 57–84.

WINKLER M A, UHER J and CEPA S (1999), 'Direct analysis and identification of *Helicobacter* and *Campylobacter* species by MALDI-TOF mass spectrometry', *Anal Chem*, **71**, 3416–19.

WOLFF S, ANTELMANN H, ALBRECHT D, BECHER D, BERNHARDT J, BRON S, ET AL. (2007), 'Towards the entire proteome of the model bacterium *Bacillus subtilis* by gel-based and gel-free approaches', *J Chrom B*, **849**, 129–40.

YOKOYAMA K, IINUMA Y, KAWANO Y, NAKANO M, KAWAGISHI M, YAMASHINO T, ET AL. (2001), 'Resolution of *Escherichia coli* O157:H7 that contaminated radish sprouts in two outbreaks by two-dimensional gel electrophoresis', *Cur Microbiol*, **43**, 311–15.

ZHANG L, SRINIVASAN U, MARRS C F, GHOSH D, GILSDORF J R and FOXMAN B (2004), 'Library on a slide for bacterial comparative genomics', *BMC Microbiol*, **4**, 12.

ZHANG W, QI W, ALBERT T J, MOTIWALA A S, ALLAND D, HYYTIA-TREES E K, ET AL. (2006), 'Probing genomic diversity and evolution of *Escherichia coli* O157 by single nucleotide polymorphisms', *Genome Res*, **16**, 757–67.

10

Development, validation and quality assurance of methods for subtyping of foodborne pathogens

E. K. Hyytia-Trees and E. M. Ribot, Centers for Disease Control and Prevention, USA

Abstract: A wide array of molecular subtyping approaches is currently being used to characterize foodborne pathogens in order to assist epidemiologists with cluster detection and trace back studies during outbreak investigations. The data generated by these methods are often used to populate reference databases or libraries. Once a new method has been developed or an old one adapted for this purpose, it must be subjected to the highest scrutiny possible to ensure that the data that will be generated meet the criteria and standard expected of a reference library. In this chapter, we will outline a typical development, validation and quality assurance process for a method to be used in the PulseNet subtyping network.

Key words: protocol development, protocol validation, standardized protocol, quality assurance/quality control program, reference library.

10.1 Introduction

A wide array of molecular subtyping approaches is currently being used to characterize foodborne pathogens in order to assist epidemiologists with cluster detection and trace back studies during outbreak investigations. In retrospect, the fusion between epidemiology and molecular biology seems obvious and natural, yet it faced many challenges in the beginning, some of which still remain today. Among these are a lack of uniformity of the techniques or protocols used between laboratories for the characterization and subtyping of pathogen(s) of interest, a lack of standardization of the methods being used and, in some cases, the use of methods that are not entirely appropriate for the intended objective (e.g., a lack of epidemiological relevance) (van Belkum *et al.*, 2007). These challenges are particularly important if a reference library or database will be populated with data generated by a network of multiple laboratories, such as PulseNet, the electronic subtyping network for foodborne pathogens (Gerner-Smidt *et al.*, 2006; Swaminathan *et al.*, 2001). In this context, an ideal subtyping method is rapid, robust, reproducible, portable, produces objective data and has high sensitivity (groups together all isolates associated with the same source) and specificity (differentiates the outbreak strain from all epidemiologically unrelated strains). Once a proper method or approach has been selected it must be subjected to the highest scrutiny possible to ensure that the data that will be generated meet the criteria and standard expected of a reference library or database. It is safe to assume that, typically, there is a direct correlation between the reproducibility and robustness of a method, and the quality of the data being stowed in the form of a reference library or database. The versatility of the method must also be established not only by its ability to produce data that can lead to the establishment of subtypes for the majority, if not all, of the strains of the pathogen being studied, but also the potential to type similar and newly emerging strains without the need to follow an entirely new approach. The method must also produce data that can be analyzed, interpreted and disseminated with a relative ease. Unlike methods developed in 'isolation' (for use by one laboratory), methods developed for the generation of data to be archived in reference libraries must be carefully tested, evaluated and validated. While the validation process for different methods may vary in the details, there are several steps that can be applied regardless of the type of method being considered. In PulseNet, the standardization of a method or a protocol consists of the following steps or phases: development, internal validation, external validation, establishment and testing of the reference library or database, along with a detailed quality assurance and quality control (QA/QC) program.

In this chapter, we outline a number of factors that should be considered when developing and validating molecular methods. The information presented herein comes from the experience the authors have in the development and validation of standardized protocols for the two approaches currently being used to subtype foodborne bacterial pathogens by PulseNet: pulsed-field gel electrophoresis (PFGE) and multi-locus variable-number tandem repeat analysis (MLVA) (Cooper *et al.*, 2006; Hyytia-Trees *et al.*, 2006; Kam *et al.*, 2008; Parsons *et al.*, 2007; Ribot *et al.*, 2006; Ribot *et al.*, 2001). PFGE is currently regarded as the 'gold standard' for subtyping foodborne bacteria, and is also the foundation of PulseNet (Gerner-Smidt *et al.*, 2006). PFGE takes advantage of infrequently cutting restriction enzymes that cut the bacterial genome in 15–35 fragments (Schwartz and Cantor, 1984). Prior to the restriction step, the DNA is prepared by embedding the bacterial cells into agarose plugs in which they undergo a chemical lysis step. After the restriction, a thin slice of the agarose plug is loaded onto an agarose gel, and the restriction fragments are separated by alternating the direction of the electric field during the electrophoresis. The gel is then stained, a digital image of the DNA banding pattern is recorded, and the image is subjected to computer assisted analysis using specialized software (in the case of PulseNet, the BioNumerics software package, Applied Maths, Kortrijk, Belgium). MLVA, a newer DNA sequence-based method, has recently been used to compliment and in some laboratories even to replace PFGE (Boxrud *et al.*, 2007; Hyytia-Trees *et al.*, 2006; Nygard *et al.*, 2007; Sperry *et al.*, 2008; Torpdahl *et al.*, 2007). MLVA targets short motif repeat units (variable-number tandem repeats or VNTRs) that are organized in tandem and occur in a single locus in the bacterial genome (van Belkum *et al.*, 1998). MLVA involves determination of the number of the repeat copy units in multiple loci using simple multiplex PCR amplification of the repeat regions followed by sizing of the fluorescent labeled amplification products using capillary electrophoresis. The sizing data are exported from the capillary electrophoresis equipment as a peak file that can be imported into analysis software (BioNumerics) for copy number (allele) determination.

10.2 Strain selection for protocol development and validation

Method development and validation are tightly intertwined and cannot be separated from each other. The number and type of isolates used in the development and validation phases must be carefully chosen during the design phase to ensure that they represent the 'typical' strains to be typed and that the method will produce data that are meaningful and meet the desired objectives. The idea is to anticipate and resolve potential problems during the development phase in order to ensure that the validation process goes smoothly. However, due to practical reasons only a limited number of isolates can be used during the development phase, hence unanticipated problems may arise during the validation phase.

The number and type of isolates needed in the development and validation depend on a number of different factors, including the type of the method and the genetic make-up of the species or serotypes to be characterized with such method. Table 10.1 summarizes the number of isolates we suggest to be used at each phase when developing and validating new PFGE and MLVA methods. Isolate selection during the development phase is critical because it is the first glance at the type of

Table 10.1 Number of isolates suggested to be used to develop, validate and quality control new PFGE and MLVA methods

Method phase	No. of isolates	
	PFGE	MLVA
Initial development	7–11	10–15
Expanded development	25–100	150–200
Internal validation	50–150	300–350
External validation		
Phase I	7–20	40–50
Phase II	25	50
QA/QC program		
Certification	4–8	6–10
Proficiency testing	1–2	1–2

data the method produces in terms of its discriminatory power and epidemiological relevance. The limited number of isolates used during the development phase should represent the population at large whereas the validation panel is an expansion of this subset and should also include more rare genotypes/phenotypes. When developing and validating a rather universal method that can be applied to multiple serotypes within one species, such as PFGE, it is always prudent to test several (three or more) strains from each of the serotypes (most common or clinically important ones) whenever possible. While it would be ideal to test at least one strain of each serotype, in practice this is not always logistically feasible. For instance, in the case of *Salmonella enterica* there are over 2500 recognized serotypes (Grimont and Weill, 2007); too many to evaluate at once. An approach to deal with this problem could consist of a ranking system to test the most commonly seen or important strains from ill subjects (human or animal) and/or environment or make the selection based on important subgroups or characteristics (severity of illness, mortality, multi-drug resistance, etc.) of the organisms being studied. MLVA assays, on the other hand, are typically serotype-specific. Only a couple of assays published so far have attempted to cover multiple serotypes and/or species (Gorge *et al.*, 2008; Lindstedt *et al.*, 2007). Since one of the objectives of the validation of a MLVA method is to get a comprehensive picture of the allelic variation seen in the target species/serotype, the number of isolates needed for validation will increase as the diversity of the genome increases. For instance, more isolates would be needed to thoroughly evaluate a MLVA method for the highly diverse *S. enterica* serotype Typhimurium than the more clonal serotype Enteritidis. The throughput of the method also has some practical implications on how many isolates can be included in the validation. It is much less time consuming to test 500 isolates with MLVA than it is with PFGE.

In order to be able to evaluate the discriminatory power of the method and the epidemiological relevance of the data produced, an ideal validation panel should contain both isolates from multiple single source outbreaks and sporadic isolates that are epidemiologically not related to each other. When the validation involves a method that targets highly variable sequences, such as MLVA, multiple (dozens) isolates from the same outbreak should be included so that the level of variation occurring during an outbreak can be evaluated. A validation panel should contain isolates from multiple geographical locations, preferably from multiple countries if the method is intended for international comparisons, in order to avoid any biases that may result from testing a few strains that might be endemic in a particular geographical area. If the validation involves a comparison to a gold standard method, the validation panel isolates should adequately capture the range of the genetic diversity seen with such method.

10.3 Protocol development

The foundation of any data repository is the quality of data that will populate such database and whether or not analysis of the data allows us to make assumptions and reach conclusions about the meaning of such information. The development phase is the first step taken toward standardization of protocols for the PulseNet system and it is, in essence, a combination of general and applied research. The general research component allows us to systematically scrutinize every step of the methodology being tested in an effort to identify the optimal conditions or parameters needed to generate a method that is robust and highly reproducible. The applied component, on the other hand, focuses on the meaning of the subtyping data being generated; for instance, is the data in agreement/concordance with the epidemiology of the organism being studied or does it establish a link between a particular phenotype and a genotype.

10.3.1 PFGE protocol development

During the early stages of PFGE protocol development for PulseNet, a relatively small set of well-characterized strains is used to test an array of conditions and reagents until high-quality and reproducible data are obtained. Typically, a set of 7–11 strains is used in this stage of the process. Development of a new PulseNet PFGE protocol typically starts with the evaluation of reagents and reaction conditions outlined in the *Escherichia coli* O157 (Ribot *et al.*, 2006) standardized protocol if the pathogen being tested is a Gram-negative bacterium or the conditions used for PFGE analysis of *Listeria monocytogenes* (Graves and Swaminathan, 2001) if the organism is Gram-positive. The steps that often require some adjustments during the development are: cell suspension concentration, lysis of the agarose-embedded cells, washes, the restriction digestion reactions and the electrophoresis conditions. Some of the factors that influence technical performance of a PFGE protocol include the type and concentration of the agarose used, the formulation of the electrophoresis buffer, the temperature the buffer is maintained at during the electrophoresis run, and the flow rate of the buffer during the electrophoresis. These parameters have been standardized for all PulseNet protocols, thus making the protocols more uniform and user friendly.

Up to this point the main goal has been to assess the protocol's ability to generate clear and reproducible data in a single laboratory with a few isolates that, in reality, may or may not represent the population at large. Once the protocol has been successfully tested with a few strains, the development phase can be expanded to include a larger number of isolates (25–100) that are tested to further assess the robustness and reproducibility of the protocol and the epidemiological relevance of the data. By testing additional isolates, one can better assess whether or not the enzymes and electrophoresis conditions are appropriate for the pathogen or if additional conditions need to be tested. The data must also be evaluated with the computer software of choice to ensure that the fingerprints generated can be easily analyzed.

10.3.2 MLVA protocol development

The development of a MLVA method begins with a screening process that involves testing of all feasible VNTRs identified in the reference genome for variability. In practice, this is done by testing each VNTR separately against a panel of 10–15 isolates. Since the goal in this phase is to achieve a maximum discriminatory power, all strains included in this initial panel should be unrelated to each other. VNTRs with very low diversity or no diversity at all are excluded from the next phase in which the promising VNTRs are screened against a larger panel (150–200) of isolates. This panel should contain both outbreak-related and sporadic isolates. The second screen will facilitate the selection of VNTRs that produce epidemiologically relevant data. After the final set of VNTRs to be included in the protocol is selected, the individual PCR reactions are multiplexed in reactions that contain usually no more than four targets each. The PCR reaction conditions are optimized using the initial set of 10–15 isolates. Different reagents, reagent concentrations and amplification conditions are tested to find the optimal combination that consistently produces reproducible and easily readable data. Finally, the multiplex reactions are evaluated by re-testing the larger panel of isolates used in the second screen. The development phase must also include a stability study for the VNTRs that are included in the final protocol. This can be accomplished in-vitro by performing a serial passage study (e.g. 40 passages) with two to three isolates. Ideally, two to three representatives of each allele detected in each VNTR should also be confirmed by sequencing to ensure that the difference seen in the fragment size was due to a difference in the repeat copy number and not due to insertions or deletions in the flanking sequences (Sperry *et al.*, 2008).

10.4 Internal validation

After the initial protocol development phase is completed, we proceed with the next phase in the process which we refer to as the 'internal validation' phase. This means that the protocol being developed has met the expected level of robustness, reproducibility, discriminatory power and quality set by the 'developers' and is now ready to be tested by others. The intention is to have unbiased or 'inexperienced' individuals test the protocol in order to obtain feedback about its overall technical performance, ease of use, and potential flaws. The information obtained from this phase is used to determine if the protocol is ready for the next phase or if further modifications are needed. In PulseNet, the internal validation phase is usually carried out by one or two individuals who are unfamiliar with the protocol (i.e. they were not involved in the protocol development phase). This step also helps us determine if the protocol instructions are clear and easy to follow.

The internal validation phase could be divided into two substages: (1) testing of the protocol by other laboratories/individuals within the agency or institution for technical performance and (2) testing of additional isolates by the 'developers' of the protocol. By subjecting a large number of isolates to analysis with the proposed protocol (50–150 additional isolates for PFGE; 300–350 for MLVA), we can further test the robustness of the protocol along with its ability to consistently produce fingerprints from all strains and generate data that are epidemiologically relevant and easy to analyze. For MLVA protocols, the additional isolates tested will help establish data interpretation guidelines and assist in building a comprehensive allele (fragment size range) table that will be used in the BioNumerics analysis (Hyytia-Trees *et al.*, 2010). If desired, the data generated during the internal validation can be used to compare the new method with a gold standard method.

The most important element of the internal validation is the analysis and review of comments provided by the staff that participated in this part of the process. If the technical comments about the protocol are positive we then move to the next step in the internal validation process: analysis of data generated with the proposed protocol by staff experienced in performing BioNumerics analysis. This step is usually carried out by members of the PulseNet USA Data Administration Team. If the raw data look acceptable to the naked eye but analysis in BioNumerics is too difficult to perform, the developers then go back to the drawing board in an attempt to resolve the issues raised by the data manager(s).

10.4.1 Evaluation of technical performance and ease of data interpretation

The evaluation of reproducibility and robustness should include comparisons between different runs, equipment, users, and reagent lots. A random subpopulation (10%) of isolates included in the validation panel should be tested independently twice to ensure the run-to-run reproducibility using the same equipment and reagent lots. The same subset of isolates should also be tested by two different technicians using the same equipment and reagent lots to evaluate the user-to-user variability. In case of MLVA, a thorough validation should also include testing the same set of isolates using two different primer lots since there are significant differences in the fluorescent intensities of primer dyes depending on their age and degradation status. The performance of a MLVA assay should also be evaluated using different types of thermocyclers. Ideally, the reproducibility of the fragment sizing in MLVA should be tested using two different capillary electrophoresis equipment of the same model. Typically, the sizing accuracy should be within one base pair between the different instruments of the same model. However, there are significant sizing differences between capillary electrophoresis equipment from different manufacturers, and also between different models of capillary electrophoresis equipment from the same manufacturer (de Valk *et al.*, 2009; Hyytia-Trees *et al.*, 2010). Since these sizing differences are not uniform among all loci, the secondary and tertiary structures of DNA fragments in each particular VNTR locus probably play some role in how each fragment migrates through the different types of gel polymers that have to be used in the different platforms. If a MLVA method is intended to be used in laboratories with multiple different capillary electrophoresis platforms, a separate protocol must be developed and validated for each platform. The validation data can be used to construct platform-specific allele tables (fragment size range table) which can be used to compensate for sizing discrepancies during the BioNumerics analysis (Hyytia-Trees *et al.*, 2010).

As additional strains are tested during this phase, a better assessment of issues affecting typeability of strains will become apparent. In the case of PFGE, an occasionally seen problem is the emergence of strains that do not generate a pattern but instead produce a smear in the lower half section of the gel (Fig. 10.1). This problem was a mystery for many years and led to the coining of the term 'PFGE untypeable' for strains that exhibit this behavior. Recent studies suggest that the smear or DNA degradation is not the result of contaminants or endonucleases released during the lysis process but that the degradation actually

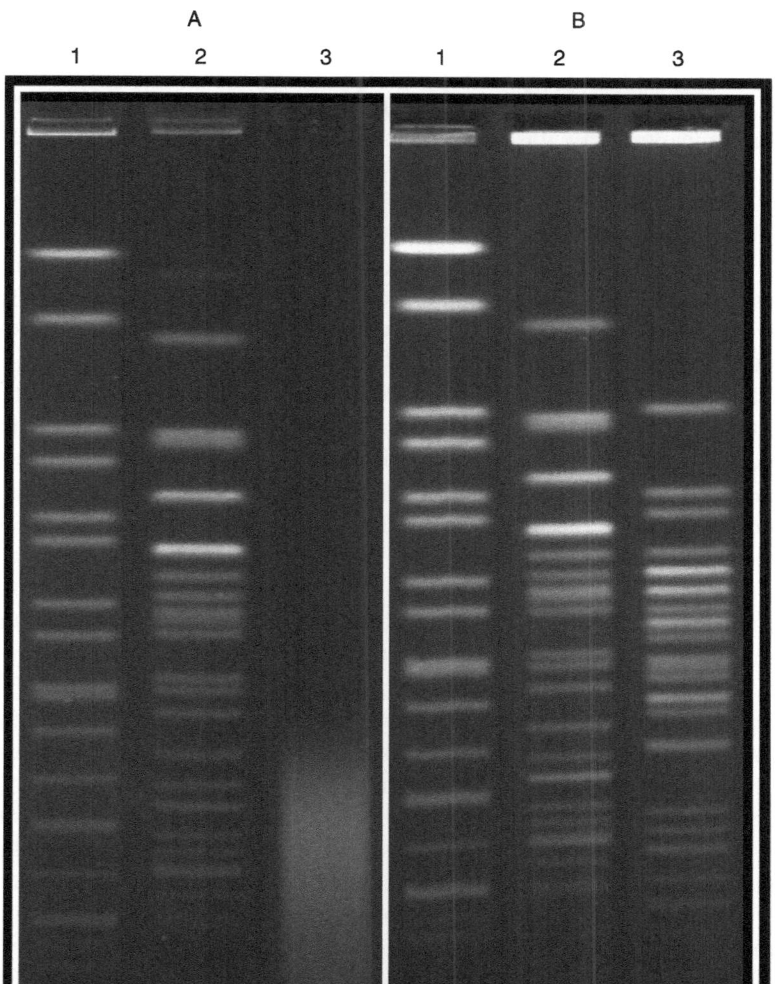

Fig. 10.1 Gel image of a *Salmonella enterica* serotype Typhimurium strain previously thought to be untypeable by PFGE (A, lane 3). Panel B (lane 3) shows the same strain electrophorized with PFGE running buffer containing a final concentration of 50 μM thiourea.

occurs during the electrophoresis step (Corkill *et al.*, 2000; Silbert *et al.*, 2003). It appears that DNA from these 'untypeable' strains is modified somehow (the actual mechanism is still unknown), and that the modifying molecule has a high affinity for the Tris radicals and superoxide molecules that are formed during the electrophoresis step. In other words, the DNA degradation is the result of a Tris-dependent strand scission reaction and not the result of nuclease activity. These studies also showed that this problem can be overcome by adding thiourea to the electrophoresis running buffer where it quenches Tris radicals thus preventing degradation of the DNA during the run.

Since MLVA protocols typically are very serotype-specific, the strain typeability is usually not a problem among that serotype. However, depending on the VNTR loci selected in the protocol, null alleles (i.e. no amplification) may occur in some isolates (Boxrud *et al.*, 2007; Hyytia-Trees *et al.*, 2006). During the internal validation phase it is highly advisable to confirm all null alleles by testing the isolates that had a null allele in a certain locus by amplifying that locus separately in order to ensure that the null allele did not occur because of suboptimal primer concentrations in the multiplex PCR.

The subjectivity of the PFGE data analysis is a well-known problem. The analysis of gel images in BioNumerics is based on an individual's ability to correctly mark all bands that were resolved on a gel. As the gel quality decreases, complexity (the number and location of bands) of the patterns increases, and so does the chance for errors in analysis and data interpretation. For this reason, it is important for any standardized PFGE protocol to consistently yield fingerprints that do not contain many areas with band compressions or too many fragments in a small area of the gel and not enough in another. As additional strains and serotypes are tested during internal validation, investigators will get a better understanding of the pattern distribution that can be observed from this organism and make any necessary adjustments to the electrophoresis conditions, such as switch and run times (Fig. 10.2). If the electrophoresis run is too short, fragments that migrate close together might not be resolved as expected. As a consequence, the analyst may mismark fragments during the analysis and obtain a wrong PFGE pattern. If the electrophoresis is allowed to run for too long and DNA fragments are lost, it would make the analysis impossible and the electrophoresis step would have to be repeated. The variations in the run length, even when a standardized protocol is used, can result from a number of factors, such as the quality of the water used to prepare the electrophoresis running buffer and the source of buffer ('home-made' versus commercial). In order to ensure uniformity in the appearance of the PFGE patterns, all PulseNet protocols recommend for the run length to be determined empirically in each laboratory by adjusting the electrophoresis run time so that the smallest DNA fragment of the size standard used migrates within 1–1.5 cm from the bottom of the gel (Ribot *et al.*, 2006).

Compared to PFGE, MLVA data are relatively easy to analyze since the data are displayed as fragment sizes. The gray areas of the MLVA data analysis include partial repeats, and multiple peaks in a single locus. Partial repeats can be problematic, particularly in loci that contain very short repeat motifs (six base pairs

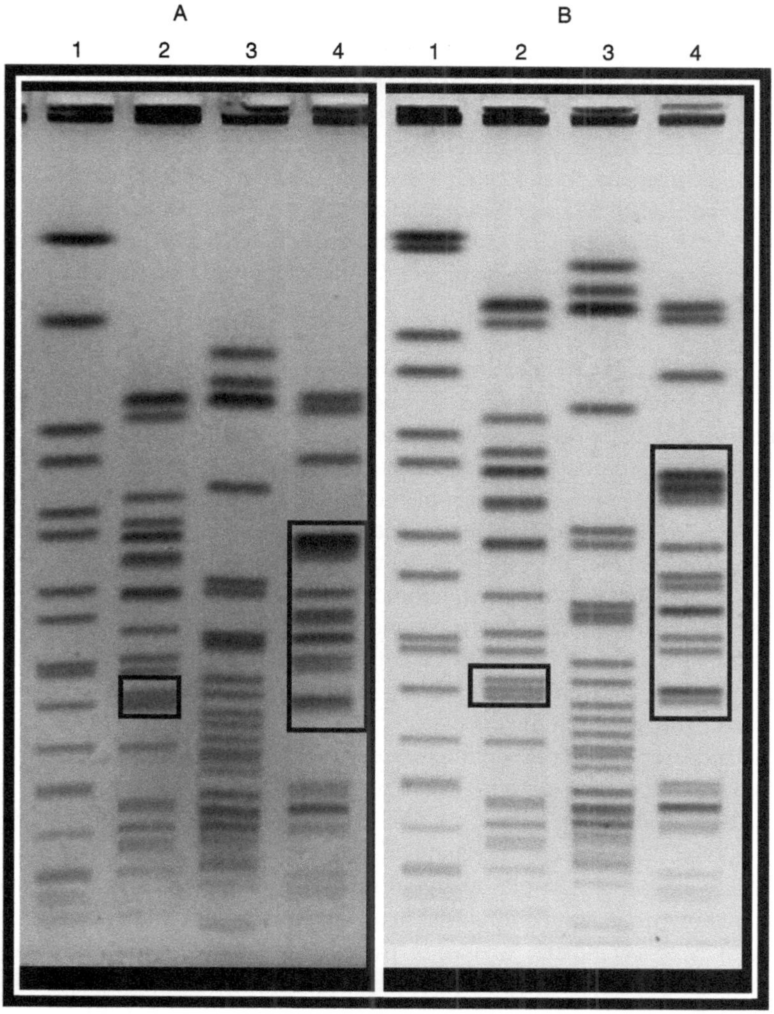

Fig. 10.2 PFGE patterns generated with two sets of electrophoresis parameters tested during the internal evaluation of the standardized PFGE protocol for non-O157 Shiga toxin-producing *Escherichia coli*. The boxed areas in lanes 2 and 4 of Panel A show band compression resulting from suboptimal electrophoresis conditions. Resolution of the fragments in areas highlighted was improved systematically (Panel B; lanes 2 and 4) by adjusting the electrophoresis conditions (initial and final switch times).

or shorter). In these cases, the copy number (allele type) should be rounded up or down to the closest complete copy number. In loci with longer repeat motifs (ten base pairs or longer), partial repeats can be handled by assigning allele types based on a difference that is less than one complete copy. For example, in a locus with a repeat motif of 12 base pairs, alleles could be spaced at every six base pairs. In either case, determining reliable and reproducible fragment size ranges for each allele (allele table) is a time consuming task that will require testing a large number of diverse isolates. The second data interpretation problem, the occurrence of multiple peaks in a single locus, arises from the fact that in some rare occasions a strain can possess two (or more) copies of the locus in its genome. Since the analysis software can only account for one peak per locus, the peak with the higher fluorescence intensity should be recorded, and the minor peak should be ignored. This does not constitute a significant problem, as long as the same peak consistently has the higher fluorescence intensity in the same strain. A locus that has multiple peaks with a higher frequency than 1.0–1.5% should be excluded from the protocol.

10.4.2 Comparison to gold standard method

When a new method is validated, its performance is often compared to that of another method that is widely used to subtype the species or serotype in question. For most foodborne bacterial organisms, this gold standard method is PFGE. Ideally, all characteristics of a new method should be equal to or better than the established gold standard method. The characteristics that most often are considered in the comparison are discriminatory power, technical demands and cost of the methods, and the epidemiological relevance of the data that the methods produce. Comparison against gold standards will also assure that the historical information in existing databases is not lost. When no widely accepted gold standard method is available, the data generated by the new method should be compared to the epidemiological data.

Discriminatory power is defined as the ability of a method to distinguish between unrelated strains. An objective measure of discriminatory power can be obtained by calculating a diversity index. Diversity is a function of subtype richness, which is defined by the number of subtypes in a population and subtype evenness, which refers to the relative distribution of individual isolates among different subtypes. A number of different mathematical algorithms are available to calculate diversity indices, and each places a slightly different emphasis on subtype richness and evenness. Dominance indices put weight towards the abundance of the most common subtype, i.e. the more even the distribution of the subtypes among the population, the higher the index. Simpson's index is the most frequently used example of a dominance index (Hunter and Gaston, 1988). It is a measure of the probability that two unrelated isolates will be characterized differently by the typing method evaluated. In contrast, another commonly used index measuring diversity, the Shannon–Weiner index, is affected by both the number of subtypes and their evenness in population (Shannon and Weaver, 1949). Shannon's formula calculates separate indices for diversity and evenness. Consequently, Shannon's diversity index is more influenced by the number of subtypes among the population. No matter which index is used to measure discriminatory power, assumptions are made that all subtypes are represented in the sample and that the sample is randomized, neither of which is true in a typical validation study. Therefore, discriminatory indices should always be interpreted in relation to the isolate collection that was used to evaluate the methods.

If the diversity indices of two methods are close the same the methods should be able to provide similar information about the isolates, i.e. they should cluster the isolates on the general level the same way. However, a complete agreement between the methods should not be expected, particularly when the methods target very different types of sequences in the bacterial genome. That is also true with PFGE and MLVA. PFGE targets infrequent cutting restriction sites in which the genetic variation happens through insertions, deletions, and recombination (Kudva *et al.*, 2002). MLVA, on the other hand, targets VNTR regions in which the variation occurs mainly through DNA polymerase slippage (van Belkum *et al.*, 1998). All these genetic events occur at varying rates in different strains. In some strains, the restriction sites may be relatively conserved, and hence we encounter PFGE types that are very common among the population. In the same set of strains, some of the VNTRs may evolve much faster and therefore MLVA may be able to break down some of the clusters identified by PFGE (Hyytia-Trees *et al.*, 2006; Torpdahl *et al.*, 2007). The opposite situation (PFGE being more discriminatory than MLVA) is also known to happen (Sperry *et al.*, 2008).

10.4.3 Comparison to epidemiological data

In most circumstances, when a new method is validated, it is more critical to have a good correlation with the epidemiological data than it is to have a good correlation with the gold standard method, unless, of course, the gold standard produces data that are in perfect agreement with the epidemiological data. This can be assessed by selecting and evaluating isolates that are known to be associated with a common source (also referred to as 'outbreak-related isolates') and epidemiologically unrelated (or 'sporadic') isolates. In fact, it is extremely important to include isolates from multiple outbreaks in the validation of a methodology. The isolates associated with a common source should all have an indistinguishable subtype that ideally should be different from the subtypes of the sporadic isolates. Each set of outbreak-related isolates should have a unique pattern that is different from any of the other outbreak-related patterns. The sporadic isolates can be used to calculate the diversity index for the method, and ideally each of them should have a unique subtype. Realistically speaking there is no single method that is capable of unequivocally identifying every single outbreak at any given time. What molecular subtyping data provide is valuable information on the likelihood that two or more strains of a given pathogen are genetically related. Depending on the evolutionary clock measured by the method of choice, one can then make interpretations as to what 'genetically related' really means. In the context of molecular epidemiology of foodborne pathogens in general, the desired timeframe is usually measured in weeks or a few months for common source outbreaks (CDC, 2006; 2007; 2008). It is important to recognize that a particular subtype may emerge multiple times within a period of months or years, but that would not necessarily mean that such subtype is responsible for a prolonged common source outbreak. As databases grow over time, it is unavoidable that common subtypes will be discovered among the population. However, if very common subtypes are already detected during the method validation phase, when only a relatively limited number (a few hundred) of isolates are used, it can be safely assumed that the method lacks sufficient discriminatory power to produce epidemiologically relevant data.

With highly variable target sequences, such as VNTRs in MLVA, some variation often occurs during an outbreak (Hyytia-Trees *et al.*, 2006). As a result, data interpretation can become complicated since the data cannot be treated as black and white anymore, i.e. should an isolate that has a pattern that is slightly different from the outbreak pattern be considered belonging to the outbreak or should it be treated as 'different'. Therefore, it is highly advisable that the validation includes a large number (dozens) of isolates from the same outbreak so that interpretation guidelines can be established on the amount of variation that can be tolerated for that organism during an outbreak. The accuracy of the epidemiological data is imperative in this process, as it will be the primary source of information used to generate the guidelines. The longer the outbreak continues, the less accurate the epidemiological data tend to be, particularly if patients are interviewed long after the exposure took place. An ideal outbreak for determining data interpretation guidelines would be one that lasted a relatively short period of time (less than two months) and from which a culture confirmed source was identified. As an example, in August–September 2006, a large *E. coli* O157 outbreak associated with consumption of raw bagged spinach took place in the United States (CDC, 2006). After subtyping over 260 isolates associated with this outbreak using the PulseNet MLVA protocol for *E. coli* O157 (Hyytia-Trees *et al.*, 2006), it was determined that variations that included up to three repeats at one locus (out of a total eight loci) or one repeat at two loci should be allowed (unpublished data). It was also determined that the ninth locus originally included in the scheme was too variable to be informative in the course of an outbreak and was consequently excluded from the final (standardized) protocol (Hyytia-Trees *et al.*, 2010). However, coming across large short-term outbreaks like the one described above may not always be feasible for pathogens with long incubation periods, such as *L. monocytogenes* (Kathariou *et al.*, 2006). It must also be noted that not all VNTRs evolve at the same rate, and in some cases, even a difference of a single repeat may be epidemiologically significant (Vogler *et al.*, 2006). An example of this kind of a situation was discovered in summer 2008 in the United States when the PulseNet MLVA protocol was used to characterize *E. coli* O157 isolates belonging to a large outbreak associated with consumption of ground beef (unpublished data). An isolate from a patient that did not seem to be epidemiologically linked to the outbreak displayed a MLVA pattern that differed from the outbreak pattern at one locus by one repeat. However, this VNTR contained a repeat motif that is relatively long compared to the other VNTRs (18 base pairs versus six to seven base pairs) included in the protocol. It is known that VNTRs containing longer repeat motifs tend to be more conserved (van Belkum *et al.*, 1998). Hence, in these types of loci more weight needs to be put even on a single repeat difference.

10.5 External validation

The external validation phase is designed to further test the robustness of the PulseNet protocol in laboratories outside the agency that developed it. The idea is to include laboratories with different levels of expertise, resources, and preferences in terms of some of the reagents and equipment needed to perform the assay. External validations have been part of PulseNet USA's protocol development strategy since the network was established in 1996 (Swaminathan *et al.*, 2001). However, it is important to mention that not all external validations are created equal. The external validation strategy to be followed may vary depending on the organism, the types of isolates available for testing (culture collection available), and the overall experience the developing laboratory and validation participants have with the organism and the method in question. Selection of laboratories that will participate in the external validation phase is achieved by carefully examining the capacity of the laboratories that volunteer to participate to ensure that they are a mixture of laboratories that are representative of the ones that are or will become part of the network. In the simplest cases, the external validation phase is conducted in three to five laboratories (Ribot *et al.*, 2001) or in as many as 10–18 laboratories. PulseNet external validations involving newer or less 'proven' methods tend to include a large number of laboratories to help identify any problems that may have gone undetected during the development and internal validation phases (Hyytịa-Trees *et al.*, 2010). Unfortunately, validations with a large number of laboratories tend to be as problematic as they are informative and are often plagued with long delays.

A typical external validation consists of the testing of a predetermined number of 'known' isolates provided by the developing laboratory; usually 7–11 isolates while others included 50 or more isolates, including duplicates used to test the reproducibility of the method. Alternatively, the external validation can be divided into two subphases; one that requires analysis of a set of 'known' isolates followed by the prospective subtyping of routine isolates received by each participating laboratory during three to four months after the first validation phase was completed or in the weeks, months or years preceding the validation study. The next step is to establish timelines for completion of the validation study. The amount of time the laboratories will have to run the isolates will depend on the number of isolates or restriction reactions (one enzyme versus two enzymes) they will have to run and the throughput of the method in question (low throughput PFGE versus high throughput MLVA). A typical PulseNet external validation study provides laboratories with a period of three to eight weeks for completion of the required analysis (running the isolates and analyzing the results in BioNumerics). While timelines should be observed strictly, sometimes extensions need to be granted to some participants (usually one to two week extensions are necessary in cases where the laboratories involved had to re-focus efforts on higher priority issues, such as outbreak support, loss of personnel or resources). When conducting validations with a large number of laboratories, it is best to reach a consensus between all participants on the actual number of isolates to be tested, the format they will be analyzed in, and the timeline for completion prior to initiating the validation study. This will increase the probability that the study will be completed successfully and without unnecessary delays.

Once the laboratories have completed their tests they are required to send both the raw data (gel image files for PFGE, peak files for MLVA) and BioNumerics bundle files (analyzed data) to the coordinating laboratory for analysis and confirmation of their results. This allows the coordinating laboratory to assess the subject's ability to generate high quality data and to perform the (computer) analysis correctly. The coordinating laboratory will also have a pre-determined amount of time (two to four weeks) to go over the data, and is charged with generating a summary report of the findings to all laboratories involved. A conference call is scheduled with all participants to discuss the results of the study and make a determination as to whether or not adjustments to the protocol are needed before proceeding to the next phase. If the results of the validation are satisfactory the next phase, establishment of a database, is initiated. If it is concluded that additional modifications are needed the protocol is revised accordingly and another external validation is conducted as described above.

10.5.1 Common problems encountered during external validation of standardized methods

The most commonly seen problems encountered during external validations of PulseNet protocols, regardless of the method being validated, usually stem from differences in reagents and equipment used by the participating laboratories (Kam *et al.*, 2008). Even though attempts are made in the protocol development phase to minimize the introduction of variables that might reduce the robustness or reproducibility of the method, it is not always possible or practical to do so. For instance, standard operating procedures (SOPs) provided to the participating laboratories contain detailed instructions about the process and the type of reagents (e.g., the brand of *Taq* polymerase) and equipment (turbidity meter or thermocyler) to be used with a particular assay. Typically, these are the equipment and reagents available to the laboratory responsible for the development of the method being evaluated. Ideally, the laboratories participating in the evaluation would use the exact same reagents and equipment listed in the SOPs. A way to approach this could be to provide all laboratories with the same equipment and reagents used in the development phase of the protocol. Unfortunately, this is not a practical approach given that it would be extremely costly to provide the infrastructure for every laboratory in a network as large and diverse as PulseNet. Furthermore, if the protocols are to be implemented in different countries or even in different regions of the world, laboratories are likely to have difficulty purchasing specific brand names, particularly if there is no authorized distributor for that product in that region. The approach PulseNet has taken is to allow a certain degree of flexibility on the use of some reagents (brand names), while explicitly stating which reagents must not be substituted when performing a given assay. In the case of PFGE, all laboratories must use a specific agarose to cast the gels (SeaKem Gold, Lonza, Rockland, ME), while laboratories can use a variety of vendors to purchase the electrophoresis running buffer, as long as its chemical composition and pH are the same as the one indicated in the SOP. Similarly, MLVA SOPs require that a specific *Taq* polymerase be used when performing the PulseNet protocols. Testing some of the most commonly used *Taq* polymerases (in the US) revealed differences between enzymes that affected fragment sizing and analysis of the MLVA products (unpublished data). Some VNTRs may also not amplify at all if more stringent *Taq* is used than what was specified in the protocol. Similarly, using deoxyribonucleotides containing dioxyuridine-triphosphate (dUTP) instead of dioxythymidine-triphosphate (dTTP) can cause sizing discrepancies. The equipment used, if different than recommended in the SOP, could also impact the quality or reproducibility of the data. PCR-based applications typically do not transfer well from one thermocycler model to another and when the feat is attempted it often results in significant differences in the amplification efficiency of the reactions or may cause the amplification reactions to fail altogether. In some thermocycler models, PCR tubes may work better than PCR plates. In other cases, poor amplification efficiency can be explained by non-optimal calibration status of the thermocycler. Different lots of PCR primers can also have an effect on the amplification efficiency. Because of all the different variables discussed above, each laboratory typically will have to re-optimize the primer concentrations in order to get an optimal amplification for all targets (Hyytia-Trees *et al.*, 2010).

Protocols developed by PulseNet are often tested with a variety of reagents and different types of equipment in order to facilitate their implementation and minimize the efforts the laboratories may have to engage in to optimize protocols for the equipment and reagents they are likely to have. In the case of PFGE protocols, restriction enzymes from the main vendors are tested to ensure that no appreciable differences are observed in the efficiency of the restriction reactions and the quality of the patterns. Different type of devices used to adjust the concentration of cell suspensions, such as turbidity meters and spectrophotometers, are also evaluated and an optimal unit range for each device is provided in all PulseNet PFGE protocols. Fortunately, the electrophoresis step is not as complicated because there is only one manufacturer of PFGE equipment with the clamp homogenous electric field (CHEF) technology used by PulseNet. For MLVA, PulseNet has tested a variety of thermocyclers, and not all perform equally well. Since public health laboratories in the USA are interested in implementing the PulseNet MLVA protocols it was important for us to develop protocols based on the capillary electrophoresis platforms most commonly used in that setting: the Genetic Analyzer 3130xl (Applied Biosystems, Foster City, CA) and the CEQ™8000 Series Genetic Analyzer System (Beckman Coulter, Fullerton, CA). As a result, PulseNet has engaged in a lengthy and thorough development and validation processes to ensure that all MLVA protocols can be performed in either platform and the data produced by both platforms can be compared in the same reference database (Hyytia-Trees *et al.*, 2010).

10.6 Establishment of reference databases and a QA/QC program

For the protocol to become an official standardized PulseNet method, it must be accompanied by a carefully crafted and fully functional national database and a QA/QC program. In order to achieve this, the BioNumerics software is customized to accommodate the proper data fields associated with each organism for which a standardized protocol exists. The functionality of the database must also be tested before it goes 'on-line' to ensure that all users are able to connect, upload, query and gather the necessary information without complications. In essence, the database must also be subjected to a validation process conducted by the PulseNet USA Database Team and the information technology support staff.

The implementation of a quality assurance program is vital to the success of any data library, in particular if the information being archived is generated by a network of laboratories. Since the data deposited in the databases/archives will be 'seen' and used by others it is imperative that such data be of the highest quality possible. PulseNet's approach to quality assurance consists of several activities including training, data curation by central database managers housed at Centers for Disease Control and Prevention (CDC, Atlanta, GA), a certification process all participants must complete successfully prior to submitting data to any of the databases, and an annual proficiency testing (PT) challenge all certified participants must pass for each organism they are certified for. A QA/QC manual listing the SOPs for all techniques implemented in PulseNet is available electronically via the internet to all participating laboratories. In addition, the QA/QC manual lists all requirements participating laboratories must comply with in the areas of data documentation, record keeping, communication and administrative policies. The certification and PT programs were implemented solely to ensure that the data generated are of the highest quality possible.

Before laboratories can upload/submit data directly to the PulseNet USA national databases they must successfully complete certification for the laboratory procedure (PFGE or MLVA) and the data analysis in BioNumerics. In the certification process, a set of well-characterized (known PFGE or MLVA patterns) isolates, typically four to eight isolates per set (Table 10.1), is provided to the laboratory by CDC. After performing the laboratory testing, the gel images (PFGE patterns) or MLVA peak files are then analyzed by the laboratory staff in BioNumerics. Once analysis of the PFGE or MLVA data is completed, the laboratory creates a file containing the raw data (a gel image for PFGE or a peak file for MLVA) and a 'bundle file' containing the results of the analysis performed in BioNumerics. The certification files are sent to CDC for evaluation. In the case of PFGE, the raw data are used to evaluate the technical skills of the staff in plug preparation, gel electrophoresis, and image acquisition (image capture). In the case of MLVA, the raw data are used to evaluate the staff's laboratory skills in PCR and fragment analysis. The analysis results included in the bundle file are used to evaluate the staff's ability to perform the computerized BioNumerics analysis accurately and consistently by comparing their results to the analysis performed by CDC staff. Laboratories that do not pass either or both of the certifications steps are provided with a report containing recommendations and troubleshooting tips on how to improve on their technique.

Individuals from PulseNet USA participating laboratories must participate in the annual PT program for the organism(s) they have been certified for. The PT challenge process is conducted much like the certification process, but only one strain per certified protocol is usually provided to the certified laboratories by CDC. The PT strain is to be tested together (on the same PFGE gel or MLVA run) with routine isolates received by the laboratory. The raw data and the analyzed files are uploaded to an on-line PT database housed at CDC and the results are evaluated and graded by the CDC staff. Laboratories that fail one PT round will receive recommendations on how to improve their technique and must submit PT results again six months later for that pathogen. Laboratories that fail two times in a row lose their certification, and as a result cannot upload data to the national database. They must successfully complete the certification process again before the full database access (privileges) can be re-instated.

10.7 Future trends

Recently, there has been a push in microbiological diagnostics to move away from culture methods to direct detection and characterization of pathogens within clinical samples. For foodborne organisms, this trend will pose an enormous challenge due to the abundant normal microflora present in clinical samples. During the method development and validation process, particular emphasis would be placed on the specificity of the assay which is typically not a concern when working with pure cultures. With PCR-based assays, possible PCR inhibition due to a high level of background DNA or any other PCR inhibitors present in clinical samples would have to be evaluated. Ideally, a validation panel should contain both naturally contaminated and spiked samples. However, an access to a wide selection of naturally contaminated samples may pose a problem since the primary samples are typically processed by hospital laboratories or commercial diagnostic companies which usually do not save the original samples.

Once the curated reference libraries, such as PulseNet databases, grow in size it will become critical to find solutions to be able to maintain the ability to easily query historical data. Currently, the BioNumerics databases have an upper size limit, and once that is exceeded, older data need to be archived, and will not be readily available for the participating laboratories. Even though the PulseNet databases are mainly used for active surveillance in the form of short-term cluster detection, it is important to recognize that these data can be utilized in many other ways that can help food safety and public health, including attribution studies (Batz *et al.*, 2005) and tracking the emergence of multi-drug resistance clones (Torpdahl *et al.*, 2007). Therefore, for detection of trends, the availability of long-term historical data is absolutely critical.

Molecular subtyping data for foodborne organisms will be increasingly used also for forensic purposes due to high-profile outbreaks that result in product recalls and class action or personal lawsuits. Individuals involved in the legal systems often want to quantify the differences between two strains, i.e. they want to get a good estimate of the probability that one strain is a derivative of the other. With non-target specific methods, such as PFGE, it is impossible to give such an estimate because there is no way of telling the reason for any band difference. However, with MLVA, estimating relationships like that is possible, but doing so accurately takes a lot of effort. You would have to determine the frequency of the different alleles in the population under study, the mutation rates in each VNTR and the frequencies of the different types of mutations (for example, did the mutation result in one versus a five-repeat difference). In other words, what is needed to accomplish this is a large unbiased database, and a comprehensive mutation study involving multiple isolates (Vogler *et al.*, 2006). It must be noted that surveillance databases, such as PulseNet databases, are heavily biased towards outbreak-related strains and therefore cannot be used to determine allele frequencies in the general population.

10.8 Sources of further information and advice

Further information can be sought from the following sources:

- Clinical and Laboratory Standards Institute (CLSI): http://www.clsi.org. 'User protocol for evaluation of qualitative test performance; approved guideline – second edition'.
- International Organization for Standardization (ISO): http://www.iso.org. 'ISO 16140:2003 Microbiology of food and animal feeding stuffs – Protocol for the validation of alternative methods'.
- Association of Analytical Communities (AOAC) International: http://www.aoac.org. 'AOAC official methods program manual'.

10.9 Disclaimer

- Use of trade names is for identification only and does not imply endorsement by the Centers for Disease Control and Prevention or by the US Department of Health and Human Services.
- The findings and conclusions in this chapter have not been formally disseminated by the Centers for Disease Control and Prevention and should not be construed to represent any agency determination or policy.

10.10 References

BATZ M B, DOYLE M P, MORRIS G, JR., PAINTER J, SINGH R, TAUXE R V, ET AL. (2005), 'Attributing illness to food', *Emerging Infectious Diseases*, **11**, 993–9.

BOXRUD D, PEDERSON-GULRUD K, WOTTON J, MEDUS C, LYSZKOWICZ E, BESSER J, ET AL. (2007), 'Comparison of multiple-locus variable-number tandem repeat analysis, pulsed-field gel electrophoresis, and phage typing for subtype analysis of *Salmonella enterica* serotype Enteritidis', *Journal of Clinical Microbiology*, **45**, 536–43.

CDC (2006), 'Ongoing multistate outbreak of *Escherichia coli* serotype O157:H7 infections associated with consumption of fresh spinach – United States, September 2006', *MMWR. Morbidity and Mortality Weekly Report*, **55**, 1045–6.

CDC (2007), 'Multistate outbreak of *Salmonella* serotype Tennessee infections associated with peanut butter – United States, 2006–2007', *MMWR. Morbidity and Mortality Weekly Report*, **56**, 521–4.

CDC (2008), 'Outbreak of *Salmonella* serotype Saintpaul infections associated with multiple raw produce items – United States, 2008', *MMWR. Morbidity and Mortality Weekly Report*, **57**, 929–34.

COOPER K L, LUEY C K, BIRD M, TERAJIMA J, NAIR G B, KAM K M, ET AL. (2006), 'Development and validation of a PulseNet standardized pulsed-field gel electrophoresis protocol for subtyping of *Vibrio cholerae*', *Foodborne Pathogens and Disease*, **3**, 51–8.

CORKILL J E, GRAHAM R, HART C A and STUBBS S (2000), 'Pulsed-field gel electrophoresis of degradation-sensitive DNAs from *Clostridium difficile* PCR ribotype 1 strains', *Journal of Clinical Microbiology*, **38**, 2791–2.

DE VALK H A, MEIS J F, BRETAGNE S, COSTA J M, LASKER B A, BALAJEE S A, ET AL. (2009), 'Interlaboratory reproducibility of a microsatellite-based typing assay for *Aspergillus fumigatus* through the use of allelic ladders: proof of concept', *Clinical Microbiology Infectious*, **15**, 180–7.

GERNER-SMIDT P, HISE K, KINCAID J, HUNTER S, ROLANDO S, HYYTIA-TREES E, ET AL. (2006), 'PulseNet USA: a five-year update', *Foodborne Pathogens and Disease*, **3**, 9–19.

GORGE O, LOPEZ S, HILAIRE V, LISANTI O, RAMISSE V and VERGNAUD G (2008), 'Selection and validation of a multilocus variable-number tandem-repeat analysis panel for typing *Shigella spp*', *Journal of Clinical Microbiology*, **46**, 1026–36.

GRAVES L M and SWAMINATHAN B (2001), 'PulseNet standardized protocol for subtyping *Listeria monocytogenes* by macrorestriction and pulsed-field gel electrophoresis', *International Journal of Food Microbiology*, **65**, 55–62.

GRIMONT P A D and WEILL F-X (2007), *Antigenic Formulae of the Salmonella Serovars*, WHO Collaborating Centre for Reference and Research on *Salmonella*, Paris

HUNTER P R and GASTON M A (1988), 'Numerical index of the discriminatory ability of typing systems: an application of Simpson's index of diversity', *Journal of Clinical Microbiology*, **26**, 2465–6.

HYYTIA-TREES E, LAFON P, VAUTERIN P and RIBOT E M (2010), 'Multilaboratory validation study of standardized multiple-locus variable-number tandem repeat analysis protocol for Shiga toxin-producing *Escherichia coli* O157: a novel approach to normalize fragment size data between capillary electrophoresis platforms', *International Journal of Food Microbiology*, **7**, 129–36.

HYYTIA-TREES E, SMOLE S C, FIELDS P A, SWAMINATHAN B and RIBOT E M (2006), 'Second generation subtyping: a proposed PulseNet protocol for multiple-locus variable-number tandem repeat analysis of Shiga toxin-producing *Escherichia coli* O157 (STEC O157)', *Foodborne Pathogens and Disease*, **3**, 118–31.

KAM K M, LUEY C K, PARSONS M B, COOPER K L, NAIR G B, ALAM M, ET AL. (2008), 'Evaluation and validation of a PulseNet standardized pulsed-field gel electrophoresis protocol for subtyping *Vibrio parahaemolyticus*: an international multicenter collaborative study', *Journal of Clinical Microbiology*, **46**, 2766–73.

KATHARIOU S, GRAVES L, BUCHRIESER C, GLASER P, SILETZKY R M and SWAMINATHAN B (2006), 'Involvement of closely related strains of a new clonal group of *Listeria monocytogenes* in the 1998–99 and 2002 multistate outbreaks of foodborne listeriosis in the United States', *Foodborne Pathogens and Disease*, **3**, 292–302.

KUDVA I T, EVANS P S, PERNA N T, BARRETT T J, AUSUBEL F M, BLATTNER F R, ET AL. (2002), 'Strains of *Escherichia coli* O157:H7 differ primarily by insertions or deletions, not single-nucleotide polymorphisms', *Journal of Bacteriology*, **184**, 1873–9.

LINDSTEDT B A, BRANDAL L T, AAS L, VARDUND T and KAPPERUD G (2007), 'Study of polymorphic variable-number of tandem repeats loci in the ECOR collection and in a set of pathogenic *Escherichia coli* and *Shigella* isolates for use in a genotyping assay', *Journal of Microbiological Methods*, **69**, 197–205.

NYGARD K, LINDSTEDT B A, WAHL W, JENSVOLL L, KJELSO C, MOLBAK K, ET AL. (2007), 'Outbreak of *Salmonella* Typhimurium infection traced to imported cured sausage using MLVA-subtyping', *Euro Surveillance*, **12**, E070315 5.

PARSONS M B, COOPER K L, KUBOTA K A, PUHR N, SIMINGTON S, CALIMLIM P S, ET AL. (2007), 'PulseNet USA standardized pulsed-field gel electrophoresis protocol for subtyping of *Vibrio parahaemolyticus*', *Foodborne Pathogens and Disease*, **4**, 285–92.

RIBOT E M, FAIR M A, GAUTOM R, CAMERON D N, HUNTER S B, SWAMINATHAN B, ET AL. (2006), 'Standardization of pulsed-field gel electrophoresis protocols for the subtyping of *Escherichia coli* O157:H7, *Salmonella*, and *Shigella* for PulseNet', *Foodborne Pathogens and Disease*, **3**, 59–67.

RIBOT E M, FITZGERALD C, KUBOTA K, SWAMINATHAN B and BARRETT T J (2001), 'Rapid pulsed-field gel electrophoresis protocol for subtyping of *Campylobacter jejuni*', *Journal of Clinical Microbiology*, **39**, 1889–94.

SCHWARTZ D C and CANTOR C R (1984), 'Separation of yeast chromosome-sized DNAs by pulsed field gradient gel electrophoresis', *Cell*, **37**, 67–75.

SHANNON C E and WEAVER W (1949), *The mathematical theory of communication*, Urbana: University of Illinois Press.

SILBERT S, BOYKEN L, HOLLIS R J and PFALLER M A (2003), 'Improving typeability of multiple bacterial species using pulsed-field gel electrophoresis and thiourea', *Diagnostic Microbiology Infectious Disease*, **47**, 619–21.

SPERRY K E, KATHARIOU S, EDWARDS J S and WOLF L A (2008), 'Multiple-locus variable-number tandem-repeat analysis as a tool for subtyping *Listeria monocytogenes* strains,' *Journal of Clinical Microbiology*, **46**, 1435–50.

SWAMINATHAN B, BARRETT T J, HUNTER S B and TAUXE R V (2001), 'PulseNet: the molecular subtyping network for foodborne bacterial disease surveillance, United States', *Emerging Infectious Disease*, **7**, 382–9.

TORPDAHL M, SORENSEN G, LINDSTEDT B A and NIELSEN E M (2007), 'Tandem repeat analysis for surveillance of human *Salmonella* Typhimurium infections', *Emerging Infectious Disease*, **13**, 388–95.

VAN BELKUM A, SCHERER S, VAN ALPHEN L and VERBRUGH H (1998), 'Short-sequence DNA repeats in prokaryotic genomes', *Microbiology and Molecular Biology Reviews*, **62**, 275–93.

VAN BELKUM A, TASSIOS P T, DIJKSHOORN L, HAEGGMAN S, COOKSON B, FRY N K, ET AL. (2007), 'Guidelines for the validation and application of typing methods for use in bacterial epidemiology', *Clinical Microbiology and Infection*, **13**, 1–46.

VOGLER A J, KEYS C, NEMOTO Y, COLMAN R E, JAY Z and KEIM P (2006), 'Effect of repeat copy number on variable-number tandem repeat mutations in *Escherichia coli* O157:H7', *Journal of Bacteriology*, **188**, 4253–63.

Part III

Molecular methods, genomics and other emerging approaches in the surveillance and study of foodborne pathogens

11

Sample preparation for the detection of foodborne pathogens by molecular biological methods

P. Rossmanith, Christian-Doppler Laboratory for Molecular Biological Food Analytics, Austria and M. Wagner, Department for Farm Animals and Public Veterinary Health, Austria

Abstract: Adequate sample preparation is the key to the successful broad-range application of molecular biological methods, such as the polymerase chain reaction in food analysis. The prerequisites for sample preparation are discussed and separation of bacterial targets from food matrices is introduced as the main methodological approach. A broad overview of several approaches to sample preparation using different separation principles is extensively reviewed. Physical, biological and chemical methods are explained and their individual pros and cons are discussed in brief including recent developments of sample preparation based on chemical digestion of the food sample which brings the matrix into aqueous solution.

Key words: sample preparation, food matrix, food pathogen detection, rapid detection methods, bacterial target separation.

11.1 Introduction

The rapid detection and subsequent identification of pathogens in foods has become a major food-science topic, driven mainly by the fact that foodstuffs are nowadays traded on a global scale. Alienated by 'food scandals' such as BSE and the impact of large outbreaks caused by foodborne microbial agents, stakeholder organizations have expressed concern about food safety worldwide. As a consequence, evaluation-driven concepts such as the risk assessment approach have been set into force to lay a scientific basis for handling the consequences of foodborne hazards in food supply chains. Quantitative detection of pathogens is increasingly important when risk assessments should be focused and integrative.

Time plays a key role in analyzing complex matrices sampled from food processing chains, which can include fecal material, swabs of food processing surfaces or carcasses and feedstuffs of animal or plant origin.

The availability of rapid detection methods is crucial in food testing and food safety management. However, the aim of an investigation will strongly influence the choice of the method used. It makes a difference whether an investigation is performed for the purpose of official food control, self-control or research-driven activities such as quantitative microbial risk assessments (QMRA).

Traditional microbiological methods are limited to cells capable of growing on nutrient-rich media. However, although these methods are reliable and widely standardised, they are time-consuming and expensive to perform, especially in presumed positive cases where species identification by biochemical profiling is warranted. Enrichments, aside from saving time, share the disadvantage of being inappropriate for non-cultivable microorganism, or microorganisms in a non-cultivable state. More rapid methods have been developed to shorten detection time by the use of chromogenic media, or advanced detection principles such as DNA hybridization or enzyme immunoassay. However, these methods still require target enrichment as the detection limit is relatively high.

DNA amplification via polymerase chain reaction (PCR) amplifies a sequence instead of a cell, and thus leads to specific detection of the target on the basis of detecting species-specific genetic signatures. Further development in other DNA-based methods has been made, and microarrays on-chip PCR and parallel sequencing in particular have established their applicability as scientific tools. Nevertheless, the advantages of these methods have stipulated new prerequisites. The theoretical performance of these methods is challenged when environmental influences or foodstuff-dependent interferences come into play. Therefore, sample preparation plays a key role in enabling the application of molecular biological methods. This, until now a neglected field, is gaining importance in connection with the application of high-tech detection methods and sophisticated analysis to the complex issue of global food safety.

11.1.1 Terms and definitions

Terminology plays a major role in the whole process of quantifying pathogens in foodstuffs – from the food sample to the results obtained by the various molecular detection methods. Exact definitions of the various steps are necessary to understand the whole process and the specific requirements of the individual steps (Fig. 11.1).

The first step within such a protocol for nearly all of the methods used involves homogenization of the food sample. The second step, which is mainly covered in this chapter, includes the separation of the target bacteria from the sample matrix. Within sample treatment, this separation of the targets is one step in a multistep process, which includes additional steps facilitating this separation process. The most frequently targeted molecule is DNA. Therefore, subsequent steps involve cleavage of the bacterial cell wall, separation of the surrounding cellular

	Workflow	Elapsed time
Matrix lysis	6.25 to 12.5 g/ml foodstuff	
	⇩	
	Homogenizing: 3 × 10 min, in matrix lysis buffer	30 min
	⇩	
	Solubilization: 30 min, 45°C, in matrix lysis buffer	60 min
	⇩	
	Separation/centrifugation: 3220 × g; 30 min	90 min
	⇩	
	Washing: 30 min, 45°C, in wash buffer	120 min
	⇩	
	Separation/centrifugation: 3220 × g; 30 min	150 min
	⇩	
	Final wash steps: 2 times in 1 × PBS, 5000 × g; 5 min	180 min
	⇩	
	Bacterial cell pellet: 5 – 200 μl	
Target detection	⇩	
	Bacterial cell wall: enzymatic digestion	> 320 min
	⇩	
	Cell disruption: nucleospin tissue kit	> 335 min
	⇩	
	DNA purification: nucleospin tissue kit	> 365 min
	⇩	
	Target detection: real-time PCR (*prfA*)	> 485 min

Fig. 11.1 Workflow of molecular biological detection of food pathogens. Sample preparation using matrix lysis is presented to illustrate the various steps and definitions involved in sample treatment (Mayrl *et al.*, 2009). The steps necessary to extract the targets from the foodstuff into the bacterial pellet after centrifugation and washing are summarized as sample treatment. The following detection steps have to be considered separately.

constituents and purification of the DNA for subsequent DNA-based molecular biological methods.

In short, sample treatment methods cover separation of the target cells from the surrounding food matrix. Downstream methods such as cell disruption, DNA purification and detection, for example, via real-time PCR, have to be taken into consideration individually while designing or choosing a suitable sample preparation method. Sample treatment is defined as the pre-detection step in the method protocol primarily necessary for reduction of the sample volume while maintaining the original target number. This concentrates the target cells and removes inhibitory substances that would otherwise hinder subsequent molecular biological detection methods. To achieve these objectives, one or more consecutive methodological steps may be necessary. The term *target* stands for the *analyte* as defined in analytical chemistry and encompasses bacterial cells, spores, viruses and other cells such as yeasts and plant cells, for example in the context of detecting genetically modified organisms (GMOs).

11.1.2 Prerequisites and challenges for sample preparation

Three challenges for molecular biological detection of pathogens from food currently exist: The required food sample volume (<25 g) contrasts with conventional sample volumes used for molecular biological methods (<250 μl), the low target copy numbers and the heterogeneous composition of food samples.

A typical 'food sample' simply does not exist. Instead, this term covers the range of all given biological matrices composed of heterogeneous chemical compounds such as fats, proteins, carbohydrates, structural compounds and others (Table 11.1). For example, in milk science foodstuffs can contain fat ranging from less than 1.5% (e.g. skim milk) to more than 80% (e.g. butter). Most molecular biological detection methods, especially real-time PCR, are *in vitro* biochemical

Table 11.1 Composition of the most frequently cited foodstuffs in sample treatment related articles

	Fat	Protein	Carbohydrates	H_2O	pH
Dairy products	%	%	%	%	
Raw milk	~ 5.4	~ 3.7	~ 4.8	~ 86.1	~ 6.5
UHT milk	~ 3.5	~ 3.3	~ 4.7	~ 88.5	~ 6.7
Hard cheese	30–50	~ 30	< 1	~ 35	–
Animal products					
Meat	1.5	21.5	< 1	76	~ 5.8
Fish	0.3–26	17–20	< 1	61–82	~ 6.2
Ice cream					
Sorbet	< 1	< 1	32	65	> 2
Ice cream	2.5	2.3	13.3	80	–

UHT: ultra high tempered

systems, which could be impaired by a range of degradation effects, fluorescence phenomena and other forms of inhibition. Real-time PCR enables, in theory, quantification down to one single nucleic acid molecule per sample (Rossmanith *et al.*, 2006). Nevertheless, these possibilities only apply to counts of pure DNA molecules in a neutral matrix. Extensive requirements must be met to achieve such a performance with a real food sample due to the inhibitory effects caused by foodstuff components on PCR (Al Soud and Radstrom, 2001; Radstrom *et al.*, 2004; Rossen *et al.*, 1992; Wilson, 1997).

The enormous variety of foodstuffs exacerbates the problem. Liquid foodstuffs are easier to work with than solids, and homogeneous food matrices are less complex than composites. For example, analyzing a piece of pizza, for example, that could be made from more than ten different food items, ranging from cheese and ham to different plant materials and herbs, illustrates this problem. The idea of a horizontal approach allowing for all sorts of foodstuffs to be tested with a single molecular technique is illusory in that context. Therefore, sample preparation methods have to reduce the impact of the food matrix to the best possible concentration and enable purification of the target organisms in a way that supports subsequent steps. Development of advanced concepts for sample preparation is challenging but nevertheless important. Interestingly, at least according to the number of publications in the field, this is somewhat underrepresented (at present) in food analytical science.

Most of the existing molecular detection methods, such as real-time PCR, require pure DNA or cell targets obtained from isolation and subsequent purification of the target molecules. These applications never lack sufficient target copy numbers as the target is multiplied to produce the necessary amounts before detection. High numbers of competing flora in a food sample are to be expected as well as very low numbers of the target pathogens, often less than one cell per gram or millilitre of food. This exacerbates the problem. Thus, the development of new concepts is necessary to ensure that the few target cells are made accessible to subsequent detection if quantification is desired.

Following a decade of attempts at implementing DNA-based food testing, conclusions are sobering. These methods have been implemented when no cultural alternatives existed, such as detection of viruses or allergens. Molecular methods for the detection of important pathogens including *Salmonella* and *Listeria* have mostly been validated by national standardization bodies such as AFNOR, AOAC or Microval. However, the development of chromogenic media has also shortened the turnover time of traditional food microbiology. Nevertheless, the problem of discriminating between viable and dead cells using DNA-based technology remains unresolved. Recent research in this area has shown conflicting outcomes (Flekna *et al.*, 2007). Therefore, in the case of litigious circumstances, PCR results are still not regarded as confirmatory in most countries as a signal does not conclusively indicate the presence of a living and thus hazardous organism. Direct molecular quantification of pathogens is still beyond our reach as long as structured research regarding integrative sampling, target extraction and target detection does not solve the problem of enrichment-independent target concentration.

Hence, a way to address the heterogeneity of the sample matrices in food analysis is to take a modular approach to sample preparation protocols derived from one basic approach and to match variants for the individual compositions of the samples routinely tested. This ensures standardized handling in the laboratory and facilitates flexibility and possible automation.

11.1.3 Separation is the core task in sample preparation

Every separation process is based on difference. In food diagnosis the targets, being bacterial cells, spores and viruses, are contained within the heterogeneous composition of the food matrices (Table 11.2). Every given physical, chemical

Table 11.2 Main target organisms for sample treatment in food surveillance

	Size	Cell wall/ capsid	Main target for direct detection	Charge	Resistance[a]
Bacteria	0.5–5 µm				
Gram +		Murein layer	Genomic DNA, surface structures, proteins	negative	++[b]
Gram –		Murein layer	Genomic DNA, surface structures, proteins	negative	+
Mycobacteria		Murein layer/ mycolic acid/ lipids	Genomic DNA, surface structures, proteins	lipophilic	++
Spores	~ 1 nm	Coat/cortex/ core	Genomic DNA, surface structures, proteins	lipophilic	++++
Viruses					
Reoviridae/ Rotavirus	~ 70 nm	Three layers of capsids/inner core	dsRNA	–	+++
Calciviridae/ Norovirus	35–39 nm	Capsid of 90 protein dimers	ssRNA/RT-PCR; viral particle/ ELISA	–	+++
Picornaviridiae / Hepatitis A	27 nm	Capsid of 60 units of proteins VP 1–4	RNA/RT-PCR; viral particle/anti HAV IgM	–	+++
Mold, yeast	3–100 µm	Cellulose, chitine, polysaccharides	Genomic DNA, surface structures, proteins	–	++

[a] Overall resistance of the cell or particle against physical and chemical impacts and enzymatic digestion.
[b] Estimation based on the composition of the particle or cell wall of the respective targets.

and biological difference can be of advantage for the principles to separate these targets. Separation parameters are defined on the basis of the characteristic of these differentiating factors.

The underlying principles of separating the target cells from the food matrix are chemical, physical, physico-chemical or biochemical-biological (Stevens and Jaykus, 2004a). Every separation process exploits features of the target that are distinct from the physical or biochemical properties of the food matrix or non-target organisms. This could be recognition of molecules on cell walls of target cells that are not present in the competing flora, separation on the basis of size, hydrophoby, conductivity, and many other principles. Separation parameters are defined on the basis of the characteristics of these differentiating factors, and separation efficiency is achieved by the extent of expression of the differentiating factors. These considerations could easily result in specific protocols that are best performed only on specified target cells/food sample categories (Table 11.3).

Also, the separation principle for a sample treatment method has also to be chosen in the context of the proposed downstream detection method. The use of harsh chemistry in combination with the downstream application of, for example, immuno-affinity based detection of bacterial cells is counter-productive. Chemical treatment of the bacterial cell wall could lead to degradation of the very epitopes necessary for antibody binding. Alternatively, if the characteristic of the bacterial cell wall is used to solubilize the foodstuff selectively, and thereby lower the viscosity of the original food homogenate, such an approach could lead to enhancement of the target-matrix difference on the basis of viscosity. Consequently, centrifugation of the bacterial cells would be improved during this physical separation process. As in this particular example, the difference between target and original food sample mixture characteristics in most cases has either to be enhanced prior to separation, or a combination of several distinct differentiating factors between target and mixture characteristics has to be recognised and exploited for successful sample treatment.

11.1.4 Criteria for evaluation of sample treatment and separation methods

The ratio of reduction of the volume of the original sample is one of the parameters to consider when a sample preparation method is to be evaluated. In addition, the absolute amount of processable foodstuff should be considered. The target recovery of the method must be considered in comparison to the reduction of the original amount of foodstuff. Methods such as aqueous two-phase systems (see Section 'Adsorption onto solid phases'), for example, show 50% recovery if no separation occurs at all and only the equal distribution of the targets within the sample is monitored (assuming a 1:1 volumetric ratio of the two phases). The reproducibility of a sample preparation method is largely mirrored in the range of values for the recovery as presented by single scientific groups or within the average number of published applications. The absolute volume of the recovered sample has to be considered with regard to the usability of subsequent methods. Splitting the sample after sample treatment to fit the requirements of subsequent

Table 11.3 A short review on target/food matrix combinations that were investigated in detail recently

Separation method	Target species	Food matrices	Limit of detection	Recovery	Authors
Direct DNA isolation	*O. oeni*	0.5 ml wine	–	–	Pinzani *et al.* (2004)
Direct DNA isolation	*L. monocytogenes*	0.1 ml Gouda cheese	3.2×10^2 CFU per g	–	Rudi *et al.* (2005)
Antibodies/specific and unspecific magnetic beads	*L. monocytogenes*	1.4 and 1.6 ml whole milk, skimmed milk, water	6–60 CFU	–	Nogva *et al.* (2000)
Antibodies/magnetic beads/ flow cytometry	*L. monocytogenes*	1 ml culture media	–	7 to 23%	Jung *et al.* (2003)
Filtration	*Yeast*	Beer, wine and beverages	–	–	Thomas *et al.* (1988)
Filtration	*L. monocytogenes*	5.1 g salmon	10 CFU per g	–	Besse *et al.* (2004)
Filtration	*E. coli*	Milk	–	10–95%	Fernandez-Astorga *et al.* (1996)
Filtration	*L. monocytogenes*	Cheese and meat	< 10 CFU per g	–	Wang *et al.* (1992)
Antibodies/magnetic beads	*L. monocytogenes*	Pure cultures enriched food	10^2 CFU per ml	–	Skjerve *et al.* (1990)
Antibodies/magnetic beads	*L. monocytogenes*	Cheese	0.1–1 CFU per g	–	Uyttendaele *et al.* (2000)
Antibodies/magnetic beads	*L. monocytogenes*	Cheese	–	5–15%	Fluit *et al.* (1993)
Antibodies/magnetic beads	*Salmonella* spp.	Poultry and beef	–	31%	Cuidjoe *et al.* (1997)
Antibodies/magnetic beads	*E. coli* O157:H7 and *Salmonella* spp.	Fish	–	–	Yu and Bruno (1996)
Antibodies/magnetic beads	*Cryptosporidium parvum*	5–100 1 tab water	10^3 CFU per ml	–	Hallier-Soullier and Guillot (1999)
Antibodies/direct colony blot	*L. monocytogenes*	0.2 ml raw milk	5×10^2 / roughly evaluated	–	Belyi *et al.* (1995)
Viral binding proteins/ magnetic beads	*L. monocytogenes*	50 ml buffer solution	> 10 CFU per 50 ml	–	Niederhauser *et al.* (1994)

Viral binding proteins/ magnetic beads	*L. monocytogenes*	Milk	10 CFU per ml	–	Amagliani *et al.* (2006)
Viral binding proteins/ magnetic beads	*L. monocytogenes*	Various rinses	< 10 CFU	< 90%	Kretzer *et al.* (2007)
Aptamer/magnetic beads	*S.* Typhimurium	Chicken rinse	$1 \times 10^2 - 1 \times 10^3$ per 25 ml	–	Joshi *et al.* (2009)
AMP/silanizised glass slides/ array based biosensor	*E. coli* and *S.* Typhimurium	25 µl bacterial suspension	$6.5 \times 10^4 - 6.8 \times 10^5$	–	Kulagina *et al.* (2005)
Aqueous two-phase system	*L. monocytogenes*	Culture media	10^4 CFU per ml	1%	Lantz *et al.* (1994)
Aqueous two-phase system	*L. monocytogenes*	0.1 ml cheese homogenate	10^6 CFU per ml	50%	Lantz *et al.* (1994)
Aqueous two-phase system	*L. monocytogenes* and *S. berta*	4 g smoked Cumberland sausage	–	8 to 95%	Pedersen *et al.* (1998)
Buoyant density gradient centrifugation/16SDNA PCR	*Y. enterocolitica*	2 ml meat juice	–	30%	Wolffs *et al.* (2004)
Buoyant density gradient centrifugation/qPCR	12 bacterial pathogenes	13 foodstuffs	–	11%	Fukoshima *et al.* (2006)
Percoll media/buoyant density gradient centrifugation	*Sh. flexneri*	0.5 ml skimmed milk	5×10^2 CFU	–	Linquist *et al.* (1997)
Percoll media/buoyant density gradient centrifugation	*Sh. flexneri*	0.5 ml 1:10 diluted blue cheese	10^4 CFU	–	Linquist *et al.* (1997)
Percoll media/buoyant density gradient centrifugation	*E. coli*	0.1 ml of 1:10 diluted beef/minced beef	1.2×10^3 to 1×10^4 CFU	–	Linquist *et al.* (1997)
Adsorption/metal hydoxides/ centrifugation	*L. monocytogenes*	Non-fat dry milk	–	65–96%	Lucore *et al.* (2000)
Adsorption/metal hydoxides	*E. coli*	Beef	–	9–99%	Berry and Siragusa (1997)
Adsorption/metal hydoxides/ centrifugation	*L. monocytogenes* and *B. cereus* spores	Non-fat dry milk	–	75–90%	Cullison *et al.* (2002)

(*Continued*)

Table 11.3 Continued

Separation method	Target species	Food matrices	Limit of detection	Recovery	Authors
Adsorption/lectines	*L. monocytogenes* and *Salmonella* spp.	Broth	–	33–215%	Patchett *et al.* (1991)
Adsorption/lectines	*L. monocytogenes*	Ground beef and milk	–	13–50%	Payne *et al.* (1992)
Dielectrophoresis	*E. coli, B. subtilis* and *Micrococcus luteus*	Broth	$> 10^7$	ca. 50% (estimated)	Markx *et al.* (1996)
Differential centrifugation/100 and 3000 × g	*L. monocytogenes*	Meat homogenates	10^3-fold increased PCR limit	–	Neiderhauser *et al.* (1992)
Differential centrifugation/500 and 10000 × g/enzymatic digest	*E. coli O157:H7*	Soft cheese	10^3 CFU	–	Meyer *et al.* (1991)
Ultrasound	*E. coli*	Culture medium	$> 10^7$	72%	Limaye and Coakley (1998)
Enzymatic digest/pronase/ PCR	*C. jejuni* and *C. Coli*	2 g cheese, yoghurt or soft cheese	50 CFU per gram	–	Wegmüller *et al.* (1993)
Enzymatic digest/pronase-lysozyme-proteinaseK/PCR	*L. monocytogenes*	40 ml raw milk and 4 g of dairy products	–	–	Allmann *et al.* (1995)
Chemical digest/guanidine thiocyanate/phenol/chloroform	*L. monocytogenes, Y. enterocolitica* and *S. enteritidis*	0,5 ml raw milk	2×10^3 CFU per ml	–	Choi and Hong (2003)
Chemical digest/diethyether-$CHCl_3$ extraction/Urea-SDS	*S. aureus* and *Y. enterocolitica*	0.4 ml of pasteurised milk	–	–	Ramesh *et al.* (2002)
Chemical digest/sodium citrate/DNAzol®BD/ filtration/centrifugation	*L. monocytogenes* and *S. enterica*	11 g of plain non-fat yoghurt or Cheddar cheese	10^0 to 10^3 CFU per g	53 to 128%	Stevens and Jaykus (2004b)
Chemical digest/Urea-SDS/ centrifugation	*L. monocytogenes*	Dairy products, egg, blood, ice cream	7.8 CFU per g	39.50%	Rossmanith *et al.* (2007)

Chemical digest/Urea-Lutensol/centrifugation	*L. monocytogenes*, *S.* Typhimurium, *E. coli*, *B. cereus*, *S. aureus*	Dairy products, egg, blood, meat, fish, ice cream	1.3 to 11.2 CFU per g	39.5 to 54.7%	Mayrl *et al.* (2009)
Chemical digest/Ionic liquids/centrifugation	*L. monocytogenes*, *S.* Typhimurium	Dairy products, egg, blood, meat, fish, ice cream	–	40–60%	Mester (pers. comm.)
ELISA	*L. monocytogenes*, *S.* Typhimurium	Pork carcasses	–	–	Fravalo *et al.* (2001)
FISH	Bacteria (16S rRNA)	Cheese	–	–	Ercolini *et al.* (2003)
Flow cytometry	*L. monocytogenes*	Seeded and naturally contaminated milk	10^2 to 10^3 CFU per ml	–	Donnely *et al.* (1986)
Chromogenic enrichment media/plate count	*Listeria* spp.	Fish, dairy products	$> 10^5$ to 10^8 CFU	86–98%	Stessl *et al.* (2009)

pers. comm.: personal communiciation

methods is not beneficial. The detection limit is important in order to estimate tolerated lower limits of pathogen contamination. The elimination of background target molecules, such as genomic DNA in the case of subsequent DNA-based detection methods, has to be taken into account too. Furthermore, secondary parameters such as costs, handling performance, and equipment are necessary considerations that should not be overlooked.

11.2 Physical separation methods used in sample preparation

11.2.1 Filtration

Filtration as a separation method is made possible by the differences in particle sizes of separate target cells and food matrix particles. This physical separation method is rapid, simple and inexpensive. Either the targets can be retained on the filter, discarded with the filtrate containing the food sample or the reverse filtrate can contain the targets. Filtration often involves sieving, which is a related but nevertheless different method. Sieving combined with filtration offers better results when used in succession and provides better separation of heterogeneous food remnants and a more homogeneous sample for filtration. Likewise, subsequent filtering using decreasing pore sizes theoretically improves precision and recovery. The recoveries of target cells demonstrated in recent publications using filtration for sample treatment vary from 10–100%. The latter recoveries have been obtained by filtration of beer, wine and other liquid, low-content beverages (Thomas *et al.*, 1988). Milk had to be treated enzymatically before filtration (Fernandez-Astorga *et al.*, 1996). Besse *et al.* (2004) described membrane filtration with subsequent enumeration of *Listeria monocytogenes* on selective media from cold-smoked salmon, whereby a volume of 5.1 g was investigated and a detection limit of 10 CFU per gram was achieved. Wang *et al.* (1992) examined the filtrates of meat and cheese samples with subsequent PCR detection and achieved a detection limit of <10 CFU for *L. monocytogenes*. However, solid foodstuffs and their homogenates are problematic due to clogging of the filter and this limits the range of processable foodstuffs (Bylund, 1995). Additionally, absorption of the bacterial targets to the filter material or captured foodstuff particles influence the performance of the method. Nevertheless, the major advantage of filtration is the nearly unlimited sample volume that can be processed.

11.2.2 Dielectrophoresis and ultrasound

Dielectrophoresis as a separation method is based on the recognition of negative charges on the target cells, which enables the movement of the targets in a high-frequency electric field (0.1–10 MHz). Dielectrophoresis has yielded approximately 50% recovery from pure bacterial cultures. A minimum concentration of 10^7 CFU per millilitre is necessary. However, the viability of the target cells is questionable and the direct application to food is limited by the various chemical reactions (e.g. oxidation and reductions on both electrodes) inherent in the food composition (Hamann and Vielstich, 2005).

Sonication of samples leads to concentration of particles in resonance nodes dependent on the wavelength and energy used. Although ultrasound has led to poor recoveries when less than 10^7 cells were applied, *Saccharomyces cerevisiae* and *Escherichia coli* have been obtained with 96% and 72% recovery respectively, using 5 ml samples of culture media (Limaye and Coakley, 1998).

11.2.3 Centrifugation

Centrifugation exploits centrifugal forces to separate particles suspended in a liquid medium. Sedimentation is basically dependent on particle density and diameter, viscosity of the liquid and the relative centrifugation force applied. Relative centrifugation force is the basis of the classification of centrifugation. However, high-speed centrifugation (>60 000 × g) is not applicable to foodstuff sample treatment due to the generation of nearly un-suspendable pellets. Thus, this application is restricted to the concentration of bacterial targets from low volumes and liquids loaded with a low amount of food particles (Fliss *et al.*, 1991; Tjhie *et al.*, 1994). Low-speed centrifugation can further be subdivided into centrifugation at less than 1000 × g and centrifugation between 1000 and 8000 × g. When forces lower than 1000 × g are applied food particles sediment out. Thereafter, when forces greater than 1000 × g are applied, bacterial cells can be isolated (Koch and Blumberg, 1976).

Differential centrifugation

Differential centrifugation works by a stepwise increase in the centrifugation speed. Lower speeds at the beginning are used to eliminate the heavier food particles from the sample, and the speed is then increased until the targets themselves are pelleted. Neiderhauser *et al.* (1992) improved the detection limits of conventional PCR by 1000-fold after combining two subsequent centrifugation steps by using 100 × g and 3000 × g spins. Detection limits of 10^3 to 10^4 have been shown from seafood and soft cheese for several bacterial targets by simple one-step centrifugation at a speed of 9000 × g (Wang *et al.*, 1997).

The advantages of centrifugation are that it is rapid, easy and inexpensive to perform once a matching centrifuge is available. Its limitations include adhesion of the targets to the food matrix and co-sedimentation of the targets with the food particles, which results in inadequate reduction of sample size.

Density gradient centrifugation

Density gradient centrifugation is reported as a tool for separation of bacteria from food matrices. The underlying principle is based on a decreasing density of the suspending solution and migration of the targets to the equilibrate portion of the sample tube during centrifugation. However, applications of this method are currently limited to liquid foodstuffs, as it is difficult to standardize and problematic with respect to some food components, especially regarding fat content (Stevens and Jaykus, 2004a). Moreover, the osmotic strength of some gradients interferes with cell viability, depending on its composition. Nevertheless, this method has become very popular over the past few years. Wolffs *et al.* (2004) used this method for quantification of *Yersinia enterocolitica* in meat juice samples and obtained a recovery of approximately 30% from 2 ml samples with a detection limit of 4.2×10^3 CFU after real-time PCR targeting of the 16S rRNA gene. Lindqvist *et al.* (1997) used Percoll media for buoyant density gradient centrifugation followed by conventional PCR and obtained a detection limit of 5×10^2 CFU of *Shigella flexneri* from 0.5 ml samples of skimmed milk and 10^4 CFU from 0.5 ml 1:10 diluted blue cheese. In another publication, this group investigated the ability of this system to extract *E. coli* from beef and minced beef homogenates (Lindqvist *et al.*, 1997). 0.1 ml aliquots of 1:10 diluted samples were processed and a detection limit from 1.2×10^3 to 1×10^4 was obtained after conventional PCR. In general recovery from this method was reported to range from 20% and 45% CFU (Stevens and Jaykus, 2004a, 2004b). Fukushima *et al.* (2007) reported a broad application of buoyant density gradient centrifugation. Twelve foodborne pathogens and 13 foodstuffs were investigated and a recovery of 11% was obtained.

11.2.4 Adsorption

Several methods for separation of food pathogens from food matrices using adsorption effects other than biochemical or biological have been reported. Adsorption is achieved by exploiting Van der Waal's forces, electrostatic interactions, hydrophobic interactions and hydrogen bonding. These physiochemical interactions mediate nonspecific adsorption to the surface of solid materials such as metal hydroxides, ion exchange resins and lectins as well as targets and these are forced to matching liquid phases. These reversible interactions are also valid in immunological binding of epitope and antibody or viral binding, respectively.

Adsorption in liquid systems

The partition of bacterial cells in one bulk phase of immiscible hydrophilic two-phase systems based on the charge of the target cells is known as the aqueous two-phase system. This nonspecific separation method is often used in combination with centrifugation. Dextran and polyethylene glycol are the most commonly used phases. Cells accumulate in either one of the bulk phases or in the interphase, depending on their surface characteristics. Additional chemical modification of the two-phase system is often performed by adding salts or phosphate buffers to improve both viability of the targets and partitioning. The aqueous two-phase system used for direct detection of *L. monocytogenes* from soft cheese was associated with recovery rates after separation and plating of less than one per cent from the bacterial culture and approximately 50% from cheese homogenate (Lantz *et al.*, 1994). A detection limit of 10^4 CFU per millilitre after subsequent conventional PCR was reported and 100 μl aliquots of the homogenate were used. Pedersen *et al.* (1998) used the same system for separation of *L. monocytogenes* and *Salmonella berta* from smoked

Cumberland sausage after processing 4 g samples. The recoveries in both phases ranged from 7.6% to 95.0%, depending on pH-value, polymers used and whether *L. monocytogenes* or *S. berta* had been separated. High detection limits and heterogeneous results are disadvantages of this method. However, the capability of processing sample volumes of 4 g is a benefit of this system. Generally, two-phase systems are sensitive to temperature and chemical composition and partitioning may be impaired by the presence of fat and other food components. Moreover, the broad range of different food matrices would necessitate a broad range of different two-phase systems; hence standardization is anticipated to be difficult.

Adsorption onto solid phases

Within the group of nonspecific adsorption methods, metal hydroxide coatings and ion exchange resins have sometimes been used to separate bacterial cells from sample matrices with varying results. Metal hydroxides, such as zirconium or hydroxyapatite were used together with centrifugation and *L. monocytogenes* and *E. coli* were removed from non-fat dry milk and beef, respectively, with variable recoveries from 9.5% to 99.0% from liquid samples and culture media (Lucore *et al.*, 2000; Berry and Siragusa, 1997; Stevens and Jaykus, 2004a). Sample pre-treatment was necessary and the method worked best on simple materials such as non-fat dry milk.

Ion exchange resins are not yet applicable to foods but they are typically used in water and protein purification techniques and provide a binding capacity of up to 10^{10} CFU per gram resin. *Pseudomonas cepacia* (*Burkholderia cepacia*) was recovered from soil with 35% efficiency (Stevens and Jaykus, 2004a). However, this nonspecific method requires pH-value manipulations to release ionic interactions for recovering the targets, which compromises the viability of the cells.

Lectins are proteins binding the *N*-acetyl glucosamine residue of the bacterial cell wall. Therefore, this group of nonspecific adsorbing molecules represents a binding link to specific immuno-separation methods. Lectins can be bound to agarose beads and used in affinity columns or used in conjunction with magnetic vehicles such as paramagnetic beads. Lectins have yielded from 13% to 50% recovery of *L. monocytogenes* from milk and ground beef (Payne *et al.*, 1992; Stevens and Jaykus, 2004a). This method is rapid and simple but requires pre-sample treatment, lacks efficient release of the bound cells and is cost intensive.

A combined adsorption–filtration method has been reported by Fumian *et al.*, (2009) providing between 5.2% and 72.3% recoveries of *Norovirus* particles from 25 g lettuce and 15 g cheese. However, these variable recoveries indicate the need for further development of this method.

11.3 Biochemical and biological separation methods used in sample preparation

Utilisation of the highly specific affinity of antibodies and viral binding proteins to separate bacterial targets from food matrices offers the advantage of a pre-selection step combined with target concentration and separation. Niche applications such as aptamers and antimicrobial peptides (AMP), both related to lectins (see Section 'Adsorption onto Solid Phases') in their nonspecific binding capacity, have also been presented. The number of publications concerned with beads and affinity binding has risen substantially over the past few years. The use of antibodies, viral binding proteins, AMP from higher plants such as magainin I and binding to the surface of bacteria have been reported in combination with magnetic beads (Amagliani *et al.*, 2006; Nogva *et al.*, 2000; Jung *et al.*, 2003; Kretzer *et al.*, 2007; Loessner *et al.*, 2002; Stevens and Jaykus, 2004a; Hallier-Soulier and Guillot, 1999; Radstrom *et al.*, 2004; Niederhauser *et al.*, 1994), silanizised glass slides (Kulagina *et al.*, 2005) and direct colony blot (Belyi *et al.*, 1995). The use of nonspecific magnetic beads has also been reported for enhanced recovery combined with real-time PCR for specific target identification (Nogva *et al.*, 2000).

11.3.1 Antibodies

Detection limits of 10^3 CFU per millilitre have been reported in general for immuno-assay separation (Stevens and Jaykus, 2004a). Skjerve *et al.* (1990) demonstrated the use of monoclonal antibodies for the separation of *L. monocytogenes* from culture media and heterogeneous suspensions. Nogva *et al.* (2000) compared four different DNA isolation bead methods on 1.4 and 1.6 ml samples of unpasteurised whole milk, skimmed milk and water and obtained detection limits after real-time PCR between 2×10^2 and 2×10^5 CFU without determining the recovery rate. A recovery rate of 7–23% from 1 ml samples of culture medium were reported by Jung *et al.* (2003) using an immunogenetic separation method combined with flow cytometry. Belyi *et al.* (1995) used *L. monocytogenes* suspended in 200 μl of raw milk for colony blotting. The detection limit of 5×10^2 CFU was roughly evaluated and no recovery rate was determined.

11.3.2 Viral binding proteins

The application of viral binding proteins combined with paramagnetic beads for concentration of bacterial targets from food has been restricted to *L. monocytogenes* so far. However, *Bacillus cereus* and *Clostridium perfringens* specific beads have demonstrated proof of principle from culture media (Kretzer *et al.*, 2007). The detection limit of these approaches is reported to be lower than that of the antibody covered beads (<10 CFU). The application of listeriolysin O coated beads has been shown for culture media and skimmed milk (Amagliani *et al.*, 2006). The phage endolysin-derived cell wall-binding domain-based magnetic separation was demonstrated by Kretzer *et al.* (2007) for *L. monocytogenes* from various naturally contaminated foodstuff samples with promising results. Recovery was not determined and an average detection limit was less than 10 CFU per millilitre for all artificially contaminated foodstuffs tested. However, solid foodstuffs were not homogenized, and thus the artificial contamination of the foodstuffs occurred on the surface only.

11.3.3 Aptamers and antimicrobial peptides

Joshi *et al.* (2009) demonstrated the use of aptamers coated to magnetic beads for processing chicken rinses. The recovery of *Salmonella enterica* subsp. *enterica* serovar Typhimurium was not determined and the detection limit ranged from 10^2 to 10^3 CFU per 25 g after quantitative real-time PCR. The AMP-based method presented by Kulagina *et al.* (2005) utilised 25 µl suspensions of *E. coli* and *S.* Typhimurium in dimethyl sulfoxide (DSMO) and was associated with a detection limit of between 6.5×10^4 and 6.8×10^5 CFU using an array-based biosensor without determination of the recovery rate.

In general, the drawbacks of this category of methods are heterogeneous detection limits and recoveries, insufficient processed volumes and overall high costs. The range of bacterial targets is limited to matching epitopes and each target requires a specific separation carrier. Cross-linking and other side effects derived from the food composition also influence the performance of these methods and washing/blocking steps reduce recovery. So far, these methods have been successfully applied mainly to pure cultures of bacteria or to liquid food matrices such as blood, skimmed milk or water, whereas their performance on solid foodstuffs remains problematic (Stevens and Jaykus, 2004a).

11.4 Chemical and enzymatic pre-separation methods for sample treatment

11.4.1 Enzymatic digestion of the food sample matrix

These methods are defined as pre-separation methods because digestion of the food matrix is always followed by a physical separation method such as centrifugation. The enzymatic digestion of food matrices is based on the specific cleavage of macromolecules by enzymes. Enzymatic digestion methods of food matrices have been described by a few groups. Wegmüller *et al.* (1993) used a pronase digest of 2 g samples of cheese or yoghurt, artificially contaminated with *Campylobacter jejuni* and *Campylobacter coli*, and obtained a detection limit of 50 CFU per gram after conventional PCR. Allmann *et al.* (1995) examined *L. monocytogenes* in 40 ml samples of raw milk and 4 g samples of dairy products using an enzymatic system combining pronase with lysozyme and proteinase K. The detection limit was not examined due to uncontrolled bacterial growth during the three-hour incubation period with pronase at 40°C. Enzymatic digestion is a widespread method in molecular biology; nevertheless this method is restricted to either low sample volumes or numbers due to the high costs of the enzymes. Moreover, most of the applications are time-consuming, as overnight incubation is usually necessary to achieve acceptable recoveries.

11.4.2 Chemical digestion of the food sample matrix

Chemical digestion or solubilization of the foodstuffs necessitates chemical cleavage of the matrix. The reaction conditions have to be selected specifically, utilising the cell wall or other target-specific characteristics. Chemical digestion or solubilization of a heterogeneous group of products such as food is difficult as the underlying reactions are influenced by pH value, temperature, salt concentration, viscosity or the main components of the foodstuff comprising fat, protein or carbohydrates. Therefore, the food composition is more important in comparison with separation techniques which are not based on digestion of the sample matrix.

A guanidine thiocyanate/phenol/chloroform extraction method has been used for direct preparation of DNA from *L. monocytogenes, Y. enterocolitica* and *Salmonella* subsp. *enterica* serovar Enteritidis by Choi and Hong (2000, 2001, 2003). They obtained a limit of detection after conventional PCR of 2×10^3 CFU of *L. monocytogenes* per millilitre, using 0.5 ml samples of artificially contaminated raw milk (Choi and Hong, 2003). Ramesh *et al.* (2002) used a diethyl ether/chloroform extraction method combined with an incubation step in 2 M urea and one per cent SDS. 0.4 ml samples of pasteurized milk artificially contaminated with *Staphylococcus aureus* and *Y. enterocolitica* were used. Stevens and Jaykus (2004b) used sodium citrate in a protocol combined with centrifugation and the DNAzol®BD reagent to produce a resulting pellet size of 500 µl when 11 g of plain non-fat yoghurt or Cheddar cheese were used. The resulting pellet size from the first step in sodium citrate was between 1 and 5 g. The Cheddar cheese was filtered after blending and the yoghurt was strained through sterile cheesecloth, thus removing a major portion of the food matrix. Bacterial recoveries after centrifugation ranged from 53 to 145 % (SD: 21.4%) based on the direct plated pellet of *L. monocytogenes* and *S.* Enteritidis in both foodstuffs. Detection limits after conventional PCR and following Southern hybridisation from 10^0 to 10^3 CFU per gram were obtained, depending on the food matrix and bacterial species.

Matrix lysis (solubilization)

Over the past few years a modular system for bacterial separation covering all relevant species and foodstuffs has been developed. This method, called 'matrix lysis', involves solubilization of the sample food matrix and subsequent separation of the target cells by centrifugation to permit isolation of their DNA and quantitative detection using real-time PCR. It is capable of reducing 6–12 g samples of the foodstuffs in 50 ml volumes down to a size that can be processed in commercial isolation kits for DNA isolation (<200 µl, (Fig. 11.2)). The first proof of principle was demonstrated by processing 12.5 ml of raw and ultra high tempered (UHT) milk samples with *L. monocytogenes* (Rossmanith *et al.*, 2007). A buffer system containing the chaotrope urea (8 M) and SDS (one per cent) as detergent was used to lyse the food matrix. Separation of the targets from the resulting solution was achieved by simple centrifugation at $3220 \times g$. A detection limit of 7.8 CFU per gram was obtained after real-time PCR, with a recovery of 39.5%. The application was broadened in a second attempt to include bacteria other than *L. monocytogenes*, especially Gram-negative species (Mayrl *et al.*, 2009). The degradation of all sorts of dairy products, egg, blood and ice cream was obtained by a further buffer system containing Lutensol™, a non-ionic

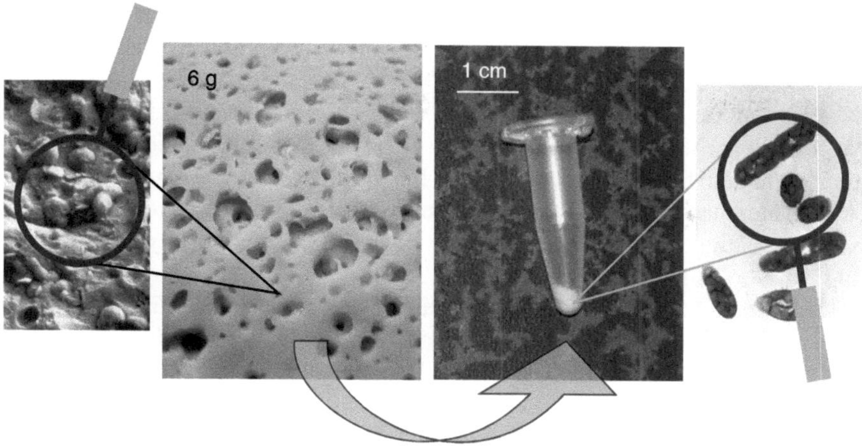

Fig. 11.2 Matrix lysis: The solid foodstuff is reduced to a pellet of < 200 μl containing mainly the bacterial content of the sample. A reduced sample size is sufficient for most downstream molecular biological detection methods and has to be achieved avoiding loss of target organisms.

detergent, instead of SDS, which allows for separation of Gram-negative bacteria. The additional use of a sucrose buffer containing an industrial protease dramatically increased the performance of the protocol when applied to meat, fish and chicken. The mean recovery of *S.* Typhimurium after real-time PCR was 54.7% from ice cream, chicken and egg with a detection limit of 11.2 CFU per gram. *S. aureus* was obtained from ice cream and milk with 43.6% recovery after real-time PCR and a detection limit of 1.3 CFU per gram. *L. monocytogenes* was demonstrated to be recoverable from blood, Gouda cheese, mozzarella cheese, chicken, salmon and ice cream with an overall 51.4% recovery. In this second development, the removal of 5-log scales of free target DNA during the protocol could be demonstrated by real-time PCR analysis. Consequently, contamination with free DNA derived from dead bacterial cells does not impair quantification of target genes. Qualitative results will be impaired with a presumptive contamination of over 100 000 dead cells, which would not be associated with growth contamination. However, this would invariably suggest the occurrence of problems within the production chain.

Starch-rich and cellulose-containing foodstuffs could not be processed with these two buffer systems and the viability of the targets was compromised due to the buffer composition. However, data yet unpublished by the authors demonstrate the successful lysis of starch-rich foodstuffs such as flour and noodles by a shift to a complementary new buffer system not based on detergents or chaotropic substances. Matrix lysis is called a modular system as the approach allows the solubilization of all food items by the use of two to three buffer systems that work under very similar conditions.

The strategy was further advanced to allow for the separation of living cells from the food matrices. Securing viability solves the problem of PCR result interpretation since such a signal would then indicate the occurrence of infective cells. Separation of *S.* Typhimurium and *L. monocytogenes* from dairy products has been demonstrated with a preliminary 60–95% recovery using ionic liquids, a completely new family of chemicals that have not yet been used in food science. The development of a modular approach currently offers a promising move towards meeting the demands of sample preparation in food analytics. A broad range of foodstuffs and target species is thereby covered with one method protocol and an altered buffer system. This could enable standardized handling, and simplify routine application of the method. Moreover, reduction of the foodstuff volume is sufficient for downstream applications, and good reproducibility of the method is achievable. The volume to be processed depends exclusively on the equipment of the laboratories and could be extended to 50 or 100 g. Concluding, matrix lysis works directly on foodstuffs, does not require more than three hours of simple incubation steps and does not require extensive skills from laboratory personnel. The matrix lysis protocol is also suitable for washing protocols to detect surface-attached targets such as bacterial pathogens on carcasses.

11.5 Related approaches and combined sample preparation and detection methods

11.5.1 Direct isolation of target DNA from food samples

Direct isolation of target DNA is based on the more-or-less selective lysis of either the bacterial cell wall or the core shell of the spore or viral envelope, without prior separation of target cells from the food matrix. Subsequent purification of the DNA from cell debris and food remnants is usually performed via affinity binding on silica gel or nitrocellulose. Overall, methods that do not separate the target cells from the rest of the sample protocols should be distinguished from methods that employ a series of steps from cell separation to DNA purification. Direct isolation of DNA is either limited by sample size or associated with impairment of subsequent PCR detection due to an enormous amount of background DNA that is usually present. Isolation methods for DNA directly from foodstuffs using commercial DNA isolation kits are widely published. Pinzani *et al.* (2004) extracted *Oenococcus oeni* DNA directly from 0.5 ml wine samples using the NucleoSpin® food kit and observed corresponding results of copy number and CFU as investigated with real-time PCR. Rudi *et al.* (2005) used the DNAeasy® tissue kit, DNA DIRECT® and the BUGS'n BEADS™ kit to isolate DNA from 100 μl volumes of Gouda cheese to detect *L. monocytogenes* directly with real-time PCR. They obtained a detection limit of 3.2×10^2 CFU per gram. Consequently, direct isolation of DNA is limited since the detection limit of these protocols is usually far from the contamination levels that can be expected in naturally contaminated samples. While pure DNA can be obtained, the method is cost intensive. In summary, this approach is restricted for basic research and should preferably not be used in routine diagnostics.

11.5.2 Target cell separation as a prerequisite for non-DNA-based detection methods

Generally, the non PCR based methods share the disadvantage that they do not proliferate, or at least not as efficiently as PCR, the target molecule to detectable concentrations.

With dye-based methods, signal strength is generated on the basis of dense loading of capture molecules with antigens/antibodies, whereby dye reporter molecules generate the measureable readout. An example is the direct detection of *Campylobacter* in faecal material by lateral flow technology (e.g. Merck, SinglePath®) using colloidal gold staining. Although many of these methods do not work without enrichment, concentration of the targets could help to improve the detection probability since it could be hypothesized that target cell proliferation is fostered when the cells are liberated from food particles prior to the enrichment step.

Chromogenic culture media that exploit biochemical pathways dependent on the presence of the target organisms have been developed to improve culture-based direct detection and quantification. Nowadays chromogenic media are not only available for most of the common pathogens tested but also for many microbial spoilage organisms and hygiene indicators. Various chromogenic plating media, such as ALOA, Rapid'L.mono®, BCM® and LIMONO-Ident agar have been successfully evaluated for the detection and enumeration of *L. monocytogenes* and some of the new media have been proposed in respective ISO methods to support the isolation of the pathogenic species from enrichments (Reissbrodt, 2004; Beumer and Hazeleger, 2003; Stessl *et al.*, 2009). Chromogenic media are of further benefit to food microbiology since their use can shorten the isolation and aid identification of the target bacteria from enrichments. However, they do not circumvent the time-consuming enrichment step. Moreover, recent research has shown that the detection probability of culture methods can be improved when the sample is deliberately manipulated. Homogenization of a >100 g portion of the foodstuff intended to be tested has increased the detection probability of *L. monocytogenes*, even when a smaller test portion was taken out of the blender and added to the tenfold volume of the enrichment medium subsequently (Danan C., pers. communication). This example demonstrates that the detection of microbial targets by culture could be improved if more emphasis was made to improve the very early steps (sample manipulation/enrichment) in the analytical chain.

In summary, despite major improvements to isolation media, culture methods are still time consuming to perform, depend on the grade of the media available, are limited in their selectivity and, particularly for chromogenic media, expensive.

Fluorescent staining of pathogens direct on to the surface of food samples has been reported using labelled antibodies recognizing bacterial surface antigens (Fravalo *et al.*, 2001) and fluorescent *in situ* hybridization (FISH) on cheese using 16S rRNA genes (Ercolini *et al.*, 2003).

Few other techniques have been established as alternatives to the above methods for direct detection and quantification of food pathogens. Changes in the optical density of salt solutions can be automatically measured and this provides the possibility for detecting *L. monocytogenes* based on hydrolysis of esculin (Peng and Shelef, 2000). These methods are alternatives to molecular and microbiological methods and likewise dependent on enrichments.

11.6 Conclusions and future trends

Food control strongly relies on standardised and validated confirmation methods that satisfy the requirements of national and international food law and codices. Until now there has been a clear objective to isolate the causative pathogen from an item that is incriminated in food contamination. This is because isolating the organism enables subsequent epidemiological investigations, and enables proof or disproof for whether the pathogen possesses the virulence traits that could explain the degree of fatality observed during outbreaks. Screening methods can support the process, but culture techniques are mostly used to confirm the screening suspicion in presumed cases of contamination.

In the case of self-control, rapid methods are preferable and molecular biological methods have already captured market segments in many food-producing countries worldwide. Food processing is expensive and therefore competitive and simpler processing steps, such as slicing or packaging, are often outsourced to places where labour costs are favorable. This necessitates that foods are shipped from one site to the other thus reducing the chance that appropriate testing can be performed. Moreover, EU food processors increasingly test product lots for the occurrence of specific pathogens before release onto retail markets. This is due to the fact that food legislation policy has conferred the responsibility for safe food production to the producer. The need for documenting an active role in securing safe production of food has led to continuous improvement and implementation of rapid tests. A limiting factor, however, is still the costs involved. Without doubt combinations of singleplex or multiplex PCR systems, immunological methods or biosensors will be the methods of choice in the future.

The picture again changes when QMRA unravels the risk that is associated with the presence of a pathogen in a foodstuff. Prevalence data are not sufficient to understand the impact of the presence of a pathogen in a respective chain. Novel biotracing concepts not only require data on numbers and growth kinetics but also strain types and distribution of marker traits within a microbial population. The ability of a pathogen to adapt to technological and environmental stresses strongly enhances the probability that consumers will be exposed to relevant numbers (Malorny *et al.*, 2008). Such a holistic approach would be required to model the flow of target organisms through the chains (http://www.biotracer.org; Barker *et al.*, in press).

Driven by accumulating knowledge and QMRAs, food legislation that has followed the precautionary principle in the past will be amended. This will lead to scenarios requiring compliance with quantitative limits. As a consequence, quantitative microbial food testing will replace qualitative testing in the future and sample treatment strategies will become essential. The intriguing problem still concerns the efficient extraction of pathogens from foods. Significant steps towards improving the efficiency of extraction of low numbers of microorganisms have been described in this chapter. However, to get such tools cost-effective and widely implemented still requires more extensive testing 'in the field' as well as lowering the cost of the approaches. We are convinced that much more research should be invested into advancing the ideas that could contribute to solving these issues.

11.7 Acknowledgments

Parts of the work presented in this chapter were funded by IP BIOTRACER (FP6-2006-FOOD-036272).

11.8 References

AL SOUD W A and RADSTROM P (2001), 'Purification and characterization of PCR-inhibitory components in blood cells', *J Clin Microbiol*, **39**, 485–93.

ALLMANN M, HOFELEIN C, KOPPEL E, LUTHY J, MEYER R, NIEDERHAUSER C, WEGMULLER B and CANDRIAN U (1995), 'Polymerase chain reaction (PCR) for detection of pathogenic microorganisms in bacteriological monitoring of dairy products', *Res Microbiol*, **146**, 85–97.

AMAGLIANI G, OMICCIOLI E, CAMPO A, BRUCE I J, BRANDI G and MAGNANI M (2006), 'Development of a magnetic capture hybridization-PCR assay for *Listeria monocytogenes* direct detection in milk samples', *J Appl Microbiol*, **100**, 375–83.

BARKER G C, GOMEZ N and SMID J (in press) 'an introduction to biotracing in food chain systems', *Trends Food Sci. Technol.*

BELYI Y F, VARFOLOMEEVA N A and TARTAKOVSKII I S (1995), 'A simple colony-blot method for identification of *Listeria* in food samples', *Med Microbiol Immunol* (Berl), **184**, 105–8.

BERRY E D and SIRAGUSA G R (1997), 'Hydroxyapatite adherence as a means to concentrate bacteria', *Appl Environ Microbiol*, **63**, 4069–74.

BESSE N G, AUDINET N, BEAUFORT A, COLIN P, CORNU M and LOMBARD B (2004), 'A contribution to the improvement of *Listeria monocytogenes* enumeration in cold-smoked salmon', *Int J Food Microbiol*, **91**, 119–27.

BEUMER R R and HAZELEGER W C (2003), '*Listeria monocytogenes*: diagnostic problems', *FEMS Immunol Med Microbiol*, **35**, 191–7.

BYLUND G (1995), *Dairy processing handbook*, Lund, Sweden, Tetra Pak Processing Systems.

CHOI W S (2000), 'Detection of *Salmonella* in milk by polymerase chain reaction', *J Food Hyg Safety*, **15**, 262–6.

CHOI W S (2001), 'Detection of *Yersinia enterocolitica* in milk by polymerase chain reaction', *Food Sci Biotechnol*, **10**, 451–4.

CHOI W S and HONG C H (2003), 'Rapid enumeration of *Listeria monocytogenes* in milk using competitive PCR', *Int J Food Microbiol*, **84**, 79–85.

CUDJOE K S and KRONA R (1997), 'Detection of *Salmonella* from raw food samples using Dynabeads anti-*Salmonella* and a conventional reference method', *Int J Food Microbiol*, **37**, 55–62.

CULLISON M A and JAYKUS L A (2002), 'Magnetized carbonyl iron and insoluble zirconium hydroxide mixture facilitates bacterial concentration and separation from nonfat dry milk', *J Food Prot*, **65**, 1806–10.

DONNELLY C W and BAIGENT G J (1986), 'Method for flow cytometric detection of *Listeria monocytogenes* in milk', *Appl Environ Microbiol*, **52**, 689–95.

ERCOLINI D, HILL P J and DODD C E (2003), 'Development of a fluorescence in situ hybridization method for cheese using a 16S rRNA probe', *J Microbiol Methods*, **52**, 267–71.

FERNANDEZ-ASTORGA A, HIJARRUBIA M J, LAZARO B and BARCINA I (1996), 'Effect of the pre-treatments for milk samples filtration on direct viable cell counts', *J Appl Bacteriol*, **80**, 511–16.

FLEKNA G, SCHNEEWEISS W, SMULDERS F J, WAGNER M and HEIN I (2007), 'Real-time PCR method with statistical analysis to compare the potential of DNA isolation methods to remove PCR inhibitors from samples for diagnostic PCR', *Mol Cell Probes*, **21**, 282–7.

FLISS I, EMOND E, SIMARD R E and PANDIAN S (1991), 'A rapid and efficient method of lysis of Listeria and other Gram-positive bacteria using mutanolysin', *Biotechniques*, **11**, 453, 456–3, 457.

FLUIT A C, TORENSMA R, VISSER M J, AARSMAN C J, POPPELIER M J, KELLER B H, KLAPWIJK P and VERHOEF J (1993), 'Detection of *Listeria monocytogenes* in cheese with the magnetic immuno-polymerase chain reaction assay', *Appl Environ Microbiol* **59**, 1289–93.

FRAVALO P, CHANET J P, MAS M, HUCHET E, QUEGUINER S and SALVAT G (2001), 'Feasibility of fluorescent detection of pathogens on pork carcasses', *Berl Munch Tierarztl Wochenschr*, **114**, 393–6.

FUKUSHIMA H, KATSUBE K, HATA Y, KISHI R and FUJIWARA S (2007), 'Rapid separation and concentration of food-borne pathogens in food samples prior to quantification by viable-cell counting and real-time PCR', *Appl Environ Microbiol* **73**, 92–100.

FUMIAN T M, LEITE J P, MARIN V A and MIAGOSTOVICH M P (2009), 'A rapid procedure for etecting noroviruses from cheese and fresh lettuce', *J Virol Methods*, **155**, 39–43.

HALLIER-SOULIER S and GUILLOT E (1999), 'An immunomagnetic separation polymerase chain reaction assay for rapid and ultra-sensitive detection of *Cryptosporidium parvum* in drinking water', *FEMS Microbiol Lett*, **176**, 285–9.

HAMANN C H and VIELSTICH W (2005), *Elektrochemie*, Weinheim, Whiley-VCH.

JOSHI R, JANAGAMA H, DWIVEDI H P, SENTHIL KUMAR T M, JAYKUS L A, SCHEFERS J and SREEVATSAN S (2009), 'Selection, characterization, and application of DNA aptamers for the capture and detection of *Salmonella enterica* serovars', *Mol Cell Probes*, **23**, 20–8.

JUNG Y S, FRANK J F and BRACKETT R E (2003), 'Evaluation of antibodies for immunomagnetic separation combined with flow cytometry detection of *Listeria monocytogenes*', *J Food Prot*, **66**, 1283–7.

KOCH A L and BLUMBERG G (1976), 'Distribution of bacteria in the velocity gradient centrifuge', *Biophys J*, **16**, 389–405.

KRETZER J W, LEHMANN R, SCHMELCHER M, BANZ M, KIM K P, KORN C and LOESSNER M J (2007), 'Use of high-affinity cell wall-binding domains of bacteriophage endolysins for immobilization and separation of bacterial cells', *Appl Environ Microbiol*, **73**, 1992–2000.

KULAGINA N V, LASSMAN M E, LIGLER F S and TAITT C R (2005), 'Antimicrobial peptides for detection of bacteria in biosensor assays', *Anal Chem*, **77**, 6504–8.

LANTZ P G, TJERNELD F, BORCH E, HAHN-HAGERDAL B and RADSTROM P (1994), 'Enhanced sensitivity in PCR detection of *Listeria monocytogenes* in soft cheese through use of an aqueous two-phase system as a sample preparation method', *App. Environ Microbiol*, **60**, 3416–18.

LIMAYE M S and COAKLEY W T (1998), 'Clarification of small-volume microbial suspensions in an ultrasonic standing wave', *J App Microbiol*, **84**, 1035–42.

LINDQVIST R (1997), 'Preparation of PCR samples from food by a rapid and simple centrifugation technique evaluated by detection of *Escherichia coli* O157:H7', *Int J Food Microbiol*, **37**, 73–82.

LINDQVIST R, NORLING B and LAMBERTZ S T (1997), 'A rapid sample preparation method for PCR detection of food pathogens based on buoyant density centrifugation', *Lett Appl Microbiol*, **24**, 306–10.

LOESSNER M J, KRAMER K, EBEL F and SCHERER S (2002), 'C-terminal domains of *Listeria monocytogenes* bacteriophage murein hydrolases determine specific recognition and high-affinity binding to bacterial cell wall carbohydrates', *Mol Microbiol*, **44**, 335–49.

LUCORE L A, CULLISON M A and JAYKUS L A (2000), 'Immobilization with metal hydroxides as a means to concentrate food-borne bacteria for detection by cultural and molecular methods', *Appl Environ Microbiol*, **66**, 1769–76.

MALORNY B, LOFSTROM C, WAGNER M, KRAMER N and HOORFAR J (2008), 'Enumeration of *Salmonella* bacteria in food and feed samples by real-time PCR for quantitative microbial risk assessment', *Appl Environ Microbiol*, **74**, 1299–304.

MARKX G H, DYDA P A and PETHIG R (1996), 'Dielectrophoretic separation of bacteria using a conductivity gradient', *J Biotechnol*, **51**, 175–80.

MAYRL E, ROEDER B, MESTER P, WAGNER M and ROSSMANITH P (2009), 'Broad range evaluation of the matrix solubilization (matrix lysis) strategy for direct enumeration of food-borne pathogens by nucleic acids technologies', *J Food Prot*, **72**, 1225–33.

MESTER P, WAGNER M and ROSSMANITH P (2010), 'Use of liquid-based extraction for recovery of *Salmonella Typhimurium* and *Listeria monocytogenes* from food matrices', *J Food Protect*, **73**, 680–7.

MEYER R, LUTHY J and CANDRIAN U (1991), 'Direct detection of polymerase chain reaction (PCR) of *Escherichia coli* in water and soft cheese and identification of enterotoxigenic strains', *Lett Appl Microbiol*, **13**, 268.

NEIDERHAUSER C, CANDRIAN U, HOFELEIN C, JERMINI M, BUHLER H P and LUTHY J (1992), 'Use of polymerase chain reaction for detection of *Listeria monocytogenes* in food', *Appl Environ Microbiol*, **58**, 1564.

NIEDERHAUSER C, HOFELEIN C, WEGMULLER B, LUTHY J and CANDRIAN U (1994), 'Reliability of PCR decontamination systems', *PCR Methods Appl*, **4**, 117–23.

NOGVA H K, RUDI K, NATERSTAD K, HOLCK A and LILLEHAUG D (2000), 'Application of 5′-nuclease PCR for quantitative detection of *Listeria monocytogenes* in pure cultures, water, skim milk, and unpasteurized whole milk', *Appl Environ Microbiol*, **66**, 4266–71.

PATCHETT R A, KELLY A F and KROLL R G (1991), 'The adsorption of bacteria to immobilized lectins', *J Appl Bacteriol*, **71**, 277–84.

PAYNE M J, CAMPBELL S, PATCHETT R A and KROLL R G (1992), 'The use of immobilized lectins in the separation of *Staphylococcus aureus, Escherichia coli, Listeria* and *Salmonella* spp. from pure cultures and foods', *J Appl Bacteriol*, **73**, 41–52.

PEDERSEN L H, SKOUBOE P, ROSSEN L and RASMUSSEN O F (1998), 'Separation of *Listeria monocytogenes* and *Salmonella berta* from a complex food matrix by aqueous polymer two-phase partitioning', *Lett Appl Microbiol*, **26**, 47–50.

PENG H and SHELEF L A (2000), 'Rapid detection of low levels of *Listeria* in foods and next-day confirmation of *L. monocytogenes*', *J Microbiol Methods*, **41**, 113–20.

PINZANI P, BONCIANI L, PAZZAGLI M, ORLANDO C, GUERRINI S and GRANCHI L (2004), 'Rapid detection of *Oenococcus oeni* in wine by real-time quantitative PCR', *Lett Appl Microbiol*, **38**, 118–24.

RÅDSTRÖM P, KNUTSSON R, WOLFFS P, LOVENKLEV M and LÖFSTRÖM C. (2004), 'Pre-PCR processing: strategies to generate PCR-compatible samples', *Mol Biotechnol*, **26**, 133–46.

RAMESH A, PADMAPRIYA B P, CHRASHEKAR A and VARADARAJ M C (2002), 'Application of a convenient DNA extraction method and multiplex PCR for the direct detection of

Staphylococcus aureus and *Yersinia enterocolitica* in milk samples', *Mol Cell Probes*, **16**, 307–14.

REISSBRODT R (2004), 'New chromogenic plating media for detection and enumeration of pathogenic Listeria spp.– an overview', *Int J Food Microbiol*, **95**, 1–9.

ROSSEN L, NORSKOV P, HOLMSTROM K and RASMUSSEN O F (1992), 'Inhibition of PCR by components of food samples, microbial diagnostic assays and DNA-extraction solutions', *Int J Food Microbiol*, **17**, 37–45.

ROSSMANITH P, KRASSNIG M, WAGNER M and HEIN I (2006), 'Detection of *Listeria monocytogenes* in food using a combined enrichment/real-time PCR method targeting the prfA gene', *Res Microbiol*, **157**, 763–71.

ROSSMANITH P, SUSS B, WAGNER M and HEIN I (2007), 'Development of matrix lysis for concentration of gram positive bacteria from food and blood', *Microbiol Methods*, **69**, 504–11.

RUDI K, NATERSTAD K, DROMTORP S M and HOLO H (2005), 'Detection of viable and dead *Listeria monocytogenes* on gouda-like cheeses by real-time PCR', *Lett Appl Microbiol*, **40**, 301–6.

SKJERVE E, RORVIK L M and OLSVIK O (1990), 'Detection of *Listeria monocytogenes* in foods by immunomagnetic separation', *Appl Environ Microbiol*, **56**, 3478–81.

STESSL B, LUF W, WAGNER M and SCHODER D (2009), 'Performance testing of six chromogenic ALOA-type media for the detection of *Listeria monocytogenes*', *J Appl Microbiol* **106**, 651–9.

STEVENS K A and JAYKUS L A (2004a), 'Bacterial separation and concentration from complex sample matrices: a review', *Crit Rev Microbiol*, **30**, 7–24.

STEVENS K A and JAYKUS L A (2004b), 'Direct detection of bacterial pathogens in representative dairy products using a combined bacterial concentration-PCR approach', *J App Microbiol*, **97**, 1115–22.

THOMAS D S (1988), 'Electropositively charged filters for the recovery of yeasts and bacteria from beverages', *J Appl Bacteriol*, **65**, 35–41.

TJHIE J H, VAN KUPPEVELD F J, ROOSENDAAL R, MELCHERS W J, GORDIJN R, MACLAREN D M, WALBOOMERS J M, MEIJER C J and VAN DEN BRULE A J (1994), 'Direct PCR enables detection of *Mycoplasma pneumoniae* in patients with respiratory tract infections', *J Clin Microbiol*, **32**, 11–16.

UYTTENDAELE M, VAN H I and DEBEVERE J (2000), 'The use of immuno-magnetic separation (IMS) as a tool in a sample preparation method for direct detection of *L. monocytogenes* in cheese', *Int J Food Microbiol*, **54**, 205–12.

WANG R F, CAO W W and CERNIGLIA C E (1997), 'A universal protocol for PCR detection of 13 species of foodborne pathogens in foods', *J Appl Microbiol*, **83**, 727–36.

WANG R F, CAO W W and JOHNSON M G (1992), '16S rRNA-based probes and polymerase chain reaction method to detect *Listeria monocytogenes* cells added to foods', *Appl Environ Microbiol*, **58**, 2827–31.

WEGMULLER B, LUTHY J and CANDRIAN U (1993), 'Direct polymerase chain reaction detection of *Campylobacter jejuni* and *Campylobacter coli* in raw milk and dairy products', *Appl Environ Microbiol*, **59**, 2161–5.

WILSON I G (1997), 'Inhibition and facilitation of nucleic acid amplification', *Appl Environ Microbiol*, **63**, 3741–51.

WOLFFS P, KNUTSSON R, NORLING B and RÅDSTRÖM P (2004), 'Rapid quantification of *Yersinia enterocolitica* in pork samples by a novel sample preparation method, flotation, prior to real-time PCR', *J Clin Microbiol*, **42**, 1042–7.

YU H and BRUNO J G (1996), 'Immunomagnetic-electrochemiluminescent detection of *Escherichia coli* O157 and *Salmonella* Typhimurium in foods and environmental water samples', *Appl Environ Microbiol*, **62**, 587–92.

12

A comparison of molecular technologies and genomotyping for tracing and strain characterization of *Campylobacter* isolates

J. van der Vossen, B. Keijser, F. Schuren, A. Nocker and R. Montijn, TNO Quality of Life, The Netherlands

Abstract: Thermophilic *Campylobacter* species are the most common cause of gastroenteritis in developed countries. *Campylobacter* spp. are ubiquitous in nature and widespread in livestock. Infections occur sporadically, and are believed to occur through consumption of contaminated meat products and from environmental sources. DNA-based techniques are essential for tracing origins of pathogens, and thereby for control over infection. This chapter provides an overview of existing technologies for characterizing *Campylobacter* isolates, in comparison to the use of pan-genomic microarrays. It is clear that microarray tools allow insight in the genetic factors important for host specificity, virulence, adaptation and antibiotic resistance.

Key words: *Campylobacter, C. jejuni*, pathogen, microarray, genomotyping, comparative genome hybridization (CGH), tracing, DNA, multi-locus sequence typing, horizontal gene transfer.

12.1 Introduction

12.1.1 *Campylobacter* spp. as a human health threat

Thermophilic *Campylobacter* spp. and in particular *C. jejuni* and *C. coli* are recognized as the most frequent cause of food-borne bacterial diarrheal disease throughout the world (Blaser, 1997; Gillespie *et al.*, 2002). An estimate of the WHO suggests the annual incidence rate of *Campylobacter* infections being as high as 1% in the population in the western nations (Humphrey *et al.*, 2007). This rate for *Campylobacter* infection also applies to the Netherlands with approximately 6200 laboratory-confirmed cases of *Campylobacter jejuni*-caused gastroenteritis (*Campylobacter* enteritis) in 2004 (Janssen *et al.*, 2006). Since *Campylobacter* enteritis is generally self-limited and patients normally recover without consulting a physician, the real incidence of *Campylobacter* enteritis in the Netherlands is estimated to be much higher with approximately 59 000 cases per year (Janssen *et al.*, 2006). Occasionally, in case of *C. jejuni* infections, the impact of illness is more dramatic due to sequela. Post-infection symptoms include reactive arthritis and the Guillain–Barré syndrome. In rare cases, the disease can even be fatal with an estimated number of 25 people who die every year from *Campylobacter* infections in the Netherlands (Kemmeren *et al.*, 2005).

Unlike many other foodborne diseases such as salmonellosis, *Campylobacter* enteritis tends to be largely sporadic rather than occurring in extensive outbreaks (Snelling *et al.*, 2005). This sporadic infection scenario is one of the reasons that make it difficult to understand the population biology and the relationships between genotype and virulence as well as sources and routes of human infection.

12.1.2 Potential sources of *Campylobacter*

Campylobacter species are zoonotic pathogens. They have a wide host range including farm animals such as poultry, pigs, cattle, and sheep as well as wildlife such as birds and free-living mammals (Petersen *et al.*, 2001; Humphrey *et al.*, 2007). In all these animals, the bacteria are able to colonize the gut. As a result, the fecal excreta of the animals may contain high doses of *Campylobacter* which can survive in dark and moist environments including surface water, coastal estuaries, and sewage (Jones, 2001). Transmission normally occurs via the fecal-oral route with the majority of transmissions being within the host-species. However, it has been shown for *C. jejuni* that these bacteria also jump frequently between hosts making use of the wide range of host species (McCarthy *et al.*, 2007).

12.1.3 Aim of this review

This chapter provides an overview of different molecular technologies applied for studying routes of transmission, epidemiology, host specificity, environmental persistence and/or virulence. Based on a comparison of the various features of these molecular methods, we advocate the potential of genomotyping for obtaining improved insight into various *Campylobacter* infection-relevant topics. Such knowledge will allow for taking adequate preventive measures to reduce the incidence of *Campylobacter* infection.

12.2 Methodologies for tracing and/or understanding strain properties

12.2.1 Multi-locus sequence typing

Multi-locus sequence typing (MLST) is a nucleotide sequence-based approach to characterize bacterial isolates. The technique involves PCR amplification of selected internal fragments of multiple housekeeping genes followed by sequence analysis and determination of allelic profiles for strain discrimination. By using MLST, McCarthy *et al.* (2007) supported evidence that *Campylobacter* lineages can acquire a host signature and potentially adapt to the host through genetic recombination events. Based on such signatures, improved prediction of host assignment appeared possible for *C. jejuni* where host-specific markers are unavailable. By using the MLST approach in a systematic study of 1231 cases of *C. jejuni* infection in the UK, Wilson *et al.* (2008) showed that probably the majority of human infections were caused by isolates associated with livestock and poultry. Similar findings were described by other researchers (Gormley *et al.*, 2008; Lévesque *et al.*, 2008). These studies are in line with previous research demonstrating host association with specific MLST types (French *et al.*, 2005; Sopwith *et al.*, 2006). In fact, MLST has been adopted by the microbial epidemiologists as the single best method for the genotyping of *Campylobacter* isolates and deciphering the epidemiologic relationships among isolates (Lévesque *et al.*, 2008). Despite this claim, there are several drawbacks of the MLST technology. The method is laborious and expensive. Moreover, the results do not provide any insight into the presence or absence of genes relevant for virulence, host specificity, antibiotic resistance or environmental survival traits.

12.2.2 Genome sequencing

The genome sequencing of *C. jejuni* strain NCTC11168 (Parkhill *et al.*, 2000) was a significant first step towards a better understanding of the species. The information was greatly expanded some years later by Fouts *et al.* (2005) who performed whole-genome sequencing on four *Campylobacter* species: *C. jejuni* strain RM1221, *C. coli*, *C. lari*, and *C. upsaliensis*. Apart from revealing species-specific differences, comparative genome analysis of the *C. jejuni* strain RM1221 (a chicken isolate) with the previously sequenced *C. jejuni* human clinical isolate NCTC11168 provided detailed insight into differences between strains from different hosts. The RM1221 strain sequence contained genes enabling colonization of chicken skin and caeca, invasion of Caco-2 cells and unique lipooligosaccharide and capsule loci as well as other open reading frames (ORF) that were shown to be absent in the clinical strain NTCT11168 (Fouts *et al.*, 2005). All in all, the comparative genome analysis revealed that many genes involved in host colonization and flagellin synthesis are conserved across species barriers, whereas species-specific variations were found for the lipooligosaccharide locus, a capsular polysaccharide locus and a putative virulence locus. Strains were also shown to vary in their metabolic profiles and resistance profiles to various antibiotics. It can be assumed that the sequence data are likely to hold additional information regarding differences in virulence and host specificity.

Comparing genome sequencing with MLST typing, the latter appears to be more useful for attributing human infection to specific *Campylobacter* strains due to the robustness of the method. Moreover, MLST is significantly more cost-effective than whole genome sequencing. However, the MLST methodology lacks more extensive information on the genetic basis of *Campylobacter* structure and function. Genome sequencing on the other hand provides detailed genetic insight in respect to efficiency of colonization and adverse human health effects, but is too expensive as a routine screening tool at this moment. The challenge for the future therefore is to develop a methodology that provides great detail, but circumvents extensive and costly sequencing of large numbers of *Campylobacter* isolates. In addition to attributing disease to specific strains, such technology should be very useful to understand *Campylobacter* epidemiology, virulence and host specificity as well as other features like survival under stress conditions.

12.2.3 Microarray hybridization technology

Microarrays are composed of hundreds to thousands of DNA probes bound to a solid surface to allow simultaneous detection of nucleic acid targets. Genes present or expressed in organisms can be detected after fluorescent labeling and hybridization to the corresponding probes on the microarray. The introduction of microarray hybridization technology for differentiation of *Campylobacter* strains represented a major technical improvement providing whole genome information without the necessity to sequence each individual genome. The technology allows the comparison of genotypic patterns obtained from different strains. The first microarray specific for *C. jejuni* was developed on the basis of cloned DNA from strain NCTC11168 (Dorrell *et al.*, 2001). The detailed gene-by-gene comparison of *C. jejuni* genomes enables conclusions about (partial) presence and absence of genes. This single-strain microarray revealed that some genes present in the sequenced strain appeared dispensable in other *C. jejuni* strains by the fact that they are absent or highly divergent in one or more of the other *C. jejuni* isolates compared with strain NCTC11168. Apparently, there is a core of sequences essential for a *Campylobacter* strain to exist and being viable. These core genes tend to attribute primarily to metabolic, biosynthetic, cellular and regulatory processes. In addition, virulence-related genes showed a high degree of conservation. Interestingly, microarray results could show that the capsule biosynthesis locus accounts for Penner serotype specificity. The authors concluded that the microarray-derived data were a solid basis for demonstrating genetic diversity among different *C. jejuni* strains and served well for identifying genes that correlate with pathogenicity. The findings correlated with observations on intraspecies variability in *C. jejuni* from follow-up studies also employing microarray-based comparative genome analysis (Leonard II *et al.*, 2003; Pearson *et al.*, 2003; Poly *et al.*, 2004; Taboada *et al.*, 2004; Parker *et al.*, 2006). Further studies were performed in the area of *C. jejuni* epidemiology and for tracing sources of human infection. Whereas On *et al.* (2006) showed how genomotyping data of different *C. jejuni* strains correlated with strain properties such as survival and toxin production, Champion *et al.* (2005) successfully demonstrated how genomotyping can add to the identification of the infection source via correlation with specific genetic markers. It has to be noted that both studies used solely the single-strain NCTC11168 containing a 1654 annotated gene sequence microarray. Further improvement can result from supplementing the microarrays with gene probes that are absent in the NCTC11168 genomes as suggested by Dorell *et al.* (2002). Offering gene complements from a wide diversity of *C. jejuni* strains isolated from different sources in the form of a pan-genomic array would remedy the drawback of the previous microarray studies using single strain-based microarrays. However, even with single-strain arrays, a comparison of different genotyping approaches could clearly show that the microarray-based approach provides higher resolution than MLST, and can therefore be considered superior for elucidating the pathogenic properties of isolates and their epidemiology (Rodin *et al.*, 2008).

12.2.4 Pan-genomic microarrays for *Campylobacter*

With the intention to meet the above-mentioned limitations, a microarray of genomic DNA fragments originating from a mixture of different strains of the same species was constructed. In contrast to a single-strain genome microarray, such pan-genomic mixed genome microarrays contain gene complements that are absent in one but present in other strains. Such an approach can be seen as an important microbiological research tool as it allows for comprehensive DNA analyses to characterize genetic diversity in a genome-wide manner. A good first example of the functionality of this extended genomotyping approach was shown for *Helicobacter pylori* by using an microarray based on ORF sequences of two distinct strains (Salama *et al.*, 2000). The array successfully identified genes shared by the two *H. pylori* strains representing a kind of 'core genome' and genes that are strain-specific and greatly helped to understand genome functionality and epidemiology. As this mixed genome array was based on known ORF sequences of only two sequenced strains, however, it can be assumed that adding ORF sequences from more than two strains will result in further refinement in terms of identification of strain-specific sequences and for making use of the full genomotyping potential of this technology.

A different approach was chosen for constructing a mixed strain *Campylobacter* genome array (Keijser and van der Vossen, Netherlands Organisation for Applied Scientific Research (TNO), unpublished). This array contains 4608 randomly cloned genetic elements derived from 12 different *C. jejuni* isolates including those originating from poultry and human gastroenteritis, arthritis and Guillain-Barré patients. Furthermore the microarray contains 1536 probes of *C. coli, C. lari*, and *C. upsaliensis*. In contrast to the single-strain arrays and those based on ORF sequences from whole-genome sequences, the TNO *Campylobacter* microarray is based on anonymous, randomly isolated genomic clones originating from a wide range of strains thus representing a random distribution of *Campylobacter* strain-derived sequences. The setup of the mixed genome array and genomotyping approach is shown in Fig. 12.1.

A collection of 350 *C. jejuni* isolates, partially selected from the CampyNet collection (Harrington *et al.*, 2003), was analyzed using this microarray. The *Campylobacter* genomotyping microarray enables efficient, high-resolution array-based typing of *C. jejuni*. The generated data furthermore revealed

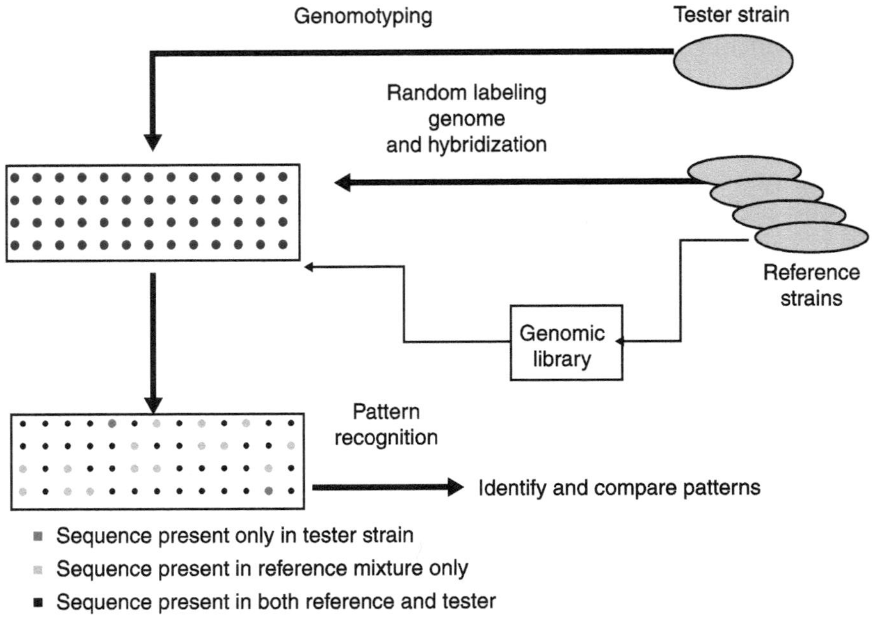

Fig. 12.1 Schematic presentation of the genomotyping approach based on mixed strain genome microarray hybridization.

information on the core and supra-genome of the *C. jejuni* species and showed clear evidence of lateral gene transfer between *Campylobacter* species. Typical hybridization results using the microarray for a limited portion of 144 strains are given in Fig. 12.2. Although only a subset of probe responses are shown (317 of a total of 6144 probes), the figure indicates that *C. jejuni* strains are highly diverse with respect to their gene complement.

By sequencing the differential displayed markers, additional *C. jejuni* related genetic information was collected. The pool of different *C. jejuni* species appeared to be composed of two main branches as observed by the distinct separation of two groups doing clustering analysis. When comparing the phylogenetic microarray data with other typing data, such as amplified fragment length polymorphism (AFLP) and Penner serotyping, an excellent correlation was found between the microarray-based typing and AFLP-based typing. The distinction between two *C. jejuni* branches directly corresponded to two main AFLP types. Applying biostatistical analysis, microarray probes could be identified that differentiated between these two branches. To elucidate the identity of the anonymous microarray probes, the corresponding genetic clones were analyzed by sequencing. The obtained data revealed that the clustering could significantly be attributed to differences in a type II restriction-modification system, present in one group and absent in the other group. Heterogeneity in the restriction-modification systems were also identified as the major differences between various *H. pylori* strains in addition to diversity in DNA methylation patterns

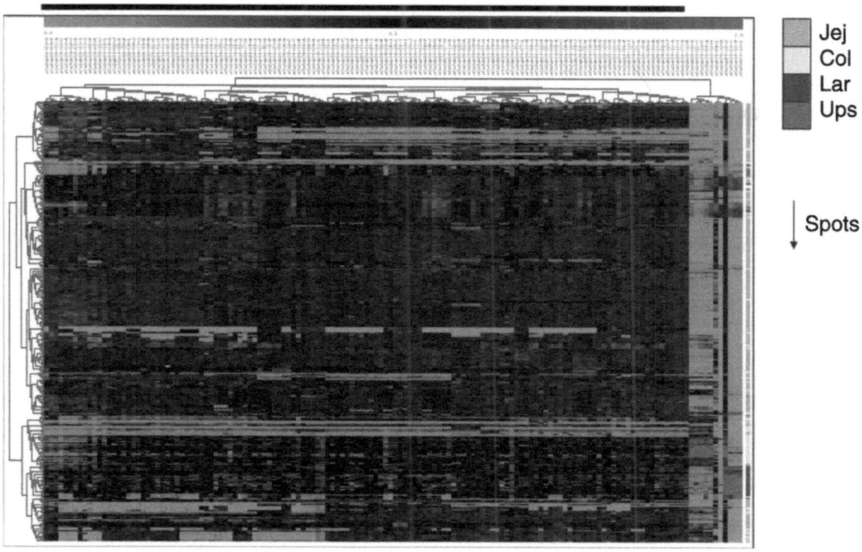

Fig. 12.2 Cluster analysis of genomotyping microarray hybridization results with a set of 144 *Campylobacter* isolates including a number of CAMPYNET strains. In the right margin, the probe spots are identified that are species-specific.

(Ando *et al.*, 2000; Xu *et al.*, 2000). Elevated divergence of these systems may be associated with the positive selection of subpopulations that contribute to genetic diversity.

Apart from yielding data regarding *C. jejuni*-specific differences, the microarray data were further utilized for selection of additional probes that will be helpful for identification of *Campylobacter* isolates at the species level based on biomarkers specific for *C. lari, C. coli, C. upsaliensis*, and *C. jejuni*. In conclusion, the *Campylobacter* genomotyping microarray is anticipated to be a good basis for a better understanding of *Campylobacter* epidemiology and identification of probes that correlate with pathogenicity, host range, antibiotic resistance, resistance to oxidative stress, and survival in specific environments.

12.2.5 Additional evidence supporting the potential of genomotyping

Since it is too early at this stage of research to draw firm conclusions with respect to the potential of microarray genomotyping for simultaneous tracing of *Campylobacter* and studying epidemiology and strain properties, further studies are needed. First steps have been taken applying the technology to study other health-threatening bacteria with clinical implications. They show up the extent to which genomotyping can assist in the understanding of epidemiology and/or the identification of virulence markers. An example is the pathogenic group A streptococci; *Streptococcus pyogenes*. These bacteria are associated with a wide variety of symptoms including mild superficial infections at one end of the scale and life threatening symptoms such as toxic shock-like syndrome (TSS) on the other end of the scale. By applying a mixed genome array, 76 clinically well-characterized strains (including strains involved in TSS) were studied with respect to genomic differences (Vlaminckx *et al.*, 2007). Previous serological information indicated that for a number of M-protein types (based on a variable surface protein) only types M1 and M3 are overrepresented in TSS. Genomic differences were studied by sequencing the corresponding differential probes. They indicated that bacteriophages are major contributors to the observed genome diversifications. In addition, novel genes were identified that are highly specific for a subset of M-types and may therefore be involved in M-type associated disease manifestations. In the M1 and M3 type strains, unique virulence factors were identified including exotoxin *spe*A, iron binding factor, collagen and fibrinogen binding factor, all of which may contribute to the potential of these strains to cause TSS. However, in contrast to the M-type specific virulence profile, a common virulence profile among strains involved in TSS irrespective of their M-type could not be identified. Therefore, it is speculated that other factors including host factors of the TSS patient are involved in the development of the syndrome.

Another example where genomotyping was successfully applied was a study aiming at understanding the evolution of *Enterococcus faecium* from a human gut commensal to a severe multidrug-resistant nosocomial pathogen. Little was known about the evolution and acquisition of drug resistance in this species. The enormous genomic diversity among different strains made whole-genome sequencing appear overwhelming. Therefore, genomotyping was performed by using a mixed-strain genome array. As expected, results revealed a high degree of strain diversity (Leavis *et al.*, 2007). Sequencing of differential probes revealed that strains had acquired sequences from other species as was observed by varying GC content of sequences as well as aberrant codon motives. The involvement of IS16 as the most prominent insert made horizontal gene transfer (HGT) appear very probable. It was hypothesized that an enterococcal subspecies evolved that became adapted to the constraints of a hospital environment via HGT events. Such information turned out to be relatively easily assessable by using genomotyping via a mixed-strain genome microarray instead of applying whole-genome sequencing of a large number of different strains.

Genomotyping was also applied on *Enterobacter cloacae* strains including strains from two outbreaks (Paauw, 2008; Paauw *et al.*, 2008). Genes for resistance, colonization, and virulence were identified that likely explain the epidemic behavior of these strains. As observed for the nosocomial *E. faecium* strains, again mobile elements turned out to be important drivers for the emergence of strains with the capacity to cause and to adapt to hospital conditions.

One step beyond understanding of epidemiology, genome plasticity, and identification of virulence genes is the selection of predictive markers. By using the genomotyping approach, a search for markers was conducted for rapid identification and discrimination of pathogenic *Legionella* strains from environmental non-pathogenic *Legionella* strains (Schuren *et al.*, in preparation).

By using the mixed genome microarray technology in combination with a large set of well-characterized *Legionella* strains from patients or the environments where disease was acquired (in total 257 samples were included), genomotyping results were further processed to identify such discriminatory biomarkers. A first biostatistical analysis revealed approximately 480 markers out of 3360 probes displaying a variation in presence. Biostatistical programing technologies (designed by B. Worzel, Genetics Squared, USA) were developed to identify genetic markers that are most predictive for discrimination between pathogenic and harmless environmental strains and to describe their interrelationships. Five markers appeared most discriminatory and predicted 100% of the pathogenic strains and 62% of the environmental strains (Schuren *et al.*, in preparation). This reliability of the outcome was further substantiated by a second independent set of strains that was not included for creating the biostatistical model. The five markers resulting from the genomotyping activity were finally translated into a rapid prognostic *Legionella* test (based on hybridization of PCR products to a user-friendly chip) that has recently become available on the market for routine testing. Application of this prognostic detection system in the drinking water industry will allow fast managerial decisions as preventive measure in the case of eminent risk of *Legionella* infection: if the detected *Legionella* strain is identified as pathogenic, immediate closure of the affected facility in combination with immediate disinfection of the water distribution system can minimize risk of disease. In case the detected *Legionella* strain turns out to be non-pathogenic, such draconic measures can be avoided and disinfection can be planned at the earliest convenient moment.

12.3 Conclusions

Although it is too early to state that genomotyping is the most appropriate methodology for tracing *Campylobacter* and studying its epidemiology and strain properties, the given examples related to other human pathogens demonstrate the power of the genomotyping approach taking advantage of the mixed genome microarray hybridization technology. Apart from whole-genome sequencing, genomotyping technology perfectly meets the demands of epidemiological studies. The technology achieves discrimination at strain level, without losing its capability to differentiate between species, a property which does not apply to many other genotyping methodologies. Information on the exchange of genes is restricted to newly inserted or deleted DNA. The technology immediately shows which gene complements are absent or present. Information on genetic exchange in terms of transmission of alleles via DNA recombination as observed by MLST typing will be invisible. In contrast to MLST typing, genomotyping allows the identification of markers correlating with host range, virulence, antibiotic resistance, etc., and is therefore far more informative and superior. For the time being, the technology is being more cost efficient than genome sequencing and will provide the microbiologist a better understanding of strain properties.

Spin-off technologies from the genomotyping approach include dedicated *Campylobacter* detection systems employing relevant biomarkers. Such rapid diagnostic tools can form the basis for better management of microbial issues in the food, feed and environmental area as well as in health care. This ability to take more targeted managerial measures, made possible by the highly predictive nature of these biomarkers, represents an essential difference with traditional molecular detection systems.

12.4 References

ANDO T, XU Q, TORRES M, KUSUGAMI K, ISRAEL D A and BLASER M J (2000), 'Restriction-modification system differences in *Helicobacter pylori* are a barrier to interstrain plasmid transfer', *Molecular Microbiology*, **37**, 1052–65.

BLASER M J (1997), The Journal of infectious Diseases, 176 (suppl 2): S103–5.

CHAMPION O L, GAUNT M W, GUNDOGDU O, ELMI A, WITNEY A A, HINDS J, ET AL. (2005), 'Comparative phylogenomics of the foodborne pathogen *Campylobacter jejuni* reveals genetic markers predictive of infection source', *Proceedings of the National Academy of Sciences of the United States of America*, **102**, 16043–8.

DORRELL N, MANGAN J A, LAING K G, HINDS J, LINTON D, AL-GHUSEIN H, ET AL. (2001), 'Whole genome comparison of *Campylobacter jejuni h*uman isolates using a low-cost microarray reveals extensive genetic diversity', *Genome Research*, **11**, 1706–15.

DUIM B, WASSENAAR T M, RIGTER A and WAGENAAR J (1999), 'High-resolution genotyping of *Campylobacter* strains isolated from poultry and humans with AFLP fingerprinting', *Applied Environmental. Microbiology*, **65**, 2369–75.

FOUTS D E, MONGODIN E F, MANDRELL R E, MILLER W G, RASKO D A, RAVEL J, ET AL. (2005), 'Major structural differences and novel potential virulence mechanisms from the genomes of multiple *campylobacter* species', *PLoS Biology*, **3**, e15. Epub.

FRENCH N, BARRIGAS M, BROWN P, RIBIERO P, WILLIAMS N, LEATHERBARROW H, ET AL. (2005), 'Spatial epidemiology and natural population structure of *Campylobacter jejuni* colonizing a farmland ecosystem', *Environmental Microbiology*, **7**(8): 1116–26.

FRIJS L M, PIN C, TAYLOR D E, PEARSON B M and WELLS J M (2007), 'A role for the *tet(O)* plasmid in maintaining *Campylobacter* plasticity', *Plasmid*, **57**, 18–28.

GORMLEY F J, MACRAE M, FORBES K J, OGDEN I D, DALLAS J F and STRACHAN N J (2008), 'Has retail chicken played a role in the decline of human *campylobacter*iosis?', *Applied Environmental Microbiology*, **74**, 383–90.

HARRINGTON C S, MORAN L, RIDLEY A M, NEWELL D G and MADDEN R H (2003), 'Inter-laboratory evaluation of three flagellin PCR/RFLP methods for typing *Campylobacter jejuni* and *Campylobacter coli*: the CAMPYNET experience', Journal of Applied Microbiology, **95**,1321–33.

HUMPHREY T, O'BRIEN S J and MADSEN M (2007), '*Campylobacter*s as zoonotic pathogens: a food production perspective', *International Journal of Food Microbiology*, **117**, 237–57.

IJZERMAN E P F (2009), *Progress in Diagnostics and Prevention of Legionnaires Disease*, Dissertation, Regional Laboratory of Public Health Kennemerland, the Netherlands.

JANSSEN R, DE JONGE R and HOEBEE B (2006), *Genetic Susceptibility to Campylobacter Infection*, RIVM report 340210002, Bilthoven.

JONES K (2001), '*Campylobacter*s in water, sewage and thee environment', *Journal of Applied Microbiology*, **90**, 68S–79S.

KEMMEREN J M, MANGEN M J, VAN DUYNHOVEN Y T and HAVELAAR A H (2005), *Priority Setting of Foodborne Pathogens*, RIVM report 330080001. Bilthoven

KOKOTOVIC B and ON S L (1999), 'High-resolution genomic fingerprinting of *Campylobacter jejuni* and *Campylobacter coli* by analysis of amplified fragment length polymorphisms', *FEMS Microbiology Letters*, **173**, 77–84.

LEAVIS H L, WILLEMS R J L, VAN WAMEL W J B, SCHUREN F H, CASPERS M P M and BONTEN M J M (2007), 'Insertion sequence-driven diversification creates a globally dispersed emerging multiresistant subspecies of *E. faecium*', *PLoS Pathogens*, **3**, 75–96.

LEONARD II E E, TAKATA T, BLASER M J, FALKOW S, TOMPKINS L S and GAYNOR E C (2003), 'Use of an open reading frame-specific *Campylobacter jejuni* DNA microarray as a new genotyping tool for studying epidemiologically related isolates', *The Journal of Infectious Diseases*, **187**, 691–4.

LÉVESQUE S, FROST E, ARBEIT R D and MICHAUD S (2008), 'Multi-locus sequence typing of *Campylobacter jejuni* isolates from humans, chickens, raw milk, and environmental water in Quebec, Canada', *Journal of Clinical Microbiology*, **46**, 3404–11.

MCCARTHY N D, COLLES F M, DINGLE K E, BAGNALL M C, MANNING G, MAIDEN M C, ET AL. (2007), 'Population genetic approaches to assigning the source of human pathogens: Host-associated genetic import in *Campylobacter jejuni*', *Emerging Infectious Diseases*, **13**, 267–72.

ON S L, DORRELL N, PETERSEN L, BANG D D, MORRIS S, FORSYTHE S J, ET AL. (2006), 'Numerical analysis of DNA microarray data of *Campylobacter jejuni* strains correlated with survival, cytolethal distending toxin and haemolysin analyses', *International Journal of Medical Microbiology*, **296**, 353–63.

PAAUW A, CASPERS M P, SCHUREN F H, LEVERSTEIN-VAN HALL M A, DELÉTOILE A, MONTIJN R C, ET AL. (2008), 'Genomic diversity within the *Enterobacter cloacae* complex', *PLoS One*, **3**(8):e3018.

PAAUW A (2008), *Enterobacter Hormaechei from Isolate to Epidemic*, PhD Dissertation, University of Utrecht, Utrecht, the Netherlands.

PARKER C T, QUIÑONES B, MILLER W G, HORN S T and MANDRELL R E (2006), 'Comparative genomic analysis of *Campylobacter jejuni* strains reveals diversity due to genomic elements similar to those present in *C. jejuni* strain RM1221', *Journal of Clinical Microbiology*, **44**, 4125–35.

PARKHILL J, WREN B W, MUNGALL K, KETLEY J M, CHURCHER C, BASHAM D, ET AL. (2003), 'Comparative genome analysis of *Campylobacter jejuni* using whole genome DNA microarrays', *FEBS Letters*, **554**, 224–30.

PENN C W, QUAIL M A, RAJANDREAM M A, RUTHERFORD K M, VAN VLIET A H, WHITEHEAD S, ET AL. (2000), 'The genome sequence of the foodborne pathogen *Campylobacter jejuni* reveals hypervariable sequences', *Nature*, **403**, 665–8.

PETERSEN L, NIELSEN E M, ENGBERG J, ON S L, DIETZ H H (2001), 'Comparison of genotypes and serotypes of *Campylobacter jejuni* isolated from Danish wild mammals and birds and from broiler flocks and humans', *Applied Environmental Microbiology*, **67**, 3115–21.

POLY F, THREADGILL D and STINTZI A (2004), 'Identification of *Campylobacter jejuni* ATCC 43431-specific genes by whole microbial genome comparisons', *Journal of Bacteriology*, **186**, 4781–95.

RODIN S, ANDERSSON A F, WIRTA V, ERIKSSON L, LJUNGSTRÖM M, BJÖRKHOLM B, ET AL. (2008), 'Performance of a 70-mer oligonucleotide microarray for genotyping of *Campylobacter jejuni*', *BMC Microbiology*, **8:** 73

SALAMA N, GUILLEMIN K, MCDANIEL T K, SHERLOCK G, TOMPKINS L and FALKOW S (2000), 'Whole-genome microarray reveals genetic diversity among *Helicobacter pylori* strains', *Proceedings of the National Academy of Sciences of the United States of America*, **97**, 14668–73.

SNELLING W J, MATSUDA M, MOORE J E, DOOLEY, J S (2005), '*Campylobacter jejuni*', *Letters in Applied Microbiology*, **41**, 297–302.

SOPWITH W, BIRTLES A, MATTHEWS M, FOX A, GEE S, PAINTER M, ET AL. (2006), '*Campylobacter jejuni* Multi-locus Sequence Types in Humans, Northwest England, 2003–2004', *Emerging Infectious Diseases*, **12**, 1500–07.

TABOADA E N, ACEDILLO R R, CARRILLO C D, FINDLAY W A, MEDEIROS D T, MYKYTCZUK O L, ET AL. (2004), 'Large-scale comparative genomics meta-analysis of *Campylobacter jejuni* isolates reveals low level of genome plasticity', *Journal of Clinical Microbiology*, **42**, 4566–76.

VLAMINCKX, B J M, SCHUREN, F H J, MONTIJN R C, CASPERS M P M, FLUIT A C, WANNET W J B, ET AL. (2007), 'Determination of the relationship between group A streptococcal genome content, M type, and toxic shock syndrome by a mixed genome microarray', Infection and Immunity. **75**, 2603–11.

WILSON D J, GABRIEL E, LEATHERBARROW A J, CHEESBROUGH J, GEE S, BOLTON E, ET AL. (2008), 'Tracing the source of *campylobacter*iosis', *PLoS Genetics*, **4**(9): e1000203.

XU Q, MORGAN R D, ROBERTS R J and BLASER M J (2000), 'Identification of type II restriction and modification systems in *Helicobacter pylori* reveals their substantial diversity among strains', *Proceedings of the National Academy of Sciences of the United States of America*, **97**, 9671–6.

13

Investigating foodborne pathogens using comparative genomics

R. A. Stabler, E. S. Nalerio, P. C. R. Strong and B. W. Wren, London School of Hygiene and Tropical Medicine, UK

Abstract: Recently, molecular typing approaches have advanced significantly through the availability of whole genome sequence data, allied with high-throughput microarray technologies and sophisticated data analysis. This has enabled us to pinpoint individual genes that may be important in survival, virulence and niche adaptation. This chapter will describe how comparative phylogenomics can be used to model relationships between strains and their genetic content, their origin (e.g. food source) and their virulence potential/disease outcome. Examples will be drawn from the established foodborne pathogens *Campylobacter jejuni*, the enteropathogenic *Yersiniae* and *Listeria monocytogenes*, as well as the suspected food-associated pathogen *Clostridium difficile*.

Key words: microarray, comparative phylogenomics, whole-genome, Bayesian.

13.1 Introduction

13.1.1 The importance of molecular typing and the emergence of problematic foodborne pathogens

The last century has seen significant advances in hygiene control and consumer knowledge, coupled with improved food treatment and food processing. However, foodborne pathogens still represent a significant threat to human health worldwide. Bacterial pathogens are particularly adept at change because of their short generation time combined with multiple mechanisms to alter their genetic repertoire. Foodborne pathogens have a further level of potential complexity as the evolutionary pressures that shape their existence are compounded by human behaviour and practices, such as changes in eating habits, ecology, agriculture and food manufacturing processes. Keeping pace with the emergence of pathogens seems intimidating, but the ready availability of whole-genome data from almost any pathogen should pave the way for new approaches to tackle our old adversaries. Furthermore, the development of complementary high-throughput genomics technologies, such as DNA microarrays, means that we are now in a position to trace and monitor pathogens, particularly those prevalent in the food chain. The era of molecular forensic microbiology is upon us.

Using case studies and examples involving foodborne pathogens, this chapter will describe how comparative phylogenomics can be used to model relationships between strains and genetic content, strains and virulence determinants, inter-strain variations, link strains and source of food contamination and strains and disease outcome. This greater understanding of strains, relationships and sources will underpin efforts that could be used to reduce pathogens in the food chain.

13.2 Molecular typing systems in tracking bacterial pathogens in the food chain

13.2.1 Traditional typing approaches

Owing to their importance, foodborne pathogens are one of the most intensively typed group of pathogens, with a myriad of typing systems applied to understand its epidemiology. This is exemplified by *Campylobacter jejuni*, the most prevalent foodborne pathogen worldwide. Traditional typing systems for *C. jejuni* have been based on phenotypes such as biotype (Hebert *et al.*, 1982), phage type (Grajewski *et al.*, 1985), antibiogram (Karmali *et al.*, 1980) and most frequently serotyping. There are two generally accepted, well-evaluated serotyping schemes. The Penner scheme is based on heat-stable (HS) antigens using a passive haemagglutination technique (Penner and Hennessy, 1980), whilst the Lior scheme is based on heat-labile (HL) antigens (Lior *et al.*, 1982) and bacterial agglutination. The major disadvantages of both techniques include the high number of 'untypable' strains together with the time-consuming and technically demanding requirements of the methods (Siemer *et al.*, 2005). Because of this, the value of serotyping techniques for national and global epidemiological studies has been restricted. For a more detailed description see Chapter 7.

13.2.2 Molecular typing approaches

Molecular typing methods which have many advantages over traditional phenotypic techniques have been developed. They generally have greater discriminatory capacity based on stable genetic elements and can be universally applied, allowing comparison between laboratories. These include pulsed-field gel electrophoresis (pulsotypes; PFGE), restriction endonuclease analysis (REA), restriction fragment length polymorphism (RFLP), toxinotyping, flagellin typing (*fla* typing), multilocus sequence typing (MLST) and PCR-ribotyping (Killgore *et al.*, 2008). A summary of genotyping techniques and the relative discriminatory power and relative cost is presented in Table 13.1. For a more detailed description see Chapter 9.

Table 13.1 Summary of genotyping techniques, relative discriminatory power and costs

Typing technique	Number of genes targeted	Relative discriminatory power	Relative cost	Reference
Comparative hybridization against array containing entire gene sequence	multiple loci	High	High	Champion *et al.* (2005)
Multilocus sequence typing (MLST)	multiple loci	Moderate to high (depends on gene choice)	Medium to High	Dingle *et al.* (2001)
Binary typing (presence/ absence of selected genes or alleles across the genome)	multiple loci	Potentially high	Medium	van Leeuwen *et al.* (1999)
Pulsed-field gel electrophoresis (PFGE)	multiple loci	Moderate to high (depends on number of bands observed)	High	Ribot *et al.* (2001)
Restriction fragment length polymorphism (RFLP)	multiple loci	Moderate to high (depends on number of bands observed)	Medium	Fujimoto *et al.* (2000)
Amplification of a single target gene specific to a pathogen	single locus	Moderate to high (depends on gene choice)	Low to medium	Eyers *et al.* (1993)
Amplified fragment length polymorphism (AFLP)	multiple loci	Moderate to high	Low to medium	Lindstedt *et al.* (2000)
Automated ribotyping	single locus	High	High	Fitzgerald *et al.* (1996)
Ribosomal RNA gel electrophoresis	single locus	Low to medium	Low	Christensen *et al.* (1999)
Plasmid profiles	single locus	Low	Low	Wachsmuth *et al.* (1991)
Restriction endonuclease of a single amplified product	single locus	High	High	Eyigor *et al.* (1999)

The lack of understanding of the population biology for many foodborne pathogens means that we do not know where these epidemic clones have emerged from and how they are continuing to evolve. To obtain this information most typing methods are of limited value, as a true population structure cannot be ascertained due to the lack of a genetic phylogeny. For many pathogens MLST has become the gold standard for assessing the population structure and genetic relatedness of a given bacterial species; provided that the species is not strictly clonal or too recombinatorial, and that an appropriate breadth of diversity of strains across the species can be tested (Maiden *et al.*, 1998). The major advantages of MLST are that it is high throughput, relatively inexpensive and portable between different research and reference laboratories. The scheme is based on sequence analysis of up to seven housekeeping genes where allele profiles define different sequence types (STs). For a more detailed description see Chapter 9.

13.3 Whole-genome approaches using microarrays

Traditional phylogenetic classification of bacteria to study evolutionary relatedness is based on the characterisation of a limited number of genes, rRNA/rDNA or signature sequences (such as MLST). However, due to the acquisition of DNA through lateral gene transfer, the differences between closely related bacterial strains can be significant. By contrast, whole-genome sequencing comparisons allow the full complement of genes to be compared in a given pathogen. Nevertheless, despite advances with next generation sequencing, whole-genome sequencing currently remains an expensive pursuit and is limited to a few institutes. DNA microarrays represent an alternative technology for whole-genome comparisons enabling a 'birds eye view' of all the genes absent or present in a given genome as compared to the reference genome on the microarray. Harnessing DNA microarray information through interrogative and robust algorithms has enabled a true 'comparative phylogenomics' approach to be developed. Recent comparative phylogenomics studies have been undertaken on increasingly large collections of diverse strains from defined origins. A common feature from many of these studies has been the unexpectedly large genetic diversity between strains within the same species, blurring our definition of species boundaries. Whole-genome comparisons typically identify sets of 'core genes' shared by all strains in a species and 'accessory genes' present in one or more strains in a species that often result from gene acquisition. It is these differences that can often be used to identify genes/genetic islands related to 'gain-of-function traits' in pathogenic strains. Uncovering the mechanisms behind this variability is fundamental in understanding and ultimately counteracting infection. Moreover, given the range of diseases associated with some bacterial pathogens and the diverse genotypic and phenotypic properties of clinical and environmental isolates, microarrays have proved to be particularly useful for determining correlates of pathogenicity (Hinchliffe *et al.*, 2003; Howard *et al.*, 2006) and in deciphering the epidemiology and ecological niches of the organism (Champion *et al.*, 2005; Stabler *et al.*, 2006). Examples of the use of microarrays for comparative genomics are given below for three well-recognised foodborne pathogens (*C. jejuni*, *Yersinia enterocolitica* and *Listeria monocytogenes*) and a further pathogen where food sources may be important in its initial transmission (*Clostridium difficile*).

13.3.1 Case study – *Campylobacter jejuni*

C. jejuni is the most frequently identified cause of acute bacterial gastroenteritis worldwide. The reported incidence of infection has markedly increased in many developed countries within the last 20 years, with the number of cases reported exceeding those of salmonellosis and shigellosis combined (Schlundt *et al.*, 2004; Rajda and Middleton, 2006). In developing countries, it is omnipresent in the stool of infants with diarrhoea as a result of contaminated food or water (Oberhelman *et al.*, 2003). Campylobacteriosis is a zoonosis – the reservoir of this disease includes wild and domestic animals, particularly birds (Nielsen *et al.*, 2006). Chickens represent the largest potential source of human infection. It is likely that the ability of these bacteria to grow at 42°C reflects their adaptation to the avian gut (Jones *et al.*, 2004). The gastrointestinal tracts of other livestock species intended for human consumption have also been shown to be frequently colonised with *C. jejuni* (Minihan *et al.*, 2004). The digestive tracts of clinically normal cattle and sheep have been demonstrated to represent a significant reservoir for a number of *Campylobacter* spp. (Colles *et al.*, 2003). A high prevalence of *Campylobacter* in pigs has been reported in numerous studies and dressed pig carcasses have been shown to be more frequently contaminated than those of either cattle or sheep (Nesbakken *et al.*, 2003). Surface and river water may be the most important reservoir of *Campylobacter* species, not only as a potential source of human infection but also as a source of infection for livestock, particularly poultry (Said *et al.*, 2003).

In recognition of the importance and prevalence of *C. jejuni*, its strain NCTC11168 was the first foodborne pathogen to be sequenced (Parkhill *et al.*, 2000). *C. jejuni* was also the first foodborne pathogen for which a microarray was constructed and used for multi-strain comparisons (Dorrell *et al.*, 2001). A number of subsequent studies have been performed generally confirming the wide diversity of this species (Leonard *et al.*, 2003; Leonard *et al.*, 2004; Keel *et al.*, 2007; Taboada *et al.*, 2007). Although these studies have been useful in identifying candidate genes that may be involved in pathogenesis and survival, it does not provide any clue to the evolutionary origins and phylogeny of *C. jejuni*. Therefore, we developed comparative phylogenomics, whole-genome comparisons of microbes using DNA microarrays combined with Bayesian-based algorithms to model the phylogeny of a given microorganism (summarised in Fig. 13.1). We applied this initially to 111 *C. jejuni* isolates from a spectrum of sources including humans (70), chickens (17), bovines (13), ovines (5) and the environment (6). Remarkably from this data, the Bayesian phylogeny of the isolates revealed two distinct clades unequivocally supported by Bayesian probabilities ($P = 1$) (Champion *et al.*, 2005). On further analysis these two clades appeared to correlate with the origin of the strains and included a livestock clade comprising 31/35 (88.6%) of the livestock isolates and a 'non-livestock' clade containing all environmental isolates. Several genes were identified as characteristic of strains in the livestock clade. The most prominent was a cluster of six genes (*cj1321* to *cj1326*) within the genetically variable flagellin glycosylation locus. Surprisingly, the initial comparative phylogenomics study

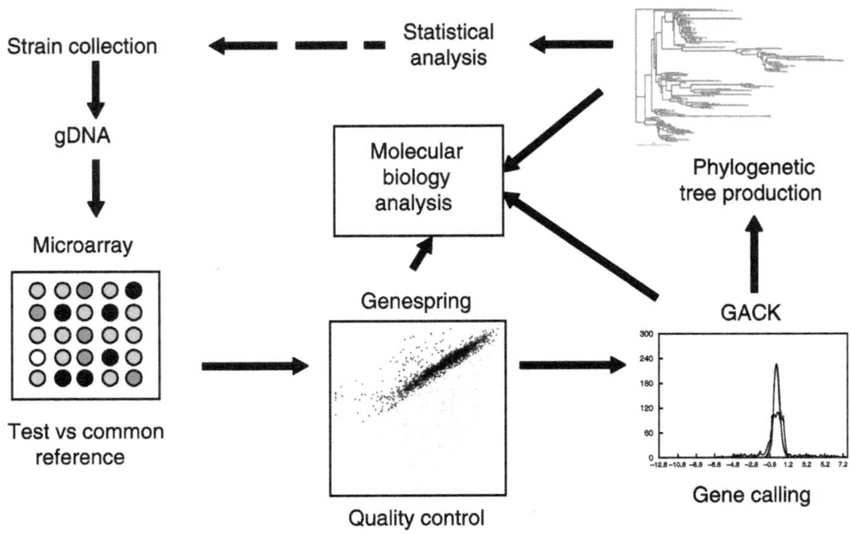

Fig. 13.1 Comparative phylogenomics pipeline.

showed that the majority 39/70 (55.7%) of *C. jejuni* human isolates were found in the non-livestock clade, suggesting that many *C. jejuni* infections may be from non-livestock (and possibly non-agricultural) sources. More recently we have studied over 230 *C. jejuni* strains from diverse origins and have further demonstrated the split of strains into two clades. Approximately half of the human isolates from this study are not associated with the livestock clade (Al-Jaberi and Wren, unpublished data). These studies may provide insight into a previously unidentified reservoir of *C. jejuni* infection that may have implications in disease control strategies.

A major advantage of the comparative genomics approach described above is that groupings of strains that relate to different origins can be traced back to the absence or presence of specific genes. Determining the function of these genes may help to explain why some strains have a distinct epidemiology, for example the identification of five genes (*Cj1321 to Cj1326*) in the *O*-linked flagellin glycosylation island that were highly prevalent in *C. jejuni* isolates from chickens. Mutagenesis, functional and structural data, has confirmed that this locus has a significant contributory role in the colonisation of *C. jejuni* in chickens (Howard *et al.*, 2009). A motile Δ*Cj1324* mutant with intact flagellae was considerably less hydrophobic and less able to autoagglutinate and form biofilms than the parent strain NCTC11168, suggesting that the surface charge of flagellae was altered in *Cj1324*-deficient strains. These physical and functional attributes were restored upon complementation. Structural analysis of flagellin protein purified from the Δ*Cj1324* mutant revealed the absence of two legionaminic acid glycan modifications that were present in the parent strain 11168H (Howard *et al.*, 2009).

These glycoform modifications were shown to be prevalent in chicken isolates and confirm that differences in the highly variable flagellin glycosylation locus can relate to strain source. The discovery of molecular mechanisms influencing persistence of *C. jejuni* in poultry aids the rational design of approaches to control this problematic pathogen in the food chain.

13.3.2 Case study – *Yersinia enterocolitica*

The enteropathogenic yersiniae – *Yersinia pseudotuberculosis* and *Y. enterocolitica* – are ubiquitous in the environment, and are a common cause of animal infections, with studies frequently isolating them from cattle, sheep, pigs, domesticated animals and avian species (Fredriksson-Ahomaa *et al.*, 1999; Fredriksson-Ahomaa *et al.*, 2001; McNally *et al.*, 2004; Wojciech *et al.*, 2004; Milnes *et al.*, 2008). Human infection with *Y. enterocolitica* is much more common than *Y. pseudotuberculosis*, and is heavily associated with consumption of contaminated or undercooked porcine products (Fredriksson-Ahomaa *et al.*, 1999; Fredriksson-Ahomaa *et al.*, 2001; Nesbakken *et al.*, 2003; McNally *et al.*, 2004; Wojciech *et al.*, 2004; Fearnley *et al.*, 2005). In Benelux countries and other areas of western and northern Europe, *Y. enterocolitica* rivals *Salmonella* as a foodborne pathogen (Doyle, 1990), and as a result is a notifiable disease. *Y. enterocolitica* comprises a biochemically and genetically heterogeneous collection of organisms that has been divided into six biogroups that are differentiated by biochemical tests (1A, 1B, 2, 3, 4 and 5). These can be placed into three lineages: a non-pathogenic group which is completely avirulent in a mouse infection model (biogroup 1A); a weakly pathogenic group that is unable to kill mice but can induce intestinal infection (biogroups 2–5) and a highly pathogenic, mouse-lethal group (biogroup 1B). Biogroup 1A lacks the pYV virulence plasmid and many of the classically defined virulence factors. Biogroup 1A strains also appear to be more distantly related to the other biogroups with molecular typing tests differentiating biogroup 1A isolates as a clear subset from all other biogroups (Fearnley *et al.*, 2005; Howard *et al.*, 2006). Despite these findings, biogroup 1A strains are ubiquitous in livestock and the environment, and in two recent UK studies biogroup 1A strains were found to be the predominant *Y. enterocolitica* biogroup isolated from livestock, and both healthy and infected humans (McNally *et al.*, 2004; Milnes *et al.*, 2008). Of the classically defined pathogenic *Y. enterocolitica*, biogroup 1B forms a geographically distinct group of strains frequently isolated in North America (so-called 'New World' strains). Biogroups 2–5 are the predominantly isolated biogroups associated with human disease cases in Europe and Japan ('Old World' strains).

Despite the separation of *Y. enterocolitica* in distinct biogroups, the genetic differences that explain variation in pathogenesis and whether different biogroups are associated with specific non-human hosts are largely unknown. Comparative phylogenomics on a diverse collection of 94 strains of *Y. enterocolitica* consisting 35 humans, 35 pigs, 15 sheeps and 9 cattle from non-pathogenic, mildly pathogenic and highly pathogenic biogroups (Fig. 13.2). The analysis confirmed three distinct statistically supported clusters comprising of a non-pathogenic clade, a mildly pathogenic clade and a highly pathogenic clade. Remarkably, there were 125 genetic differences present in all highly pathogenic strains but absent in the other clades. These included several previously uncharacterised predicted coding sequences (CDSs) that may encode novel virulence determinants including a hemolysin, metalloprotease and type III secretion effector protein. Additionally, 27 CDSs were identified which were present in all 47 low pathogenicity strains but absent in all non-pathogenic 1A isolates. Further analysis revealed that *Y. enterocolitica* does not cluster according to source (host), and suggests pigs as a major reservoir for human infection (Fig. 13.3).

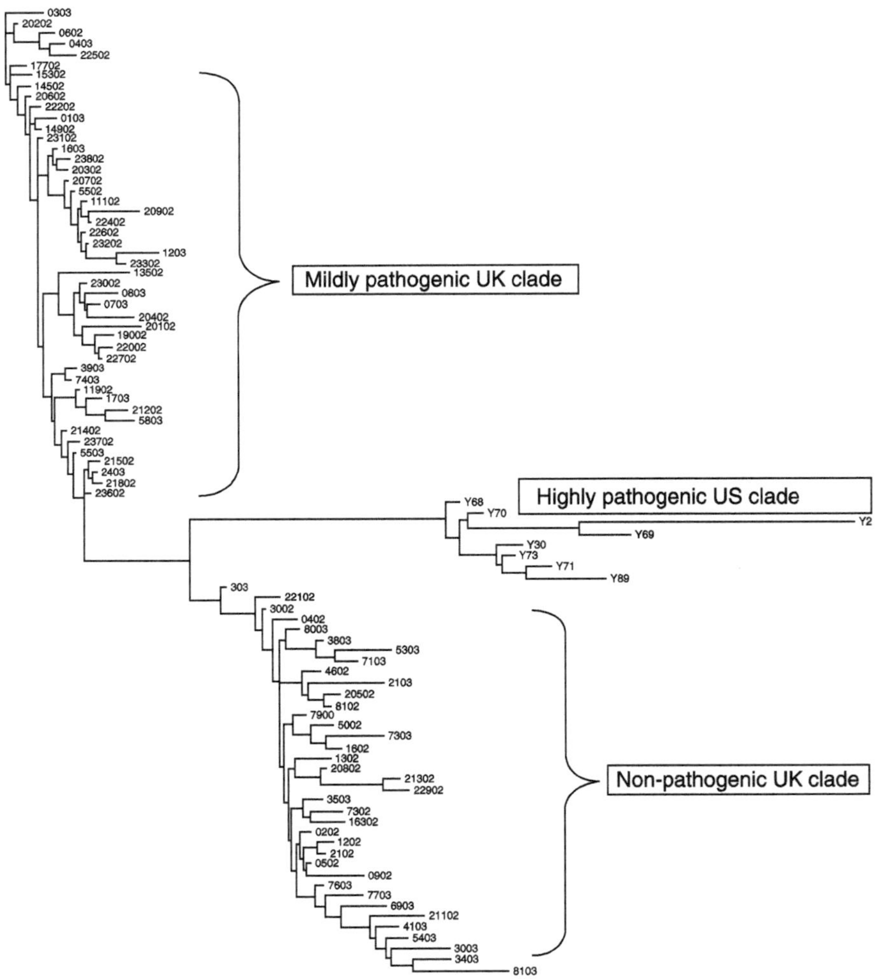

Fig. 13.2 Example of *Yersinia enterocolitica phylogenetic* tree. Comparative phylogenomics using Bayesian algorithm analysis of microarray data showed three main clusters which correlate with known biotypes.

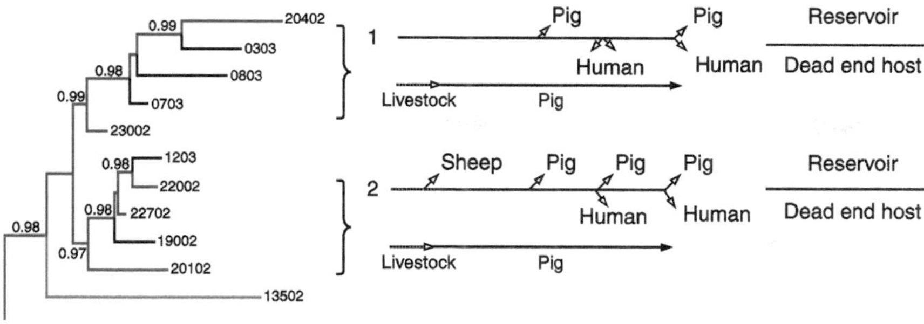

Fig. 13.3 *Yersinia* tree showing animal transmission to humans. Analysis of two comparative phylogeny subclades (1 and 2) identifies animals as reservoirs of infection with occasional transmission to humans. No human-to-human spread indicates that this is a dead end for transmission.

13.3.3 Case study – *Listeria monocytogenes*

L. monocytogenes is the causative agent of listeriosis, which can cause diverse clinical symptoms from gastroenteritis to meningitis and possible death in humans and in over 40 animal species (Kathariou, 2002). The infection is usually acquired by the consumption of dairy and meat products (Autio *et al.*, 2002). High-risk groups for infection include infants, elderly, immunocompromised individuals and pregnant women; however, healthy individuals can also be infected (Schlech, 1998). Despite the fact that listeriosis has a low incidence compared to other foodborne diseases, it is a great public health concern due its mortality rate which is around 20–30%. In the USA, there are approximately 2500 cases of listeriosis every year, with a hospitalisation rate of 91% and approximately 500 deaths (Mead *et al.*, 1999). Importantly, the epidemiology of listeriosis is changing and the incidence is increasing (Goulet *et al.*, 2008). In the last five years, the incidence has increased 59% in the European countries, with increases being reported in Belgium, Denmark, Germany, Finland, France and Switzerland (Goulet *et al.*, 2008). In England and Wales the morbidity rate has increased to 44% (Gillespie *et al.*, 2006). According to Cairns and Payne (2009), the sudden increase of listeriosis in those countries can be associated with changes in food processing and production, and also modification of consumption of the food could be the root of the problem.

L. monocytogenes is another pathogen found ubiquitous in the environment and has been isolated from fruits, vegetables and animals (Seeliger, 1961; Farber and Peterkin, 1991; Rocourt and Cossart, 1997). Animals acquire *L. monocytogenes* during feeding with contaminated food/silage, which can result in asymptomatic carrier status and shedding of bacteria in their faeces (Rocourt and Cossart, 1997). Therefore, food processing plants can be contaminated from raw food materials; *L. monocytogenes* has been isolated from food processing plant environments worldwide (Miettinen *et al.*, 2001; Chasseignaux *et al.*, 2002; Rorvik *et al.*, 2003; Vitas and Garcia-Jalon, 2004; Gudmundsdottir *et al.*, 2005). Contamination can also be through human activity since 2–3% of humans are asymptomatic carriers of this pathogen (Rocourt *et al.*, 2000).

Although this microorganism does not form spores and has no capsule, *L. monocytogenes* can survive in adverse conditions in the environment and food chain (Vazquez-Boland *et al.*, 2001). *L. monocytogenes* has been shown to grow between 0°C and 45°C and a broad range of pH (4.5–9.0); additionally, *L. monocytogenes* can survive at high salt concentrations (up to 10% NaCl) (Bell and Kyriakides, 1998; Lado and Yousef, 1999). These attributes enable *L. monocytogenes* to survive the environmental conditions typically found in the industrial food chain, including refrigeration. If food becomes contaminated, refrigeration temperatures that inactivate or even eliminate other bacteria will not inhibit *L. monocytogenes* multiplication (Rocourt *et al.*, 2000).

Although nearly all *L. monocytogenes* strains have been considered pathogenic, it appears that not all strains are capable of developing disease. In fact, disease outcome can be related to strain serotype. There are 13 recognised *L. monocytogenes* serotypes, but only three (4b, 1/2a and 1/2b) have been associated with approximately 95% of cases and listeriosis outbreaks (Schuchat *et al.*, 1991; Kathariou, 2002). Serotype 4b strains are over-represented in clinical cases and outbreaks and potentially possess unique virulence properties (Doumith *et al.*, 2004b). Similarly, serotype 1/2a is the most isolated in food and food processing plants.

Molecular analyses using PFGE demonstrated that *L. monocytogenes* is a highly diverse species which can be grouped into three lineages (Brosch *et al.*, 1994). Lineage I (consisting of serogroups 1/2a, 3a, 1/2c and 3c), lineage II (1/2b, 3b, 4b, 4d, 4e and 7) and lineage III (4a and 4c) (Doumith *et al.*, 2004a, Nightingale *et al.*, 2005). This classification has been subdivided in six phylogenetic groups, lineage I subtype 1 (serotypes 1/2a and 3a), I.2 (1/2c and 3c), II.1 (4b, 4d and 4e), II.2 (1/2b, 3b and 7), III.1 (4a) and III.2 (4c) (Doumith *et al.*, 2004b). Furthermore, it has been suggested that lineage I represents an environmentally adapted lineage, whilst lineage II strains represent a human host adapted lineage (Wiedmann, 2002).

Comparative phylogenomics of *L. monocytogenes* isolates from Europe and Brazil have verified the three main lineages (Stabler *et al.*, unpublished) confirming that the lineages represent isolates with similar genetic content. Additionally, there was evidence that isolates found to persist in food production environments but from different countries were also very similar, possibly showing a transcontinental spread of persister isolates or co-evolution of persistent genotypes. This study provides an understanding in the molecular phylogeny and evolution of *L. monocytogenes* strains giving information about transmission, pathogenesis and niche adaptation.

13.3.4 Case study – *Clostridium difficile*

C. difficile is a unique pathogen that often predominates in the bowel microflora as a result of the microbial compositional changes following antibiotic treatment.

The hospital environment and patients undergoing antibiotic treatment provide a discrete ecosystem where *C. difficile* persists and where virulent clones thrive. The continued rise of *C. difficile* infection (CDI) worldwide has been accompanied by the rapid emergence and transcontinental spread of highly virulent clones designated PCR-ribotypes 017, 027 and 078. These strains have risen from obscurity to become the most frequently isolated *C. difficile* strain types. Additionally, patients infected with these strains often experience more severe diarrhoea, more recurrent episodes and higher mortality. Although *C. difficile* appears to be evolving to occupy the hospital niche, community-acquired CDI is also on the increase. Equally, changes in human activity are likely to be responsible for creating the microenvironment for *C. difficile* to thrive. The rapid worldwide spread of the 017 and 027 clones of *C. difficile* provides a valuable opportunity to study the very recent emergence of a bacterial pathogen.

Recently, epidemiology studies have suggested a possible zoonotic or foodborne link between human acquisition of *C. difficile* and its subsequent spread within a hospital environment (Rupnik, 2007). This illustrates the importance of choosing isolates from wide sources, to be able to identify the potential reservoir of a given pathogen. To date, most *C. difficile* epidemiological studies have focused on human isolates, which probably miss the bigger picture. Other recently emerging strains include PCR-ribotypes 053 and 078 (Keel *et al.*, 2007), of which 078 is particularly interesting as it has been found in both animals and humans, forming the major PCR-ribotype in both calves and pigs (Keel *et al.*, 2007). The increased incidence of community-acquired infection with ribotype 078 (Goorhuis *et al.*, 2008) could be linked to the fact that this isolate is found in important food animals and has been passed thorough processed meats in the food chain.

We applied comparative phylogenomics to model the phylogeny of *C. difficile*, including 75 diverse isolates comprising hypervirulent, toxin variable and animal strains. The analysis identified four distinct statistically supported clusters comprising a hypervirulent clade, a toxin A–B+ clade and two clades with human and animal isolates. The human and animal II clade was mainly dominated with ribotype 078, confirming that both human and animal 078 isolates are genetically similar. Genetic differences among clades revealed several genetic islands relating to virulence and niche adaptation, including antibiotic resistance, motility, adhesion and enteric metabolism. This study has provided insight into the possible origins of *C. difficile* and its evolution that may have implications in disease control strategies.

13.4 Conclusions and future trends

The bedrock of infectious disease prevention and control is high-quality microbiology and surveillance that allows outbreaks to be anticipated and prevented. Traditionally, this has relied on phenotypic markers, such as serotyping and phage typing, and limited genotypic methods, such as PFGE. More recently,

MLST has been used effectively for retrospective population genetic studies and for determining clonality of strains of diverse origin. All of these methods suffer from providing limited information. In an ideal world, the complete genome sequence of every problem pathogen would be determined as and when outbreaks occurred. Such information would enable us to determine if the genetic information from a given pathogen is changing both in terms of gene acquisition and gene loss.

The complete genome sequence of any bacterium is the definitive standard to fully distinguish between strains and with the advent of next generation sequencing and reduced sequencing costs, this will become a reality. However, it will take considerable time to be used on a routine basis, but may be used in reference laboratories in the near future. Alternatively, next-generation sequencing, which has been developed extensively for tracking Single Nucleotide Polymorphisms (SNPs) in human genetic studies, may have wider applications in bacterial genetic studies. Such SNP analysis has recently been applied to provide the detailed epidemiological distinction and micro-evolution of the clonal bacterial pathogen *Salmonella typhi* (Holt *et al.*, 2008). This can be scaled up for 96-well microtitre plate analysis using Illumina Golden Gateway SNP CHIP technology. This could be usefully applied to study fine genetic differences between clonal lineages, such as the *C. difficile* 027, 017, and 078 PCR-ribotypes to define their evolutionary origin and to monitor how they will evolve in the future. These concerted efforts within a population genetic framework will provide the basis for rational strategies to reduce the burden of pathogens in the food chain.

In the future, DNA microarrays will continue to be used. Microarray analysis is a powerful enabling technology that allows global comparative analysis of the gene content between different strains in a given species. Comparative genomics using microarrays from a range of bacterial pathogens, such as *H. pylori*, *C. jejuni*, *E. coli*, *L. monocytogenes*, *C. difficile* and the enteropathogenic *Yersinia*, clearly demonstrate the diversity and adaptability of these specialised groups of organisms. Studies have revealed much evidence of lateral gene transfer and recombination. This supports an evolutionary scenario involving vertical diversification by mutagenesis, punctuated by frequent lateral gene transfer resulting in a global mosaic genome structure. All microarray analyses are limited by the genetic information on the microarray. However, as more genomes are sequenced and the capacity of microarrays is increased further, information on microarrays can be used to interrogate a given bacterial genome or mixture of genomes. Another stumbling block to reaping the benefits of DNA microarrays is insufficient bioinformatics tools to analyse data. In the future, sustained improvements in software, algorithms, computing speed and information storage will dramatically increase the scale of problems we tackle to understand the basic biology, evolution and traceability of foodborne pathogens. The development of a database of nucleotide differences among strains should allow the design of a universal microbial pathogen microarray, which would have wide application in studying the epidemiology, population genetics, molecular phylogeny and evolution of microbial pathogens, as well as diagnostic applications. A 'lateral gene transfer' microarray consisting of genes from mobile elements, such as pathogenicity islands, phage and plasmid sequences, may have multiple applications. For example, it could be used in active microbial surveillance as an early warning system to alert public health officials to the presence or potential emergence of a more virulent pathogen. The development and application of such a prototype microarray, termed an 'Active Surveillance of Pathogens' (ASP) microarray, have recently been published (Stabler *et al.*, 2008).

No doubt that the knowledge garnered from these studies will be applied to well-designed intervention strategies to reduce the burden of infectious disease. Currently, we have only just begun to scratch the surface in terms of the potential applications of DNA microarrays and next-generation sequencing. The next few years will be a voyage of discovery in terms of developing our understanding of how pathogens persist and evolve in the food chain.

13.5 References

AUTIO T, LUNDEN J, FREDRIKSSON-AHOMAA M, BJORKROTH J, SJOBERG A M & KORKEALA H (2002), Similar *Listeria monocytogenes* pulsotypes detected in several foods originating from different sources. *Int J Food Microbiol*, **77**, 83–90.

BELL C & KYRIAKIDES A (1998), *Listeria* – Una aproximación prática al microorganismo y su control en los alimientos. *Zaragoza*, 173.

BROSCH R, CHEN, J & LUCHANSKY J B (1994), Pulsed-field fingerprinting of listeriae: identification of genomic divisions for *Listeria monocytogenes* and their correlation with serovar. *Appl Environ Microbiol*, **60**, 2584–92.

CAIRNS B J & PAYNE R J (2009), Sudden increases in listeriosis rates in England and Wales, 2001 and 2003. *Emerg Infect Dis*, **15**, 465–8.

CHAMPION O L, GAUNT M W, GUNDOGDU O, ELMI A, WITNEY A A, HINDS J, DORRELL N & WREN B W (2005), Comparative phylogenomics of the foodborne pathogen Campylobacter jejuni reveals genetic markers predictive of infection source. *Proc Natl Acad Sci USA*, **102**, 16043–8.

CHASSEIGNAUX E, GERAULT P, TOQUIN M T, SALVAT G, COLIN P. & ERMEL G (2002), Ecology of *Listeria monocytogenes* in the environment of raw poultry meat and raw pork meat processing plants. *FEMS Microbiol Lett*, **210**, 271–5.

CHRISTENSEN H, JORGENSEN K & OLSEN J E (1999), Differentiation of *Campylobacter coli* and *C. jejuni* by length and DNA sequence of the 16S-23S rRNA internal spacer region. *Microbiology*, **145**(1), 99–105.

COLLES F M, JONES K, HARDING R M & MAIDEN M C (2003), Genetic diversity of *Campylobacter jejuni* isolates from farm animals and the farm environment. *Appl Environ Microbiol*, **69**, 7409–13.

DINGLE K E, COLLES F M, WAREING D R, URE R, FOX A J, BOLTON F E, ET AL. (2001), Multilocus sequence typing system for *Campylobacter jejuni*. *J Clin Microbiol*, **39**, 14–23.

DORRELL N, MANGAN J A, LAING K G, HINDS J, LINTON D, AL-GHUSEIN H, BARRELL B G, PARKHILL J, STOKER N G, KARLYSHEV A V, BUTCHER P D & WREN B W (2001), Whole genome comparison of *Campylobacter jejuni* human isolates using a low-cost microarray reveals extensive genetic diversity. *Genome Res*, **11**, 1706–15.

DOUMITH M, BUCHRIESER C, GLASER P, JACQUET C & MARTIN P (2004a), Differentiation of the major *Listeria monocytogenes* serovars by multiplex PCR. *J Clin Microbiol*, **42**, 3819–22.

DOUMITH M, CAZALET C, SIMOES N, FRANGEUL L, JACQUET C, KUNST F, MARTIN P, COSSART P, GLASER P & BUCHRIESER C (2004b), New aspects regarding evolution and virulence of *Listeria monocytogenes* revealed by comparative genomics and DNA arrays. *Infect Immun*, **72**, 1072–83.

DOYLE M P (1990), Pathogenic *Escherichia coli, Yersinia enterocolitica*, and *Vibrio parahaemolyticus*. *Lancet*, **336**, 1111–15.

EYERS M, CHAPELLE S, VAN CAMP G, GOOSSENS H & DE WACHTER, R (1993), Discrimination among thermophilic *Campylobacter* species by polymerase chain reaction amplification of 23S rRNA gene fragments. *J Clin Microbiol*, **31**, 3340–3.

EYIGOR A, DAWSON K A, LANGLOIS B E & PICKETT C L (1999), Cytolethal distending toxin genes in *Campylobacter jejuni* and *Campylobacter coli* isolates: detection and analysis by PCR. *J Clin Microbiol*, **37**, 1646–50.

FARBER J M & PETERKIN P I (1991), *Listeria monocytogenes*, a foodborne pathogen. *Microbiol Rev*, **55**, 476–511.

FEARNLEY C ON, S L KOKOTOVIC B, MANNING G, CHEASTY T & NEWELL D G (2005), Application of fluorescent amplified fragment length polymorphism for comparison of human and animal isolates of *Yersinia enterocolitica. Appl Environ Microbiol*, **71**, 4960–5.

FITZGERALD C, OWEN R J & STANLEY J (1996), Comprehensive ribotyping scheme for heat-stable serotypes of *Campylobacter jejuni. J Clin Microbiol*, **34**, 265–9.

FREDRIKSSON-AHOMAA M, BUCHER M, HANK C, STOLLE A & KORKEALA H (2001), High prevalence of *Yersinia enterocolitica* 4:O3 on pig offal in southern Germany: a slaughtering technique problem. *Syst Appl Microbiol*, **24**, 457–63.

FREDRIKSSON-AHOMAA M, HIELM S & KORKEALA H (1999), High prevalence of yadA-positive *Yersinia enterocolitica* in pig tongues and minced meat at the retail level in Finland. *J Food Prot*, **62**, 123–7.

FUJIMOTO S, UMENE K, SAITO M, HORIKAWA K & BLASER M J (2000), Restriction fragment length polymorphism analysis using random chromosomal gene probes for epidemiological analysis of *Campylobacter jejuni* infections. *J Clin Microbiol*, **38**, 1664–7.

GILLESPIE I A, MCLAUCHLIN J, GRANT K A, LITTLE C L, MITHANI V, PENMAN C, LANE, C & REGAN M (2006), Changing pattern of human listeriosis, England and Wales, 2001–2004. *Emerg Infect Dis*, **12**, 1361–6.

GOORHUIS A, DEBAST S B, VAN LEENGOED L A, HARMANUS C, NOTERMANS D W, BERGWERFF A A & KUIJPER E J (2008), *Clostridium difficile* PCR ribotype 078: an emerging strain in humans and in pigs? *J Clin Microbiol*, **46**, 1157–8.

GOULET V, HEDBERG C, LE MONNIER A & DE VALK H (2008), Increasing incidence of listeriosis in France and other European countries. *Emerg Infect Dis*, **14**, 734–40.

GRAJEWSKI B A, KUSEK J W & GELFAND H M (1985), Development of a bacteriophage typing system for *Campylobacter jejuni* and *Campylobacter coli. J Clin Microbiol*, **22**, 13–18.

GUDMUNDSDOTTIR S, GUDBJORNSDOTTIR B, LAUZON H L, EINARSSON H, KRISTINSSON K G & KRISTJANSSON M (2005), Tracing *Listeria monocytogenes* isolates from cold-smoked salmon and its processing environment in Iceland using pulsed-field gel electrophoresis. *Int J Food Microbiol*, **101**, 41–51.

HEBERT G A, HOLLIS D G, WEAVER R E, LAMBERT M A, BLASER M J & MOSS C W (1982), 30 years of campylobacters: biochemical characteristics and a biotyping proposal for *Campylobacter jejuni. J Clin Microbiol*, **15**, 1065–73.

HINCHLIFFE S J, ISHERWOOD K E, STABLER R A, PRENTICE M B, RAKIN A, NICHOLS R A, OYSTON P C, HINDS J, TITBALL R W & WREN B W (2003), Application of DNA microarrays to study the evolutionary genomics of *Yersinia pestis* and *Yersinia pseudotuberculosis. Genome Res*, **13**, 2018–29.

HOLT K E, PARKHILL J, MAZZONI C J, ROUMAGNAC P, WEILL F X, GOODHEAD I, RANCE R, BAKER S, MASKELL D J, WAIN J, DOLECEK C, ACHTMAN M & DOUGAN G (2008),

High-throughput sequencing provides insights into genome variation and evolution in *Salmonella* Typhi. *Nat Genet*, **40**, 987–93.

HOWARD S L, GAUNT M W, HINDS J, WITNEY A A, STABLER R & WREN B W (2006), Application of comparative phylogenomics to study the evolution of *Yersinia enterocolitica* and to identify genetic differences relating to pathogenicity. *J Bacteriol*, **188**, 3645–53.

HOWARD S L, JAGANNATHAN A, SOO E C, HUI J P, AUBRY A J, AHMED I, KARLYSHEV A, KELLY J F, JONES M A, STEVENS M P, LOGAN S M & WREN B W (2009), A *Campylobacter jejuni* glycosylation island important in cell charge, legionaminic acid biosynthesis and colonisation of chickens. *Infect Immun*, **77**(6): 2544–56.

JONES M A, MARSTON K L, WOODALL C A, MASKELL D J, LINTON D, KARLYSHEV A V, DORRELL N, WREN B W & BARROW P A (2004), Adaptation of *Campylobacter jejuni* NCTC11168 to high-level colonization of the avian gastrointestinal tract. *Infect Immun*, **72**, 3769–76.

KARMALI M A, DE GRANDIS S & FLEMING P C (1980), Antimicrobial susceptibility of *Campylobacter jejuni* and *Campylobacter fetus* subsp. fetus to eight cephalosporins with special reference to species differentiation. *Antimicrob Agents Chemother*, **18**, 948–51.

KATHARIOU S (2002), *Listeria monocytogenes* virulence and pathogenicity, a food safety perspective. *J Food Prot*, **65**, 1811–29.

KEEL K, BRAZIER J S, POST K W, WEESE S & SONGER J G (2007), Prevalence of PCR ribotypes among *Clostridium difficile* isolates from pigs, calves, and other species. *J Clin Microbiol*, **45**, 1963–4.

KILLGORE G, THOMPSON A, JOHNSON S, BRAZIER J, KUIJPER E, PEPIN J, FROST E H, SAVELKOUL P, NICHOLSON B, VAN DEN BERG R J, KATO H, SAMBOL S P, ZUKOWSKI W, WOODS C, LIMBAGO B, GERDING D N & MCDONALD L C (2008), Comparison of seven techniques for typing international epidemic strains of *Clostridium difficile*: restriction endonuclease analysis, pulsed-field gel electrophoresis, PCR-ribotyping, multilocus sequence typing, multilocus variable-number tandem-repeat analysis, amplified fragment length polymorphism, and surface layer protein A gene sequence typing. *J Clin Microbiol*, **46**, 431–7.

LADO B H & YOUSEF A E (1999), Characteristics of *Listeria monocytogenes* important to food processors. In Ryser, E. T. & Marth, E. H. (Eds) *Listeria, Listeriosis, and Food Safety*, 3rd ed., CRC Press, Boca Raton, pp. 158–98.

LEONARD E E, TAKATA T, BLASER M J, FALKOW S, TOMPKINS L S & GAYNOR E C (2003), Use of an open-reading frame-specific *Campylobacter jejuni* DNA microarray as a new genotyping tool for studying epidemiologically related isolates. *J Infect Dis*, **187**, 691–4.

LEONARD E E, TOMPKINS L S, FALKOW S & NACHAMKIN I (2004), Comparison of *Campylobacter jejuni* isolates implicated in Guillain–Barre syndrome and strains that cause enteritis by a DNA microarray. *Infect Immun*, **72**, 1199–203.

LINDSTEDT B A, HEIR E, VARDUND T & KAPPERUD G (2000), A variation of the amplified-fragment length polymorphism (AFLP) technique using three restriction endonucleases, and assessment of the enzyme combination BglII-MfeI for AFLP analysis of *Salmonella enterica* subsp. enterica isolates. *FEMS Microbiol Lett*, **189**, 19–24.

LIOR H, WOODWARD D L, EDGAR J A, LAROCHE L J & GILL P (1982), Serotyping of *Campylobacter jejuni* by slide agglutination based on heat-labile antigenic factors. *J Clin Microbiol*, **15**, 761–8.

MAIDEN M C, BYGRAVES J A, FEIL E, MORELLI G, RUSSELL J E, URWIN R, ZHANG Q, ZHOU J, ZURTH K, CAUGANT D A, FEAVERS I M, ACHTMAN M & SPRATT B G (1998), Multilocus sequence typing: a portable approach to the identification of clones within populations of pathogenic microorganisms. *Proc Natl Acad Sci USA*, **95**, 3140–5.

MCNALLY A, CHEASTY T, FEARNLEY C, DALZIEL R W, PAIBA G A, MANNING G & NEWELL D G (2004), Comparison of the biotypes of *Yersinia enterocolitica* isolated from pigs, cattle and sheep at slaughter and from humans with yersiniosis in Great Britain during 1999–2000. *Lett Appl Microbiol*, **39**, 103–8.

MEAD P S, SLUTSKER L, DIETZ V, MCCAIG L F, BRESEE J S, SHAPIRO C, GRIFFIN P M & TAUXE R V (1999), Food-related illness and death in the United States. *Emerg Infect Dis*, **5**, 607–25.

MIETTINEN M K, PALMU L, BJORKROTH K J & KORKEALA H (2001), Prevalence of *Listeria monocytogenes* in broilers at the abattoir, processing plant, and retail level. *J Food Prot*, **64**, 994–9.

MILNES A S, STEWART I, CLIFTON-HADLEY F A, DAVIES R H, NEWELL D G, SAYERS A R, CHEASTY T, CASSAR C, RIDLEY A, COOK A J, EVANS S J, TEALE C J, SMITH R P, MCNALLY A, TOSZEGHY M, FUTTER R, KAY A & PAIBA G A (2008), Intestinal carriage of verocytotoxigenic *Escherichia coli* O157, *Salmonella*, thermophilic *Campylobacter* and *Yersinia enterocolitica*, in cattle, sheep and pigs at slaughter in Great Britain during 2003. *Epidemiol Infect*, **136**, 739–51.

MINIHAN D, WHYTE P, O'MAHONY M, FANNING S, MCGILL K & COLLINS J D (2004), *Campylobacter* spp. in Irish feedlot cattle: A longitudinal study involving pre-harvest and harvest phases of the food chain. *Journal of Veterinary Medicine Series B*, **51**, 28–33.

NESBAKKEN T, ECKNER K, HOIDAL H K & ROTTERUD O J (2003), Occurrence of *Yersinia enterocolitica* and *Campylobacter* spp. in slaughter pigs and consequences for meat inspection, slaughtering, and dressing procedures. *Int J Food Microbiol*, **80**, 231–40.

NIELSEN E M, FUSSING V, ENGBERG J, NIELSEN N L & NEIMANN J (2006), Most *Campylobacter* subtypes from sporadic infections can be found in retail poultry products and food animals. *Epidemiol Infect*, **134**, 758–67.

NIGHTINGALE K K, WINDHAM K & WIEDMANN M (2005), Evolution and molecular phylogeny of *Listeria monocytogenes* isolated from human and animal listeriosis cases and foods. *J Bacteriol*, **187**, 5537–51.

OBERHELMAN R A, GILMAN R H, SHEEN P, CORDOVA J, TAYLOR D N, ZIMIC M, MEZA R, PEREZ J, LEBRON C, CABRERA L, RODGERS F G, WOODWARD D L & PRICE L J (2003), *Campylobacter* transmission in a Peruvian shantytown: a longitudinal study using strain typing of campylobacter isolates from chickens and humans in household clusters. *J Infect Dis*, **187**, 260–9.

PARKHILL J, WREN B W, MUNGALL K, KETLEY J M, CHURCHER C, BASHAM D, CHILLINGWORTH T, DAVIES R M, FELTWELL T, HOLROYD S, JAGELS K, KARLYSHEV A V, MOULE S, PALLEN M J, PENN C W, QUAIL M A, RAJANDREAM M A, RUTHERFORD K M, VAN VLIET A H, WHITEHEAD S & BARRELL B G (2000), The genome sequence of the foodborne pathogen *Campylobacter jejuni* reveals hypervariable sequences. *Nature*, **403**, 665–8.

PENNER J L & HENNESSY J N (1980), Passive hemagglutination technique for serotyping *Campylobacter fetus* subsp. jejuni on the basis of soluble heat-stable antigens. *J Clin Microbiol*, **12**, 732–7.

RAJDA Z & MIDDLETON D (2006), Descriptive epidemiology of enteric illness for selected reportable diseases in Ontario, 2003. *Can Commun Dis Rep*, **32**, 275–85.

RIBOT E M, FITZGERALD C, KUBOTA K, SWAMINATHAN B & BARRETT T J (2001). Rapid pulsed-field gel electrophoresis protocol for subtyping of *Campylobacter jejuni*. *J Clin Microbiol*, **39**, 1889–94.

ROCOURT J & COSSART P (1997), *Listeria monocytogenes*. In Doyle, M. P., Beucaht, L. R. & Montville, T. J. (Eds) *Food Microbiology Fundamental and Frontiers* (pp. 337–52). Washington, DC: ASM Press.

ROCOURT J, JACQUET C & REILLY A (2000), Epidemiology of human listeriosis and seafoods. *Int J Food Microbiol*, **62**, 197–209.

RORVIK L M, AASE B, ALVESTAD T & CAUGANT D A (2003), Molecular epidemiological survey of *Listeria monocytogenes* in broilers and poultry products. *J Appl Microbiol*, **94**, 633–40.

RUPNIK M (2007), Is *Clostridium difficile*-associated infection a potentially zoonotic and foodborne disease? *Clin Microbiol Infect*, **13**, 457–9.

SAID B, WRIGHT F, NICHOLS G L, REACHER M & RUTTER M (2003), Outbreaks of infectious disease associated with private drinking water supplies in England and Wales 1970–2000. *Epidemiol Infect*, **130**, 469–79.

SCHLECH W F (1998), Foodborne listeriosis. *Clin Infect Dis*, **31**, 770–5.

SCHLUNDT J, TOYOFUKU H, JANSEN J & HERBST S A (2004), Emerging foodborne zoonoses. *Rev Sci Tech*, **23**, 513–33.

SCHUCHAT A, SWAMINATHAN B & BROOME C V (1991), Epidemiology of human listeriosis. *Clin Microbiol Rev*, **4**, 169–83.

SEELIGER H P R (1961), *Listeriosis*. Basel: Karger.

SIEMER B L, NIELSEN E M & ON S L (2005), Identification and molecular epidemiology of *Campylobacter coli* isolates from human gastroenteritis, food, and animal sources by amplified fragment length polymorphism analysis and Penner serotyping. *Appl Environ Microbiol*, **71**, 1953–8.

STABLER R A, DAWSON L F, OYSTON P C, TITBALL R W, WADE J, HINDS J, WITNEY A A & WREN B W (2008), Development and application of the active surveillance of pathogens microarray to monitor bacterial gene flux. *BMC Microbiol*, **8**, 177.

STABLER R A, GERDING D N, SONGER J G, DRUDY D, BRAZIER J S, TRINH H T, WITNEY A A, HINDS J & WREN B W (2006), Comparative phylogenomics of *Clostridium difficile* reveals clade specificity and microevolution of hypervirulent strains. *J Bacteriol*, **188**, 7297–305.

TABOADA E N, VAN BELKUM A, YUKI N, ACEDILLO R R, GODSCHALK P C, KOGA M, ENDTZ H P, GILBERT M & NASH J H (2007), Comparative genomic analysis of *Campylobacter jejuni* associated with Guillain-Barre and Miller Fisher syndromes: neuropathogenic and enteritis-associated isolates can share high levels of genomic similarity. *BMC Genomics*, **8**, 359.

VAN LEEUWEN W, VERBRUGH H, VAN DER VELDEN J, VAN LEEUWEN N, HECK M & VAN BELKUM A (1999), Validation of binary typing for *Staphylococcus aureus* strains. *J Clin Microbiol*, **37**, 664–74.

VAZQUEZ-BOLAND J A, KUHN M, BERCHE P, CHAKRABORTY T, DOMINGUEZ-BERNAL G, GOEBEL W, GONZALEZ-ZORN B, WEHLAND J & KREFT J (2001), *Listeria* pathogenesis and molecular virulence determinants. *Clin Microbiol Rev*, **14**, 584–640.

VITAS A I & GARCIA-JALON V A (2004), Occurrence of *Listeria monocytogenes* in fresh and processed foods in Navarra (Spain). *Int J Food Microbiol*, **90**, 349–56.

WACHSMUTH I K, KIEHLBAUCH J A, BOPP C A, CAMERON D N, STROCKBINE N A, WELLS J G, ET AL. (1991), The use of plasmid profiles and nucleic acid probes in epidemiologic investigations of foodborne, diarrheal diseases. *Int J Food Microbiol*, **12**, 77–89.

WIEDMANN M (2002), Molecular subtyping methods for *Listeria monocytogenes*. *J AOAC Int*, **85**, 524–31.

WOJCIECH L, STARONIEWICZ Z, JAKUBCZAK A & UGORSKI M (2004), Typing of *Yersinia Enterocolitica* Isolates by ITS profiling, REP- and ERIC-PCR. *J Vet Med B Infect Dis Vet Public Health*, **51**, 238–44.

14

Protein-based analysis and other new and emerging non-nucleic acid based methods for tracing and investigating foodborne pathogens

J. P. Bowman, University of Tasmania, Australia

Abstract: Understanding the activity and viability of bacterial populations provides an insight into their physiology, reactions to stress and response to environmental change. As the key goals in food safety involve the surveillance, control and risk reduction of foodborne pathogens, physiologically oriented studies are of great relevance, allowing for detection and prediction of growth and activity. This review examines the progress, utilization and pitfalls of innovative and emerging non-nucleic acid based methods that can be linked with data generated through genome sequencing projects to provide an enhanced understanding of foodborne pathogen physiology and activity.

Key words: foodborne pathogens, methods, visability assays, Fourier transform spectroscopy, Nano-SIMS, Raman spectroscopy, electrophysiology, proteomics, gel-free proteomics, metabolomics.

14.1 Introduction

Using molecular methods, it is now possible to gain quite detailed information about individual cells through to entire microbial communities that belong or occur within any given ecosystem or setting. The progress in building up information on simple and complex systems is inherently iterative, driven by technology and by the type of questions that can be asked and answered as a result. The science of microbiology has now entered an increasingly information-rich historical phase. Integrating several forms of information is critical in developing more complete conceptualizations of biological systems (Brul, 2007). Understanding foodborne pathogens in complex human-modified systems is an example where combining different forms of knowledge may pay dividends in the form of enhanced surveillance, control and risk reduction.

14.2 Distinguishing live from dead cells: viability and pH-sensitive stains for assessing cell physiology

One of the interesting aspects of microbial life is that there is often tendency to describe bacteria as simply 'alive and dead'. It is a simplification (and an avoidance) of that fact that defining viability in bacteria is actually a rather ambiguous area in many circumstances. To overcome this, various approaches have been developed for assessment of viability that do not involve cultivation in or on laboratory growth media. Many forms of inactivating stress result in loss of culturability, as opposed to just viability. This process is often generally referred to and assumed as 'killing'. Culturability involves many complex processes occurring in the cell. For example, formation of the septal ring for cell division is a complex process consisting of several proteins that operates to condense and partition the chromosome, control the synthesis of cell wall polymers (peptidoglycan, etc.), and is strictly monitored by cytoskeletal-like proteins controlling cell shape and septum formation (Cabeen and Jacobs-Wagner, 2007). Several forms of stress can interrupt this relatively delicate process as well as other processes in the cell, for example high hydrostatic pressure, heat, acid, etc. (Holtje, 1998). Thus, loss of culturability is not necessarily associated with death but rather, in a more general sense, involves cell injury, and viable but not culturable (VBNC) cells are assumed to be injured or dormant (e.g. Sachidanandham and Gin, 2009). Insights into basic aspects of cell physiology, such as inactivation, injury, repair and death, should increasingly be illuminated with genome-based approaches that in combination with other techniques provide more complete views on cell physiological features and states. Thus, it is a consideration that not only genomic methods be discussed but also other relevant non-nucleic methodologies, since the integration of these techniques, are more likely to be useful in revealing new insights. Genomic data in the end still need to be connected with phenotypes and mechanisms.

Various compounds, typically with a fluorescent moiety allowing detection either by epifluorescent microscopy or via flow cytometry, can be applied to assess cellular viability. Such approaches are especially useful when standard methods for assessing culturability are not practical or absent from samples. Compounds such as carboxyfluoroscein diacetate are hydrolyzed by broad-spectrum cytoplasmic enzymes while others bind to nucleic acids, for example the cyanine dye SYTO Green (Fig. 14.1), or are reduced by metabolic redox reactions in the cell, for example 5-cyano-2,3-ditolyl tetrazolium chloride (CTC); all require uptake into the cell by active transport (Blackburn and McCarthy, 2000). Membrane potential, the voltage difference between the inside and the outside the cell, a general indicator of cell integrity, can also be assessed in the same ways, for example using bis-oxonol [DiBAC(4)(3)] (Lopez Amoros *et al.*, 1997; Nebe-von

Caron *et al.*, 1998; Papadimitriou *et al.*, 2006). Counterstains, such as propidium iodide (Fig. 14.1), TOTO-1 and propidium azide, reveal membrane-permeable cells that can be assumed to be either injured or dead depending on the experimental stress conditions applied (Nocker *et al.*, 2006). Similarly, dyes that are ratiometric indicators of pH [e.g. 2',7'-bis-(2-carboxyethyl)-5,6-carboxyfluorescein, 1-N-phenylnaphthylamine] can also be used to assess cell viability since a compromised cell wall leads to deviations in internal pH levels (e.g. detected using fluorescence ratio imaging microscopy or FRIM) and can be used to assess single cells (Helander and Mattila-Sandholm, 2000; Siegumfeldt *et al.*, 2000). Usage of viability stains in a practical sense has been considered, for example, to assess potential of cross-contamination (Wilks *et al.*, 2006) and to verify activity of starter cultures and probiotic cultures (Bunthof and Abee, 2002), and in that respect has a role in providing rapid information regarding food-related microorganisms. Rapid kit-based methodology is available for both microscopic and cytometric approaches, for example ChemChrome and LIVE/DEAD BacLight Viability Kit (Fig. 14.1), and has been proven to be convenient for use in many applications (Maukonen *et al.*, 2006).

Viability stain-based approaches have also been utilized to assess foodborne pathogens in a number of stress scenarios, such as starvation, and the outcomes of inimical stress treatments (e.g. Besnard *et al.*, 2002; Ritz *et al.*, 2006; Tangwatcharin *et al.*, 2006; Asakura *et al.*, 2007; Dreux *et al.*, 2007; Falcioni *et al.*, 2008). In the late 1990s, the debate over the ambiguity of the VBNC state peaked (e.g. Kell *et al.*, 1998; McDougald *et al.*, 1999). The current literature still has not provided any real answers to what exactly involve VBNC states and what processes or conditions govern the bonafide ability for cells to repair injuries and resuscitate from the VBNC or less defined dormant states (Fig. 14.1). Clearly, viable stains provide evidence of cellular activity, but other more advanced methods are needed to provide further answers, for example quantitative polymerase chain reaction (PCR) based on mRNA (Gonzalez-Escalona *et al.*, 2006). VBNC cells have been slated a food safety concern due to their difficulty to capture in traditional 'shake and plate' procedures and potential to confound standard PCR-, probe- or immunological-based approaches. Uncertainty in the safety of exported food products can lead to negative market perceptions (Davies, 1997), so it is a concern that can become economical. The use of integrated approaches, such as transcriptomics and proteomics, would be clearly valuable in further developing concepts in the area of non-culturable bacterial cells and provide a possible means to develop more discriminatory tracking procedures for industry.

14.3 Rapid sample scanning: fluorescent *in situ* hybridization coupled to secondary ion mass spectrometry (SIMS), Fourier transform and Raman spectroscopy

Fluorescent *in situ* hybridization (FISH) involves using fluorescently labelled oligonucleotide probes, targeting small or large subunit ribosomal RNA, which are applied to permeabilized cells and typically visualized by epifluorescence microscopy or flow cytometry. Recent variations to the method include catalyzed reporter deposition-FISH (CARD-FISH) (Eschenhagen *et al.*, 2008) and halogen *in situ* hybridization-secondary ion mass spectroscopy (HISH-SIMS) (see below) that offer improved sensitivity and utility. The general approach offers a means to detect bacteria without cultivation in complex samples. In general, FISH has had limited application for food-based matrices (Moreno *et al.*, 2001; Fang *et al.*, 2003; Gunasekera *et al.*, 2003; Wilks *et al.*, 2006), as certain technical problems mainly associated with food sampling and autofluorescence of various food components limit the applicability of the approach. Importantly, it cannot be made 'high-throughput' without the development of some form of automation, such as light-emitting diode (LED) detector technology and the 'Bioplorer' (Shimakita *et al.*, 2006; Nishimura *et al.*, 2008). The necessary efforts required to make this technology feasible are currently ongoing, for example streamlining the sample processing time (Ootsubo *et al.*, 2003; Laflamme *et al.*, 2009) and improving detection sensitivity, for example the CARD-FISH methodological variations (Fuchizawa *et al.*, 2008). Such improvements, if coupled with instrumental advances, may give FISH-type approaches a push towards greater feasibility in large-scale raw and processed food testing (Cenciarini-Borde *et al.*, 2009). The approach would be very useful for assessing shelf-life of products and factory hygiene. Integrated technological approaches could also include greater utility, for example detection plus verification of visualized targets such as the rapid identification of bacterial cells in food samples. Possible future nanotechnological developments could couple FISH-oriented approaches to SIMS technology. Nano-SIMS was originally developed to detect comparatively stable isotopes in extraterrestrial samples at the sub-micron scale (Stadermann *et al.*, 1999; Zinner *et al.*, 2003). On the basis of this beginning, high-resolution nano-SIMS has been used for investigating biological material, including modern clinical and ancient archaeological samples all the way to detecting supernovae-derived isotopic signatures (Guerquin-Kern *et al.*, 2005; Quintana *et al.*, 2006; Wacey *et al.*, 2008).

Essentially, the technology works by using a laser to ablate the surface of a sample a few molecules deep. The atoms and clusters of atoms removed in this process ('primary ions') may ionize, forming 'secondary ions'. These ions are funnelled into a mass spectrometer and are analysed on the basis of mass using a high-resolution ion probe. By scanning over the surface using the ion probe, a false colour image can be created that indicates the distribution of stable isotopes over the sample surface. A prototype device was developed that was adapted for analysis of ratios of stable isotopes of carbon and nitrogen, and basically, it was demonstrated that the distributions of $^{12}C/^{13}C$ and $^{14}N/^{15}N$ could be measured at the subcellular level (Peteranderl and Lechene, 2004). An instrument, the Nano-SIMS 50, manufactured by French company Cameca, has the capacity to record up to seven atomic mass images simultaneously at a resolution level of 50 nm. This can be done in combination with FISH using oligonucleotide probes that are halogenated, for example $^{127}I^-$, $^{19}F^-$, in this respect, creating a new method named HISH-SIMS. An example of HISH-SIMS in action involved individual cells of a mixed population of *Escherichia coli* and *Azoarcus* sp., visualized using specific phylogenetic probes and simultaneously information on relative uptake of carbon and nitrogen established on the basis of the image analysis of $^{13}C/^{12}C$ and $^{13}C + {}^{14}N/^{12}C + {}^{14}N$ ratios (Li, Wu, *et al.*, 2008). Higher ratios of the heavier stable isotopes compared to the lighter much more abundant isotopes were found to be indicative of higher relative uptake rates. Differential uptake of C and N was demonstrated to occur in photosynthetic bacterial cells collected from lakewater showing that some cell populations are physiologically heterogeneous, with some cells being much more active than others (Musat *et al.*, 2008). The method thus potentially provides a way to gauge how active a cellular population is and what species are the main drivers in a given sample. Essentially, SIMS-based methods provide a higher resolution than offered by standard microscopic or electrophysiological approaches such as micro-ion flux estimation (MIFE) (see below), which integrate information of a cellular population. Rather, SIMS provides a view into heterogeneity within a population. A limitation is the ablation and subsequent mass analysis restricts what can be analysed; however, the potential for new developments in the mass spectrometry (MS) field may widen the utility of SIMS-like interrogative approaches.

Fourier transform (FT) and Raman spectroscopy are emerging as other means to obtain information on bacteria either directly or for bacteria in contact with or in association with biological or non-biological matrices. In this approach, spectra, based on vibrational time-domain measurements of reflected or scattered (in the case of Raman spectroscopy) radiation or energy (which can be light, ultraviolet, near-infrared, infrared, nuclear magnetic or electron spin resonance), are acquired and can provide visualizable information on the chemical and, indirectly, the biological properties of a sample. Recent advances in FT/Raman spectroscopy include higher resolution spectral analysis that can be coupled to microscopic scanning, allowing broad-scale sample sweeps through to single-cell and macromolecular level detection (Harz *et al.*, 2009). Raman spectroscopy and its many variations offer rapid and high-resolution analysis of molecules with potential for specific identification. Distinctive two-dimensional Raman resonance spectral features in the ultraviolet region could be useful for rapid, automated identification of microorganisms (Grun *et al.*, 2007). Incorporation into convenient scanning-type instruments may in the future benefit surveillance and auditing procedures in food production systems. FT infrared spectroscopy (FTIR) has been recently applied to scan for bacteria in situations that also have applicability to hygiene surveillance, and the approach is inherently useful for analysis of bulk volumes or large-scale surfaces. For example, the presence of biological films or deposits on surfaces that either comprise a biofilm or provide potential for bacterial growth can be readily detected using attenuated total reflectance (ATR)-FTIR (Oulahal *et al.*, 2009). Such an application has been proposed as a way to investigate the efficacy of sanitizing biocide solutions (Fernandez-Saiz *et al.*, 2009), antibacterial coatings (Suci *et al.*, 1998; Amalric *et al.*, 2009) or explore food or beverage contamination scenarios (Noack *et al.*, 2008). Spectral information analysed with advanced approaches, such as neural network analysis

(Gupta *et al.*, 2006), can currently resolve bacteria within food matrices to certain limits. For example, Nicolaou and Goodacre (2008) determined that ATR-FTIR was able to detect down to approximately 10^3 colony-forming units (CFUs) in whole and skimmed milk based on metabolic influence of the bacteria on milk proteins and carbohydrates. Spectral features, dependent on cell surface polymer properties, may be potentially useful in distinguishing different bacterial groups or even subgroups (Irudayaraj *et al.*, 2002) such as different serotypes of *Salmonella* (Baldauf *et al.*, 2007) or *Listeria monocytogenes* (Rebuffo-Scheer *et al.*, 2007). For the purpose of differentiating under undefined conditions, FTIR-based approaches are still rather limited, being dependent on general phenotypic broad traits (e.g. oxidized metabolites, storage polymers) (Garip *et al.*, 2009).

14.4 Electrophysiology

A wealth of physiological information can be obtained by studying the transport of molecules into and out of cells. Such information provides indicators of microbial viability, metabolic activity, responses to imposed stresses and environmental change, and also capacity for biochemical transformations. Genome sequence data have provided an indication of the immense capacity of bacterial transport and also indicated large knowledge gaps (Ren and Paulsen, 2007). Methods used to study transport have included detectors, (bio)sensors and (bio)chemical assays. The main limitations of many of these systems are that they are invasive and/or destructive, do not have sufficient selectivity or sensitivity or both, are subject to the vagaries of hysteresis, and usually only characterize bulk liquid phenomena since they are usually based on steady-state concentrations of target molecules (Shabala, Ross, *et al.*, 2006; McLamore *et al.*, 2009). Ion-selective electrodes (ISEs) have been used for many years as a way to obtain information on the distribution of ions in or deriving from complex samples, including stratified ecosystems of biological assemblages such as sediments and biofilms.

Electrophysiology is still a relatively undeveloped area for studying bacterial physiology (Shabala *et al.*, 2001), though it is based upon ideas that go back to the 1940s. Subsequently, single and dual microelectrode probes for directly measuring ion fluxes on cell surfaces (patch clamping) were developed (e.g. vibrating probe, Jaffe and Nuccitelli, 1974) to study animal, plant and fungal (i.e. 'large') cells. The desire to extend electrophysiological analyses to bacteria, especially those of public health concern, relates to investigating biophysical phenomena surrounding the cell membrane. This is due to the simple reason that bacteria, like other lifeforms, need the cytoplasmic membrane for many aspects of survival. Most importantly, these include intracellular pH homeostasis; osmotic potential; nutrient acquisition through membrane-bound transporters; and removal of organic and inorganic toxins and metabolites using membrane-bound efflux pumps. Controlling membrane processes are sensory systems embedded in membranes (e.g. two-component histidine kinase and response regulator systems) that elicit responses within seconds to external stimuli. These include factors constantly influencing bacterial growth: light, redox, oxygen (these have common links in protein structure, e.g. PAS domains; Davis *et al.*, 1999; Taylor and Zhulin, 1999; Vuillet *et al.*, 2007); nutrients, toxins, etc. that induce a chemotactic response (Wadhams and Armitage, 2004); effects that mechanically stretch or compress membranes such as temperature and osmotica (Braun *et al.*, 2007); and quorums (Taga and Bassler, 2003). Thus, within very short time frames, bacteria have the capacity to respond to a mélange of environmental changes and depending on the physiological state of the cell and the change itself, adjust phenotypically to maximize growth or attempt to avoid inactivation. The initial events that lead to the eventual gene expression, protein translation and phenotype alterations occur rapidly enough such that very high time resolution techniques are needed to capture information.

New or variant approaches to study membrane functionality continue to be developed mainly in an attempt to circumvent limitations associated with existing methodologies and to avoid issues associated with the invasiveness of the approach or hysteresis since they may affect the end conclusions. Relatively, non-invasiveness methodologies include nuclear magnetic resonance (NMR) spectroscopy, fluorescence imaging approaches (discussed above) and patch clamping (MacDonald and Martinac, 2005; Nguyen *et al.*, 2005; Sotomayor *et al.*, 2007; Tang *et al.*, 2009). NMR-based approaches have been mainly used to assess cytoplasmic pH, with a focus towards human disease/cancer systems (Gillies *et al.*, 2004). Other methods have appeared more recently, including ISEs, self-referencing probes and dielectrophoresis.

14.4.1 Non-invasive ion flux measurements using microelectrodes

Measurement of ion fluxes in a way that otherwise does not disturb a cell or cell population can be done on the basis of a very simple principle. When cells either extrude or take up ions, the concentrations of ions are lower as one moves away from cell surface since ions diffuse along a concentration gradient. Ions also move down the electrical potential gradient; thus, by measuring the electrical potential at two different points of a gradient, the rate at which the ions move can be determined. By using two separate ISEs, one close to the cell surface ('near pole') and another farther away ('far pole'), and slowly oscillating them (to form a square-wave signal), the concentration of specific ions can be determined, while the voltage difference between these ISEs can be used to determine the ion mobility and direction. Assuming a simple scenario of a nearly complete flat surface, that is a monolayer of bacterial cells or biofilm, the Nernst slope can be used to directly calculate the *net ion flux* (as described in Newman, 2001; Shabala, McMeekin, *et al.*, 2006).

On the basis of this principle, the MIFE system (Fig. 14.1) was developed using an inverted microscope, with ISEs controlled by micromanipulators (Newman, 2001). The ISEs are connected to an amplifier and electrometers to measure voltages and voltage gradients. The microscope and probes are poised over an open experimental chamber in which biological specimens can be placed and exposed to different ionic solutions introduced by a peristaltic pump. In the latest manifestations of MIFE (Sergey Shabala, Svetlana Shabala, Tracey Cuin,

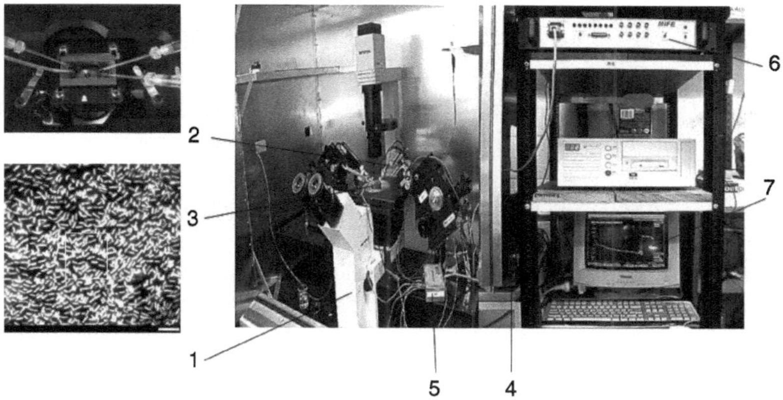

The MIFE setup: microscope (1); an electrode holder (2); a measuring chamber (3); a stepper motor and controller (4); a pre-amplifier (5); a main amplifier and controller (6) connected to a personal computer (7).

Fig. 14.1 The components of the micro ion flux estimation (MIFE) system are shown here. The MIFE system provides non-invasive simultaneous electrophysiological measurements of cell transporter activity, using ion–ion-selective self-referencing microelectrodes operated at a high temporal (ca. five seconds) and spatial (micron scale) resolution over sustained periods of time (hours to days).

personal communication), the system incorporates a temperature conductor stage that allows controlled cooling and heating of samples. Shabala, Ross, *et al.* (2006) provide more details on microelectrode fabrication, calibration and what is required for maximizing signal-to-noise ratios in the experiments.

Owing to the size of commercial ISEs in order to study bacteria, 'monolayers' of cells need to be created on glass coverslips using poly-L-lysine to adhere cells (Shabala *et al.*, 2001), with measurements performed at several locations of the layer to account for variations in the thickness and sensitivity of the cell layer. Mature biofilms have also been investigated (McLamore *et al.*, 2009) using an adaptation of MIFE in which rapid oscillation of the ISEs and self-referencing software are used to correct for noise occurring between the ISE at near and far poles on the basis of Fick's law of diffusion (Smith *et al.*, 1999). A downside of this approach is that time resolution is reduced somewhat compared to MIFE in which electrode detection of different ions relies on the ISEs being filled with ion-selective ionophore in a back-filling solution (LIX). LIX has the advantage of being more rapidly responsive (on the scale of seconds) to the presence of ions than standard glass microelectrodes. Regarding MIFE, net fluxes that have been experimentally measured reliably include H^+, Cl^-, K^+, Na^+, Ca^{2+}, Mg^{2+}, Cd^{2+}, Zn^{2+}, Cu^{2+}, $NH4^+$ and NO_3^-. Theoretically, this list could be extended to other analytes relevant to physiological and metabolic aspects of bacterial cells (e.g. CO_2, CO_3^{-2}, PO_4^{-3}, S^{-2}, CN, NOx) through experimentation with other LIX reagents that are commercially available or determined through empirical experimentation (Mikhel'son, 2008).

Applications of MIFE with respect to examination of stress-induced changes to ion fluxes have been, at the present, geared towards foodborne pathogens, though any type of bacteria could be studied theoretically. Research since 2001 has revealed some fundamental information on ion fluxes in *E. coli* and other species exposed to various forms of stress, growth condition variations and nutrient additions (Shabala, McMeekin, *et al.*, 2006; Shabala, Ross, *et al.*, 2006; Shabala *et al.*, 2009). Since ion flux correlates with metabolic activity (Shabala, Ross, *et al.*, 2006), MIFE has potential in providing information on bacterial responses under many different scenarios, including establishing viability. Thus, MIFE could also be used to test the effects of antimicrobials, other bioactive compounds, as well as other forms of biotic and abiotic stress on microbial gene expression and metabolism, and also determine the effects on cell viability over different levels of time resolution and could be coupled with specific bacterial capture technologies (e.g. Koo *et al.*, 2009). Overall, MIFE has a distinct advantage in observing physiological phenomena that change over short time scales for extended periods of time.

14.5 Proteomics

Proteins are fundamental central features of life, and most cellular functions require proteins acting in a myriad of roles. Like all other lifeforms, foodborne pathogenic species have a complement of proteins that are highly conserved, which are involved in the basic processes of the cell and for the most part are essential (Gerdes *et al.*, 2003). These include proteins involved in or associated with DNA transcription, protein translation, protein folding and export, cell wall polymer biosynthesis, and central metabolism. These proteins tend to be relatively abundant in the cell (Ishihama *et al.*, 2008). Many proteins are species- or strain-specific and can include proteins involved in virulence, specific metabolism (e.g. carbon source uptake and catabolism) as well as resistance (e.g. antibiotic resistance) and stress defense. The copy numbers of these proteins vary to a much greater degree, with numbers dependent on cell physiology and circumstances. This complement of both conserved abundant proteins and more abundant variable proteins makes a more comprehensive survey of protein abundance useful in understanding the physiology of the cell as well for other applications such as identification and potentially direct detection. Analysis of the transcriptome of cells comprising the mRNA formed during gene transcription can infer to some degree of the proteins formed in the cell under different situations. However, such analyses make large assumptions due to post-transcriptional modifications that may take place subsequent to mRNA synthesis.

In order to investigate proteins in cells relevant to problems related to foodborne pathogens and the food production chain environment, proteomics approaches have rapidly developed to a point at which a large proportion of the proteome (whole-cell protein complement) (Wilkins *et al.*, 1996; James, 1997) can be assayed and identified. This is possible using a combination of liquid chromatographic separation and high-resolution MS, with comparison to databases that include genome sequence-derived data. Thus, like microarray analysis, highly detailed information can be obtained to examine microbial physiology. A major limitation, however, is to arrive at reliable abundance determinations for individual proteins. For microarray analysis, this is easier due to probes being arrayed massively in parallel and the specific binding of labelled RNA samples producing fluorescent signal that can be directly measured. In the case of proteomics, proteins need to be identified and quantified simultaneously. For bacteria, even for *E. coli*, we still do not have a comprehensive database on the relative copy numbers of proteins in the cells under specified growth conditions. Such databases are available for *Saccharomyces cerevisiae* arrived at laboriously by the use of a series of methodologies, culminating in a flow cytometric investigation of large green fluorescent protein (GFP)-tagged libraries that allowed single-cell observations (Newman *et al.*, 2006). In order to achieve a similar outcome for bacteria and other organisms, proteomics appears to be the only feasible approach. Methods in the last few years have resorted to stable isotope labelling for calculating protein abundances, but this represents a major limitation (and expense) in performing experiments. However, in the last two years, significant advances have occurred in the development of label-free methods.

14.5.1 One- and two-dimensional gel electrophoresis

Proteins can be separated on polyacrylamide gels on the basis of isoelectric point (using isoelectric focusing), protein molecular weight or as protein complexes where they exist in the native state. In their simplest form, proteins can be separated in one dimension, usually on the basis of size, yielding a series of bands (Shapiro *et al.*, 1967). Sodium dodecyl sulfate (SDS)-polyacrylamide gel electrophoresis (SDS-PAGE) has been used for various applications in bacteriology including identification of bacteria by comparing banding patterns (Priest and Austin, 1993); alternatively, the gel containing the protein is flooded with a colorimetric substrate that leads to visualization of enzyme bands (zymogram) (Lantz and Ciborowski, 1994) and, with the advent of mass spectrometric procedures, identification of individual proteins (see below). SDS-PAGE of proteins is also used in western blotting procedures using antisera to detect specific proteins amongst a background of other proteins (Burnette, 1981). Separation of proteins can also be done in two dimensions, with the goal to identify protein abundances with greater resolution and the intention to determine changes in protein abundance. Proteins are usually initially separated by net charge using isoelectric point focusing and then on the basis of mass.

For two-dimensional gel electrophoresis (2-DE) applications, protein extracts are usually fractionated on the basis of pH gradient in order to better resolve acidic, neutral or basic proteins; one consequence of this is that only a limited portion of the proteome can be visualized any one time. Proteins are visualized by the use of colour stains (comassie blue, silver staining, etc.) or fluorescent dyes (e.g. SYPRO-ruby) for subsequent image analysis. Images are either captured using standard scanning or by use of a phosphorimager (Fig. 14.2a), and the latter offers better resolution and less smearing of spots. Image analysis using dedicated software (e.g. Delta2D, ImageMaster, Melanie, PDQuest, Progenesis, REDFIN) is a critical aspect of gel-based proteomics providing the means to match spots and validate changes in abundance of proteins (different methods have been extensively assessed by Nishihara and Champion, 2002) (Fig. 14.2b). Unfortunately, the procedure is met with a number of serious problems mainly with spot separation and signal strength, contamination leading to 'ghost spots', inconsistencies in positions of proteins due to variations in the performance of the

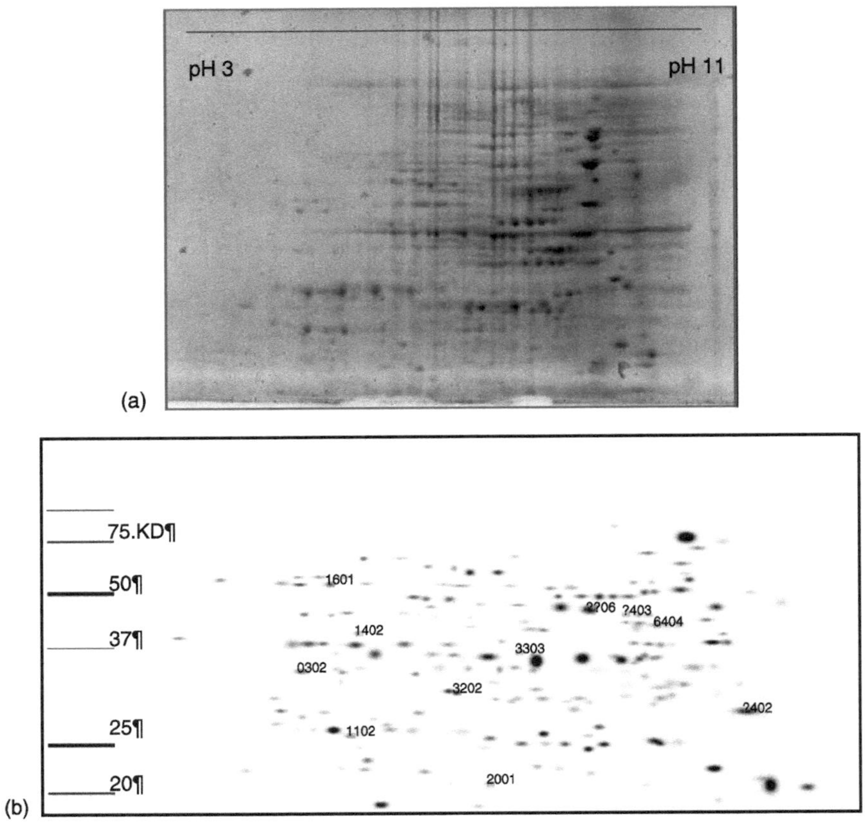

Fig. 14.2 Example of a 2-DE protein gel. (a) Raw image (gel stained with Sypro Ruby and visualized using a phosphorimager). (b) Image of a master gel (integrated and digitized from multiple gel raw images using PDQuest, BioRad) based on lysate obtained from *E. coli* OH111:H in lag phase (three hours incubation, 25°C) in nutrient broth acidified to pH 4.6. The first dimension separation was based on isoelectric point focusing, between pH 11 and pH 3. Labelled spots (image B) were analysed by in-gel trypsin digestion followed by MALDI-TOF mass fingeprinting identification (with permission from H. Haines).

electrophoresis leading to protein mismatching or lack of matching, problems with quantification, and inconsistent standards for quantification. The main way to resolve this has been the necessity of performing a large number of replicate gels, typically as many as six to ten per experimental condition.

As there is a paramount desire by scientists to obtain quantitative and informative information regarding protein abundance changes, especially relating to complex phenomena such as that related to stress physiology, various approaches have been used to try and improve the resolution of differentially translated proteins and avoid inter-gel comparison issues. The most widespread approach applied successfully is difference gel electrophoresis (DIGE) (Unlu *et al*., 1997), which involves labelling of extracted proteins with Cy3, Cy5 or Cy2 and simultaneously running up to three different labelled proteins together on a gel. Dedicated software is available for analysis of differential gel electrophoresis (DIGE) gels (e.g. DeCyder, Delta2D, Progenesis, REDFIN). Overall, gel electrophoresis methods still remain a popular and powerful tool to directly investigate bacterial physiology (Hecker *et al*., 2008). Improvements in separation technologies, however, should eventually replace gel-based techniques by providing vastly improved information capture depth and rapidity of analysis.

14.5.2 Peptide mass spectrometric analysis

One major advance in gel-based proteomics has been the development of the means to identify peptide amino acid sequences using MS (Aebersold and Mann, 2003). This advance has allowed determination of proteins directly from gels, ushering in the concept of proteomics. The process of protein identification involves cutting protein spots from gels followed by in-gel digestion that comprises de-staining, alkylation, reduction and then trypsin digestion steps. Trypsin has become highly favoured in proteomics applications as it has a very well-defined and consistent specificity (Olsen *et al*., 2004). The peptides yielded are identified using MS. Analytical applications involving proteins (and other biopolymers) include the development of MS technology such as matrix-assisted laser desorption/ionization (MALDI). In the case of MALDI, soft ionization of the peptides is done by sequential pulsation of a nitrogen laser. The peptides are solubilized in a matrix solution that promotes adsorption of the laser energy and encourages ionization (e.g. 3,5-dimethoxy-4-hydroxycinnamate, α-cyano-4-hydroxycinnamate or 2,5-dihydroxybenzoate). The mass and abundance of the ions generated through MALDI are typically determined using a time-of-flight (TOF) spectrometer owing to its applicability to emergent ion pulses rather than continuous streams (Tanaka *et al*., 1988). Other popular mass spectrometers that are useful for proteomics include ion trap mass spectrometers (Laiko *et al*., 2000) or MS systems utilizing electrospray ionization (Fenn *et al*., 1989). Recently developed or improved mass spectrometers include FT MS, providing large improvements in resolution (Shen *et al*., 2001), enhanced TOF-TOF with attomole detection limits (Suckau *et al*., 2003) and the linear trap quadrapole (LTQ)-Orbitrap (Hu *et al*., 2005; Makarov *et al*., 2006) – the main instrumental

MS technology for shotgun proteomics. The latter instruments have the necessary resolution, sensitivity and accuracy of peptide mass analysis (to four significant places) such that more than 1000 proteins can be potentially identified in a single replicate in complex mixtures, such as bacterial lysates. Increasing replicates increases information capture.

14.5.3 Peptide identification against genome sequence information

MS-based peptide identification benefits naturally from the burgeoning richness in genome sequence data, and accurate protein identification is increasingly easier to obtain. The most common problem with peptide identification is that the segments of amino acids can be replaced by other amino acid segments with approximately the same masses leading to erroneous identifications. The incidence of this issue has been reduced using *de novo* sequencing software such as SPIDER, in which short amino acid sequences (sequence tags) are cross-referenced against known MS-derived interpretation errors (Han *et al.*, 2005). In general, peptide sequence match searches can be performed against genomic sequence databases using dedicated software, typically involving cross-correlation or hypergeometric analysis. This type of software has been initially proprietary, such as Sequest (Eng *et al.*, 1994), Mascot (Perkins *et al.*, 1999; Koenig *et al.*, 2008) and Phenyx (Colinge *et al.*, 2003), but open access software is now available, such as X!TANDEM (Craig and Beavis, 2004), OMSSA (Geer *et al.*, 2004), MyriMatch (Tabb *et al.*, 2007) and greylag (greylag.org; Stowers Institute for Medical Research, Kansas City, MO, USA). It is important to note that statistical validation of the protein identification based on the quality of the mass spectra (Zhang *et al.*, 2006) is an important consideration, especially in analysis of peptides derived from unknown or poorly characterized sources, for example unsequenced bacterial strains, and whole communities ('metaproteomics') (Maron *et al.*, 2007).

14.5.4 Bacterial protein extraction and fractionation

Soluble proteins can be readily extracted and trypsin digested from cell lysates, though predominantly only cytosolic proteins are detectable (see below). Several commercial kits are available in the market to rapidly extract proteins for various proteomic applications. All use detergents, for example 3-[3-(cholamidopropyl)-dimethyl-amonio]-1-propanesulfonate (CHAPS), amidosulfobetaine-14 (ASB-14), N-decyl-N,N'-dimethyl-3-ammonio-1-propanesulfonate (trade name SB-10), to ensure that proteins are solubilized usually in a solution also containing water-breaking compounds, mainly urea and/or thiourea. In the case of high-resolution analysis of complex protein mixtures directly by liquid chromatography (LC)/MS (see below), the detergents can represent a technical issue, causing column blockages. New proprietary acid-labile surfactants are available in kit form that avoids this issue (e.g. Protea Biosciences). Development of new rapid methods for protein extraction that will allow deep proteomic analysis of smaller and smaller samples is ongoing (e.g. Thompson *et al.*, 2008). Also, the application of proteomics to environmental samples is increasing (e.g. Park *et al.*, 2008). Such developments in the near future may eventually provide for direct analysis of food samples, including using proteomics to detect bacteria either separated from food matrices or after enrichment. Adaptations of MALDI-TOF, such as surface-enhanced laser desorption/ionization (SELDI)-TOF (often referred to as 'protein or peptide chips'), provide improvements in complex proteome analysis (Merchant and Weinberger, 2000). In this case, chip surfaces functionalized with spots of different forms of substances with a capacity to bind proteins (cation exchange resins, antibodies, etc.) can selectively bind proteins from a complex sample. The chip is then treated with a matrix compound as done with MALDI (see above), and soft ionization of the chip spot to a TOF MS yields mass data that can be digitized into a signal. In this respect, fingerprint patterns can be assembled for samples containing many proteins such as clinical samples (for which the technology has been targeted by the commercial owners BioRad). With respect to microbiological applications, SELDI has been used to differentiate bacterial strains (Yang *et al.*, 2009) and preliminary analyses have been applied to food-based systems (Rawel *et al.*, 2005).

14.5.5 Proteomic analysis of cell wall and membrane fractions

Many aspects of the virulence of foodborne pathogens involve proteins associated with the cell wall and membranes, including toxins, host cell interactions, specific nutrient acquisition, immunological responses and protective responses. Unfortunately, solubilization of cell wall and membrane-associated proteins is a significant challenge as proteins are often tightly associated with peptidoglycan or lipid membranes and are also often much more hydrophobic. In addition, many cell wall and membrane proteins have large, low-complexity regions consisting of repeated transmembrane domains that can make identification difficult (Weiner and Li, 2008). This is a particular issue for Gram-positive bacteria such as *L. monocytogenes* and *Staphylococcus aureus* in which many proteins are anchored to peptidoglycan (Pucciarelli *et al.*, 2005). A cell wall proteome has been established for *L. monocytogenes*, but this relied on laborious high salt precipitation extractions followed by MS identification using one-dimensional gels (Schaumburg *et al.*, 2004). Calvo *et al.* (2005) used trypsin digests of peptidoglycan fractions from *L. monocytogenes* and subsequent LC/MS analysis following gel fractionation to more directly detect cell wall proteins. LC/MS has also been used successfully to more extensively investigate the cell membrane subproteome of *L. monocytogenes* (Wehmhoner *et al.*, 2005). Mujahid *et al.* (2007) found that good solubilization for cell wall-associated proteins from *L. monocytogenes* could be achieved using mutanolysin and ASB-14.

There are several strategies for fractionation of complex protein mixtures for protein identification. This includes several modes in which high-performance liquid chromatography (HPLC) fractionation of protein samples is followed by trypsin digestion of sub-fractions followed by LC/MS analysis. Alternatively, protein mixtures can be separated in one-dimensional gels and sliced and subjected to in-gel trypsin digestion followed by LC/MS analysis. Shotgun proteomic approaches (see below) increasingly are moving to direct trypsin digestion of the initial protein mixture before separation and MS analysis. The separation methods in this case can include various forms of HPLC/ultra performance liquid chromatography (UPLC) separation including strong cation exchange chromatography, performed either offline or online and mixed-mode pH reverse phase chromatography. A comparison of these methods revealed that the best recovery of total proteins was with mixed-mode reverse phase chromatography of directly trypsin-digested mixtures, yielding almost twice as many proteins as a gel slicing-based approach performed in parallel (Dowell *et al.*, 2008).

14.5.6 Gel-free proteomics using LC-MS/MS approaches

Gel-free (shotgun) approaches to proteomics involve adaptations of different forms of relatively new separation technologies using HPLC or UPLC, operating in a single or two-dimensional gradient mode, or capillary electrophoresis (CE). In the case of the single-gradient LC modes, MS using soft ionization (MALDI-TOF) can provide initial resolution and is coupled with tandem MS analysis for further resolution (i.e. LC-MS/MS). HPLC/UPLC technology development has galvanized proteomics advances, including improvements in separation resolution and, in particular, sensitivity through development of new separation techniques, columns and new detectors (Mitulović and Mechtler, 2006). This includes mixed-mode, two-dimensional phase separation using either modified reverse phase or monolithic column packing. Peptides eluting from the HPLC separation phase are ionized and the ions analysed typically by ion trap spectroscopy. Linear trap quadrupole (LTQ) Orbitrap mass spectrometers (i.e. ThermoFinnegan), for example, offer a start-of-the-art combination of resolution, dynamic range and mass accuracy. The main limitation of LC/MS has been related to quantification since this form of data is not directly determinable from the mass spectra. Another limitation is sensitivity (see below for practical outcomes) due to the current state of HPLC/UPLC column technology. Peak resolution is generally more adequate and improvements have been shown with the use of UPLC systems (Wilson *et al.*, 2005). The shotgun proteomics process is essentially a semi-random process; thus, due to the large difference in protein abundances (on a scale of up to 10^5-fold), a certain proportion of a given proteome will lie beyond detectability, even with the best current technology.

14.5.7 Stable isotope (and metabolic) labelling

Until recently, the main way of resolving quantification issues for proteins in mixtures has been to use stable isotope probe labelling of samples. Several methods in this regard are available, including isotope-coded affinity tags (ICAT), isobaric tag for relative and absolute quantitation (iTRAQ) and metabolic labelling approaches (Gygi *et al.*, 1999; Ross *et al.*, 2004; Shadforth *et al.*, 2005).

ICAT involves labelling peptides using chemical tagging reagents that consist of a chemical that is reactive to a particular amino acid side chain (e.g. cysteine modification by iodoacetamide), a linker compound that is either isotopically light or heavy and an affinity-reactive tag such as biotin. The modified light or heavy peptides are purified using affinity chromatography and subsequently analysed using LC-MS. The heavy peptides and light peptides can be compared allowing determination of ratios in peptide abundance. The iTRAQ method involves a similar simpler principle and depends on N-terminal amine-reactive isobaric tags. These tags (theoretically) allow for multiplex determination of protein samples as up to four different tags can be used simultaneously. The main limitation of ICAT and iTRAQ is the relative high cost imposed per sample. ICAT is also limited as the amino acid to be derivatized, such as cysteine, has to be present in the peptide. Metabolic labelling relies on the growth of a test organism or microorganism in a medium containing a compound comprising a heavy elemental isotope that can be assimilated and incorporated efficiently into protein, for example $^{13}C/^{2}H$-arginine (SILAC method) (Ong *et al.*, 2002), ^{18}O (Mirgorodskaya *et al.*, 2000) or $^{15}NH_4^+$ (Conrads *et al.*, 2001). These labels offer a comparatively cheaper approach to ratio determination but are limited by the need to grow test cultures directly in defined media. A major overall limitation, regardless of the method type, is that though stable isotopes provide ratio determinations for proteins, comparisons in abundance between different proteins is not directly possible. More recently, inherent limitations are being overcome with the switch to label-free approaches, possibly a consequence of growing confidence in the technology.

14.5.8 Label-free approaches

Several label-free strategies have been applied to obtain absolute abundance determination of peptides. The simplest approach involved spiking peptide samples with an internal standard of known concentration (Gerber *et al.*, 2003). Another approach, termed the xPAI (extracted ion density-based protein abundance index) method, calculates abundance by averaging ion intensities of the three most intense peptides for each protein (Forner *et al.*, 2006). Although xPAI correlates well with abundance determined by gel-based imaging, it is not applicable to complex mixtures since the proportion of proteins possessing three peptide ion matches or more is generally quite low. Fractionated peptide samples also cannot be easily investigated due to carry-over of proteins separated into different fractions. An approach generally termed PAI (protein abundance index) uses certain mass spectra-determined criteria to estimate absolute abundance of proteins. In this respect, different researchers have used different criteria to calculate PAI, including the number of actually identified and the theoretical number of tryptic peptides per protein (Rappsilber *et al.*, 2002), the known molecular weight of the protein identified (Sanders *et al.*, 2002), and the protein length (Zybailov *et al.*, 2006). The criterion based on detection of parent ions was used to define an exponentially modified PAI (see below), referred to as emPAI, which was shown to be as accurate as gel-based quantification (Ishihama *et al.*, 2005) and has been validated using several different methodological variations, performed in separate laboratories (Ishihama *et al.*, 2008):

$$\text{emPAI} = 10^{\text{PAI}} - 1,$$

where PAI is the number of observed MS/MS spectra per protein over the total number of observable MS/MS spectra per protein. Certain high-abundance proteins may saturate emPAI signals leading to potential underestimation of abundance. The saturation correlates with frequency of detection of the parent ions of the protein; some proteins, due to their small size (e.g. ribosomal proteins), are likely to have high over-representation of ions in analyses compared to most other proteins (Ishihama *et al.*, 2008). The method would also be compatible with metabolic labelling and other simple, stable isotope-labeling procedures (such as iTRAQ) since they do not introduce biases into abundance measurements.

Essentially, the PAI-based methods work on a signal-to-noise ratio approach; however, due to reduced replication, the accuracy of these approaches is not clear. Development of better statistical approaches, including significance testing and false discovery analysis, is useful for evaluating and filtering proteomic data (Old *et al.*, 2005; Zhang *et al.*, 2006), including large and complex datasets now being generated by label-free shotgun approaches (Choi *et al.*, 2008; Li and Roxas, 2009). Another strategy used is to measure protein abundance from spectral count data, essentially the number of peptide MS/MS spectra determined for a given protein (Old *et al.*, 2005). Spectral counts correlate extremely well with samples in which control proteins are spiked; indeed, it has been found that spectral count can be more reliable than spiking samples. This value is usually filtered to remove peptides of uncertain protein affiliation. For example, this can be done through X!TANDEM searches and Protein Prophet analyses. The fold abundance (as a log ratio to the base of 2) of given proteins can be compared between sample sets (in which peptide count data of replicates are pooled), taking into account variations in depth of sampling (effectively normalization) and the situation occurring in a sample set where a given protein is not detected (Old *et al.*, 2005):

$$\text{RSC} = \log_2[(n_2 + f)/(n_1 + f)] + \log_2[(t_1 - n_2 + f)/(t_2 - n_1 + f)],$$

where n_1, n_2 are the spectral counts for a given protein in samples 1 and 2; t_1, t_2 are the total spectral counts (accounting for sampling depth) for samples 1 and 2; and f is a pseudospectral count used to avoid comparisons of 0 log values (different values have been used, e.g. 1.25 by Old *et al.*, 2005, or 0.5 by Zhang *et al.*, 2006). The validity of the fold change can be tested by an adaptation of the *G*-test with significance examined on the basis of chi-square distributions (Zhang *et al.*, 2006).

One issue with spectral count comparisons is that they tend to vary greatly for low-abundance proteins. The only way to circumvent this is to increase sampling depth and biological replication (which are one and the same since the whole process involves semi-random sampling). One method used to take into account low-abundance issues, as well as provide a means to statistically test assumptions on what is a significant difference between peptide sample sets, involves calculation of a spectral index (SI) for which protein is detected in a given peptide sample set:

$$SI(P) = [\{S_A/(S_A + S_B)\}(N_A^D/N_A^T)] - [\{S_B/(S_A + S_B)\}(N_B^D/N_B^T)],$$

where S_A, S_B are the total spectral counts of a given protein in sets of replicates A and B; N_A^D, N_B^D, are the number of replicates in which the peptide of a given protein was recorded; N_A^T, N_B^T, are the total number of replicates for given samples A and B. The derived SI values fall between 1 and –1, with 1 indicating that the protein is found in sample A but not in B, while –1 represents the opposite possible outcome. Permutation analysis using the Bayesian *t*-statistic can be used to test at what SI indices are needed to record whether a protein is enriched in sample A or B at an $\alpha = 0.05$ level. For example, in the study involving the broncheoaleveolar lavage fluid of patients with cystis fibrosis compared to healthy individuals, Fu *et al.* (2008) found SI indices of –0.75 and 0.75 were indicative of statistically significant protein enrichments at a 0.5% significance level (0.67/–0.67 at a 5% level).

A significant bottleneck facing proteomics-based analyses is that not all proteins that can be produced by a cell, be it bacterial, archaeal or eukaryotic, can be detected in a complex mixture and that sizable differences are needed between estimated protein abundances to reliably detect expression changes. Using a sensible combination of gel-based and gel-free methods, theoretically 70–80% of proteins can be detected at any one time in a typical mid-sized bacterial proteome (e.g. Zhang *et al.*, 2006; Dowell *et al.*, 2008; Hecker *et al.*, 2008). Since the proteomics technology area is in a vigorous state of development (a useful overview has been provided by Egas and Wirth, 2008), further improvements are likely in the near future.

One aspect that proteomic investigations of complex mixtures, such as bacterial lysates, cannot be ignored is that sample fractionation is inevitable. The easily solubilized protein fraction will primarily capture information for cytosolic proteins and for some more abundant membrane-associated proteins. In a highly comprehensive study of the *E. coli* cytosolic protein fraction, produced by isotonic lysis of biomass, about 29% of proteins potentially present in the cytoplasm (cytosolic proteins and precursors of membrane proteins) were detected and abundance determined using the emPAI method (Ishihama *et al.*, 2008). If the protein has transmembrane domains, the detectability (proportion of proteins of a given type) declined significantly, from about 18% for proteins with one transmembrane domain to only 3–8% for proteins with two or more domains.

An example of gel-free proteomic experiment (utilizing 2D LC/LTQ Orbitrap MS) with a *L. monocytogenes* factory-persistent strain (Porteus *et al.*, 2010, unpublished data) is shown in Fig. 14.3(a). Data comprising four replicates of approximately 100 μg extracted protein each, approximately 33% of the total proteome (947 proteins), were detected, based on an analysis of 33 552 peptides. More sampling noise is observed with low spectral counts (low abundance or poorly extracted proteins) and 28% of the proteins were only detected from a single peptide identification. Due to the low sampling level, the return on protein identification

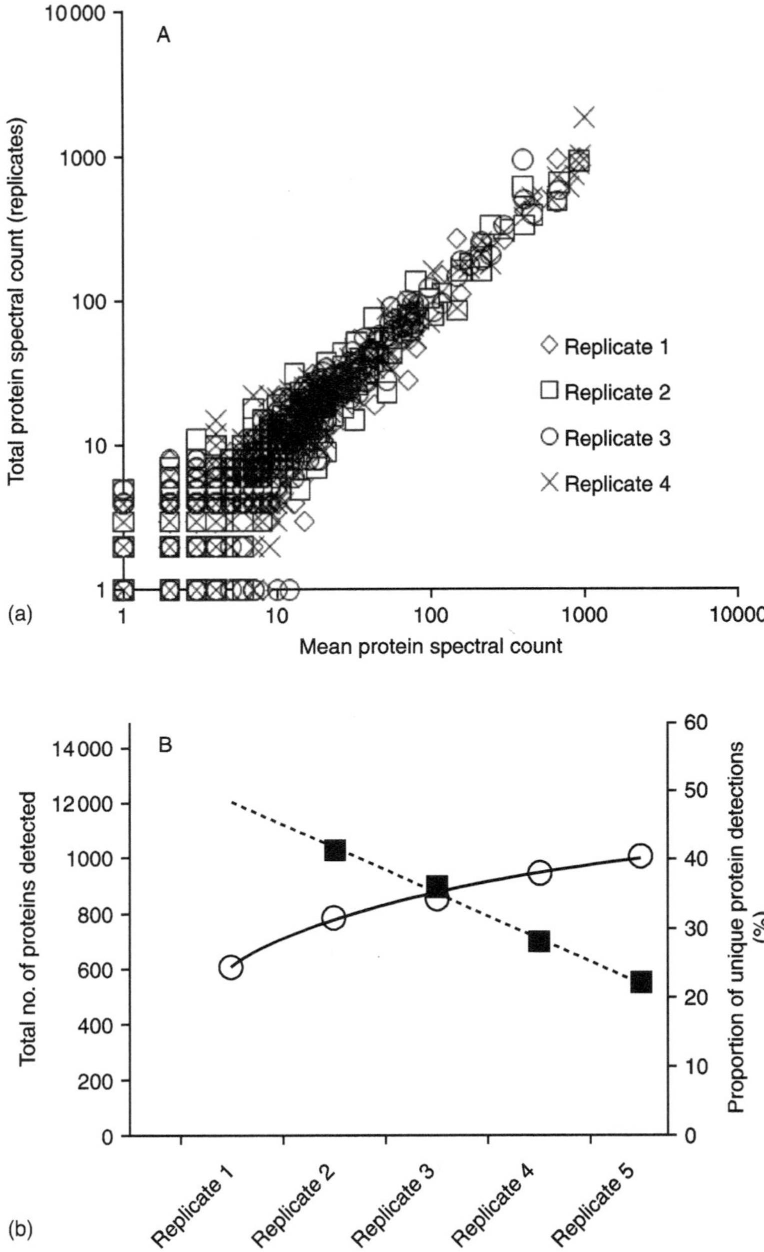

Fig. 14.3 (a) Protein abundance as measured by spectral counting of MS spectra variation between biological replicates of *Listeria monoctytogenes* D-81 (processed meat factory persistent strain, serotype 1/2b) showing increased variation for low spectral count values. (b) Comparison of number of proteins detected in *L. monoctytogenes* ScottA with each additional replicate in comparison to the incidence of proteins identified only from a single peptide.

probability match scores is predominantly less than 0.1 (such low probabilities are normally removed by data filtering). Increasing the replicates to five, the number of proteins found increases only by about 9% (Fig. 14.3b); however, the proportion of unique detections declines and also improves discrimination of protein abundances between comparative datasets. From these data, the spectral count data can be compiled using gene/protein ontological-based approaches (e.g. http://www.geneontology.org) to examine broad comparisons of data (Fig. 14.4) and, as mentioned above, be compared to examine whether significant changes occur. An example of this type of data is shown in Table 14.1.

14.5.9 Correlations of protein abundance with gene expression and essentiality

Most integrated transcriptomic and proteomic studies tend to find poor coupling of datasets. The reason for this is likely attributed to biological reasons, such as post-transcriptional events, but also could be due to the inadequacy or simply lack of statistical tools to compensate for biases in the quite different data collection approaches (Nie *et al.*, 2007). Much of this limitation has to do with the gulf in knowledge in how transcription links in with translation. Some advances in this area have been possible due to better *en masse* quantification of proteins. Knowing how abundant proteins are in the cell will eventually allow for better correlations to be found with transcriptome data; however, this is likely to require a large effort and will also need improvements in existing LC/MS technology. The in-depth analysis of the *E. coli* (strain MC4100) cytosol proteome (Ishihama *et al.*, 2008) revealed that the median number of copies of each protein detected was approximately 500 per cell, with the detection limit for proteins at about 50 copies per cell. Some proteins had abundances exceeding 10^4 copies per cell and 179 proteins were present at more than 2000 copies per cell. The highly abundant proteins tend to be shorter (median 252 vs. 327 amino acids) and less hydrophobic on average (–0.24 vs. –0.21 on the Kyte–Doolittle scale). Less abundant proteins may be prone to aggregation due to greater frequency of hydrophobic stretches and isoelectric points being closer to neutrality. An abundance bias was also noted on the basis of protein topology, which also relates to protein functionality. An interesting observation is that higher abundance of proteins correlates with their increased importance to the cell (Gerdes *et al.*, 2003). This essentiality relationship is also found for highly expressed genes in yeast (Jansen and Gerstein, 2000). Protein abundance generally follows data obtained from gene expression studies, with correlations made with codon usage biases (Karlin *et al.*, 2001; Greenbaum *et al.*, 2002). In yeast, abundant mRNAs are coded by genes with a consistent codon usage bias leading to definition of the codon adaptation index (Sharp and Li, 1987; Jansen *et al.*, 2003). From the study of Ishihama *et al.* (2008), about 10% of the detected proteins were found to deviate from codon usage correlations and they suggested that other mechanisms may independently control either mRNA or protein abundance. These could involve small RNAs or specific peptidases.

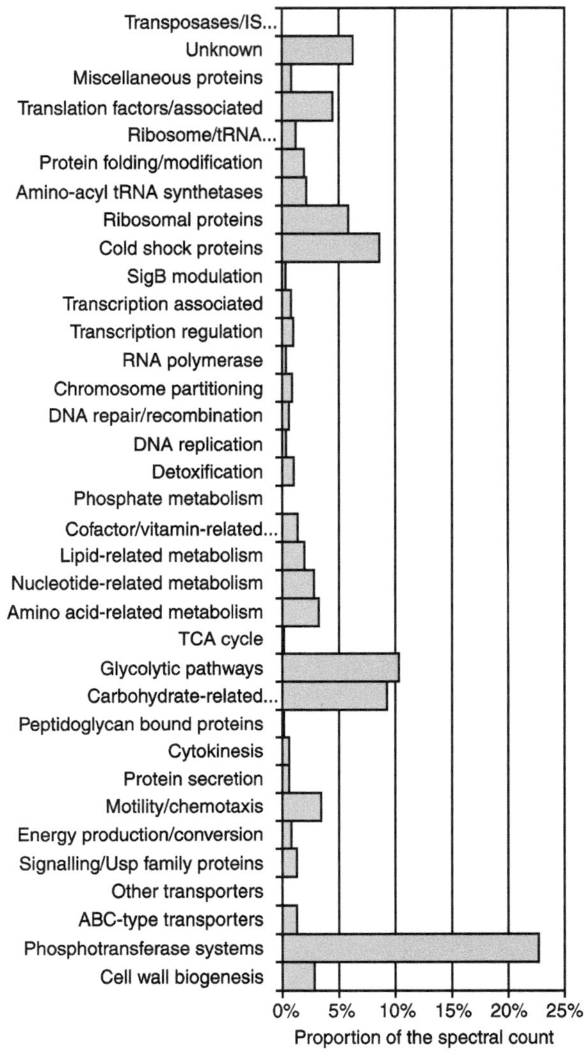

Fig. 14.4 Proportion of spectral counts (pooled from four biological replicates, see Fig. 14.3) for different groups of proteins defined on the basis of functionality, obtained by HPLC/LTQ Orbitrap MS analysis of cell lysate proteins of *L. monoctytogenes* at mid-exponential growth phase at 25°C in a carbohydrate-containing medium (Porteus *et al.*, 2009). There is likely to be an over-representation of some protein groups compared to others due to differing extraction efficiencies; however, the data are useful for a preliminary assessment of protein abundance and can also be used in comparative experiments, including manipulated culture conditions, stress, comparison of different strains, etc.

Table 14.1 Example dataset showing the use of a gel-free proteomics approach to analyse the phenotypic disposition of cells exposed to different conditions. In this example proteins of persistent *L. monoctytogenes* strain DS-81 are shown that were determined to have the greatest changes in abundance (*ca.* >50-fold) following heat shock (50°C for 1 h). See section 1.5.8 for analysis methods and explanation for fold change (RSC) and spectral index derived from spectral count data

ORF no.*	Protein name	Function	Control	Heat-shocked	Fold change (RSC)	Spectral Index
			Normalized spectral counts:			
lmo2641	IspB	polyprenyl synthetase	1	151	123	0.997
lmo2248		phosphate transport regulator, putative	3	360	111	0.990
lmo0342	TktB	transketolase	2	209	94	0.991
lmo0906	Gor	glutathione reductase	3	255	79	0.985
lmo1571	PfkA	6-phosphofructokinase	3	243	75	0.985
lmo1530	Tgt	queuine tRNA-ribosyltransferase	2	163	74	0.989
lmo0345	RpiB	ribose 5-phosphate isomerase B	1	77	64	0.994
lmo1288	LuxS	S-ribosylhomocysteine lyase	1	68	56	0.993
lmo2398	LtrC	putative phosphotidylglycerophosphatase	2	104	47	0.982
lmo0399	FrwB	PTS system, fructose-like-specific IIB component	381	1	–124	–0.999
lmo0782	ManY	PTS system, mannose-specific IIC component	301	1	–99	–0.998
lmo2666		PTS system, galacitol-specific IIB component	1679	15	–86	–0.982
lmo1267	Tig	trigger factor (prolyl isomerase)	201	1	–66	–0.998
lmo1322	NusA	transcription elongation factor	181	1	–60	–0.997
lmo1657	Tsf	elongation factor EF-Ts	161	1	–53	–0.997
lmo0398	FrwA	PTS system, fructose-like-specific IIA component	281	3	–51	–0.985
lmo1808	FabD	malonyl CoA-acyl carrier protein transacylase	281	3	–51	–0.985
lmo2577		Cof subfamily of IIB subfamily of haloacid dehalogenase superfamily	145	1	–48	–0.997

* ORF reference no. is based on the genome of *L. monoctytogenes* EGD-e (http://www.listilist.org).

14.6 Applications of proteomics for detection of foodborne pathogens

Given that the technology for protein detection and quantification is still in development, an acceleration of studies examining microbiological issues with proteomic approaches is evident. A growing number of studies on foodborne pathogens that involve proteomics are reviewed here. 2-DE studies coupled to gel-spot identification, usually by MALDI-TOF, still vastly predominate as the method of choice, but LC/MS-based shotgun studies are increasingly emerging since 2008.

Proteomics may provide opportunities for pathogen detection in food sample safety and quality diagnostic analysis. Such complex samples could include food samples, enrichment cultures and clinical samples. Although perhaps a long way off for routine diagnostic application, the use of advanced, miniaturized proteomics-based systems holds appeal. A 'bottom-up' approach depends on new developments in separation and MS technology to improve sensitivity and portability for detection of specific proteins in a complex sample. A comprehensive review of the current possibilities for MS-based technology for rapid and accurate identification of bacteria has been provided by Demirev and Fenselau (2008).

Approaches that are in development have also used proteomics as a 'top-down' tool to profile microorganisms to obtain a unique fingerprint. This can involve direct analysis of cell lysates and database comparison of peptide sequences (Fagerquist, 2007) or chemical profiling by examining peptides derived from enzymes activated on exposure to an array of specific substrates. This approach would be of interest in the concurrent development of phenotypic arrays (Bochner *et al.*, 2008; Bochner, 2009), for example matching proteins to functions determined for bacterial strains that have no genome sequence data. Some examples of recent attempts to devise such systems include CE-MS/MS being used to detect and identify pathogenic bacteria-specific peptides directly from clinical sample digests (pus, blood, sputum and urine) by matching the peptides detected against protein databases (Hu *et al.*, 2006). A chemical proteomics profiling approach was developed based on an array of 18 specific glycosidase enzyme substrates and subsequent EI/MS to identify the activated enzymes marker peptides to produce a fingerprint (Yu *et al.*, 2008). Substantial research is now underway to develop microfluidic chip-based electrophoresis platforms that could be used to streamline protein extraction and separation for subsequent MS analyses (Pizarro *et al.*, 2007).

Proteomics has been used extensively for strain-level comparisons. Such comparisons can better establish patterns of genetic diversity, potentially identify unique markers that could be used in detection of pathogenic genotypes (perhaps by other methods such as real-time PCR), and also assess strain variation with respect to physiology and stress responses. Attribution of proteins that are unique to a given strain or genotype should ideally be based on other data such as genome sequences. This reduces the chances of protein misidentification as well as providing fore-knowledge of proteins that may or may not be present. Good control of growth conditions for strain comparison experiments is required, especially for bacterial species with significant strain-level diversity. Differences caused by physiological variations in growth conditions can be readily observed with proteomics, even between genetically closely related strains (Vijayendran *et al.*, 2007). Proteomics-based studies that have been performed, which are relevant to identification-based and strain comparative analyses of foodborne pathogens, are indicated below.

At the current time, CE/MS-based and other separation/MS methods are also being vigorously researched for applications for food-based systems in order to rapidly detect bacterial toxins, mycotoxins, plant toxins, immunogenic proteins and various other proteinaceous contaminants. This area has been reviewed recently by Garcia-Canas and Cifuentes (2008).

14.6.1 Virulence, stress and biofilm-based studies on foodborne pathogens utilizing proteomics

Responses to a diverse range of physical- and chemical-derived stress, as well as interesting recovery from stress responses, have been investigated using other 2-DE approaches in various foodborne pathogens. In most cases, experiments have been performed using laboratory media and conditions that are growth permissive or partially inactivating. Further determination of the fates or outcomes of non-growth permissive situations would be useful information as it would provide basic information on proteins that are critical for survival, pass the point at which growth is halted and inactivation begins to occur. The lack of complete coverage of proteins, since most studies involve 2-DE, unfortunately prevents much in the way of in-depth comparisons for many studies, especially when more complex experimental conditions are utilized. Typically, only 50–100 proteins are identified simply because the process of protein identification is laborious and relatively expensive. The lack of quantification is also a major limitation and results tend to be rather descriptive and rely heavily on prior published information, tapping into online database resources such as BioCyc/EcoCyc. However, it is expected that much larger and comprehensive datasets will become available in the literature over the next several years due to the uptake of shotgun-based proteomics analysis approaches. Like microarray analysis, some form of standardization of methodology and database storage of datasets will be necessary to allow data, especially large datasets, to be extractable and potentially comparable (Nie *et al.*, 2007).

14.6.2 *Escherichia coli*

Certain *E. coli* serotypes, such as O157:H7, O111 and O26, are food or waterborne human pathogens causing infections with varying severity. Strains have large differences in genomes due to infiltration of phage-introduced genes, integrons and integrated plasmids, which include toxin-encoding genes (Shiga toxins) and other virulence factors. These introduced extra genomic sections can comprise pathogenicity islands, such as the important locus of enterocyte effacement. Plasmids also contribute virulence, and proteomics has been used to detect and characterize plasmid-encoded proteins that have roles in survival and pathogenicity of *E. coli* O157:H7 strains (Lim *et al.*, 2007). Thus, with advances in proteomics technologies and with greater capacity to capture information, studies are increasingly analysing toxin-producing serotypes in addition to non-pathogenic strains. Given the difference between these strains, involving perhaps as many as 1000 genes, proteomics data need to be compared with care. Proteomics provides a means to directly compare *E. coli* strains and can be potentially used to discover proteins that can differentiate strains in other applications such as surveillance of food and environmental samples (Xia *et al.*, 2008).

Not surprisingly, proteomics has been vigorously applied to study *E. coli*. This not only owes to it being the most genetically characterized species (e.g. http://www.ecocyc.org), but the species has also been used to assess and develop new proteomics technology. Proteomics has been used more recently to study virulence and quorum-sensing mechanisms in *E. coli*, helped along by the wealth of genomic data now available (> ten finished genomes, http://www.genome.jp/KEGG/). Studies have been performed in many cases in relation to food ecosystems (e.g. raw meat isolates or clinical strains derived from patients with foodborne infections), though few, if any, studies have explored the proteome of *E. coli* strains within or on actual food matrices.

Stress-related studies of *E. coli* are of general interest to the field of microbial physiology, as they often involve proteins found in many bacteria. Such proteomics-based studies, however, have been largely performed on non-toxigenic strains (such as K12 variants), so connections to virulence-related responses are often not considered. Studies to date have investigated responses to acidic and alkaline pH (Slonczewski *et al.*, 1999; Stancik *et al.*, 2002; Padan *et al.*, 2005; Huang *et al.*, 2007; Wu *et al.*, 2009), disinfectants (Bore *et al.*, 2007), oxidizing conditions and oxidizing compounds (Turlin *et al.*, 2005; Bore *et al.*, 2007; Han *et al.*, 2008; Leichert *et al.*, 2008), protein denaturants and heat shock (Rosen and Ron, 2002; Han *et al.*, 2008; Kwon *et al.*, 2008), low temperature (Kim *et al.*, 2005; Strocchi *et al.*, 2006; Jones *et al.*, 2008), osmotic pressure (Weber *et al.*, 2006), cells forced into VBNC/dormant states (Asakura *et al.*, 2007), and antibiotics exposure (Xu *et al.*, 2006; Camara and Hays, 2007; Li, Wang, *et al.*, 2008; Lin *et al.*, 2008; Zhang *et al.*, 2008). In this respect, innovative approaches to studying proteins using proteomics often utilize *E. coli* as a model species. Strain K12 has been a platform used in integrating different data (Reed *et al.*, 2003), only limited by the fact that instruments have been in rapid development in the last few years. Examples include defining an 'alkaline proteome' (Wang, Zhu, *et al.*, 2006), which provided information on proteins involved in cell death processes. Another example includes the study of oxidative stress involving discovery of proteins that are sensitive and protective under oxidizing conditions, made possible by using a novel thiol-trapping procedure (Leichert *et al.*, 2008). On the basis of these studies, similar approaches can be applied to other bacterial species, some of which have been performed recently with respect to other foodborne pathogens (see below).

The role that proteins play in *E. coli* attachment and development of biofilms is a promising area of research in which proteomics-based approaches are useful. The ability to attach to animal cells and biological surfaces is an important aspect in the distribution and spread of *E. coli*, for example in its capacity to cross-contaminate food during early food processing stages, as well in its inherent virulence capacity (Rivas *et al*., 2008). The role that autoinducers play in helping *E. coli* establish infections, or at least the potential to do so, has been explored using a number of scenarios with proteomics-based approaches (Tremoulet *et al*., 2002; Collet *et al*., 2007; Kim, Oh, *et al*., 2007; Soni *et al*., 2007). For example, the addition of quorum-sensing autoinducer-2 directly to cultures was examined with 2-DE. The addition increases the virulence of an O157:H7 serotype strain, and this change appears to be attributed to an increased expression of flagella-related proteins and certain colonization factors (Kim, Oh, *et al*., 2007). *E. coli* grown in fresh bovine milk increased translation of the LuxS protein, suggesting that bacterial quorum-sensing mechanisms are activated in some food matrices. This was also suggested by an increased expression of flagellar proteins and colonization factors that are known to be LuxS regulated (Lippolis *et al*., 2009). An interesting counterpoint to this study was the use of proteomics to determine responses of bovine hosts to *E. coli* mastitis infections by examining proteins appearing in raw milk samples (that had been depleted of major milk proteins by immunoglobulin adsorption). Proteins detected included host-derived antimicrobial peptides such as cathelicidin, indolicidin and bactenecin (Boehmer *et al*., 2008). The combined application of proteomics with other methods can provide a powerful approach to examine pathogenesis mechanisms; doing so may allow discovery of novel intervention strategies or avenues for targeted drug development (Cash, 2008).

14.6.3 *Salmonella*

Salmonella strains, mainly those defined within *Salmonella enterica* subspecies 1 (which includes Typhi and Typhimurium serovars), are a major cause of gastroenteritis and more serious illness (such as typhoid fever), due to consumption of contaminated foodstuffs and water and direct colonization of animals (e.g. poultry, reptiles, etc.; Bhan *et al*., 2005). A lot of current interest related to *Salmonella* revolves around improved strain comparisons (Encheva *et al*., 2007; Ansong *et al*., 2008); targeted drug development and understanding of drug resistance, a major problem in dealing with *Salmonella* (Becker *et al*., 2006; Coldham *et al*., 2006; Hu *et al*., 2007; Sriramulu, 2008; Webber *et al*., 2008); its adept adaptability (e.g. quorum sensing, Soni *et al*., 2008; acquisition of chemical/antimicrobial resistance, Randall *et al*., 2007; Karatzas *et al*., 2008; enhanced virulence in zero gravity, Wilson *et al*., 2007; intracellular survival, Shi *et al*., 2006; Manes *et al*., 2007; acid habituation, Berk *et al*., 2005; Lee *et al*., 2007); and better understanding of its pathogenesis mechanisms. *Salmonella* strains, again mainly belonging to the Typhi and Typhimurium serovars, have been subjected to large-scale proteomics (Ansong *et al*., 2008). Rodland *et al*. (2008) made the conclusion of these studies in a review that high-throughput mass spectrometric analysis of proteins 'provides a new view of host-pathogen interactions, emphasizing the protein products and defining how protein interactions determine the outcome of infection'. Responses specific to host cell growth and survival were simulated *in vitro* and assessed using a variation of the DIGE technique and showed good coupling to data from microarray studies (Sonck *et al.*, 2009); however, from more detailed proteomic-based surveys, it was found that of proteins present in *Salmonella*, as many as 20% show indication of post-transcriptional modification. Novel applications involving proteomics-based techniques have been used on *Salmonella* Typhimurium to study protein–protein interactions relevant to pathogenesis mechanisms. This involved using His-biotin/His-tagged proteins that had been cross-linked and further identified by tandem MS/MS and LC/MS analysis. The advantage of such a combined approach is that it reduces noise from non-specific background cross-reactions (Chowdhury *et al.*, 2009). This could be useful in studying specific protein–protein interactions in other bacteria for a whole host of goals, including understanding regulatory networks. These studies have found unexpected protein–protein interactions and exemplify the complexity of genetic mechanisms within highly evolved bacteria (Rodland *et al.*, 2008).

14.6.4 *Yersinia enterocolitica*

Yersinia enterocolitica can cause occasionally severe illness in humans and animals (termed yersiniosis) due to consumption of animal offal. Like its more foreboding relative *Yersinia pestis*, the virulence of *Y. enterocolitica* is dependent on secretion of various chromosomal and plasmid-encoded effector proteins that work against host cells (Chromy *et al.*, 2004). Proteomics approaches have been used to unravel the secretion and effector systems, revealing them to be quite complex (Matsumoto and Young, 2006; Axler-DiPerte *et al.*, 2009). Higher resolution proteomics approaches could be useful in further determining both bacterial and host responses during yersiniosis. Substantial opportunity exists to further understand the physiology and survival of this pathogen and determine its potential as an emerging food pathogenic risk using proteomics and other genome data-dependent approaches.

14.6.5 *Enterobacter sakazakii*

An opportunistic enteric, pathogen *Enterobacter sakazakii*, has been found to cause serious infections in neonates derived from consumption of food, especially desiccated food products such as powdered milk (Skovgaard, 2007). Proteomics has been used so far to study stress responses in different *E. sakazakii* strains, including the ability to withstand osmotic stress and to learn more about its thermal tolerance, factors important for its survival in dry foods. 2-DE/MALDI-TOF studies along with other analyses of physically desiccated and NaCl-exposed cultures suggested *E. sakazakii* shifts from an active growth metabolism to a less motile, quiescent physiological state in which certain stress-protective and structural proteins were more evident (Riedel and Lehner, 2007). A protein, identified using LC/MS, with a putative role in thermal tolerance was found to be discriminatory between several different *E. sakazakii* strains (Williams *et al.*, 2005).

14.6.6 *Vibrio* species

Vibrio parahaemolyticus and *Vibrio vulnifus*, occurring in sufficient numbers in seafood products (especially mollusks), can elicit gastrointestinal illnesses and wound infections. *V. parahaemolyticus* foodborne illness is the most common due to production of a thermostable haemolysin. The profile of *V. parahaemolyticus*, as a ubiquitous, animal-associated species, has made it of interest in ecophysiological studies. Proteomics has only been applied in limited applications so far but would clearly be of use to better define the ecophysiology of these species. So far, proteomics (mainly 2-DE/MALDI-TOF) has been used to define targets for tracking pathogenic (and pandemic) strains (Williams *et al.*, 2004). Another application is to better understand modes of survival outside and within animal hosts, for example survival mechanisms within the host (Lee *et al.*, 2006; Oh *et al.*, 2008; Miyamoto *et al.*, 2009). Another area of research has been investigations on how *Vibrio* spp. transit between and survive in estuarine and marine ecosystems, for example modulating salt-sensitive proteins (Xu *et al.*, 2004) and biofilms (Kim, Lee, *et al.*, 2007; Shin *et al.*, 2007). The latter studies are significant due to the tendency of *Vibrio* spp. to become residents in aquaculture systems causing problems with husbandry of the fish stocks that can lead to downstream public health issues.

14.6.7 *Campylobacter jejuni* (and other relatives)

Campylobacter jejuni is one of the most common causes of traveller's diarrhoea and food poisoning and is often associated with consumption of contaminated poultry products. It can cause chronic disease in some patients, including Guillain–Barré syndrome and Miller Fisher syndrome. Proteomics and a sub-specialization called glycoproteomics have been used in the last few years to improve strain and species identification for virulence characterization, vaccine development and to study stress responses relevant to disease.

2-DE/MALDI-TOF has been used to investigate proteomes of *C. jejuni* at the strain level and also determine proteomes of different *Campylobacter* species, with the goal to develop new ways to distinguish virulent strains. Peptide analysis indicated that certain abundant proteins (e.g. HU, IF-1, TrxA, GroES) could be used to discriminate strains due the presence of distinctive amino acid substitutions (Fagerquist, 2007; Fagerquist *et al.*, 2006, 2007). The membrane proteins in different *C. jejuni* strains have been identified and compared using a combination of 2-DE/MALDI-TOF and two-dimension HPLC/ion trap MS analysis. A substantial number of strain-specific proteins have been identified using this combined approach, which allows the capture of a higher number of proteins (Cordwell *et al.*, 2008).

2-DE analysis was used to investigate proteins specifically associated with *C. jejuni* in the biofilm mode of growth. The results indicated differential expression of certain surface-associated and stress-related proteins; however, it was also found that biofilm cells were not as stress tolerant as planktonic cells (Dykes *et al.*, 2003). This is in keeping with *C. jejuni* not possessing a well-evolved stress defense system (Murphy *et al.*, 2006). *C. jejuni* does have a defense against bile, one reason for its significance in gastroenteritis. The use of 2-DE and MS analysis revealed a comprehensive list of proteins involved in protection against bile, including ferritin and various chaperones (Fox *et al.*, 2007).

Comparisons of virulent/hypovirulent and good colonizer/poor colonizer variants of *C. jejuni* NCTC 11168 using microarray and 2-DE/MALDI-TOF-based approaches have improved details of virulence mechanisms. These studies have indicated that regulation of flagella biosynthesis and flagella themselves plays a major role in colonization and subsequent virulence (Carrillo *et al.*, 2004; Seal *et al.*, 2007). The glycosylation of flagella with novel carbohydrates determined using a metabolomics approach (see below) has been shown to be very important for virulence (Logan *et al.*, 2009). Using 1-DE and MS analysis, surface-associated proteins from *C. jejuni* expressed in *E. coli* as vaccine candidates have been analysed and identified (Prokhorova *et al.*, 2006), a process that could be used for vaccine development of other bacterial pathogens.

Glycoproteomics (reviewed by Hitchen and Dell, 2006) has been used to extensively study *C. jejuni* strains. Using lectin-based affinity chromatography, glycoproteins from lysates or cell wall preparations have been analysed by a variety of approaches, starting off with 2-DE and MS analysis (Young *et al.*, 2002). NMR spectroscopic techniques have been used to identify the glycosyl residues attached to the proteins from the same samples. A rapid version glycoproteomics method was devised to provide more accurate screening of isolates associated with Guillain–Barré syndrome and Miller Fisher syndrome (Dzieciatkowska *et al.*, 2008).

One of the few obtained protein interaction maps is available for *C. jejuni*, created using a high-throughput yeast two-hybrid screening approach (Parrish *et al.*, 2007). This interactive map could be valuable for downstream proteomic and other investigations involving this pathogenic species.

14.6.8 *Listeria monocytogenes*

The causative agent of listeriosis *L. monocytogenes* is a Gram-positive bacterium related peripherally to the lactic acid bacteria and *Staphylococcus* and its relatives in the phylum *Firmicutes*. The main use of proteomics to study this species has been to gain improved (and new) insights into its stress physiology, compare strains with different virulence potential, and for more specific investigations of gene and protein functionality. The first published proteomics experiments investigated acid stress with identification of several proteins including those associated with oxidative stress, for example Thioredoxin TrxA (Phan-Thanh and Mahouin, 1999). A later study investigated proteins associated with biofilm and planktonic growth (Tremoulet *et al.*, 2002), demonstrating that about 30 proteins showed differential translation when cultures were in a biofilm-like mode. Another study investigating biofilms also examined carbon starvation (Helloin *et al.*, 2003). Global analysis of the proteome in cells transiting from the exponential growth phase to the stationary growth phase has been investigated using 2-DE approaches (Folio *et al.*, 2004; Weeks *et al.*, 2004). A high proportion of proteins detected showed changes over a seven to nine-hour period of growth, indicating that the occurrence of growth phase transition involves substantial cellular adjustments.

One of the few direct food-based proteomic studies involved growth of a serotype 4b strain (ATCC 19115) on deli turkey meat with comparison to growth on brain heart infusion agar at 15°C (Mujahid *et al.*, 2008). On agar, some proteins specifically expressed are relevant to osmotic or oxidative stress (e.g. OsmC/Ohr), while on the meat surface, several metabolism-related proteins show an increased translation (Mujahid *et al.*, 2008). Experiments have been performed to examine post-stress inactivation in order to investigate cell repair, which remains a relatively poorly understood process. 2-DE analysis of *L. monocytogenes* and other *Firmicutes* following high-pressure processing treatment and possibly present in a state of recovery was found to express quite different proteins, some associated with stress responses. It was considered that the relative fitness of the strains could influence the types of proteins differentially expressed (Jofre *et al.*, 2007). The inhibitory activity of liquid smoke (relevant to cold smoke preservation) was investigated by 2-DE, but known stress protein, besides ClpP, was not observed after exposure; however, it was found that haemolysis was inhibited, though listeriolysin could be readily detected, suggesting an undefined effect on protein functionality by liquid smoke (Guilbaud *et al.*, 2008).

A proteome reference prototype map was developed in order to allow comparison of different *L. monocytogenes* strains isolated from food; this was of interest, given the general association of certain strain serotypes with listeriosis. In the study, 33 reference proteins were identified by 2-DE under normal growth conditions for genome sequenced strain EGD-e. About 13% of the more prominent protein spots could not be detected in other food isolates (Ramnath *et al.*, 2003). A similar study (using 2-DE/MALDI-TOF) compared *L. monocytogenes* strains of serotypes 1/2a, 1/2b and 4b, including a strain connected with a listeriosis epidemic (Dumas *et al.*, 2008).

The analyses could demonstrate statistically significant differences in protein patterns (determined by calculating estimated relative variances, factorial discrimination and principal coordinate analyses) on the basis of virulence and serotype, but strains could be separated on the basis of strain origin. A few proteins were serotype or strain-specific, including serine proteases found only in serotype 4b strain. Proteins found to be different between the strains mainly belong to the core genome and thus are likely being expressed at different levels and not consistently detected on 2-DE gels. Trost *et al.* (2005) more specifically used both 2-DE/MALDI-TOF and LC/MS to examine the secretosome of

L. monocytogenes and *Listeria innocua* strains and suggested that secreted proteins could include useful discriminatory marker proteins between pathogenic and non-pathogenic *Listeria* strains.

14.6.9 *Staphylococcus aureus*

S. aureus is a facultatively anaerobic member of phylum *Firmicutes* often implicated in human opportunistic infections, including toxic shock syndrome and foodborne toxigenesis owing to formation of enterotoxin, secreted exotoxins and superantigenic proteins (Ziebandt *et al.*, 2001). Since much of *S. aureus*' virulence is tied to surface and secreted proteins, proteomics provides a means to examine virulence-related processes in more detail, possibly leading to a more targeted approach for drug discovery (Brotz-Oesterhelt *et al.*, 2005; Heinemann *et al.*, 2005; Wang, White, *et al.*, 2006). Proteomics has been applied to define and better understand growth phase transition (Kohler *et al.*, 2005; Scherl *et al.*, 2005); regulation of secreted superantigenic proteins and understand the secretosome in general (Rogasch *et al.*, 2006; Sibbald *et al.*, 2006); and to create a reference map of proteins (Hecker *et al.*, 2003), including cell wall proteins associated with virulent and avirulent strains (Nandakumar *et al.*, 2005; Gatlin *et al.*, 2006; Taverna *et al.*, 2007).

S. aureus often causes problems due to its tendency to form biofilms and then elicit virulence activity by secretion of extracellular toxin enzymes such as alpha-toxin (a phospholipase C). Proteomics has been applied with the consideration that proteins formed specifically in the biofilm mode of growth could be targeted (i.e. vaccines or equivalent) to prevent *S. aureus* colonization and contamination (Brady *et al.*, 2006; Bernard *et al.*, 2009). Integrated analyses with transcriptomics of biofilm-derived protein and RNA extracts showed reasonable coupling of data (Resch *et al.*, 2006). These comparative experiments demonstrate the value of incorporating proteomics in a more a holistic approach to understanding bacterial physiology. This is also demonstrated in studies of *S. aureus* in which various in-frame deletion mutants have been compared with wild-types, for example examining proteins important in regulating anaerobic metabolism (Throup *et al.*, 2001; Kohler *et al.*, 2003; Fuchs *et al.*, 2007; Schlag *et al.*, 2008).

Drug discovery is an important activity in relation to *S. aureus*, and proteomics is being used extensively to examine response of this species to antimicrobials. This includes studying the effects of antibiotics that induce the cell wall stimulon and cause membrane depolarization (Singh *et al.*, 2001; Muthaiyan *et al.*, 2008). Better understanding of the acquisition of vancomycin (glycopeptides) resistance (Pieper *et al.*, 2006; Scherl *et al.*, 2006; Drummelsmith *et al.*, 2007), response to oxidative stress (Wolf *et al.*, 2008) and examining staphylococcal phage characteristics (Eyer *et al.*, 2007) have also utilized the 2-DE/MALDI-TOF approach.

14.6.10 *Clostridium perfringens* and related species

Clostridium species are anaerobic, fermentative, spore-forming Gram-positive bacteria belonging to the phylum *Firmicutes*. Toxin-producing species can cause mild-to-fatal food poisoning, most famously *Clostridium botulinum* and *Clostridium perfringens*. Several other *Clostridium* species can cause meat spoilage. It was recognized that using proteomics and comparative genomics would be valuable in learning more about the virulence mechanisms and microbiology of *C. perfringens*, the most commonly encountered problematic *Clostridium* in food systems (Sawires and Songer, 2006). To date, proteomics-based research has been minimal and 2-DE/MS has been used to identify proteins that were cross-reactive between *Clostridium tetani* and *C. perfringens* and thus could be used in a multi-species vaccine (Alam *et al.*, 2008).

14.6.11 *Bacillus cereus*

Strains of *Bacillus cereus*, a spore-forming soil-associated Gram-positive species, may form toxins that can cause acute food poisoning, termed 'Fried Rice Syndrome'. Toxins can lead either to diarrhoeal-type or emetic-type symptoms. Proteomics has been mainly used for as a possible way of identifying strains, owing to *B. cereus* similarity to other species of medical or agricultural interest, namely *Bacillus anthracis* and *Bacillus thuriengiensis*. Phenotypically, these species cannot be distinguished. MALDI-TOF identification of trypsin-derived peptides from whole-cell proteins (Warscheid and Fenselau, 2004) and spore-associated proteins (Demirev *et al.*, 2005) has been used to distinguish different *Bacillus* species, including *B. cereus*.

Other studies of *B. cereus* utilizing proteomics-based approaches relevant to food-based systems include investigations of responses to nutrient availability and biofilm formation. Vilain and Brozel (2006) used extensive 2-DE analysis (quantifying >800 protein spots) to investigate the biofilm-specific proteome of *B. cereus* and revealed that it had a distinctly different array of translated proteins compared to the planktonic state. Luo *et al.* (2007) examined the responses of *B. cereus* in soil liquid extracts compared to normal nutrient broth. The proteins identified from 2-DE gels suggested that *B. cereus* can readily adjust to consume several different soil-associated substrates. Using 2-DE/MS analysis, beta-lactamase was found in the spores of antibiotic-resistant *B. cereus* strains (Fenselau *et al.*, 2008).

14.7 Metabolomics

Using a variety of approaches, it is possible to obtain data for the many metabolites that occur in bacterial cells (Holms, 1996), referred collectively as the 'metabolome'. Essentially, the metabolome brings cell physiology 'full circle', making it possible to connect genomics (information) and proteomics (function) and thus build sounder systems biology (Brul, 2007). Metabolomics relies, like proteomics, on relatively recent advances in separation technology and MS (Makarov *et al.*, 2006). Identification of the hundreds of metabolites that may occur in bacteria at any particular time has been performed using LC (Edwards and Palsson, 2000), CE or gas chromatography followed by some sort of mass spectral analysis (Dunn *et al.*, 2005; Kiefer *et al.*, 2008). Alternatively, NMR spectroscopy-based approaches, mainly using high-resolution magic angle-spinning spectra, can be used to discriminate different samples (using ordination analysis of spectra) or identify specific compounds in complex biological mixtures (Dunn *et al.*, 2005; Bolten *et al.*, 2007).

With respect to foodborne pathogens [other than studies on generic non-Shiga toxin (Verotoxin) producing *E.coli* (STEC/VTEC) strains], metabolomics has so far had limited application. A study has investigated the effect of foodborne pathogens on the metabolome of *Brassica rapa*, detecting increases in various metabolites occurring in the presence of different foodborne pathogens; for example, Gram-negative pathogens (*E. coli, S. typhimurium*, *Shigella flexneri*) stimulated production of various plant alkaloids, certain amino acids, alcohols, carbohydrates and phenols as discriminating metabolites between different pathogen species on brassica (Jahangir *et al.*, 2008). Metabolomics has been used specifically to study distinctive flagella-associated glycans formed by *C. jejuni* and by analysis of deletion mutants that allowed identification of biosynthetic genes (Logan *et al.*, 2009). More generally, NMR-based metabolomics has been shown to be useful in distinguishing different types of *E. coli* and *B. cereus*, though the different factors that contribute to these differences have still yet to be elucidated in detail (Bundy *et al.*, 2005; Tian *et al.*, 2008).

Although it is relatively straightforward to obtain different patterns of metabolites in metabolomes of different samples or strains, the main challenges to metabolomics are dominated by quantification issues. This relates to metabolites being diverse in terms of physicochemical properties as well as in inherent chemical stability, rapid turnover rates in cells and various limitations with respect to mass spectrophotometric technology (Kiefer *et al.*, 2008). When growth conditions are kept constant, metabolomes seem to be quite reproducible (approximately 20% variation between biological replicates for abundant proteins) (Winder *et al.*, 2008); however, in studies in which *E. coli* was grown at different growth rates in chemostats under controlled levels of glucose limitation, it was found that metabolome components (e.g. glycolysis intermediates) varied considerably (Schaub and Reuss, 2008), providing issues with data comparability. This is in keeping with the cell metabolic flux networks being rather asymmetric (Almaas *et al.*, 2004). This issue could be compounded by the inconsistency in cultivation parameters, so very well-defined systems will be needed to be utilized to fully realize the development of metabolomics data analysis and interpretability (Gjersing *et al.*, 2007). Novel methods are being developed, such as stable isotope dilution approaches, which can be useful for resolving specific differences between organisms such as uropathogenic *E. coli* (Henderson *et al.*, 2009). It is clear that there is also a need to use metabolomic approaches under a wider range of growth conditions to also better define overall metabolic flux networks, that is 'to test the limits'. For example, metabolite tracking was useful in understanding the physiological behaviour of *C. jejuni* on transition to stationary growth phase, including motility, metabolite production and nutrient preferences (Wright *et al.*, 2009). It seems that for metabolomics to become more broadly used, a better foundation of data is needed (as is the case for protein abundance determinations in proteomics analyses) and it is assumed that increasing improvements in quantification and data analysis will help this endeavour. Since the technology is still very much in a state of development, metabolomics, as a routine tool, is some way off. Nevertheless, it has the potential to provide significant advances to the study of the physiology and biochemistry of foodborne pathogens and along with integration of gene and protein-based approaches further deepens the sophistication and value of systems biology.

14.8 Sources of further information and advice

For viability staining of bacteria, sources of kits and dyes useful for epifluorescence can be obtained from Invitrogen (http://www.invitrogen.com, MolecularProbes brand) and Aes-Chemunex (http://www.aeschemunex.com, ChemChrome brand). SIMS-based technology is currently being marketed by Cameca (http://www.cameca.com; Nano SIMS 50/50L). The MIFE system is commercially available (at a cottage industry scale!) from the University of Tasmania (http://www.phys.utas.edu.au/physics/biophys/mife.htm). An extremely useful beginning point for those venturing into FTIR and Raman spectroscopic chemometrics and informatics (news service, publications, tools, education, forums, etc.) is http://www.spectroscopyNOW.com. There are an extraordinarily large number of online resources for proteomics-based research. A useful consolidated form for proteomics for systems biology (and thus microbiology) is based at http://www.systemsbiology.org/, which supports the Trans Proteomic Pipeline, including X!Tandem and SEQUEST search engines. Rapid and comprehensive analysis and compilation of shotgun proteomic MS/MS data (Protein Prophet, Spectral counts, etc.) is available through MS2 Dashboard server software supported by LabKey Software Foundation (http://www.labkey.org) as well as other online resources. Numerous online tools are available for proteomics analysis usually interoperable with nucleotide/protein databases (e.g. http://expasy.org/; www.ncbi.nlm.nih.gov/Tools/; http://ncrr.pnl.gov/). Educational and news-based resources for proteomics are also easy to find (some useful examples include http://www.ionsource.com/links/proteolinks.htm; http://www.hsls.pitt.edu/guides/genetics/obrc/proteomics; http://www.genome.duke.edu/cores/proteomics/education/presentations.php; http://www.lubw.com/tutorial-on-protein-quantification-using-spectral-counting-g-test.html).

A good introduction to the more technical aspects of metabolomics can be obtained at the Scripps Centre for Mass Spectrometry (http://masspec.scripps.edu/metabo_science/index.php). Several software tools are also available online once: MathDamp (http://mathdamp.iab.keio.ac.jp/); MZmine 2 (http://mzmine.sourceforge.net/index.shtml); SIEVE (find on http://www.thermo.com); and MetaAlign (http://www.metalign.wur.nl/).

14.9 References

AEBERSOLD R and MANN M (2003), 'Mass spectrometry-based proteomics', *Nature*, **422**, 198–207.

ALAM S I, BANSOD S and SINGH L (2008), 'Immunization against *Clostridium perfringens* cells elicits protection against *Clostridium tetani* in mouse model: identification of cross-reactive proteins using proteomic methodologies', *BMC Microbiol*, **8**, Article no. 194.

ALMAAS E, KOVACS B, VICSEK T, OLTVAI Z N and BARABASI A L (2004), 'Global organization of metabolic fluxes in the bacterium *Escherichia coli*', *Nature*, **427**, 839–43.

AMALRIC J, MUTIN P H, GUERRERO G, PONCHE A, SOTTO A and LAVIGNE J P (2009), 'Phosphonate monolayers functionalized by silver thiolate species as antibacterial nanocoatings on titanium and stainless steel', *J Mater Chem*, **19**, 141–9.

ANSONG C, YOON H, NORBECK A D, GUSTIN J K, MCDERMOTT J E, MOTTAZ H M, ET AL. (2008), 'Proteomics analysis of the causative agent of typhoid fever', *J Proteome Res*, **7**, 546–57.

ASAKURA H, PANUTDAPORN N, KAWAMOTO K, IGIMI S, YAMAMOTO S and MAKIN S (2007), 'Proteomic characterization of enterohernorrhagic *Escherichia coli* O157: H7 in the oxidation-induced viable but non-culturable state', *Microbiol Immunol*, **51**, 875–81.

AXLER-DIPERTE G L, HINCHLIFFE S J, WREN B W and DARWIN A J (2009), 'YtxR Acts as an overriding transcriptional off-switch for the *Yersinia enterocolitica* Ysc-Yop Type 3 secretion system', *J Bacteriol*, **191**, 514–24.

BALDAUF N A, RODRIGUEZ-ROMO L A, MANNIG A, YOUSEF A E and RODRIGUEZ-SAONA L E (2007), 'Effect of selective growth media on the differentiation of *Salmonella* enterica serovars by Fourier-transform mid-infrared spectroscopy', *J Microbiol Meth*, **68**, 106–14.

BECKER D, SELBACH M, ROLLENHAGEN C, BALLMAIER M, MEYER T F, MANN M, ET AL. (2006), 'Robust *Salmonella* metabolism limits possibilities for new antimicrobials', *Nature*, **440**, 303–7.

BERK P A, DE JONGE R, ZWIETERING M H, ABEE T and KIEBOOM J (2005), 'Acid resistance variability among isolates of *Salmonella enterica* serovar Typhimurium DT104', *J Appl Microbiol*, **99**, 859–66.

BERNARD L, LITZLER P Y, COSETTE P, LEMELAND J F, JOUENNE T and JUNTER G A (2009), 'Proteomic analysis of *Staphylococcus aureus* biofilms grown in vitro on mechanical heart valve leaflets', *J Biomed Mat Res Part A*, 88A: 1069–78.

BESNARD V, FEDERIGHI M, DECLERQ E, JUGIAU F and CAPPELIER J M (2002), 'Environmental and physico-chemical factors induce VBNC state in *Listeria monocytogenes*', *Vet Res*, **33**, 359–70.

BHAN M K, BAHL R and BHATNAGAR S (2005), 'Typhoid and paratyphoid fever', *Lancet*, **366**, 749–62.

BLACKBURN C D and MCCARTHY J D (2000), 'Modifications to methods for the enumeration and detection of injured *Escherichia coli* O157:H7 in foods', *Int J Food Microbiol*, **55**, 285–90.

BOCHNER B R (2009), 'Global phenotypic characterization of bacteria', *FEMS Microbiol Rev*, **33**, 191–205.

BOCHNER B R, GIOVANNETTI L and VITI C (2008), 'Important discoveries from analysing bacterial phenotypes', *Mol Microbiol*, **70**, 274–80.

BOEHMER J L, BANNERMAN D D, SHEFCHECK K and WARD J L (2008), 'Proteomic analysis of differentially expressed proteins in bovine milk during experimentally induced *Escherichia coli* mastitis', *J Dairy Sci*, **91**, 4206–18.

BOLTEN C J, KIEFER P, LETISSE F, PORTAIS, J-C and WITTMANN, C (2007), 'Sampling for metabolome analysis of microorganisms', *Anal Chem*, **79**, 3843–9.

BORE E, HEBRAUD M, CHAFSEY I, CHAMBON C, SKJAERET C, MOEN B, ET AL. (2007), 'Adapted tolerance to benzalkonium chloride in *Escherichia coli* K-12 studied by transcriptome and proteome analyses', *Microbiology*, **153**, 935–46.

BRADY R A, LEID J G, CAMPER A K, COSTERTON J W and SHIRTLIFF M E (2006), 'Identification of *Staphylococcus aureus* proteins recognized by the antibody-mediated immune response to a biofilm infection', *Infect Immun*, **74**, 3415–26.

BRAUN Y, SMIRNOVA A V, WEINGART H, SCHENK A and ULLRICH M S (2007), 'A temperature-sensing histidine kinase – function, genetics, and membrane topology. Two-component signaling systems', *Meth Enzymol*, **423**, 222–49.

BROTZ-OESTERHELT H, BANDOW J E and LABISCHINSKI H (2005), 'Bacterial proteomics and its role in antibacterial drug discovery', *Mass Spec Rev*, **24**, 549–65.

BRUL S (2007), 'Systems biology and good microbiology', in Brul S, van Gerwen, S. and Zwietering, M.H. (Eds): *Modelling Microorganisms in Food*, Woodhead Publishing Ltd., Cambridge, pp. 250–88.

BUNDY J G, WILLEY T L, CASTELL R S, ELLAR D J and BRINDLE K M (2005), 'Discrimination of pathogenic clinical isolates and laboratory strains of *Bacillus cereus* by NMR-based metabolomic', *FEMS Microbiol Lett*, **242**, 127–36.

BUNTHOF C J and ABEE T (2002), 'Development of a flow cytometric method to analyze subpopulations of bacteria in probiotic products and dairy starters', *Appl Environ Microbiol*, **68**, 2934–42.

BURNETTE W N (1981), 'Western blotting: electrophoretic transfer of proteins from sodium dodecyl sulfate polyacrylamide gels to unmodified nitrocellulose and radiographic detection with antibody and radioiodinated protein', *Anal Biochem*, **112**, 195–203.

CABEEN, M T and JACOBS-WAGNER C (2007), 'Skin and bones: the bacterial cytoskeleton, cell wall, and cell morphogenesis', *J Cell Biol*, **179**, 381–7.

CALVO E, PUCCIARELLI M G, BIERNE H, COSSART P, ALBAR, J P and GARCIA-DEL PORTILLA, F (2005), 'Analysis of the *Listeria* cell wall proteome by two-dimensional nanoliquid chromatography coupled to mass spectrometry', *Proteomics*, **5**, 433–43.

CAMARA J E and HAYS F A (2007), 'Discrimination between wild-type and ampicillin-resistant *Escherichia coli* by matrix-assisted laser desorption/ionization time-of-flight mass spectrometry', *Anal Bioanal Chem*, **389**, 1633–8.

CARRILLO C D, TABOADA E, NASH J H E, LANTHIER P, KELLY J, LAU P C, ET AL. (2004), 'Genome-wide expression analyses of *Campylobacter jejuni* NCTC11168 reveals coordinate regulation of motility and virulence by flhA', *J Biol Chem*, **279**, 20327–38.

CASH P (2008), 'Proteomics of bacterial pathogens', *Exp Opin Drug Discov*, **3**, 461–73.

CENCIARINI-BORDE C, COURTOIS S and LA SCOLA B (2009), 'Nucleic acids as viability markers for bacteria detection using molecular tools', *Future Microbiol*, **4**, 45–64.

CHOI H, FERMIN D and NESVIZHSKII A I (2008), 'Significance analysis of spectral count data in label-free shotgun proteomics', *Mol Cell Proteomics*, **7**, 2373–85.

CHOWDHURY S M, SHI L, YOON H J, ANSONG C, ROMMEREIM L M, NORBECK A D, ET AL. (2009), 'Method for investigating protein-protein interactions related to *Salmonella* Typhimurium pathogenesis', *J Proteome Res*, **8**, 1504–14.

CHROMY B A, PERKINS J, HEIDBRINK J L, GONZALES A D, MURPHY G A, FITCH J P, ET AL. (2004), 'Proteomic characterization of host response to *Yersinia pestis* and near neighbours', *Biochem Biophys Res Comm*, **320**, 474–9.

COLDHAM N G, RANDALL L P, PIDDOCK L J V and WOODWARD M J (2006), 'Targeted protein degradation by *Salmonella* under phagosome-mimicking culture conditions investigated using comparative peptidomics', *J Antimicrob Chemoth*, **58**, 1145–53.

COLINGE J, MASSELOT A, GIRON M, DESSINGY T and MAGNIN J (2003), 'OLAV: towards high-throughput tandem mass spectrometry data identification', *Proteomics*, **3**, 1454–63.

COLLET A, VILAIN S, COSETTE P, JUNTER G A, JOUENNE T, PHILLIPS R S, ET AL. (2007), 'Protein expression in *Escherichia coli* S17-1 biofilms: impact of indole', *Anton Leeuw Int J G*, **91**, 71–85.

CONRADS T P, ALVING K, VEENSTRA T D, BELOV M E, ANDERSON G A, ANDERSON D J, ET AL. (2001), 'Quantitative analysis of bacterial and mammalian proteomes using a combination of cysteine affinity tags and ^{15}N-metabolic labeling', *Anal Chem*, **73**, 2132–9.

CORDWELL S J, ALICE C L L, TOUMA R G, SCOTT N E, FALCONER L, JONES D, ET AL. (2008), 'Identification of membrane-associated proteins from *Campylobacter jejuni* strains using complementary proteomics technologies', *Proteomics*, **8**, 122–39.

CRAIG R and BEAVIS R C (2004), 'TANDEM: matching proteins with mass spectra', *Bioinformatics*, **20**, 1466–7.

DAVIES P (1997), 'Food safety and its impact on domestic and export markets', *Swine Health Prod*, **5**, 13–20.

DAVIS S J, VENER A V and VIERSTRA R D (1999), 'Bacteriophytochromes: phytochrome-like photoreceptors from non-photosynthetic eubacteria', *Science*, **286**, 2517–20.

DEMIREV P A, FELDMAN A B, KOWALSKI P and LIN J S (2005), 'Top-down proteomics for rapid identification of intact microorganisms', *Anal Chem*, **77**, 7455–61.

DEMIREV P A and FENSELAU C (2008), 'Mass spectrometry for rapid characterization of microorganisms', *Ann Rev Anal Chem*, **1**, 71–93.

DOWELL J A, FROST D C, ZHANG J and LI L (2008), 'Comparison of two-dimensional fractionation techniques for shotgun proteomics', *Anal Chem*, **80**, 6715–23.

DREUX N, ALBAGNAC C, FEDERIGHI M, CARLIN F, MORRIS, C E and NGUYEN-THE C (2007), 'Viable but non-culturable *Listeria monocytogenes* on parsley leaves and absence of recovery to a culturable state', *J Appl Microbiol*, **103**, 1272–81.

DRUMMELSMITH J, WINSTALL E, BERGERON M G, POIRIER G G and OUELLETTE M (2007), 'Comparative proteomics analyses reveal a potential biomarker for the detection of vancomycin-intermediate *Staphylococcus aureus* strains', *J Proteome Res*, **6**, 4690–702.

DUMAS E, MEUNIER B, BERDAGUE J-L, CHAMBON C, SESVAUX M and HEBRAUD M (2008), 'Comparative analysis of extracellular and intracellular proteomes of *Listeria monocytogenes* strain reveals a correlation between protein expression and serovar', *Appl Environ Microbiol*, **74**, 7399–409.

DUNN W B, BAILEY N J C and JOHNSON H E (2005), 'Measuring the metabolome: current analytical technologies', *Analyst*, **130**, 606–25.

DYKES G A, SAMPATHKUMAR B and KORBER D R (2003), 'Planktonic or biofilm growth affects survival, hydrophobicity and protein expression patterns of a pathogenic *Campylobacter jejuni* strain', *Int J Food Microbiol*, **89**, 1–10.

DZIECIATKOWSKA M, LIU X, HEIKEMA A P, HOULISTON R S, VAN BELKUM A, SCHWEDA E K H, ET AL. (2008), 'Rapid method for sensitive screening of oligosaccharide epitopes in the lipooligosaccharide from *Campylobacter jejuni* strains isolated from Guillain–Barré syndrome and Miller Fisher syndrome patients', *J Clin Microbiol*, **46**, 3429–36.

EDWARDS J S and PALSSON B (2000), 'Metabolic flux balance analysis and the *in silico* analysis of *Escherichia coli* K12 gene deletions', *BMC Bioinform*, 1: Article no. 1.

EGAS D A and WIRTH M J (2008), 'Fundamentals of protein separations: 50 years of nanotechnology, and growing', *Ann Rev Anal Chem*, **1**, 833–55.

ENCHEVA V, WAIT R, BEGUM S, GHARBIA S E and SHAH H N (2007), 'Protein expression diversity amongst serovars of *Salmonella enterica*', *Microbiology*, **153**, 4183–93.

ENG J K, MCCORMACK, A L and YATES J R III (1994), 'An approach to correlate tandem mass spectral data of peptides with amino acid sequences in a protein database', *J Am Soc Mass Spectrom*, **5**, 976–89.

ESCHENHAGEN U, ESCHENHAGEN M, LUDWIG F, KIESSLING A, SYMANK K, BOSCHKE E, ET AL. (2008), 'In situ hybridization of microcolonies using catalyzed reporter deposition with tetramethylbenzidine: a method for detecting low numbers of bacterial cells in drinking water', *Eur Food Res Technol*, **227**, 995–9.

EYER L, PANTUCEK R, ZDRAHAL Z, KONECNA H, KASPAREK P, RUZICKOVA V, ET AL. (2007), 'Structural protein analysis of the polyvalent staphylococcal bacteriophage 812', *Proteomics*, **7**, 64–72.

FAGERQUIST C K (2007), 'Amino acid sequence determination of protein biomarkers of *Campylobacter upsaliensis* and *C. helveticus* by "composite" sequence proteomic analysis', *J Proteome Res*, **6**, 2539–49.

FAGERQUIST C K, BATES A H, HEATH S, KING B C, GARBUS B R, HARDEN L A, ET AL. (2006), 'Sub-speciating *Campylobacter jejuni* by proteomic analysis of its protein biomarkers and their post-translational modifications', *J Proteome Res*, **5**, 2527–38.

FAGERQUIST C K, YEE E and MILLER W G (2007), 'Composite sequence proteomic analysis of protein biomarkers of *Campylobacter coli*, *C. lari* and *C. concisus* for bacterial identification', *Analyst*, **132**, 1010–23.

FALCIONI T, PAPA S, CAMPANA R, MANTI A, BATTISTELLI M and BAFFONE W (2008), 'State transitions of *Vibrio parahaemolyticus* VBNC cells evaluated by flow cytometry', *Cytometry*, 74B: 272–81.

FANG Q, BROCKMANN S, BOTZENHART K and WIEDENMANN A (2003), 'Improved detection of *Salmonella* spp. in foods by fluorescent in situ hybridization with 23S rRNA probes: a comparison with conventional culture methods', *J Food Prot*, **66**, 723–31.

FENN J B, MANN M, MENG C K, WONG S F and WHITEHOUSE C M (1989), 'Electrospray ionization for mass spectrometry of large biomolecules', *Science*, **246**, 64–71.

FENSELAU C, HAVEY C, TEERAKULKITTIPONG N, SWATKOSKI S, LAINE O and EDWARDS N (2008), 'Identification of beta-lactamase in antibiotic-resistant *Bacillus cereus* spores', *Appl Environ Microbiol*, **74**, 904–6.

FERNANDEZ-SAIZ P, LAGARON J M and OCIO M J (2009), 'Optimization of the biocide properties of chitosan for its application in the design of active films of interest in the food area', *Food Hydrocolloid*, **23**, 913–21.

FOLIO P, CHAVANT P, CHAFSEY I, BELKORCHIA A, CHAMBON C and HEBRAUD M (2004), 'Two-dimensional electrophoresis database of *Listeria monocytogenes* EGDe proteome and proteomic analysis of mid-log and stationary growth phase cells', *Proteomics*, **10**, 3187–201.

FORNER F, FOSTER L J, CAMPANARO S, VALLE G and MANN M (2006), 'Quantitative proteomic comparison of rat mitochondria from muscle, heart, and liver', *Mol Cell Proteomics*, **5**, 144–56.

FOX E M, RAFTERY M, GOODCHILD A and MENDZ G L (2007), '*Campylobacter jejuni* response to ox-bile stress', *FEMS Immunol Med Microbiol*, **49**, 165–72.

FU X, GHARIB S A, GREN P S, AITKEN M L, FRAZER D A, PARK D R, ET AL. (2008), 'Spectral index for assessment of differential protein expression in shotgun protomics', *J Proteome Res*, **7**, 845–54.

FUCHIZAWA I, SHIMIZU S, KAWAI Y and YAMAZAKI K (2008), 'Specific detection and quantitative enumeration of *Listeria* spp. using fluorescent in situ hybridization in combination with filter cultivation (FISHFC)', *J Appl Microbiol*, **105**, 502–9.

FUCHS S, PANE-FARRE J, KOHLER C, HECKER M and ENGELMANN S (2007), 'Anaerobic gene expression in *Staphylococcus aureus*', *J Bacteriol*, **189**, 4275–89.

GARCIA-CANAS V and CIFUENTES A (2008), 'Recent advances in the application of capillary electromigration methods for food analysis', *Electrophoresis*, **29**, 294–309.

GARIP S, GOZEN A C and SEVERCAN F (2009), 'Use of Fourier transform infrared spectroscopy for rapid comparative analysis of *Bacillus* and *Micrococcus* isolates', *Food Chem*, **113**, 1301–7.

GATLIN C L, PIEPER R, HUANG S T, MONGODIN E, GEBREGEORGIS E, PARMAR P P, ET AL. (2006), 'Proteomic profiling of cell envelope-associated proteins from *Staphylococcus aureus*', *Proteomics*, **6**, 1530–49.

GEER L Y, MARKEY S P, KOWALAK J A, WAGNER L, XU M, MAYNARD D M, ET AL. (2004), 'Open mass spectrometry search algorithm', *J Proteomics Res*, **3**, 958–64.

GERBER S A, RUSH J, STEMMAN O, KIRSCHNER M W and GYGI S P (2003), 'Absolute quantification of proteins and phosphoproteins from cell lysates by tandem MS', *Proc Natl Acad Sci USA*, **100**, 6940–5.

GERDES S Y, SCHOLLE M D, CAMPBELL J W, BALAZSI G, RAVASZ E, DAUGHERTY M D, ET AL. (2003), 'Experimental determination and system level analysis of essential genes in *Escherichia coli* MG1655', *J Bacteriol*, **185**, 5673–84.

GILLIES R J, RAGHUNAND N, GARCIA-MARTIN M L and GATENBY R A (2004), 'pH imaging', *IEEE Eng Med Biol*, **23**, 57–64.

GJERSING E L, HERBERG J L, HORN J, SCHALDACH C M and MAXWELL R S (2007), 'NMR metabolomics of planktonic and biofilm modes of growth in *Pseudomonas aeruginosa*', *Anal Chem*, **79**, 8037–45.

GONZALEZ-ESCALONA N, FEY A, HOFLE M G, ESPEJO R T and GUZMAN C A (2006), 'Quantitative reverse transcription polymerase chain reaction analysis of *Vibrio cholerae* cells entering the viable but non-culturable state and starvation in response to cold shock', *Environ Microbiol*, **8**, 658–66.

GREENBAUM D, JANSEN R and GERSTEIN M (2002), 'Analysis of mRNA expression and protein abundance data: an approach for the comparison of the enrichment of features in the cellular population of proteins and transcripts', *Bioinformatics*, **18**, 585–96.

GRUN J, MANKA C K, NIKITIN S, ZABETAKIS D, COMANESCU G, GILLIS D, ET AL. (2007), 'Identification of bacteria from two-dimensional resonant-Raman spectra', *Anal Chem*, **79**, 5489–93.

GUERQUIN-KERN J L, WU T D, QUINTANA C and CROISY A (2005), 'Progress in analytical imaging of the cell by dynamic secondary ion mass spectrometry (SIMS microscopy)', *BBA-Gen Subjects*, **1724**, 228–38.

GUILBAUD M, CHAFSEY I, PILET M F, LEROI F, PREVOST H, HEBRAUD M, ET AL. (2008), 'Response of *Listeria monocytogenes* to liquid smoke', *J Appl Microbiol*, **104**, 1744–53.

GUNASEKERA T S, VEAL D A and ATTFIELD P V (2003), 'Potential for broad applications of flow cytometry and fluorescence techniques in microbiological and somatic cell analyses of milk', *Int J Food Microbiol*, **85**, 269–79.

GUPTA M J, IRUDAYARAJ J M, SCHMILOVITCH Z and MIZRACH A (2006), 'Identification and quantification of foodborne pathogens in different food matrices using FTIR spectroscopy and artificial neural networks', *T ASAE*, **49**, 1249–55.

GYGI S P, RIST B, GERBER S A, TURECEK F, GELB M H and AEBERSOLD R (1999), 'Quantitative analysis of complex protein mixtures using isotope-coded affinity tags', *Nature Biotechnol*, **17**, 994–99.

HAN K Y, PARK J S, SEO H S, AHN K Y and LEE J (2008), 'Multiple stressor-induced proteome responses of *Escherichia coli* BL21(DE3)', *J Proteome Res*, **7**, 1891–903.

HAN Y, MA B and ZHANG K (2005), 'SPIDER: software for protein identification from sequence tags containing de novo sequencing error', *J Bioinform Comput Biol*, **3**, 697–716.

HARZ A, ROSCH P and POPP J (2009), 'Vibrational spectroscopy – a powerful tool for the rapid identification of microbial cells at the single-cell level', *Cytometry*, 75A: 104–13.

HECKER M, ANTELMANN H, BUETTNER K and BERNHARDT J (2008), 'Gel-based proteomics of Gram-positive bacteria: a powerful tool to address physiological questions', *Proteomics*, **8**, 4958–75.

HECKER M, ENGELMANN S and CORDWELL S J (2003), 'Proteomics of *Staphylococcus aureus* – current state and future challenges', *J Chromatogr*, **787**, 179–95.

HEINEMANN M, KUMMEL A, RUINATSCHA R and PANKE S (2005), 'In silico genome-scale reconstruction and validation of the *Staphylococcus aureus* metabolic network', *Biotechnol Bioeng*, **92**, 850–64.

HELANDER, I M and MATTILA-SANDHOLM T (2000), 'Permeability barrier of the Gram-negative bacterial outer membrane with special reference to nisin', *Int J Food Microbiol*, **60**, 153–61.

HELLOIN E, JANSCH L and PHAN-THANH L (2003), 'Carbon starvation survival of *Listeria monocytogenes* in planktonic state and in biofilm: a proteomic study', *Proteomics*, **3**, 2052–64.

HENDERSON J P, CROWLEY J R, PINKNER J S, WALKER J N, TSUKAYAMA P, STAMM W E, ET AL. (2009), 'Quantitative metabolomics reveals an epigenetic blueprint for iron acquisition in uropathogenic *Escherichia coli*', *PLoS Path*, **5**, e1000305.

HITCHEN P G and DELL A (2006), 'Bacterial glycoproteomics', *Microbiology*, **152**, 1575–80.

HOLMS H (1996), 'Flux analysis and control of the central metabolic pathways in *Escherichia coli*', *FEMS Microbiol Rev*, **19**, 85–116.

HOLTJE J V (1998), 'Growth of the stress-bearing and shape-maintaining murein sacculus of *Escherichia coli*', *Microbiol Mol Biol Rev*, **62**, 181–203.

HU Q, NOLL R J, LI H, MAKAROV A, HARDMAN M and GRAHAM COOKS R (2005), 'The Orbitrap: a new mass spectrometer', J *Mass Spectrom*, **40**, 430–43.

HU S, LOO J A and WONG D T (2006), 'Human body fluid proteome analysis', *Proteomics*, **6**, 6326–53.

HU W S, LIN Y H and SHIH C C (2007), 'A proteomic approach to study *Salmonella enterica* serovar Typhimurium putative transporter associated with ceftriaxone resistance', *Biochem Biophys Res Comm*, **361**, 694–9.

HUANG Y J, TSAI T Y and PAN T M (2007), 'Physiological response and protein expression under acid stress of *Escherichia coli* O157:H7 TWC01 isolated from Taiwan', *J Agric Food Chem*, **55**, 7182–91.

IRUDAYARAJ J, YANG H and SAKHAMURI S (2002), 'Differentiation and detection of microorganisms using Fourier transform infrared photoacoustic spectroscopy', *J Mol Struct*, **606**, 181–8.

ISHIHAMA Y, ODA Y, TABATA T, SATO T, NAGASU T, RAPPSILBER J, ET AL. (2005), 'Exponentially modified protein abundnace index (emPAI) for estiation of absolute protein amount in proteomics by the number of sequenced peptides per protein', *Mol Cell Proteomics*, **4**, 1265–72.

ISHIHAMA Y, SCHMIDT T, RAPPSILBER J, MANN M, ULRICH HARTL F, KERNER M J, ET AL. (2008), 'Protein abundance profiling of the *Escherichia coli* cytosol', *BMC Genomics*, **9**, Article no. 102.

JAFFE L F and NUCCITELLI R (1974), 'An ultrasensitive vibrating probe for measuring steady state extracellular currents', *J Cell Biol*, **63**, 614–28.

JAHANGIR M, KIM H K, CHOI Y and VERPOORTE R (2008), 'Metabolomic response of *Brassica rapa* submitted to pre-harvest bacterial contamination', *Food Chem*, **107**, 362–8.

JAMES P (1997), 'Protein identification in the post-genome era: the rapid rise of proteomics', *Q Rev Biophys*, **30**, 279–331.

JANSEN R, BUSSEMAKER H J and GERSTEIN M (2003), 'Revisting the codon adaptation index from a whole-genome perspective: analyzing the relationship between gene expression and codon occurrence in yeat using a variety of models', *Nucl Acids Res*, **31**, 2242–51.

JANSEN R and GERSTEIN M (2000), 'Analysis of the yeast transcriptome with structural and functional categories: characterizing highly expressed proteins', *Nucleic Acids Res*, **28**, 1481–8.

JOFRE A, CHAMPOMIER-VERGES M, ANGLADE P, BARAIGE F, MARTIN B, GARRIGA M, ET AL. (2007), 'Protein synthesis in lactic acid and pathogenic bacteria during recovery from a high-pressure treatment', *Res Microbiol*, **158**, 512–20.

JONES T H, JOHNS M W and GILL C O (2008), 'Changes in the proteome of *Escherichia coli* during growth at 15°C after incubation at 2, 6 or 8°C for 4 days', *Int J Food Microbiol*, **124**, 299–302.

KARATZAS, K A G, RANDALL L P, WEBBER M, PIDDOCK L J V, HUMPHREY T J, WOODWARD M J, ET AL. (2008), 'Phenotypic and proteomic characterization of multiply antibiotic-resistant variants of *Salmonella entetica* serovar typhimurium selected following exposure to disinfectants', *Appl Environ Microb*, **74**, 1508–16.

KARLIN S, MRAZEK J, CAMPBELL A and KAISER D (2001), 'Characterizations of highly expressed genes of four fast-growing bacteria', *J Bacteriol*, **183**, 5025–40.

KELL D B, KAPRELYANTS A S, WEICHART D H, HARWOOD C R and BARER M R (1998), 'Viability and activity in readily culturable bacteria: a review and discussion of the practical issues', *Anton Leeuwen Int J G*, **73**, 169–87.

KIEFER P, PORTAIS, J-C and VORHOLT J A (2008), 'Quantitative metabolome analysis using liquid chromatography-high-resolution mass spectrometry', *Anal Biochem*, **382**, 94–100.

KIM H S, LEE M A, CHUN S J, PARK S J and LEE K H (2007), 'Role of NtrC in biofilm formation via controlling expression of the gene encoding an ADP-glycero-manno-heptose-6-epimerase in the pathogenic bacterium *Vibrio vulnificus*', *Mol Microbiol*, **63**, 559–74.

KIM Y, OH S, AHN E Y, IMM J Y, OH S, PARK S, ET AL. (2007), 'Proteome analysis of virulence factor regulated by autoinducer-2-like activity in *Escherichia coli* O157:H7', *J Food Prot*, **70**, 300–7.

KIM Y H, HAN K Y, LEE K and LEE J (2005), 'Proteome response of *Escherichia coli* fed-batch culture to temperature downshift', *Appl Microbiol Biotechnol*, **68**, 786–93.

KOENIG T, MENZE B H, KIRCHNER M, MONIGATTI F, PARKER K, PATTERSON T, ET AL. (2008), 'Robust prediction of the MASCOT score for an improved quality assessment in mass spectrometric proteomics', *J Proteome Res*, **7**, 3708–17.

KOHLER C, VON EIFF C, PETERS G, PROCTOR R A, HECKER M and ENGELMANN S (2003), 'Physiological characterization of a heme-deficient mutant of *Staphylococcus aureus* by a proteomic approach', *J Bacteriol*, **185**, 6928–37.

KOHLER C, WOLFF S, ALBRECHT D, FUCHS S, BECHER D, BUTTNER K, ET AL. (2005), 'Proteome analyses of *Staphylococcus aureus* in growing and non-growing cells: a physiological approach', *Int J Med Microbiol*, **295**, 547–65.

KOO O K, LIU Y S, SHUAIB S, BHATTACHARYA S, LADISCH M R, BASHIR R, ET AL. (2009), 'Targeted capture of pathogenic bacteria using a mammalian cell receptor coupled with dielectrophoresis on a biochip', *Anal Chem*, **81**, 3094–101.

KWON S, JUNG Y and LIM D (2008), 'Proteomic analysis of heat-stable proteins in *Escherichia coli*', *BMB Reports*, **41**, 108–11.

LAFLAMME C, GENDRON L, TURGEON N, FILION G, HO J and DUCHAINE C (2009), 'Rapid detection of germinating *Bacillus cereus* cells using fluorescent in situ hybridization', *J Rapid Meth Aut Mic*, **17**, 80–102.

LAIKO V V, MOYER S C and COTTER R J (2000), 'Atmospheric pressure MALDI/ion trap mass spectrometry', *Anal Chem*, **72**, 5239–43.

LANTZ M S and CIBOROWSKI P (1994), 'Zymographic techniques for detection and characterization of microbial proteases', Methods Enzymol, **235**, 563–94.

LEE A Y, PARK S G, JANG M, CHO S, MYUNG P K, KIM Y R, ET AL. (2006), 'Proteomic analysis of pathogenic bacterium *Vibrio vulnificus*', *Proteomics*, **6**, 1283–9.

LEE Y H, KIM B H, KIM J H, YOON W S, BANG S H and PARK Y K (2007), 'CadC has a global translational effect during acid adaptation in *Salmonella enterica* serovar Typhimurium', *J Bacteriol*, **189**, 2417–25.

LEICHERT L I, GEHRKE F, GUDISEVA H V, BLACKWELL T, ILBERT M, WALKER A K, ET AL. (2008), 'Quantifying changes in the thiol redox proteome upon oxidative stress in vivo', *Proc Natl Acad Sci U S A*, **105**, 8197–202.

LI H, WANG B C, XU W J, LIN X M and PENG X X (2008), 'Identification and network of outer membrane proteins regulating streptomysin resistance in *Escherichia coli*', *J Proteome Res*, **7**, 4040–9.

LI, Q B and ROXAS B A P (2009), 'An assessment of false discovery rates and statistical significance in label-free quantitative proteomics with combined filters', *BMC Bioinform*, **19**, Article no. 10.

LI T, WU T D, MAZÉAS L, TOFFIN L, GUERQUIN-KERN J L, LEBLON G, ET AL. (2008), 'Simultaneous analysis of microbial identity and function using NanoSIMS', *Environ Microbiol*, **10**, 580–8.

LIM J Y, SHENG H Q, SEO K S, PARK Y H and HOVDE C J (2007), 'Characterization of an *Escherichia coli* O157:H7 plasmid O157 deletion mutant and its survival and persistence in cattle', *Appl Environ Microb*, **73**, 2037–47.

LIN X M, LI H, WANG C and PENG X X (2008), 'Proteomic analysis of nalidixic acid resistance in *Escherichia coli*: Identification and functional characterization of OM proteins', *J Proteome Res*, **7**, 2399–405.

LIPPOLIS J D, BAYLES D O and REINHARDT T A (2009), 'Proteomic changes in *Escherichia coli* when grown in fresh milk versus laboratory media', *J Proteome Res*, **8**, 149–58.

LOGAN S M, HUI, J P M, VINOGRADOV E, AUBRY A J, MELANSON J E, KELLY J F, ET AL. (2009), 'Identification of novel carbohydrate modifications on *Campylobacter jejuni* 11168 flagellin using metabolomics-based approaches', *FEBS J*, **276**, 1014–23.

LOPEZ AMOROS R, CASTEL S, COMAS RIU J and VIVES REGO J (1997), 'Assessment of *E. coli* and *Salmonella* viability and starvation by confocal laser microscopy and flow cytometry using rhodamine 123, DiBAC4(3), propidium iodide, and CTC', *Cytometry*, **29**, 298–305.

LUO Y, VILAIN S, VOIGT B, ALBRECHT D, HECKER M and BROZEL V S (2007), 'Proteomic analysis of *Bacillus cereus* growing in liquid soil organic matter', *FEMS Microbiol Lett*, **271**, 40–7.

MACDONALD A G and MARTINAC B (2005), 'Effect of high hydrostatic pressure on the bacterial mechanosensitive channel MscS', *Eur Biophys J Biophy*, **34**, 434–41.

MAKAROV A, DENISOV E, KHOLOMEEV A, BALSCHUN W, LANGE O, STRUPAT K, ET AL. (2006), 'Performance evaluation of a hybrid linear ion trap/Orbitrap mass spectrometer', *Anal Chem*, **78**, 2113–20.

MANES N P, GUSTIN J K, RUE J, MOTTAZ H M, PURVINE S O, NORBECK A D, ET AL. (2007), 'Effect of fluoroquinolone exposure on the proteome of *Salmonella enterica* serovar Typhimurium', *Mol Cell Proteomics*, **6**, 717–27.

MARON P A, RANJARD L, MOUGEL C and LEMANCEAU P (2007), 'Metaproteomics: a new approach for studying functional microbial ecology', *Microb Ecol*, **53**, 486–93.

MATSUMOTO H and YOUNG G M (2006), 'Proteomic and functional analysis of the suite of Ysp proteins exported by the Ysa type III secretion system of *Yersinia enterocolitica* Biovar 1B', *Mol Microbiol*, **59**, 689–706.

MAUKONEN J, ALAKOMI H L, NOHYNEK L, HALLAMAA K, LEPPAMAKI S, MATTO J, ET AL. (2006), 'Suitability of the fluorescent techniques for the enumeration of probiotic bacteria in commercial non-dairy drinks and in pharmaceutical products', *Food Res Int*, **39**, 22–32.

MCDOUGALD D, RICE S A, WEICHART D and KJELLEBERG S (1998), 'Nonculturability: adaptation or debilitation?', *FEMS Microbiol Ecol*, **25**, 1–9.

MCLAMORE E S, PORTERFIELD D M and BANKS M K (2009), 'Non-invasive self-referencing electrochemical sensors for quantifying real-time biofilm analyte flux', *Biotechnol Bioeng*, **102**, 791–9.

MERCHANT M and WEINBERGER S R (2000), 'Recent advancements in surface-enhanced laser desorption/ionization time-of-flight mass spectrometry', *Electrophoresis*, **21**, 1164–77.

MIKHEL'SON K N (2008), 'Electrochemical sensors based on ionophores: current state, trends, and prospects', *Russ J Gen Chem*, **78**, 2445–54.

MIRGORODSKAYA O A, KOZMIN Y P, TITOV M I, KORNER R, SONKSEN C P and ROEPSTORFF P (2000) 'Quantification of peptides and proteins by matrix-assisted laser desorption/ionization mass spectrometry using (18),O-labeled internal standards', *Rapid Commun Mass Spectrom*, **14**, 1226–32.

MITULOVIĆ G and MECHTLER K (2006), 'HPLC techniques for proteomics analysis – a short overview of latest developments', *Brief Funct Genomics Proteomics*, **5**, 249–60.

MIYAMOTO K, KOSAKAI K, IKEBAYASHI S, TSUCHIYA T, YAMAMOTO S and TSUJIBO H (2009), 'Proteomic analysis of *Vibrio vulnificus* M2799 grown under iron-repleted and iron-depleted condition', *Microb Pathogenesis*, **46**, 171–7.

MORENO Y, HERNANDEZ M, FERRUS M A, ALONSO J L, BOTELLA S, MONTES R, ET AL. (2001), 'Direct detection of thermotolerant campylobacters in chicken products by PCR and in situ hybridization', *Res Microbiol*, **152**, 577–82.

MUJAHID S, PECHAN T and WANG C L (2007), 'Improved solubilization of surface proteins from *Listeria monocytogenes* for 2-DE', *Electrophoresis*, **28**, 3998–4007.

MUJAHID S, PECHAN T and WANG C L (2008), 'Protein expression by *Listeria monocytogenes* grown on a RTE-meat matrix', *Int J Food Microbiol*, **128**, 203–11.

MURPHY C, CARROLL C and JORDAN K N (2006), 'Environmental survival mechanisms of the foodborne pathogen *Campylobacter jejuni*', *J Appl Microbiol*, **100**, 623–32.

MUSAT N, HALM H, WINTERHOLLER B, HOPPE P, PEDUZZI S, HILLION F, ET AL. (2008), 'A single-cell view on the ecophysiology of anaerobic phototrophic bacteria', *Proc Natl Acad Sci U S A*, **105**, 17861–6.

MUTHAIYAN A, SILVERMAN J A, JAYASWAL R K and WILKINSON B J (2008), 'Transcriptional profiling reveals that daptomycin induces the *Staphylococcus aureus* cell wall stress stimulon and genes responsive to membrane depolarization', *Antimicrob Agents Chemoth*, **52**, 980–90.

NANDAKUMAR R, NANDAKUMAR M P, MARTEN M R and ROSS J M (2005), 'Proteome analysis of membrane and cell wall associated proteins from *Staphylococcus aureus*', *J Proteome Res*, **4**, 250–7.

NEBE-VON CARON G, STEPHENS P and BADLEY R A (1998), 'Assessment of bacterial viability status by flow cytometry and single cell sorting', *J Appl Microbiol*, **84**, 988–98.

NEWMAN I A (2001), 'Ion transport in roots: meaurement of fluxes using ion-selective microelectrodes to characterize transporter function', *Plant Cell Environ*, **24**, 1–14.

NEWMAN J R, GHAEMMAGHAMI S, IHMELS J, BRESLOW D K, NOBLE M, DERISI J L, ET AL. (2006), 'Single-cell proteomic analysis of *S. cerevisiae* reveals the architecture of biological noise', *Nature*, **441**, 840–6.

NGUYEN T, CLARE B, GUO W and MARTINAC B (2005), 'The effects of parabens on the mechanosensitive channels of *E. coli*', *Eur Biophys J Biophys*, **34**, 389–95.

NICOLAOU N and GOODACRE R (2008), 'Rapid and quantitative detection of the microbial spoilage in milk using Fourier transform infrared spectroscopy and chemometrics', *Analyst*, **133**, 1424–31.

NIE L, WU G, CULLEY D E, SCHOLTEN J C M and ZHANG W (2007), 'Integrative analysis of transcriptomic and proteomic data: challenges, solutions and applications', *Crit Rev Biotechnol*, **27**, 63–75.

NISHIHARA J C and CHAMPION K M (2002), 'Quantitative evaluation of proteins in one- and two-dimensional polyacrylamide gels using a fluorescent stain', *Electrophoresis*, **23**, 2203–15.

NISHIMURA M, SHIMAKITA T, MATSUZAKI T, TASHIRO Y and KOGURE K (2008), 'Automatic counting of FISH-labeled microbes by an LED illuminated detecting apparatus', *Fisheries Sci*, **74**, 405–10.

NOACK D, KNODL C and LACHENMEIER D W (2008), 'Evaluation of fluorescence-marked gene probes and Fourier transform infrared spectroscopy as novel methods to detect beer spoilage bacteria', *Deutsche Lebensm-Rundsch*, **104**, 65–9.

NOCKER A, CHEUNG C Y and CAMPERA A K (2006), 'Comparison of propidium monoazide with ethidium monoazide for differentiation of live vs. dead bacteria by selective removal of DNA from dead cells', *J Microbiol Meth*, **67**, 310–20.

OH M H, JEONG H G and CHOI S H (2008), 'Proteomic identification and characterization of *Vibrio vulnificus* proteins induced upon exposure to INT-407 intestinal epithelial cells', *J Microbiol Biotechnol*, **18**, 968–74.

OLD W M, MEYER-ARENDT K, AVELINE-WOLF L, PIERCE K G, MENDOZA A, SEVINSKY J R, ET AL. (2005), 'Comparison of label-free methods for quantifying human proteins by shotgun proteomics', *Mol Cell Proteomics*, **4**, 1487–1502.

OLSEN J V, ONG S E and MANN M (2004), 'Trypsin cleaves exclusively C-terminal to arginine and lysine residues', *Mol Cell Proteomics*, **3**, 608–14.

ONG S E, BLAGOEV B, KRATCHMAROVA I, KRISTENSEN D B, STEEN H, PANDEY A, ET AL. (2002), 'Stable isotope labeling by amino acids in cell culture, SILAC, as a simple and accurate approach to expression proteomics', *Mol Cell Proteomics*, **1**, 376–86.

OOTSUBO M, SHIMIZU T, TANAKA R, SAWABE T, TAJIMA K and EZURA Y (2003), 'Seven-hour fluorescence in situ hybridization technique for enumeration of *Enterobacteriaceae* in food and environmental water sample', *J Appl Microbiol*, **95**, 1182–90.

OULAHAL N, ADT I, MARIANI C, CARNET-PANTEIZ A, NOTZ E and DEGRAEVE P (2009), 'Examination of wooden shelves used in the ripening of a raw milk smear cheese by FTIR spectroscopy', *Food Control*, **20**, 658–63.

PADAN E, BIBI E, ITO M and KRULWICH T A (2005), 'Alkaline pH homeostasis in bacteria: new insights', *BBA-Biomembranes*, **1717**, 67–88.

PAPADIMITRIOU K, PRATSINIS H, NEBE-VON-CARON G, KLETSAS D and TSAKALIDOU E (2006), 'Rapid assessment of the physiological status of *Streptococcus macedonicus* by flow cytometry and fluorescence probes', *Int J Food Microbiol*, **111**, 197–205.

PARK C, NOVAK J T, HELM R F, AHN Y O and ESEN A (2008), 'Evaluation of the extracellular proteins in full-scale activated sludges', *Water Res*, **42**, 3879–89.

PARRISH J R, YU J, LIU G, HINES J A, CHAN J E, MANGIOLA B A, ET AL. (2007), 'A proteome-wide protein interaction map for *Campylobacter jejuni*', *Genome Biol*, **8**, Article R130.

PERKINS D N, PAPPIN D J, CREASY D M and COTTRELL J S (1999), 'Probability-based protein identification by searching sequence databases using mass spectrometry data', *Electrophoresis*, **20**, 3551–67.

PETERANDERL R and LECHENE C (2004), 'Measurement of carbon and nitrogen stable isotope ratios in cultured cells', *J Am Soc Mass Spectr*, **15**, 478–85.

PHAN-THANH L and MAHOUIN F (1999), 'A proteomic approach to study the acid response in *Listeria monocytogenes*', *Electrophoresis*, **20**, 2214–24.

PIEPER R, GATLIN-BUNAI C L, MONGODIN E F, PARMAR P P, HUANG S T, CLARK D J, ET AL. (2006), 'Comparative proteomic analysis of *Staphylococcus aureus* strains with differences in resistance to the cell wall-targeting antibiotic vancomycin', *Proteomics*, **6**, 4246–58.

PIZARRO S A, LANE P, LANE T W, CRUZ E, HAROLDSEN B and VANDERNOOT V A (2007), 'Bacterial characterization using protein profiling in a microchip separations platform', *Electrophoresis*, **28**, 4697–704.

PRIEST F G and AUSTIN B (EDS) (1993), 'Chemosystematics and molecular biology II: proteins, lipids, carbohydrates and whole cell', in *Modern Bacterial Taxonomy*, Chapman & Hall, London, pp. 95–110.

PROKHOROVA T A, NIELSEN P N, PETERSEN J, KOFOED T, CRAWFORD J S, MORSCZECK C, ET AL. (2006), 'Novel surface polypeptides of *Campylobacter jejuni* as traveller's diarrhoea vaccine candidates discovered by proteomics', *Vaccine*, **24**, 6446–55.

PUCCIARELLI M G, CALVO E, SABET C, BIERNE H, COSSART P and GARDIA-DEL PORTILLO F (2005), 'Identification of substrates of the *Listeria monocytogenes* sortases A and B by a non-gel proteomic analysis', *Proteomics*, **5**, 4808–17.

QUINTANA C, BELLEFQIH S, LAVAL J Y, GUERQUIN-KERN J L, WU T D, AVILA J, ET AL. (2006), 'Study of the localization of iron, ferritin, and hemosiderin in Alzheimer's disease hippocampus by analytical microscopy at the subcellular level', *J Struct Biol*, **153**, 42–54.

RAMNATH M, RECHINGER K B, JANSCH L, HASTINGS J W, KNOCHEL S and GRAVESEN A (2003), 'Development of a *Listeria monocytogenes* EGDe partial proteome reference map and comparison with the protein profiles of food isolates', *Appl Environ Microb*, **69**, 3368–76.

RANDALL L P, COOLES S W, COLDHAM N G, PENUELA E G, MOTT A C, WOODWARD M J, ET AL. (2007), 'Commonly used farm disinfectants can select for mutant *Salmonella enterica* serovar Typhimurium with decreased susceptibility to biocides and antibiotics without compromising virulence', *J Antimicrob Chemoth*, **60**, 1273–80.

RAPPSILBER J, RYDER U, LAMOND A I and MANN M (2002), 'Large-scale proteomic analysis of the human spliceosome', *Genome Res*, **12**, 1231–45.

RAWEL H M, ROHN S, KROLL J and SCHWEIGERT F J (2005), 'Surface enhanced laser desorption ionization time-of-flight mass spectrometry analysis in complex food and biological systems', *Mol Nutrit Food Res*, **49**, 1104–11.

REBUFFO-SCHEER C A, SCHMITT J and SCHERER S (2007), 'Differentiation of *Listeria monocytogenes* serovars by using artificial neural network analysis of Fourier-transformed infrared spectra', *Appl Environ Microb*, **73**, 1036–40.

REED J L, VO T D, SCHILLING C H and PALSSON B O (2003), 'An expanded genome-scale model of *Escherichia coli* K-12 (iJR904 GSM/GPR)', *Genome Biol*, **4**, Article no. R54.

REN Q and PAULSEN I T (2007), 'Large-scale comparative genomic analyses of cytoplasmic membrane transport systems in prokaryotes', *J Mol Microbiol Biotechnol*, **12**, 165–79.

RESCH A, LEICHT S, SARIC M, PASZTOR L, JAKOB A, GOTZ F, ET AL. (2006), 'Comparative proteome analysis of *Staphylococcus aureus* biofilm and planktonic cells and correlation with transcriptome profiling', *Proteomics*, **6**, 1867–77.

RIEDEL K and LEHNER A (2007), 'Identification of proteins involved in osmotic stress response in *Enterobacter sakazakii* by proteomics', *Proteomics*, **7**, 1217–31.

RITZ M, PILET M F, JUGIAU F, RAMA F and FEDERIGHI M (2006), 'Inactivation of *Salmonella* Typhimurium and *Listeria monocytogenes* using high-pressure treatments: destruction or sublethal stress?' *Lett Appl Microbiol*, **42**, 357–62.

RIVAS L, FEGAN N and DYKES G A (2008), 'Expression and putative roles in attachment of outer membrane proteins of *Escherichia coli* O157 from planktonic and sessile culture', *Foodborne Path Disease*, **5**, 155–64.

RODLAND K D, ADKINS J N, ANSONG C, CHOWDHURY S, MANES N P, SHI L, ET AL. (2008), 'Use of high-throughput mass spectrometry to elucidate host-pathogen interactions in *Salmonella*', *Future Microbiol*, **3**, 625–34.

ROGASCH K, RUHMLING V, PANE-FARRE J, HOPER D, WEINBERG C, FUCHS S, ET AL. (2006), 'Influence of the two-component system SaeRS on global gene expression in two different *Staphylococcus aureus* strains', *J Bacteriol*, **188**, 7742–58.

ROSEN R and RON E Z (2002), 'Proteome analysis in the study of the bacterial heat-shock response', *Mass Spectrom Rev*, **21**, 244–65.

ROSS P L, HUANG, Y L N, MARCHESE J N, WILLIAMSON B, PARKER K, HATTAN S, ET AL. (2004), 'Multiplexed protein quantitation in *Saccharomyces cerevisiae* using amine-reactive isobaric tagging reagents', *Mol Cell Proteomics*, **3:** 1154–69.

SACHIDANANDHAM R AND GIN, K Y H (2009), 'A dormancy state in nonspore-forming bacteria', *Appl Microbiol Biotechnol*, **81**, 927–41.

SANDERS S L, JENNINGS J, CANUTESCU A, LINK A J and WEIL P A (2002), 'Proteomics of the eukaryotic transcription machinery: identification of proteins assocated with components of yeast TFIID by multidimensional mass spectrometry', *Mol Cell Biol*, **22**, 4723–38.

SAWIRES Y S and SONGER J G (2006), '*Clostridium perfringens*: insight into virulence evolution and population structure', *Anaerobe*, **12**, 23–43.

SCHAUB J and REUSS M (2008), 'In vivo dynamics of glycolysis in *Escherichia coli* shows need for growth-rate dependent metabolome analysis', *Biotechnol Prog*, **24**, 1402–7.

SCHAUMBURG J, DIEKMANN O, HAGENDORFF P, BERGMANN S, ROHDE M, HAMMERSCHMIDT S, ET AL. (2004), 'The cell wall subproteome of *Listeria monocytogenes*', *Proteomics*, **4**, 2991–3006.

SCHERL A, FRANCOIS P, BENTO M, DESHUSSES J M, CHARBONNIER Y, CONVERSET W, ET AL. (2005), 'Correlation of proteomic and transcriptomic profiles of *Staphylococcus aureus* during the post-exponential phase of growth', *J Microbiol Meth*, **60**, 247–57.

SCHERL A, FRANCOIS P, CHARBONNIER Y, DESHUSSES J M, KOESSLER T, HUYGHE A, ET AL. (2006), 'Exploring glycopeptide-resistance in *Staphylococcus aureus*: a combined proteomics and transcriptomics approach for the identification of resistance-related markers', *BMC Genomics*, **7**, Article no. 296.

SCHLAG S, FUCHS S, NERZ C, GAUPP R, ENGELMANN S, LIEBEKE M, ET AL. (2008), 'Characterization of the oxygen-responsive NreABC regulon of *Staphylococcus aureus*', *J Bacteriol*, **190**, 7847–58.

SEAL B S, HIETT K L, KUNTZ R L, WOOLSEY R, SCHEGG K M, ARD M, ET AL. (2007), 'Proteomic analyses of a robust versus a poor chicken gastrointestinal colonizing isolate of *Campylobacter jejuni*', *J Proteome Res*, **6**, 4582–91.

SHABALA L, BOWMAN J, BROWN J, ROSS T, MCMEEKIN T and SHABALA S (2009), 'Ion transport and osmotic adjustment in *Escherichia coli* in response to ionic and non-ionic osmotica', *Environ Microbiol*, **11**, 137–48.

SHABALA L, MCMEEKIN T, BUDDE B and SIEGUMFELDT H (2006), '*Listeria innocua* and *Lactobacillus delbrueckii* subsp. *bulgaricus* employ different strategies to cope with acid stress', *Int J Food Microbiol*, **110**, 1–7.

SHABALA L, ROSS T, MCMEEKIN T and SHABALA S (2006), 'Non-invasive microelectrode ion flux meaurements to study adpative responses of microorganisms to the environment', *FEMS Microbiol Rev*, **30**, 472–86.

SHABALA L, ROSS T, NEWMAN I, MCMEEKIN T and SHABALA S (2001), 'Measurements of net fluxes and extracellular changes of H^+, Ca^{2+}, K^+, and NH_4^+ in *Escherichia coli* using ion-selective microelectrodes', *J Microbiol Meth*, **46**, 119–29.

SHADFORTH I P, DUNNLEY P J, LILLEY K S and BESSANT C (2005), 'i-Tracker: for quantitative proteomics using iTRAQ', *BMC Genomics*, **6**, Article no. 145.

SHAPIRO A L, VIÑUELA E and MAIZEL J V JR (1967), 'Molecular weight estimation of polypeptide chains by electrophoresis in SDS-polyacrylamide gels', *Biochem Biophys Res Comm*, **28**, 815–20.

SHARP P M and LI W H (1987), 'The codon adaptation index – a measure of directional synonymous codon usage bias, and its potential applications', *Nucl Acids Res*, **15**, 1281–95.

SHEN Y F, ZHAO R, BELOV M E, CONRADS T P, ANDERSON G A, TANG K Q, ET AL. (2001), 'Packed capillary reversed-phase liquid chromatography with high-performance electrospray ionization Fourier transform ion cyclotron resonance mass spectrometry for proteomics', *Anal Chem*, **73**, 1766–75.

SHI L, ADKINS J N, COLEMAN J R, SCHEPMOES A A, DOHNKOVA A, MOTTAZ H M, ET AL. (2006), 'Proteomic analysis of *Salmonella enterica* serovar Typhimurium isolated from RAW 264.7 macrophages – identification of a novel protein that contributes to the replication of serovar Typhimurium inside macrophages', *J Biol Chem*, **281**, 29131–40.

SHIMAKITA T, TASHIRO Y, KATSUYA A, SAITO M and MATSUOKA H (2006), 'Rapid separation and counting of viable microbial cells in food by nonculturable method with Bioplorer, a focusing-free microscopic apparatus with a novel cell separation unit', *J Food Prot*, **69**, 170–6.

SHIN N R, LEE D Y and YOO H S (2007), 'Identification of quorum sensing-related regulons in *Vibrio vulnificus* by two-dimensional gel electrophoresis and differentially displayed reverse transcriptase PCR', *FEMS Immunol Med Microbiol* **50**, 94–103.

SIBBALD, M J J B, ZIEBANDT A K, ENGELMANN S, HECKER M, DE JONG A, HARMSEN H J A, ET AL. (2006), 'Mapping the pathways to staphylococcal pathogenesis by comparative secretomics', *Microbiol Mol Biol R*, **70**, 755–88.

SIEGUMFELDT H, RECHINGER K B and JAKOBSEN M (2000), 'Dynamic changes of intracellular pH in individual lactic acid bacterium cells in response to a rapid drop in extracellular pH', *Appl Environ Microbiol*, **66**, 2330–5.

SINGH V K, JAYASWAL R K and WILKINSON B J (2001), 'Cell wall-active antibiotic induced proteins of *Staphylococcus aureus* identified using a proteomic approach', *FEMS Microbiol Lett*, **199**, 79–84.

SKOVGAARD N (2007), 'New trends in emerging pathogens', *Int J Food Microbiol*, **120**, 217–24.

SLONCZEWSKI J L, BLANKENHORN D, FOSTER J W, MATIN A, BOOTH I R, STOCK J B, ET AL. (1999), 'Acid and base regulation in the proteome of *Escherichia coli*. Bacterial Response to pH', *Novart Fdn Symp*, **221**, 75–92.

SMITH, P J S, HAMMAR K, PORETRFIELD D M, SNAGER R H and TRIMARCHI J R (1999), 'Self-referencing, non-invasive, ion selective electrode for single cell detection of trans-plasma membrane calcium flux', *Microsc Res Tech*, **46**, 398–417.

SONCK, K A J, KINT G, SCHOOFS G, VANDER WAUVEN C, VANDERLEYDEN J, DE KEERSMAECKER, S C J, ET AL. (2009), 'The proteome of *Salmonella* Typhimurium grown under in vivo-mimicking conditions', *Proteomics*, **9**, 565–79.

SONI K, JESUDHASAN P, CEPEDA M, WILLIAMS B, HUME M, RUSSELL W K, ET AL. (2007), 'Proteomic analysis to identify the role of LuxS/Al-2 mediated protein expression in *Escherichia coli* O157:H7', *Foodborne Path Disease*, **4**, 464–71.

SONI K A, JESUDHASAN P R, CEPEDA M, WILLIAMS B, HUME M, RUSSELL W K, ET AL. (2008), 'Autoinducer AI-2 is involved in regulating a variety of cellular processes in *Salmonella* Typhimurium', *Foodborne Path Disease*, **5**, 147–53.

SOTOMAYOR, M, VASQUEZ V, PEROZO E and SCHULTEN K (2007), 'Ion conduction through MscS as determined by electrophysiology and simulation', *Biophys J*, **92**, 886–902.

SRIRAMULU D D (2008), 'Adaptive expression of foreign genes in the clonal variants of bacteria: from proteomics to clinical application', *Proteomics*, **8**, 882–92.

STADERMANN F J, WALKER R M and ZINNER E (1999), 'The next generation ion probe for the microanalysis of extraterrestrial material', *Meteorits Planet Sci*, **34**, A111–12.

STANCIK L M, STANCIK D M, SCHMIDT B, BARNHART D M, YONCHEVA Y N and SLONCZEWSKI J L (2002), 'pH-dependent expression of periplasmic proteins and amino acid catabolism in *Escherichia coli*', *J Bacteriol*, **184**, 4246–58.

STROCCHI M, FERRER M, TIMMIS K N and GOLYSHIN P N (2006), 'Low temperature-induced systems failure in *Escherichia coli*: Insights from rescue by cold-adapted chaperones', *Proteomics*, **6**, 193–206.

SUCI P A, VRANY J D and MITTELMAN M W (1998), 'Investigation of interactions between antimicrobial agents and bacterial biofilms using attenuated total reflection Fourier transform infrared spectroscopy', *Biomaterials*, **19**, 327–39.

SUCKAU D, RESEMANN A, SCHUERENBERG M, HUFNAGEL P, FRANZEN J and HOLLE A (2003), 'A novel MALDI LIFT-TOF/TOF mass spectrometer for proteomics', *Anal Bioanal Chem*, **376**, 952–65.

TABB D L, FERNANDO C G and CHAMBERS M C (2007), 'MyriMatch: highly accurate tandem mass spectral peptide identification by multivariate hypergeometric analysis', *J Proteomics Res*, **6**, 654–61.

TAGA M E and BASSLER B L (2003), 'Chemical communication among bacteria', *Proc Natl Acd Sci U S A*, **100**, 14549–54.

TANAKA K, WAKI H, IDO Y, AKITA S, YOSHIDA Y and YOSHIDA T (1988), 'Protein and polymer analyses up to m/z 100 000 by laser ionization time-of-flight mass spectrometry', *Rapid Commun Mass Spectrom*, **2**, 151–3.

TANG Y, CHEN X, YOO J, YETHIRAJ A and CUI Q (2009), 'Numerical simulation of nanoindentation and patch clamp experiments on mechanosensitive channels of large conductance in *Escherichia coli*', *Exp Mech*, **49**, 35–46.

TANGWATCHARIN P, CHANTHACHUM S, KHOPAIBOOL P and GRIFFITHS M W (2006), 'Morphological and physiological responses of *Campylobacter jejuni* to stress', *J Food Prot*, **69**, 2747–53.

TAVERNA F, NEGRI A, PICCININI R, ZECCONI A, NONNIS S, RONCHI S, ET AL. (2007), 'Characterization of cell wall associated proteins of a *Staphylococcus aureus* isolated from bovine mastitis case by a proteomic approach', *Vet Microbiol*, **119**, 240–7.

TAYLOR B L and ZHULIN I B (1999), 'PAS domains: Internal sensors of oxygen, redox potential, and light', *Microbiol Mol Biol R*, **63**, 479–506.

THOMPSON M R, CHOUREY K, FROELICH J M, ERICKSON B K, BERKMOES N C and HETTICH R L (2008), 'Experimental approach for deep proteome measurements from small-scale microbial biomass samples', *Anal Chem*, **80**, 9517–25.

THROUP J P, ZAPPACOSTA F, LUNSFORD R D, ANNAN R S, CARR S A, LONSDALE J T, ET AL. (2001), 'The *srhSR* gene pair from *Staphylococcus aureus*: genomic and proteomic approaches to the identification and characterization of gene function', *Biochemistry*, **40**, 10392–401.

TIAN J, SHI C Y, GAO P, YUAN K L, YANG D W, LU X, ET AL. (2008), 'Phenotype differentiation of three *E. coli* strains by GC-FID and GC-MS based metabolomics', *J Chromatogr B*, **871**, 220–6.

TREMOULET F, DUCHE O, NAMANE A, MARTINIE B and LABADIE J C (2002), 'A proteomic study of *Escherichia coli* O157:H7 NCTC 12900 cultivated in biofilm or in planktonic growth mode', *FEMS Microbiol Lett*, **215**, 7–14.

TROST M, WEHMHONER D, KARST U, DIETERICH G, WEHLAND J and JANSCH L (2005), 'Comparative proteome analysis of secretory proteins from pathogenic and non-pathogenic *Listeria* species', *Proteomics*, **5**, 1544–57.

TURLIN E, SISMEIRO O, LE CAER J P, LABAS V, DANCHIN A and BIVILLE F (2005), '3-phenylpropionate catabolism and the *Escherichia coli* oxidative stress response', *Res Microbiol*, **156**, 312–21.

UNLU M, MORGAN M E and MINDEN J S (1997), 'Difference gel electrophoresis: a single-gel method for detecting changes in protein extracts', *Electrophoresis*, **18**, 2071–7.

VIJAYENDRAN C, POLEN T, WENDISCH V F, FRIEHS K, NIEHAUS K and FLASCHEL E (2007), 'The plasticity of global proteome and genome expression analyzed in closely related W3110 and MG1655 strains of a well-studied model organism, *Escherichia coli* K12', *J Biotechnol*, **128**, 747–61.

VILAIN S and BROZEL V S (2006), 'Multivariate approach to comparing whole-cell proteomes of *Bacillus cereus* indicates a biofilm-specific proteome', *J Proteome Res*, **5**, 1924–30.

VUILLET L, KOJADINOVIC M, ZAPPA S, JAUBERT M, ADRIANO J M, FARDOUX J, ET AL. (2007), 'Evolution of a bacteriophytochrome from light to redox sensor', *Embo J*, **26**, 3322–31.

WACEY D, KILBURN M R, MCLOUGHLIN N, PARNELL J, STOAKES C A, GROVENOR C R M, ET AL. (2008), 'Use of NanoSIMS in the search for early life on Earth: ambient inclusion trails in a c. 3400 Ma sandstone', *J Geol Soc*, **165**, 43–53.

WADHAMS G H and ARMITAGE J P (2004), 'Making sense of it all: bacterial chemotaxis', *Nat Rev Mol Cell Biol*, **5**, 1024–37.

WANG S Y, ZHU R, PENG B, LIU M M, LOU Y, YE X T, ET AL. (2006), 'Identification of alkaline proteins that are differentially expressed in an overgrowth-mediated growth arrest and cell death of *Escherichia coli* by proteomic methodologies', *Proteomics*, **6**, 5212–20.

WANG W, WHITE R and YUAN Z Y (2006), 'Proteomic study of peptide deformylase inhibition in *Streptococcus pneumoniae* and *Staphylococcus aureus*', *Antimicrob Agents Chemoth*, **50**, 1656–63.

WARSCHEID B and FENSELAU C (2004), 'A targeted proteomics approach to the rapid identification of bacterial cell mixtures by matrix-assisted laser desorption/ionization mass spectrometry', *Proteomics*, **4**, 2877–92.

WEBBER M A, COLDHAM N G, WOODWARD M J and PIDDOCK L J V (2008), 'Proteomic analysis of triclosan resistance in *Salmonella enterica* serovar Typhimurium', *J Antimicrob Chemoth*, **62**, 92–7.

WEBER A, KOGL S A and JUNG K (2006), 'Time-dependent proteome alterations under osmotic stress during aerobic and anaerobic growth in *Escherichia coli*', *J Bacteriol*, **188**, 7165–75.

WEEKS M E, JAMES D C, ROBINSON G K and SMALES C M (2004), 'Global changes in gene expression observed at the transition from growth to stationary phase in *Listeria monocytogenes* ScottA batch culture', *Proteomics*, **4**, 123–35.

WEHMHONER D, DIETERICH G, FISCHER E, BAUMGARTNER M, WEHLAND J and JANSCH L (2005), '"LANESPECTOR", a tool for membrane proteome profiling based on sodium dodecyl sulfate-polyacrylamide gel electrophoresis/liquid chromatography – tandem mass spectrometry analysis: application to *Listeria monocytogenes* membrane proteins', *Electrophoresis*, **26**, 2450–60.

WEINER J H and LI L (2008), 'Proteome of the *Escherichia coli* envelope and technological challenges in membrane proteome analysis', *BBA-Biomembranes*, **1778**, 1698–713.

WILKINS M R, PASQUALI C, APPEL R D, OU K, GOLAZ O, SANCHEZ J-C, ET AL. (1996), 'From proteins to proteomes: large scale protein identification by two-dimensional electrophoresis and amino acid analysis', *Nat Biotechnol*, **14**, 61–5.

WILKS S A, MICHELS H T and KEEVIL C W (2006), 'Survival of *Listeria monocytogenes* Scott A on metal surfaces: implications for cross-contamination', *Int J Food Microbiol*, **111**, 93–8.

WILLIAMS T L, MONDAY S R, EDELSON-MAMMEL S, BUCHANAN R and MUSSER S M (2005), 'A top-down proteomics approach for differentiating thermal resistant strains of *Enterobacter sakazakii*', *Proteomics*, **5**, 4161–9.

WILLIAMS T L, MUSSER S M, NORDSTROM J L, DEPAOLA A and MONDAY S R (2004), 'Identification of a protein biomarker unique to the pandemic O3:K6 clone of *Vibrio parahaemolyticus*', *J Clin Microbiol*, **42**, 1657–65.

WILSON I D, NICHOLSON J K, CASTRO-PEREZ J, GRANGER J H, JOHNSON K A, SMITH B W, ET AL. (2005), 'High resolution ultra performance liquid chromatography coupled to oa-TOF mass spectrometry as a tool for differential metabolic pathway profiling in functional genomic studies', *J Proteome Res*, **4**, 591–8.

WILSON J W, OTT C M, BENTRUP K H Z, RAMAMURTHY R, QUICK L, PORWOLLIK S, ET AL. (2007), 'Space flight alters bacterial gene expression and virulence and reveals a role for global regulator Hfq', *Proc Natl Acad Sci U S A*, **104**, 16299–304.

WINDER C L, DUNN W B, SCHULER S, BROADHURST D, JARVIS R, STEPHENS G M, ET AL. (2008), 'Global metabolic profiling of *Escherichia coli* cultures: an evaluation of methods for quenching and extraction of intracellular metabolites', *Anal Chem*, **80**, 2939–48.

WOLF C, HOCHGRAFE F, KUSCH H, ALBRECHT D, HECKER M and ENGELMANN S (2008), 'Proteomic analysis of antioxidant strategies of *Staphylococcus aureus*: diverse responses to different oxidants', *Proteomics*, **8**, 3139–53.

WRIGHT J A, GRANT A J, HURD D, HARRISON M, GUCCIONE E J, KELLY D J, ET AL. (2009), 'Metabolite and transcriptome analysis of *Campylobacter jejuni* in vitro growth reveals a stationary-phase physiological switch', *Microbiology*, **155**, 80–94.

WU L, LIN X M and PENG X X (2009), 'From proteome to genome for functional characterization of pH-dependent outer membrane proteins in *Escherichia coli*', *J Proteome Res*, **8**, 1059–70.

XIA X X, HAN M J, LEE S Y and YOO J S (2008), 'Comparison of the extracellular proteomes of *Escherichia coli* B and K-12 strains during high cell density cultivation', *Proteomics*, **8**, 2089–103.

XU C X, REN H X, WANG S Y and PENG X X (2004), 'Proteomic analysis of salt-sensitive outer membrane proteins of *Vibrio parahaemolyticus*', *Res Microbiol*, **155**, 835–42.

XU C X, LIN X M, REN H X, ZHANG Y L, WANG S Y and PENG X X (2006), 'Analysis of outer membrane proteome of *Escherichia coli* related to resistance to ampicillin and tetracycline', *Proteomics*, **6**, 462–73.

YANG Y C, YU H, XIAO D W, LIU H, HU Q, HUANG B, ET AL. (2009), 'Rapid identification of *Staphylococcus aureus* by surface enhanced laser desorption and ionization time of flight mass spectrometry', *J Microbiol Meth*, **77**, 202–6.

YOUNG N M, BRISSON J R, KELLY J, WATSON D C, TESSIER L, LANTHIER P H, ET AL. (2002), 'Structure of the N-linked glycan present on multiple glycoproteins in the Gram-negative bacterium, *Campylobacter jejuni*', *J Biol Chem*, **277**, 42530–9.

YU Y, MIZANUR R M and POHL N L (2008), 'Glycosidase activity profiling for bacterial identification by a chemical proteomics approach', *Biocat Biotransform*, **26**, 25–31.

ZHANG B, VERBERKMOES N C, LANGSTON M A, UBERBACHER E, HETTICH R L and SAMATOVA N F (2006), 'Detecting differential and correlated protein expression in label-free shotgun proteomics', *J Proteome Res*, **5**, 2909–18.

ZHANG D F, JIANG B, XIANG Z M and WANG S Y (2008), 'Functional characterisation of altered outer membrane proteins for tetracycline resistance in *Escherichia coli*', *Int J Antimicrob Agents*, **32**, 315–19.

ZIEBANDT A K, WEBER H, RUDOLPH J, SCHMID R, HOPER D, ENGELMANN S, ET AL. (2001), 'Extracellular proteins of *Staphylococcus aureus* and the role of SarA and sigma(B)', *Proteomics*, **1**, 480–93.

ZINNER E, AMARI S, GUINNESS R, NGUYEN A, STADERMANN F J, WALKER R M, ET AL. (2003), 'Presolar spinel grains from the Murray and Murchison carbonaceous chondrites', *Geochim Cosmochim Acta*, **67**, 5083–95.

ZYBAILOV B, MOSLEY A L, SARDIU M E, COLEMAN M K, FLORENS L and WASHBURN M P (2006), 'Statistical analysis of membrane proteome expression changes in *Saccharomyces cerevisiae*', *J Proteome Res*, **5**, 2339–47.

15

Virulotyping of foodborne pathogens

T. M. Wassenaar, Molecular Microbiology and Genomic Consultants, Germany

Abstract: Virulotyping is a special case of genotyping where the bacterial DNA that is to be detected or characterized contains virulence genes. This requires previous knowledge of the genes that determine the virulence of a given pathogen, which again depends on the definition of virulence and on the methods of how virulence genes are being identified. In the first part of this chapter, the definition of virulence and the identification of virulence genes are discussed, and theoretical advantages and disadvantages of virulotyping over other genotyping methods are outlined. In the second part, genotyping methods that are based on described virulence genes are summarized for the most important foodborne pathogens.

Key words: virulence factor, virulence gene, detection, virulotyping, genotyping.

15.1 Introduction

Genotyping is used to differentiate isolates within a species. Genotyping can characterize a single allele, e.g., when the presence or absence of a gene is assessed, in which case the outcome is binary. Alternatively, a single allele can be assessed for nature or sequence, so that the outcome is variable, as for instance a nucleotide sequence or a banding pattern. Multiple alleles can be combined, again as binary outcomes but more frequently as DNA sequences, as is the case for Multilocus Sequence Typing (MLST), or banding patterns, as in Pulsed-Field Gel Electrophoresis (PFGE). Banding patterns can be generated from multiple, variable loci of which the exact gene content is irrelevant, as with PFGE or Amplified Fragment Length Polymorphism (AFLP) (see Chapter 8). Virulotyping is a specific form of genotyping whereby the allele(s) of interest are virulence genes. The term was coined relatively recently (Timothy *et al.*, 2008), although the variation existing within and between virulence genes of bacterial populations has been exploited since the early days of PCR-dependent genotyping, which was first applied to streptococci (Gardiner *et al.*, 1995). In this chapter, the term virulotyping is restricted to those cases where the target sequences are known and specifically chosen, coding for factors with a proven or assumed role in virulence.

Genotyping and virulotyping are used for various reasons, such as outbreak investigations (are given cases of infection related to a common source?), source tracking (do particular genotypes enter the food chain through a particular animal species?), clinical microbiology (are particular genotypes typically causing more severe infections?) and risk analysis (should all isolates of a given species or subtype be regarded with equal risks?). In all these given examples of applications it is of key importance to recognize isolates that are 'equal' and differentiate these from isolates that are 'different'. The difficulty is, though, to know how equal isolates must be to be screened as such (Wassenaar, 2003).

Bacterial isolates of a common ancestor can be considered 'equal' as long as genetic events have not changed the genetic repertoire of the offspring. Such genetic events, or mutations, can be DNA insertions, deletions, transitions including inversions or single point mutations. As long as such mutations have not taken place, all offspring of a common ancestor should be determined as being equal and identical to the ancestor, and ideally be distinguishable from isolates of a different ancestor. In case a typing method fails to differentiate populations from distinct ancestors, the applied method does not have sufficient discrimination power to discriminate clonal lineages, although it may still be sufficiently discriminatory to address specific questions. When, on the other hand, differences (due to mutations) are detected within a population derived from a common ancestor, whilst these mutations are not considered significant for the question to be addressed, the discrimination could be considered too high, as commonness would be missed. This is illustrated in Fig. 15.1.

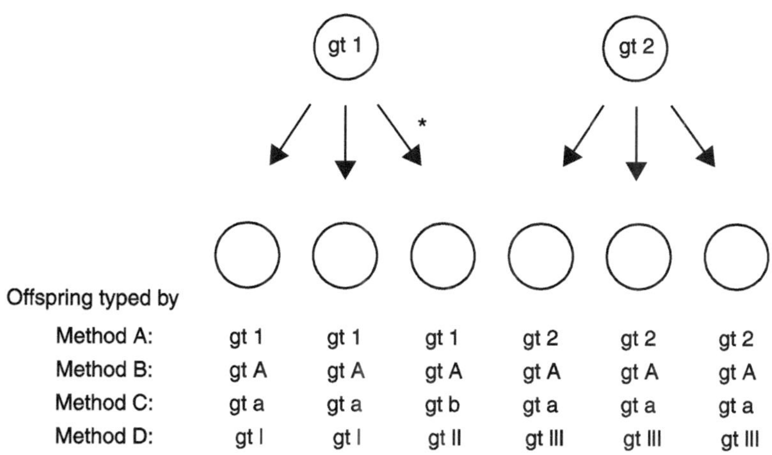

Fig. 15.1 Schematic representation of two bacterial populations, derived from two different ancestors ('gt 1' and 'gt 2'), whose offspring is typed by four different hypothetical methods. The asterisk indicates a genetic mutation introduced in this offspring. Method A recognizes two genotypes but misses the introduced genetic mutation. Method B lacks discriminatory power and detects only one genotype. Method C also detects two genotypes but the grouping based on this result would be incorrect, whereas Method D has such a high discrimination power that it misses the relatedness within lineage gt 1.

False grouping can occur when a gene or allele is subjected to horizontal gene transfer, especially when this gene is the only typing target. These difficulties challenge any genotyping method, and virulotyping is no exception. Virulotyping has particular advantages over other methods of genotyping, but there are also difficulties to overcome, and even disadvantages, as will be discussed in this chapter. Before considering what genes should be targeted for virulotyping of specific food pathogens, we need to realize how virulence is defined and how virulence genes are identified.

15.2 Defining and identifying virulence genes

The definition of virulence genes builds on the definition of virulence, which was proposed by Robert Koch in his famous postulates of 1890. For application of experimental genetic methodology, Finlay and Falkow (1989) modified Koch's postulates into four new molecular postulates to define 'virulence genes':

1 The phenotype or property under investigation should be associated with pathogenic members of a genus or pathogenic species of a strain.
2 Specific inactivation of the gene(s) associated with the expected virulence trait should lead to a measurable loss in pathogenicity or virulence.
3 Reversion or allelic replacement of the mutated gene should lead to restoration of pathogenicity.
4 If such genetic manipulations are not possible, specific antibodies against the product(s) of the investigated gene(s) should neutralize pathogenicity.

These postulates are easy to envisage for a hypothetical pathogen that relies on its virulence by means of toxin production: the toxin gene would be found in that pathogen only, inactivating the toxin gene would result in loss of virulence and repairing the inactivation in the knock-out mutant would restore virulence. In case inactivation of the toxin gene would be lethal to the bacteria, antibodies that neutralize the toxin could give evidence for the fourth postulate. Although these requirements thus seem quite appropriate to define and identify virulence genes, their strict application can easily result in both false-positive and false-negative findings. Moreover, genes are nowadays more frequently identified from genome sequences than from tedious wet lab experimental work, in which case sequence homology is interpreted to imply similarity in function. In a novel genome sequence, a predicted gene that shows a significant similarity to a previously described virulence gene (of the same or of another species) would be annotated as a 'putative virulence gene'. This leads to many genes in sequence databases to be annotated as functional in virulence by inferred function, but it ignores the fact that only a specific combination of virulence genes, many of which need to be expressed under specific regulation, would typically lead to pathogenicity (Falkow, 1997). Moreover, experimental horizontal transfer of a particular gene can result in unexpected phenotypes (e.g., Susanna *et al.*, 2006) and likewise, unpredictable functional changes may also result from natural transfer of genes.

Therefore, prediction of gene function based on sequence similarity across species remains tricky even without the additional effects of evolutionary mutations and adaptation of genes to fit novel roles and functions. It is risky to assume conservation of function if two proteins share a degree of sequence similarity without further experimental evidence.

As an example of a false-positive outcome of the molecular Koch's postulates, consider a gene whose inactivation results in loss of virulence (the mutant can no longer cause disease) because it is no longer adequately equipped for its niche in the host where it causes disease. Examples would be inactivation of regulatory genes, attachment genes, metabolic genes, etc. There are numerous examples where 'virulence genes', identified by gene inactivation (e.g., via transposon mutagenesis) that resulted in a decrease in virulence, were later recognized as regulatory genes that (co)regulate true virulence genes (e.g., *virF* in *Shigella*, Mills *et al.*, 1992). This author would prefer the more precise descriptive function 'virulence regulatory gene' in such a case, and to reserve the term 'virulence gene' for those genes encoding factors that interact with, and are directly responsible for the damage to the host (Wassenaar and Gaastra, 2001).

False-negative results (true virulence genes are missed in an experimental approach) could result from pathogenic mechanisms that are multifactorial and coded by redundant genes: inactivation of one gene would not necessarily impair virulence properties in such a case. The evolutionary advantage of a virulent phenotype can be high, so that a degree of redundancy has frequently arisen as a result of selective pressure. The recently reported cytotoxic effects observed in *Campylobacter jejuni*, that were independent of cytolethal distending toxin (CDT), exemplify such redundancy (Kalischuk *et al.*, 2007). In cases like these, experimental identification of genes with a role in virulence can be problematic (Craven and Neidle, 2007; Wyckoff *et al.*, 2007; Seleem *et al.*, 2008). The requirement of complementation (postulate 3) is also imprecise. For example, even under simplified *in vitro* conditions, a presumably straightforward process such as bacterial invasion is driven and regulated by several genes and genetic loci. The destruction of a single gene may diminish virulence but its transfer into a non-virulent organism is unlikely to induce invasiveness; in most cases large DNA segments containing multiple genes are required to transfer virulence properties to a non-virulent recipient (e.g., O'Gorman *et al.*, 1996).

Concentrating on immunological methods (postulate 4) does not necessarily lead to clear-cut results either: many virulence factors are immunogenic but the argument cannot be reversed. Immunogenic surface structures frequently do not play a role in virulence, in that they are not involved in the damage the organism does to its host. For instance, the major outer membrane protein of *C. jejuni* is immunogenic (Huang *et al.*, 2007) but functions as a porin and not as a virulence factor (Khan *et al.*, 2005). Nor are all virulence factors immunogenic, as some may be shielded from antibodies by external membrane structures or by their specific composition (e.g., Pfleger *et al.*, 2008).

Ignoring the above-mentioned complication of presumed functional conservation inferred by sequence similarity, how valid is a finding of significant homology to a gene in a database annotated as a virulence gene? Can we trust that annotation? As already mentioned, annotated functions frequently have not been tested in the lab, but are based on homology to other genes. GenBank contains many of so-called 'putative' virulence genes that were annotated based on long trails of similarity. Without further investigation and experimental evidence such annotation should be interpreted with care. In general, a GenBank entry that is based on wet lab evidence is seen to be most reliable, but even in these files mistakes can be present. There are numerous examples of genes annotated on the wrong strand (in case the reading frame was open in both directions), or even of 'wrong DNA' being sequenced (for instance vector DNA instead of the cloned insert), and when such mistakes are not corrected they can 'contaminate' the database with wrongly annotated sequences. Although these problems are recognized, a recent proposal to allow community-driven corrections in GenBank was not welcomed (Pennisi, 2008).

Another complication arises if a suitable animal model to mimic disease is absent. In that case *in vitro* models provide a substitute model system but that may be unsatisfactory. Taking *C. jejuni* as an example, multiple *in vitro* models have been applied to measure invasiveness or toxicity of these bacteria to tissue culture cells, but there is no clear evidence that either mechanism is of importance in clinical manifestations. For example, the majority of clinical isolates are only weakly invasive when tested *in vitro* (Fearnley *et al.*, 2008). Non-invasive or non-toxigenic *C. jejuni* are found amongst clinical isolates, although it is possible that these pathogenic mechanisms might be mutually exclusive in particular isolates.

As a result of these various approaches, the label 'virulence' is very commonly added to genes and this inflated the term somewhat. As previously proposed (Wassenaar and Gaastra, 2001), it would be helpful to distinguish true virulence genes from virulence-associated genes in the following way: *true virulence genes* are present only in pathogenic bacteria. Their products have to be involved in interactions with the host and be directly responsible for the damage to that host. They must be absent in commensal bacteria. *Virulence-associated genes* specifically regulate, modify, secrete or otherwise support the activity and/or expression of virulence genes or their products. An example would be type III secretion genes, which encode a secretion machinery that can specifically secrete true virulence factors. Virulence-associated genes can have counterparts in commensal bacteria, where they may regulate or support expression of genes other than true virulence genes (for instance a type III secretion system is present in an insect endosymbiont; Dale *et al.*, 2005).

Opportunistic pathogens live as commensals in a healthy host, but only cause infections under specific conditions, such as a breakdown in immune defense, when underlying diseases are present (cystic fibrosis), or when they are introduced to unnatural sites (intubation). Under those conditions they cause disease, and rely on genes to do so. Under the strict definition given above, opportunistic pathogens cannot contain true virulence genes; instead, the genes involved in virulence under specific circumstances only could be called *virulence life-style genes*, which form a third class in our division. This class would also contain genes whose product help pathogens to colonize the site where they can cause damage, as for instance specific fimbriae, that they may share with commensal bacteria. One could argue, from a clinical point of view, that clinically relevant antibiotic resistance genes are also part of the virulence life-style gene complement, as such genes would, for instance, tell a Methicillin-resistant *Staphylococcus aureus* (MRSA) apart from a susceptible *S. aureus*, which is much easier to treat. This example illustrates how genes that are not generally described as virulence genes can have a major effect on the outcome of infection and the severity of disease, and further illustrates that it is useful to distinguish virulence life-style genes from true virulence genes.

As the opportunistic pathogens exemplify, disease is one of several possible outcomes of an interplay between bacteria and host. The host species or its immune status can determine whether colonization progresses into disease, or the infection remains asymptomatic. A clinical microbiologist would consider *E. coli* O157:H7 a pathogen, but to a veterinarian it is a commensal to cattle (up to 74% of dairy cattle can shed *E. coli* O157:H7 asymptomatically, Hussein and Sakuma, 2005). Similarly, *C. jejuni* is pathogenic to humans but it is a commensal to poultry, and *Campylobacter coli* does not cause disease in pigs where it is frequently detected. It is as yet unknown whether the same genes are involved in asymptomatic carriage in one host and symptomatic disease in another, or whether different genes are at play; if the first applies, this further complicates the labeling of genes related to virulence, as the host species in which the bacteria multiply then defines whether a gene is a virulence gene or not.

In conclusion, we can safely state that the definition of virulence genes is controversial and definitions or means of identification used are far from universal. Nevertheless, there are a large number of genes in the public domain called 'virulence genes' and most, many or some of these (depending on the personal preferences of the used definition) could be targets for virulotyping.

15.3 Virulotyping: advantages and disadvantages

The advantages of virulotyping seem obvious: since there is a direct correlation between the detected genes and the virulence potential of the organism, it would be straightforward to identify where the biggest risk is for human health. This could allow more accurate source attribution, more effective control measures or more relevant microbial risk analysis. At present, our knowledge on bacterial populations that persist in the food chain is biased: particular selection pressures applied during food processing potentially enrich the bacterial population, but an unknown fraction of these are (more) virulent, as the two properties (persistence and virulence) may not be linked. Detection of the virulent members of a population gives more reliable information on the true impact on human health and since non-pathogenic isolates remain undetected, the outcome of detection would be more relevant.

Virulotyping could potentially revolutionize bacteriological detection, when virulence genes rather than virulent organisms are detected. When virulence genes are determined by PCR or other methods that specifically detect DNA, bacterial culture is principally not necessary and detection can be fast and sensitive. In practice, due to the presence of inhibitory factors, PCR is not very sensitive when directly applied to food matrices. Enrichment procedures are often necessary to improve detection limits, decrease inhibitor concentrations and prevent detection of dead bacteria, but they reduce the presumed speed of PCR detection considerably. On the other hand, without a bacteriological culture, the species from which detected virulence genes were derived cannot always be precisely determined, antibiograms cannot be carried out and reference material cannot be stored for future experiments. Such considerations have largely prevented the routine application of PCR for detection of DNA without bacteriological culture. Nevertheless, promising examples of PCR detection of virulence genes directly on food (following enrichment) have been described, for instance to detect botulinum (De Medici *et al.*, 2009) or Verotoxin in beef by LAMP (Hara-Kudo *et al.*, 2007).

Virulotyping can be advantageous over other methods of genotyping, but this is not necessarily so: it depends on the questions addressed, and also on the type of organism under study. If the presence or absence of virulence genes correlates neatly with other markers that can be targeted by genotyping using previously standardized, portable, easy-to-use or faster/cheaper methods, there is little benefit to switch to virulotyping. Most Shiga-toxin (Verotoxin) producing *E. coli* (STEC) strains are of the serotype O157 (although both non-toxigenic O157 *E. coli* and verotoxigenic non-O157 strains exist) which forms characteristic colorless colonies on sorbitol-MacConkey agar, due to a lack of sorbitol fermentation. This characteristic is frequently used as a relatively simple predictor of the O157 serotype, and indirectly of potential virulence (though sorbitol-fermenting non-O157 STEC would thus be missed). Zooming in at the virulence genes of STEC, independent of their serotype, however, has revealed a large variation in their presence or absence, although common patterns could be recognized between the O157 and the, presumably independently evolved, other STEC serotypes (Ogura *et al.*, 2007). This observed diversity would only complicate identification of a virulent sample, if that is all that were needed; a simpler test for presence of one telltale gene, such as *tir*, could then suffice (Bono, 2009). Assessing the diversity of virulence genes could, however, provide a valuable tool to differentiate outbreak cases from background sporadic cases, or to identify the relative attribution of various food sources. Phage typing is still being applied to *Salmonella enterica* because its outcome is highly predictive of virulent potential, and different serotypes of this organism correlate neatly with either virulence or animal source or both. In such cases, virulotyping would only be needed if more information is essential. Various successive *Salmonella* serotypes or phagetypes have predominated in the human population over time (e.g., Enteritidis, Typhimurium DT104 or the recently emerged monophasic serotype 4,5,12:i:-), and although their virulence gene repertoire was found variable to some degree within a given type, the phenotypic differentiation proved suitable to define intervention strategies in a farm-to-fork approach. Whether a more refined picture, based on virulence genes, would have led to more effective intervention strategies is disputable.

Virulotyping also has disadvantages. Presence of a gene does not guarantee it is actually expressed, and thus non-expressed virulence genes can result in false-positive results. A virulence gene can be present but not expressed due to mutations that inactivate the gene (whereby the gene but not the mutation is detected by PCR). For instance, occasionally an isolate of *C. jejuni* is found to contain, but not express, CDT (Eyigor *et al.*, 1999). Virulence genes are usually heterogeneous within a bacterial species and generally have a relatively high mutation rate, as they are under high selective pressure. The discriminatory power of virulotyping can be so high that offspring of a single ancestor may no longer be recognized as such (see Fig. 15.1). When increased discrimination power breaks down sought correlations, high resolution is no longer beneficial. An additional disadvantage is that in case a bacterial population is replaced by a population with a different genetic repertoire, the detection methods based on virulence genes may have to be altered as well. Furthermore, the detection of virulence genes without further knowledge of the organism is risky, as many of these genes are known to be subject to horizontal gene transfer. The genetic context in which a transferred virulence gene is accepted may be different from that of the original donor which, without further knowledge of the genomic background, could remain unnoticed – with possible consequences to the virulence phenotype. Finally, virulence genes are sometimes found present in non-virulent organisms, even if detection is restricted to 'true virulence genes'. For instance, several virulence factors of pathogenic *E. coli* were found in stools of healthy children and their presence and expression fluctuated with time (Schlager *et al.*, 2003).

Most virulotyping methods are based on either multiplex PCR or microarray analysis. Either approach usually produces a binary outcome for multiple genes, which is interpreted as indicative of presence or absence (divergence is usually ignored) of that particular gene. For the interpretation of the results, binary outcomes of individual genes are usually combined and equally weighed when analyzed. This is an over-simplification, as the presence or absence of different virulence genes is frequently linked, so that their individual detection does not report truly independent events. Microarray DNA analysis can reveal information on gene divergence as well as on gene presence/absence, provided sufficient probes are represented on the chip to assess the true biological diversity of that gene in the examined population. Although microarray RNA analysis overcomes detection of non-transcribed genes, the outcome now depends on the growth conditions applied to the culture from which RNA was isolated. It is assumed that all virulent isolates of a given species respond in a similar manner to external signals, so that standardized growth conditions should produce standardized results. In view of the large variation existing in virulence gene presence and their regulation, it remains to be seen whether this assumption is correct.

15.4 Examples of specific pathogens

15.4.1 *Escherichia coli*

A classical example of a bacterial pathogen in need of virulotyping is *E. coli*, where commensals need to be distinguished from serious pathogens. In particular, Shiga-toxin producing *E. coli* (STEC, alternatively called VTEC for Verotoxin *E. coli* or EHEC for Enteroheamorhagic *E. coli*) is implicated with serious bloody diarrhea and hemolytic-uremic syndrome (HUS), a potentially lethal complication. EHEC is mostly transmitted through food, in particular (but not exclusively) through contaminated beef. Outbreaks are frequent, and sometimes of an extensive scale. EHEC is frequently of serotype O157:H7, though over 430 EHEC serotypes have been isolated from cattle (Gyles, 2007). Even within the major O157:H7 serotype there is a wide variation in genetic background (Gyles, 2007). Some of that variation can be linked to virulence, for instance which of the two toxin (*stx*) genes is present, and which subtype thereof (Erickson and Doyle, 2007). However, the variation within O157:H7 is much more substantial than just a few alleles. This variation can actually be used for strain identification in outbreak situations by a number of genotyping techniques (reviewed in Gyles, 2007). Moreover, for food safety management and clinical predictions typing of the virulence genes directly would be advantageous over serotyping or genotyping.

The major virulence genes of EHEC are well characterized. Stx production is essential but not sufficient for EHEC virulence. In addition, the locus of enterocyte effacement (LEE) pathogenicity island, which contains a type III secretion system as well as the genes responsible for the typical 'attaching and effacing' cell lesions, is needed. The large plasmid carrying the alpha hemolysin *(hlyA)* locus is a further requirement (Caprioli *et al.*, 2005). Ideally, virulence typing of EHEC would thus at least target *hlyA*, LEE genes, such as *eae*, and the *stx* genes. When detection of these and other genes was compared by PCR and hybridization, the latter performed better for genes present on plasmids (Gerrish *et al.*, 2007). When PCR detection fails but hybridization detects a gene, diversity in the PCR primer sequences is the most likely cause. This illustrates the difficulty in designing a simple, rapid but also reliable test for detection of highly variable virulence genes.

15.4.2 *Salmonella enterica*

In case of *Salmonella* spp., there is no need to differentiate commensal from pathogenic bacteria, but many serovars of *Salmonella* that can be detected in food are relatively apathogenic to humans (Foley and Lynne, 2008). Moreover, amongst the pathogenic serotypes, the diversity in genetic repertoire is high, and maybe virulotyping of this organism is more likely to complicate and confuse matters than to improve current practices. Within the given serovar *S. enterica* Typhimurium, five *Salmonella* Pathogenicity Islands (SPIs) have been recognized, of which SPI-1 and SPI-2 have been described to correlate with virulence (Gerlach and Hensel, 2007; Rychlik *et al.*, 2009). Extensive variation in virulence genes exists within the species, for instance 20 virulence genes were found to vary widely between 114 *Salmonella* isolates that covered 38 serovars (Khoo *et al.*, 2009).

These target genes were all chromosomal, but virulence genes can also reside on plasmids (e.g., *spv*), which frequently also carry antimicrobial resistance genes that can be of clinical relevance (Herrero *et al.*, 2009). Imaginably, the genetic variation of plasmids can be even larger than that detected with chromosomal markers (Baggesen *et al.*, 2000). Moreover, these plasmids can be highly transferable (Guerra *et al.*, 2002). To complicate things further, chromosomal *Salmonella* Genome Island 1 (SGI1), responsible for the resistance phenotype ACSSuT, can be mobilized by a plasmid, so that its presence may no longer correlate with serotype or phage type (Doublet *et al.*, 2005). Detection of the resistance type ACSSuT, which is indicative for the presence of SGI1, combined with detection of the *spv* virulence locus would most likely point to serovar Typhimurium phage type DT104, but other phage types can carry these genes, too (Chiu *et al.*, 2006).

Recently, it was shown that despite the mobile nature of these genes, presence of particular resistance or virulence genes is distributed in a non-random manner over MLST sequence types (Wiesner *et al.*, 2009). The observed non-random distribution of these 'accessory genes' illustrates that presence or absence of such genes, when assessed by virulotyping, should not be interpreted as completely independent events.

To put this observed degree of variation into perspective, recent virulotyping work by a European collaboration must be mentioned here (Huehn *et al.*, 2010). They investigated gene presence by PCR for five genes indicative of the five SGIs, three independent prophage genes, one virulence plasmid gene and a fimbrial cluster marker, in a large set of strains. Variation within the serovar Typhimurium was observed with two of the three prophages and with the virulence plasmid gene. Variation within Enteritidis was found for only one prophage gene and the plasmid gene. Only two Infantis and one 'Virchow' strain displayed variation in SPI-1. Analysis of this gene set was carried out under the assumption that their presence was unlinked, which in this choice of selected genes is a safe assumption. However, subsequently 275 genes were analyzed in depth by microarray hybridization and the results were analyzed by UPGMA. It is questionable if that algorithm is suitable to analyze presence or absence of signals that may (in part) be linked. The presented results showed that strains belonging to the same serovar grouped together, indicating that the broader virulence-associated gene complement corresponded with the serovar (Huehn *et al.*, 2010). The achieved increased resolution within serovars, as obtained by this virulotyping, might improve, or largely obfuscate the epidemiological insights into *Salmonella*, which time will tell.

15.4.3 *Campylobacter jejuni* and *C. coli*

The species *C. jejuni* and *C. coli* are pathogenic to humans; exceptions to this rule are as yet not recognized. Although some MLST genotypes (e.g., from wild birds) are hardly ever observed in human clinical cases (Sheppard *et al*., 2009), this may be caused by lack of exposure routes, rather than lack of virulence – though the latter cannot be excluded. It is currently not known if non-pathogenic *C. jejuni/coli* (*Campylobacter* for short) exists, nor is it known if every *Campylobacter* has the potential to cause the whole disease spectrum that can be observed. Analogous to the situation in many other food pathogens, it could be expected that gene content correlates, to some extent, with virulence potential, though the evidence to support this view is still inconclusive. If it turns out to be the case, virulotyping should differentiate more virulent types from less, or non-virulent types. If, however, all *Campylobacters* would be able to cause disease (whereby the outcome would then presumably be due to differences in doses and host-related factors), virulotyping is no longer advantageous, and differentiation could be obtained with any variable gene or genes of choice.

Several publications investigated a possible correlation between *in vitro* determined virulence properties (adhesion, invasion, translocation, toxin production) and isolation source (clinical versus animal or environmental isolates), with conflicting results that may partly be explained by low numbers or biased selection of investigated isolates. In a study including 113 isolates, clinical isolates were not found more invasive or toxigenic on average than isolates from other sources, though strongly invasive strains were over-represented in clinical isolates (Fearnley *et al*., 2008). There is also no clear correlation between *in vitro* virulence and genotype (determined by whatever method), though again conflicting observations have been published on this subject.

The clearest correlation between bacterial properties and clinical outcome is found for the post-infectious neuropathies, such as Guillain–Barré Syndrome (GBS) and Miller Fisher Syndrome (MFS), which correlate with some subtypes of *C. jejuni*. Although originally a correlation with specific serotypes (e.g., O19) was observed, this could not be confirmed in larger studies (Endtz *et al*., 2000). More precisely, particular types of lipooligosaccharide (LOS) can immunologically mimic neural gangliosides, resulting in pathological cross-reactivity. In this case, genetic detection of these traits could have predictive clinical relevance, though it should be noted that GBS or MFS only result in a small minority of infections caused by Campylobacters expressing these LOS types. Most likely, both bacterial and host factors determine whether an infection results in post-infectious neuropathies. Moreover, LOS is not directly coded by genes, but produced by enzymes in a multi-step process, which makes it harder to identify those genes responsible for the formation of a particular GBS-inducing LOS. Nevertheless, combinations of particular LOS biosynthesis genes were found to correlate with *in vitro* models of GBS, such as reaction to GBS-patient's sera (Yuki and Odaka, 2005).

A correlation between LOS-producing genes and other *in vitro* virulence models was investigated by Habib and co-workers who assessed the variation in LOS genes by PCR and compared these to *in vitro* invasiveness (Habib *et al*., 2009). The investigated isolates were sorted in four LOS categories that were shown to differ in *in vitro* invasiveness, and the LOS type correlated to their

MLST clonal complex (Habib *et al.*, 2009). Using microarray hybridization, multiple genes can be detected at the same time, and with the use of a 70-mer chip, *C. jejuni, C. coli* and *Arcobacter butzleri* could be detected and characterized without PCR amplification (Quiñones *et al.*, 2007). The chip included probes for the highly variable LOS biosynthesis genes and the surprising finding was that many poultry isolates were identified as LOS type B, which is implied in GBS and MFS. The high detection rate of this LOS type contradicts the low frequency of these sequelae, illustrating that these clinical outcomes are multifactorial.

In Unicomb *et al.* (2008), clinical outcome was evaluated in a case-case comparison whereby strains were typed for their flagellin gene. A genotype designated '*flaA-2* type' correlated with more serious cases. Since flagellin is described as a virulence factor (non-motile bacteria do not colonize and are not invasive), it would be tempting to assume that this particular flagellin is the basis of this increased virulence. However, this flagellin type could be a marker for a subpopulation whose increased virulence is due to other genes, a possibility that was not discussed by the authors.

15.4.4 *Listeria monocytogenes*

Human foodborne listeriosis cases occur at a lower frequency than *Salmonella, E. coli* and *Campylobacter* infections, and occur both as sporadic cases and as outbreaks. Invasive listeriosis, though rare, has a high fatality. The geographical scattering of cases, combined with a large within-species diversity of both pathogenic and non-pathogenic strains, challenges reliable source attribution and identification of transmission routes. Many virulence-associated genes have been described for *L. monocytogenes*, but the differentiation between virulent and non-virulent isolates remains problematic. A multiplex PCR for detection of the internalin genes *inlA, inlC* and *inlJ* has been developed (Liu *et al.*, 2007). A set of five different genes (*plcA*, *prfA*, *hlyA*, *actA* and *iap*) were analyzed by PCR from bovine mastitis-derived samples, but the number of analyzed samples was too low to draw conclusions (Rawool *et al.*, 2007). Although the virulence regulatory gene *prfA* is specific for the species and essential for virulence (Freitag *et al.*, 2009), it can also be found in strains with a low virulence phenotype, in which case it is either not expressed or mutated (Roche *et al.*, 2009). Either way, a simple PCR detection of *pfrA* is not sufficient as a marker for virulence, and molecular detection of virulence genes regulated by *pfrA* (*inlABC*, *actA*, *plcAB*, *hly* and others) will not improve this predictive value. It seems that despite a reasonable understanding of its pathogenicity, reliable virulotyping of *L. monocytogenes* is not yet feasible.

15.5 Future trends

Currently, virulotyping has the most potential when carried out by microarray analysis, as this can analyze multiple genes in one step, but the technique is costly and elaborate. Modern chips can easily cover the genetic content of several bacterial genomes, but their production is costly, the hybridization is a technical challenge and the interpretation of the resulting data is a profession on its own. A cheap and simple alternative is multiplex PCR, with or without sequencing of the product. Here, conservation of primer sequences is essential for detection, which limits its application.

In the future, sequencing complete genomes may become available, when the cost of sequencing continues to decrease and throughput continues to improve. Consider the differentiation that will be possible if every bacterial isolate can be completely sequenced in approximately 30 minutes, at a cost of two dollar or so. When technical development continues the way it does, this may be envisaged within the next decade (Anonymous, 2008).

The differentiation we could then apply to bacterial isolates is so enormous that a whole novel picture would arise. The genetic variation within a species can be quite extensive, and is not restricted to a few point mutations: here we are dealing with variation in gene absence, presence and divergence. The bacterial lineages within a species as recognized by MLST will be challenged when whole genome sequences will show the true diversity between isolates sharing identical MLST types, as preliminary data are already indicating (T.M. Wassenaar and D.W. Ussery, unpublished observations). A bacterial 'species pan-genome' that is artificially constituted of any gene that can possibly be detected in a particular isolate of that species was estimated ten times bigger than the number of genes per individual genome (Snipen *et al.*, 2009). With this degree of variation it will be difficult if not impossible to choose which genes should be regarded as key importance, in terms of virulence, and which can safely be ignored. Our current view of virulence may well be an extreme over-simplification, due to our limited knowledge of the true genetic diversity within a species. In ten years time from now, we may have dropped the concept of virulence genes completely, and instead may be assessing gene combinations instead of individual genes.

Whether the concept of virulence genes remains steadfast or might be completely overhauled, our insight of bacterial virulence is likely to improve over the next decade. The outcome of ongoing research may have direct therapeutic consequences. The development of novel antimicrobial drugs that specifically target virulence genes is a promising approach (Lynch and Wiener-Kronish, 2008). The possibility to design antibiotics that might kill pathogens and leave commensals undisturbed is very promising. This future application is all the more reason to define virulence genes with care.

15.6 References

ANONYMOUS (2008), 'Prepare for the deluge', *Nat Biotechnol*, **26**: 1099.

BAGGESEN D L, SANDVANG D and AARESTRUP F M (2000), 'Characterization of *Salmonella enterica* serovar typhimurium DT104 isolated from Denmark and comparison with isolates from Europe and the United States', *J Clin Microbiol*, **38**(4): 1581–6.

BONO J L (2009), 'Genotyping *Escherichia coli* O157:H7 for its ability to cause disease in humans', *Curr Protoc Microbiol*, Chapter 5: Unit 5A.3.

CAPRIOLI A, MORABITO S, BRUGÈRE H and OSWALD E (2005), 'Enterohaemorrhagic *Escherichia coli*: emerging issues on virulence and modes of transmission', *Vet Res*, **36**(3): 289–311.

CHIU C H, SU L H, CHU C H, WANG M H, YEH C M, WEILL F X and CHU C (2006), 'Detection of multidrug-resistant *Salmonella enterica* serovar Typhimurium phage types DT102, DT104, and U302 by multiplex PCR', *J Clin Microbiol*, **44**(7): 2354–8.

CRAVEN S H and NEIDLE E L (2007), 'Double trouble: medical implications of genetic duplication and amplification in bacteria', *Future Microbiol*, **2**: 309–21.

DALE C, JONES T and PONTES M (2005), 'Degenerative evolution and functional diversification of type-III secretion systems in the insect endosymbiont *Sodalis glossinidius*', *Mol Biol Evol*, **22**(3): 758–66.

DE MEDICI D, ANNIBALLI F, WYATT G M, LINDSTRÖM M, MESSELHÄUSSER U, ALDUS C F, DELIBATO E, KORKEALA H, PECK M W and FENICIA L (2009), 'Multiplex PCR for detection of botulinum neurotoxin-producing clostridia in clinical, food, and environmental samples', *Appl Environ Microbiol*, **75**(20): 6457–61.

DOUBLET B, BOYD D, MULVEY M R and CLOECKAERT A (2005), 'The *Salmonella* genomic island 1 is an integrative mobilizable element', *Mol Microbiol*, **55**(6): 1911–24.

ENDTZ H P, ANG C W, VAN DEN BRAAK N, DUIM B, RIGTER A, PRICE LJ, WOODWARD D L, RODGERS F G, JOHNSON W M, WAGENAAR J A, JACOBS B C, VERBRUGH H A and VAN BELKUM A (2000), 'Molecular characterization of *Campylobacter jejuni* from patients with Guillain–Barré and Miller Fisher syndromes', *J Clin Microbiol*, **38**(6): 2297–301.

ERICKSON M C and DOYLE M P (2007), 'Food as a vehicle for transmission of Shiga toxin-producing *Escherichia coli*', *J Food Prot*, **70**(10): 2426–49.

EYIGOR A, DAWSON K A, LANGLOIS B E and PICKETT C L (1999), 'Detection of cytolethal distending toxin activity and *cdt* genes in *Campylobacter* spp. isolated from chicken carcasses', *Appl Environ Microbiol*, **65**(4): 1501–5.

FALKOW S (1997), 'Perspectives series: host/pathogen interactions. Invasion and intracellular sorting of bacteria: searching for bacterial genes expressed during host/pathogen interactions', *J Clin Invest*, **100**(2): 239–43.

FEARNLEY C, MANNING G, BAGNALL M, JAVED M A, WASSENAAR T M and NEWELL D G (2008), 'Identification of hyperinvasive *Campylobacter jejuni* strains isolated from poultry and human clinical sources', *J Med Microbiol*, **57**(5): 570–80.

FINLAY B B and FALKOW S (1989), 'Common themes in microbial pathogenicity', *Microbiol Rev*, **53**(2): 210–30.

FOLEY S L and LYNNE A M (2008), 'Food animal-associated *Salmonella* challenges: pathogenicity and antimicrobial resistance', *J Anim Sci*, **86**(14): E173–87.

FREITAG N E, PORT G C and MINER M D (2009), '*Listeria monocytogenes* – from saprophyte to intracellular pathogen', *Nature Rev Microbiol*, **7**(9): 623–8.

GARDINER D, HARTAS J, CURRIE B, MATHEWS J D, KEMP D J and SRIPRAKASH K S (1995), 'Vir typing: a long-PCR typing method for group A streptococci', *PCR Methods Appl*, **4**(5): 288–93.

GERLACH R G and HENSEL M (2007), 'Salmonella pathogenicity islands in host specificity, host pathogen-interactions and antibiotics resistance of *Salmonella enterica*', *Berl Munch Tierarztl Wochenschr*, **120**(7/8): 317–27.

GERRISH R S, LEE J E, REED J, WILLIAMS J, FARRELL L D, SPIEGEL K M, SHERIDAN P P and SHIELDS M S (2007), 'PCR versus hybridization for detecting virulence genes of enterohemorrhagic *Escherichia coli*', *Emerg Infect Dis*, **13**(8): 1253–5.

GUERRA B, SOTO S, HELMUTH R and MENDOZA M C (2002), 'Characterization of a self-transferable plasmid from *Salmonella enterica* serotype typhimurium clinical isolates carrying two integron-borne gene cassettes together with virulence and drug resistance genes', *Antimicrob Agents Chemother*, **46**(9): 2977–81.

GYLES C L (2007), 'Shiga toxin-producing *Escherichia coli*: an overview', *J Anim Sci*, **85**(13): E45–62.

HABIB I, LOUWEN R, UYTTENDAELE M, HOUF K, VANDENBERG O, NIEUWENHUIS E E, MILLER WG, VAN BELKUM A and DE ZUTTER L (2009), 'Correlation between genotypic diversity, lipoologosaccharide locus class variation and Caco-2 invasion potential of *Campylobacter jejuni* from chicken meat and human origin: a contribution to virulotyping', *Appl Environ Microbiol*, **75**(13): 4277–88.

HARA-KUDO Y, NEMOTO J, OHTSUKA K, SEGAWA Y, TAKATORI K, KOJIMA T and I KEDO M (2007), 'Sensitive and rapid detection of Verotoxin-producing *Escherichia coli* using loop-mediated isothermal amplification', *J Med Microbiol*, **56**(3): 398–406.

HERRERO A, MENDOZA M C, THRELFALL E J and RODICIO M R (2009), 'Detection of *Salmonella enterica* serovar Typhimurium with pUO-StVR2-like virulence-resistance hybrid plasmids in the United Kingdom', *Eur J Clin Microbiol Infect Dis*, **228**(9): 1087–93.

HUANG S, SAHIN O and ZHANG Q (2007), 'Infection-induced antibodies against the major outer membrane protein of *Campylobacter jejuni* mainly recognize conformational epitopes', *FEMS Microbiol Lett*, **272**(2): 137–43.

HUEHN S, LA RAGIONE R M, ANJUM M, SAUNDERS M, WOODWARD M J, BUNGE C, HELMUTH R, HAUSER E, GUERRA B, BEUTLICH J, BRISABOIS A, PETERS T, SVENSSON L, MADAJCZAK G, LITRUP E, IMRE A, HERRERA-LEON S, MEVIUS D, NEWELL D G and MALORNY B (2010), 'Virulotyping and antimicrobial resistance typing of *Salmonella enterica* serovars relevant to human health in Europe', *Foodborne Pathog Dis*, **7**(5): 523–35.

HUSSEIN H S and SAKUMA T (2005), 'Prevalence of shiga toxin-producing *Escherichia coli* in dairy cattle and their products', *J Dairy Sci*, **88**(2): 450–65.

KALISCHUK L D, INGLIS G D and BURET A G (2007), 'Strain-dependent induction of epithelial cell oncosis by *Campylobacter jejuni* is correlated with invasion ability and is independent of cytolethal distending toxin', *Microbiology*, **153**(9): 2952–63.

KHAN I, ADLER B, HARIDAS S and ALBERT M J (2005), 'PorA protein of *Campylobacter jejuni* is not a cytotoxin mediating inflammatory diarrhoea', *Microbes Infect*, **7**(5/6): 853–9.

KHOO C H, CHEAH Y K, LEE L H, SIM J H, SALLEH N A, SIDIK S M, RADU S, and SUKARDI S (2009), 'Virulotyping of *Salmonella enterica* subsp. enterica isolated from indigenous vegetables and poultry meat in Malaysia using multiplex-PCR', *Antonie Van Leeuwenhoek*, **96**(4): 441–57.

LIU D, LAWRENCE M L, AUSTIN F W and AINSWORTH A J (2007), 'A multiplex PCR for species- and virulence-specific determination of *Listeria monocytogenes*', *J Microbiol Methods*, **71**: 133–40.

LYNCH S V AND WIENER-KRONISH J P (2008), 'Novel strategies to combat bacterial virulence', *Curr Opin Crit Care*, **14**(5): 593–99.

MILLS J A, VENKATESAN M M, BARON L S and BUYSSE J M (1992), 'Spontaneous insertion of an IS1-like element into the *virF* gene is responsible for avirulence in opaque colonial variants of *Shigella flexneri* 2a', *Infect Immun*, **60**(1): 175–82.

OGURA Y, OOKA T, ASADULGHANI, TERAJIMA J, NOUGAYRÈDE J P, KUROKAWA K, TASHIRO K, TOBE T, NAKAYAMA K, KUHARA S, OSWALD E, WATANABE H and HAYASHI T (2007), 'Extensive genomic diversity and selective conservation of virulence-determinants in enterohemorrhagic *Escherichia coli* strains of O157 and non-O157 serotypes', *Genome Biol*, **8**(7): R138.

O'GORMAN L E, KREJANY E O, BENNETT-WOOD V R and ROBINS-BROWNE R M (1996), 'Transfer of attaching and effacing from a strain of enteropathogenic *Escherichia coli* to *E. coli* K-12', *Microbiol Res*, **151**(4): 379–85.

PENNISI E (2008), 'DNA data: proposal to "Wikify" GenBank meets stiff resistance', *Science*, **319**: 1598–9.

PFLEGER B F, KIM Y, NUSCA T D, MALTSEVA N, LEE J Y, RATH C M, SCAGLIONE J B, JANES B K, ANDERSON E C, BERGMAN N H, HANNA P C, JOACHIMIAK A and SHERMAN

D H (2008), 'Structural and functional analysis of AsbF: origin of the stealth 3,4-dihydroxybenzoic acid subunit for petrobactin biosynthesis', *Proc Natl Acad Sci USA*, **105**(44): 17133–8.

QUIÑONES B, PARKER C T, JANDA J M JR., MILLER W G and MANDRELL R E (2007), 'Detection and genotyping of *Arcobacter* and *Campylobacter* isolates from retail chicken samples by use of DNA oligonucleotide arrays', *Appl Environ Microbiol*, **73**(11): 3645–55.

RAWOOL D B, MALIK S V, SHAKUNTALA I, SAHARE A M and BARBUDDHE S B (2007), 'Detection of multiple virulence-associated genes in *Listeria monocytogenes* isolated from bovine mastitis cases', *Int J Food Microbiol*, **113**(2): 201–7.

ROCHE S M, GRACIEUX P and VELGE P (2009), 'Poor detection of low-virulence field strains of *L. monocytogenes* is related to selective agents in selective media and is unrelated to PrfA', *Food Microbiol*, **26**(1): 21–26.

RYCHLIK I, KARASOVA D, SEBKOVA A, VOLF J, SISAK F, HAVLICKOVA H, KUMMER V, IMRE A, SZMOLKA A and NAGY B (2009), 'Virulence potential of five major pathogenicity islands (SPI-1 to SPI-5) of *Salmonella enterica* serovar Enteritidis for chickens', *BMC Microbiol*, **9**: 268.

SCHLAGER T A, WHITTAM T S, HENDLEY J O, BHANG J L, WOBBE C L and STAPLETON A (2003), 'Variation in frequency of the virulence-factor gene in *Escherichia coli* clones colonizing the stools and urinary tracts of healthy prepubertal girls', *J Infect Dis*, **188**(7): 1059–64.

SELEEM M N, BOYLE S M and SRIRANGANATHAN N (2008), '*Brucella*: a pathogen without classic virulence genes', *Vet Microbiol*, **129**(1/2): 1–14.

SHEPPARD S K, DALLAS J F, MACRAE M, MCCARTHY N D, SPROSTON E L, GORMLEY F J, STRACHAN N J, OGDEN I D, MAIDEN M C and KEN J F (2009), '*Campylobacter* genotypes from food animals, environmental sources and clinical disease in Scotland 2005/6', *Int J Food Microbiol*, **134**(1/2): 96–103.

SNIPEN L, ALMØY T and USSERY D W (2009), 'Microbial comparative pan-genomics using binomial mixture models', *BMC Genomics*, **10**: 385.

SUSANNA K A, DEN HENGST C D, HAMOEN L W and KUIPERS O P (2006), 'Expression of transcription activator *ComK* of *Bacillus subtilis* in the heterologous host *Lactococcus lactis* leads to a genome-wide repression pattern: a case study of horizontal gene transfer', *Appl Environ Microbiol*, **72**(1): 404–11.

TIMOTHY S, SHAFI K, LEATHERBARROW A H, JORDAN F T and WIGLEY P (2008), 'Molecular epidemiology of a reproductive tract-associated colibacillosis outbreak in a layer breeder flock associated with atypical avian pathogenic *Escherichia coli*', *Avian Pathol*, **37**(4): 375–8.

UNICOMB L E, O'REILLY L C, KIRK M D, STAFFORD R J, SMITH H V, BECKER N G, PATEL M S and GILBERT G L (2008), 'Risk factors for *Campylobacter jejuni flaA* genotypes', *Epidemiol Infect*, **136**(11): 1480–91.

WASSENAAR T M (2003), 'Molecular typing of pathogens', *Berl Munch Tierarztl Wochenschr*, **116**(11/12): 447–453.

WASSENAAR T M and GAASTRA W (2001), 'Bacterial virulence: can we draw the line?', *FEMS Microbiol Lett*, **201**(1): 1–7.

WIESNER M, ZAIDI M B, CALVA E, FERNÁNDEZ-MORA M, CALVA J J and SILVA C (2009), 'Association of virulence plasmid and antibiotic resistance determinants with chromosomal multilocus genotypes in Mexican *Salmonella enterica* serovar Typhimurium strains', *BMC Microbiol*, **9**: 131.

WYCKOFF E E, MEY A R and PAYNE S M (2007), 'Iron acquisition in *Vibrio cholerae*', *Biometals*, **20**(3/4): 405–16.

YUKI N and ODAKA M (2005), 'Ganglioside mimicry as a cause of Guillain–Barré syndrome', *Curr Opin Neurol*, **18**(5): 557–61.

16

Using ribotyping to trace foodborne aerobic spore-forming bacteria in the factory: a case study

A. C. M. van Zuijlen, Unilever R&D Vlaardingen, The Netherlands

Abstract: Spores from mesophilic aerobic spore-forming bacteria (*Bacillus*) sometimes are able to survive the thermal process required for commercially sterile products and cause spoilage or food poisoning. Highly heat-resistant spores are a growing concern in the production of heat-preserved foods. These spores are introduced into the product either through the ingredients with a high spore load or through growth of bacterial spores during processing. This chapter describes a case study in which relevant aerobic spore-formers are identified by ribotyping to track their origin and obtain information about their dispersal. This information is subsequently used to control the levels of spore-formers in the product. In the future, rapid methods based on genomics should provide almost in-line information about the levels of heat-resistant mesophilic aerobic spore-formers.

Key words: mesophilic bacterial spores, spore load, *Bacillus sporothermodurans*, ribotyping.

16.1 Introduction

This chapter approaches tracing of spore-forming bacteria in the factory as a case study on thermoresistant *Bacillus* spores in sterilised soups. The techniques used in analysing the bacteria, typing them and using typing data for tracing can be applied to any other group of microorganisms in food environments.

Historically the majority of soups and sauces have been produced directly for the consumer in small cans or glass jars. Soups in larger packaging formats used in catering services often were either dried or frozen. In large scale production of these products, using batch systems (retort) or continuous sterilisation systems (cooker-coolers, continuous hydrostats), safe thermal processes are based on inactivation of *Clostridium botulinum* (botulinum cook). But in practice, for low-acid soups and sauces (pH > 4.70) the processes are generally increased to F0 5 or 8 for inactivation of more heat-resistant (*Bacillus*) spore-formers.

In general, Bacilli are considered to be harmless micro-organisms, with a few well-known exceptions in food, including *Bacillus cereus* (McKillip, 2000). However, it is commonly found that the more common spoilage Bacilli, for instance *B. subtilis* and *B. licheniformis*, can also cause food poisoning. Salkinga-Salonen, *et al.* (1999) reported *B. licheniformis* in relation to food poisoning and Mosupye, *et al.* (2002) described the potential risk due to toxin production by both *B. subtilis* and *B. licheniformis*. Phelps and McKillip (2002) mention a.o. *B. amyloliquefaciens* as one of the *Bacillus* outside the cereus group being able to produce toxin.

The thermal processes applied in practice, which result in commercial sterile products, will not eliminate all bacterial spores, but will result in ambient stable products at temperatures up to 37°C. Within the given shelf life of the products no outgrowth of (mesophilic) spores will occur. However, in the last decade, especially in UHT milk, highly heat-resistant spores were able to survive the process and cause spoilage. Petterson (1996) described a new species *Bacillus sporothermodurans* that produces highly heat-resistant spores. Scheldemans (2006) mentions the growing concern about tolerance and resistance of spores in his report on highly heat-resistant spores of *B. sporothermodurans* in UHT-treated milk. In the past Westhof (1981) already reported the occurrence of *B. cereus* and *B. subtilis* in UHT processed, spoiled milk.

In general, thermal processes for sterilised foods are based on worst case approaches, meaning that for thermal process setting the maximum spore load, the lowest initial temperature and the coldest point in both packaging and heat equipment have to be taken into consideration. As a result of this approach processes are set very conservatively, resulting in rather long processes and, as a consequence, destroying the organoleptical product quality.

For this reason there is a constant search for quality improvement, possibly coming from:

- Preparing products in a different way in the production phase, resulting in more freshness.
- Make change in the composition to improve the taste: for example add less salt.
- Finding alternative ways to heat process the products: an example is rotational sterilisation, which will shorten thermal processes and improve the quality.
- Finding alternative packaging methods, with adapted thermal processes resulting in superior quality when compared with cans and jars. Examples are soups and sauces in small pouches for the consumer and big pillow pouches aiming at the catering market.

All these changes will affect the thermal process and consequently can also affect possible survival of spores and outgrowth at ambient storage temperatures.

The work used as a case study for this chapter was initialised after problems occurred when regular low-acid soups were filled in large pillow bags instead of in cans or glass jars. Using the same thermal process setting conditions (F0 = 5 minutes in the coldest point) very low numbers of mesophilic spores survived the process and were able to grow at ambient storage. Reasons for this were complex, because measurements revealed a large spread in calculated lethality from thermocouples. The calculated F0 values in one thermal process could vary between F0 5 and 21 for one type of soup or between F0 6 and 34 for another soup.

On top of that it was estimated that the total amount of heat given to these pouches was far less than that given to cans or glass jars. This increases the probability of spores getting less heat, surviving the process and consequently growing out.

Increasing the thermal processes to F0 = 8 minutes for low-acid soups still proved to be insufficient to obtain the required stability. Based on a large number of F0 calculations from thermocouple measurements in processes it was estimated that with a one per cent probability of the F0 being < 5 minutes, the process should be set at a minimum F0 of 10 minutes. Those 10 minutes should be achieved in the position with the lowest integrated lethal effect in the bag. This increase in the thermal process proved to be successful, though survival and growth of spores still occurred in one particular Indian curry soup.

Table 16.1 lists the soups with their defect rate after three weeks of incubation at 33°C, together with the minimum calculated F0 from a thermocouple at the coldest point during the thermal process.

There are in principle two possible causes for survival and outgrowth of bacterial spores in the sterilised product:

1. Too high numbers of spores entering the process through the raw materials. Some of these spores could survive the process, especially when they are highly heat resistant.
2. Spore-formers grow in the production line and produce spores during production. Because these spores are usually produced at elevated temperatures, they can become more heat resistant as reported by Palop *et al.* (1999).

Both possibilities are illustrated with examples in this chapter.

Table 16.1 List with sterilised soups, the measured F0 and the defect rate after incubation of bags for 3 weeks at 33°C

Soup	Production code	Measured F0	% of samples with growth
Chicken soup	L90303	> 5	0.17
Chicken base	L90510	> 5	3.0
Creamy curry	L92010	> 5	20.0
Creamy curry	L90311	> 5	50.0
Asparagus soup	L01701	> 8	1.3
Chicken soup	L01901	> 8	1.3
Mushroom soup	L02001	> 8	3.6
Curry soup	L02501	> 8	1.3
Creamy curry	L02801	> 8	1.3
Indian curry soup	L01105	> 10	23.0
Indian curry soup	L03006	> 10	23.0

16.2 Ingredients as a source of bacterial spores

It is an assumption that all spores introduced in the process are coming from the raw materials. Both the prerequisite and HACCP-programme must guarantee that cleaning procedures are effective and under control. Some of the raw materials used in soup production, for instance spices, are often notoriously highly contaminated with bacterial spores. But also natural ingredients, like garlic, onion and red pepper often have very high spore counts. Some ingredients with high spore loads are listed in Table 16.2. These show that very high amounts of mesophilic and thermophilic aerobic spores can sometimes occur. By performing spore counts after different heat treatments of the ingredient the load with highly heat-resistant spores can be estimated. In this example the thermoresistance of spores in some ingredients was very high and in some cases the mesosphilic spores appeared to be more heat resistant than the thermophilic spores.

Dispersal of many different bacterial spores into a production environment such as in this case study is very likely to happen. More and more ingredients are sourced worldwide and any possible contamination with microorganism is likely to occur. In the case of spices, contamination with bacterial spores is mainly influenced by growth and harvest conditions. But during specific production steps, such as for instance drying, moisture content and drying temperature will be important in allowing for possible growth and sporulation of high heat-resistant bacterial spores. For ingredients such as garlic and onion, the levels of bacterial spores and their heat resistance are known to be very variable following different harvest conditions.

It is therefore very important that analytical data for relevant target microorganisms are known before ingredients are introduced into a product or into a production line.

The process of tracing spore-formers usually starts by analysing all ingredients that are used in the recipe of the product. The method used to analyse samples for mesophilic spores is as follows. Samples of the ingredients are homogenised in Pepton water to obtain a 1:10 suspension. After pasteurisation at 80°C for 15 minutes the suspension is plated on nutrient agar (Oxoid CM3) and incubated for two days 37°C.

Table 16.2 Examples of highly contaminated ingredients with bacterial spores

Ingredient	Mesophilic aerobic spores 15 min. 80°C	Mesophilic aerobic spores 15 min. 100°C	Thermophilic aerobic spores 15 min. 80°C	Thermophilic aerobic spores 15 min. 100°C
Turmeric	6.68	5.99	6.00	3.52
Coriander	5.62	5.29	6.08	4.81
Cumin	5.24	3.32	3.59	2.20
Chili	6.38	<1.00	4.19	<1.00
Ginger	6.90	<1.00	3.88	<1.00
Black pepper	7.10	6.23	4.28	4.16

Numbers in log per gram ingredient

However, this does not give any information about the real hazard for mesophilic thermoresistant spores. If this is required, the suspension of the ingredient should also be analysed for spores after pasteurisation at 100°C or even 110°C.

This enables a better understanding of the real hazard of the spore load for the thermal process of the soup under examination. An example of such a spore load calculation is given in Table 16.3. Using the spore counts of the ingredients, here at 80° and 100°C, and the weight of each ingredient in the recipe, the contribution

Table 16.3 Analysis for spore load calculation of ingredients for Indian curry soup

Ingredient	% of ingredient in recipe	*Bacillus* spores 15'/80°C log CFU/g	*Bacillus* spores 15'/100°C log CFU/g	Spore load at 100°C log/g
Chicken broth	3.9	3.86	3.87	2.45
Ginger pieces	0.9	–	–	
EFP salt	0.7	–	–	
Sodium glutamate	0.3	–	–	
Tomato paste CB	0.9	–	–	
Fresh bean sprouts	6.0	1.78	–	0.56
Onion frozen	5.0	–	–	
Water chestnut	4.0	–	–	
Red pepper frozen	3.0	1.30	–	
Green leek frozen	2.0	1.70	–	
Black fungus frozen	2.0	3.02	2.81	1.11
Chicken meat frozen	5.0	–	–	
Binding flower	2.2	2.64	2.74	1.08
Colflo starch	2.5	–	–	
Meat extract	0.2	1.30	–	
Curry powder aroma	0.5	2.86	–	
Curry powder ster.	0.25	–	–	
Sodium ascorbate	0.12	–	–	
Curafos phosphate	0.14	–	–	
Skim milk powder	1.4	2.76	2.73	0.88
Chicken fat	2.8	3.69	2.15	0.60
Rapeseed oil	2.6	–	–	
Water	53.2	–	–	
White pepper extract	0.04	–	–	
Nutmeg extract	0.02	–	–	
Fried onion aroma	0.08	3.34	1.48	
Garlic powder	0.08	4.70	3.79	0.64
Coriander extract	0.06	–	–	
Pepper extract	0.02	–	–	
Cumin	0.04	3.88	1.95	
Chili extract	0.004	1.95	–	
Cinnamon extract	0.01	–	–	
Soup before sterilisation		2.72	2.43	2.52

–: below the detection level of log 1.0 CFU/g

in spores of each ingredient in the recipe can be calculated. The total of these contributions will give the total spore load for the specific recipe.

To illustrate this, the example of Indian curry soup is used. Samples of all ingredients used in the recipe were analysed for *Bacillus* spores after pasteurisation of the 1:10 suspension for 15′/80°C and for 15′/100°C. The results are given in Table 16.3 and show that five ingredients mainly contribute to the spore load of this recipe, being chicken broth, black fungus, binding flour, skim milk powder and garlic powder. The final soup has a spore load of approximately log 2.5 per gram with heat-resistant spores, which is considered to be quite high.

Strains were isolated from those ingredients with a significant spore load after pasteurisation of 15'/100°C for further typing (16.5).

During a next production of the same curry soup, based on the knowledge of the previous run, a selected number of ingredients were analysed again for thermoresistant spores after pasteurisation of the 1:10 suspension of the ingredient for 15 minutes at 100°C and 15 minutes at 110°C.

This analysis shows that results can differ considerably between two separate sampling times, in spite of the fact that there were only two months between the sampling times (see Table 16.4). Now only two ingredients, black fungus and skim milk powder, showed high numbers of heat-resistant spores at 100°C. Most of the spores in black fungus also were able to survive the 15 minutes at 110°C.

All the colonies from black fungus were isolated for further typing. The same was done with all the colonies from the plates of the other ingredients originating from the 15 minutes at 100°C treatment. The large numbers of isolates from black fungus and milk powder were typed because of the high numbers of heat-resistant spores in the previous analysis.

The findings in these investigation resulted in applying a different approach in monitoring the spore load of ingredients. To start with, all samples to monitor

Table 16.4 Microbiological analysis for high heat-resistant spores

Ingredient	*Bacillus* spores 15 min. 100°C log CFU/g	*Bacillus* spores 15 min. 110°C log CFU/g
Black fungus	3.01	2.65
Chicken meat (cooked)	–	–
Binding flour	1.30	–
Curry flavour	1.00	–
Curry HTST	–	–
Skim milkpowder	2.60	–
Chicken fat	–	–
White pepper	–	–
Fried onion flavour	1.95	–
Garlic powder	1.85	–
Cumin	1.70	–

–: below the detection level of log 1.00 CFU/g

microbiological quality of incoming raw materials, are analysed for the presence of mesophilic spores after a pasteurisation of 15 minutes at 80°C. When spore counts are high, (> 100/gram), the sample is analysed again for spore counts, but now after pasteurisation for 15 minutes at 100°C and 15 minutes at 110°C.

In practice, the maximum target level of thermoresistant spores should be well below 100 per gram of the final recipe. Experience has shown that levels above 1000 per gram can result in survival and successive growth of mesophilic aerobic spores in the product stored at ambient temperatures.

16.3 Growth of bacterial spores in production

In production of canned soups and sauces unknown numbers of *Bacillus* spores are always introduced into the product through the raw materials. Production and cleaning regimes should prevent germination, growth and possibly sporulation during the allowed maximum production time. The HACCP-plan for such a line should include sufficient preventive measures to control the formation of numbers of spores that might survive the thermal process.

Historically it was assumed that keeping the product temperature well above 60°C during processing was sufficient to prevent germination and subsequently the outgrowth and spore forming of aerobic and anaerobic spores. In practice this was supported by having a long history of commercial sterility in canned, sterilised soups and sauces, when using the generally applied thermal processes in retorts and in continuous sterilisation equipment. Examinations showed that during production at these elevated temperatures of above 60°C, sometimes growth of thermophilic Bacilli, more specific of *Bacillus stearothermophilus* was observed. When, according to the GMP, the sterilised product is cooled immediately after sterilisation to below 42°C, outgrowth of these spores is successfully prevented. These products are, of course, not intended for tropical countries, because this would require more severe sterilisation programs, in excess of F0 = 15 minutes.

As stated in the introduction, non-sterility problems caused by outgrowth of Bacillus spores occurred when new ways of packaging and sterilisation were introduced. These introductions were accompanied by upscaling soup production into larger, more automated production lines, with the intention of producing longer runs of the same recipe without intermediate rinsing or cleaning.

To establish maximum holding times in the line a number of soups were tested, originally at 50°C because this was assumed to be the critical temperature for outgrowth of aerobic and anaerobic spores. Analyses for mesophilic aerobic spores were performed after pasteurisation of the soup for 15 minutes at 80°C; the plates were incubated at 40°C. Figure 16.1 gives an example of what frequently occurs when soup is waiting in the line to be filled. Especially in low-acid soups, with a pH > 5.50, rapid growth can occur of the total aerobic plate count, often already exceeding 10^7 CFU/g after four hours. Increase of anaerobic spores was never observed, but at 50°C the aerobic spore counts increased from 10–50 CFU/g to 1000–5000 CFU/g.

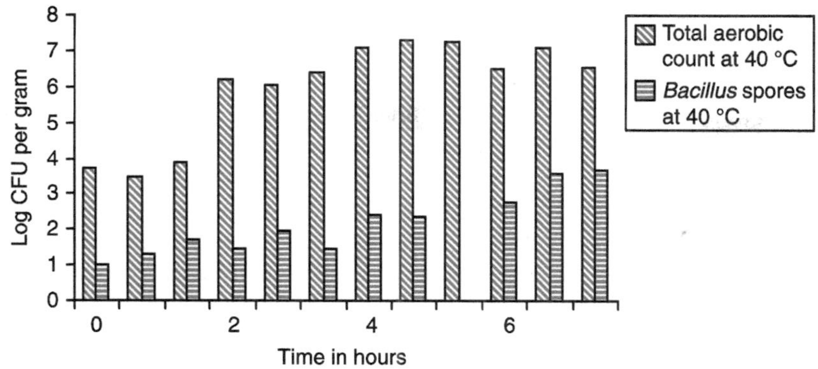

Fig. 16.1 Growth of *Bacillus* spores in soup at 50 °C.

To establish maximum production runs as part of the line HACCP-plan samples were taken from the line during runs lasting longer than eight hours. Samples were collected from equipment throughout the whole line, starting with the soup preparation (weighing, mixing and cooking) and finishing in the filling department.

Most samples were taken from places where the product visibly collected and remained seated for longer times, such as from the wall and shaft of the stirrer in the vessels, from the lobes of a pump and from the upper part of the wall in the fillers. Because of the high temperatures here, 70°C–80°C, aerobic plate counts and spore counts were performed at 55°C.

Surprisingly, especially from the hot part of the line, starting with the pump after the cooking vessel, rapid increase in total aerobic plate count and in aerobic spores was observed.

The example in Fig. 16.2 shows rapid growth of the total aerobic plate count already after six hours, resulting in counts above 10^7 CFU/gram after 12 hours

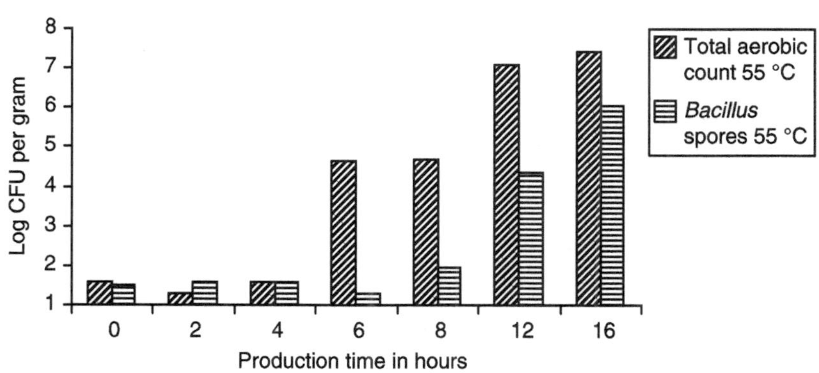

Fig. 16.2 Growth of *Bacillus* spores in equipment during production.

production. After approximately eight hours the aerobic spores started to increase and after 16 hours have reached a level of 10^6 CFU/g. These numbers are far too high to be able to assure safety or stability of the sterilised product during its shelf life. Generally it is assumed that the thermal process should be able to either kill or sufficiently damage 1000 spores per gram to assure ambient stability.

In the example here it was shown that possibly germination, but certainly growth and sporulation of aerobic spore-formers, could take place at high production temperatures. To be able to assess the probability of this to occur in soup during production, soup was taken from the final filler in production and incubated at controlled temperatures at 60, 70 and 80°C for a maximum of 16 hours. Total aerobic plate counts and aerobic spores (15 minutes, 80°C) were analysed and plates were incubated at 55°C.

Figure 16.3 shows results of one these incubation trials. This chicken soup was taken from the filler, which is the stage immediately before sterilisation. At 60°C growth of total aerobic count and aerobic spores was much faster than at 55°C. Already after four hours the total aerobic count reached its maximum level of 10^6 CFU/g and after eight hours the aerobic spores also reached the same level of 10^6 CFU/g. Extending the incubation time to 16 hours did not increase these levels any more.

Also at 70°C growth occurred of both the total aerobic plate count and of aerobic spores, though the levels remained lower that at 60°C. After eight hours incubation the total aerobic plate count reached its maximum level of a little over 10^5 CFU/g and after 12 hours the aerobic spores reached its maximum level of just under 10^3 CFU/g.

At 80°C no growth of aerobic bacteria and aerobic spores was observed any more. The conclusion from these experiments therefore was that up to 70°C, outgrowth of aerobic spore-forming bacteria followed by formation of spores was possible.

From different soups, either sampled in the production line, or from incubated samples at 55, 60 or 70°C, a number of strains were isolated from the plates for typing with the RiboPrinter. All isolates were first checked for growth at 37°C, making sure that only potential mesophilic spore-formers were typed. As stated earlier, thermophilic spores are regarded to be unimportant for these ambient stable products.

16.4 Identifying relevant spore-formers

Ribotyping has been widely used for typing and characterisation of bacteria, for instance for *Salmonella* (Oscar, 1998), for *Clostridium botulinum* (Skinner *et al.*, 2000) and for *Listeria monocytogenes* (Gendel and Ulaszek, 2000). Guillaume *et al.* (2002) reported about using ribotyping for demonstrating genetic heterogenity in *B. sporothermodurans*.

All isolates from soups and ingredients were sent to Unilever Research in Colworth, UK, to be typed with the DuPont RiboPrinter. Before sending, it was

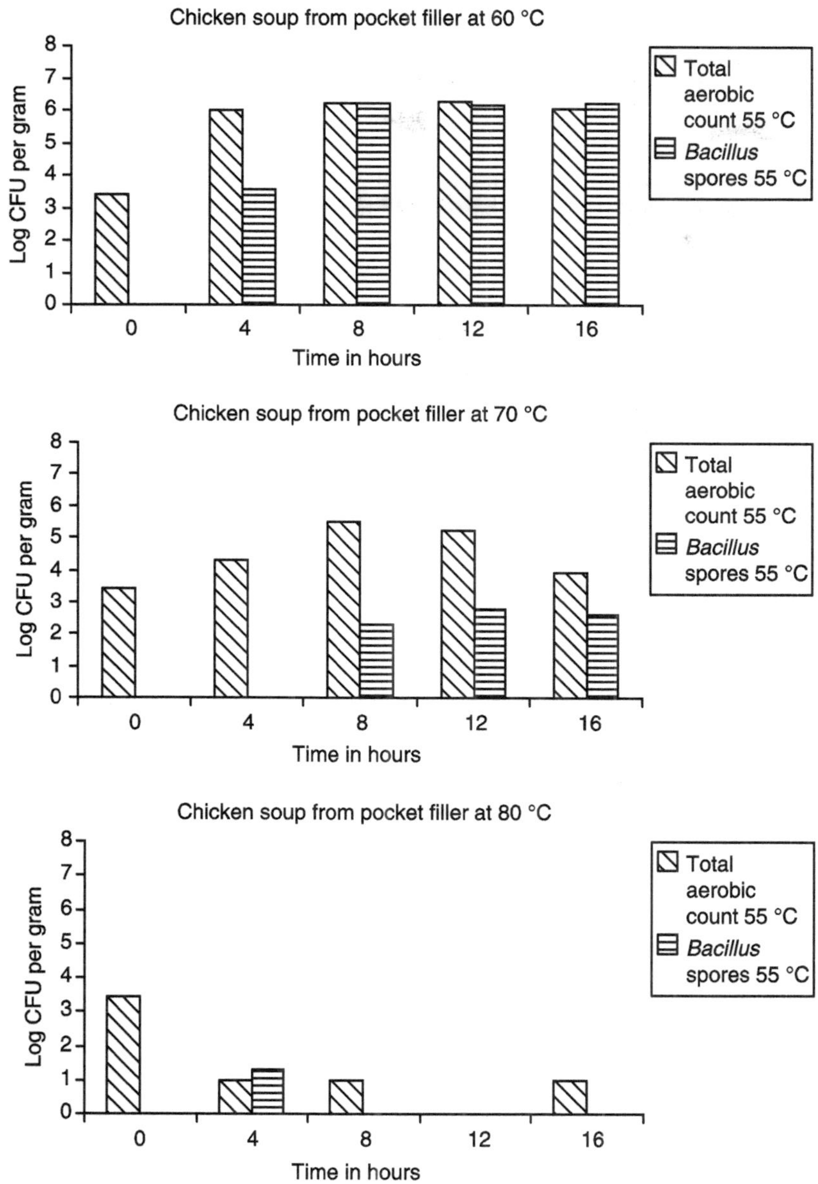

Fig. 16.3 Growth and sporulation of Bacilli in soup at 60 °C, 70 °C and 80 °C.

confirmed that all isolates were aerobic spore-forming rods. A large number of isolates, approximately 400, were typed in order to try to relate thermal-resistant spores from the ingredients with spores that had survived the thermal process.

The DuPont Qualicon RiboPrinter™ is an automated ribotyping instrument for characterising and identifying bacteria. The RiboPrint® patterns are generated from restriction fragments of the ribosomal RNA genes. In the standard process of the RiboPrinter the DNA is digested with the endonuclease E*co*RI. The restriction fragments are separated by electrophoresis and transferred directly to nylon membranes by Southern blotting. Only particular restriction fragments associated with specific sequences are detected. The RiboPrint® pattern is then created by hybridisation of the rRNA information with a chemiluminescent-labelled DNA probe containing the rRNA operon from *E. coli*. The chemiluminescent patterns are imaged electronically with a CCD camera and stored in a database

The resulting EcoRIrRNA operon fingerprint is compared to a dynamic database for comparison with other patterns and with a fixed identification database for classical taxonomy. A taxonomic name is assigned when the new RiboPrint® pattern has >85% similarity with a pattern in the identification database.

Similarity between new and existing patterns can also be used for grouping patterns within a specific similarity range in RiboGroups. For this, any given pattern is compared with the patterns in the database. There is a match when the patterns are undistinguishable from each other, having a similarity of more than 90%. In that case the patterns are assigned to the same RiboGroup. This RiboGroup code has the format:

RIBO1 162 – 249 – S-1

RIBO1 = the name of the database built with the restriction enzyme EcoR1
249 = the batch number of a run of up to 8 samples per run
S-1 = the sample number or the lane in the gel that the sample was positioned.

When a sample is tested on the RiboPrinter and produces a series of bands, the instrument immediately checks the library to test if the band has been seen before. If that is not the case a new RiboGroup is produced. Following the assignment of groups, each RiboGroup therefore is named after the sample that first generated the band pattern.

At a certain moment it became apparent that it was not possible to do an in-depth analysis for similarity with the software of the RiboPrinter. To be able to do this, more advanced assistance was found at the NIZO-institute in the Netherlands. All results from the RiboPrinter were added to a data bank in BioNumerics computer software (Applied Maths, St. Martens-Latem Belgium). With this software, dendrograms were generated by means of product moment/ UPGMA cluster analysis. With BioNumerics it was possible to do the required in-depth analysis for similarity between the RiboGroup patterns in the data bank. Figure 16.4 shows an example of a small part of the isolates that were grouped in this way. Using BioNumerics made it possible to recognise more precisely similarities between heat-resistant soup isolates and isolates from ingredients.

16.5 Tracking sources of relevant spore-formers

To be able to trace specific Bacilli that survived the thermal process it was necessary to have these typed to at least the species level. In total, 40 isolates that

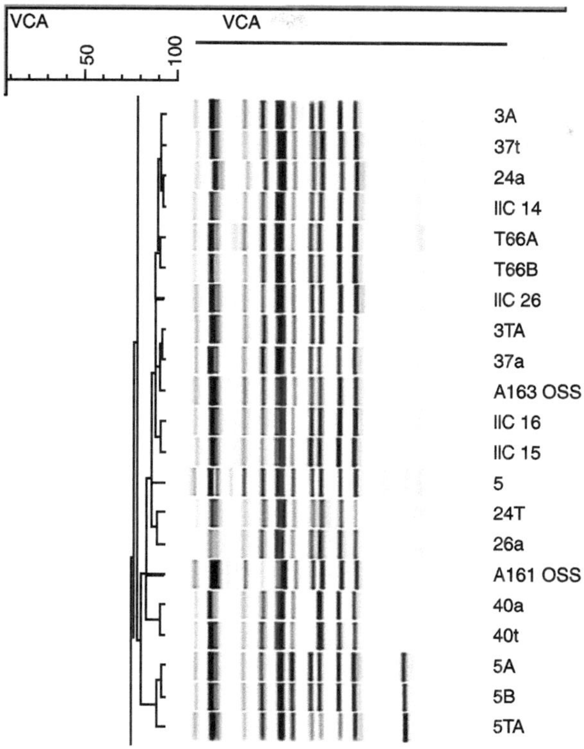

Fig. 16.4 Part of the BioNumerics database showing some differentiation within *Bacillus subtilis*.

grew in sterilised soups after incubation for three weeks at 33°C were typed using the RiboPrinter. Most of these could be identified to the species level by comparison with reference strains or with earlier isolates. As shown in Table 16.5, there were five species of *Bacillus* isolated, *B. subtilis*, *B. cereus*, *B. sporothermodurans*, *B. licheniformis* and *B. amyloliquefaciens*. Special attention was given to the tracing of *B. sporothermodurans* that survived in Indian curry soup, because this Bacillus proved to be very heat resistant. Depending on the way it was sporulated it had a D121°C between 2.5 and 6 minutes. Characterisation of this organism, together with other Bacilli from low acid canned products, was reported by Oomes (2007).

By typing over 300 isolates from ingredients and comparing the RiboPrinter profiles through BioNumerics, it was possible to list ingredients with *Bacillus* species that closely matched the soup isolates. This list with, in total, 109 isolates is given in Table 16.6. Notice that the high numbers in black fungus and milk powder are the results of the selection of a large number of isolates from the same ingredient of the same production date. For black fungus this was done because it was the only ingredient with spores surviving 15 minutes at 110°C. For milk powder, the reason was that dairy products are notorious sources for *Bacillus*

Table 16.5 Observed survival of *Bacillus* spores in commercially sterilised soups

Bacillus	Soup	Number of production runs	Number of isolates
subtilis	Curry cream	3	12
	Indian curry	1	1
	Asparagus	1	1
cereus	Indian curry	1	2
	Curry cream	1	1
	Pea soup	1	1
sporothermodurans	Indian curry	2	12
licheniformis	Mushroom	1	2
	Chicken	1	6
	Mushroom cream	1	1
amyloliquefaciens	Creamy chicken	1	5

Table 16.6 Incidence of heat-resistant spores in ingredients, resembling isolates from sterilised soups

Bacillus	Ingredient	Number of batches	Number of isolates
subtilis	White pepper	1	2
	Garlic powder	1	4
	Turmeric	1	6
	Black fungus	2	43[a]
	Cumin	2	3
	Curry aroma	1	1
	Binding flour	1	2
	Chicken fat	1	1
cereus	Chicken fat	1	3
sporothermodurans	Black fungus	1	1
licheniformis	Curry powder spec.	1	2
	Milk powder	2	20[b]
	Cumin	1	2
	Garlic powder	2	3
	White pepper	1	3
amyloliquefaciens	Binding flour	2	5

[a] all isolates from the same batch
[b] 18 isolates from the same batch

spores. Scheldeman *et al*. (2000) reported heat-resistant *B. licheniformis* in raw milk and milking equipment.

B. subtilis and *B. licheniformis* were most frequently found in a range of ingredients. It must be emphasised here that especially for *B. subtilis* and *B. licheniformis* the RiboPrinter profiles showed a large variation within the species, most likely referring to subspecies. The other relevant *Bacillus* species in this study were only found in one single ingredient, *B. cereus* in chicken fat, *B. sporothermodurans* in black fungus and *B. amyloliquefaciens* in binding flour.

As indicated earlier, a lot of effort was spent on trying to trace *B. sporthermodurans* in one of the ingredients. Based on the analysis of the ingredients (16.2) black fungus (frozen) was the main suspect for being the source. The results of typing the isolates of this black fungus emphasises the difficulties one can encounter in trying to trace a single species in an ingredient. Of the 40 isolates typed the vast majority proved to be *B. subtilis* and only one isolate was typed as *B. sporothermodurans* (isolate IIC65). This isolate however matched perfectly with isolates taken from the sterilised Indian curry soup (IC32), as is shown in Fig. 16.5.

This was the only *B. sporothermodurans* encountered in the ingredients examined in this study, although it has been widely reported in the dairy industry. Herman (2000) describes methods for isolation and detection of this microorganism and also Da Silva *et al.* (1998) reported the detection and isolation. Vaerewijck *et al.* (2001) reported on tracing *B. sporothermodurans* in feed for dairy cattle, stating that feed can be a source of contamination of milk with spores, including *B. sporothermodurans*.

Although *B. licheniformis* was found frequently in ingredients, there were no close matches with one of the isolates from the sterilised soups. There was some similarity between the isolates from milk powder and those from chicken soup, but the isolates clearly were not the same subspecies and, on top of that, milk powder is not used in chicken soup.

For *B. subtilis* there were more ingredient isolates that were closely similar to isolates from sterilised soups. The closest similarity was found between an isolate from binding flour (IIC14) and an isolate from curry cream soup, as is shown in Fig. 16.6. These isolates also matched very close to another well-typed isolate (*B. subtilis* A163) that was isolated from sterilised canned soups many years earlier.

A last example of a good match is given for *B. cereus* in Fig. 16.7, showing a very close similarity between an isolate from milk powder (IIC99a) and an isolate from Indian curry soup (31a and 32a).

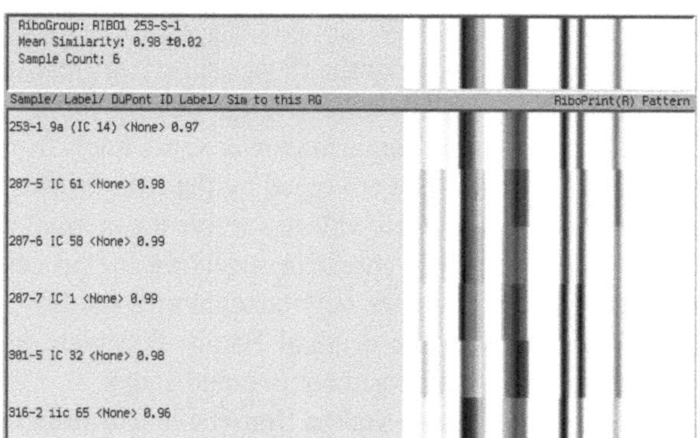

Fig. 16.5 Matching profiles for *B. sporothermodurans* isolated from sterilised Indian curry soup and from black fungus.

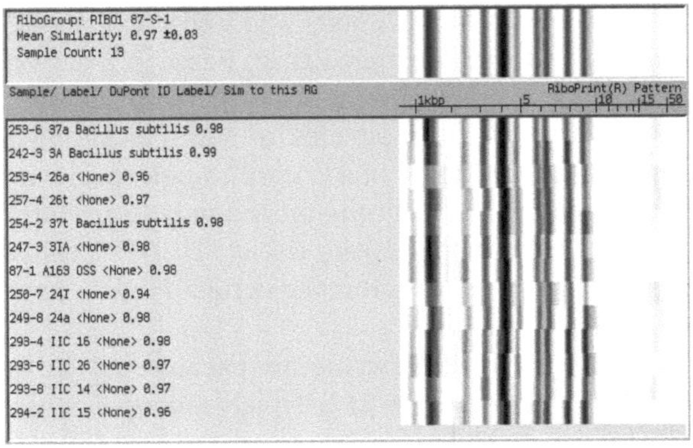

Fig. 16.6 Matching profiles for *B. subtilis* from sterilised curry cream soup and from binding flour.

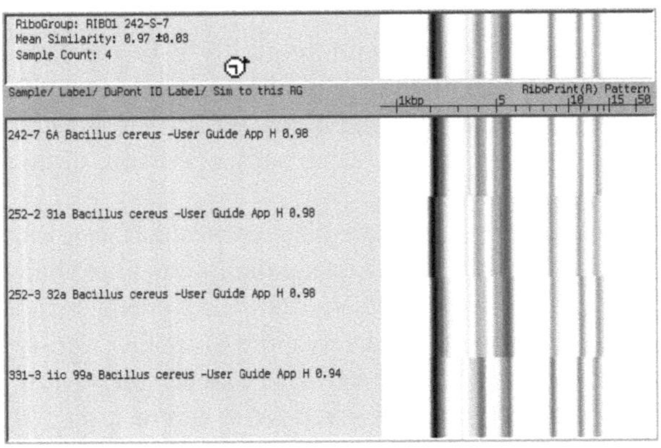

Fig. 16.7 Matching profiles for *B. cereus* for isolates from Indian curry soup and from milk powder.

These similarities show the origin of the most riskful heat-resistant spores and enable the producer to take measures to reduce the risk by simply replacing this ingredient with an alternative or by acquiring lower spore loads from the supplier.

In this study many other isolates were typed by the RiboPrinter that could not be linked to any of the species that were able to survive the thermal process. These isolates could in principle also be a threat in surviving the process, because of their high heat resistance. This chapter only concentrates on those isolates from ingredients that matched closely the original Bacilli that were known to have survived the thermal process of sterilised commercial soups.

As could be expected, most isolates taken from the line or from incubated soup at 55, 60 and 70°C were typed as *Geobacillus stearothermophilis* or other Geobacilli. Especially in the soups that were challenged by incubation at 60 and 70°C, *G. stearothermophilus* was the predominant organism. The only other

Bacillus that was occasionally encountered in different soups was *B. licheniformis*. This proves at least that it is possible for mesophilic Bacilli to grow and sporulate at temperatures as high as 70°C in a production line. This was confirmed in practice when samples were taken from the equipment at the end of production runs that on a few occasion contained high spore counts. Some of these isolates were also typed as *B. licheniformis*. This fact has to be taken into account when, for instance, maximum production times on a line have to be established. The most vulnerable soups, for example chicken soup, should be checked for possible growth after incubation for the maximum time allowed at realistic temperatures. This study shows that mesophilic spores are able to germinate, grow and subsequently sporulate at temperatures as high as 70°C.

16.6 Controlling levels of spore-formers in production

As indicated in this chapter there are two major causes for the survival and possible growth of highly heat-resistant mesophilic aerobic spores in a sterilised product. These are contamination with too high levels of highly heat-resistance spores through the ingredients and/or growth of and sporulation of these spores during production. For both of these causes preventive measures have to be implemented in the process and must be part of the HACCP-plan.

1. Reducing the numbers of heat-resistant spores entering the production by controlling the raw material quality. There are a number of measures that together should control the spore load entering the process:

   - Raw material specifications, making sure that the right levels of spores are agreed with the suppliers.
   - Supplier selection, especially in those ingredients where high spore loads can be expected.
   - Monitoring of the microbiological quality, either by the supplier or through an intake control program. This should take into account the risks of highly heat-resistant spores, resulting in an increased sampling frequency when risk is high.
   - Always test new ingredients or ingredients obtained from a new supplier for the spore load before accepting these for production. In some cases, e.g. garlic and onion, analyses should also be performed when newly harvested material is delivered. Sometimes there is a significant difference in spore load depending on harvest conditions.

   Make sure that there is sufficient knowledge about the ingredients, especially with regard to highly heat-resistant spores. As shown in this chapter, sometimes ingredients with a low spore load can still be the cause of survival and growth of spores in the sterilised product when the spores are very heat resistant. Examples of such ingredients are flours, milk powder and mushrooms. For such ingredients the best practice is to perform spore counts after pasteurisation not only after heating at 80°C, but also perform counts after heating at 100°C and 110°C.

2. Prevention of possible outgrowth and subsequent sporulation of Bacilli in the production line.

   A HACCP-plan for the soup production line must include maximum production times for the different parts of the line if this is required. It must also include maximum waiting times and temperatures for ingredients, mixes and final product before sterilisation.

   Both maximum production times and waiting times should be checked before they are implemented in the operational procedures. This can be done by sampling and analysing at the end of runs at the end of the maximum run time. But it is also possible, as described in this chapter, to simulate production and waiting times by incubating samples at realistic temperatures and analyse for growth of microorganisms and for possible growth of spores. If necessary, the process should be redesigned to avoid critical temperatures for growth and sporulation.

   Based on the growth and sporulation rate, cleaning and rinsing frequencies are established to control numbers of spores in the line. Because this will mainly depend on temperature, different cleaning or rinsing frequencies may occur for the different equipment in the line. All of this must be implemented in the operational procedures, including checks that cleaning and rinsing was performed after the prescribed time.

16.7 Future trends

As stated in the introduction, thermally processed foods are often over-processed taking into account unknown heat resistance of the spores present and also because of the worst case approach in establishing thermal processes.

Knowledge about which spores, and how many are present in the ingredients or in the product immediately before sterilisation, could help in reducing the thermal process.

Rapid tests identifying possible 'problem organisms' before the thermal process would make it possible to achieve this. PCR based methods to characterise mesophilic *Bacillus* spore from UHT milk have already been described by Hammer *et al.* (2000). It is possible to identify specific organisms, such as *B. sporothermodurans*, as reported by Scheldeman *et al.* (2002) and Montanari *et al.* (2004).

Rapid detection of *Bacillus* spores in the ingredients and in the products was part of a Dutch subsidised project, named EET (Environment, Energy and Technology). Relevant Bacilli were selected based on the similarity analyses performed with BioNumerics, described in this chapter. Markers for species and sometimes subspecies, in the case of *B. subtilis*, were identified by means of molecular methods, first by using 16S-rDNA sequences, later by genomotyping. Attempts were made to find specific markers for the group of heat-resistant spores. Furthermore, a specific marker for the genus *Bacillus* is necessary to make it possible to identify strains that were not included yet in the set of markers for the selected Bacilli.

To perform the tests DNA has to be extracted from a great variety of samples, varying from all kinds of ingredients to many different finished products. For this purpose adapted DNA-extraction techniques are developed, to be able to extract DNA with the highest possible yield from any given sample.

After pre-treatment of the sample, all DNA-fragments are simultaneously amplified (multiplex PCR amplification) and subsequently tested on a multi-array system, identifying on the species level those Bacilli that match the markers on the multi-array. With the present status of these techniques it is possible to detect approximately 200 cells/spores in the sample.

Identifying spores that could survive the thermal process could be used as an early-warning system in the production of heat-preserved foods. But identifying these spores already in the raw materials creates an opportunity to prevent dispersal into the production environment. Nowadays ingredients are resourced from all over the world, leading to possible ingress of almost any type of spore-forming bacteria. The example in this case study is the introduction of *Bacillus sporothermodurans* through black fungus mushrooms from an Asian country. The products in this study will at any stage of their production be susceptible to the germination and growth of almost any microorganisms. The techniques described here that are developed now, would even be able to identify typically heat-resistant bacterial spore-formers. This knowledge can be used to avoid these ingredients from being used in those production environments where they are a hazard for the safety or stability of the heat-treated products.

With a suitable rapid test system any sample can be screened for possible problematic Bacilli from raw materials (mainly dry ingredients and spices), samples in the production line, where growth of these organisms is suspected, and finally from the products after the thermal process.

The present status of these techniques does not allow this last step at the moment, because surviving spore-formers will be well below the present detection level of 200 cells/spores per gram. This would require a detection level of one spore per packaging unit.

The results from screening with a rapid test will enable the producer to apply the optimum thermal process without having to take into account maximum spore load. Reduced thermal processes will result in better quality products, while still assuring ambient stability.

16.8 References

DE SILVA S, PETTERSSON B, AQUINO DE MURO M A and PRIEST F G (1998), 'A DNA probe for the detection and identification of *Bacillus sporothermodurans* using the 16S–23S rDNA spacer region and phylogenetic analysis of some field isolates of *Bacillus* which form highly heat resistant spores', *Systematic and Applied Microbiology*, **21**, 398–407.

GENDEL S M and ULASZEK J (2000), 'Ribotype analysis of strain distribution in *Listeria monocytogenes*', *Journal of Food Protection*, **63**, 179–85.

GUILLAUME-GENTIL O, SCHELDEMAN P, MARUGG J, HERMAN L, JOOSTEN H and HEYNDRICKX M (2002), 'Genetic heterogeneity in *Bacillus sporothermodurans* as demonstrated by ribotyping and repetitive extragenic palindromic PCR fingerprinting', *Applied and Environmental Microbiology*, **68**, 4216–24.

HAMMER P, LEMBKE F, SUHREN G and HEESCHEN W (1995), 'Characterization of a heat resistant mesophilic *Bacillus* species affecting quality of UHT milk – a preliminary report', *Kieler Milchwirtschaftliche Forschungsberichte*, **47**, 303–11.

HERMAN L, HEYNDRICKX M, VAEREWIJCK M and N KLIJN N (2000), '*Bacillus sporothermodurans* – a *Bacillus* forming highly heat-resistant spores. 3. Isolation and methods of detection', *Bulletin of the International Dairy Federation*, **357**, 9–14.

MCKILLIP J L (2000), 'Prevalence and expression of enterotoxins in *Bacillus cereus* and other *Bacillus* spp., a literature review', *Antonie van Leeuwenhoek*, **77**, 393–9.

MONTANARI G, BORSARI A, CHIAVARI C, FERRI G, ZAMBONELLI C and GRAZIA L (2004), 'Morphological and phenotypical characterization of *Bacillus sporothermodurans*', *Journal of Applied Microbiology*, **97**, 802–9.

MOSUPYE F M, LINDSAY D, DAMELIN, L H and VONA HOLY A (2002), 'Cytotoxicity assessment of *Bacillus* strains ilsolated from street-vended foods in Johannesburg, South Africa', *Journal of Food Safety*, **22**, 95–105.

OOMES S J C M, VAN ZUIJLEN A C M, HEHENKAMP J O, WITSENBOER H, VAN DER VOSSEN, J M B M and BRUL S (2007), 'The characterisation of *Bacillus* spores occurring in the manufacturing of (low-acid) canned products', *International Journal of Food Microbiology*, **120**, 85–94.

OSCAR T P (1998), 'Identification and characterization of *Salmonella* isolates by automated Ribotyping', *Journal of Food Protection*, **61**(5), 519–24.

PALOP A, MANAS P and CONDON S (1999), 'Sporulation temperature and heat resistance of *Bacillus* spores: a review', *Journal of Food Safety*, **19**, 57–72.

PETTERSSON B, LEMBKE F, HAMMER P, STACKEBRANDT E and PRIEST F G (1996), '*Bacillus sporothermodurans*, a new species producing highly heat-resistant endospores', *International Journal of Systematic Bacteriology*, **46**, 759–64.

PHELPS R J and MCKILLIP J L (2002), 'Enterotoxin production in natural isolates of Bacillaceae outside the *Bacillus cereus* group', *Applied and Environmental Microbiology*, June, 3147–51.

SALKINOJA-SALONEN M S, VUORIO R, ANDERSSON M A, KÄMPFER P, ANDERSSON M C, HONKANEN-BUZALSKI T and SCOGING A C (1999), 'Toxigenic strains of *Bacillus licheniformis* related to food poisoning', *Applied and Environmental Microbiology*, Oct., 4637–45.

SCHELDEMAN P, HERMAN L, FOSTER L S and HEYNDRICKX M (2006), '*Bacillus sporothermodurans* and other highly heat-resistant spore-formers in milk', *Journal of Applied Microbiology*, **101**, 542–55.

SCHELDEMAN P, PIL A, HERMAN L, DE VOS P and HEYNDRICKX M (2005), 'Incidence and Diversity of Potentially Highly Heat-Resistant Spores Isolated at Dairy Farms', *Applied and Environmental Microbiology*, **71**, 1480–94.

SCHELDEMAN P, HERMAN L, GORIS J, DE VOS P and HEYNDRICKX M (2002), 'Polymerase chain reaction identification of *Bacillus sporothermodurans* from dairy sources', *Journal of Applied Microbiology*, **92**, 983–91.

SKINNER G E (2000), 'Differentiation between types and strains of *Clostridium botulinum* by Riboprinting', *Journal of Food Protection*, **63**(10), 1347–52.

SKINNER G E, GENDEL S M, FINGERHUT G A, SOLOMON H A and ULASZEK J (2000), 'Differentiation between types and strains of *Clostridium botulinum* by riboprinting', *Journal of Food Protection*, **63**, 1347–52.

VAEREWIJCK M J M, DE VOS P, LEBBE L, SCHELDEMAN P, HOSTE B and HEYNDRICKX M (2001), 'Occurrence of *Bacillus sporothermodurans* and other aerobic spore-forming species in feed concentrate for dairy cattle', *Journal of Applied Microbiology*, **91**, 1074–84.

WESTHOFF D C and DOUGHERTY S L (1981), 'Characterization of *Bacillus* species isolated from spoiled ultra-high temperature processed milk', *Journal of Dairy Science*, **64**, 572–80.

17

Biotracing: a novel concept in food safety integrating microbiology knowledge, complex systems approaches and probabilistic modelling

J. Hoorfar, Technical University of Denmark, Denmark, M. Wagner, Department of Farm Animal and Veterinary Public Health, Austria, K. Jordan, Teagasc, Ireland and G. C. Barker, Institute of Food Research, UK

Abstract: The interface between pathogen testing and tracing systems in food production is crucial if we are to improve tracing the origin of accidental or deliberate microbial contamination of feed and food. As part of that, it is essential to model the development of contamination from the point of entry to the point of detection, and beyond. Modelling is important not only to trace the origin of the contamination, but also to inform the food/feed producers of the appropriate corrective actions to protect the consumer. A complete chain approach is needed, beginning with sampling and ending with recommendations for control measures, well-justified and targeted product recalls, timely activation of rapid alert systems and proper emergency responses. Using mathematical modelling of contamination, biotracing can help food producers to know exactly what is coming into the factory and what is leaving the factory. The large European project BIOTRACER has taken up this approach.

Key words: pathogen monitoring, contamination modelling, rapid methods, decision support, predictive modelling.

17.1 What is BIOTRACER?

BIOTRACER is one of the largest Integrated Projects launched by the European Union to support research in improved food safety (http://www.biotracer.org). It has 46 project partners from 24 countries, including countries that supply feed to Europe such as Brazil, Indonesia, Russia and South Africa. The main objective of

BIOTRACER is to create tools and models that improve tracing of both accidental and deliberate microbial contamination of feed and food (including bottled water). The work is focused on several pathogen-food/feed chain matrices:

1 Quantitative chain modelling for *Staphylococcus aureus* in milk and *Salmonella* in pork.
2 Traceability of mycotoxins and *Salmonella* in the feed chain.
3 Traceability of *Campylobacter jejuni* in the chicken chain and *Salmonella* in the pork chain.
4 Traceability of *Listeria monocytogenes* and *S. aureus* in the dairy chain.
5 Virtual contamination scenarios (simulating bio-terror activity) for *Bacillus anthracis*, *Clostridium botulinum* and norovirus.

BIOTRACER is a multi-disciplinary project that includes experts in modelling, database developers, software companies, risk assessors, risk managers, system biologists, food and molecular microbiologists and food/feed companies. The ambition is that all these will work together in 'biotraceability'. In order to further application of the results of BIOTRACER, and ensure a positive impact on tracking/tracing pathogens, it will be possible for companies and researchers to become Associate Members of BIOTRACER, through the website http://www.biotracer.org.

The BIOTRACER Integrated Project takes one step further by defining a new research discipline called biotracing and establishing principles that might support the use of new food chain information sets to drive improved food safety by addressing source level propositions.

17.2 Definition of biotracing

Biotracing is integrated tracing (backward) and tracking (forward) of biological contamination along entire food production chains. Importantly, tracking and tracing relate to the agents, not the food, and this separates biotraceability from traceability.

In any process or production chain, there are access points along the line at which contaminants may enter the chain. In food and feed production chains, these access points can occur on the farm, in transportation, production, slaughter processes and storage, among others.

What biotracing is trying to achieve is to identify (using advanced microbiological methods, information supplies and mathematical models) where and when the contamination happened and what effect it may have further down the chain (Fig. 1). Biotracing promotes a better understanding of pathogen physiology

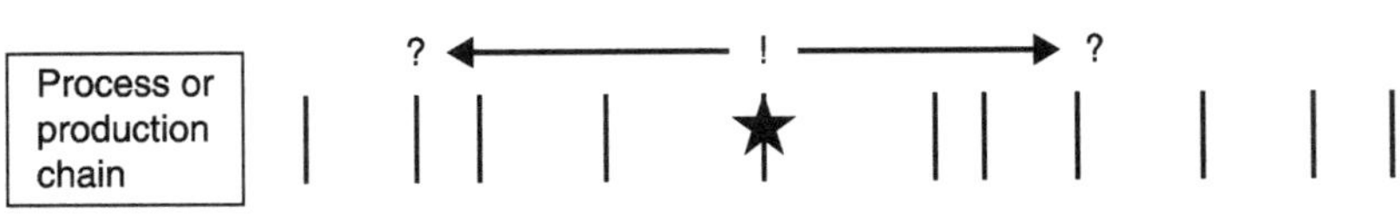

Fig. 17.1 Any process or production chain has access points where contamination or unexpected influences can enter the process.

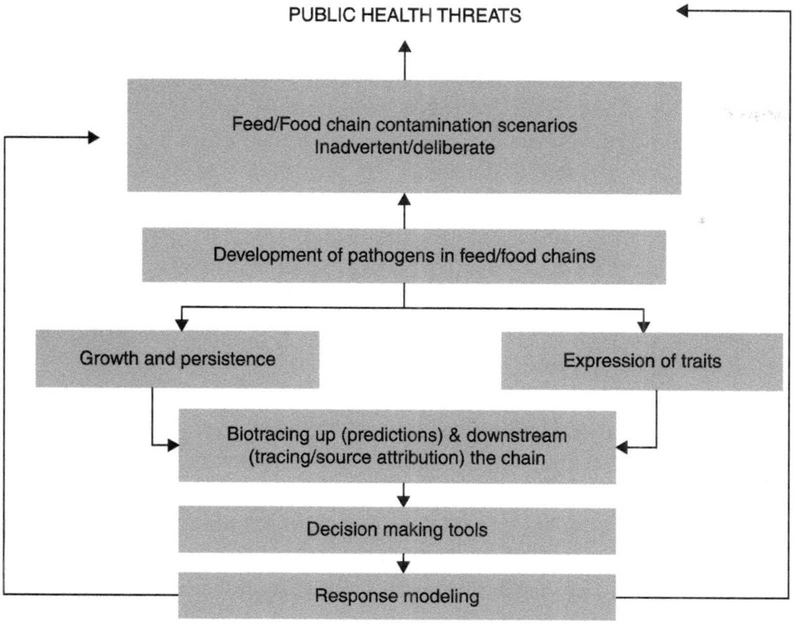

Fig. 17.2 The conceptual food chain modelling of biotracing.

and detection, combined with a domain modelling approach in which many variables and their dependencies are identified and the relevant information flows are established. In this regard, probabilistic models and statistical inference are appropriate mathematical tools – a Bayesian framework is particularly powerful. The phrase 'farm-to-fork' is commonly used in 'traceability' issues. For biotracing, the relevant time sequence goes both ways from 'farm-to-fork' and from 'fork-to-farm' in order to establish not only when the contamination happened (past), but also how it might affect the production process and the consumer (future).

Biotracing can be considered a new concept because it combines the knowledge and technology of food microbiology with complex systems approaches and probabilistic modelling (Fig. 17.2). The food microbiology aspect is used to answer questions such as what strain of pathogen has contaminated the chain, can we improve detection capabilities and what are the growth or no-growth expectations of that pathogen in the various process or production points of the chain (in response to temperature, light, oxygen effects from processes in the production chain, etc.). The mathematical modelling aspect is used to identify a probable time and place of contamination using information about the chain and data from in-line monitoring but, additionally, domain models support improved predictions for downstream progression of the agent population. This allows informed decision-making in relation to remedial actions, interventions and recalls in a cost-efficient way.

However, this may lead one to ask, is not biotracing already part of the process for production chains? The answer is no – at least not consistently. While tracking and tracing are sometimes achieved, biotracing aims to bring a new level of integration, incorporating decision support tools based on limited sampling and using modern techniques for pathogen detection and modelling.

A progressive approach to biotracing is important in the EU and worldwide because food production is crossing country borders within the EU and beyond. The pork and chicken eaten in Denmark or elsewhere may not come from Denmark or even another EU country. Globalisation of food production and trading will become more and more of a food safety issue (Lynch *et al.*, 2009). Import of feed products from beyond EU borders is at a high level. How does one know what the farmer fed the chicken you are cooking for dinner tonight? Similar questions establish biotracing as important not only to the food production industry, but also to the consumer.

At a practical level, biotracing will combine finite amounts of sampling with improved detection methods and inference models to establish biotraceability as the ability to use downstream information to point to materials, processes or actions within a particular food chain that can be identified as a source of undesirable agents. Biotracing models support rational decision making and prioritise actions (for tracking) and explanations (for tracing) in complex scenarios. Biotracing is a driver for improved food safety rather than a mechanism for establishing liability for a particular event.

17.3 Fundamental concepts of biotracing

Biotracing is a new concept in data collection, improved detection and modelling; in this sense biotracing goes beyond traditional risk assessment or predictive modelling. Biotracing aims to complement established food safety assessment schemes by combining data flows with mathematical models to give quantitative information about the likely sources of particular hazards or potential problems. Biotracing is different because it involves the integration of laboratory technologies into modelling of actual food chains to trace back and track forward contamination in food or feed. Biotracing aims to operate within food chain systems in such a way that responses to safety signals are increasingly targeted at particular sources, so are less intrusive and are ultimately more efficient. Biotracing is a whole-chain approach to providing tools, both microbiological and mathematical, for tracing contamination in food/feed. The combination of laboratory experiments with chain-related data and chain information flows leads to an embracing nature of biotracing; for example, a biotrace might include transport data, process failure data, as well as microbiological data, into a consistent scheme that provides quantitative information about source level propositions. In this way, biotracing can overcome the problematic predictions obtained from complex food/feed chain scenarios where there may be gaps in the available knowledge. Established quantitative risk assessment methodologies address questions that relate, predominantly, to identified endpoint measures. Biotracing aims to supplement this analysis with an inferential scheme to quantify beliefs about source events such as inline contamination of process failures.

The biotracing idea has been driven by the rapidly increasing amount of data that is now routinely collected in relation to many food safety and public health issues and also by a desire to identify particular elements, such as materials or processes, rather than whole chains for ongoing attention or remedial actions in relation to food safety management. The aim is to harness the information supply to create an effective driver for overall improved food safety, increased efficiency and less waste. More about biotracing is included in a recently published viewpoint by Barker *et al.* (2009).

When the BIOTRACER project began, tracing ideas were influenced by the same thinking as forensics and transfer of liability where endpoint measures are used to indicate failure at a precise space/time location with almost certainty. However, in food chain scenarios, this approach might not be best, because it does little to improve food safety for the consumer and it is often initiated too late in the chain to identify responsible sources with precision. The evolution of biotracing will be a driving force to improve food safety for the consumer in a global fashion based on sound scientific information that supports good process management decisions.

Most recently, it has become possible to identify a significant element of scientific literature that both indicates a desire for biotracing and establishes some of the tools and techniques that can support the initiative. Wilson *et al.* (2008) indicate how multilocus sequence typing (MLST) can be used to trace human *Campylobacter* infections to animal or environmental sources while Viera-Pinto *et al.* (2006) use pulsed-field gel electrophoresis (PFGE) to identify contamination sources and dissemination routes for *Salmonella* in slaughtered pigs. Similarly, Lomonaco *et al.* (2009) use PFGE patterns to select pasteurisation failure or cross-contamination as the probable source of *L. monocytogenes* in Gorgonzola cheese, and Andre *et al.* (2008) employ PFGE and antimicrobial resistance markers to trace the origin of *S. aureus* in cheese to either raw milk or cross-contamination. Gudmundsdottir *et al.* (2005) use electrophoresis to trace the sources of *L. monocytogenes* in Icelandic smoked salmon, and Nakamura *et al.* (2006) pursued similar objectives in the smoked fish chain in Japan using PFGE and polymerase chain reaction (PCR)-based methods. Jarman *et al.* (2008) indicate that chromatography can be combined with spectroscopic data to determine the source of *Bacillus* spores in a non-food chain scenario.

However, inspection shows that these reports, and increasingly others, are incomplete because they do not include full quantification of the trace. That is, they do not provide an apparatus by which posterior probability can be assigned over a list of possible sources. The absence of this assignment is crucial in decision making as, apart from exceptional cases, the source identification is not categorical. Probabilities assigned to potential sources provide the means by which uncertainties can be included into the decision-making process; most food chain scenarios include significant uncertainties.

17.4 Why is biotraceability needed?

In the EU, food safety policy is governed by the White Paper on Food Safety (COM, 1999, 719 Final), which calls for 'a comprehensive integrated approach' to 'ensure a high level of human health and consumer protection'. This policy has been enacted into legislation by the General Food Law of 1 January 2006 (Commission Regulation 178/2002/EC). This legislation places responsibility for traceability on the food/feed producer (Section 4, Articles 18–20). The legislation obliges each company to provide their own records to food safety authorities on request. However, there are no standards on the delivery media or the format of the records. Additionally, there are no indications how the traceability concept, which operates widely and very successfully, can be linked to biotraceability, i.e. to inferential consideration of the passage of foodborne agents rather than actual food units.

When dealing with food contamination, the traditional approach has focused on retrospective testing of food samples to ensure freedom from pathogens. In the case of positive results, tracing systems are usually activated to withdraw the whole production batch from the market or, in more serious cases, activate consumer alert systems. Insufficient attention has been paid to the interface between food testing and (digitalised) traceability systems, i.e. how do we translate retrospective laboratory data and chain information into tracing the source of contamination, and tracking contaminated units of a product that has been already released for consumption.

Increasingly, the ability to trace the origins of contamination in food is recognised, by science and regulation, as a step forwards in safety management. The role of traceability as an incentive for food safety is outlined, in a mathematical model, by Pouliot and Sumner (2008). In their review of risks from contaminated produce, Lynch *et al.* (2009) remark that 'whenever possible, outbreak investigations need to include rapid and detailed traceback and exposure assessments so that likely sources of contamination can be identified as far back as the field of production'. In a recent review of Staphylococci in dairy food, Berry (2008) clearly states when considering threats from resistant strains 'the origin of isolates was not discussed and is of importance'. In a recent report from the American Academy of Microbiology relating to biocrime, Keim (2003) addresses questions concerning the way that evidence can be used to identify the source of identified microorganisms and concludes that 'some pathogen attributes that are unimportant in protecting public health may provide clues in a forensic investigation'. Biotracing applies similar reasoning to investigation of foodborne hazards and to improved food safety.

17.5 What are the gaps in biotraceability?

The feed hygiene regulation (Commission Regulation 183/2005/EC, which became effective on 1 January 2006) lays down a set of measures on hygiene, traceability, conditions and arrangements for feed production. The focus of the regulation is on the safety of animal feed, which is the primary responsibility of the operators in the feed chain. Some of the most important aspects of this regulation are the registration of establishments and rules for record keeping and improved traceability throughout the whole feed chain. Thus, feed is recognised as a possible entry point of contamination into the food production chain. Feed research is, however, largely lagging behind food research in terms of research funds and experts, who usually consider food research as more prestigious. The bulk nature of many feed operations means that biotracing within the feed chain is particularly challenging (e.g. Sapkota *et al.*, 2007).

Second, most microbial models for predicting the behaviour of microorganisms in food environments are based on laboratory studies (e.g. Combase and Micromodel). Microorganisms may behave differently in food, so there are extended uncertainties related to this type of modelling (Schvartzman *et al.* 2010). In order to be able to estimate their persistence and virulence and to predict their behaviour in particular food types at a particular stage of production (Dorrell *et al.*, 2005), biotraceability requires deeper knowledge of the microorganisms' physiology and behaviour when present in food.

Third, the sequencing of microbial genomes (http://www.jcvi.org) has made it possible to construct whole-genome DNA microarrays and to construct tailor made PCR typing methods that can be used to give useful information about the genetic composition of bacterial strains. However, to date, these advanced technologies, which can yield valuable information in a single step, have not been widely applied to tracing and tracking of pathogens in the food chain (Champion *et al.*, 2005), or to model development. PFGE, MLST and many other typing frameworks have been applied to biotracing; particularly, in the form of classification systems that align types with natural reservoirs for biological agents. However, currently, this information is not integrated into food chain models and it is not sufficiently quantified to support strong decision making. The conversion of classification information into matching information, based on uncertain statistics for types, is a target of new biotracing methods, but identification of the appropriate markers and of sufficient analyses is currently lacking. Recent advances in microbial forensics and in molecular epidemiology indicate potential progressions for food chain biotracing (e.g. Keim, 2003).

Finally, there is a shortage of reliable standardised test systems for fast and precise characterisation of pathogenic microorganisms (Malorny *et al.*, 2003). Current methods for microbial analysis of foodborne pathogens are time consuming, have a limited sample throughput and do not provide all of the information needed to assess virulence potential. Even molecular genotyping, which usually involves some form of electrophoresis following culture, is labour intensive, time consuming, not easily automated and the results are not generally comparable between laboratories. Thus, a major focus of BIOTRACER is the improvement of tools for earlier tracking of pathogens (Fig. 17.2).

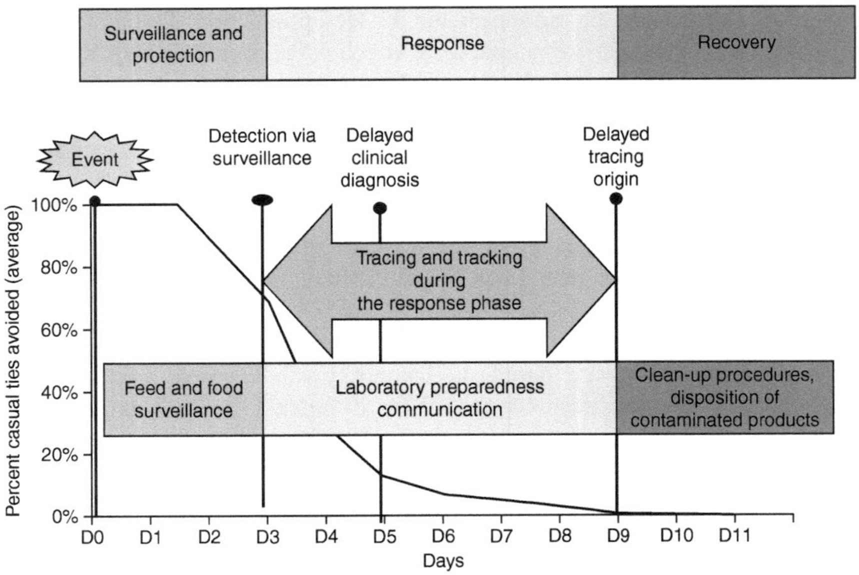

Fig. 17.3 The potential impact of BIOTRACER on early tracing actions and rapid responses. The aim of BIOTRACER is to shift the entire graph to the left by improving detection, increasing preparedness and communication, improving tracking and tracing and therefore avoiding casualties.

17.6 How can these gaps be closed?

There is a need for an integrated approach, providing a set of advanced detection tools and models that collectively identify and estimate the most relevant contamination scenarios and critical points within the whole food chain. Through food and feed chain models, these can be used in order to quantify the likely impact on food safety.

The BIOTRACER project is developing novel frontier technologies and exploiting them for tracking and tracing of microorganisms in selected food and feed chains. The technologies provide data to support models for the behaviour of pathogens in these environments. This information will be made available through regional and national databases, to promote a better understanding of how microbes are transmitted over borders.

17.7 What are the achievements so far?

With just three years completed, BIOTRACER has already made some major advances.

17.7.1 Setting up the biotracing concept

After numerous meetings, the concept of biotracing has been developed in theory and the pros and cons are highlighted (Barker *et al.*, 2009). Relevant databases, such as Combase and Symprevious, were used to present the status of knowledge on selected pathogen/matrix combinations. Other activities dealt with the development of concepts, such as how to integrate biotracing in the Laboratory Information Management System (LIMS) environments, which are in place at many institutions worldwide.

17.7.2 Established proof of principle in simple Bayesian Network models

Simple models, without the complications built into real food chains, have been established to illustrate the inferential element of a biotrace based on (1) Bayesian inference combining food chain information and endpoint observation into posterior belief about sources and (2) inference based on type matching in systems with uncertain statistics for sources and types.

17.7.3 Available literature sources and databases were screened to identify knowledge gaps

A core activity of BIOTRACER is to develop modelling to stages beyond simple data mining within currently accessible databases. Most data in these databases are accumulated from scientific literature without a systematic approach. An illustrative example is that many growth curves are available at temperature conditions complying with laboratory requirements, but not referring to the 'chill chain' as found in reality. By using the data available, we could identify data gaps. These data gaps serve as templates for further research activities, but also as templates for considerations of how to study the impact of these data gaps on the chain model.

17.7.4 Biotracing concepts implemented

Models allowing the study of the impact of chain factors on the sources and the fate of a pathogen upstream and downstream in the chain (so-called biotracing questions) were developed for selected pathogen/food matrix combinations in dairy (*S. aureus* in milk), meat (*Salmonella* in pork) and feed (*Salmonella* in feed) chains.

17.7.5 Understanding the impact of food chain on the physiology, persistence and proliferation of pathogenic microorganisms

Emphasis for the physiological studies was directed at the most pivotal food chain elements such as the slaughtering and post-slaughtering carcass contamination with *Salmonella* (Eijkelkamp *et al.*, 2009), growth capacity of *Campylobacter* at post-harvest levels, adaptation and survival of *Listeria* in plant environments (Belessi *et al.*, 2008) and food dependent toxin gene expression in *S. aureus*.

17.7.6 A greater microbiological toolbox of methods needed for answering biotracing questions

We have developed methods to detect organisms that could be used as targets for deliberate or accidental contamination scenarios. PCR methods were developed both for the detection of classical foodborne pathogens, such as *Salmonella* in feed and pork (Malorny *et al.*, 2008), and also for organisms that could serve as bioweapons (Norovirus and spore-formers such as *B. anthracis* and *C. botulinum*; Fach *et al.*, 2009). The toolbox was further expanded by developing molecular methods useful for studying target cell quantities and tracing or tracking of epidemiologically relevant marker traits or even whole genomes of pathogens (genomotyping).

17.7.7 Exploitation of molecular concepts to study and measure the dependence of pathogens from autochthonous microbial communities

A particular activity of BIOTRACER deals with the development of methods to study whole microbial communities. Most growth data refer to spiking experiments of target microorganisms in food, or in unrelated systems like laboratory media, without taking the important influence of microbial flora present in food into consideration. Efficient research on the interaction between autochthonous flora and contaminant demands methods that elucidate the complete population structure of the most prevalent species.

17.8 Specific achievements to date

BIOTRACER is arranged in five different research areas. Each research area focuses on a particular aspect of biotracing, or on specific pathogen-food/feed matrices. Within each research area, specific advances have been made.

17.8.1 Research area 1: modelling

- We have created the possibility to discuss 'the needs' for operational biotracing from an industrial perspective.
- Current sampling protocols have been re-addressed from a scientific viewpoint and tested as potential evidence streams for biotracing.
- The opportunity for modelling experts to have access to samples actually taken in industry has been generated. Modelling has highlighted both analytical and computational generalities leading to transferable elements in biotracing and has planned the way forward in terms of software implementation (Barker *et al.*, 2009).

17.8.2 Research area 2: feed chain

- New methods to purify the mycotoxins, ochratoxin and aflatoxin have been developed.
- New sampling strategies and integration of new analytical tools, such as use of sol-gel purification and immunoultrafiltration have been used on mycotoxins.

(Andersson and Haggblom, 2009; Koyuncu and Haggblom, 2009; Reiter *et al.*, 2009)

17.8.3 Research area 3: meat chain

- In *Salmonella*, specific genes have been identified that are associated with growth in a matrix resembling the structured matrix of a food.
- Phylogenomics is being transformed into a typing tool such as MLST. This will allow rapid analysis of population structure.

(Josefsen *et al.*, 2007; Malorny *et al.*, 2008; Löfström *et al.*, 2009a; Löfström *et al.*, 2009b; Olsen *et al.*, 2009)

17.8.4 Research area 4: dairy chain

- The safety level of each step involved in the production of pasteurised milk and soft, semi-hard and hard cheeses has been characterised.
- Scientific information has been submitted to international databases, such as PulseNet and Combase.
- The behaviour of the isolated pathogens with regard to the stresses involved in food processing, and with regard to persistence in the industrial environment, has been evaluated.
- Information on molecular, prevalence and phenotypic data of *L. monocytogenes* and *S. aureus* in cheese and milk has been integrated into domain models using advanced statistical and predictive modelling tools.

(Fox *et al.*, 2009; Reichert-Schwillinsky *et al.*, 2009; Tinganitas *et al.*, 2009)

17.8.5 Research area 5: bio-terror

- A PCR assay has been developed for *C. botulinum* using the GeneDisc system.
- A new multiplex PCR assay for *B. anthracis* has been developed.
- An integrated lab-on-a-chip (ILOC) detection system for *B. anthracis* has been demonstrated.
- Whole genome sequencing of *B. anthracis* and *C. botulinum* using the 454-technology has been performed.
- New sampling methods have been developed for Norovirus.

(Fach *et al.*, 2009; Perelle *et al.*, 2009)

Modelling studies relating to different parts of the food/feed chain will result in increased awareness of the risk factors involved in production and what measures can be used to reduce the identified risks (Wein and Liu, 2005). Integration of large volumes of expression data, and data from other high throughput techniques, with information from biological and physiological sources promotes a representation that includes superior beliefs and one that can be used to address new queries that relate to actual conditions experienced in food production.

Sophisticated data-mining algorithms and inference schemes can be used to analyse the data relating to when, where and how products are transported within food chain systems, where and when contamination most likely occurred, in addition to information on the primary level of contamination. Model development will be highly interactive, involving many stakeholders, and in many cases, will be used to develop user-friendly tools (software, expert systems) for operational biotracing.

17.9 Future trends

Strategically, BIOTRACER strives towards the following:

1. *An extension of risk management concepts*: One of the goals of BIOTRACER is to foster decision making based on resilient fundamentals. Resilience in this context means a combination of scientific soundness with a high conceptual flexibility to cover different feed and food contamination scenarios. One vision is that biotracing supports food chain management options, beyond those achieved by hazard analyses or risk assessments, that target follow-on activities in an unobtrusive fashion and so lead to a steady improvement in food safety integrated over a full chain.
2. *Integration of experts of various fields and disciplines*: Most importantly, total food chain approaches require an integration of research along the food chain. Feed specialists were segregated from food specialists for decades, and method developers usually have scientific backgrounds distinct from the expertise of other food-related professions. Thus, from the very beginning, BIOTRACER was devoted to creating a strong interaction among the partners by launching workshops and discussion meetings.
3. *Development of a European science on traceability*: BIOTRACER is only one initiative (EU-funded research project) among others dealing with traceability in a wider context. The coordinators of these projects have agreed on a common understanding to exchange knowledge for the sake of bringing together vital European players in this area. Furthermore, BIOTRACER has developed strong partner dissemination by supporting exchange of ideas and knowledge with young scientists through a mobility programme.
4. *Stakeholder involvement and dissemination*: Since the biotracing concept is a completely new concept, a major focus, and considerable effort, is being put into communicating the innovation to the food/feed industry, stakeholders and interested parties outside the consortium.

17.10 Acknowledgements

BIOTRACER is funded by the EU 6th Framework, Contract #036272.

17.11 References

ANDERSSON G and HAGGBLOM P (2009), 'Sampling for contaminants in feed – estimating the concentration of mycotoxins found in bulk batches of feed can be a complicated process', *Feed International*, **30**, 16–19.

ANDRE M C D P B, CAMPOS M R H, BORGES L J, KIPNIS A, PIMENTA F C and SERAFINI A B (2008), 'Comparison of *Staphylococcus aureus* isolates from food handlers, raw bovine milk,

Minas Frescal cheese by antibiogram and pulsed-field gel electrophoresis following SmaI digestion', *Food Control*, **19**, 200–7.

BARKER G C, GOMEZ N and SMID J (2009), 'An introduction to biotracing in food chain systems', *Trends Food Sci. Technol.*, **20**, 220–6.

BELESSI C I, PAPANIKOLAOU S, DROSINOS E H and SKANDAMIS P N (2008), 'Survival and acid resistance of *Listeria innocua* in Feta cheese and yoghurt, in the presence or absence of fungi', *J. Food Prot.*, **71**, 742–9.

BERRY E (2008), 'Methecillin-Resistant Staphylococci and dairy food', *IDF Animal Health Newsletter*, **2**, 12.

CHAMPION O L, GAUNT M W, GUNDOGDU O, ELMI A, WITNEY A A, HINDS J, DORRELL N and WREN B W (2005), 'Comparative phylogenomics of the foodborne pathogen *Campylobacter jejuni* reveals genetic markers predictive of infection source', *Proc. Natl. Acad. Sci.*, **102**, 16043–8.

COM (1999), '719 final, 12 January 2000', *E.C.* White Paper on Food Safety. Available from: http://ec.europa.eu/dgs/health_consumer/library/pub/pub06_en.pdf (accessed 23 February 2009).

Commission Regulation (EC) No 178/2002 of the European Parliament and of the Council of 28 January 2002 laying down the general principles and requirements of food law, establishing the European Food Safety Authority and laying down procedures in matters of food safety. OJ L31, 01.02.2002, 1–24.

Commission Regulation (EC) No 183/2005 of the European Parliament and of the Council for feed business establishments manufacturing or placing on the market feed additives of the category coccidiostats and histomonostats, OJ L35, 08.02.2005, 1–22.

DORRELL N, HINCHLIFFE S J and WREN B W (2005), 'Comparative phylogenomics of pathogenic bacteria by microarray analysis', *Curr. Opin. Microbiol.*, **8**, 620–6.

EIJKELKAMP J M, AARTS H J M and VAN DER FELS-KLERX, H J (2009), 'Suitability of rapid detection methods for *Salmonella* in poultry slaughterhouses', *Food Anal. Methods*, **2**, 1–13.

FACH P, MICHEAU P, MAZUET C, PERELLE S and POPOFF M (2009), 'Development of real-time PCR tests for detecting botulinum neurotoxins A, B, E, F producing *Clostridium botulinum*, *Clostridium baratii* and *Clostridium butyricum*', *J. Appl. Microbiol.*, **107**(2), 465–73.

FOX E, O'MAHONY T, CLANCY M, DEMPSEY R, O'BRIEN M and JORDAN K (2009), '*Listeria monocytogenes* in the Irish dairy farm environment', *J. Food Prot.*, **72**, 1450–6.

GUDMUNDSDOTTIR S, GUDBJORNSDOTTIR B, LAUZON HL, EINARSSON H, KRISTINSSON KG and KRISTJANSSON M (2005), 'Tracing *Listeria monocytogenes* isolates from cold-smoked salmon and its processing environment in Iceland using pulsed-field gel electrophoresis', *Int. J. Food Micro.*, **101**, 41–51.

JARMAN K H, KREUZER-MARTIN H W, WUNSCHEL D S, VALENTINE N B, CLIFF J B, PEARSON C E, COLBURN H A and WAHL K L (2008), 'Bayesian-integrated microbial forensics', *Appl. Environ. Microbiol.*, **74**, 3573–82.

JOSEFSEN M H, KRAUSE M, HANSEN F and HOORFAR J (2007), 'Optimization of a 12-hour TaqMan PCR-based method for detection of *Salmonella* in meat', *Appl. Environ. Microbiol.*, **73**, 3040–8.

KEIM P (2003), *Microbial Forensics: A Scientific Assessment*, American Academy of Microbiology, Washington, DC.

KOYUNCU S and HAGGBLOM P (2009), 'A comparative study of cultural methods for the detection of *Salmonella* in feed and feed ingredients', *BMC Vet. Res.* **5**, 6.

LÖFSTRÖM C, HOORFAR J, SCHELIN J, RADSTROM P and MALORNY B (2009a), '*Salmonella*', in *Molecular Detection of Foodborne Pathogens*, ed. Dongyou Liu, CRC Press, Boca Raton, FL, 495–512.

LÖFSTRÖM C, KRAUSE M, JOSEFSEN MH, HANSEN F and HOORFAR J (2009b), 'Validation of a same-day real-time PCR method for screening of meat and carcass swabs for *Salmonella*', *BMC Microbiology*, **9**, 85.

LOMONACO S, DECASTELLI L, NUCERA D, GALLINA S, BIANCHI DM and CIVERA T (2009), '*Listeria monocytogenes* in Gorgonzola: subtypes, diversity and persistence over time', *Int. J. Food Micro.*, **128**, 516–20.

LYNCH M F, TAUXE R V and HEDBERG C W (2009), 'The growing burden of foodborne outbreaks due to contaminated fresh produce: risks and opportunities', *Epidemiol. Infect.*, **137**, 307–15

MALORNY B, LÖFSTRÖM C, WAGNER M, KRÄMER N and HOORFAR, J (2008), 'Enumeration of *Salmonella* in food and feed samples by real-time PCR for quantitative microbial risk assessments', *Appl. Environ. Microbiol.*, **74**, 1299–304.

MALORNY B, TASSIOS P, RÅDSTRÖM P, COOK N, WAGNER M and HOORFAR J (2003), 'Standardization of diagnostic PCR for the detection of foodborne pathogens', *Int. J. Food Micro.*, **83**, 39–48.

NAKAMURA H, TOKUDA Y, SONO A, KOYAMA T, OGASAWARA J, HASE A, HARULI K and NISHIKAWA Y (2006), 'Molecular typing to trace *Listeria monocytogenes* isolated from cold smoked fish to a contamination source in a processing plant', *J. Food Prot.*, **69**, 835–41.

OLSEN K N, LUND M, SKOV J, CHRISTENSEN L S and HOORFAR J (2009), 'Towards real-time monitoring of broiler flocks: detection of *Campylobacter* in air samples for continuous monitoring of *Campylobacter* colonization in broiler flocks', *Appl. Environ. Microbiol.*, **75**, 2074–8.

PERELLE S, CAVELLINI L, BURGER C, BLAISE-BOISSEAU S, HENNECHART-COLLETTE C, MERLE G and FACH P (2009), 'Use of a robotic RNA purification protocol based on the NucliSens R easyMAGtm for real-time RT-PCR detection of hepatitis A virus in bottled water', *J. Virol. Methods*, **157**, 80–3.

POULIOT S and SUMNER D A (2008), 'Traceability, liability and incentives for food safety and quality', *Am. J. Agric. Econ.*, **90**, 15–27.

REICHERT-SCHWILLINSKY F, PIN C, DZIECIOL M, WAGNER M and HEIN I (2009), 'Stress and growth-rate-related differences between plate count and real-time PCR data during growth of *Listeria monocytogenes*', *Appl. Environ. Microbiol.*, **75**(7), 2132–8.

REITER E V, ZENTEK J and RAZZAZI-FAZELI E (2009), 'Review on sample preparation strategies and methods used for the analysis of aflatoxins in food and feed', *Mol. Nutr. Food Res.*, **53**, 508–24.

SAPKOTA A R, LEFFERTS L Y, MCKENZIE S and WALKER P (2007), 'What do we feed to food production animals? A review of animal feed ingredients and their potential impacts on human health', *Environ. Health Perspect.*, **115**(5), 663–70.

SCHVARTZMAN M S, BELESSI X, BUTLER F, SKANDAMIS P and JORDAN K (2010) 'Comparison of growth limits of *Listeria monocytogenes* in milk, broth and cheese, *Journal of Applied Microbiology*, in press.

TIGANITAS A, ZEAKI N, GOUNADAKI A S, DROSINOS E H and SKANDAMIS P N (2009), 'Study of the effect of lethal and sublethal pH and aw stresses on the inactivation or growth of *Listeria monocytogenes* and *Salmonella Typhimurium*', *Int. J. Food Micro.*, **134**(1/2), 104–12.

VIERA-PINTO M, TENREIRO R and MARTINS C (2006), 'Unveiling contamination sources and dissemination routes of *Salmonella* sp. in pigs at a Portuguese slaughter house through macrorestriction profiling by pulsed field get electrophoresis', *Int. J. Food Micro.*, **110**, 77–84.

WEIN L M and LIU Y (2005), 'Analysing a bioterror attack on the food supply: the case of botulinum toxin in milk', *Proc. Natl. Acad. Sci.*, **102**, 9984–9.

WILSON D J, GABRIEL E, LEATHERBARROW A J H, CHEESBROUGH J, GEE S, BOLTON E, FOX A, FEARNHEAD P, HART CA and DIGGLE P J (2008), 'Tracing the source of Campylobacteriosis', *PloS Genet.*, **4**(9), e1000203.

Part IV

Tracing pathogens in particular food chains

18

Tracing pathogens in red meat and game production chains and at the abattoir

P. Whyte, S. Fanning, S. O'Brien, L. O'Grady and K. Solomon, University College Dublin, Ireland

Abstract: Meat can be contaminated by a range of pathogenic microorganisms, including *Salmonella*, *E. coli* O157 and *Campylobacter* which impact significantly on public health in terms of morbidity and mortality. This may be further exacerbated by the emergence of resistance to a number of clinically relevant antimicrobials. In recent years, it has become apparent that an integrated approach to meat safety is required which assesses where hazards may be introduced or where risks increase in the entire meat chain along with the implementation of targeted preventive interventions ('farm to fork'). This approach has facilitated the implementation of sequential risk reduction along the entire food chain using the HACCP concept. By incorporating appropriate controls at the pre-harvest level, including, for example, good husbandry practice, herd health and disease management programmes and prudent drug use, along with good hygiene practices at slaughter level and greater food chain surveillance, adequate levels of safety assurance can be achieved.

Key words: pathogens, verocytotoxigenic *E. coli*, *Salmonella*, *Campylobacter*, control, risk, meat.

18.1 Introduction

18.1.1 General background

Red meat is a regularly consumed commodity in Western diets and increasingly in populations in developing countries around the world. Indeed lean meat is accepted as a wholesome and healthy source of protein, iron and other essential nutrients. Global demand for red meat, including beef, has increased in recent years and this trend is expected to continue with growth particularly in developing countries such as China, India and Brazil as well as Russia.

Average annual per capita consumption of meat (including bovine, ovine, porcine and poultry meat) in the world in 2007 was estimated at 42.1 kg, ranging from 31.2 to 82.3 kg for developing and developed countries respectively (FAO, 2008). In the United States the average per capita consumption of red meat in 2007 was estimated at circa 53 kg per person of which beef accounted for 29.5 kg (USDA, 2007). Data within the European Union for the year 2007 revealed that per capita beef consumption varied widely between member states where information was available with ranges from 6.6 to 32.5 kg reported (Eurostat, 2008).

In a global context, the red meat industry is a vital component of modern agriculture and plays a significant role in national and international trade and economic activity. Estimates of cattle populations globally range from approximately 1 to 1.5 billion, with for example large populations in India (281 million), Brazil (187 million), China (140 million), the United States (97 million) and the European Union (90 million) (FAO, 2005; Cattlenetwork, 2008; USDA, 2008a, 2008b). The largest exporters of beef are Australia, Brazil, the United States and the European Union (Chatellier *et al*., 2003; Horchner *et al*., 2006; USDA, 2008a). Total global meat production was reported at 278.5 million tonnes for 2007 of which 66.4 million tonnes was of bovine origin (FAO, 2008). The monetary value of the red meat industry annually is several hundreds of billions of Euro/dollars with for example beef production in the United States alone worth $36.1 billion in 2007 (USDA, 2008b, 2008c).

18.1.2 Foodborne disease and red meat

The magnitude of the burden of foodborne disease in humans is enormous and has significant associated economic costs (USDA, 2009). For example, in the United States it is estimated that approximately 76 million illnesses occur each year, resulting in 325 000 hospitalisations and 5000 deaths (Mead *et al*., 1999). Red meat has frequently been reported to be contaminated with pathogenic microorganisms resulting in significant disease and morbidity in human populations (Rangel *et al*., 2005; Doyle and Erickson, 2006). Beef was identified as the vehicle of transmission in up to 3.4% of foodborne outbreaks reported in the United States between 1993 and 1997 (CDC, 2000). In 2006 in the European Union, 10.3% of reported outbreaks where a vehicle was identified were attributed to the consumption of contaminated red meat/meat products (EFSA, 2007). An extensive study analysing international outbreak data for source attribution by Greig and Ravel (2009) reports that 12.2% of outbreaks reviewed were caused by beef or beef products, with a variety of pathogens implicated (Greig and Ravel, 2009).

18.1.3 Evolution of food safety in meat production

Towards the end of the nineteenth century, with advances in veterinary medicine and other related disciplines, it became apparent that contaminated meat was responsible for widespread morbidity in humans and the need for regulated public health controls was recognised. During the 1890s Robert von Ostertag, a

German veterinarian, began to develop an organoleptic approach to assessing the suitability of meat as fit for human consumption (Bulling and Schonberg, 1999). The principles of his inspection-based approach were enshrined in legislation in Germany in 1900 and he is acknowledged as 'the father' of traditional *ante-* and *post-mortem* veterinary meat inspection. This approach has been adopted in most developed countries since then, including for example, the United States as summarised by the Food and Drug Administration (Young, 1981). These controls were based around the principles of visual inspection, palpation and incision of lymph nodes, various muscle tissues and organs of food animal carcases to detect pathological evidence of parasitic infestation or microbiological infection. This system has to date achieved considerable success in identifying conditions such as tuberculosis or *Cysticercus bovis* in bovines and removal of high-risk carcases or tissues from the human food chain. It has improved public health protection and facilitated commerce through trade in animals, meat and meat products by providing some level of safety and quality assurance (Berends, 1998).

However, it is now universally recognised that the risk of transmission of such pathogens through meat is substantially lower in developed countries due to modern husbandry methods and the implementation of effective pre-harvest eradication or control measures in production animals thus reducing the risk of exposure in consumers. Furthermore, in recent years, based on epidemiological and attribution studies associated with human foodborne disease surveillance of outbreaks and cases, it is obvious that a range of pathogens including haemorrhagic *Escherichia coli* (EHEC), *Campylobacter* or salmonellae are now the most significant hazards transmitted through foods of animal origin. These zoonotic agents are commonly carried by animals asymptomatically and indeed their spread within and between herds can be facilitated by modern animal production systems. Conventional meat inspection procedures were not designed to detect these meatborne hazards and indeed certain aspects, for example incision, can even exacerbate contamination and have a detrimental effect on meat safety or quality. As a consequence, it has been noted that traditional meat inspection techniques alone are inadequate in attaining the required levels of public health protection and that additional risk-based measures are required.

In recent years veterinarians and other public health professionals have identified the need for a more integrated approach to food safety management. This is based on concepts and tools which include longitudinal integration, risk assessment, sequential risk reduction from farm to fork, process control, hazard monitoring and surveillance, food chain information and traceability. These objectives can be achieved through for example the implementation of good farming practices (GFPs), good manufacturing practices (GMPs), good hygienic practices (GHPs) and the application of hazard analysis critical control point (HACCP) systems. Those principles have now been enshrined in the legislation of many developed countries, including for example the United States (USDA, 1996) and the European Union (Anonymous, 2004a, 2004b, 2004c).

18.1.4 Monitoring and surveillance in the red meat chain

Monitoring and surveillance of hazards forms an essential component underpinning any modern food safety management system. Integrated approaches are required to produce data which establishes prevalences of the more significant microbial pathogens in feeds, food animal populations and in foods of animal origin at all pertinent stages of the food chain. Such information allows trends in zoonotic agent prevalences to be compared over time in animals at herd, regional, national or international levels and enables policy makers to more effectively assess risks and ascertain if, when and at what points in a food chain interventions may be required. Well-designed surveillance (both passive and active) would ensure timely intervention and ultimately protect food chains. Risk managers also need to consider the necessity of vigilance in relation to identifying and assessing threats of new, emerging or re-emerging food safety hazards (Whyte *et al.*, 2005; Doyle and Erickson, 2006). Robust surveillance of foodborne disease in human populations is also critical so that risks associated with relevant hazards can be assessed and rankings made to assist with the prioritisation and mobilisation of resources where the public health impact can be maximised.

Such information is also vital in order to protect trade in animals and meat by providing the necessary reassurances. Furthermore, subsets of pathogens isolated from these sources can be further characterised and sub-typed using both phenotypic and genotypic methods. This could provide relevant public health professionals and management within food business operations with information to enable them to more accurately assess the risks and significance of detecting potential hazards in foodstuffs. (This area will be discussed in more detail in Section 18.3).

18.2 Foodborne pathogens in red meat and their public health significance

18.2.1 Verocytoxigenic *E. coli* (EHEC)

Pathogenic *E. coli* are categorised by their virulence characteristics, epidemiological and clinical effects, mechanism(s) of pathogenicity, distinct O:H serogroups and mode(s) of attachment to the host cells (Doyle *et al.*, 1997). These categories include enteropathogenic *E. coli* (EPEC) strains, enteroinvasive *E. coli* (EIEC) strains, enterotoxigenic *E. coli* (ETEC) strains, enteroaggregative *E. coli* (EAEC) strains, diffuse-adhering *E. coli* (DAEC) strains and enterohaemorrhagic *E. coli* (EHEC) strains.

Over the last three decades the EHEC group has become of considerable importance as causative agents in major food poisoning outbreaks, causing severe and potentially life-threatening illness. EHEC strains were first identified as human pathogens in 1982, when *E. coli* O157:H7 was associated with two separate outbreaks of severe bloody diarrhoeal syndrome in the USA (Riley *et al.*, 1983). Other serotypes identified from the EHEC group as major causes for concern include: O26:H11, O103:H2, O111:H8, O145:H28 and O157:NM (Non-Motile) (Doyle *et al.*, 2001; Mora *et al.*, 2007).

EHEC strains are characterised by the expression of verotoxins (vt1 and vt2), the presence of a 60 MDa plasmid involved in pathogenesis and a 90 kDa outer membrane protein, intimin, known to have a major part in the formation of attachment and effacement (AE) lesions. The AE property is chromosomally encoded by the 41 genes that are present on a 'pathogenicity island' referred to as the locus of enterocyte effacement (LEE) (Saunders *et al.*, 1999). LEE encodes the adhesion protein called intimin (*eaeA gene*) and the effector molecules (EspA, EspB and EspD) of a type III secretion system that are responsible for the genesis of the AE lesion (Morabito *et al.*, 2003). The pathogen expresses the receptor (intimin), which is translocated into the host epithelial cells. This allows the pathogens to attach to the epithelial cells of the large intestine, resulting in the effacement of underlying microvilli followed by the formation of characteristic AE lesions (Doyle *et al.*, 1997; Ludwig and Muller-Wiefel, 1998).

Infection by EHEC can have a number of consequences, from symptom-free carriage (Swerdlow and Griffin, 1997) through to mild uncomplicated diarrhoea, to severe haemolytic colitis (HC) and haemolytic uremic syndrome (HUS) (Heuvelink, 2000). The incubation period from ingestion of the pathogen to the appearance of the first symptoms can vary widely, that is between 1 and 8 days. Generally the illness begins with abdominal cramps and non-bloody diarrhoea, which may develop into bloody diarrhoea in 2–3 days (Griffin, 1995; Mead and Griffin, 1998). The illness may progress into HC, HUS and sometimes thrombotic thrombocytopenic purpura (TTP) (a variant form of HUS where patients become more febrile with a marked neurological involvement) (Coia, 1998; Karmali, 1989; Morrison *et al.*, 1986). Treatment for serious cases of EHEC infection is mainly supportive care. The mainstay of therapy for colitis is fluid and electrolyte management, monitoring for blood loss and observation for the progression to HUS. Again, while management of HUS is mainly supportive, it may include dialysis, hemofiltration and transfusion of packed electrolytes (Kaplan *et al.*, 1998; Neild, 1998).

It has been suggested that the infectious dose of EHEC O157:H7 may be as low as 10 organisms (Coia, 1998). Such suggestions are supported by the findings of an outbreak study reporting infection following consumption of food products containing as few as two cells per 25 g (Willshaw *et al.*, 1994). The infectious dose of other EHEC may also be very low making their survival and growth in food or the environment a significant public health risk.

The incidences of EHEC infections are highest amongst children and can vary greatly between geographical region: 0.7 cases per 100 000 in American children <15 years of age (WHO, 2005a), 3.4 per 100 000 in Scottish children and 1.54 per 100 000 in English children <5 years of age (Lynn *et al.*, 2005). Up to 85% of cases are as a result of exposure to the pathogen through food (WHO, 2005a). Furthermore, EHEC accounts for approximately 90% of all HUS in childhood (Taylor, 2008).

EHEC strains can be transmitted by food, water, person-to-person or animal-to-person contact and occasionally through occupational exposure (e.g. laboratory workers). The gastrointestinal tract of cattle is the major reservoir of EHEC with most infections being caused by the ingestion of contaminated food of bovine origin, particularly undercooked minced beef (Meng and Doyle, 1998; Tozzi *et al.*, 2001). Meat most likely becomes contaminated during slaughter, with dissemination of such contamination during further processing and mixing of meats from different animals and batches. The problem becomes compounded when meat is minced and reformed, as the pathogen can be introduced to the interior of the reformed products such as burgers, where it may be partially protected from lethal surface temperatures during inadequate cooking (Heuvelink, 2000). Other foods of bovine origin such as raw (unpasteurised) milk and raw milk products (i.e. cheese and yoghurt) have been implicated as sources and vehicles of EHEC infection (Ansay and Kaspar, 1997; Coia *et al.*, 2001). Apart from these primary sources of contamination the dispersion of untreated manure in the environment can spread EHEC to secondary vehicles of human infection (Coia, 1998). These include other animals in contact with ruminants (e.g. horses and rabbits), fruits (e.g. unpasteurised apple juice), cider, vegetables (e.g. sprouts, potatoes and lettuce) fertilised with contaminated manure and contaminated drinking or surface waters. Several reports identified petting farms, open zoos and domestic pets as sources for EHEC O157 infections, implicating exposure to infected calves, horses, goats, cats and dogs (Chapman *et al.*, 2000; CDC, 2005a; Bentancor *et al.*, 2007).

18.2.2 *Salmonella*

Salmonella spp. are the aetiological agents of typhoid fever, a systemic infection, and a generally self-limiting infection of the intestinal epithelium, known as gastroenteritis, both of which are major public health threats. Salmonellosis is one of the most common and widely distributed foodborne diseases. It constitutes a major public health burden and represents a significant cost in many countries. In the United States, an estimated 1.4 million non-typhoidal *Salmonella* infections, resulting in 168 000 visits to physicians, 15 000 hospitalisations and 580 deaths are reported annually. The total cost associated with *Salmonella* is estimated at US$ 3 billion annually in the United States (WHO, 2005b).

Animal infection with *Salmonella* may occur in all red meat production species and is present worldwide. Infection is *via* faecal oral route. Clinically these infections present commonly as septicaemia, enteritis and abortion. Other, less common, manifestations include arthritis, dry gangrene and osteomyelitis. Subclinical chronic infections are also an important part in the epidemiology of *Salmonella* infections. Shedding from subclinical carriers is variable with increased shedding associated with periods of stress.

The transmission of *Salmonella* between farms is *via* direct or indirect means. Infected animals (often subclinical) are the main source of infection. These animals may either spread infection in a herd *via* direct contact or indirectly *via* environmental contamination with faeces. *Salmonella* may persist for up to 6 years in dried bovine faeces. Other potential vectors of transmission include contaminated feed or water supplies, milk from infected animals, infected wildlife and vermin or *via* contaminated vehicles and farm visitors.

Persistence of carrier status in infected animals, resistant environmental contamination and potential wildlife/vermin reservoirs make the elimination from infected premises very challenging. Currently more than 2500 serotypes of *Salmonella enterica* are known. The antigenic formulae of *Salmonella* serotypes are defined and maintained by WHO Collaborating Centre for Reference and Research on *Salmonella* at the Pasteur Institute, France, and new serotypes are listed in annual updates of the Kauffmann-White Scheme (Brenner *et al.*, 2000) – a scheme based on antigenic variation in the outer membrane lipopolysaccharide (O), phase 1 (H1) and phase 2 (H2) flagella and a capsular (Vi) antigen that occurs only in serotypes Typhi, Paratyphi and Dublin (D'Aoust, 2007).

The ubiquity of salmonellae in the natural environment, coupled with intensive animal husbandry practices used in the meat and fish industries and the recycling of offal and inedible raw materials into animal feeds, has favoured the continued prominence of this pathogen in the global food chain. Mammals, birds, rodents, reptiles, amphibians and insects contribute to the environmental reservoir of salmonellae and can contaminate humans by direct contact and indirectly by faecal contamination of waters (D'Aoust, 2007).

Human infection is initiated by the oral route either directly following the consumption or handling of contaminated food including red meat, poultry and others. Epidemiological links between the hosts, along with an ability of the organism to survive in the environment, have led to complex cycles of infection with humans being the final host.

Contaminated poultry products are widely recognised as an important source of *Salmonella* infection (Cogan and Humphrey, 2003). In addition, the aquaculture industry is particularly vulnerable to contamination as fish and shellfish are reared in unprotected earthen ponds in third world countries. Furthermore, the use of wash waters of uncertain microbiological quality, non-specialised workers with poor personal hygiene and the use of untreated human and animal wastes and of *Salmonella*-contaminated feeds increase the hazard to humans as these products are frequently consumed raw or lightly cooked (D'Aoust, 2007). It has also been well documented that raw milk from ovine, bovine and caprine sources are potential vehicles of human infections with *Salmonella*. Also the fresh fruit and vegetable industries encounter potentially hazardous practices such as the use of contaminated waters, fertilisation with raw or partially composted animal manure and sewage sludge with infected wildlife contaminating crops (D'Aoust, 2007).

Improving the microbiological quality of foods alone is insufficient to reduce the considerable mortality and morbidity throughout the world caused by *Salmonella* since food-processing technologies cannot guarantee its absence. Therefore, efforts must be made to adhere strictly to hygiene measures in food processing technologies by following GHPs, GMPs and stringently implementing HACCPs along the complete food chain. At present, it is widely acknowledged that the most cost effective method of controlling foodborne hazards from farm to fork is the HACCP system (Panisello *et al.*, 2000).

18.2.3 *Campylobacter*

Campylobacters are now the leading cause of bacterial gastroenteritis in humans in many developed countries (Tauxe, 2002, Vierikko *et al.*, 2004). *Campylobacter jejuni* and to a lesser extent *Campylobacter coli* are the main species associated with disease in humans (Moore *et al.*, 2005; Foley and McKeown, 2007). Most cases of campylobacteriosis in humans result in a self-limiting enterocolitis with diarrhoea lasting up to 24 hours, while others produce more severe symptoms including bloody diarrhoea, abdominal cramps and vomiting lasting sometimes in excess of 10 days (FSAI, 2002). The high incidence of disease associated with these pathogens, their low infectious dose in humans and potentially severe sequelae (including Guillain–Barré syndrome), make it a very significant public health issue. A substantial proportion of cases are transmitted through foods and have been associated with the consumption or handling of contaminated meat, particularly poultry (Mead *et al.*, 1999; Smerdon *et al.*, 2001; Little *et al.*, 2008).

In the United States it has been estimated that up to 2.4 million human cases annually are attributable to *C. jejuni* infection from a total of 5.2 million cases of bacterial foodborne disease (Mead *et al.*, 1999; CDC, 2005b). In the European Union, a total of 176 013 cases of campylobacteriosis were reported in 2006 with a mean crude incidence rate (CIR) of 46.1 per 100 000 population (EFSA, 2007). Reported CIRs between member states varied highly from 0.3 to 220.2 cases per 100 000 population. These observed variations may be attributable to real differences in exposure and subsequent human cases of disease or differences in laboratory and reporting structures used in each country. Furthermore, there were substantive differences between member states in the reported origin of cases (imported/domestic) of campylobacteriosis (EFSA, 2007). It is also widely accepted that there is gross underreporting of disease and that the true incidence of campylobacteriosis could be anything from 8 to 100 times greater than the actual numbers of cases reported in developed countries (Madsen, 2007).

The reservoir of campylobacters is the gastrointestinal tract of warm-blooded animals, as this zoonotic agent has been recovered from feral animals, wild birds as well as pets such as cats and dogs (Pacha *et al.*, 1998; Wahlstrom *et al.*, 2003; Acke *et al.*, 2009). In the vast majority of instances, intestinal carriage of *C. jejuni* or *C. coli* in animals is asymptomatic. Although contaminated poultry meat is thought to be the most significant risk factor associated with *Campylobacter* infection, these pathogens display a complex epidemiology and are extensively found in a wide range of animal populations and environmental sources. As a result, other routes of transmission to humans cannot be overlooked (Humphrey *et al.*, 2007).

Other domestic animals, including cattle, pigs and sheep, are frequently colonised by *Campylobacter* species, and raw meat derived from these species may be contaminated with these pathogens (Stanley *et al.*, 1998; Stanley and Jones, 2003; Payot *et al.*, 2004). A total of 58% of Irish feedlot cattle were found to be carriers of *Campylobacter* species (Minihan *et al.*, 2004). Humphrey *et al.* (2007) in a review article found that mean percentage prevalences of campylobacters recovered from dairy and beef cattle were 30% (range 6–64%) and 62.2% (range 42–83%) respectively from data obtained from 21 countries.

Data from the same review reported prevalences in sheep and pigs at levels of 31.1% (range 18–44%) and 61% (range 50–69%) respectively. These data clearly demonstrate the potential risks associated with these food animal species.

It has been estimated that up to 50% of human cases of campylobacteriosis may be explained by poultry meat as a source, thus leaving a further 50% whose origin is more difficult to determine (Madsen, 2007). Raw red meat has been identified as one such potential source of infection, although prevalences are in general lower but can vary widely (Humphrey and Jorgensen, 2006; Little *et al.*, 2008). Whyte *et al.* (2004) reported prevalences of campylobacters in retail samples of beef, pork and lamb at rates of 3.2%, 5.1% and 11.8% respectively. A review of prevalence data in retail samples from 21 countries reported mean rates of campylobacters in beef, pork and lamb of 2.7% (range 0–9.8%), 2% (range 0–5.1%) and 6% (range 0–12.2%) respectively (Humphrey *et al.*, 2007). Higher prevalences and concentrations have been detected in edible offal (e.g. kidney, liver, heart) and consumption and handling of these may pose a far greater risk to public health (Madsen, 2007).

18.3 Potential amplification steps and control of enteropathogens in red meat and game production chains

18.3.1 Entry and amplification – pre-harvest

The infection with or carriage of enteric pathogens such as *Salmonella, Campylobacter* and enterohaemorrhagic *E. coli* in red meat animals represents a significant risk to public health (Humphrey and Jorgensen, 2006). Indeed foodborne transmission of these agents is often the most likely route of exposure to humans as a result of the consumption or handling of contaminated meat derived from these animals. A pre-requisite for the identification of any effective measures to control these pathogens in animal populations is a comprehensive understanding of their aetiology and epidemiology at this stage in the food chain. Unfortunately substantial data gaps remain in relation to the reservoirs, modes of transmission and dissemination of some of these pathogens, thus making meaningful intervention at primary production level difficult. These challenges are further compounded by the fact that most infection in animals by such organisms is asymptomatic.

Numerous opportunities exist in food animal production for the introduction and spread of hazards of public health significance. Furthermore transport and slaughter of animals and processing of carcases may result in exposure to, and contamination by, enteric pathogens, thereby compromising the safety of end products. A number of relevant stages exist where enteric zoonotics can be introduced and concentrations may be amplified.

At the pre-harvest level, livestock may be either intensively or extensively raised depending on a number of factors, including local practices, economics and consumer preferences to mention a few. As a consequence, the on-farm epidemiology associated with foodborne enteric pathogens is often highly complex, variable and pathogen-specific. In general, a range of potential risk factors exist pertaining to the colonisation of livestock by human enteropathogens; these include ingestion of contaminated feedstuffs or water, contact with a contaminated environment, direct contact with infected animals within a cohort or introduction *via* vectors (Humphrey and Beckett, 1987; Minihan, Whyte, O'Mahony, Clegg, *et al.*, 2003; Humprey and Jorgensen, 2006). Often, once introduced, pathogens can spread rapidly within groups resulting in high proportions of animals being infected. Infection may be transient or prolonged and numbers of infected animals within a cohort may increase or decrease over time. These epidemiological features are not fully understood but a number of possibilities have been suggested including seasonality altering the level of challenge animals are exposed to or development of host immune responses (Minihan, Whyte, O'Mahony, Clegg, *et al.*, 2003; LeJeune *et al.*, 2004). In addition, concentration of pathogens carried in the intestinal tract and consequently shed in faecal material can vary markedly between individuals or herds and is dependent on both the host and pathogen species. For example, reported faecal shedding rates for *E. coli* O157 in positive beef cattle can vary from <100 cfu/g to >10^7 cfu/g (Laegreid *et al.*, 1999; Omisakin *et al.*, 2003; Ogden *et al.*, 2004). Reasons for these observed differences may include age of the animal, commensal gut flora, immune status, initial level of pathogen exposure, seasonality, underlying host susceptibility (genotypic) or the pathogens' ability to colonise as well as husbandry methods used (van Donkersgoed *et al.*, 1999; Nielsen *et al.*, 2002).

As cattle and sheep are generally extensively reared, attaining stringent levels of control is difficult and in many instances is neither practical nor feasible for several of these commensal enteropathogens. Also, there is always potential for repeated exposure (Humphrey and Jorgensen, 2006; Madsen, 2007). However, implementation of good farming and husbandry practices, along with other pathogen-specific risk management measures, is recommended to minimise the risk of introduction or to control the spread, particularly where clinical disease is evident (for example salmonellosis in cattle) so that cases are detected as early as possible, removed from the herd and treated appropriately (Kabagambe *et al.*, 2000; Warnick *et al.*, 2001).

High levels of faecal soiling on cattle hides is a significant risk as this can increase the pathogen load on cattle presented for slaughter (O'Brien *et al.*, 2005). The type of farming system (e.g. cattle kept predominantly indoor vs. outdoor), bedding, climatic conditions and diet can also influence levels of contamination on hides (McGee *et al.*, 2004; Kuhnert *et al.*, 2005). The hides of cattle may also become excessively dirty during transport and holding at the abattoir, for example due to inadequate cleaning of vehicles or pens in the lairage, overstocking of animals and increased stress. Ultimately, the additional challenge posed by excessively dirty cattle at slaughter can increase the risk of bacteria, including zoonotic enteropathogens, being transferred to the dressed carcase and meat during the dehiding procedures (McEvoy *et al.*, 2000; Nastasijevic *et al.*, 2008, 2009).

18.3.2 Entry and amplification – harvest

Fresh meat is an ideal substrate to support microbial growth and consequently is susceptible to rapid spoilage or pathogen growth unless some of its properties are modified or it is stored in such a manner so as to retard microbial activity and reproduction (Jackson *et al.*, 2001). The water availability (a_w) of fresh muscle is high at 0.99 with a total water content of approximately 75% (Kraft, 1992). It also contains other essential components including peptides and amino acids, fats, carbohydrates (glycogen), soluble phosphorous and trace elements necessary to support microbial growth.

The main sources of microbial contamination of meat are the skin/hide, the intestinal tract, the nasopharyngeal cavities and the urogenital tract (O'Brien *et al.*, 2005). Carcases and meat may also become contaminated from external sources such as the slaughter plant environment, including reservoirs in the lairage, slaughter hall and equipment, chills, deboning areas and staff. Grau (1986) suggested that meat animals could be regarded as a source of edible tissue sandwiched between two layers that are heavily contaminated with microorganisms. Muscle tissue in live animals on occasion may also be contaminated with bacteria, if for example the animal is clinically infected or bacteraemic at time of slaughter.

From an operational perspective, bacteria may initially be introduced into an animal during the sticking/bleeding stage. As this procedure initially requires penetration of the skin, external contaminants can be transferred into blood vessels which are severed to facilitate bleeding. Although the animal is desensitised, the heart remains beating and contaminants may enter the circulatory system and be distributed to edible muscle tissues around the body.

Dehiding of cattle is one of the most significant points of entry of pathogens on to dressed carcases and subsequently onto meat. This is particularly important in relation to contamination of carcases with VTEC *E. coli* given the severity of illness and the low infectious dose often associated with these organisms. A number of high-risk operations are carried out before and during hide removal that can result in the transfer of bacteria onto the underlying carcase. These include the making of a number of incisions in the hide, the generation of aerosols and dust, contact with operators hands, use of dirty knives, contact between the carcase and the outer surfaces of the hide and contact between the carcase and abattoir equipment.

Evisceration procedures are also a potential and significant entry point for the introduction of enteric pathogens to carcases. Inadequate care while removing the intestinal tract can result in accidental piercing or rupture causing leakage of faecal material on to the carcase. Leakage of intestinal contents may also occur from either the oesophagus or rectum of the animal. Such scenarios can lead to gross contamination of the carcase with potentially very high concentrations of pathogenic organisms. Bacteria can also be transferred between carcases when the sternum/brisket is cut using a saw to facilitate removal of offal from the thoracic cavity. The processing environment itself may also be a source of contamination such as walls, floors, food contact surfaces, knives, operators, air or scald water (in the case of pigs).

Once evisceration is complete and the carcases have been split, carcases are normally washed to remove any traces of blood or bone dust using cold or warm water. This step has little impact on reducing bacteria and indeed can result in their redistribution on carcase surfaces. Carcase sides and offal are then chilled to prevent or retard the growth of microorganisms. In Europe, beef carcases must be chilled to ≤7°C while offal must be chilled and maintained at a temperature of ≤3°C due to these organs typically containing higher levels of contamination and possessing more favourable attributes to support microbial growth (Anonymous, 2004b). Inadequate chilling or failure to maintain product at the correct chill temperatures during subsequent deboning, packing and all stages of distribution can result in the proliferation of pathogens and increase the risks to public health. The deboning and cutting into subprimal and retail cuts also represents an opportunity for bacteria to be transferred from the outer carcase surfaces onto newly exposed meat surfaces. This is particularly important in the case of minced or comminuted meat where microbes may be dispersed throughout the product due to the process significantly increasing the surface area available.

18.3.3 Strategies to control pathogens in the red meat chain

The asymptomatic colonisation of livestock with zoonotic pathogens poses a significant challenge in modern agriculture as currently few measures have been identified which effectively control these hazards (Besser *et al.*, 2001; Bach *et al.*, 2005; Humphrey and Jorgensen, 2006). Exposure of animals to pathogens through contact with infected animals within cohorts, contamination of the environment or emergence of other environmental reservoirs along with contamination of feedstuffs and water can all contribute to their dissemination and persistence. This deficiency in the availability of practical and effective interventions at pre-harvest level has resulted in efforts to control pathogens such as *E. coli* O157, campylobacters or asymptomatically carried salmonellae being applied later in the food chain (Madsen, 2007). However, implementation of good farming practices (including provision of quality feed and water and prevention of their contamination through for example hygienic design of water and feed troughs, general farm hygiene, waste management, husbandry and herd health approaches, etc.) is essential to try and minimise the risks of introducing or spreading zoonotic agents within or between herds.

With live cattle, considerable efforts should be made to ensure that hides are not excessively soiled at time of slaughter. Clean livestock policies have been implemented in several countries at abattoir level, including Ireland by the Department of Agriculture, Fisheries and Food, and the United Kingdom by the Food Standards Agency, where by cattle are visually graded to ensure that only acceptable categories of animals are slaughtered (FSA, 2009; DAFF, 2009). All relevant stakeholders, including farmers, animal transporting contractors, abattoir owners and official veterinarians, have responsibilities in the implementation and enforcement of these food hygiene policies. In Ireland for example, where animals are deemed to be excessively dirty or wet, clipping coats or holding until sufficiently dry is recommended, and any associated costs are charged to the farmer to incentivise farmers to present clean cattle for slaughter. This is an important component of the *ante-mortem* examination (AME) of cattle in the lairage, along with clinical assessment to detect and exclude any visibly diseased animals and verification of cattle identification and other essential food chain information.

Once accepted for slaughter, stringent hygienic practices must be observed, particularly where any hide incisions are made to facilitate bleeding and hide removal. Operators at these process stages should have a number of knives and have access to a sterilisation bath containing water at a minimum of 82°C. A freshly sterilised knife should initially be used to make a small incision in the hide. A second sterile knife is then used to enlarge the incision by using a spear cut technique (cutting from inside to out). This technique avoids the transfer of external hide contaminants to the underlying carcase surfaces. Furthermore, the operators completing the high-risk steps of dehiding should ensure that the hand in contact with the hide (dirty) is not allowed to make contact with any part of the carcase where the hide has been removed. Similarly, the clean hand (i.e. the knife-holding hand) should not come in contact with the external surfaces of the hide, again to avoid the spread of contamination. It is also important to hygienically remove the skin from the tail, release and seal off the rectum by placing a plastic bag over it and holding it in place by means of a rubber band (Buncic, 2006a). This prevents any leakage of faecal material from the rectum during dehiding and evisceration. The use of disposable paper towels is also recommended where shackling and dehiding equipment are attached to the hocks of the carcase to prevent equipment to carcase contamination. Depending on the type of hide puller used, skin is released manually from the abdomen or back as well as the fore and hind legs. In females, the udder must be removed carefully and intact from the carcase to avoid spillage of milk which can often contain pathogens. Prior to evisceration, the oesophagus should be separated from the trachea and sealed with a rubber band using a special tool (rodding). Again, this measure prevents leakage of rumen contents onto the carcase. Once the abdominal cavity has been opened, the entire intestinal tract can then be removed intact and any edible tissues hygienically separated. The liver and kidneys can also be removed from the abdominal cavity at this time and other red offal detached when the thoracic cavity is opened.

Carcases should be visually inspected prior to rinsing with hot or cold water and subsequent chilling. One of the main purposes of this inspection is to detect any traces of faecal contamination. If detected, these should be removed by trimming any affected parts of the carcase. Once all carcase dressing procedures have been completed, it is essential that they are removed to chill rooms immediately. The benefits of refrigeration of meat on shelf life and product preservation are well understood and documented. Chilling has been carried out under commercial conditions in the beef industry for over a century. The objective of refrigeration is to reduce the temperature of meat from body temperature to a final temperature of 4–7°C or less (Anonymous, 2004b; Buncic, 2006a). For red meat, chilled air is used to reduce carcase temperature. From a microbiological perspective, it would be best to achieve this final chill temperature as quickly as possible, however, in order to prevent a deterioration in the quality of red meat as a result of 'cold shortening' the rate of chilling is normally controlled so that carcase temperatures do not drop below *circa* 10°C until rigor is complete (usually 10 hours for beef). It is also important that the chill provides good air circulation, that carcases are not touching and that carcases are protected from condensate drip. The humidity of air must also be carefully controlled during chilling and storage of carcases. If humidity levels are excessive, carcase surfaces do not adequately dry and are more susceptible to microbial growth. Maintenance of chilled temperatures of product and air in boning halls is also imperative to inhibit and retard microbial growth.

It is also essential that abattoirs implement effective food safety management systems using the HACCP principles. These risk management systems should be supported with robust procedures covering good hygienic and manufacturing practices, hygiene and environmental monitoring and other ongoing microbiological surveillance. It is also essential that personnel in the abattoir are adequately trained and supervised so that they can carry out their responsibilities effectively and hygienically. This is particularly important for high-risk process stages where high concentrations of pathogens could be introduced.

18.3.4 Strategies to control pathogens in game

Game animals can be categorised based on whether they are farmed or wild. With wild animals, species can include for example deer, kangaroo and wild boar. These species can also be farmed along with others including bison, crocodiles, alligators, ostriches and emus. Wild game may be found in either controlled or wild populations (e.g. national parks, estates or reserves), while farmed game are intentionally reared to produce meat and hides. Both scenarios present unique food safety challenges as game animals may be colonised with a variety of zoonotic bacteria and parasites. In general, similar food safety principles should be applied in game animal production as domestic livestock. It has been reported that the incidence of enteric pathogens in meat from wild or farmed game animals can be less than those for meat from intensively reared domestic animals when carcases are dressed at suitable abattoirs (Gill, 2007).

Likelihood of exposure and colonisation by specific zoonotic agents is dependent on species, geographical location and prevalence in the environment and prevalence in other wildlife reservoirs. Crocodiles, for example, have been frequently found to be colonised with *Salmonella*, while the reported prevalence of *E. coli* O157 and campylobacters in deer is high (Wahlstrom *et al.*, 2003; Thomas, 1999). Interestingly, *Salmonella* has rarely been detected in wild deer, while colonisation may occur in farmed deer where exposure was attributed to contamination by domestic livestock species (Deutz *et al.*, 2000; Paulsen and Winkelmayer, 2004; Branham *et al.*, 2005; Atanassova *et al.*, 2008). This may be due to lower stocking densities used in game farming or found in the wild resulting in lower risk of spread between animals. Also in the wild there may be increased selective pressures exerted where symptomatically infected animals may not survive. How animals are killed and dressed in the field can impact significantly the microflora of game meat. In general, game animals are killed and eviscerated in the field, with visibly damaged or contaminated tissue removed by trimming with dehiding and butchering often carried out at a remote location (Gill, 2007; Atanassova *et al*., 2008). Evisceration of a carcase within two hours of killing is recommended as the intestines swell over time, increasing the likelihood of rupture or piercing during subsequent removal (Deutz *et al*., 2000; Gill, 2007).

Formal controls currently exist in many countries, including member states of the European Union for game meat intended for sale in the market (Anonymous, 2002, 2004b). Under this legislation, such hunters are defined as food business operators and are therefore responsible for food safety and traceability. The Regulation outlines competencies that at least one 'trained' person within a hunting party must possess, including basics of anatomy and pathology of wild game (including normal and abnormal changes), handling of wild game and meat after hunting and the ability to undertake an initial examination of wild game on the spot. Furthermore, they are required to possess adequate knowledge of hygiene rules and techniques for handling, transportation and evisceration of game animals after killing. Large game must be chilled as soon as possible after killing to ≤7°C and inspection completed by the competent authority at a game-handling establishment. Based on previously published data, such controls appear to provide equivalent levels of assurance to retailers, restaurateurs and consumers alike in relation to the bacterial status of game meat when compared with meat from farmed domestic animals. However, game animals may be more susceptible to infection with certain parasites which could be transmitted to humans, and this area warrants further investigation to fully assess potential risks to public health (Gill, 2007).

18.3.5 Monitoring strategies along the red meat and game chains

Several laboratory methods have been described to detect and identify a range of zoonotic foodborne pathogens (e.g. *Salmonella*, pathogenic *Escherichia coli*, *Campylobacter* and others) at points along the farm-to-fork continuum. One of the most commonly used approaches is the application of conventional microbiological techniques, including pre-enrichment, to recover sub-lethally injured cells, followed by a second enrichment step and then selective plating onto agar media formulated to support the growth of a specific bacterium (e.g. cefixime-tellurite sorbitol MacConkey [CT-SMAC] for *Escherichia coli* O157 or xylose lysine deoxycholate [XLD] for *Salmonella*) (see Table 18.1). On occasion these approaches also support the growth of non-target organisms, but usually these are sufficiently different in presentation that they can be easily distinguished. Importantly, as some of these pathogens are infective in low numbers, along with the fact that they retain their virulence characteristics, reliable recovery from the red meat and game food chain is essential.

Table 18.1 Some selective media applied to detect zoonotic foodborne pathogens

Bacterium	Enrichment media	Selective agar	Biochemical confirmation	Molecular detection
Campylobacter jejuni	– Preston broth – Bolton broth	– modified charcoal cefperazone deoxycholate selective agar (CCDA) – Colombia blood agar – Preston agar	Hippurate hydrolysis test (+), Indoxyl acetate test (+), Urease test (–) API *Campy*	16S rRNA *hipO*
Escherichia coli O157 (VTEC)	– modified tryptone soya broth (mTSA) containing novobiocin	– serotype-specific immunomagnetic separation – sorbitol MacConkey agar containing cefixime & tellurite (CT-SMAC) – [tryptone bile X-glucuronide agar (TBGA) for O26 & O111 serotypes]	API 20E	stx-1 stx-2 eae hlyA
Salmonella Typhimurium	– first enrichment in buffered peptone water (18 h at 37°C) – second enrichment in Rappaport-Vasiliadis Soya (RVS) peptone broth (6 h at 41.5°C) & Selenite broth	– xylose desoxycholate agar (XLD) – brilliant green agar (BGA)	API 20E	*fliC* *floR* (DT104) 16S rRNA

The composition of selective media is usually based on unique physiological traits or phenotypes associated with individual pathogens. For example, in the case of verocytotoxigenic *E. coli* O26, these bacteria do not ferment the sugar rhamnose and can utilise sorbitol as a carbon source. Further, the culture media can be supplemented with antimicrobial compounds cefixime and tellurite to which *E. coli* O26 is resistant and which will suppress background flora. In the case of other bacteria, selection criteria will be different. Generally, once the culture-based protocols have produced a presumptive colony, this is investigated further to make a definitive identification. Additional steps may include the use of biochemical galleries to further identify the organism (e.g. API20E). Application of molecular-based protocols may also be helpful, especially when characterising gene markers associated with pathogenicity (see Section 18.5.4). Figure 18.1 provides a general scheme to include the broad outline of the food chain (in the left panel) and a general summary of the methods of detection (modified from Murinda and Oliver, 2006). Molecular-based methods with carefully designed primers can provide further information about an isolate previously identified, such as the identification of virulence markers (e.g. vt-1 and/or vt-2 in the case of verocytotoxigenic *Escherichia coli*).

Animals are reservoirs for many of these human pathogens. On-farm control measures including hygiene and biosecurity will facilitate a reduction in the microbial load and reduce the risk of transmission further along the food chain. Microbiological assessment of animal feed, water and the environment is essential to understand the microbial ecology and identify points along the production chain that could pose problems. These measures could also be further supported during the pre-harvest phase by suitable vaccination programmes, targeting particular zoonotic bacteria along with the use of competitive exclusion and/or bacteriophage cocktails to reduce the microbial load in animals and their environments (see Section 18.5.3). Management of animal waste is also an important measure to prevent the re-cycling of the microbial ecology. Further along the production chain similar approaches can be applied. Such measures may also complement traditional meat inspection approaches and provide a novel approach towards the control of microbiological hazards (Fosse *et al.*, 2008).

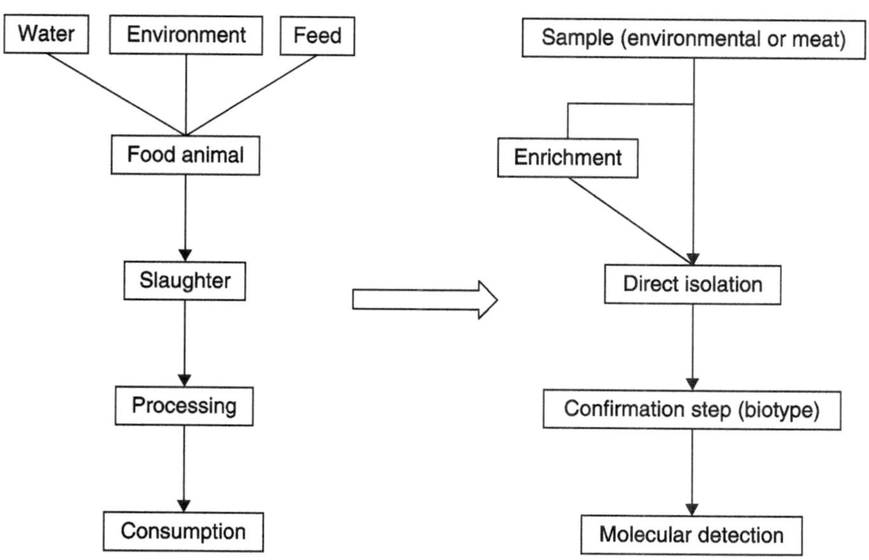

Fig. 18.1 A schematic illustration of the broad production chain linked to the microbiological analysis of points along this food chain.

18.4 Antimicrobial resistance in red meat pathogens

18.4.1 Emergence of resistance in microorganisms

The relative extent to which veterinary clinical treatments contribute to the selection of antimicrobial resistance in foodborne pathogens remains an issue of much debate. However, increasingly the scientific evidence indicates that the use of antibiotics in food animals leads to the development of resistant pathogenic microorganisms which are capable of dissemination along the food chain and infection of the consumer (Van Looveren *et al.*, 2001; Levy, 2002). When pathogens acquire antimicrobial drug resistance to clinically important drugs, early empirical treatment may be less effective and the choices of treatment may be limited (Mølbak, 2005).

It is clear that antibiotics provide significant selective pressure on susceptible microbial populations, where cells that are less sensitive are at a great advantage to survive and multiply (Wassenaar and Silley, 2008). Also, a number of studies have shown that various types of food and environmental sources harbour bacteria that are resistant to one or more antimicrobial drugs used in veterinary medicine and in food-producing animals (Anderson *et al.*, 2003; Schroeder *et al.*, 2004).

Genes encoding for resistance are often located on mobile genetic elements (i.e. bacteriophages, plasmids, transposons and integrons), which may be horizontally transferred by conjugation, transduction and transformation to other bacteria. However, bacteria can also be intrinsically resistant to antibiotics (Fig. 18.2).

Salmonella Typhimurium DT104 has proven to be very adept at obtaining multiple antibotic resistance (MAR), generally having resistance to ampicillin, chloramphenicol, streptomycin, sulfonamides and tetracyclines (ACSSuT). Unusually, the spectrum of resistance is encoded on the chromosome (*Salmonella* Genomic Island 1) instead of being mediated by a mobile genetic element. This may reduce the fitness cost generally associated with antibiotic resistance as scientific evidence suggests that this organism is resistant to commonly used control measures and is more persistent in the environment (Walsh, Kennedy, *et al.*, 2008; Vernozy-Rozand, 2007). Studies have shown Campylobacters are capable of acquiring antimicrobial resistance genes from both Gram-negative and Gram-positive microorganisms (Werner *et al.*, 2001; Pinto-Alphandary *et al.*, 1990). A recent European study into *Campylobacter* in food animals found that a significant number of isolates recovered from cattle and pigs were resistant to ciprofloxacin, erythromycin, tetracycline (EFSA, 2007; Piddock *et al.*, 2000). To date relatively little research on antibiotic resistance in EHEC has been carried out, possibly due to the fact that antibiotic therapy is not recommended for the treatment of the infection. However, the development of antibiotic resistance in these zoonotic pathogens may result in preferential colonisation in food-producing animals and improved survival in the environment, highlighting the need for constant surveillance (Walsh, Kennedy, *et al.*, 2008).

The development of antibiotic resistance in pathogens raises some important public health concerns (Fig. 18.3). The scientific community is constantly working

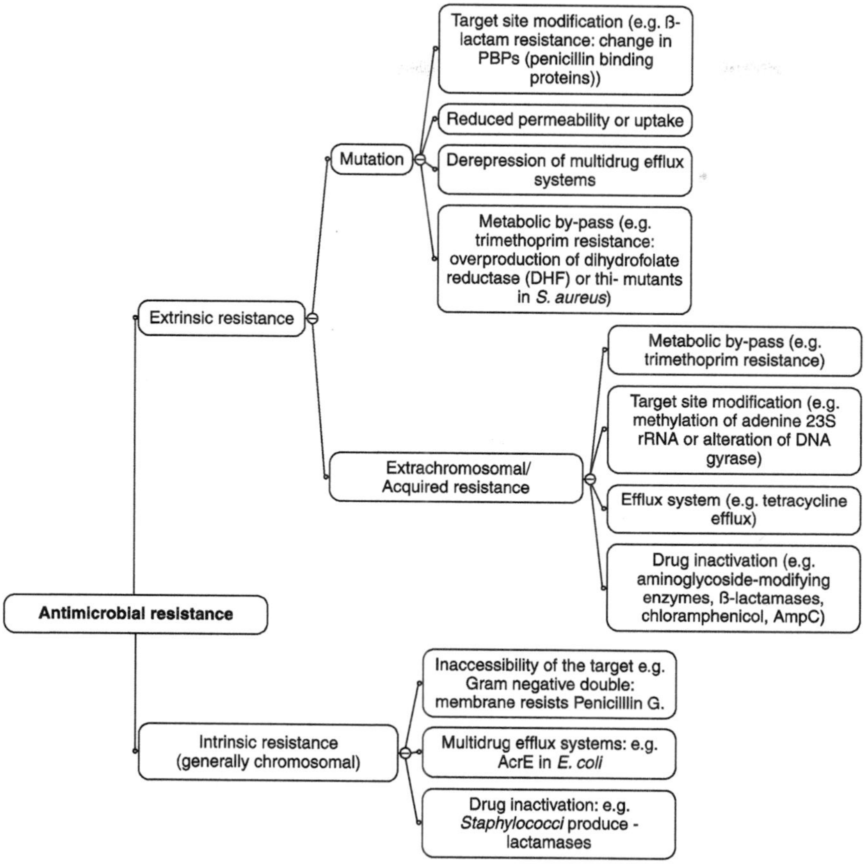

Fig. 18.2 Pathways of intrinsic and extrinsic antibiotic resistance in bacteria.

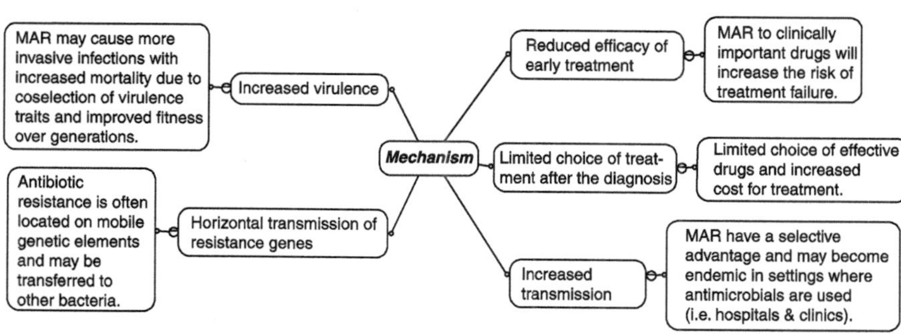

Fig. 18.3 Effects on public health caused by the emergence of antimicrobial drug resistance in foodborne pathogens (based on a table by Mølbak [2005]).

on new strategies for tackling antibiotic resistance. While the development of new antibiotic analogues for the treatment of antibiotic resistant pathogens will save lives in the short-term, more research is needed to discover novel gene candidates from bacterial genomes, as antibacterial targets (Mills, 2006). Other novel strategies currently being examined include the development of inhibitors for resistance genes and antibiotic efflux, as well as novel compounds to disrupt intercellular communication (Wright and Sutherland, 2007).

18.4.2 Use of antimicrobials in animal medicine

Antimicrobial drugs have been used in agriculture (including livestock and growth promotion) since the 1950s to treat, control and prevent infectious disease in high-risk situations and to improve feed efficiency (Angulo *et al.*, 2004; Walsh, Duffy, *et al.*, 2008). Administration of antimicrobial drugs at sub-therapeutic levels to food animals for the purposes of growth promotion is increasingly difficult to justify in the context of a growing antimicrobial resistance (AMR – defined as the failure of an antimicrobial compound to inhibit the growth of a bacterium) problem and has been largely discontinued in developed societies. Therapeutic administration to food animals with infectious diseases is acceptable practice, in addition to being an economic necessity. In the case of large groups of intensively produced food animals, the differentiation between therapeutic, prophylactic and metaphylactic administration may be more difficult to discern.

Several bodies including the Institute of Medicine (IOM) have attempted to quantify the use of antimicrobial drugs in agriculture, by various methods of analysis (IOM, 1989; Mellon and Benbrook, 2001). For example, in 1999, 22.9 million pounds of antibiotics were used for non-therapeutic purposes, as estimated by the Union for Concerned Scientists (UCS) in the USA (Jones and Ricke, 2003).

Development of resistance is an inevitable consequence of the selective pressure imposed by indiscriminate use of these valuable therapeutic agents, in both animal and human medicine. Over reliance and misuse of antimicrobial drugs in both animal production and human medicine has led to an increase in resistant strains being isolated from specimens, further increasing the risk of therapeutic failure in future outbreaks of severe infection, where the administration of a drug is indicated. Consequently, higher rates of morbidity and mortality are being reported leading to increased health-economic impacts (WHO, 2005b).

Two types of resistance to antimicrobial drugs arise:

- *Intrinsic resistance:* is associated with the physiological properties of the organism. Gram-negative bacteria are generally regarded to be more resistant to antimicrobial drugs than Gram-positive organisms (Russell and Day, 1996). For instance Gram-negative bacteria are resistant to penicillin G by virtue of their double-membrane structure. All of these features are generally (but not always) chromosomally encoded and therefore persist even in the absence of selective pressure.

- *Extrinsic resistance:* arises from a mutation or through the acquisition of a resistant determinant following horizontal DNA transfer. Mutations in housekeeping structural or regulatory genes are reported to arise once in every billion cell divisions (Courvalin and Trieu-Cuot, 2001).

Rates of mutation can differ between organisms, with some demonstrating a hypermutable phenotype, wherein mutations arise at a faster rate, facilitating an increased rate of adaptation to a selected environment. In contrast, horizontal gene transfer can occur in a range of different environments (see examples below). This transfer can also occur between pathogens and commensal or non-pathogenic organisms and between bacteria of different genera (Walsh, Duffy, *et al.*, 2008; Walsh and Fanning, 2008). Increasingly, commensal *Escherichia coli* colonising the intestine are being recognised as an important reservoir of resistance genes (Oppegaard *et al.*, 2001). The term 'resistome' was recently coined to reflect the fact that antibiotic resistance genes are present not only in pathogenic bacteria, but in a wider environment and transfer from these domains is probably ongoing (Wright, 2007).

18.4.3 Antimicrobial resistance and impact on animal health

An increasing prevalence of antimicrobial resistance has important consequences for animal health and welfare. These consequences manifest as two potential threats: (1) increased treatment failures, and (2) increased selection of pathogenic organisms by decreased competition. Treatment failure has immediate consequences for the individual animal and both consequences may result in increased morbidity at a herd level *via* increased exposure/challenge.

18.4.4 Antimicrobial resistance and impact on human health

The increased prevalence of antimicrobial drug-resistance in red meat pathogens such as *Campylobacter* and *Salmonella* has important public health implications. The majority of *Campylobacter* and *Salmonella* infections are mild, self-limiting and not requiring treatment, but antibiotic treatment may be used in immunocompromised patients or in those with prolonged or worsening symptoms (Allos and Blaser, 1995). *Campylobacter* infections are 2–4.5 times more common than *Salmonella* infections (Wheeler *et al.*, 1999; Blaser *et al.*, 1983) and ciprofloxacin resistance is found in ~10% of *Campylobacter* cases with *C. coli* isolates more likely to exhibit antimicrobial drug resistance than *C. jejuni* (Dingle *et al.*, 2005; The *Campylobacter* Sentinel Surveillance Scheme Collaborators, 2002).

Evidence suggests that patients with antibiotic-resistant *Campylobacter* infection are up to six times more likely to suffer more prolonged illnesses (Helms *et al.*, 2005). This can lead to post-infectious complications including Guillain–Barré syndrome following severe immunoreactivity (Taylor *et al.*, 1987; Travers and Barza, 2002) or life-threatening systemic disease as a result of invasive illness after therapeutic failure (Moore *et al.*, 2005; Moore *et al.*, 2006).

Adverse clinical outcomes associated with antimicrobial drug resistance in *Salmonella* are more common (Helms *et al*., 2005; Martin *et al*., 2004; Varma *et al*., 2005) than with *Campylobacter* (Helms *et al*., 2003). A multi-resistant strain of *Salmonella* Typhimurium DT 104, displaying resistance to up to five commonly used antimicrobial drugs, emerged in the late 1980s in the UK and has spread to Europe, the USA and several other countries worldwide, gaining significant clinical importance (Threlfall, 2000; Threlfall *et al*., 2000). Complications of infection with *Salmonella* Typhimurium DT 104 include intestinal perforation, peritonitis and fatal bloodstream infection (Mølbak *et al*., 2000). It is important to note that multi-resistance is not an absolute requirement, as *Salmonella* strains resistant to just one clinically relevant antibiotic are also frequently associated with clinical complications (Varma *et al*., 2005).

For both pathogens described above, an association between drug-resistance and an increase in other virulence factors has been suggested, as an increased incidence of adverse clinical events could not be attributed to therapeutic failure alone (Helms *et al*., 2005). The use of antibiotics in both the food chain and as therapeutic agents would select for the presence of these factors, ensuring that antimicrobial-resistant pathogens remain a considerable public health concern.

18.4.5 Surveillance of antimicrobial resistance and intervention

Antimicrobial resistance has been attributed to the overuse and/or misuse of antimicrobials in humans, animals and horticulture (Levy, 2002). Therapeutic failure in cases of severe bacterial infection, resulting in higher rates of morbidity and mortality, remains a significant concern. Until recently, the addition of antimicrobials to feed and water in order to promote growth and prevent bacterial infection in food producing animals was common practice. In 2003, the Food and Agriculture Organisation (FAO), the Organization International Epizooties (OIE) and the World Health Organisation (WHO) concluded that non-human usage of antimicrobials (contributing to the emergence of resistance) had serious human health consequences. The public health risk associated with the development of antimicrobial resistance resulting from this process was deemed 'significant' by WHO (2008). Subsequently, the use of antimicrobials for growth promotion purposes was banned in Europe.

Concerns related to the emergence of foodborne zoonotic pathogens with a multidrug-resistant (MDR) phenotype highlighted the need to safeguard the food chain as an avenue for the dissemination of antimicrobial resistance. This problem remains a real concern. MDR to clinically significant antibiotics including (fluoro) quinolones, extended spectrum β-lactamases and cephalosporins are being isolated from food and food-producing animals. In addition, mechanisms facilitating the accumulation, expression and transfer of antibiotic resistance-encoding genes have been described. Between 40 and 70% of Gram-negative bacteria (cultured from clinical and livestock sources) are known to carry class 1 integrons. These genetic elements have the ability to capture and express one or more resistant genes. MDR represents a direct hazard to the food industry contributing to the emergence of MAR pathogens, which can cause serious clinical problems in humans.

In more general terms, it has been speculated that typical food processing conditions or common practices such as disinfection can play a role in the development and dissemination of resistant genes. Concern exists that the stresses commonly encountered by bacteria in the food industry increase rates of plasmid transfer (Walsh, Duffy, *et al.*, 2008) or even induce the multiple antibiotic resistance (*mar*) operon, giving rise to a MDR phenotype. These stresses can also cause changes in membrane permeability and nucleic acid damage, which can promote gene transfer by natural transformation mutations (Lado and Yousef, 2002). Alternatively, there are concerns that antibiotic resistance genes can confer cross-resistance to food-related stresses, making the elimination of pathogens from food more difficult (Walsh and Fanning, 2008). Increased resistance of bacteria in low pH environments may compromise the stability and safety of some food products. For a wide range of heat-treated foods, in particular meat products, reduced susceptibility to thermal stress would be a major problem for food processors. In the case of low infectious dose pathogens (such as *E. coli* O157:H7), even a minor adaptation to current food processing techniques would need to be considered in the calculation of processing parameters. Furthermore, factors which augment the development of antimicrobial resistance within the food chain should be addressed, in order to safeguard development and dissemination of MAR in foodborne pathogens.

Monitoring antimicrobial drug resistance is essential to support measures aimed at controlling the emergence of resistance to these agents in both animal and human populations. Several countries have set up antimicrobial resistance monitoring systems, two examples of which include DANMAP established in 1995 (Aarestrup *et al.*, 1998) and in the USA, the National Antimicrobial Resistance Monitoring System (NARMS) based at the Centres for Disease Control and Prevention was set up in 1996 as a collaborative effort between several organisations. In future the establishment of a coordinated global system based upon sharing data in existing and newly established national agency databases would be a welcome development. This would have global impacts on animal and public health, and international trade.

18.5 Future trends

18.5.1 Herd health

The traditional role of herd health is to control or eliminate diseases and management inefficiencies that impact on animal welfare or limit productivity. The framework of herd health control is risk monitoring and assessment in conjunction with best farming practices. The implementation of herd health control must be in a practical and economically feasible way. The reduction in pathogen load on the farm will reduce the initial challenge to food safety management systems further along the food chain. Improved health will lead to the reduced need for antimicrobial treatments therefore slowing the development of antimicrobial resistance.

The delivery of herd health programmes may take different forms: (1) implementation and accreditation of general best farming practice, (2) specific disease control programmes, or (3) broader risk reduction programmes, such as HACCP programmes. Of these, broad risk reduction programmes, coupled with specific disease reduction strategies, are likely to prove most beneficial to reducing the zoonotic pathogen load originating from the farm. Potential benefits for good farming practices such as improved bio-security may also help to reduce the impact of national zoonotic disease outbreaks.

At harvesting stages of food production the integration of herd level monitoring data allows for the potential of risk categorisation of farms based on disease risk. This would facilitate potential control measures to be focused at the time of harvest and further along the food chain.

Challenges that currently exist for the development and integration of herd health into food safety are a lack of coordinated pathogen surveillance and agreed control plans that include the farm. Also, although anecdotally the control of disease at farm level is beneficial in reducing foodborne zoonoses, evidence is lacking as to what level of reduction in pathogen load at farm level is required. In addition, the ultimate effects of disease reduction on the farm on many foodborne zoonoses are unclear. It is conceivable that contamination during further processing may negate any efforts at farm level. Finally, there may be conflict in the economic benefits at the farm level and the balance of cost sharing will need to be agreed. For example, controlling infections such as *E. coli* O157, where there are no gains in terms of animal health or productivity, the economic benefit at a farm level may need to lie in increased market access or increase product value.

18.5.2 Risk-based inspection systems

In recent years, it has become apparent that traditional meat inspection systems were not designed to control many of the meatborne pathogens such as *Salmonella*, *E. coli O157*, *Campylobacter* and *Listeria monocytogenes*, which are now the cause of significant morbidity and mortality in human populations in developed countries (Hathaway and Richards, 1993; Berends, 1998; USDA, 2007; Leps and Fries, 2009). Indeed, there is now a consensus that an integrated risk-based approach to meat safety is required which concentrates on the implementation of preventive controls along the food chain with ongoing monitoring of processes, in addition to surveillance systems to detect hazards and other microbiological testing to verify that the required levels of assurance are achieved. Much progress has been made through the implementation of food safety management systems using HACCP principles and ensuring that responsibility for ensuring adequate levels of safety assurance is shared between the food business operator and relevant regulatory agencies (USDA, 1996; Anonymous, 2004a, 2004b, 2004c). The completion of food safety audits, in the form of internal, regulatory or third-party (customer) audits are also now frequently used to provide additional assurances and verification that food safety management systems and official controls are adequately implemented.

Improvements in food chain information are also required, in particular at the pre-harvest level where essential information pertaining to animal and herd histories can be provided and used to assess risk and subsequent levels of inspection required at the harvest level. These might include for example, feed information, animal identification and history, animal/herd disease and pathogen monitoring status and records of medicines used (Sofos, 2008). The collection and ongoing analysis of data at harvest level, such as *ante-* and *post-mortem* inspection information would also provide valuable insights and again facilitate the completion of more meaningful risk assessments. Traceability throughout all phases of the red meat chain is also a pre-requisite in any risk-based management system so that effective and timely interventions can be made if required.

Moving forward this risk-based approach is essential to adequately protect public health as global consumption and international trade in meat increases. This would facilitate harmonisation of regulatory and inspection systems both nationally and internationally (Sofos, 2008). Implementation of such an approach would also allow specific meatborne hazards to be prioritised and resources mobilised to implement specific measures to manage associated risks at all pertinent stages of the food chain, for example at pre-harvest level by preventing/reducing colonisation in livestock or controlling contamination in the harvest phase.

The evolution and implementation of risk-based inspection systems is currently being considered or underway in many countries, and the process is dynamic as it will require ongoing review and modification as dictated by risk analysis arising from surveillance of pathogens and other hazards (both current and emerging) in animals, food of animal origin and humans (Berends, 1998; Anonymous, 2004a, 2004b, 2004c; USDA, 2006). In meat processing plants in the United States and the United Kingdom, for example, it is envisaged that inspection resources will be allocated based on the risk profile of each establishment. These profiles will be compiled based on ongoing performance evaluations of each plant, with increased inspection activities devoted to historically higher risk facilities (Schnirring, 2007; El Amin, 2007).

Moreover, the development and implementation of robust microbiological testing criteria for foods and process hygiene evaluation has complimented the new risk-based inspection systems. These are often supported by regulatory enforcement and reviewed by the relevant food safety inspectorate with non-compliances resulting in official investigation and intervention. A further tool commonly used to verify the effectiveness of food safety management systems and official controls is the conducting of periodic audits.

18.5.3 Recent and novel technologies

Pre-harvest approaches

Control of zoonotic foodborne pathogens is an essential step towards reducing the impact of bacterial disease on public health. Several approaches, some novel, have attracted attention including the use of vaccines, bacteriophages which specifically target bacteria, the inclusion of probiotics in animal feeds and the application of naturally derived antimicrobial peptides, derived from milk and other proteins.

Bacteriophages. Use of bacteriophages to control infectious diseases dates back to the beginning of the last century (Debarbieux, 2008; Dublanchet and Patey, 2008). Phages can specifically attack bacteria and are an attractive alternative to antimicrobials to control zoonotic pathogens. Bacteriophages have been applied to control foodborne pathogens including *Campylobacter* (Wagenaar *et al*., 2005; Atterbury *et al*., 2003) and *E. coli* O157:H7 (Callaway *et al*., 2008). Usually, application involves the use of a cocktail of defined phages that limits the development of resistance. A three-phage cocktail was applied to reduce *E. coli* O157:H7 on meat samples and was effective in reducing initial numbers of approximately 3 $\log_{10}$ CFU/ml to undetectable levels over a two-hour period (O'Flynn *et al*., 2004). Importantly, an understanding of the phage-bacteria kinetics may lead to improvements in the application of this form of biocontrol (Cairns *et al*., 2009), providing a realistic pre-harvest control protocol. In addition to the use of phages, endolysins have also attracted attention. These phage-encoded enzymes breakdown the bacterial peptidoglycan layer and are considered to be as effective as antibtioics, with important applications for pathogen control (Hermoso *et al*., 2007).

The United States Department of Agriculture has approved the use of a commercial bacteriophage for hide washing. OmniLytics is applied to the hides of live animals prior to slaughter and results in the reduction of *E. coli* O157 (http://www.omnilytics.com).

Bacteriocins. Bacteriocins are ribosomally synthesised antimicrobial peptides (AMPs) capable of inhibiting or killing other bacteria either within the same species (narrow spectrum activity) or across genera (broad spectrum activity) (Cotter *et al*., 2005). These antimicrobial peptides are produced by Gram-positive and Gram-negative bacteria. Since lactic acid bacteria (LAB) are generally regarded as safe (GRAS), these organisms have been studied in greater detail (Chen and Hoover, 2003). Several studies have been reported describing the application of bacteriocins to meat aimed at controlling *Listeria* (Aasen *et al*., 2003; Ananou *et al*., 2005; Castellano and Vignolo, 2006). Despite the fact that most reported bacteriocins are aimed at controlling Gram-positive pathogens, *Lactobacillus salivarius* NRRL B-30514 was identified which produced a bacteriocin OR-7 effective against *Campylobacter jejuni* (Stern *et al*., 2006). More recent reports identified bacteriocins with a broader spectrum of activity against Gram-negative pathogens including *E. coli*, *Salmonella* and *Campylobacter* (Line *et al*., 2008; Svetoch *et al*., 2008).

Bacteriocin producing bacteria have also been added to animal feed and this strategy has been shown to reduce pathogenic populations of bacteria. Colicins are bacteriocins produced by *E. coli* and these peptides have been applied as a pre-harvest pathogen control measure aimed at reducing *E. coli* O157 in cattle populations (Diez-Gonzalez, 2007; Schamberger *et al*., 2004). Colicin-producing

E. coli yielded a reduction of 1.1 log10 CFU/g in the population of *E. coli* O157:H7. An advantage of this approach is the inherent specificity of the AMP to selectively eliminate only *E. coli* O157:H7, leaving the remaining bacterial flora undisturbed (Schamberger *et al*., 2004).

Vaccines. Food producing animals are a source of zoonotic pathogens including *E. coli* O157:H7 and others that infect humans through either direct contact or *via* the food chain following consumption of contaminated meat. In the former case, the type III secretion system (T3SS) facilitates the colonisation of the intestinal mucosa. Typically this involves secreted effectors being delivered to the host including bacterial proteins such as Tir, EspA, EspB, intimin and others. These proteins are also required for colonisation in cattle and have become targets for the development of vaccines. An efficient vaccine could reduce the carriage of this pathogen in food-producing animals (Rogan *et al*., 2009). In this case a commercial vaccine reduced the numbers of days of shedding by one-third, with a 2.28 $\log_{10}$ reduction in the bacterial count in the faeces. Similarly, novel fusion of proteins using sub-units of the Stx2B toxin gene were used to develop vaccines. This vaccine induced a high-level humoral IgG response in mice (Gao *et al*., 2009). A negative-marker vaccine was developed aimed at controlling *Salmonella* Typhimurium (Selke *et al*., 2007). Using sequence-tagged mutagenesis, a number of vaccine candidates were also identified in *Salmonella* Typhimurium (Carnell *et al*., 2007).

Antimicrobial peptides. These are short (12–100 residues) cationic peptides produced by both plants and animals with a broad spectrum of activity against both Gram-positive and Gram-negative bacteria, fungi, viruses and protozoa. These peptides are part of the host immune system (Hancock, 1997) and can also modulate the immune response in higher organisms (Bowdish *et al*., 2005). Since the majority of AMPs kill microorganisms quickly it is unlikely that resistance would develop. Given their small size these can be chemically synthesised. Their mode of action is not completely understood but thought to act *via* permeabilisation of the bacterial cell membrane (Hancock and Rozek, 2002). Studies have demonstrated the efficacy of AMPs against a number of foodborne pathogens including *Salmonella*, *E. coli* O157:H7 and *Listeria* (Anderson and Yu, 2005; Higgs *et al*., 2007).

Harvest and post-harvest approaches

In recent years a plethora of antimicrobial technologies have been explored as possible means of reducing levels of microbial contaminants on carcases and meat. The intervention technologies may be either physical or chemical and can be applied to clean animals before hide removal, decontaminate dressed carcases following dehiding or subsequently on meat products to reduce, eliminate or control the growth of pathogens. In addition, these technologies can be used individually or in combination. Examples of physical decontamination techniques include animal cleaning, hair removal, knife trimming of any soiling on carcases, washing/rinsing of carcases with hot or cold water and the use of steam pasteurisation or steam-vacuuming prior to carcase chilling (Sofos, 2008). Using such techniques, substantial reductions in bacterial populations can be achieved on dressed carcases and meat products (Phebus *et al.*, 1997; Minihan, Whyte, O'Mahony, and Collins, 2003; Arthur *et al.*, 2007).

The use of chemicals to decontaminate hides and carcases has also been investigated. Significant reductions in populations of *E. coli* O157 and Enterobacteriaceae were reported in cattle hides washed with bactericidal solutions (Bosilevac *et al.*, 2004). A range of organic acid solutions (e.g. acetic, lactic and citric) and other chemicals (e.g. chlorine, peroxyacids and trisodium phosphate) have also been used to rinse carcases resulting in substantial reductions in microbial populations, including zoonotic pathogens (Castillo *et al.*, 1999; Bolton *et al.*, 2001; Hugas and Tsigarida, 2008).

In the United States, the use of approved chemicals and the application of physical technologies for hide/carcase decontamination are permitted as part of a pathogen reduction strategy (Koohmaraie *et al.*, 2007). Within the European Union, physical decontamination techniques have been permitted for use in slaughter plants, and with the introduction of the new food hygiene package in 2006, the use of substances other than potable water for carcase decontamination is permitted once an appropriate risk assessment has been completed and approval sanctioned by the European Commission (Anonymous, 2004b; Hugas and Tsigarida, 2008). However, either chemical or physical decontamination techniques should only be applied as part of an integrated risk management system where the emphasis is placed on preventing microbial contamination in the first instance through the application of GHPs and HACCP. This approach is imperative as none of these techniques can guarantee elimination of pathogens (Huffman, 2002; Minihan, Whyte, O'Mahony, and Collins, 2003).

A number of other technologies have been applied to either decontaminate meat products or prevent/retard the proliferation of pathogenic and spoilage organisms in meat. Advances in packaging technology have enabled the successful application of a range of systems that alter the gaseous atmosphere under which meat is stored (e.g. vacuum or modified atmosphere packaging). Such systems, when used in combination with effective cold line control, have resulted in substantial increases in the shelf life of fresh red meat while simultaneously controlling pathogen populations (O'Brien *et al.*, 2005). Other novel decontamination technologies for decontaminating fresh meat include the application of pulsed electric fields, ultrasonication and thermosonication, high hydrostatic pressure, irradiation, ohmic heating, bacteriophage and bacteriocin treatments, and their potential for improving red meat safety should be explored further (Aymerich *et al.*, 2008; Sofos, 2008).

18.5.4 Advances in pathogen detection and surveillance

In applying traditional approaches to pathogen detection, two limitations are obvious: the time taken to recover and grow the organism and the labour input necessary to complete the manual steps. Research over the past few years has focused on the development of detection methods that are faster and require less manual input. These approaches are slowly gaining acceptance in the food industry. In particular, attention has been focused on the application of molecular-based approaches to pathogen detection. Conventional PCR has been applied for detection of food pathogens (Rijpens and Herman, 2002; Josefsen *et al.*, 2003; Naravaneni and Jamil, 2005; Neubauer and Hess, 2006) and more recently the use of real-time PCR applications has begun to emerge (O'Regan *et al.*, 2008). In these assays, detection is based on specific marker genes unique to a foodborne pathogen (e.g. *invA* which is commonly used for the detection of *Salmonella; stx1/stx2* in *E. coli* O157). With the continued advances in DNA microarray technology, pathogen detection arrays known as 'DNA chips' are beginning to emerge which provide a high-throughput capacity to identify several pathogens simultaneously. Some of these approaches are based on the use of taxonomic markers including the 16S rDNA genes, wherein an array is constructed, containing unique oligonucleotide probes to detect specific bacteria. Some arrays can discriminate up to 204 strains of bacteria from pure culture with a sensitivity of 100 CFU (Wang *et al.*, 2007) whereas others have used a whole-genome approach to compare individual genomes one-against-another, to reveal unique markers which are then incorporated into an array and tested for their utility (Kim *et al.*, 2008). The latter can be applied to accurately detect and identify foodborne pathogens of interest. Furthermore, these new protocols will provide useful epidemiological data, which can be applied to support HACCP protocols, thereby improving the microbiological quality of the food matrix.

In comparing the traditional and modern approaches to detection of pathogens above, it is obvious that the same broad steps are included in each protocol, but the speed at which final results are obtained is very different (see Fig. 18.4, which shows a comparison of conventional vs. molecular-based approaches applied to detect *Salmonella*). As these technologies continue to develop, it is to be expected that it will be possible to evaluate the microbiological quality of meat carcasses directly on the slaughter line. Detection, identification and quantitation will be possible, using modern molecular-based methods that are superior compared with the traditional culture-based approaches. (A detailed description of molecular methods of detection is presented in Part I of this book).

18.6 Sources of further information and advice

The following websites are useful sources of information on the topics covered in this chapter.

1 World Health Organization http://www.who.int
2 Food and Agriculture Organization http://www.fao.org
3 Office International des Epizooties http://www.oie.int
4 European Union Legislation http://eur-lex.europa.eu/
5 Central Statistics Office Ireland http://www.cso.ie
6 European Statistics http://epp.eurostat.cec.eu.int/pls/portal

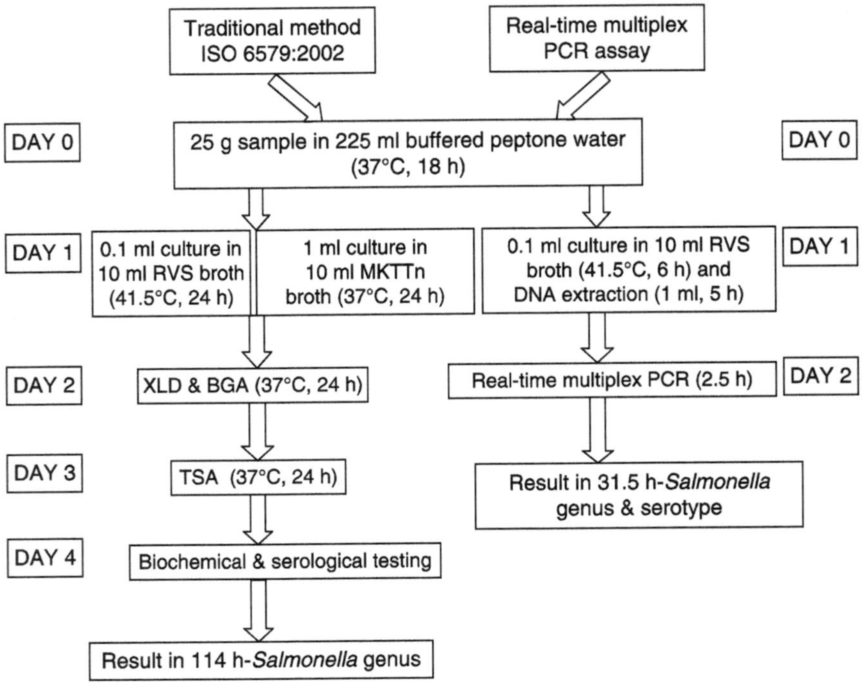

Fig. 18.4 Timelines for traditional culture and RT-PCR detection methods for *Salmonella* isolation.

7 Food Safety Authority of Ireland http://www.fsai.ie
8 Irish Department of Agriculture, Fisheries and Food http://www.agriculture.gov.ie
9 Health Protection Surveillance Centre Ireland http://www.hpsc.ie
10 Food Safety Promotion Board http://www.safefoodonline.com
11 Food Standards Agency United Kingdom http://www.food.gov.uk
12 Health Protection Agency http://www.hpa.org.uk
13 United States Department of Agriculture http://www.usda.gov
14 Food Safety and Inspection Service http://www.fsis.usda.gov
15 Centers for Disease Control and Prevention http://www.cdc.gov
16 USDA Economic Research Service http://www.ers.usda.gov
17 European Food Safety Authority http://www.efsa.eu.int

18.7 References

AARESTRUP F M, BAGER F, MADSEN M, JENSEN N E, MEYLING A and WEGENER H C (1998), 'Surveillance of antimicrobial resistance in bacteria isolated from food animals to antimicrobial growth promoters and related therapeutic agents in Denmark', *APMIS*, **10**, 606–22.

AASEN I M, MARKUSSEN S, MORETRO T, KATLA T, AXELSSON L and NATERSTAD K (2003), 'Interactions of the bacteriocins sakacin P and nisin with food constituents', *Int J Food Microbiol*, **87**, 35–43.

ACKE E, MCGILL K, GOLDEN O, JONES B R, FANNING S and WHYTE P (2009), 'Prevalence of thermophilic *Campylobacter* species in household cats and dogs in Ireland', *Vet Rec*, **164**, 44–7.

ALLOS B M and BLASER M J (1995), '*Campylobacter jejuni* and the expanding spectrum of related infections', *Clin Infect Dis*, **20**, 1092–101.

ANANOU S, GARRIGA M, HUGAS M, MAQUEDA M, MARTINEZ-BUENO M, GALVEZ A, ET AL. (2005), 'Control of *Listeria monocytogenes* in model sausages by enterocin AS-48', *Int J Food Microbiol*, **103**, 179–90.

ANDERSON A D, NELSON J M, ROSSITER S and ANGULO F J (2003), 'Public health consequences of use of antimicrobial agents in food animals in the United States', *Microb Drug Resist*, **9**, 373–9.

ANDERSON R C and YU P L (2005), 'Factors affecting the antimicrobial activity of ovine-derived cathelicidins against *E. coli* O157:H7', *Int J Antimicrobial Agents*, **25**, 205–10.

ANGULO F J, BAKER N L, OLSEN S J, ANDERSON A and BARRETT T J (2004), 'Antimicrobial use in agriculture: controlling the transfer of antimicrobial resistance to humans', *Semin Pediatr Infect Dis*, **15**, 78–85.

ANONYMOUS (2002), 'Regulation (EC) No 178/2004 of the European Parliament and of the Council of 28 January 2002 laying down the general principles and requirements of food law, establishing the European Food Safety Authority and laying down procedures in matters of food safety', *Official J European Union*, L31/1.

ANONYMOUS (2004a), 'regulation (EC) No 852/2004 of the European Parliament and of the Council of 29 April 2004 on the hygiene of foodstuffs', *Official J European Union*, L139/1.

ANONYMOUS (2004b), 'regulation (EC) No 853/2004 of the European Parliament and of the Council of 29 April 2004 laying down specific hygiene rules for the hygiene of foodstuffs', *Official J European Union*, L139/55.

ANONYMOUS (2004c), 'regulation (EC) No 854/2004 of the European Parliament and of the Council of 29 April 2004 laying down specific rules for the organisation of official controls on products of animal origin intended for human consumption', *Official J European Union*, L139/206.

ANSAY S E and KASPAR C W (1997), 'Survey of retail cheeses, dairy processing environments and raw milk for *Escherichia coli* O157:H7', *Lett Appl Micro*, **25**, 131–4.

ARTHUR T M, BOSILEVAC J M, BRICHTA-HARHAY D M, KALCHAYANAND N, SHACKELFORD S D, WHEELER T L, ET AL. (2007), 'Effects of a minimal hide wash cabinet on the levels and prevalence of *E. coli* O157:H7 and *Salmonella* on the hides of beef cattle at slaughter', *J Food Prot*, **70**, 1076–9.

ATANASSOVA V, APELT J, REICH G and KLEIN G (2008), 'Microbiological quality of freshly shot game in Germany', *Meat Sci*, **78**, 414–19.

ATTERBURY R J, CONNERTON P L, DODD C E, REES C E and CONNERTON I F (2003), 'Application of host-specific bacteriophages to the surface of chicken skin leads to a reduction in recovery of *Campylobacter jejuni*', *Appl Environ Microbiol*, **69**, 6302–6.

AYMERICH T, PICOUET P A and MONFORT J M (2008), 'Decontamination technologies for meat products', *Meat Sci*, **78**, 114–29.

BACH S J, SELINGER L J, STANFORD K and MCALLISTER T A (2005), 'Effect of supplementing corn- or barley-based feedlot diets with canola oil on faecal shedding of *Escherichia coli* O157:H7 by steers', *J Appl Microbiol*, **98**, 464–75.

BENTANCOR A, RUMI M V, GENTILINI M V, SARDOY C, IRINO K, AGOSTINI A, ET AL. (2007), 'Shiga toxin-producing and attaching and effacing *Escherichia coli* in cats and dogs in a high hemolytic uremic syndrome incidence region in Argentina', *FEMS Microbiol Lett*, **267**, 251–6.

BERENDS B R (1998), *A Risk Assessment Approach to the Modernization of Meat Safety Assurance*, Printpartners, Enschede, The Netherlands.

BESSER T E, RICHARDS B L, RICE D H and HANCOCK D D (2001), '*Escherichia coli* O157:H7 infection of calves: infectious dose and direct contact transmission', *Epidemiol Infect*, **127**, 555–60.

BLASER M J, WELLS J G, FELDMAN R A, POLLARD R A and ALLEN J R (1983), '*Campylobacter* enteritis in the United States: a multicenter study', *Ann Intern Med*, **98**, 360–5.

BOLTON J D, DOHERTY A M and SHERIDAN J J (2001), 'Beef HACCP: intervention and non-intervention systems', *Int J Food Microbiol*, **66**, 119–29.

BOSILEVAC J M, ARTHUR T M, WHEELER T L, SHACKELFORD S D, ROSSMAN M, REAGAN J O, ET AL. (2004), 'Prevalence of *Escherichia coli* O157 and levels of aerobic bacteria and Enterobacteriaceae are reduced when hides are washed and treated with cetylpyridinium chloride at a commercial beef processing plant', *J Food Prot*, **67**, 646–50.

BOWDISH D M, DAVIDSON D J and HANCOCK R E (2005), 'A re-evaluation of the role of host defense peptides in mammalian immunity', *Curr Protein Pept Sci*, **6**, 35–51.

BRANHAM L A, CARR M A, SCOTT C B and CALLAWAY T R (2005), '*E. coli* O157 and *Salmonella* spp. in white tailed deer and livestock', *Curr Issues Intest Microbiol*, **6**, 25–9.

BRENNER F W, VILLAR R G, ANGULO F J, TAUXE R and SWAMINATHAN B (2000), '*Salmonella* nomenclature,' *J Clin Microbiol*, **38**, 2465–7.

BULLING E and SCHONBERG A (1999), 'Robert von Ostertag (1864–1940) a veterinarian contemporary with R. Virchow and R. Koch,' *Hist Vet Med*, **24**, 96–119.

BUNCIC, S (2006a), 'slaughter and dressing', *integrated food safety and Veterinary Public Health*, Cabi, Oxfordshire, pp. 139–78.

BUNCIC, S (2006b), 'meat preservation and processing', *integrated food Safety and Veterinary Public Health*, Cabi, Oxfordshire, pp. 213–61.

CAIRNS B J, TIMMS A R, JANSEN V A, CONNERTON I F and PAYNE R J (2009), 'Quantitative models of in vitro bacteriophage-host dynamics and their application to phage therapy', *PLoS Pathog*, 5, e1000253.

CALLAWAY T R, EDRINGTON T S, BRABBAN A D, ANDERSON R C, ROSSMAN M L, ENGLER M J, ET AL. (2008), 'Bacteriophage isolated from feedlot cattle can reduce *Escherichia coli* O157:H7 populations in ruminant gastrointestinal tracts', *Foodborne Pathog Dis*, **5**, 183–91.

CARNELL S C, BOWEN A, MORGAN E, MASKELL D J, WALLIS T S and STEVENS M P (2007), 'Role in virulence and protective efficacy in pigs of *Salmonella* enterica serovar Typhimurium secreted components identified by signature-tagged mutagenesis', *Microbiol*, **153**, 1940–52.

CASTELLANO P and VIGNOLO G (2006), 'Inhibition of *Listeria innocua* and *Brochotrix thermosphacta* in vacuum-packed meat by the addition of bacteriocinogenic *Lactobacillus curvatus* CRL705 and its bacteriocins', *Lett Appl Microbiol*, **43**, 194–9.

CASTILLO A, LUCIA L M, GOODSON K J, SAVELL J W and ACUFF G R (1999), 'Decontamination of beef carcass surface tissue by steam vacuuming alone and combined with hot water and lactic acid sprays', *J Food Prot*, **62**, 146–51.

CATTLENETWORK (2008), *World report: cattle population by country*. Lenexa, Kansas. Available from: http://www.cattlenetwork.com/content.asp?contentid=226025 [Accessed 19 January 2009].

CDC (CENTERS FOR DISEASE CONTROL AND PREVENTION) (2000), 'Surveillance for foodborne outbreaks – United States, 1993–1997', *MMWR Morb Mortal Wkl Rep*, **49**(No.SS-1), 1–66.

CDC (2005a), 'outbreaks of *escherichia coli* O157:H7 associated with petting zoos – North Carolina, Florida, and Arizona, 2004 and 2005', *MMWR Morb Mortal Wkly Rep*, **54**, 1277–80.

CDC (2005b), Campylobacter *infections*. Available from: http://www.cdc.gov/ncidod/dbmd/diseaseinfo/campylobacter_g.htm [Accessed 28 January 2009].

CHAPMAN P A, CORNELL J and GREEN C (2000), 'Infection with verocytotoxin-producing *Escherichia coli* O157 during a visit to an inner city open farm', *Epidemiol Infect*, **125**, 531–6.

CHATELLIER H, GUYOMARD, H and LE BRIS K (2003), 'La production et les echanges de viande bovine dans le monde et dans l'Union europeenne', *INRA Prod Anim*, **16**, 365–80.

CHEN H and HOOVER D G (2003), 'Bacteriocins and their food applications', *Comp Rev Food Sci Food Saf*, **2**, 82–100.

COGAN T A and HUMPHREY T J (2003), 'The rise and fall of *Salmonella* Enteritidis in the UK', *J Appl Microbiol*, **94** (Suppl), S114–119.

COIA J (1998), 'Clinical, microbiological and epidemiological aspects of *Escherichia coli* O157 infection', *FEMS Immun Med Micro*, **20**, 1–9.

COIA J, JOHNSON Y, STEERS N J and HANSON M F (2001), 'A survey of the prevalence of *Escherichia coli* O157 in raw meats, raw cow's milk and raw-milk cheeses in the south-east Scotland', *Int J Food Micro*, **66**, 63–9.

COTTER P D, HILL C and ROSS R P (2005), 'Bacteriocins: developing innate immunity for food', *Nat Rev Microbiol*, **3**, 777–88.

COURVALIN, P and TRIEU-CUOT P (2001), 'Minimizing potential resistance: the molecular view', *Clin Infect Dis*, **33**(Suppl 3), S138–46.

DAFF (IRISH DEPARTMENT OF AGRICULTURE, FISHERIES AND FOOD) (2009), *Clean Livestock Policy*. Available from: http://www.agriculture.irlgov.ie/printindex.jsp?file=publicat/sch00/cattle.xml [Accessed 10 February 2009].

D'AOUST J (2007), 'Salmonella', in Storrs, M. (Ed.): *Food Safety Handbook – Microbiological Challenges*, Biomerieux Editions, France, pp. 128–41.

DEBARBIEUX L (2008), 'Experimental phage therapy in the beginning of the 21st century', *Med Mal Infect*, **38**, 421–5.

DEUTZ A, FUCHS K, PLESS P, DEUTZ-PIEBER U and KOFER J (2000), 'Hygienerisiken bei wildfleisch-oberflachenkeimgehalte und human pathogene keime', *Fleischwirtschaft*, **80**, 106–8.

DIEZ-GONZALEZ F (2007), 'Applications of bacteriocins in livestock', *Curr Issues Intest Microbiol*, **8**, 15–24.

DINGLE K E, CLARKE, L and BOWLER, I C J W (2005), 'Ciprofloxacin resistance among human *Campylobacter* isolates 1991–2004: an update', *J Antimicrob Chemother*, **56**, 435–7.

DOYLE M P, BEUCHAT L R and MONTVILLE T J (2001), *Food Microbiology: Fundamentals and Frontiers*, 2nd ed., ASM Press, Washington, DC, pp. 171–91.

DOYLE M P and ERICKSON M C (2006), 'Emerging microbiological food safety issues related to meat', *Meat Sci*, **74**, 98–112.

DOYLE M P, ZHAO T, MENG J and ZHAO S (1997), '*Escherichia coli* O157:H7', in Doyle M.P., Beuchat, L.R. and Montville, T.J. (Eds): *Food Microbiology: Fundamentals and Frontiers*, ASM Press, Washington, pp. 171–91.

DUBLANCHET A and PATEY O (2008), 'Bacterial infection-update on phage therapy', *Med Mal Infect*, 38; 407–9.

EFSA (EUROPEAN FOOD SAFETY AUTHORITY) (2007), 'The community summary report on trends and sources of zoonoses, zoonotic agents, antimicrobial resistance and foodborne outbreaks in the European Union in 2006', *EFSA Journal*, **130**, 3–352.

EL AMIN A (2007), *UK moves to Risk-based Inspection System*. Available from: http://www.foodproductiondaily.com/quality-safety/uk-moves-to-risk-based-inspection-system [Accessed 18 February 2009].

EUROSTAT (2008), *Food: From Farm to Fork Statistics*. European Commission. Available from: http://epp.eurostat.ec.europa.eu/portal/page?_pageid=1090,30070682,1090_33076576&_dad=portal&_schema=PORTAL [Accessed 20 January 2009].

FAO (FOOD AND AGRICULTURE ORGANIZATION) (2005), *Summary of World Food and Agricultural Statistics 2005, ESS/ESSA/DGS-4*. FAO Rome, Italy. Available from: http://www.fao.org/es/ess/sumfas/sumfas_en_web.pdf [Accessed 19 January 2009].

FAO (2008), *Food Outlook – Global Market Analysis*, November 2008. Available from: http://www.fao.org/docrep/011/ai474e/ai474e09.htm [Accessed 19 January 2009].

FOLEY B and MCKEOWN P (2007), 'Campylobacteriosis in Ireland, 2005', *Epi-Insight*, **8**, 2–3.

FOOD STANDARDS AGENCY (2009), *Clean Livestock*. Available from: http://www.food.gov.uk/foodindustry/farmingfood/cleancattleandmeatsafety/#h_3 [Accessed 10 February 2009].

FOSSE J, SEEGERS H and MAGRAS C (2008), 'Foodborne zoonoses due to meat: a quantitative approach for a comparative risk assessment applied to pig slaughtering in Europe', *Vet Res*, **39**, 1.

FSAI (FOOD SAFETY AUTHORITY OF IRELAND) (2002), 'Control of *Campylobacter* species in the food chain', FSAI report, Dublin.

GAO X, CAI K, SHI J, LIU H, HOU X, TU W, ET AL. (2009), 'Immunogenicity of a novel Stx2b-Stx1b fusion protein in a mice model of Enterohemorrhagic *Escherichia coli* O157:H7 infection', *Vaccine*, **24**, 2070–6.

GILL C O (2007), 'Microbiological conditions of meats from large game animals and birds', *Meat Sci*, **77**, 149–60.

GRAU F H (1986), 'Microbial ecology of meat and poultry', in Pearson, A.M. and Dutson, T.R. (Eds): *Advances in Meat Research, Vol. 2, Meat and Poultry Microbiology*, AVI Publishing, Westport, CT, pp. 1–47.

GREIG J D and RAVEL A (2009), 'Analysis of foodborne outbreak data reported internationally for source attribution', *Int J Food Microbiol*, **130**, 77–87.

GRIFFIN P M (1995), '*Escherichia coli* O157:H7 and other enterohemorrhagic *Escherichia coli*', in Blaser M.J., Smith P.D., Ravdin J.I., Greenberg, H.B. and Guerrant, R.L., *Infections of the Gastrointestinal Tract*, Raven Press, New York, pp. 739–61.

HANCOCK R E (1997), 'Peptide antibiotics', *Lancet*, **349**, 418–22.

HANCOCK R E and ROZEK A (2002), 'Role of membranes in the activities of antmicrobial cationic peptides', *FEMS Microbiol Lett*, **206**, 143–9.

HATHAWAY S C and RICHARDS M S (1993), 'Determination of the performance attributes of post-mortem meat inspection procedures', *Prev Vet Med*, **16**, 119–31.

HELMS M, SIMONSEN J, OLSEN K E and MØLBAK K (2005), 'Adverse health events associated with antimicrobial drug resistance in *Campylobacter* species: a registry-based cohort study', *J Infect Dis*, **191**, 1050–5.

HELMS M, VASTRUP P, GERNER-SMIDT, P and MØLBAK K (2003), 'Short and long-term mortality associated with foodborne bacterial gastrointestinal infections: registry-based study', *BMJ*, **326**, 357.

HERMOSO J A, GARCIA J L and GARCIA P (2007), 'Taking aim on bacterial pathogens: from phage therapy to enzybiotics', *Curr Opin Microbiol*, **10**, 461–72.

HEUVELINK A E (2000), *Verocytotoxin-Producing Escherichia Coli in Humans and the Food Chain*, PhD Thesis, University Hospital Nijmegen, The Netherlands.

HIGGS R, LYNN D, CAHALANE S, ALANA I, HEWAGE C, JAMES T, ET AL. (2007), 'Modification of chicken avian b-defensin-8 at positivity selected amino acid sites enhnces specific antmicrobial activity', *Immunogenetics*, **59**, 573–80.

HORCHNER P M, BRETT D, GORMLEY B, JENSON I and POINTON A M (2006), 'HACCP-based approach to the derivation of an on-farm food safety program for the Australian red meat industry', *Food Cont*, **17**, 497–510.

HUFFMAN R D (2002), 'Current and future technologies for the decontamination of carcasses and fresh meat', *Meat Sci*, 62; 285–94.

HUGAS M and TSIGARIDA E (2008), 'Pros and cons of carcass decontamination: the role of the European Food Safety Authority', *Meat Sci*, **78**, 43–52.

HUMPHREY T and JORGENSEN F (2006), 'Pathogens on meat and infection in animals – establishing a relationship using *Campylobacter* and *Salmonella* as examples', *Meat Sci*, **74**, 89–97.

HUMPHREY T, O'BRIEN S and MADSEN M (2007), 'Campylobacters as zoonotic pathogens: A food production perspective', *Int J Food Microbiol*, **117**, 237–57.

HUMPHREY T J and BECKETT P (1987), '*Campylobacter jejuni* in dairy cows and raw milk', *Epidemiol Infect*, **98**, 263–9.

IOM (INSTITUTE OF MEDICINE) (1989), *Human Health Risks with the Subtherapeutic Use of Penicillin or Tetracyclines in Animal Feeds*, National Academy Press, Institute of Medicine, Washington, DC.

JACKSON T C, MARSHALL D L, ACUFF G R and DICKSON J S (2001), 'Meat, poultry and seafood', in Doyle M.P., Beuchat, L.R. and Montville, T.J., *Food Microbiology Fundamentals and Frontiers*, 2nd ed., ASM Press, Washington, pp. 91–110.

JONES F T and RICKE S C (2003), 'Observations on the history of the development of antimicrobials and their use in poultry feeds', *Poult Sci*, **82**, 613–7.

JOSEFSEN M H, LAMBERTZ S T, JENSEN S and HOORFAR J (2003), 'Food-PCR. Validation and standardization of diagnostic PCR for detection of *Yersinia enterocolitica* and other foodborne pathogens', *Adv Exp Med Biol*, **529**, 443–9.

KABAGAMBE E K, WELLS S J, GARBER L P, SALMAN M D, WAGNER B and FEDORKA-CRAY P J (2000), 'Risk factors for fecal shedding of *Salmonella* in 91 US dairy herds in 1996', *Prev Vet Med*, **43**, 177–94.

KAPLAN B S, MEYERS K E and SCHULMAN S L (1998), 'The pathogenesis and treatment of hemolytic uremic syndrome', *J Am Soc Nephrol*, **9**, 1126–33.

KARMALI M A (1989), 'Infection by verocytotoxin-producing *Escherichia coli*', *Clin Micro Rev*, **2**, 15–38.

KIM H J, PARK S H, LEE T H, NAHM B H, KIM Y R and KIM H Y (2008), 'Microarray detection of foodborne pathogens using specific probes prepared by comparative genomics', *Biosens Bioelectron*, **24**, 238–46.

KOOHMARAIE M, ARTHUR T M, BOSILEVAC J M, BRICHTA-HARHAY D M, KALCHAYANAND N, SHACKELFORD S D, ET AL. (2007), 'Interventions to reduce/eliminate *Escherichia coli* O157:H7 in ground beef', *Meat Sci*, **77**, 90–6.

KRAFT A A (1992), *Psychrotrophic Bacteria in Foods: Diseases and Spoilage*, CRC Press, Florida.

KUHNERT P, DUBBOSSON C R, ROESCH M, HOMFIELD E, DOHERR M G and BLUM J W (2005), 'Prevalence and risk-factor analysis of Shiga toxigenic *Escherichia coli* in faecal samples of organically and conventionally farmed dairy cattle', *Vet Microbiol*, **109**, 37–45.

LADO B and YOUSEF A (2002), 'Alternative food preservation technologies: efficacy and mechanisms', *Microbes Infect*, **4**, 433–40.

LAEGREID W W, ELDER R O and KEEN J E (1999), 'Prevalence of *Escherichia coli* O157:H7 in range beef calves at weaning', *Epidemiol Infect*, **123**, 291–8.

LEJEUNE J T, BESSER T E, RICE D H, BERG J L, STILBORN R P and HANCOCK D D (2004), 'Longitudinal study of faecal shedding of *Escherichia coli* O157:H7 in feedlot cattle: predominance and persistence of specific clonal types despite massive cattle population turnover', *Appl Environ Microbiol*, **70**, 377–84.

LEPS J and FRIES R (2009), 'Incision of the heart during meat inspection of fattening pigs – a risk-profile approach', *Meat Sci*, **81**, 22–7.

LEVY S B (2002), 'Active efflux, a common mechanism for biocide and antibiotic resistance', *J Appl Mirco*, **92**, 65–71.

LINE J E, SVETOCH E A, ERUSLANOV B V, PERELYGIN V V, MITSEVICH E V, MITSEVICH I P, ET AL. (2008), 'Isolation and purification of enterocin E-760 with broad antimicrobial activity against Gram-positive and Gram-negative bacteria', *Antimicrob Agents Chemother*, **52**, 1094–100.

LITTLE C L, RICHARDSON J F, OWEN R J, DE PINNA E and THRELFALL E J (2008), '*Campylobacter* and *Salmonella* in raw red meats in the United Kingdom: prevalence, characterization and antimicrobial resistance pattern, 2003–2005', *Food Microbiol*, **25**, 538–43.

LUDWIG K and MULLER-WIEFEL D E (1998), 'Pathomechanisms in the haemolytic-uraemic syndrome', *Nephrol Diay Transplant*, **13**, 23–7.

LYNN R, O'BRIEN S J, TAYLOR C M, ADAK G K, CHART H, CHEASTY T, ET AL. (2005), 'Childhood hemolytic uremic syndrome, United Kingdom and Ireland', *Emerg Infect Dis*, **11**, 5990–6.

MADSEN M (2007), '*Campylobacter*', in Storrs, M. (Ed.): *Food Safety Handbook-Microbiological Challenges*, Biomerieux Editions, France, pp. 20–37.

MARTIN L J, FYFE M, DORÉ K, BUXTON J A, POLLARI F, HENRY B, ET AL. (2004), 'Increased burden of illness associated with antimicrobial-resistant *Salmonella enterica* serotype Typhimurium infections', *J Infect Dis*, **189**, 377–84.

MCEVOY J M, DOHERTY A M, FINNERTY M, SHERIDAN J J, MCGUIRE L, BLAIR I S, ET AL. (2000), 'The relationship between hide cleanliness and bacterial numbers on beef carcasses at a commercial abattoir', *Lett Appl Microbiol*, **30**, 390–5.

MCGEE P, SCOTT L, SHERIDAN J J, EARLEY B and LEONARD N (2004), 'Horizontal transmission of *Escherichia coli* O157:H7 during cattle housing', *J Food Prot*, **67**, 2651–6.

MEAD P and GRIFFIN P (1998), '*Escherichia coli* O157:H7', *Lancet*, **352**, 1207–12.

MEAD P S, SLUTSKER L, DIETZ V, MCCAIG L F, BRESEE J S, SHAPIRO C, ET AL. (1999), 'Food-related illness and death in the Unites States', *Emerg Infect Dis*, **5**, 607–25.

MELLON M B C and BENBROOK K L (2001), *Hogging It: Estimates of Antimicrobial Abuse in Livestock*, UCS Publications, Cambridge.

MENG J and DOYLE M P (1998), 'Microbiology of shiga-toxin producing *Escherichia coli* in foods', in Kaper, J.B. and O'Brien, A.D. (Eds): Escherichia coli *O157:H7 and other Shiga toxin-producing* E. coli *strains*, American Society for Microbiology, Washington, DC, pp. 92–108.

MILLS S D (2006), 'When will the genomics investment pay off for antibacterial discovery', *Biochem Pharmacol*, **71**, 1096–102.

MINIHAN D, WHYTE P, O'MAHONY M, CLEGG T and COLLINS J D (2003), '*Escherichia coli* O157 in Irish feedlot cattle: a longitudinal study involving preharvest and harvest phases of the food chain', *J Food Safety*, 167–78.

MINIHAN D, WHYTE P, O'MAHONY M and COLLINS J D (2003), 'The effect of commercial steam pasteurization on the levels of Enterobacteriaceae and *Escherichia coli* on naturally contaminated beef carcasses', *J Vet Med B*, **50**, 352–6.

MINIHAN D, WHYTE P, O'MAHONY M, FANNING S, DOYLE M and COLLINS J D (2004), 'An investigation of transport, lairage and hide cleanliness on *Campylobacter* prevalence in feedlot cattle and dressed carcasses', *J Food Safety*, **24**, 37–52.

MØLBAK, K (2005), 'Human health consequences of antimicrobial drug-resistant *Salmonella* and other foodborne pathogens', *Clin Infect Dis*, **41**, 1613–20.

MØLBAK K, BAGGESEN D L, AARESTRUP F M, EBBESEN J M, ENGBERG J, FRYDENDAHL K, ET AL. (2000), 'An outbreak of multidrug-resistant, quinolone-resistant *Salmonella* enterica serotype typhimurium DT104', *N Engl J Med*, **342**, 661.

MOORE J E, BARTON M D, BLAIR I S, CORCORAN D, DOOLEY J S, FANNING S, ET AL. (2006), 'The epidemiology of antibiotic resistance in *Campylobacter*', *Microbes Infect*, **8**, 1955–66.

MOORE J E, CORCORAN D, DOOLEY J S G, FANNING S, LUCEY B, MATSUDA M, ET AL. (2005), '*Campylobacter*', *Vet Res*, **36**, 351–82.

MORA A, BLANCO M, BLANCO J E, DAHBI G, LÓPEZ C, JUSTEL P, ET AL. (2007), 'Serotypes, virulence genes and intimin types of Shiga toxin (verocytotoxin)-producing *Escherichia coli* isolates from minced beef in Lugo (Spain) from 1995 through 2003', *BMC Microbiol*, **7**, 13.

MORABITO S, TOZZOLI R, OSWALD E and CAPRIOLI A (2003), 'A mosaic pathogenicity island made up of the locus of enterocyte effacement and a pathogenicity island of *Escherichia coli* O157:H7 is frequently present in attaching and effacing *E. coli*', *Infect Immun*, **71**, 3343–8.

MORRISON D M, TYRELL D L, JEWELL, L D (1986), 'Colonic biopsy in verotoxin-induced hemorrhagic colitis and thrombotic thrombocytopenic purpura (TTP)', *Am J Clin Pathol*, **86**, 108–12.

MURINDA S E and OLIVER S P (2006), 'Physiologic and molecular markers for detection of shiga toxin-producing *Escherichia coli* serotype O26 strains', *Foodborne Path Dis*, **3**, 163–77.

NARAVANENI R and JAMIL K (2005), 'Rapid detection of foodborne pathogens by using molecular techniques', *J Med Microbiol*, **54**, 51–4.

NASTASIJEVIC I, MITROVIC R and BUNCIC S (2008), 'Occurrence of *Escherichia coli* O157 on hides of slaughtered cattle', *Lett Appl Microbiol*, **46**, 126–31.

NASTASIJEVIC I, MITROVIC R and BUNCIC S (2009), 'The occurrence of *Escherichia coli* O157 in/on faeces, carcases and fresh meats from cattle', *Meat Sci*, **82**(1), 101–5.

NEILD G H (1998), 'Hemolytic uremic syndrome/thrombotic thrombocytopenic purpura: pathophysiology and treatment', *Kidney Int Suppl*, **64**, 45–9.

NEUBAUER C and HESS M (2006), 'Detection and identification of foodborne pathogens of the genera *Campylobacter, Arcobacter* and *Helicobacter* by multiplex PCR in poultry and poultry products', *J Vet Med B Infect Dis Vet Public Health*, **53**, 376–81.

NIELSEN E M, TEGTMEIER C, ANDERSEN H J, GRONBAEK C and ANDERSEN J S (2002), 'Influence of age, sex and herd characteristics on the occurrence of verocytotoxin-producing *Escherichia coli* O157 in Danish dairy farms', *Vet Microbiol*, **88**, 245–57.

O'BRIEN S B, DUFFY G, CARNEY E, SHERIDAN J J, MCDOWELL D A and BLAIR I S (2005), 'Prevalence and numbers of *Escherichia coli* O157 on bovine hides at a beef slaughter plant', *J Food Prot*, **68**, 660–5.

O'FLYNN G, ROSS R P, FITZGERALD G F and COFFEY A (2004), 'Evaluation of a cocktail of three bacteriophages for biocontrol of *Escherichia coli* O157:H7', *Appl Environ Microbiol*, **70**, 3417–24.

OGDEN I D, MACRAE M and STRACHEN N J C (2004), 'Is the prevalence and shedding concentrations of *E. coli* O157 in beef cattle seasonal', *FEMS Microbiol Lett*, **233**, 297–300.

OMISAKIN F, MACRAE M, OGDEN I D and STRACHAN N J C (2003), 'Concentration and prevalence of *Escherichia coli* O157 in cattle faeces at slaughter', *Appl Environ Microbiol*, **69**, 2444–7.

OPPEGAARD H, STEINUM T M and WASTESON Y (2001), 'Horizontal transfer of a multi-drug resistance plasmid between coliform bacteria of human and bovine origin in a farm environment', *Appl Environ Microbiol*, **67**, 3732–4.

O'REGAN E, MCCABE E, BURGESS C, MCGUINNESS S, BARRY T, DUFFY G, ET AL. (2008), 'Development of a real-time multiplex PCR assay for the detection of multiple *Salmonella* serotypes in chicken samples', *BMC Microbiology*, **8**, 156.

PACHA R E, CLARK G W, WILLIAMS E A and CARTER A M (1998), 'Migratory birds of central Washington as reservoirs of *Campylobacter jejuni*', *Can J Microbiol*, **34**, 80–2.

PANISELLO P J, ROONEY R, QUANTICK, P C and STANWELL-SMITH R (2000), 'Application of foodborne disease outbreak data in the development and maintenance of HACCP systems', *Int J Food Microbiol*, **59**, 221–34.

PAULSEN P and WINKELMAYER R (2004), 'Seasonal variation in the microbial contamination of game carcasses in an Austrian hunting area', *Eur J Wildlife Res*, **50**, 157–9.

PAYOT S, DRIDI S, LAROCHE M, FEDERIGHI M and MAGRAS C (2004), 'Prevalence and antimicrobial resistance of *Campylobacter coli* isolated from fattening pigs in France', *Vet Microbiol*, **101**, 91–9.

PHEBUS R K, NUTSCH A L, SCHAFER R, WILSON M J, RIEMANN J D, LEISING C L, ET AL. (1997), 'Comparison of steam pasteurization and other methods for reduction of pathogens on surfaces of freshly slaughtered beef', *J Food Prot*, **60**, 476–84.

PIDDOCK L J, WHITE D G, GENSBERG K, PUMBWE L and GRIGGS D J (2000), 'Evidence for an efflux pump mediating multiple antibiotic resistance in *Salmonella* enterica serovar Typhimurium', *Antimicrob Agents Chemother*, **44**(11), 3118–21.

PINTO-ALPHANDARY H, MABILAT C and COURVALIN P (1990), 'Emergence of aminoglycoside resistance genes aadA and aadE in the genus *Campylobacter*', *Antimicrob Agents Chemother*, **34**, 1294–6.

RANGEL J M, SPARLING P H, CROWE C, GRIFFIN P M and SWERDLOW D L (2005), 'Epidemiology of *Escherichia coli* O157:H7 outbreaks, United States, 1982–2002', *Emerg Infect Dis*, **11**, 603–9.

RIJPENS N P and HERMAN L M (2002), 'Molecular methods for identification and detection of bacterial food pathogens', *JAOAC Int*, **85**, 984–95.

RILEY L W, REMIS R S, HELGERSON S D, MCGEE H B, WELLS J G, DAVIS B R, ET AL. (1983), 'Hemorrhagic colitis associated with a rare *Escherichia coli* serotype', *N Engl J Med*, **308**, 681–5.

ROGAN D R, SMITH D R, MOXLEY R A, POTTER A A and STRAUSS C E (2009), 'Vaccination with type III secretion proteins reduces *E. coli* O157:H7 shedding and contamination in cattle', *Vet Immunol Immunopathol*, **128**, 334.

RUSSELL A D and DAY M J (1996), 'Antibiotic and biocide resistance in bacteria', *Microbios*, **85**, 45–65.

SAUNDERS J R, SERGEANT M J, MCCARTHY A J, MOBBS K J, HART C A, MARKS T S, ET AL. (1999), 'Genetics and molecular ecology of *Escherichia coli* O157', in Steward, C.S. and Flint, H.J. (Eds): *E. Coli O157 in Farm Animals*, CAB International, Oxfordshire, UK, pp.1–25.

SCHAMBERGER G P, PHILLIPS R L, JACOBS J L and DIEZ-GONZALEZ F (2004), 'Reduction of *Escherichia coli* O157:H7 populations in cattle by feeding colicin E7-producing *E. coli*', *Appl Environ Microbiol*, **70**, 6053–60.

SCHNIRRING L (2007), *USDA Plans Risk-based Meat Inspection System*. Centre for Infectious Disease Research and Policy, University of Minnesota. Available from: http://www.cidrap.umn.edu/cidrap/contents/fs/food-disease/news/mar0707inspection.html [Accessed 18 February 2009].

SCHROEDER C M, WHITE D G and MENG J (2004), 'Retail meat and poultry as a reservoir of antimicrobial-resistant *Escherichia coli*', *Food Microbiol*, **21**, 249–55.

SELKE M, MEENS J, SPRINGER S, FRANK R and GERLACH G F (2007), 'Immunization of pigs to prevent disease in humans: construction and protective efficacy of a *Salmonella enterica* serovar Typhimurium live negative-marker vaccine', *Infect Immun*, **75**, 2476–83.

SMERDON W J, ADAK G K, O'BRIEN S J, GILLESPIE I A and REACHER M (2001), 'General outbreaks of infectious intestinal disease linked with red meat, England and Wales, 1992–1999', *Commun Dis Public Health*, **4**, 259–67.

SOFOS J N (2008), 'Challenges to meat safety in the 21st century', *Meat Sci*, **78**, 3–13.

STANLEY K and JONES K (2003), 'Cattle and sheep farms as reservoirs of *Campylobacter*', *J Appl Microbiol*, **94**, 104s–13s.

STANLEY K, WALLACE J S, CURRIE J E, DIGGLE P J and JONES K (1998), 'The seasonal variation of thermophilic campylobacters in beef cattle, dairy cattle and calves', *J Appl Microbiol*, **85**, 472–80.

STERN N J, SVETOCH E A, ERUSLANOV B V, PERELYGIN V V, MITSEVICH E V, MITSEVICH I P, ET AL. (2006), 'Isolation of a *Lactobacillus salivarius* strain and purification of its bacteriocin, which is inhibitory to *Campylobacter jejuni* in the chicken gastrointestinal system', *Antimicrob Agents Chemother*, **50**, 3111–16.

SVETOCH E A, ERUSLANOV B V, PERELYGIN V V, MITSEVICH E V, MITSEVICH I P, BORZENKOV V N, ET AL. (2008), 'Diverse antimicrobial killing by *Enterococcus faecium* E 50–52 bacteriocin', *J Agric Food Chem*, **56**, 1942–8.

SWERDLOW D L and GRIFFIN P M (1997), 'Duration of faecal shedding of *Escherichia coli* O157:H7 among children in day-care centres', *Lancet*, **347**, 745–6.

TAUXE R V (2002), 'Emerging foodborne pathogens', *Int J Food Microbiol*, **78**, 31–41.

TAYLOR C M (2008), 'Enterohaemorrhagic *Escherichia coli* and *Shigella dysenteriae* type 1-induced haemolytic uraemic syndrome', *Pediatr Nephrol*, **23**, 1425–31.

TAYLOR D N, BLASER M J, ECHEVERRIA P, PITARANGSI C, BODHIDATTA L and WANG W L (1987), 'Erythromycin-resistant *Campylobacter* infections in Thailand', *Antimicrob Agents Chemother*, **31**, 438–42.

THE *CAMPYLOBACTER* SENTINEL SURVEILLANCE SCHEME COLLABORATORS (2000), 'Ciprofloxacin resistance in *Campylobacter jejuni*: case–case analysis as a tool for elucidating risks at home and abroad', *J Antimicrob Chemother*, **50**, 561–8.

THOMAS B (1999), 'Nachweis von verotoxinbildenden *Escherichia coli* in rehfleisch', *Arch fur Lebensmittelhygiene*, **50**, 52–4.

THRELFALL E J (2000), 'Epidemic *Salmonella* Typhimurium DT104 – a truly international multiresistant clone', *J Antimicrob Chemother*, **46**, 7–10.

TOZZI A E, GORIETTI S and CAPRIOLI A (2001), 'Epidemiology of human infections by *Escherichia coli* O157 and other verocytotoxin-producing *E. coli*', in Duffy G, Garvey P, McDowell DA, *Verocytotoxigenic E. coli*', Food and Nutrition Press, Connecticut, pp. 161–80.

TRAVERS K and BARZA M (2002), 'Morbidity of infections caused by antimicrobial-resistant bacteria', *Clin Infect Dis*, **34**(Suppl 3): S131–4.

USDA (UNITED STATES DEPARTMENT OF AGRICULTURE) (1996), 'Pathogen reduction: Hazard Analysis Critical Control Point (HACCP) systems: Final rule', *Federal Register*, **61**, 38806–988.

USDA (2006), 'Risk-based inspection system Docket No. FSIS-2006-0028', *Federal Register*, **71**, 56470–1.

USDA (FOOD SAFETY INSPECTION SERVICE) (2007), *The Evolution of Risk-based Inspection.* Available from: http://www.fsis.usda.gov/PDF/evolution_of_RBI_022007.pdf [Accessed18 February 2009].

USDA (FOREIGN AGRICULTURE SERVICE) (2008a), *Livestock and Poultry: World Markets and Trade.* Available from: http://www.fas.usda.gov/psdonline/circulars/livestock_poultry.pdf [Accessed 20 January 2009].

USDA (ECONOMIC RESEARCH SERVICE) (2008b), *Agricultural Projections to 2017.* Available from: http://www.ers.usda.gov/Publications/OCE081/OCE20081fm.pdf [Accessed 20 January 2009].

USDA (ECONOMIC RESEARCH SERVICE) (2008c), *US Beef and Cattle Industry: Background Statistics and Information.* Available from: http://www.ers.usda.gov.news/BSECoverage.htm [Accessed 14 January 2009].

USDA (ECONOMIC RESEARCH SERVICE) (2009), *Food Safety: Economic Costs of Foodborne Illness.* Available from: http://www.ers.usda.gov/briefing/foodsafety/economic.htm [Accessed 26 January 2009].

VAN DONKERSGOED J, GRAHAM T and GANNON V (1999), 'The prevalence of verotoxins, *Escherichia coli* O157:H7, and *Salmonella* in the feces and rumen of cattle at processing', *Can Vet J*, **40**, 332–8.

VAN LOOVEREN M, DAUBE G, DE ZUTTER L, DUMONT J M, LAMMENS C, WIJDOOGHE P, ET AL. (2001), 'Antimicrobial susceptibilities of *Campylobacter* strains isolated from food animals in Belgium', *J Antimicrob Chemother*, **48**, 235–40.

VARMA J K, MØLBAK K, BARRETT T J, BEEBE J L, JONES T F, RABATSKY-EHR T, ET AL. (2005), 'Antimicrobial-resistant nontyphoidal *Salmonella* is associated with excess bloodstream infections and hospitalizations', *J Infect Dis*, **191**, 554–61.

VERNOZY-ROZAND C (2007), '*Escherichia coli*/EHEC', in Storrs M. (Ed.): *Food Safety Handbook-Microbiological Challenges*, BioMerieux, France, p. 142.

VIERIKKO A, HANNINEN M, SIITONEN A, RUUTU P and RAUTELIN H (2004), 'Domestically acquired *Campylobacter* infections in Finland' *Emerg Infect Dis*, **10**, 127–30.

WAGENAAR J A, VAN BERGEN M A, MUELLER M A, WASSENAAR T M and CARLTON R M (2005), 'Phage therapy reduces *Campylobacter jejuni* colonization in broilers', *Vet Microbiol*, **109**, 275–83.

WAHLSTROM H, TYSEN E, ENGVALL O, BRANDSTROM E, ERIKSSON E and MORNER T (2003), 'Survey of *Campylobacter* species, VTEC O157 and *Salmonella* species in Swedish wildlife', *Veterinary Record*, **153**, 74–80.

WALSH C, DUFFY G, NALLY P, O'MAHONY R, MCDOWELL D A and FANNING S (2008), 'Transfer of ampicillin resistance from *Salmonella Typhimurium* DT104 to *Escherichia coli* K12 in food', *Lett Appl Microbiol*, **46**, 210–15.

WALSH C and FANNING S (2008), 'Antimicrobial resistance in foodborne pathogens – a cause for concern?', *Curr Drug Targets*, **9**, 808–15.

WALSH C, KENNEDY J and FANNING S (2008), 'Antimicrobial resistance in bacteria from food', in *Research advances in Applied and Environmental Microbiology 1*, Global Research Network, pp. 9–31.

WANG X W, ZHANG L, JIN L Q, JIN, M, SHEN Z Q, AN S, ET AL. (2007), 'Development and application of an oligonucleotide microarray for the detection of food-borne bacterial pathogens', *Appl Microbiol Biotechnol*, **76**, 225–33.

WARNICK L D, CROFTON L M, PELZER K D and HAWKINS M J (2001), 'Risk factors for clinical salmonellosis in Virginia, USA cattle herds', *Prev Vet Med*, **49**, 259–75.

WASSENAAR T M and SILLEY P (2008), 'Antimicrobial resistance in zoonotic bacteria: lessons learned from host-specific pathogens', *Animal Health Res Rev*, **9**, 177–86.

WERNER G, HILDEBRANDT B and WITTE W (2001), 'Aminoglycoside-streptothricin resistance gene cluster aadE-sat4-aphA-3 disseminated among multiresistant isolates of *Enterococcus faecium*', *Antimicrob Agents Chemother*, **45**, 3267–9.

WHEELER J G, SETHI D, COWDEN J M, WALL P G, RODRIGUES L C, TOMPKINS D S, ET AL. (1999), 'Study of infectious intestinal disease in England, rates in the community, presenting to General Practice and reported to national surveillance', *BMJ*, **318**, 1046–50.

WHO (WORLD HEALTH ORGANIZATION) (2005a), *Enterohaemorrhagic* Escherichia coli *(EHEC)*. Available from: http://www.who.int/mediacentre/factsheets/fs125/en/ [Accessed March 26 2009].

WHO (2005B), *Drug-Resistant* Salmonella. Fact Sheet No 139.

WHO (2008), International Food Safety Authorities Network (INFOSAN) Information note no. 2/2008, 7 March 2008.

WHYTE P, FANNING S, O'MAHONY M, O'MAHONY R and DRUDY D (2005), 'Public health and economic burden of infectious food-borne diseases in Ireland', *Irish Vet J*, **58**, 279–83.

WHYTE P, MCGILL K, COWLEY D, MADDEN R H, MORAN L, SCATES P, ET AL. (2004), 'Occurrence of *Campylobacter* in retail foods in Ireland', *Int J Food Microbiol*, **95**, 111–18.

WILLSHAW G A, THIRLWELL J, JONES A P, PARRY S, SALMON R L and HICKEY M (1994), 'Verocytotoxin-producing *Escherichia coli* O157 in beefburgers linked to an outbreak of diarrhoea, haemorrhagic colitis and haemolytic uraemic syndrome in Britain', *Lett Appl Micro*, **19**, 304–7.

WRIGHT G D (2007), 'The antibiotic resistome: the nexus of chemical and genetic diversity,' *Nat Rev Microbiol*, **5**, 175–86.

WRIGHT G D and SUTHERLAND A D (2007), 'New strategies for combating multidrug-resistant bacteria', *Trends Mol Med*, **13**, 260–7.

YOUNG J H (1981), *The Long Struggle for the 1906 Law, FDA Consumer*. Available from: http://www.foodsafety.gov/~Ird/history2.html [Accessed 26 January 2009].

19

Tracing pathogens in fish production chains

B. T. Lunestad and A. Levsen, NIFES, Norway and J. T. Rosnes, NOFIMA Norconserv, Norway

Abstract: If found in seafood, microorganisms and parasites may negatively affect the product in two ways: by reducing the quality or by compromising safety through intoxications, gastrointestinal infections or allergies. This chapter discusses how bacteria, fungi and parasites or their products may affect the safety of fish and fishery products throughout the production chains. Furthermore, possible strategies of tracking the sources, reservoirs, survival and potential amplification of pathogens in the fish production chains are discussed. The chapter also deals with control strategies, including new preservation methods, hazard analysis and critical control point (HACCP) analysis and microbial modelling.

Key words: seafood, fish, microorganisms, bacteria, fungi, parasites, production chains.

19.1 Introduction

In 2005, the world fisheries and aquaculture production reached an estimated total volume of 141.6 million metric tons. Of this, capture and aquaculture represented 93.8 million and 47.8 million tons, respectively (FAO, 2007). It is predicted that aquaculture production will continue to expand, whereas marine fisheries seem to have reached a ceiling. The patterns of food consumption are also changing throughout the world, shifting towards higher-quality and more expensive foods such as meat, fish and dairy products. A growing and generally richer world population (http://www.gapminder.org) will have increasing demands for animal protein including fish, necessitating an increased focus on production of safe seafood products of high quality.

19.2 Foodborne pathogens in the fish production chains

Microorganisms and parasites in fish products may negatively influence the product in two ways, either by reducing the quality as food or compromising the safety due to the potential to cause intoxications, infections or allergies.

Lynch *et al.* (2006) have reported on foodborne disease outbreaks in the United States in the period from 1998 to 2002. In this period, a total of 6647 outbreaks were reported, and in 33% of these cases, the ethological agent was determined. These outbreaks caused a reported 128 370 persons to become ill. In this material, finfish was responsible for 337 outbreaks (5.1%), 1692 diseased persons (1.3%) and one fatal case (1.1%).

In this chapter, we will discuss how microorganisms and parasites, or their products, may affect the safety of fish and fish products throughout the production chains.

19.3 Bacteria

Table 19.1 gives an overview of growth characteristics with respect to temperature, pH, water activity (a_w) and presence of oxygen for the pathogenic bacteria covered in this chapter.

19.3.1 *Listeria monocytogenes*

The genus *Listeria* includes six species at present, *Listeria monocytogenes*, *Listeria ivanovii*, *Listeria innocua*, *Listeria welshimeri*, *Listeria seeligeri* and *Listeria grayi*. Of these, only *L. monocytogenes* is considered a problem for food safety. *L. monocytogenes* is able to grow under both aerobic and anaerobic conditions, at temperatures normally found during refrigeration and at rather high concentrations of salt (10%). The robustness of this bacterium, together with its wide distribution, makes it a special challenge in the production of lightly preserved ready-to-eat foods. In seafood produced under temperate and cold climatic conditions, *L. monocytogenes* is the main bacterium of concern with respect to fish and fish product safety.

The ubiquity of the *L. monocytogenes* suggests that human exposure to the bacterium must be frequent. *L. monocytogenes* has been shown to cause the disease listeriosis in animals and man. Even though the number of listeriosis cases is relatively low, the fatality rate is as high as 20–30% (Paoli *et al.*, 2005). Listeriosis in humans has been associated with the consumption of a number of different foods including green salad, cheese, cold cuts, mussels, cold smoked fish and fermented fish and meat products. Some cases of listeriosis with fish as the most likely source have been reported (Facinelli *et al.*, 1989; Ericsson *et al.*, 1997; Miettinen *et al.*, 1999).

The habit of eating raw or lightly processed fish, such as cold smoked salmon, is not new. However, a recent trend is the increased consumption of raw seafood

Table 19.1 Growth characteristics of pathogenic bacteria relevant for fish and fish products included in this chapter

Microorganism	Minimum temp. °C	Minimum pH	Minimum a_w	Aerobic/ Anaerobic	Typical food item and environmental reservoirs
Bacillus cereus	4	4.3	0.95	Facultative	Rice, spices, vegetables, eggs, dairy products, heat treated fish products
Clostridium botulinum (mesophilic, proteolytic)	10	4.6	0.93	Anaerobic	Meat, fish, vegetables, soil, sediments
C. botulinum (psychrotrophic, non-proteoloytic)	3	5.0	0.97 (or ≥ 5.5% NaCl)	Anaerobic	Seafood (type E), meat (type B, F)
C. perfringens	12	5.0	0.95		Heat treated meat and fish products, soil, aquatic sediments
Escherichia coli	5	3.8	0.95	Facultative	Meat and fish products intestine of warm-blooded animals, faecal contaminated water and soil
Listeria monocytogenes	0	4.3	0.92	Facultative	Seafood, meat, vegetables, non pasteurised diary products, soil, water, plants, sewage drain
Salmonella sp.	5	3.8	0.94	Facultative	Poultry, egg, spices, animal feeds, dried ingredients
Staphylococcus aureus	6 (10 for toxin)	4.0 (4.5 for toxin)	0.83 (0.9 for toxin)	Facultative	Recontaminated heat treated foods, humans and warm blooded-animals
Vibrio cholerae	10	5.0	0.97	Facultative	Seafood, human intestine, faecal polluted water
V. parahaemolyticus	5	4.8	0.94 (Halophile)	Facultative	Seafood, coastal and brackish water
Aeromonas hydrophila	0	4.0	0.97 > 5% NaCl	Facultative	Fish and shellfish, some red meats (beef, pork, lamb) and poultry
Plesiomonas shigelloides	8	4.5	0.97 > 5% NaCl	Facultative	Seafood, coastal and brackish water

Source: Adapted from European Chilled Food Federation (ECFF, November 1996). In: Martins, T., Harmonization of safety criteria for minimally processed foods. Inventory report FAIR concerted action FAIR CT96.1020, September 1997, and updated with information from Adams and Moss, 2008.

as sashimi, often referred to as sushi. This may lead to new safety challenges, especially concerning parasites. Regarding *L. monocytogenes*, sashimi is not considered a product of increased risk since it has a short shelf-life that is not sufficient to allow multiplication of the bacterium to high numbers.

Survey programs for *L. monocytogenes* in foods are performed on a regular basis by many national food control authorities. The findings from such examinations show large year-to-year variations even for the same product and country. The general tendency, however, is that the proportion of positive samples and the number of bacteria in positive samples increase during processing of products with no or minimal preservation.

Several international bodies that deal with food safety issues have addressed *L. monocytogenes* and works to establish acceptable numbers in different types of foods are currently undertaken. It has been documented that products containing less than a certain number of bacteria (100–1000 bacteria per gram) are not involved in disease among healthy adults. Some national food control authorities have given recommendations for susceptible groups of consumers (children, pregnant and immunocompromised persons) to avoid products where there is a risk for *L. monocytogenes* contamination. Such products include smoked vacuumed fish products and cold cuts.

In a recent study, 75 isolates of *L. monocytogenes* from pelagic and farmed fish and seafood processing environments have been typed by multiple-locus variable number of tandem repeat analysis (MLVA; Lunestad *et al.*, 2009). This typing method is based on the identification of repeated sequences of highly polymorphic DNA from multiple genome loci in bacteria. The MLVA typing method has previously been applied for the typing of several foodborne pathogens, including *L. monocytogenes* (Lindstedt *et al.*, 2004a, 2004b; Lindstedt, 2005; Lindstedt *et al.*, 2007; Lindstedt *et al.*, 2008). This typing method divided the 75 isolates into 26 distinct MLVA profiles, 14 of which were found to match profiles from human clinical isolates. Even though the result from this study demonstrates a possible epidemiological link to seafood, it cannot on the basis of this study be concluded that humans have been infected by *L. monocytogenes* through the consumption of fish.

19.3.2 *Salmonella*

Bacteria in the genus *Salmonella* belong to the family Enterobacteriaceae, and are Gram-negative and oxidase-negative rods. The range of foods involved in transmission of salmonelosis is large, and also includes seafood such as fish, shellfish and crustaceans (Huss *et al.*, 2004). Due to the heterothermic properties of non-mammalian marine animals, *Salmonella* is not a constituent of the normal flora of these animals. The presence of *Salmonella* in seafood must therefore be a result of contamination via polluted rearing water, food handlers or from cross-contamination during catch, production or transport. Compared to other foods, fish is not considered an important vehicle for *Salmonella* transmission even though fish feed may occasionally harbour *Salmonella* (Lunestad *et al.*, 2007).

19.3.3 *Aeromonas* and *Plesiomonas*

The genera *Aeromonas* and *Plesiomonas* comprise Gram-negative rods that are commonly found in aquatic environments. *Aeromonas* and *Plesiomonas* species have attracted increased attention due to their ability to grow under refrigerated conditions, possibly giving increased challenges as consumption of chilled foods increase. The reported time of generation for *A. hydrophila* at 4–5°C is in the range of 9–14 hours (Adams and Moss, 2008). Common symptoms seen for foodborne infections with *Aeromonas* and *Plesiomonas* are watery or sometimes bloody mucoid diarrhoea, stomach pain and nausea. In some cases, invasive infections may be seen, and is then often accompanied by fever. In children, vomiting may also be seen. As for most other infections, young, old and immunocompromised persons are those that most often contract infections.

Aeromonads are widespread and may be detected in raw meat, unpasteurised milk, fish, shellfish and vegetables. These bacteria are not considered a normal component of the human intestinal flora, but may sometimes be detected in asymptomatic carriers especially in the tropical areas. *Aeromonas* strains associated with disease in humans seems to be only a subset of those found in the environment, and it has been proposed that certain strains of aeromonads are particularly adapted to a rapid multiplication in foods. The species most often isolated during gastrointestinal disease in humans are *Aeromonas hydrophila, A. caviae* and *A. sobria*. The foods most frequently linked to outbreaks by aeromonads are water, seafood and beef, pork or chicken meat.

In contrast to the genus *Aeromonas*, *Plesiomonas* consists of one homogenous species, *P. shigelloides*. This organism has been implicated in at least two waterborne outbreaks involving several thousand persons. Other outbreaks with *P. shigelloides* have been linked to consumption of crab, oysters, fish and shrimp, showing that this organism is of special relevance to seafood.

Plesiomonas shigelloides and the aeromonads will not survive even mild heating, and do not tolerate reduced pH, or salt concentrations above 5% in food products.

19.3.4 *Vibrio cholerae, Vibrio parahaemolyticus* and *Vibrio vulnificus*

Bacteria belonging to the genus *Vibrio* are Gram-negative, oxidase-positive, non-spore forming and typically appear as slightly curved rods. In contrast to most other bacteria of importance for seafood safety, members of this group are widespread in aquatic habitats at different salinities. These bacteria are common in marine and estuarine environments, and on the surfaces of marine plants and animals (Baumann *et al.*, 1984). They may also occur naturally in the intestinal content of marine animals (Sakata, 1990), and some *Vibrio* species are also found in freshwater (West, 1989). There is no correlation between the occurrence of vibrios and bacteria of mammalian faecal origin. Therefore, common indicator organisms for faecal pollution do not give information on the presence of pathogenic vibrios.

The genus *Vibrio* contains at present 63 species, of which more than 20 *Vibrio* species have been described as being able to cause disease in animals, while twelve species have so far been reported to be pathogenic to humans. The latter species are *V. cholerae*, *V. parahaemolyticus*, *V. vulnificus*, *V. fluvialis*, *V. alginolyticus*, *V. damsela*, *V. furnissii*, *V. hollisae*, *V. mimicus*, *V. cincinnatiensis*, *V. metschnikovii* and *V. carchariae*. Eight of these species have been reported as associated with foodborne infections, of which *V. cholerae*, *V. vulnificus* and *V. parahaemolyticus* are considered to be predominant as human pathogens. The pathogenicity of *V. cholerae* and *V. parahaemolyticus* is highly correlated to the presence of specific virulence factor genes. An overview of the pathogenic *Vibrio* species associated with human infections is given in Table 19.2.

The incidence and density of human pathogenic vibrios in the environment and consequently in seafood products are highly dependent on the ambient water temperatures. These bacteria are occurring at increased numbers at elevated seawater temperatures (Baffone *et al.*, 2000; Høi *et al.*, 1998; Oliver and Kaper, 2001; O'Neil *et al.*, 1992; West, 1989; Dalsgaard, 2002). Pathogenic vibrios may also be detected in seafood organisms in temperate waters, as in the Nordic countries during the summer months, but their abundance is comparatively low. Foodborne infections with vibrios are most common in Asian countries, and less common in Europe especially countries having a cold climate (Sutherland and Varnam, 2002).

Epidemic *V. cholerae* has historically been associated with large outbreaks involving several countries and a large number of cases, i.e. pandemics. The most predominant symptom is excessive watery diarrhoea, leading to circulation failure and possibly death. The largest recent cholera pandemic occurred in South America in the early 1990s and involved more than 400 000 cases. This outbreak was at least partially caused by a contaminated ceviche, a raw fish product (Huss *et al.*, 2004).

V. parahaemolyticus typically gives disease characterized by rapidly emerging abdominal pain, nausea and explosive diarrhoea, sometimes followed by fever,

Table 19.2 *Vibrio*-species associated with infections in humans. The table gives an overview of *Vibrio* species reported in the literature and their site of infection. GI tract: gastro intestinal tract; ++ : most common site of infection; + : other sites of infection; (+) : rare sites of infection; ? : infection remains to be established

Vibrio sp.	Site of infection						
	GI tract	Wound	Ear	Primary septicaemia	Bacteraemia	Lung	Menings
1. *V. cholerae* O1/O139	++	+	?	?	?	?	?
V. cholerae non O1/O139	++	+	+	(+)	+	?	(+)
2. *V. parahaemolyticus*	++	+	+	?	(+)	(+)	(+)
3. *V. vulnificus*	+	++	?	++	+	(+)	(+)
4. *V. fluvialis*	++	?	?	?	?	?	?
5. *V. alginolyticus*	?	++	+	?	(+)	?	?
6. *V. damsela*	?	++	?	?	?	?	?
7. *V. furnissii*	+	?	?	?	?	?	?
8. *V. hollisae*	++	?	?	(+)	?	?	?
9. *V. mimicus*	++	+	+	?	?	?	?
10. *V. metschnikovii*	(+)	?	?	(+)	?	?	?
11. *V. cincinnatiensis*	?	?	?	?	(+)	?	(+)
12. *V. carchariae*	?	++	?	?	?	?	?

Source: From: West, 1989; Oliver and Kaper, 2001 and Adams and Moss, 2008.

vomiting and headache. Infections have exclusively been linked to consumption of seafood, especially raw or undercooked products. As for *V. cholerae*, the dose of infection is high, and are typically reported as 10^6 cells per gram food. The time of incubation lasts from some few to 24 hours, and the duration of the disease is usually 2–3 days. The fatality rate is generally low, even though some cases of infection by *V. parahaemolyticus* may lead to death.

19.3.5 *Escherichia coli*

Escherichia coli belong to the family of Enterobacteriaceae, and are a typical species in the group commonly designated as coliform bacteria. *E. coli*, as the other coliform bacteria, are facultative anaerobic, Gram-negative rods capable to ferment lactose with gas production within 48 hours at 30–37°C.

Most of the *E. coli* strains are considered harmless commensals in the intestine, but some strains are pathogenic and can cause disease. Pathogenic *E. coli* are currently divided into groups depending on virulence and clinical symptoms. Important groups are: enteropathogenic (EPEC) and enteroaggregative *E. coli* (EaggEC) both involved in diarrhoea among children, enterotoxigenic *E. coli* (ETEC) that give the common traveller's diarrhoea, enteroinvasive *E. coli* (EIEC), diffusely-adherent *E. coli* (DAEC), and enterohemorrhagic *E. coli* (EHEC) that may cause haemorrhagic colitis and, in some cases among children, haemolytic uremic syndrome (HUS; Smith and Fratamico, 2005).

Outbreaks of EHEC have been connected to many food items including meat, milk, vegetables, fruit juices and fish (Jay *et al.*, 2005). Neither of these *E. coli* types are commonly found in temperate waters or on seafood products from temperate waters.

Poor hygiene, cross-contamination by food handlers or contaminated water may, however, transfer the organism. The use of good hygienic practice (GHP) with emphasis on clean water and good personnel hygiene will help in the control these organisms. As all are sensitive to heating, the GHP-programme must be particularly strict when ready-to-eat foods are processed.

19.3.6 *Clostridium botulinum*

Bacteria in the genus *Clostridium botulinum* are anaerobic spore-formers of great importance for food safety. They are naturally found in soil, sediments and water on a world-wide basis (Fach *et al.*, 2002; Hielm *et al.*, 1998; Huss, 1980; Gram, 2001). The bacterium has even been detected in sea-salt intended for fish salting (Fenicia *et al.*, 2002). Most strains within the species of *C. botulinum* are able to produce very potent proteineous toxins during growth, and the presence of such bacteria is consequently of great concern in the assessment of food safety.

The Gram-positive rod-shaped *C. botulinum* can be divided into seven types (A to G) based on the serology of the toxins produced. Clostrids producing the G toxin are currently re-designated to the spesies *C. argentinence*. The amount of toxin necessary to cause human illness is very small, and may be measured in nanograms. Thus, if an environment favourable to germination and growth of spores from any of the seven types of *C. botulinum* known to produce neurotoxins exists in fish before or during processing, there is a potential risk of illness. These toxins are thermally unstable, and will generally be inactivated at temperatures above 85°C for 5 minutes. Proper reheating instructions, ensuring this thermal load on chilled ready meals, could add an extra safety margin against botulinum toxins, although the consumer handling is not reliable. Bacterial toxins in general, including botulinum toxin are stable at high salt concentrations and low pH (Huss and Rye Pedersen, 1980). Any toxin present or preformed in the raw material could be carried over to the final product, and strict growth control from harvest to consumption is necessary (Huss *et al.*, 2004). Human botulism, i.e. infections or intoxications associated with *C. botulinum*, is in the vast majority of cases associated with the types A, B, E and rarely F (Austin, 2001).

The type E *C. botulinum* is together with the types B and F classified as psychrotrophic, and thereby able to grow at relatively low temperatures. At otherwise optimal conditions *C. botulinum* type E may grow and produce toxins at temperatures down to 3°C (Graham *et al.*, 1997), and in products containing up to 5% NaCl (Gram, 2001). Optimal temperature of growth is reported to be 18–25°C, and the minimum water activity (a_w) required is 0.97.

Several authors have reported on the prevalence of *C. botulinum* in seafood (Cann *et al.*, 1966; Fach *et al.*, 2002; Gram, 2001; Hielm *et al.*, 1998; Huss, 1980; Hyytiä *et al.*, 1998; Hyytiä-Trees, 1999). When found in seafood products from cold-water areas such as Scandinavia, Canada, Alaska, Russia and some parts of Japan, *C. botulinum* type E is reported to be the most prevalent type (Huss, 1994). Based on the widespread distribution reported in the present publication, the author considers *C. botulinum* type E to be a true aquatic organism.

Eklund *et al.* (1982) cautioned that the non-proteolytic nature of type E organisms would not result in the development of odours indicative of spoilage. Thus toxin could be formed with little evidence to the consumer that the fish was spoiled and possibly unsafe. In a 1963 outbreak described by these authors, only 3 of 16 affected people reported any unusual flavours or off odours.

Many cases of botulism are linked to the consumption traditional preserved fish products, used for many centuries. An example is non-vacuumed, lightly salted and fermented trout from freshwater sources ('rakfisk'), or from home produced lightly salted and dried ham. Rakfisk is produced on a small-scale local basis, and has at present no export value.

The sodium chloride concentration is an important factor in controlling the outgrowth and toxin production of *C. botulinum*. It is generally accepted that the inhibitory concentration for non-proteolytic *C. botulinum* is 5.0% water-phase salt (WPS), equivalent to a water activity of 0.97. A 10% WPS concentration is necessary to inhibit the proteolytic strains. The hurdle principle will apply in such a situation, and it should be stressed that these NaCl limits apply under optimal conditions for bacterial growth, and that other factors as the presence of oxygen, low temperature and high or low pH will reduce the salt tolerance of *C. botulinum* in a practical situation. Huss (1994) concluded that in fish products stored at temperatures below 10°C, and at a WPS concentration of 3 % would be sufficient to inhibit the growth of *C. botulinum* for at least 30 days. In the traditional salting process applied for herring in Norway, the salt content will within one week reach at least 12–13% in the water phase through the fish, giving a narrow time interval where growth could occur, if the temperature were favourable for bacterial growth.

The Codex Alimentarius Commission made an evaluation of traditionally prepared salted herring and sprat, and concluded that botulism from such products does not seem to be a public health issue. Even if spores of *C. botulinum* should be detected in such fish, the temperature during the brining operation and the required concentrations of NaCl is sufficient to arrest the growth and toxin production even in uneviscerated fish (Anon., 2003).

19.3.7 *Clostridium perfringens*

Clostridium perfringens is widely distributed in the environment, and may, for example, be found in soil at levels from 10^3 to 10^4 per gram. These anaerobic, Gram-positive spore-forming bacteria are also isolated from water and sediments, from faeces of healthy human individuals and from the gut of cod (Aschfalk and Müller, 2002) and different freshwater fishes (Cai *et al.*, 2008).

C. perfringens does not grow at chill temperatures, and the minimum growth temperatures are often referred to as 12°C, and the growth rate is slow below 20°C. In general, *C. perfringens* type A food poisoning ranks as one of the most common causes of foodborne diseases. This is promoted by two main characteristics: a rapid doubling time (<10 min for vegetative cells) at favourable temperatures and very heat stable spores (tolerate boiling for > 1 h). The vegetative cells are sensitive to acid (minimum pH of 5), salt (maximum 6%), and they do not grow at water activities below 0.95. Suitable growth conditions in chilled fish products are therefore only sporadically found, and growth control in seafood should not be difficult to maintain. Controlling proper time and temperature conditions and avoiding cross-contamination to heat-treated foods is essential. *C. perfringens* inoculated in fish fillets of jack mackerel subsequently packaged under a controlled carbon dioxide gas atmosphere showed no increase in numbers at 15°C in 3 days, when all samples were considered microbiologically spoiled (Kimura *et al.*, 1996).

Approximately seven annual outbreaks of *C. perfringens* are reported in the US with links to seafood and it is estimated that approximately 200 cases linked to seafood occur every year (Feldhusen, 2000). Food poisoning may occur if high levels of cells are found in the food, making it likely that a certain number of cells may survive the gut passage, sporulate in the small intestine and produce an enterotoxin. Resulting symptoms occurring 8–24 hours after ingestion may be nausea, abdominal pain, diarrhoea and sometimes vomiting.

19.3.8 *Bacillus cereus*

The genus *Bacillus* comprise facultative anaerobic, Gram-positive spore-forming bacteria widely distributed in the environment. They may be isolated from a wide variety of foods, especially of plant origin, but also from meat, fish and dairy products. The most important pathogenic species of relevance to food is *B. cereus*, but also other *Bacillus* species, such as *B. subtilis*, *B. licheniformis* and *B. pumilis* have been shown to cause food poisoning. *B. cereus* may cause different types of intoxications. The two types of food poisoning, emetic and diarrhoeal disease are caused by very different virulence factors.

Most strains are mesophilic and are able to grow in low-acid foods at temperatures down to 15°C and up to 55°C, with an optimum of 30–40°C. During the last decade, psychrotrophic strains of *B. cereus* that are able to grow at temperatures down to 4°C and 6°C, have been recognized. Complex fish products composed of several ingredients, e.g. vegetables, spices, milk and flour, may contain *Bacillus* spp. Many products only receive a mild-heat treatment, sufficient to safeguard the product against spores of *Clostridium botulinum*, but not for the more heat resistant spores of *Bacillus cereus*. *Bacillus* species has been isolated from sous vide cod fillets (Ben Embarek, 1994) and in many other sous vide products (Nissen *et al.*, 2002). Food containing more than 10^4 *B. cereus* cells per gram may not be safe for consumption. Control of *Bacillus cereus* is efficiently obtained by chilling, except for the few psychrotrophic strains.

19.3.9 *Staphylococcus aureus*

The genus *Staphylococcus* comprises several species of which *S. aureus* is the most important associated with foodborne disease. *S. aureus* has a lower growth limit of 6°C and a limit for toxin production of 10°C. *S. aureus* are rarely found on fresh fish from temperate waters, but may be isolated from newly caught fish in warm waters. The most likely contamination route for *S. aureus*, is from food production personnel and handlers with infections on the skin or in mucosal linings.

Investigations have shown that many food handlers, 6% and 12% according to hand and nasal sampling, carry enterotoxigenic *S. aureus*. This is documented in that some investigations have found 2–10% *S. aureus* in fresh fish and bivalves, while 24–52% are found in some vulnerable cooked products (Jablonski and Bohach, 1997).

The staphylococci are Gram-positive cocci with their primary habitat in the skin, glands and mucous membranes of warm-blooded animals including humans. Infected sores and scratches are often harbourage sites for *S. aureus*. Outside the body, *S. aureus* is one of the most resistant non-spore forming human pathogens and can survive for extended periods in a dry state. Staphylococcal food poisoning ranks as one of the most prevalent foodborne causes of gastroenteritis worldwide. It results from ingestion of one or more preformed staphylococcal enterotoxins in contaminated food. Due to the mesophilic growth some sort of temperature abuse must precede intoxication, and a level of about 10^6 CFU per gram is needed to form toxic doses. The bacterium is very salt tolerant and levels and toxins may be produced in up to 10–15 % NaCl. Growth and toxin production may occur in products handled by man such as cooked and peeled crustaceans, warm smoked fish or fish products which are ripened, like sardines (Arkoudelos *et al.*, 2003) and 'rakad fish'.

Prevention and control of staphylococcal toxin food poisoning include proper chilling and avoidance of cross-contamination of both raw and cooked products. The toxins are heat stable and may resist boiling temperatures.

19.4 Biogenic amines

Biogenic amines are decomposition products from bacterial degradation of substances found in the food. The most important biogenic amine of relevance for seafood is histamine. Histamine is the result of microbial decarboxylation of the amino acid histidine, which is found in high levels in certain fish species. Such species belong to the Scombridae and Scomberesocidae families, such as tuna, bonito and mackerel, and it is therefore commonly referred to as scombroid poisoning. Non-scombroid fish such as sardines and herring have also been implicated in histamine formation. Histamine is remarkably heat stable, and is not destroyed by common preparation and preservation techniques. Fortunately, scombroid poisoning is usually a mild intoxication and it is not a significant cause of death. Histamine-forming bacteria isolated from scombroid fish involved in food poisoning are *Morganella* (formerly *Proteus*) *morganii*, *Klebsiella pneumonia* and *Hafnia alvei*. Furthermore, *Proteus vulgaris*, *Proteus mirabilis*, *Clostridium perfringens*, *Enterobacter aerogenes* and *Vibrio alginolyticus* are isolated from fish and shown able to decarboxylate amino acids (Shalaby, 1996). In addition to these, the following bacteria have been found to have histamine decarboxylase activity in fish: *Acinetobacter lwoffi*, *Shewanella putrefaciens*, *Aeromonas hydrophila* and *Plesiomonas shigelloides*. If fish are not properly refrigerated, there is greater risk of unacceptable histamine levels to accumulate before sensory rejection. Low-temperature storage of fish at temperatures under +5°C at all times is generally the most effective way to avoid histamine production. However, special concern is paid to histamine production by psychrotolerant bacteria as *Morganella psychrotolerans* and *Photobacterium phosphoreum* (Emborg and Dalgaard, 2008). For these bacteria, chilling of fish to 0–5°C alone is not sufficient to prevent formation of histamine.

19.5 Parasites

Virtually all of the commercially exploited fish species are infected with a variety of different parasites. However, while some parasites may have a significant impact on the fish's health or reduce the aesthetical appearance of the product, only a few but widely distributed fish parasites are of direct consumer health concern. The currently increasing popularity of Asian-inspired seafood based on semi-processed or even raw fish meat underlines the importance of providing the fish processing industry with up-to-date knowledge on the detection and control of the parasites of fish most relevant to food hygiene.

The larvae of several species of roundworms (nematodes) commonly occur in most commercially exploited fish stocks in temperate seas around the globe. Besides the considerable quality reducing effect of these parasites, they are of direct human health concern, especially regarding the consumption of undercooked, lightly brined, marinated or raw fish meat. Additionally, the potential of these worms to induce hypersensitivity reactions in humans has recently received increased attention. Thus, this group of roundworms will be used as a model to illustrate the most commonly applied monitoring strategies regarding human pathogenic parasites in fish production chains.

In parasitic disease associated with the consumption of seafood, the by far most frequently involved roundworms belong to the genus *Anisakis*. The best known and most intensively studied species is *A. simplex*, commonly known as the herring or whale worm. As inferred from its vernacular name, the worms' life cycle involves various whale species as definitive host, mainly krill as intermediate host, and numerous fish species including herring as transport host, transferring the larvae from krill to the whales. In a fish, the majority of *A. simplex* larvae are typically encapsulated as flat tight spirals, measuring 4–5 mm in cross section, in and on the visceral organs. However, a smaller number of larvae may migrate from the abdominal cavity into the flesh (Fig. 19.1). This behaviour eventually results in the presence of worms in the fish fillets, which again may draw the attention of consumers and food safety authorities. Most of the flesh-invading larvae seem to reside in the belly flaps – some, however, may penetrate deeply into the dorsal musculature of their fish host. When liberated from the capsule the worm, 20–30 mm long, moves vigorously.

The third-stage larvae of another anisakid nematode species of food hygienic concern, *Pseudoterranova decipiens*, also known as the cod or seal worm due to

Fig. 19.1 *Anisakis simplex* third-stage larva partially bored into the flesh of Atlantic herring (*Clupea harrengus*).

its respective host preferences, frequently occur in the liver and flesh of various fish species from coastal areas where seals are common. Compared to *Anisakis* however, the larvae of *Pseudoterranova* are not coiled and considerably larger, reaching 50 mm in length, and hence may be easier detected and removed during filleting and processing.

Anisakidosis, i.e. human infection with live *Anisakis* larvae, is reported most frequently from East Asia and some southern European countries where raw or lightly salted or marinated fish is part of the everyday diet. Resulting from an immediate immune reaction, the clinical manifestations of acute anisakidosis include epigastic pain, nausea, vomiting and diarrhoea. The condition may resemble several other gastrointestinal disorders, such as gastric ulcer, tumours or acute appendicitis (Bouree *et al.*, 1995). Moreover, the results of various studies indicate that *A. simplex* larvae, both dead and alive, may cause allergic reactions after consumption of infected seafood (Audicana *et al.*, 2002), or even at indirect contact with the parasite during occupational activities such as filleting or cooking (Nieuwenhuizen *et al.*, 2006). The allergic reactions were shown to be triggered by an IgE mediated response. So far at least five allergens have been identified, but there is reason to believe that several other exist (Kobayashi *et al.*, 2007).

19.6 Fungi and mycotoxins

A diverse range of moulds is able to produce toxic secondary metabolites of unknown function in the fungal cell. Three genera are especially important in feeds and foods: *Aspergillus*, *Penicillium* and *Fusarium* (Kuiper-Goodman, 2005; Moss, 2002). In terms of acute toxicity the mycotoxins most commonly encountered in feed and food are much less potent than the botulinum toxins and many of the algal toxins. However, long-term chronic exposure is of special concern, since several of the mycotoxins are carcinogenic, affect the foetus and may give multiorgan toxicity or influences the immune system. Most moulds can grow at a_w down to 0.6 and are thus able to grow at high salt concentrations or in dry products.

Traditional smoked fish, e.g. salmon and mackerel, may be vulnerable to mould growth if stored under suboptimal conditions over time. The surface of dried fish may be colonized by certain moulds, in particular *Wallemia sebi*. This mould produces the mycotoxin walleminol, but possible health effects of this toxin on humans are largely unknown.

19.7 Tracking the sources, reservoirs, survival and potential amplification steps of human pathogens in fish production chains

19.7.1 Bacteria

The sources and reservoirs of relevant pathogens in fish production chains are of diverse origin. On the one hand, we find the aeromonads and vibrios, having the aquatic environment as the natural habitat. On the other hand, *L. monocytogenes* may be found in the production environment and also on the raw materials. And finally *Salmonella*, that may be found in fish feed, samples from the production environment, among the personnel or in rodents and birds with access to production facilities.

For *L. monocytogenes*, several authors have demonstrated that *L. monocytogenes* are able to colonise fish production facilities by site-specific clones, constituting a reservoir for contamination of products during processing (Klæboe *et al.*, 2006; Rørvik *et al.*, 1995; Rørvik *et al.*, 1997). Moreover, there is a general tendency to observe increasing concentrations of the bacterium during increased level of processing of fish product in question.

Even though *Salmonella* sp. represents typical intestinal bacteria, they may survive and even multiply in food or feed if the conditions are favourable. *Salmonella* may grow both aerobically and aerobically in the temperature span between 5°C and 46°C, and at a pH ranging from 3.8 to 9.5. As for other bacteria in the Enterobacteriaceae family, the optimal growth temperature is 37°C. The heat resistance of salmonellae is strongly influenced by the surroundings in which the bacterium is found, and increases by decreasing water content or increasing fat content in the product in question. In frozen or dried products and products rich in fats such as fish feed, *Salmonella* sp. may survive for months, as has been shown for oil for fish feed production (Lunestad and Borlaug, 2009). Feed materials and compound feed ready for use may sometimes be contaminated by certain types of *Salmonella.* In feed materials intended for aquaculture species, certain types of *Salmonella* dominate. Common *Salmonella enterica* serotypes in vegetable or animal components for production of fish feed or in ready-to-use feed are Agona, Senftenberg, Montevideo, Livingstone, Bloemfontain, Johannesburg, Lexington, Anatum, Cerro, Worthington, Lille and Oranienburg (Lunestad *et al.*, 2007). These authors summarised the prevalence of *Salmonella* in samples of compound feed, feed materials and environmental samples. The authors reported the prevalence of *Salmonella* in ready-to-use compound fish feed to be 0.3%, and the most common serovars were Senftenberg, Agona, Montevideo and Kentucky. The prevalence in feed materials varied from 0.14 to 0.33%, and the prevalence in environmental samples were found to be 3.78%. The predominant serovars found in fish meal were Senftenberg and Montevideo. The same serovars were dominating in isolates from the production environment, and could in these cases be considered 'house strains'. Under natural rearing conditions for farmed Atlantic salmon, the risk of transmission to humans via fish products is considered minimal (Nesse *et al.*, 2005). This assumption is supported by the finding that the most common serovars in fish feed ingredients, fish feed and fish feed factories counts for a minor proportion of clinical Salmonella isolates from humans.

Although *Salmonella* sp. has not been shown to establish or multiply in the intestine of cold water fish, the safety for handlers of fish feed and the possible dissemination of new *Salmonella* serotypes among wild homoeothermic animals such as birds and rodents have been of some concern.

Vibrios are generally not robust bacteria and are sensitive to heating, freezing, drying and several other common preservation techniques, as well as the low pH in the stomach of humans. Some of the vibrios may demonstrate very rapid growth under optimal conditions. For instance, provided that the cooling of previously heat-treated and recontaminated seafood products is inadequate, the numbers of *V. parahaemolyticus* may increase drastically during storage (Huss *et al.*, 2004). An obvious problem in this respect seems to be the cooling of heat-treated seafood products with non treated seawater during production.

The coliform bacteria, as *E. coli* are common in the gut and intestine of warm-blooded animals and humans, and are therefore often used as an indicator for faecal contamination. Fish in their natural habitat do not harbour coliform bacteria (Austin, 1988; Hovda *et al.*, 2007), and when such bacteria are detected it is considered to indicate contamination of faecal origin. The coliforms include several geni and some stains are psychrotrophic capable to survive and even grow at chilled temperatures (4°C).

In the case of foodborne intoxications, preformed toxins in the products may give symptoms when ingested. Representatives from several bacterial geni, such as *S. aureus*, *B. cereus* and *C. botulinum* are of relevance as toxin producers. Furthermore, some microbes may produce food degradation products with a toxic potential, such as histamine. The microbial production of such toxins is due to improper handling and storage, with time and temperature as key elements. The temperature stability of toxins varies greatly. *S. aureus* toxins, histamine and some *B. cereus* toxins are heat stabile, and will not be inactivated by boiling or other normal food preparation techniques involving heat. On the other hand, toxins from *C. botulinum* and other types of *B. cereus* toxins are shown to be heat labile. However, in practical food preparation one cannot rely on heating as a means of inactivating toxins if found in foods.

19.7.2 Parasites

Anisakis larvae show considerable tolerance to salt (NaCl) and acetic acid (vinegar), both major ingredients in brines and marinades. For example, the survival of *Anisakis* larvae in freshly marinated herring fillets increased from 35 days to more than 119 days when the salt content in the water phase of the fish tissue was reduced from 9% (w/w) to 4.3% (w/w) while keeping the concentration of acetic acid constant at 2.6% (Karl *et al.*, 1995). Moreover, some major allergens originating from *Anisakis simplex* 3rd stage larvae seem to be resistant to heat, low pH (gastric acid) and short-time freezing (Valls *et al.*, 2005), which could account for gastroallergic symptoms that may appear after eating poached or canned fish. However, prolonged deep-freezing of the actual products seems currently the best alternative to lower the IgE-mediated allergic response in sensitized persons/consumers.

19.7.3 Fungi and mycotoxins

When feed contaminated by mycotoxins is given to farmed fish, there is a possibility for transmission of these components. However, in annual surveillance programs for Norwegian fish feed and farmed fish, fungi are regularly detected in these products. Mycotoxins above established acceptable limits, when available, has so far not been seen.

Mycotoxins are generally considered stable during storage. The heat stability for mycotoxins are unknown, but a reduction of the concentration of certain toxins is observed during heat treatments as the extruding of fish feed (Castelles *et al.*, 2005).

Fish feed has traditionally been based on the marine ingredients fish oil and fish meal. To reduce feed cost and improve the fish-farming sustainability, a recent development is the partial replacement of marine ingredients with vegetable ingredients. Fungi are not abundant in the marine environment, and it is a tendency that vegetable components have higher counts of fungi, and higher concentrations of mycotoxins (Moss, 2002). The replacement of marine animal feed ingredients with vegetable counterparts may lead to increased mycotoxin load on farmed fish.

19.8 Pathogen monitoring and control strategies

19.8.1 Bacteria

Food safety management relies on the interplay between several elements, and both knowledge on microbiology and tools to perform risk assessments, as the HACCP, have to be enforced.

At the centre lies the provision of safe food defined by food safety objectives (FSO), which is a statement of the frequency or maximum concentration of a microbiological hazard in a food when considering it acceptable for consumer protection. The FSO is set by application of a number of systems, which have been adopted by the food industry and are used in an integrated fashion. These include good manufacturing practice (GMP), GHP and the HACCP.

According to the recommendations of the FAO/WHO, criteria should be established according to the principles outlined by the Codex Alimentarius. The sampling plans for the relevant microorganisms, pathogens and indicator, as well as of the analytical methods have to be an integrated part of these criteria. Details on proposed sampling plans for finished products have been discussed by the International Commission on Microbial Specifications for Foods (ICMSF) (www.icmsf.iit.edu). Sampling and testing of finished fish products cannot provide more than a certain level of confidence, and is not an absolute guarantee for the absence of the pathogen. This is the case even for the most stringent sampling plans at the very high level of 60 samples of 25 g. In the case of food manufacturers, microbiological testing of end products is mainly performed to verify and confirm the efficiency of the implemented preventive measures outlined above. A complete sampling program has therefore to include as well analysis of raw materials, samples from food contact surfaces providing information on possible pathogen build up, and environmental samples from the processing environment.

The limitations of testing of product samples alone are related to the distribution of microorganisms in a sample. In general, distribution is presumed to be random but frequently cells occur in clumps or aggregates and do not follow a Poisson distribution.

The numbers of pathogenic microorganisms in most raw materials and food products are usually low and analysis on single samples may provide little information of use for the implementation and maintenance of GMP and HACCP systems. Instead, the enumeration of so-called 'indicator organisms' has an important role. Indicator organisms are groups of microorganisms that are indicative for the possible presence of pathogens. Although there is not necessarily a relationship between indicator and pathogen numbers, it can be generally assumed that the numbers of pathogen are less than the numbers of the organisms indicating its presence. It can also be assumed that reduction in the numbers of the indicator organisms will give a similar reduction in the numbers of any corresponding pathogen (Brown *et al.*, 2000). For the same reasons indicator organisms can also provide a measure of post-pasteurisation contamination by pathogens.

The importance of the sampling procedure itself as has been illustrated by a recent study on samples of Atlantic salmon (*Salmo salar L.*). In this study, three different sampling strategies from each individual fish were compared. From a total of 50 fish, samples were withdrawn as either muscle and skin (25 g and 25 cm^2), gill material (25 g) or swabbing by a sterile sponge of the skin on one side of the fish. In this study, none of the samples of muscle and skin was found positive, three of the 50 samples of gill material were positive and two of the 50 swab samples of the skin were positive for *L. monocytogenes*. The results of this limited trial indicate that the chances of detecting the bacterium on a fish are higher when using the gills as sample material, followed by swabbing, and finally sampling of muscle and skin by the conventional method. Similar findings have also been reported by Miettinen and Wirtanen (2005). This demonstrates that attention should be paid to the choice of sampling method, and that these differences should be kept in mind when comparing results obtained after dissimilar sampling procedures.

To efficiently control *L. monocytogenes*, every potential route of entry and cross-contamination should be taken into consideration (Gravani, 1999). In addition, GMPs need to be followed (Blanchfield, 2005). The state of the production facilities is important, as plants in a good state of repair had a lower risk of *L. monocytogenes* contamination than those with moderate or heavy signs of wear and tear (Rørvik *et al.*, 1997). In addition, the facilities should be designed or arranged to restrict or eliminate the transmission by personnel, equipment and conveyance between raw, processing, packaging and shipping areas, as the cross-contamination between raw and finished product areas is a major source of contamination (Lappi *et al.*, 2004a). Well-designed lines and facilities enable good working routines and effective departmentalisation of facilities help in controlling both employees and product traffic (Autio *et al.*, 2004). The employee movement between departments, due to rotation of the assigned duties, was found to be a risk factor for *L. monocytogenes* contamination of final products, especially if limited or no precautions were taken to avoid the spread of bacteria (Rørvik *et al.*, 1997). Cross-contamination by employees and the environment of the processing plants may represent important causes of contamination of the finished product (Thimothe *et al.*, 2004). Separate equipment, including tools employed by maintenance persons, should be available for use in raw and finished product areas (Gravani, 1999). In addition, all equipment within the factory should be designed to facilitate cleaning and minimise cross-contamination of products (Dauphin *et al.*, 2001). The slicing and brining processes of smoked salmon were the most critical steps of the fish production line mainly due to difficulties with cleaning the equipment thoroughly (Autio *et al.*, 1999; Johansson *et al.*, 1999). In addition, *L. monocytogenes* was extremely difficult to eliminate using routine disinfection procedures in a meat-bone separator which constitutes complex equipment (Lappi *et al.*, 2004a). The prevention of *L. monocytogenes* contamination in products is based on avoiding the colonisation of the processing environment and equipment with *L. monocytogenes*. The hygienic aspects should be stressed when selecting new processing equipment as the complex construction of equipment often hampers the cleaning and disinfection (Autio *et al.*, 2004).

Listeria spp., including *L. monocytogenes*, have been most frequently isolated from floor drains and floors, suggesting that these areas may function as one of several reservoirs in food processing facilities (Rørvik *et al.*, 1997; Norton *et al.*, 2001; Thimothe *et al.*, 2004). These should be thoroughly cleansed and disinfected daily without high-pressure hosing, since such practices readily promote the spread of *Listeria* to nearby equipment and other areas of the factory through splashing and the generation of aerosols (Gravani, 1999). Clean and especially dry floors are important for the control of *Listeria* in processing plants (Thimothe *et al.*, 2004).

Disinfection is the final step in eliminating *L. monocytogenes* and other foodborne pathogens as well as the myriad of spoilage organisms present in the production environment. Since the presence of organic debris readily decreases the effect of disinfecting agents against microbes (Best *et al.*, 1990), it is important to make sure that every item must first be thoroughly cleaned before it is disinfected. Disinfection with hot water is not advised, since sufficiently high water temperature cannot be easily maintained (Gravani, 1999). In some cases, however, the use of hot water (80°C), heating in oven (80°C) and treating with gas flame or a hot steam treatment has proven efficient in eradication of *L. monocytogenes* (Autio *et al.*, 1999; Lappi *et al.*, 2004b). In addition to inadequate separation between raw and finished products resulting from non-optimal factory design, indifferent attitudes of employees towards proper cleaning and disinfecting has been most frequently cited as factors that promote post-processing contamination. Effective cleaning and disinfection programs for standard operating procedures for every factory job along with master schedules listing the frequency of cleaning and disinfection procedures are needed (Gravani, 1999). The effectiveness of the cleaning and disinfection programs should be verified through daily microbiological analysis of both product and environmental samples gathered from all areas of facility. During environmental sampling, the efficacy of cleaning and disinfection procedures can be easily determined through the use of ATP bioluminescence monitoring systems. Routine testing of environmental samples for *Listeria* spp. remains, however, a critical component of any disinfection verification program (Gravani, 1999). Development of focused clean-up and disinfection procedures as well as implementation of the HACCP programme are of great importance in the prevention of *L. monocytogenes* colonisation (Autio *et al.*, 2004).

Consistent monitoring of *L. monocytogenes* contamination over time should be a component of every *L. monocytogenes* control strategy (Hoffman *et al.*, 2003). The prevalence data and subtyping of *L. monocytogenes* provide a base for implementing effective cleaning and disinfecting procedures focusing on *L. monocytogenes* niches and transmission pathways (Thimothe *et al.*, 2004). A plant-specific *Listeria* control program should also include strategies to minimise both the raw material and the environmental contaminations, procedures to prevent cross-contamination and employee training (Lappi *et al.*, 2004a). Even when handled under the best possible conditions, raw seafood or processing environment will probably never be completely free from *L. monocytogenes* (Gravani, 1999; Fonnesbech Vogel *et al.*, 2001; Autio *et al.*, 2004).

The procedures for sampling may affect the outcome when testing for pathogenic microorganisms, as have been seen for analysis with respect to *L. monocytogenes*. Sampling procedures may also be important when new preservation methods are introduced. For example, microwave (MW) heating is an uneven heating process, creating hot and cold spots, and challenge testing with selected pathogens may be necessary to validate that a suitable pasteurisation value is achieved.

19.8.2 Parasites

Due to their comparatively small size and transparency, most *Anisakis* larvae in the fish flesh may remain undetected during industrial processing, and are hence still present when the final products reach the market. Some fishery products that have been implicated in human anisakidosis are sashimi, ceviche, raw fish roe, green herring, undercooked or grilled fish and cold smoked fish. Attempts to kill or inactivate the larvae in fresh fillets by irradiation (Loaharanu and Murrell, 1994) or high hydrostatic pressure (Molina Garcia and Sanz, 2002) failed due to an irreversible negative effect on the texture or taste of the final product.

A set of international rules and regulations aims to minimise the consumer health risk inflicted by the possible presence of parasites in fish and fishery products. For example, according to the EC regulations 853/2004 (Anon., 2004), any fishery product which is to be consumed raw or almost raw must be frozen at a core temperature of at least 20°C for not less than 24 hours. Moreover, processing plant operators must ensure that fishery products have been subjected to a visual inspection in order to detect any visible parasites before being placed on the market. In this respect, candling, i.e. a brief visual inspection on a light table, is the recommended method for processing plants. However, Levsen *et al.* (2005) showed that only 7–10% of the *Anisakis* larvae that are actually present in the flesh of herring, Atlantic mackerel and blue whiting are detected by candling under industrial conditions. In white-fleshed fish species such as rockfish (*Sebastes* sp.) and Pacific cod, the efficiency of candling appears to be higher since Adams *et al.* (1997) detected between 43% and 76% of the *Anisakis* larvae present. Several other nematode detection methods including multispectral imaging and imaging spectroscopy (Wold *et al.*, 2001) have been investigated with respect to their applicability under industrial conditions, but proved unsuitable.

19.8.3 Fungi and mycotoxins

The prevention of mycotoxin production must be based on good agricultural practice and GMP aiming to control the growth of moulds at any stage in the line for fish feed production and storage.

19.9 New preservation strategies

Traditional methods for acceptable preservation of foods include heating, chilling, freezing, drying, curing, salting, preservation with sugar, acidification, fermentation, modified atmosphere packaging and smoking (Leistner, 1992). Recent consumer demands have focused on fresh tasting; high-quality foods and the numbers of ready meals have increased greatly in the industrialised countries. These meals are also preferred to be free of preservatives, have a high degree of convenience and require minimal preparation time. A great challenge is therefore to ensure the safety of such products. During the last 10 years, many new preservation technologies have been investigated. Combinations of several factors that secure microbial safety, stability and the sensory quality have been regarded as a basis of food preservation. Some of these combinations are still in the exploratory and developmental stages, whereas other methods have obtained regulatory approval and been introduced in the marketplace. Combined preservation has been well described and the principle is often referred to as 'hurdle technology' (Leistner, 1992; Leistner, 1995; Leistner, 2002). The intelligent selection of hurdles in terms of the number required, the intensity of each and the sequence of applications to achieve a specified outcome, are supposed to have significant potential for the future (McMeekin *et al.*, 2002).

Heat processing of food may induce desirable and undesirable changes. Useful changes are texture softening, starch swelling, protein coagulation and formation of taste and aroma components. Loss of vitamins and minerals, formation of thermal reaction components of biopolymers and, loss of fresh appearance, flavour and texture are among the undesirable changes (Ohlsson, 2002). During the last years, new thermal methods are tested in order to overcome or at least minimise undesirable quality changes. Examples of such methods are high-temperature short-time, ohmic heating, infrared heating, high-frequency or radio-frequency heating and micro wave (MW) heating. MW and radiofrequency heating refers to the use of electromagnetic waves of certain frequencies to generate heat.

MW heating of food in pouches or ready meals in trays has been widely investigated and is also in large-scale industrial use. The practical problems in achieving uniform and reproducible heating by MW heating have been addressed by carefully designed MW cavities, by application of shielding or field directing films onto certain areas of the packages, by complicated power-control schemes to direct controlled amounts of MW power to defined areas on food trays or by the use of multi-mode cavity tunnels, where the scattered MW fields are expected to give a fairly uniform coupling of the MW energy into the food packages. Still, the limited penetration of MW fields into large volume packages of high-moisture food, edge and corner overheating effects and the power intensity patterns of the standing (interference) waves in many MW applicators pose difficulties that have limited the industrial acceptance of the technique.

Combinations of modified atmosphere packaging and post-packaging thermal treatments are rarely seen. This is due to the fact that headspace gas in thermally processed products has an insulating effect in classic heat transfer. With non-contact dielectric heating this is not an issue, but with dielectric heating, e.g. MW or radiofrequency heating, a rapid non-contact heating is possible. Still one has to deal with steam that is generated inside the package with the risk that it may explode. Counter pressure or venting is therefore necessary to prevent over-pressure. A number of new valves are now available that release steam during thermal processing, and during chilling the condensation of the vapour will generate a partial vacuum in the processed package. Although not yet commercially available, technologies may also be developed to reintroduce preserving gas mixtures after processing, if found commercially viable.

Light pulses have been used successfully as a new technique for the inactivation of bacteria and fungi on the surface of food products when the major composition of the emitted spectrum is UV (Marquenie *et al.*, 2002). High UV light doses can cause damage to the treated tissue, e.g. of fruits and berries. The possibility of decreasing the treatment intensity by combining two or more treatments to preserve the food quality without decreasing the inactivation properties therefore appears promising. Microbial reductions of up to 2 log-units have been reported by combination of UV and modified atmosphere. High-pressure processing (HPP) at low temperatures combined with modified atmosphere packaging has been used for the preservation of salmon (Amanatidou *et al.*, 2000). When salmon had been subjected to HPP treatment in the presence of 50% O_2, the threshold value for microbial spoilage of salmon (7.0–7.2 log CFU per gram) was not reached for at least 18 days at 5°C. Spoilage microorganisms (lactic acid bacteria, *Shewanella putrefaciens*) as well as pathogens (*Listeria monocytogenes*, *Salmonella* Typhimurium) spiked on salmon prior to the treatment, were susceptible to HPP in the presence of 50% O_2 and 50% CO_2.

Interest in developing biopreservation techniques in lightly preserved food has increased in recent decades, and has focused on minimizing the presence of and/or suppressing the growth of *L. monocytogenes*. In cold smoked salmon, biopreservation techniques have involved the introduction of a competitive lactic acid bacteria (LAB) microflora for their protective effects, including bacteriocin-producing LAB, and purified antilisterial bacteriocins (Hugas and Monfort, 1997). It has been demonstrated that LAB comprises the dominant microflora in vacuum packed cold smoked salmon (Gonzales-Rodrigues *et al.*, 2002). Most of them are able to hamper the growth of *L. monocytogenes* due to the production of diverse compounds including antimicrobials produced at high cell density, or peptides called bacteriocins (Nilsson *et al.*, 1999; Nilsson *et al.*, 2004; Brillet *et al.*, 2005).

19.10 Hazard analysis and critical control point

Over the past decade, the HACCP system has become internationally recognized with respect to the control of food hazards. The evolution of HACCP from concept to an international standard has been relatively rapid. In some countries the fishery industry adopted the system early compared to other food industries, mainly because those countries had extensive export and international trade. It is important to recognise that HACCP has been in a constant state of development. A basic principle has been that HACCP must be an industry-driven program, with the role of the regulatory agency being that of approval of the processing plant's basic plan design, onsite verification, and inspector training. The HACCP can be separated in two parts; the first part focuses on defining the nature of the product and developing a flow diagram that details each operational step in the process. Understanding the nature of the product is essential to determine the potential food safety hazards, and it includes knowledge of the consumers' use, distribution and marketing in the retail chain. The second part consists of applying the seven HACCP principles: (1) conduct a hazard analysis, (2) determine the critical control points, (3) establish critical limits, (4) establish monitoring procedures, (5) establish a plan for corrective actions, (6) establish verification procedures and finally (7) establish record-keeping and documentation procedures. An extended review of the HACCP principles and applications is presented in Chapter 22 of this book.

Foodborne diseases are of major concern to consumers, producers and authorities. Despite increased awareness, there still seem to be a high number of outbreaks, and these numbers are recognised as being significantly underestimated since reporting is limited.

19.11 Microbial modelling

19.11.1 Predictive models in seafood

Application of mathematical modelling for shelf-life prediction requires sufficient knowledge of the product spoilage mechanisms (Koutsoumanis and Nychas, 2001). In the case of fish and fish products, spoilage is caused by a fraction of the total fish microflora, known as the specific spoilage organisms (SSO; Gram and Huss, 1996). Since temperature is one of the most important factors influencing microbial growth, modelling the growth of the SSO as a function of temperature is essential in shelf life prediction. Although a large number of models for the prediction of growth of spoilage organisms at various temperatures have been developed, many of these studies have been carried out under constant conditions (McClure *et al.*, 1993; Dalgaard 1995). However, unlike other factors affecting microbial growth (e.g. pH and water activity), temperature may vary extensively throughout the complete production and distribution chain. In practice, foods are frequently exposed to significant temperature fluctuations during transport and storage before delivery to the consumer.

The objective of predictive food microbiology is to mathematically describe the growth or reduction in numbers of foodborne microbes or spoilage bacteria under specific environmental conditions and thereby make it possible to predict the safety and shelf life of products. It is assumed that growth or decline of a particular microorganism is governed by the environment it experiences. This environment includes intrinsic factors as pH, and water activity (a_W) and extrinsic factors such as temperature and atmosphere. A large number of factors affect the microorganism, however in most foods only a few exert most of the growth-limiting activity. Several assumptions are made during modelling, e.g. that the effect of one factor is assumed to be independent of whether the microbe is found in a broth or present in food.

19.11.2 Safety models

The earliest predictive models for food safety were the ones developed for canned products during the 1920s. These models described thermal processes sufficient to destroy 10^{12} spores of *Clostridium botulinum* type A, chosen as a model organism. Such treatment have later been referred to as a 'botulinum cook' (Whiting and Buchanan, 2001; McMeekin *et al.*, 2002). The model described production processes with a very large safety margin and, whilst that probably accounts for its continued use, it perhaps also inhibited its widespread recognition as a predictive model.

The origin of 'modern' predictive microbiology can be traced to the 1960s and 1970s when kinetic models were used to address food spoilage problems (Spencer and Baines, 1964; Olley and Ratkowsky, 1973a, b), followed by the use of probability models to address food poisoning problems, particularly botulism and other intoxications (Genigeorgis, 1981; Roberts *et al.*, 1981).

More precise models have been developed during the later years, and they are often classified based on the population behaviour that they describe, e.g. growth models, limit of growth models or inactivation models. Predictive microbiology provides invaluable information for the production of safe food with adequate shelf life, and it can predict effects of several environmental factors combined.

For food safety evaluation, Pathogen Modelling Programme (Buchanan, 1993) and Food Micromodel (Anonymous, 1997) have illustrated the potential of predictive microbiology to wide groups of users. Application software with kinetic models for growth of food spoilage microorganisms have also been developed (Zwietering *et al.*, 1992; Schellekens *et al.*, 1994; Avery *et al.*, 1996; Anonymous, 1998; Wijtzes *et al.*, 1998). Growth models for *Brochothrix thermosphacta*, *Lactobacillus plantarum*, *Saccharomyces cerevisiae* and *Zygosaccharomyces bailii* are included in Food Micromodel (Anonymous, 1997) and the Food Spoilage Predictor (Anonymous, 1998), which is commercially available application software, contains a model for growth of psychrotolerant pseudomonads (Neumeyer *et al.*, 1997). The remaining programmes with kinetic spoilage models do not seem available, and software with models for growth of important seafood spoilage bacteria like *Shewanella putrefaciens* and *Photobacterium phosphoreum* has not been developed yet.

The objectives of several papers have been to describe the development and distribution of the Seafood Spoilage Predictor (SSP) software as well as approaches used in modelling of seafood spoilage (Dalgaard *et al.*, 2002). The SSP software contains both kinetic models for growth of SSO and empirical relative rates of spoilage models. The software has been distributed via the internet and experiences with this type of software distribution are reported.

19.12 Future trends

19.12.1 Trends among consumers and the implications for seafood safety

As consumers, we have high demands to the food products we buy. The product should be of high quality, cheap, easy to prepare, be of high nutritional value and, of cause, safe. The enlightened consumer is also concerned about sustainable production, fair trade and animal welfare. These demands are often in conflict with each other.

A general trend for seafood, as well as other products, is to favour products that are mildly preserved and thereby appearing as close to the fresh original as possible. On the end of the scale regarding freshness is sashimi, often referred to as sushi. These products are based on raw seafood products, and give special challenges with respect to parasites and microorganisms.

Consumers expect parasite-free fish and fishery products. However, large-scale harvested and processed wild fish appears to be the only industrially produced food which is at risk to regularly carry parasites when put on the market. This is largely due to the fact that both economic and technological constrains impede the detection and subsequent removal of the parasites that might be present in the fish flesh. However, several comprehensive studies have shown that farmed fish, especially net-pen reared Atlantic and Pacific salmon, are free from nematode larvae of known human infective potential (Angot and Brasseur, 1993; Deardorff and Kent, 1989; Lunestad, 2003). This apparent absence is due to the widespread use of industrially produced dry-feed which does not contain viable nematode larvae. Although the main components of such fish-feed are of marine origin, any larvae possibly present in the raw material are killed during the processing of the feed. Thus, whenever unprocessed, raw marine fish products are to be consumed, farmed fish should be preferred. However, the use of fresh unprocessed pelagic fish (e.g. herring or capelin) for feed in capture-based marine aquaculture may result in transmission of nematode larvae from the feed to the actual cultured fish.

19.12.2 Emerging pathogens relevant to seafood

The predicted shift in the consumption patterns towards more diary, egg, meat and fish products will inevitably make the foodborne pathogens associated with these products of higher importance.

19.12.3 Bacterial and parasital resistance to therapeutics

In all situations where antibacterial agents are applied, as in hospitals, in the community during treatment outside a hospital, in animal farms or during treatment of pet animals, bacterial resistance is proven to be a problem (Teale, 2002). Many authors are concerned about the negative effects of the use of antibiotics in agriculture and aquaculture (Bogaard and Stobberingh, 2000; Willis, 2000; Russel, 2002; Aarestup, 2006). Infections mediated by bacteria may necessitate the use of antibiotics in aquaculture, and such agents are today applied on a world-wide basis in the farming of fish and other aquatic organisms. Human diseases, caused by multi-resistant microorganisms have so far not been reported to be linked directly to consumption of seafood. However, an epidemic outbreak of cholera in Ecuador with multi-drug resistant *Vibrio cholerae* indicates an influence from aquaculture (Weber *et al*., 1994). When assessing microbial food safety, it seems to be a trend to not be concerned only with pathogenic microorganisms, but also with genes that mediate the transfer of resistance among microbes. Since the methods involved are laborious and necessitate specific equipment, these analyses are not performed on a routine basis.

19.12.4 Increased use of vegetable components in fish feed and burden of mycotoxins

Fish feed has traditionally been based on the marine ingredients fish oil and fish meal. To reduce feed cost and improve the fish farming sustainability, a recent development is the partial replacement of marine ingredients of animal origin (fish oil and meal) with vegetable ingredients. It is a general observation that vegetable feed ingredients are more susceptible to growth of fungi, due to intrinsic or storage and transport-related factors, compared to feed ingredients from an animal origin (Moss, 2002). When the proportion of vegetable ingredients increase in fish feed, one could expect problems with mycotoxins in fish to increase.

19.13 Sources of further information and advice

The following books and websites are recommended for general information on foodborne pathogens, predictive food and seafood microbiology or seafood related human pathogens.

ADAMS, MR and MOSS, MO, 2008. *Food microbiology*, Guildford, UK, RSC Publishing, ISBN-978-0-85404-284-5.

BUHINA, AK 2008. *Foodborne microbial pathogens – mechanism and pathogenesis*, New York, Springer Science and Business media, ISBN: 978-0-387-74536-7.

HUSS, HH, L. ABABOUCH and L GRAM., 2004. Assessment and management of seafood safety and quality, FAO Fisheries Technical Paper No. 444, Rome. Also available free of charge at the web pages of FAO: http://www.fao.org/docrep/

LEVSEN, A, LUNESTAD, BT and BERLAND, B, 2008. Parasites in farmed fish and fishery products, in: Lie, Ø. (Ed.), *Improving farmed fish for the consumer*, Woodhead publishing, ISBN: 978-1-84569-299-5.

LUNESTAD, BT and ROSNES, JT, 2008. Microbiological quality and safety of farmed fish, in: Lie, Ø. (Ed.), *Improving farmed fish for the consumer*, Woodhead publishing, ISBN: 978-1-84569-299-5.

19.14 References

AARESTUP F M (2006), *Antimicrobial resistance in bacteria of animal origin*, Washington, DC: American Society for Microbiology Press, ISBN: 9781555813062.

ADAMS A M, MURRELL K D and CROSS J H (1997), Parasites of fish and risk to public health. *Rev. Sci. Tech. Off. Int. Epizoot.*, **16**, 652–60.

ADAMS M R and MOSS M O (2008), *Food microbiology*, Guildford, UK, RSC Publishing, ISBN-978-0-85404-284-5.

AMANATIDOU A, SCHLUTER O, LEMKAU K, GORRIS L G M, SMID E J and KNORR D (2000), Effect of combined application of high-pressure treatment and modified atmospheres on the shelf life of fresh Atlantic salmon. *Innovat. Food Sci. Emerg. Tech.*, **1**, 87–98.

ANGOT V and P BRASSEUR (1993), European farmed Atlantic salmon (*Salmo salar* L.) are safe from anisakid larvae. *Aquaculture*, **118**, 339–44.

ANONYMOUS (1998), *FSP, Food Spoilage Predictor Tool. Hastings Data Loggers*, Port Macquarie, New South Wales, Australia.

ANONYMOUS (2003), *CODEX – Risk assessment on* Clostridium botulinum *in salted Atlantic herring and sprat*. Available from: //ftp.fao.org/codex/ccffp26/fp0303be.pdf.

ANONYMOUS (2004), *EU Regulation (EC) No 853/2004 of the European Parliament and of the Council of 29 April 2004 laying down specific hygiene rules for food of animal origin*. Available from: http://www.fsai.ie/uploadedFiles/Reg853_2004(1).pdf.

ARKOUDELOS J S, SAMARAS F J and TASSOU C C (2003), Survival of *Staphylococcus aureus* and *Salmonella enteritidis* on salted sardines (*Sardina pilchardus*) during ripening. *J. Food. Prot.*, **66**, 1479–81.

ASCHFALK A and MÜLLER W (2002), *Clostridium perfringens* toxin types from wild-caught Atlantic cod (*Gadus morhua* L.) determined by PCR and Elisa. *Canadian J. Microbiol.*, **48**(4), 365–8.

AUDICANA M T, ANSOTEGUI I J, CORRES D E and KENNEDY M W (2002), *Anisakis simplex*: dangerous – dead and alive? *Trends in Parasitol.*, **18**, 20–25.

AUSTIN B (1988), *Marine microbiology*. Cambridge University Press, Cambridge, UK. ISBN:0 521 31130 6.

AUSTIN J W (2001), *Clostridium botulinum*. In: Doyle M.P., Beuchat, L.R. and Montville, T.J. (eds.). *Food Microbiology, fundamentals and frontiers*, ASM Press, Washington, pp. 329–50.

AUTIO T, HIELM S, MIETTINEN M, SJÖBERG A, AARNISALO K, BJÖRKROTH J, ET AL. (1999), Sources of *Listeria monocytogenes* contamination in a cold smoked rainbow trout processing plant detected by Pulsed-Field Gel Electrophoresis typing. *Appl. Environ. Microbiol.*, **65**(1), 150–5.

AUTIO T J, LINDSTRÖM M K and KORKEALA H J (2004), Research update on major pathogens associated with fish products and processing of fish. In: Smulders, J.M. and Collins, J.D. (eds.). *Food safety assurance and veterinary public health, vol. 2, Safety assurance during food processing*, Wageningen Academic Publishers, the Netherlands, pp. 115–34.

AVERY S M, HUDSON J A and PHILLIPS D M (1996), Use of response surface models to predict bacterial growth from time/temperature histories. *Food Control*, **7**, 121–8.

BAFFONE W, PIANETTI A, BRUSCOLINI F, BARBIERI E and CITTERIO B (2000), Occurrence and expression of virulence-related properties of *Vibrio* species isolated from widely consumed seafood products. *Int. J. Food Microb.*, **54**, 9–18.

BAUMANN P, FURNISS A L and LEE J V (1984), Facultatively anaerobic Gram-negative rods. Genus I Vibrio Pacini 1854, 411 AL. In: *Bergey's Manual of Systematic Bacteriology*, Vol. 1, N. Krieg and J.G. Holt, Williams and Williams, Baltimore, pp. 518–38.

BEN EMBAREK P K (1994), *Microbial safety and spoilage of sous vide fish products. 1994.* Thesis, Technological Laboratory of the Danish Ministry of Agriculture and Fisheries & the Royal Veterinary and Agricultural University, Denmark.

BEST M, KENNEDY M E and COATES F (1990), Efficacy of a variety of disinfectants against *Listeria* spp. *Appl. Environ. Microbiol.*, **56**, 377–80.

BLANCHFIELD J R (2005), Good manufacturing practice (GMP) in the food industry. In: Lelieveld, H.L.M., Mostert, M.A. and Holah, J. (eds.), *Handbook of hygiene control in the food industry*, Woodhead Publishing Limited, Cambridge, UK, pp. 324–47.

BOGAARD A E V D and STOBBERINGH E E (2000), Epidemiology of resistance to antibiotics. Links between animals and humans. *Int. J. Antimicrobial Agents*, **14**, 327–35.

BOUREE P, A PAUGAM and J C PETITHORY (1995), Anisakidosis: Report of 25 cases and review of the literature. *Comp. Immun. Microbiol. Infect. Dis.*, **18**(2), 75–84.

BRILLET A, LILET M, PRÉVOST H, CARDINAL M and LEROI F (2005), Effect of inocululation of *Carnobacterium divergens* V41, a biopreservative strain against *Listeria monocytogenes* risk, on the microbiological, chemical and sensory quality of cold-smoked salmon. *Int. J. Food Microbiol.*, **104**, 309–24.

BROWN M H, GILL C O, HOLLINGSWORTH J, NICKELSON R, SEWARD S, SHERIDAN J J, ET AL. (2000), The role of microbiological testing in systems for assuring the safety of beef. *Int. J. Food Microbiol.*, **62**, 7–16.

BUCHANAN R L (1993), Developing and distributing user-friendly application software. *J. Ind. Microbiol.*, **12**, 251–5.

CAI Y, GAO J, WANG X, CHAI T, ZHANG X, DUAN H, ET AL. (2008), *Clostridium perfringens* toxin types from freshwater fishes in one water reservoir of Shangdong Province of China, determined by PCR. *Deutsche tierärztliche Wochenschrift*, **115**(8), 292–7.

CANN D C, WILSON B B, SHEWAN, J M and HOBBS G (1966), Incidence of *Clostridium botulinum* type E in fish products in the United Kingdom. *Nature*, **211**, 205–6.

CASTELLS M, MARIN S, SANCHIS V and RAMOS J (2005), Fate of mycotoxins in cereals during extrusion cooking: A rewiev. *Food Additives and Contam.*, **22**(2), 150–7.

DALGAARD P (1995), Modelling of microbial activity and prediction of shelf life of packed fresh fish. *Int. J. Food Microbiol.*, **19**, 305–17.

DALGAARD P, BUCH P and SILBERG S (2002), Seafood Spoilage Predictor – development and distribution of a product specific application software. *Int. J. Food Microbiol.*, **73**, 343–9.

DALSGAARD A (2002), *Vibrio vulnificus*. *Culture*, **23**(2), 5–8.

DAUPHIN G, RAGIMBEAU C and MALLE P (2001), Use of PFGE typing for tracing contamination with *Listeria monocytogenes* in three cold-smoked salmon processing plants. *Int. J. Food Microbiol.*, **64**, 51–61.

DEARDORFF T L and M L KENT (1989), Prevalence of larval *Anisakis simplex* in pen-reared and wild-caught salmon (Salmonidae) from Pudget Sound, Washington. *J. Wildl. Dis.*, **25**, 416–19.

ECFF (EUROPEAN CHILLED FOOD FEDERATION) (1996, November) *In*, Martins T, Harmonization of safety criteria for minimally processed foods. Inventory report FAIR concerted action FAIR CT96.1020, September 1997.

EKLUND M W, PELROY G A, PARANJPYE R, PETERSON M E and TEENT F M (1982), Inhibition of *Clostridium botulinum* types A and E toxin production by liquid smoke and NaCl in hot-process smoke-flavoured fish. *J. Food. Protect.*, **45**, 935–41.

EMBORG, J and DALGAARD P (2008), Growth, inactivation and histamine formation of *Morganella psychrotolerans* and *Morganella morganii* – development and evaluation of predictive models. *Int. J. Food Sci.*, **128**, 234–43.

ERICSSON H, EKLOW W, DANIELSON-THAM M L, LONCAREVIC S, MENTZING L O, UNNERSTAD H, ET AL. (1997), An outbreak of listeriosis suspected to have been caused by rainbow trout. *J. Clin. Microbiol.*, **35**, 2904–7.

FACH P, PERELLE S, DILLASSER F, GROUT J, DARGAIGNARATZ C, BOTELLA L, ET AL. (2002), Detection by PCR-enzyme-linked-immunosorbet assay of *Clostridium botulinum* in fish and environmental samples from a coastal area in Northern France. *Appl. Environ. Microbiol.*, **68**(12), 5870–6.

FACINELLI B, VARALDO P E, TONI M, CASOLARI, C and FABIO U (1989), Ignorance about Listeria. *Br. Med. J.*, **299**, 738.

FAO 2007. *The state of world fisheries and aquaculture, 2006*. Available from: http://www.fao.org

FELDHUSEN F (2000), The role of seafood in bacterial foodborne diseases. *Microbes Infect*, **2**, 1651–60.

FENICIA L, ANNIBALLI F, POUSHABAN M, FRANCIOSA, G and AURELI P (2002), Presence of *Clostridium botulinum* spores in sea-salt in Italy. Poster presented at the *18th International ICFHM Symposium, Food Microbiology*, Lillehammer, Norway, 18–23 August 2002.

FONNESBECH VOGEL B, HUSS H H, OJENIYI B, AHRENS, P and GRAM L (2001), Elucidation of *Listeria monocytogenes* contamination routes in cold-smoked salmon processing plants detected by DNA-based typing methods. *Appl. Environ. Microbiol.*, **67**, 2586–95.

GENIGEORGIS C A (1981), Factors affecting the probability of growth of pathogenic microorganisms in foods. *J. American Veterinary Medicine Association*, **179**, 1410–17.

GONZÁLEZ-RODRIGUES M N, SANZ J J, SANTOS J A, OTERO, A and GARCIA-LÓPEZ M L (2002), Numbers and types of microorganisms in vacuum-packaged cold-smoked freshwater fish at the retail level. *Int. J. Food Microbiol.*, **77**, 161–8.

GRAHAM A F, MASON D R, MAXWELL, F J and PECK M W (1997), Effect of pH and NaCl on growth from spores of nonproteolytic *Clostridium botulinum* at chill temperature. *Lett. Appl. Microbiol.*, **24**, 95–100.

GRAM L (2001), Potential hazards in cold-smoked fish: *Clostridium botulinum* type E. *J. Food Sci.*, Suppl. **66**, 1082–7.

GRAM L and HUSS H H (1996), Microbiological spoilage of fish and fish products. *Int. J. Food Microbiol.*, **33**, 121–37.

GRAVANI R (1999), Incidence and control of *Listeria monocytogenes* in food processing facilities. In: Ryser, E.T. and Marth, E.H. (eds.). *Listeria, Listeriosis and Food Safety*, 2 ed. Marcel Dekker, New York, USA, pp. 657–709.

HIELM S, HYYTIÄ E, ANDERSIN A B and KORKEALA H (1998), High prevalence of *Clostridium botulinum* type E in Finnish freshwater and Baltic Sea sediment samples. *J. Appl. Microbiol.*, **84**, 133–7.

HOFFMAN A, GALL K L, NORTON D M and WIEDMANN M (2003), *Listeria monocytogenes* contamination patterns for the smoked fish processing environment and for raw fish. *J. Food Prot.*, **66**, 52–60.

HØI L, LARSEN J L, DALSGAARD I and DALSGAARD A (1998), Occurrence of *Vibrio vulnificus* biotypes in Danish marine environments. *Appl. Environ. Microbiol.*, **64**(1), 7–13.

HOVDA M B, LUNESTAD B T, FONTANILLAS R and ROSNES J T (2007), Molecular characterisation of the intestinal microbiota of farmed Atlantic salmon (*Salmo salar* L.). *Aquaculture*, **272**, 581–8.

HUGAS M and MONFORT J M (1997), Bacterial starter cultures for meat fermentation. *Food Chemistry*, **59**(4), 547–54.

HUSS H H (1980), Distribution of *Clostridium botulinum*. *Appl. Environ. Microbiol.*, **39**, 764–9.

HUSS H H and RYE PEDERSEN E (1980), *Clostridium botulinum* in fish. *Scan. J. Vet. Med.*, **31**, 214–21.

HUSS H H (1994), *Assurance of seafood safety*. FAO Fisheries technical papers No. 334, pp. 8–26, FAO, Rome.

HUSS H H, ABABOUCH L and GRAM L (2004), *Assessment and management of seafood safety and quality*, FAO Fisheries Technical Paper No. 444, Rome.

HYYTIÄ-TREES E (1999), *Prevalence, molecular epidemiology and growth of Clostridium botulinum type E in fish and fishery products*. Dissertation, Department of Food and Environmental Hygiene, Faculty of Veterinary Medicine, University of Helsinki, Finland, ISBN 952-45-8684-0.

HYYTIÄ E, HIELM S and KORKEALA H (1998), Prevalence of *Clostridium botulinum* type E in Finnish fish and fisheries products. *Epidemiology Infect.*, **120**, 245–50.

JABLONSKI L M and BOHACH G A (1997), *Staphylococcus aureus*. In: Doyle M.E., Beuchat, L.R. and Montville, T.J (Eds.), ASM Press, Washington, DC, pp. 353–75.

JAY J M, LOESSNER M J, GOLDEN D A (2005), *Modern food microbiology*, Springer, New York.

JOHANSSON T, RANTALA L, PALMU L and HONKANEN-BUZALSKI T (1999), Occurrence and typing of *Listeria monocytogenes* strains in retail vacuum-packed fish products and in a production plant. *Int. J. Food Microbiol.*, **47**, 111–19.

KARL H, ROEPSTORFF A, HUSS H H and BLOEMSMA B (1995), Survival of *Anisakis* larvae in marinated herring fillets. *Int. J. Food Sci. Tech.*, **29**, 661–70.

KIMURA B, KURODA S, MURAKAMI M, FUJII T (1996), Growth of *Clostridium perfringens* in fish fillets packaged with a controlled carbon dioxide atmosphere at abuse temperatures. *J. Food Protect.*, **59**(7), 704–10.

KLÆBOE H, ROSEF O, FORTES E and WIEDMANN M (2006), Ribotype diversity of *Listeria monocytogenes* isolates from two salmon processing plants in Norway. *Int. J. Environ. Health Res.*, **16**(5), 375–83.

KOBAYASHI Y, ISHIZAKI S, SHIMAKURA K, NAGASHIMA Y and SHIOMI K (2007), Molecular cloning and expression of two new allergens from *Anisakis simplex*. *Parasitol. Res.*, **100**, 1233–41.

KOUTSOUMANIS K and NYCHAS G J E (2001), Application of a systematic experimental procedure to develop a microbial model for rapid fish shelf life prediction. *Int. J. Food Microbiol.*, **60**, 174–84.

KUIPER-GOODMAN T (2005), Risk assessment and risk management of mycotoxins in food. In: Magan, N. and Olsen, M. (eds.), *Mycotoxins in food*, Woodhead Publishing, CRC Press, Cambridge, pp. 4–31.

LAPPI V R, THIMOTHE J, KERR NIGHTINGALE K, GALL K, SCOTT V N and WIEDMANN M (2004a), Longitudinal studies on *Listeria* in smoked fish plants: impact of intervention strategies on contamination patterns. *J. Food Prot.*, **67**, 2500–14.

LAPPI V R, THIMOTHE J, WALKER J, BELL J, GALL K, MOODY M W, ET AL. (2004b), Impact of intervention strategies on *Listeria* contamination patterns in crawfish processing plants: a longitudinal study. *J. Food Prot.*, **67**, 1163–9.

LEISTNER L (1992), Food preservation by combined methods. *Food Res. Int.*, **25**, 151–8.

LEISTNER L (1995), Use of hurdle technology in food processing: recent advances. In: Barbosa-Cánovas, G.V. and Welti-Chanes, J. (eds.), *Food preservation by moisture control*, Technomic Publishing Company, Basel, Switzerland, pp. 377–96.

LEISTNER L (2002), Hurdle technology. In: Juneja, V.K. and Sofos, J.N (eds.), *Control of foodborne microorganisms*. Marcel Dekker, Basel, Switzerland, pp. 493–508.

LEVSEN A, LUNESTAD B T and BERLAND B (2005), Low detection efficiency of candling as a commonly recommended inspection method for nematode larvae in the flesh of pelagic fish. *J. Food Prot.*, **68**, 828–32.

LINDSTEDT B A, VARDUND T and KAPPERUD G (2004a), Multiple-Locus Variable Number of Tandem Repeat Analysis of *Escherichia coli* O157 using PCR multiplexing and multicolour capillary electrophoresis. *J. Microbiol. Meth.*, **58**, 213–22.

LINDSTEDT B A, VARDUND T, AAS L and KAPPERUD G (2004b), Multiple-locus variable number tandem repeats analysis of *Salmonella enterica* subsp. enterica serovar

Typhimurium using PCR multiplexing and multicolour capillary electrophoresis. *J. Microbiol. Meth.*, **59**, 163–72.

LINDSTEDT B A (2005), Multiple-locus variable number tandem repeats analysis for genetic fingerprinting of pathogenic bacteria. *Electrophoresis*, **26**, 2567–82.

LINDSTEDT B A, BRANDAL L T, AAS E, VARDUND, T and KAPPERUD G (2007), Study of polymorphic variable number of tandem repeats loci in the ECO collection and in a set of pathogenic *Escherichia coli* and *Shigella* isolates for use in a genotyping assay. *J. Microbiol. Meth.*, **69**, 197–205.

LINDSTEDT B A, THAM W, DANIELSSON-THAM M L, VARDUND T, HELMERSSON, S and KAPPERUD G (2008), Multiple-locus variable number tandem repeats analysis of *Listeria monocytogenes* using multicolor capillary electrophoresis and comparison with pulsed-field gel electrophoresis typing. *J. Microbiol. Meth.*, **72**(2), 141–8.

LOAHARANU P and MURRELL D (1994), A role of irradiation in the control of foodborne parasites. *Tr. Food Sci. Technol.*, **5**, 190–5.

LUNESTAD B T (2003), Absence of nematodes in farmed Atlantic salmon (*Salmo salar* L.) in Norway. *J. Food Prot.*, **66**, 122–4.

LUNESTAD B T, NESSE L, LASSEN J, SVIHUS B, NESBAKKEN T, FOSSUM K, ET AL. (2007), *Salmonella* in fish feed: occurrence and implications for fish and human health in Norway. *Aquaculture*, **165**, 1–8.

LUNESTAD B T, THI TRUONG T T, BORLAUG K, GALLUZZI T, VARDUND T and LINDSTEDT B A (2009), *Listeria monocytogenes* from farmed fish and processing environments – prevalence and typing by multiple-locus variable number of tandem repeat analysis. Paper presented at the International Meeting on Emerging Diseases and Surveillance (IMED), 13–16 February, Vienna, Austria, http://imed.isid.org/

LUNESTAD B T and BORLAUG K (2009), Persistence of *Salmonella enterica* serovar Agona in oil for fish feed production. *J. Aquaculture Feed Sci. and Nutr.*, **1**(3), 73–7.

LYNCH M, PAINTER J, WOODRUFF, R and BRADEN C (2006), Surveillance for foodborne-disease outbreaks – United States, 1998–2002. *Morbidity and Mortality Weekly Report*, Vol. 55, CDC.

MCCLURE P J, BARANYI J, BOOGARD E, KELLY and ROBERTS T A (1993), A predictive model for the combined effect of pH, sodium chloride and storage temperature on the growth of *Brochothrix thermosphacta. Int. J. Food Microbiol.*, **19**, 161–78.

MCMEEKIN T A, OLLEY J, RATKOWSKY D A and ROSS T (2002), Predictive microbiology: towards the interface and beyond. *Int. J. Food Microbiol.*, **73**, 395–407.

MARQUENIE D, MICHIELS C W, GEERAERD A H, SCHENK A, SOONTJENS C, VAN IMPE J F, ET AL. (2002), Using survival analysis to investigate the effect of UV-C and heat treatment on storage rot of strawberry and sweet cherry. *Int. J. Food Microbiol.*, **73**, 187–96.

MIETTINEN M K, SIITONEN A, HEISKANEN P, HAAJANEN H, BJORKROTH, K J and KORKEALA H J (1999), Molecular epidemiology of an outbreak of febrile gastroenteritis caused by *Listeria monocytogenes* in cold-smoked rainbow trout. *J. Clin. Microbiol.*, **37**, 2358–60.

MIETTINEN H and WIRTANEN G (2005), Prevalence and location of *Listeria monocytogenes* in farmed rainbow trout. *Int. J. Food Microbiol.*, **104**, 135–43.

MOLINA-GARCIA A D and SANZ P D (2002), Anisakis simplex larva killed by high-hydrostatic-pressure processing. *J. Food Prot.*, **65**, 383–8.

MOSS M (2002), Toxogenic fungi. In: C.W. Blackburn and P.J. McClure (eds.), *Foodborne pathogens*, Woodhead Publishing, CRC Press, Cambridge, pp. 479–88.

NESSE L L, LOVOLD T, BERGSJO B, NORDBY K, WALLACE, C and HOLSTAD G (2005), Persistence of orally administered *Salmonella enterica* serovars Agona and Montevideo in Atlantic salmon (*Salmo salar* L.). *J. Food Prot.*, **68**, 1336–9.

NEUMEYER K, ROSS, T and MCMEEKIN T A (1997), Development of a predictive model to describe the effect of temperature and water activity on the growth of spoilage pseudomonads. *Int. J. Food Microbiol.*, **38**, 45–54.

NIEUWENHUIZEN N, LOPATA A L, JEEBHAY M F, HERBERT D R and ROBINS T G (2006), Exposure to fish parasite *Anisakis* causes allergic airway hyperactivity and dermatitis. *J. Allergy Clin. Immunol.*, **117**, 1098–105.

NILSSON L, GRAM L and HUSS H H (1999), Growth control of *Listeria monocytogenes* on cold-smoked salmon using competitive lactic acid bacteria flora. *J. Food Protect.*, **62**, 336–342.

NILSSON L, NG Y Y, CHRISTIANSEN J N, JORGENSEN B L, GROTINUM D and GRAM L (2004), The contribution of bacteriocin to the inhibition of *Listeria monocytogenes* by *Carnobacteriaum piscicola* strains in cold-smoked salmon system. *J. Appl. Microbiol.*, **96**, 133–43.

NISSEN H, ROSNES J T, BRENDEHAUG J and KLEIBERG G H (2002), Safety evaluation of sous vide-processed ready meals. *Lett. Appl. Microbiol.*, **35**, 433–8.

NORTON D M, MCCAMEY M A, GALL K L, SCARLETT J M, BOOR K J and WIEDMANN M (2001), Molecular studies on the ecology of *Listeria monocytogenes* in the smoked fish processing industry. *Appl. Environ. Microbiol.*, **67**, 198–205.

OHLSSON T (2002), Minimal processing of foods with thermal methods. In: Ohlsson, T. and Bengtsson, N. (eds.). *Minimal processing technologies in the food industry*, Cambridge, UK, Woodhead Publ. Ltd.

OLIVER J D and KAPER J B (2001), *Vibrio* species. In: Doyle M.P., Beuchat, L.R. and Montville, T.J. (eds.), *Food microbiology: fundamentals and frontiers*, ASM Press, Washington, DC, pp. 263–300.

OLLEY J and RATKOWSKY D A (1973a), Temperature function integration and its importance in the storage and distribution of flesh foods above the freezing point. *Food Technology in Australia*, **25**, 66–73.

OLLEY J and RATKOWSKY D A (1973b). The role of temperature integration in monitoring fish spoilage. *Food Technology in New Zealand*, **8**, 147–53.

O'NEIL K R, JONES S H and GRIMES D J (1992), Seasonal incidence of *Vibrio vulnificus* in the Great Bay Estuary of New Hampshire and Maine. *Appl. Environ. Microbiol.*, **58**(10), 3257–62.

PAOLI G C, BHUINA A K and BAYLES D O (2005), *Listeria monocytogenes*. In: Fratamico, P.M, Bhuina, A.K. and Smith, J.L. (eds.), *Foodborne pathogens, microbiology and molecular biology*, Caister Academic Press, Norfolk, UK, pp. 295–325.

ROBERTS T A, GIBSEN A M, ROBINSON A (1981), Prediction of toxin production by *Clostridium botulinum* in pasteurized pork slurry. *J. Food Technol.*, **16**, 337–55.

RØRVIK L M, CAUGANT D A and YNDESTAD M (1995), Contamination pattern of *Listeria monocytogenes* and other *Listeria* spp. in a salmon slaughterhouse and smoked salmon processing plant. *Int. J. Food Microbiol.*, **25**(1), 19–27.

RØRVIK L M, SKJERVE E, KNUDSEN B R and YNDESTAD M (1997), Risk factors for contamination of smoked salmon with *Listeria monocytogenes* during processing. *Int. J. Food Microbiol.*, **37**(2–3), 215–19.

RUSSEL A D (2002), Introduction of biocides into clinical practice and the impact on antibiotic-resistant bacteria. *Symp. Ser. Soc. Appl. Microbiol.*, **92**, S121–35.

SAKATA T (1990), Microflora of the digestive tract of fish and shellfish. In: R. Lesel (ed.). *Microbiology in poecilotherms*, Elsevier Science Publ. B.V., Biomedical Division, Amsterdam, The Netherlands, pp. 171–6.

SCHELLEKENS M, MARTENS T, ROBERTS T A, MACKEY B M, NICOLAI B M, VAN IMPE J F, ET AL. (1994), Computer aided microbial safety design of food processes. *Int. J. Food Microbiol.*, **24**, 1–19.

SHALABY A R (1996), Significance of biogenic amines to food safety and human health. *Food Res. Int.*, **29**, 675–90.

SMITH J L and FRATAMICO P M (2005), In: Fratamico, P.M, Bhuina, A.K. and Smith, J.L. (eds.), *Foodborne pathogens, microbiology and molecular biology*, Caister Academic Press, Norfolk, UK, pp. 357–82.

SPENCER R and BAINES C R (1964), The effect of temperature on spoilage of wet fish: I. Storage at constant temperature between –1 and 25°C. *Food Technology Champaign*, **18**, 769–72.

SUTHERLAND J and VARNAM A (2002), Enterotoxin-producing *Staphylococcus, Shigella, Yersinia, Vibrio, Aeromonas* and *Plesiomonas*. In: Blackburn, C.W. and McClure, P. (eds.), *Foorborn pathogenes*, Woodhead Publishing, Cambridge, UK, ISBN 1 85573 454 0.

TEALE C J (2002), Antimicroblial resistance and the food chain. *J. Appl. Microbiol. Symposium Suppl.*, **92**, 85–9.

THIMOTHE J, NIGHTINGALE K K, GALL K, SCOTT V N and WIEDMANN M (2004), Tracking of *Listeria monocytogenes* in smoked fish processing plants. *J. Food Prot.*, **67**(2), 328–41.

VALLS A, PASCUAL C Y and MARTIN ESTEBAN M (2005), *Anisakis* allergy: an update. *Revue francaise d'allergologie et d'immunologie clinique*, **45**, 108–13.

WEBER J T, MINTZ E D, CANIZARES R, SEMIGLIA A, GOMEZ I, SEMPERTEGUI R, ET AL. (1994), Epidemic cholera in Ecuador: multidrug-resistance and transmission by water and seafood. *Epidem. Inf.*, **112**, 1–11.

WEST P A (1989), The human pathogenic vibrios – a public health update with environmental perspectives. *Epidem Inf*, **103**, 1–34.

WHITING R C and BUCHANAN R L (2001), Predictive microbiology and risk assessment. In: Doyle M.P., Beauchat L.R., Montville, T.J. (eds.), *Food microbiology: fundamentals and frontiers*, 2nd ed. American Society for Microbiology Press, Washington, DC, pp. 813–31.

WIJTZES T, VAN'T RIET K, HUIS IN'T VELD J H J and ZWIETERING M H (1998), A decision support system for the prediction of microbial food safety and food quality. *Int. J. Food Microbiol.*, **24**, 1–19.

WILLIS C (2000), Antibiotics in the food chain: their impact on the consumer. *Rev. Med. Microbiol.*, **11**(3), 153–60.

WOLD J P, WESTAD F and HEIA K (2001), Detection of parasites in cod fillets by using SIMCA classification in multispectral images in the visible and NIR region. *Appl. Spectroscopy*, **55**, 1025–34.

ZWIETERING M H, WIJTZES T, DE WIT J C, VAN'T RIET K (1992), A decision support system for prediction of the microbial spoilage in foods. *J. Food Prot.*, **55**, 973–79.

20

Tracing pathogens in poultry and egg production and at the abattoir

K. L. Hiett, United States Department of Agriculture, USA

Abstract: One of the most rapidly growing categories of food commodities consists of poultry meat and hen eggs. The rise in the production and consumption of these commodities has resulted in increased concerns regarding their safety to the consumer. This chapter presents a summary of the most common zoonotic bacterial pathogens associated with poultry meat and hen egg production, as well as a review of various intervention strategies aimed at the reduction of these pathogens on/in the products.

Key words: poultry, eggs, zoonotic bacterial pathogens, food safety.

20.1 Introduction

Access to a nutritionally adequate and safe food supply should be considered a basic human right and thus a priority for the global community. Among the varieties of food products, one of the most rapidly growing categories consists of poultry meat and hen eggs (Scanes, 2007). The Food and Agricultural Organization of the United Nations (FAO, 2007a) reported that between 1995 and 2005, the global consumption of chicken meat and hen eggs increased by approximately 53% and 39%, respectively (FAO, 2007a, 2007b). With respect to the United States and Europe, chicken meat production increased by 38% and 30%, respectively, with a 2% (USA) and 6% (Europe) increase for hen egg production (FAO, 2007a, 2007b). The observed increase in the production and consumption of these two commodities has been attributed to several factors including the nutritional value (low fat content of chicken meat, nutrient dense, and high-quality protein sources of both chicken meat and hen eggs), as well as the relatively low price of the two products (Mead *et al.*, 1999; Scanes, 2007).

The rise in the production and consumption of poultry meat and hen eggs has resulted in increased concerns regarding the safety of these products for consumers

(Mead *et al.*, 1999; Mbata, 2005; Humphrey *et al.*, 2007; Juneja and Sofos, 2010). While bacterial contamination of the raw chicken and egg products occurs, the general consensus is that most poultry- and egg-related human illnesses arise as a result of either consumer or retail outlet mishandling of the raw products (cross-contamination to other foods), improper cooking or improper storage of products after preparation (Bryan and Doyle, 1995; Mbata, 2005; Humphrey *et al.*, 2007; Todd *et al.*, 2007; de Jong *et al.*, 2008; Luber, 2009). These observations have generated a discussion as to which intervention strategy is best applied to achieve a reduction in human food-borne illness: a reduction in the prevalence of zoonotic bacterial pathogenson/in the products, or the development and application of improved risk communication aimed at educating the consumer and thus changing or improving their food preparation habits (Luber, 2009). A combination of the two strategies is likely to be the most efficacious approach. An additional concern is that as technology advances, our understanding of bacterial contamination of these products greatly increases. Newly detected or 'emerging' bacterial pathogens are now being recognized as causative agents of food-borne illness related to the consumption of poultry meat. Lastly, the complexity of poultry meat and egg production/processing systems is a significant challenge for reducing the prevalence of pathogens on the final products. In major poultry-producing countries, broiler meat production and processing systems are generally intensive and vertically integrated. A single company is often responsible for management of all activities involved, including grand-parent and parent breeder maintenance, hatching of broiler chicks, feed production, broiler rearing (grow-out), catching, transport, processing, packaging, production of 'ready-to-eat' (RTE) products, and distribution. A similar vertically integrated system is also in place for table egg production. Each stage of poultry meat and egg production and processing systems presents its own unique challenges regarding pathogen contamination and control. As a result of the complex nature of the production and processing environments (farm to fork), pathogen contamination can occur at any level. Thus, it is quite likely that a 'multi-hurdle' approach is the best strategy for pathogen reduction and subsequent elimination.

20.2 Pathogens associated with broiler meat

The two zoonotic bacterial pathogens most often associated with chicken meat are *Campylobacter* spp., more specifically *C. jejuni* and *C. coli*, and *Salmonella* spp. (Ekdahl *et al.*, 2005; Mbata, 2005; Anonymous, 2007). Additional zoonotic pathogens occasionally associated with poultry meat include *Clostridium perfringens* and *Listeria monocytogenes*. Recently, however, newly emerging (or newly detected/recognized) pathogens such as *Arcobacter* spp. (de Boer *et al.*, 1996; Mbata, 2005; Ho *et al.*, 2006; Quiñones *et al.*, 2007; Lipman *et al.*, 2008), *Helicobacter* spp. (Fox, 1997; Mbata, 2005; Ceelen *et al.*, 2007), and verotoxigenic *Eschericia coli* (Mbata, 2005) have been recovered from poultry products and reported as causative agents of human gastroenteritis.

20.2.1 *Campylobacter* spp.

Taxonomy

Campylobacter enteritis was first reported by Theodor Escherich (1886). In several articles, Escherich described finding, in the feces of children with diarrhea, spirally curved bacteria that could not be cultured on solid media (Goossens and Butzler, 1992). By the early 1900s, as cultural methods improved, isolates of bacteria associated with ovine and bovine abortions were identified; in 1918, Smith and Taylor described infectious abortion in cattle due to a bacterium classified as 'vibrio' or 'spirilla'. A more definitive link to human infection was established in 1957 when King observed bacteria in blood cultures taken from humans with entercolitis. These bacteria were designated 'related vibrios'. Morphological and physiological differences between the abortion associated strains and the known vibrios became apparent in 1963, whereupon a separate genus, *Campylobacter*, was proposed by Sebald and Veron (Kaijser, 1988; Griffiths and Park, 1990). Currently, *Campylobacter* spp. are in the class of Epsilon Proteobacteria which include the family *Campylobacteracea*, comprised of the genera *Campylobacter*, *Arcobacter*, and *Sulfurospirillum*, and the family *Helicobateraceae*, comprised of the genera *Helicobacter*, *Wolinella*, and *Thiovulum* (Kabeya *et al.* 2004; On, 2001; Vandamme *et al.*, 2000, http://www.gbif.net/and http://www.uniprot.org/).

Characteristics

Bacteria of the genus *Campylobacter* are Gram-negative, non-spore forming, highly motile, spirally curved rods, 0.5–8 μm in length and 0.2–0.5 μm in width; coccoid forms may be present in older cultures (Ketley, 1995; Penner, 1988; Griffiths and Park, 1990). The bacteria generally possess a single polar unsheathed flagellum, referred to as monotrichous, or a flagellum at each end, referred to as amphitrichous; however, aflagellated forms may appear spontaneously and reversibly in culture (Caldwell, 1985; Ketley, 1997). *Campylobacter* spp. grow optimally between 30°C and 42°C, and are microaerophilic, requiring an O_2 concentration of 3–5%, and a CO_2 concentration of 3–10%, with a nitrogen balance (Ketley, 1997). However, several of the newly recognized *Campylobacter* species require hydrogen for optimal growth (Vandamme *et al.*, 1991; Lastovica *et al.*, 1993).

Clinical manifestations and significance to public health

Campylobacter spp. are presently considered to be leading bacterial etiological agents of acute gastroenteritis in industrialized human populations (Tauxe, 1992; Friedman *et al.*, 2004; Humphrey *et al.*, 2007). The economic burden of campylobacteriosis is significant. A conservative estimate in the United Kingdom is £65 million yearly with a true cost estimate of £500 million. In the United States, the annual estimated cost during the 1990s was approximately $4.3 billion USD (Buzby and Roberts, 1997). These figures do not take into consideration long-term sequelae or loss of productivity.

Campylobacter enteritis is generally caused by two closely related species, *C. jejuni* and *C. coli*; however, *C. jejuni* is the dominant causal species, responsible for 80–90% of all infections in the developed world (Skirrow, 1991; Altekruse *et al.*, 1994). The total number of *Campylobacter* spp. enteritis cases in the United States is estimated at 2.4 million per year, or approximately 1–2% of the population per year, similar to that of the United Kingdom (Blaser and Reller, 1981; Skirrow, 1991; Tauxe, 1992; Taylor, 1992; Slutsker *et al.*, 1998). The vast majority of cases occur as isolated, sporadic events, rather than as recognized outbreaks. Additionally, many cases go undiagnosed or unreported (CDC). Ingestion of as few as 100–500 organisms can result in disease (Anonymous, 2005). The majority of *C. jejuni* cases are enteric with most episodes confined to local acute gastroenteritis characterized by nausea, abdominal cramps, fever, diarrhea, and fatigue. Infections are generally self-limited and are resolved within several days (7–10) after initial onset (Chowdhury, 1984; Peterson, 1994), although, relapses have been reported in a small number of patients (Allos and Blaser, 1995). Antibiotic treatment is rarely indicated for clinical infections; this is especially important as antimicrobial resistance to clinically important drugs is increasing in *Campylobacter* spp. Helms *et al.* (2005) reported an increased severity of disease (invasiveness or death) when patients were infected with antibiotic-resistant *C. jejuni* isolates.

Campylobacter spp. infections have also been associated with unnecessary appendectomies, reactive arthritis, Miller Fisher Syndrome, and development of Guillain–Barré Syndrome (GBS; Butzler and Skirrow, 1979; Bokkenheuser and Sutter, 1981; Walker *et al.*, 1986; Bryan and Doyle, 1995). *C. jejuni*-associated reactive arthritis is an oligoarticular arthritis that is asymmetric and primarily affects the ankles, knees or wrists (Keat, 1983; Peterson, 1994). The development of reactive arthritis in patients previously infected with *C. jejuni* has been associated with the presence of the human leucocyte antigen-B27 (HLA-B27 antigen) (Kaslow *et al.*, 1984; Butzler, 1991). GBS is an acute disease of the peripheral nerves that is characterized by segmental demyelination resulting in a rapidly ascending paralysis that can lead to respiratory muscle compromise and ultimately death (Allos, 2001; Humphrey *et al.*, 2007). In the United States, a disproportionate number of GBS-associated *C. jejuni* isolates belong to the relatively uncommon O:19 and O:41 Penner serotypes (lipopolysaccharide) (Allos and Blaser, 1995; Allos, 1998; Nachamkin *et al.*, 2002; Takahashi *et al.*, 2005). Molecular mimicry by the *C. jejuni* lipooligosaccharide outer core results in the production of cross-reactive antibodies by certain individuals whereupon the gangliosides in the nervous tissue are targeted and undergo segmental demyelination (Nachamkin *et al.*, 1998; Godschalk, 2004).

20.2.2 *Salmonella* serotypes

Taxonomy

The genus *Salmonella* was named after Daniel E. Salmon when in 1885, Salmon and Smith (1885) first described an organism, then termed *Bacillus cholerae-suis*, from pigs (Andrews and Bäumler, 2005). Later, Smith delineated *Salmonella* as a separate genus. Subsequently, the history of *Salmonella* nomenclature can best be described as complex. The current classification scheme recognizes only two species, *Salmonella bongori* and *Salmonella enterica* (comprised of six subspecies, I–VI) (Reeves *et al.*, 1989; Brenner *et al.*, 2000). Members of the genus are classified based upon the expression of two surface antigens: the O antigen (LPS) and the H antigen (flagellar). The subspecies are divided into over 50 sero groups based on O antigen classification. These sero groups are further divided into over 2300 serotypes based on H antigen classification. There are approximately 20 serotypes of *S. bongori*, while *S. enterica* is comprised of over 2443 serotypes (Andrews and Bäumler, 2005).

Characteristics

Bacteria of the genus *Salmonella* are Gram-negative rod-shaped, non-spore forming, predominantly motile enterobacteria, that are approximately 2–5 μm in length and 0.7–1.5 μm in diameter (Giannella, 1996). *Salmonella* isolates generally possess peritrichous flagella; however, aflagellated variants, such as *S. enterica* serovar Pullorum and *S. enterica* serovar Gallinarum, can occur. The bacteria are facultative anaerobes, and can metabolize nutrients by respiratory and fermentative pathways. Optimum growth temperatures range from 35°C to 40°C (Giannella, 1996).

Clinical manifestations and significance to public health

The human and economic burden of *Salmonella*-associated entercolitis is significant. In the United States, the Centers for Disease Control and Prevention (CDC) estimates that there are approximately 40 000 culture-confirmed cases of non-typhoidal *Salmonella* infections annually (Mead *et al.*, 1999). However, the actual number of cases is likely to be roughly 1.4 million annually (Voetsch *et al.*, 2004) resulting in 15 000 hospitalizations and 400 deaths. Cost estimates are believed to approach $2.3 billion USD annually (Frenzen *et al.*, 1999).

Salmonella isolates vary greatly in their ability to cause disease in different hosts. With respect to human illness, it is estimated that 99% of clinical cases can be attributed to members of *S. enteritidis* subspecies I; more specifically, human disease associated with poultry meat and egg consumption is generally attributed to *S.* Enteritidis and *S.* Typhimurium (Aleksic *et al.*, 1996). Most episodes occur within 12–72 hours after infection, and are confined to acute gastroenteritis characterized by abdominal cramps, fever and diarrhea. Infections are generally self-limited and are resolved within several days (4–7) after initial onset (Bailey, 1988). Antibiotic treatment is rarely indicated; exceptions include the elderly, infants, and immunocompromised individuals.

20.2.3 *Clostridium perfringens*

Taxonomy

Clostridia species are members of the Firmicutes division including *Clostridium* and other similar genera. The species most associated with food production and processing is *C. perfringens*; however *C. botulinum* is a significant cause for concern in food storage. *C. perfringens* is classified into one of five types (A, B,

C, D or E) based on toxin production of four extracellular toxins, alpha, beta epsilon, and iota. These extracellular toxins are not associated with food poisoning (Meer *et al.*, 1997; Petit *et al.*, 1999; Sarker *et al.*, 2000; Brynestad and Granum, 2002). Rather, an enterotoxin (*cpe*), produced by types A and C is responsible for food poisoning. The cpe enterotoxin can be carried on a plasmid or chromosomal (Cornillot *et al.*, 1995).

Characteristics

C. perfringens is a Gram-positive, anaerobic, non-motile, spore forming, rod-shaped bacterium (Novak *et al.*, 2005). In the United Kingdom and United States, *C. perfringens* bacteria are the third most common cause of food-borne illness, with improperly prepared poultry meat as a primary source (Warrell *et al.*, 2003). Many areas of poultry processing plants were determined contaminated with *C. perfringens* (Craven *et al.*, 2001). The *C. perfringens* enterotoxin (CPE) which is responsible for enteritis is heat-labile at 74°C and can be detected in contaminated food and feces (Murray *et al.*, 2009). The incidence of *C. perfringens* in the intestinal tract of poultry has been reported to be as high as 80% (Tschirdewahn *et al.*, 1991). Numbers of *C. perfringens* in the feces were $>10^4$/g in 31% of chickens. The percentage of *C. perfringens*-positive samples from tested hatcheries demonstrated an overall incidence of 20% (Craven *et al.*, 2001). However, the overall prevalence of enterotoxigenic, i.e., capable of causing human disease, isolates in chicken feces has been reported at 40–60% (Miwa *et al.*, 1997).

Clinical manifestations and significance to public health

The number of *C. perfringens* related human illness in the United States is estimated at 248 520 cases per year with an estimated cost of $200.00 per case (Mead *et al.*, 1999). The estimated time from ingestion to onset of symptoms is between 6 and 24 hours (Mead *et al.*, 1999). Symptoms typically include abdominal cramping and diarrhea, however vomiting and fever are unusual (Meer *et al.*, 1997). To date, efforts directed at preventing *C. perfringens* food-borne disease outbreaks have concentrated on developing guidelines for food handling and preparation techniques that prevent multiplication of this organism in the contaminated product (Bryan and McKinley, 1974). Despite these guidelines, outbreaks involving *C. perfringens* continue (Brynestad and Granum, 2002).

20.2.4 *Listeria monocytogenes*

Taxonomy

L. monocytogenes was first described in 1926, by E.G.D. Murray as the etiologic agent of a septic illness in laboratory rabbits (Murray, 1926). Murray originally referred to the organism as *Bacterium monocytogenes* before J.H. Harvey Pirie changed the genus name to *Listeria* in 1940 (Pirie, 1940). The *Listeria* genus includes six distinct species, but only *L. monocytogenes* is consistently associated with human illness. Currently, there are 13 serotypes of *L. monocytogenes* associated with human illness, however only three of these serotypes, 1/2a, 1/2b, and 4b, are predominantly associated with human food-borne illness (Kathariou, 2002).

L. monocytogenes has three distinct lineages with differing evolutionary histories and pathogenic potentials (Jeffers *et al.*, 2001). Lineage I isolates are most associated with human clinical cases, but are underrepresented in animal clinical isolates (Gray *et al.*, 2004; den Bakker *et al.*, 2008). Lineage II strains are more commonly isolated in environmental, food, and animal samples (Nightingale *et al.*, 2005). Isolation of Lineage III strains is rare and most often found in animal samples (Gray *et al.*, 2004; den Bakker *et al.*, 2008).

Characteristics

Listeria monocytogenes is a Gram-positive, non-spore forming, motile, aerobic or facultative anaerobic, rod-shaped bacterium. *L. monocytogenes* is actively motile (tumbling), via peritrichous flagella, at room temperature (20–25°C), however, the organism does not synthesize flagella at 37°C (Paoli *et al.*, 2005). The optimum growth temperature is 30–37°C; however, growth has been reported between the temperatures of –0.4°C and 45°C (Rowan and Anderson, 1998). *L. monocytogenes* are prevalent in the environment, and are often introduced into raw or post-processed food through incidental contact with contaminated surfaces; therefore contamination is primarily a concern for RTE foods (Meng and Doyle, 1997; CFSAN, 2003). Additionally, since *L. monocytogenes* can grow at typical refrigeration temperatures, the organism can outcompete other bacteria and multiply in most RTE food environments. The recent strengthening of regulatory standards by the Food Safety and Inspection Service (FSIS) and the FDA allow 'zero tolerance' for *L. monocytogenes* on RTE meat and poultry products. These standards and the USDA's October 2003 'Interim Final Rule on Control of *Listeria monocytogenes*' have focused attention on the need to control this threat to food safety in the processing environment (US Department of Agriculture, FSIS. 9 CFR Part 430; USDA, 2003).

Clinical manifestations and significance to public health

While *L. monocytogenes* may account for only 2% (2500 individuals) of all food-borne related illnesses per year in the United States, the organism is responsible for 4% (2300 individuals) of hospitalizations, and 28% (500 individuals) of food-borne related deaths (Mead *et al.*, 1999). Symptoms range from influenza-like (persistent fever and muscle aches), to gastrointestinal, septicemia, meningitis, encephalitis, cornela ulcers, and intrauterine infections of pregnant women that may result in spontaneous abortions during the 2nd or 3rd trimester or stillbirth (Holland *et al.*, 1987; Whitelock-Jones *et al.*, 1989; Armstrong and Fung, 1993; Vásquez-Boland *et al.*, 2001). The elderly, neonates, and immunosuppressed individuals are at a greater risk for severe illness (i.e. meningitis, encephalitis, and septicemia) or death from *L. monocytogenes*-contaminated food (Mead *et al.*, 1999). The time from ingestion to onset of gastrointestinal symptoms is estimated at approximately 12 hours, while the onset of more serious forms of listeriosis is believed to range from a few days to three weeks (Posfay-Barbe and Wald, 2009).

20.3 Source tracking

As mentioned previously, the intensive nature of poultry meat and table egg production/processing coupled with consumer-demand for high-quality products for the lowest price makes reducing and eventually eliminating zoonotic bacterial pathogens from the final commodity problematic. Methods for rapid detection and identification of these pathogens from a variety of environments (source tracking) and complex matrices will greatly facilitate the identification of reservoirs of these pathogens. Additionally, the ability to determine the relatedness of pathogens in agricultural settings is becoming increasingly important for epidemiological investigations and for the subsequent development of risk assessment models and intervention strategies to eliminate these pathogens from the food supply.

20.3.1 Traditional microbiological/cultural methods

Traditional microbiological/cultural techniques have been useful as they have been extensively tested for accuracy and are often considered the 'gold standard' by regulatory agencies. Additionally, cultural techniques provide isolates that can be stored for long periods and used in subsequent investigations. However, culture-based methods are labor intensive and often require extended times for detection. Additionally, with some organisms such as *Campylobacter* spp., traditionally accepted cultivation media, such as Campy-Cefex (Stern *et al.*, 1995) and modified cefoperazone charcoal deoxycholate (mCCDA; Bolton *et al.*, 1983) may possess an inherent bias for recovery for specific species (*C. jejuni* and *C. coli*) as well as not provide for the detection of newly emerging organisms (Aspinall *et al.*, 1996; Bourke *et al.*, 1998; Byrne *et al.*, 2001). As different species are recovered or detected using newer technologies such as metagenomics analyses coupled with pyrosequencing, it has become evident that additional *Campylobacter* spp., as well as other members of the Epsilon Proteobacteria, may be more prevalent than previously thought (Tauxe, 2002; Schlundt *et al.*, 2004; Skovgaard, 2007). Furthermore, if distinct *Campylobacter* spp., as well as other members of the Epsilon Proteobacteria, arise or emerge, current recovery technologies are inadequate to detect this trend. These observations can likely be applied to other zoonotic pathogens as well. Excellent sources for culture-based recovery techniques are texts such as the *Compendium of Methods for the Microbiological Examination of Foods* (Vanderzant and Splittstoesser, 1992), *Official Methods of Analysis of the Association of Official Analytical Chemists* (AOAC, 1990), the *Bacteriological Analytical Manual* (FDA, 1992), *Standard Methods for the Examination of Dairy Products* (APHA, 1985), and *Modern Food Microbiology* (Jay, 1992).

20.3.2 Antibody-based/immunological methods

Several technologies that employ immunologic capture mechanisms are being used for the rapid detection of zoonotic pathogens. These detection methods rely on antibody-based probe and affinity probe specificity to capture the organism of interest. Antibodies (monovalent and divalent antibody fragments and single chain variable regions) recombinant antibodies, bacteriphage display antibody probes, aptamers (nucleic acid ligands) and peptide ligands are all examples of ligands used for capture (Tuerk and Gold, 1990; Goldschmidt, 2006; Fipula, 2007). Disadvantages associated with this set of technologies include difficulties in replicating or maintaining physical structures, methods for synthesis can be tedious, and long-term stability.

20.3.3 Molecular typing and tracking methods

Molecular typing and tracking of pathogens allows investigators to determine the chain of infection/contamination, through detection of the reservoir of the agent of interest, by uncovering the mode of transmission to a susceptible host, and through the determination of the means of entry into the host (Levin *et al.*, 1999). Additionally, information acquired from molecular typing and tracking of pathogens can (1) facilitate the detection of outbreaks within flocks or herds, (2) facilitate the detection of emerging isolates with increased virulence properties, (3) assist in the assessment of the effectiveness of current control measures, (4) assist in the establishment of risk reduction strategies, and (5) assist in evaluating the effectiveness of food safety programs.

There are several parameters to consider when choosing a typing method. One of the most critical considerations is the clonal relatedness of the organism of interest (Riley, 2004). When investigating organisms that are not strictly clonal, the careful choice of an appropriate target (one that is not affected by genetic instability) for subtype analyses can circumvent subsequent analytical concerns. A second factor to take into account when choosing a subtyping method is the discriminatory power of the method for the particular organism of interest (Maslow *et al.*, 1993). A typing method should be able to distinguish between truly different isolates; however, the typing method should not type all isolates differently (Tenover *et al.*, 1994, 1997). Upon developing and applying a new typing method to an organism, the method should initially be tested in an epidemiologic background to assess the discriminatory power (Swaminathan and Matar, 1993; Riley, 2004). If it is determined that a specific technique does not provide adequate discrimination, a second technique can often be performed to provide more information. It is also important to keep in mind that one particular subtyping method may not be adequate for all organisms of interest. Often, different methodologies must be employed for different organisms. A third consideration for choosing a typing method is the simplicity in performing the technique (Riley, 2004). Related to simplicity is a fourth consideration, reproducibility. A fifth concern is the throughput potential associated with a technique. A final consideration is the cost associated with performing a technology; a cost analysis should include both the initial investment for equipment and the subsequent cost of reagents and consumables. Application of the same technique, with minor modifications, to several organisms, is one means to offset a large initial equipment expense.

In general, the classification and subtyping of organisms can be divided into two categories, phenotypic based technologies or genotypic based technologies. Phenotypic methods are based upon the detection of characteristics that are expressed by an organism, whereas genotypic methods are based upon the analysis of nucleic acids and sequence polymorphisms within an organism (Maslow *et al.*, 1993; Tenover *et al.*, 1997; Lipuma, 1998). Traditionally, phenotypic methods have been used for differentiation of organisms. However, recent advances in the development of molecular techniques have led to the widespread use of these methods. Often, both phenotypic and genotypic technologies can be combined to provide the most useful information. Below are brief descriptions of the most commonly used typing methods in each category.

20.4 Phenotypic based tracking methods

20.4.1 Biochemical characteristics (biotyping)

Biotyping is based upon the analysis of metabolic activities of distinct organisms. Isolates of interest are subjected to a panel of biochemical tests and differentiated by the responses to those tests. In general, the greater the number of biochemical tests applied to a panel of organisms, the greater the discriminatory power of the technique (Tenover *et al.*, 1994). Consequently, conducting this technique in an optimal manner can be costly in terms of both labor and money. Additionally, because biotypes are dependent upon growth conditions, reproducibility is often a concern.

20.4.2 Serotyping

Serotyping is based upon the analysis of antigenic expression of an organism. The antigens used for analysis can be comprised of proteins, polysaccharides, and lipopolysaccharides (Edwards and Ewing, 1972; Penner, 1988; Popoff and Le Minor, 1997). Most often, the somatic antigen (O polysaccharide) and/or the flagellar protein (H antigen) are used in combination for serotype analyses. The greatest disadvantage associated with serotyping is the expense and limited availability of quality-controlled reagents. The differences in antiserum preparation often lead to typing discrepancies both within a laboratory and between laboratories. Additionally, visual determination of agglutination can lead to inconsistent typing. Other considerations include the possibility of different species expressing antigens that cross-react (false-positives) and the lack of expression of some antigens under certain growth conditions (false-negatives).

20.4.3 Phage typing

Phage typing is a method that classifies bacteria on the basis of their susceptibility to lysis by bacteriophage. An important parameter to consider when performing phage type analysis is that initially the technique must be developed and specifically designed for each individual species of bacteria (Grajewski *et al.*, 1985; Ward *et al.*, 1987). Additionally, the maintenance of stocks of typing phages for all species of bacteria can be cumbersome.

20.5 Nucleic acid based methods

20.5.1 Randomly amplified polymorphic DNA (RAPD)

Randomly amplified polymorphic DNA (RAPD) analysis is a whole-genome typing method where a single arbitrary oligonucleotide primer (generally 9–10 bp in length) is used for a PCR amplification. Because only one primer is used, the primer must be able to bind to opposite strands of the target DNA approximately 100–3000 bp in distance from one another for amplification to be achieved. After the PCR is completed, agarose electrophoresis is performed to resolve the resulting bands. Computer algorithms are then used for analysis and comparison of fingerprints (Williams *et al.*, 1990; Welsh and McClelland, 1991; Micheli and Bova, 1997). RAPD is a technically simple technique that is relatively inexpensive and subject to automation (high throughput). The primary disadvantage is the lack of reproducibility; factors such as the ratio of DNA template concentration to primer concentration, model of thermocycler, Mg^{+2} concentration, and brand of Taq™ polymerase used can greatly affect banding patterns (Meunier and Grimont, 1993; Berg *et al.*, 1994; Tyler *et al.*, 1997).

20.5.2 Repetitive element polymerase chain reaction (rep-PCR)

Repetitive element sequence based polymerase chain reaction (rep-PCR) is a PCR-based method that targets known, conserved, repetitive DNA sequences, that are usually present in multiple copies in bacterial genomes (Versalovic *et al.*, 1991; Lupski and Weinstock, 1992). Examples of repetitive DNA motifs include repetitive extragenic palindromes (REP), enterobacterial repetitive consensus elements (ERIC), BOX elements, and RepMP3 (Stern *et al.*, 1984; Wenzel and Herrmann, 1988; Hulton *et al.*, 1991; Martin *et al.*, 1992). Rep-PCR is a discriminatory typing method that provides a whole-genome analysis of an organism; because the primer(s) sequences are defined, reproducibility is enhanced relative to RAPD (Tyler *et al.*, 1997). Additionally, rep-PCR is becoming increasingly technically straightforward due to recent automation of the PCR amplification, band resolution, and analysis steps. A disadvantage associated with automation is that the reagents and the specialized equipment can be costly.

20.5.3 Polymerase chain reaction-restriction fragment length polymorphism (PCR-RFLP)

Polymerase chain reaction-restriction fragment length polymorphism (PCR-RFLP) involves PCR amplification of a target sequence, restriction digestion (using an enzyme with a 4 bp recognition site) of the resulting amplicon, electrophoretic resolution of the digested fragments, and analysis of the resulting banding pattern (Williams *et al.*, 1990; Tornieporth *et al.*, 1995; Gonzalez *et al.*, 1997). PCR-RFLP is technically simple, inexpensive, and allows for rapid processing of samples. A disadvantage of the technique is that PCR-RFLP examines only a small section of the total genome. Genomic instability of an organism will not be detected using this method.

20.5.4 Pulsed-field gel electrophoresis (PFGE)

Pulsed-field gel electrophoresis (PFGE) is a macro-restriction profiling technique based on the digestion of DNA with restriction enzymes that cut DNA infrequently ('rare-cutters' of 6 or 8 bp recognition sites) to produce large fragments of DNA. Initially, whole cells are embedded in agarose plugs and lysed using detergent and enzymes. All treatments are performed on the plug in an effort to prevent shearing of the DNA. The digested gel plugs are then loaded into an agarose gel and resolved using a special gel apparatus where various pulsed-field techniques can be applied. One such technique is field-inversion gel electrophoresis, used to resolve DNA fragments from ~10–2000 kb. This method employs a periodic inversion of the electric field in forward and reverse directions. Resolution of larger DNA fragments requires the use of a field-angle alternation electrophoresis such as a contour-clamped homogenous electric field (CHEF) (Chu *et al.*, 1986). The CHEF method employs a uniform electric field that constantly alternates angles 120° to one another for varying time intervals.

PFGE is considered to be one of the most discriminatory typing methods currently available as it provides a whole-genome analysis of an organism. Presently, PFGE is the typing method employed by the US Centers for Disease Control PulseNet program (The National Molecular Subtyping Network for Foodborne Disease Surveillance) (http://www.cdc.gov/pulsenet/) (Swaminathan *et al.*, 2001). An additional benefit is that PFGE can be performed without prior knowledge of nucleotide sequence. However, there are several disadvantages associated with PFGE. Preparation of the agarose plugs and the subsequent restriction digest of those plugs are time-consuming and tedious procedures. A second disadvantage is that the reagents and the specialized equipment are costly. However, once a laboratory has made the initial investment in equipment, it can be used in the analysis of several different organisms.

20.5.5 Amplified fragment length polymorphism (AFLP)

Amplified fragment length polymorphism (AFLP) is a whole-genome analysis technique based on the selective PCR amplification of restriction fragments generated from a total digest of genomic DNA (Blears *et al.*, 1998). Typically, 50–100 fragments in the size range of 50–550 bp are amplified and detected using this method (Blears *et al.*, 1998, Duim *et al.*, 1999). AFLP is considered to be a highly discriminatory typing method that provides a whole-genome analysis of an

organism. Additionally, this technique can be performed without prior knowledge of the organism's nucleotide sequence. The primary parameters that require consideration include the G+C content and the size of the organism's genome. A principal disadvantage associated with AFLP is that the initial capital investment in equipment is expensive. A second major disadvantage is that the final data output is complex, thus interpretation can be difficult.

20.5.6 Variable-number tandem repeats (VNTR)

Variable-number tandem repeat (VNTR) analysis is a technique based on the selective PCR amplification, and subsequent agarose gel resolution, of regions of a prokaryotic genome that contain either minisatellite DNA or microsatellite DNA. Prokaryotic minisatellites and microsatellites are tandem repeats of DNA that range in size from 10 to 100 base pairs or from 1 to 10 base pairs, respectively, and are arranged in a head-to-tail configuration. The size of the specific repeat and the number of times the repeat occurs in tandem define the subtype (van Belkum, 1999; Yeramian and Buc, 1999). A primary disadvantage associated with VNTR analysis is that a prior knowledge of the whole genome of an organism is required for the development of an appropriate target that provides accurate epidemiologic information.

20.5.7 DNA sequence analysis

Direct DNA sequence analysis is a typing technique where specific genomic DNA fragments are generated by a targeted PCR and subsequently sequenced and analyzed (Mullis and Faloona, 1987). This technique can provide very detailed information to allow for comprehensive comparison and tracking of organisms. Additionally, as technology advances, automation of the production of target PCR amplicons and the automation of DNA sequencing permits for faster processing of a larger number of isolates. However, a number of disadvantages are associated with this technique. Initial capital expense for the equipment, a fluorescent automated DNA sequencer, is steep. This expense can be circumvented as several core laboratories and commercial laboratories have been established that perform DNA sequencing for a fee. A second disadvantage is that, often, a great deal of prior knowledge regarding the organism of interest is required to develop an appropriate target that provides accurate epidemiologic information. A third disadvantage is that this technique samples only a small portion of the organism's total genome.

20.5.8 Multilocus enzyme electrophoresis (MLEE) and multilocus sequence typing (MLST)

Multilocus sequence typing (MLST) is an extension of multilocus enzyme electrophoresis (MLEE), a phenotypic typing method (Milkman, 1973; Selander *et al.*, 1986). In MLEE, the electrophoretic mobilities of several proteins are analyzed in starch gels and assigned as allozymes or electrophorectic types based on observed differential mobilities as a result of amino acid substitutions. The enzymes chosen for analysis generally are under low selective pressure for variability. Additionally, the genes encoding the selected proteins are located throughout the genome; thus the total genome is sampled using this method. Execution of MLEE and analysis of the resulting data can be complex, so an extension of the procedure, MLST, was developed. MLST consists of the direct nucleotide sequencing of internal regions, approximately 450 bp, of 'housekeeping genes' that encode the proteins analyzed in MLEE (Maiden *et al.*, 1998). Generally, seven or eight genes are analyzed; each different DNA sequence for an individual gene is assigned as a distinct allele. A seven digit allelic profile is obtained for that isolate where the combination of all allelic profiles is referred to as a sequence type. MLST is a highly discriminatory typing scheme that has proven to be most useful in studies on population genetics of organisms. With the automation of PCR amplification and sequencing, MLST is rapid and technically straightforward. Additionally, the technique is highly reproducible between laboratories. The primary disadvantage is the initial capital expense for the equipment, a fluorescent automated DNA sequencer. Again, this expense can be circumvented as several core laboratories and commercial laboratories have been established that perform DNA sequencing for a fee.

20.5.9 Automated ribotyping

Ribotyping involves analysis of genomic DNA restriction fragments that contain all or a part of the genes that code for 16S and 23S ribosomal RNA (rRNA) (Grimont and Grimont, 1986; Stull *et al.*, 1988). The only parameter to consider for optimization is the choice of the appropriate restriction enzyme(s). Commonly enzymes that possess 6 bp recognition sites are used. In general, organisms contain several copies of ribosomal genes, therefore ribotyping provides sufficient information for adequate discrimination between isolates. The discriminatory power of ribotyping can be increased by using a combination of restriction enzymes for analyses. An additional strength of this method is that rRNA genes are located at different positions of the chromosome, therefore, the total genome is sampled with this method. Fully automated systems have been developed that make this technique technically easy as well as reproducible (Bruce, 1996). However, the automated equipment, as well as the consumables, are relatively expensive. Additionally, the automated system allows for only a small number of samples to be analyzed at one time.

20.5.10 The future

Recent advances in DNA sequencing and computational power have provided new tools for subtype analysis of organisms. The emerging field of comparative genomics is one outcome of the complete genome sequencing of organisms. The comparative analysis of whole genomes has provided information such that new paradigms regarding evolutionary relationships between organisms are emerging (Blattner *et al.*, 1997; Perna *et al.*, 2001). As more information is obtained, subtyping technology is being refined leading to more informative analyses. Another outcome of comparative genomics is the 'in silico' identification of improved targets for subtype analyses, with subsequent testing for effectiveness on epidemiologic defined isolates.

Whole-genome sequence analysis has also led to the development of microarray technology. Microarray analysis has the potential to address such questions as (1) gross identification of bacteria or viruses, (2) identification of genes involved in antimicrobial resistance, (3) identification of distinct isolates based on total DNA hybridization patterns, (4) identification of distinct isolates based on differential gene expression (RNA), (5) identification of novel genes from related isolates, and (6) identification of nucleotide differences in target genes. While microarray analysis offers a promising future, there are currently several limitations associated with the technology. Reproducibility, cost (both initial and consumable expenses), technical difficulty, and interpretation are all concerns. However, as advances are made regarding this technology, its usefulness as a tool for subtype analysis should increase.

20.6 Reservoirs and potential amplification steps of human pathogens in poultry production chains

Each stage of poultry meat and egg production and processing systems presents its own unique challenges regarding pathogen contamination and control. As a result of the complex nature of the production and processing environments (farm to fork), pathogen contamination can occur at any level. Thus, it is quite likely that a 'multi-hurdle' approach will serve as the best strategy for pathogen reduction and subsequent elimination.

20.6.1 Poultry meat

Pre-harvest

The zoonotic pathogens of greatest concern are *Campylobacter* spp. and *Salmonella* spp. While *C. perfringens* is currently considered more of an animal health issue (the etiologic agent of necrotic enteritis and gas gangrene in the bird; Songer, 1996), it will likely become more of a concern if, like the European Union, the United States bans the use of subtherapeutic antimicrobials.

Campylobacter spp. are ubiquitous in the production environment, but are rarely recovered, using traditional culture techniques, until the third or fourth week of production. This phenomenon has been attributed to the presence of maternal antibodies, however it could be due to insufficient culture-based methods (Aspinall *et al.*, 1996; Bourke *et al.*, 1996; Byrne *et al.*, 2001; Newell *et al.*, 2001; Sahin *et al.*, 2001). Several suspected sources or vectors of contamination have been studied and include exposure of birds to contaminated water, a previously contaminated rearing environment, hatchery pads, litter, feed, personnel, small animals on the farm, flies, and rodents (Genigeorgis *et al.*, 1986; Lindblom *et al.*, 1986; Hood *et al.*, 1988; Kazwala *et al.*, 1990; Pearson *et al.*, 1993, 1996).

A survey conducted by the United States FSIS confirmed the presence of *Salmonella* spp. in 11.4% of broiler chickens tested during production and in 45.0% of ground chicken products (http://www.fsis.usda.gov/Science/Progress_Report_Salmonella_Testing_Table/index.asp). Regarding *Salmonella* spp., a significant mode of transmission of the organism is through vertical transmission and pseudovertical (fecal contamination on egg shell) transmission from the breeder hen to the broiler offspring (Cox *et al.*, 1990; Berchieri *et al.*, 2001; Hensel and Neubauer, 2002). Vaccination and hygienic measures have considerably reduced vertical transmission of the infection from the parent flocks to the broiler birds (Van Immerseel *et al.*, 2004); however, if contaminated chicks are placed, the rapid spread of *Salmonella* spp. throughout the house is often observed (Heyndrickx *et al.*, 2002, 2007; Marin *et al.*, 2009). A second significant source implicated in *Salmonella* spp. contamination associated with production is the feed (Durand *et al.*, 1990; Schluter *et al.*, 1994; Bailey *et al.*, 1999; Ranta and Maijala, 2002). Contamination of the finished feed is believed to originate from the incorporation of previously contaminated vegetable ingredients (Davies *et al.*, 2001). Other vectors investigated, but not believed to be as significant of a concern include wildlife, insects, and rodents (Davies *et al.*, 2001; Garber *et al.*, 2003).

Catching and transport

The stresses associated with feed-withdrawal, catch, loading, and transport from production facilities (farms) to processing facilities are believed to increase both *Campylobacter* spp. and *Salmonella* spp. contamination of birds entering the processing facilities (Mulder, 1995; Hiett *et al.*, 2002; Slader *et al.*, 2002; Hansson *et al.*, 2005; Allen *et al.*, 2008; Burkholder *et al.*, 2008). Notably, transport to the slaughterhouse in contaminated transport coops is a great concern as it has been reported that *Salmonella* spp. can be introduced into *Salmonella*-free flocks during this practice (Heyndrickx *et al.*, 2002; Slader *et al.*, 2002; Cooksey, 2005; Marin *et al.*, 2009). Additionally, Stern *et al.* (1995) observed that, with respect to *Campylobacter* spp., the mean total counts from birds prior to transport to processed carcasses increased from 2.71 to 5.15 log cfu, respectively.

Post-harvest/processing/ready-to-eat

Once *Campylobacter* spp. and *Salmonella* spp. contaminated poultry enters the processing plant, cross-contamination to negative flocks is a concern (Hiett *et al.*, 2002; Whyte *et al.*, 2002). Processing areas implicated in contamination include the neck-cutting knife blade, scald tank, defeathering, and the immersion chill tank (Oosterom *et al.*, 1983; Carramiñana *et al.*, 1997; Geornaras *et al.*, 1997; Sarlin *et al.*, 1998). Inadequate input of fresh water and an increased level of organic material (fecal contamination) in the scald tank and during immersion chilling are cited as significant contributors to cross-contamination during processing (Byrd *et al.*, 1999).

An increasing demand from consumers is the subsequent preparation of processed poultry meat into RTE products. Concomitant with this post-processing/ preparation associated environment is the emergence of a niche for subsequent contamination with the zoonotic pathogens *L. monocytogens* and *C. perfringens* (Bell and Kyriakides, 2005). The psychrotrophic nature of *L. monocytogenes* allows the organism to proliferate in foods and persist in the processing plant environment, thus making it a difficult pathogen to control (Swaminathan, 2001). With respect to *C. perfringens*, the inefficient cooling of large batches of cooked food allows rapid germination or the organism that results in the outgrowth of other organisms (Juneja *et al.*, 1994, 1999). Additionally, temperature abuse of prepared foods is a significant risk factor for *C. perfringens* contamination (Angulo *et al.*, 1998). The three sources considered most critical for RTE product contamination are the contamination status of food entering the area, environmental sources (such as counters, tools, and utensils), and the food handlers themselves (USDHHS-FDA-CFSAN, 1999).

20.6.2 Table eggs

While a variety of zoonotic pathogens can be recovered from the shells of table eggs, the overwhelming majority of disease outbreaks associated with table eggs can be attributed to *S. enterica* serovar Enteritidis (SE) (Angulo and Swerdlow, 1999; van de Giessen *et al.*, 1999; Wall and Ward, 1999). The organism is most likely introduced to laying hens through oral ingestion, whereupon the organism spreads and colonizes the crop and ceca (Turnbull and Snoeyenbos, 1974). The subsequent invasion of mucosal epithelial cells leads to a systemic dissemination to a variety of internal organs, of particular interest, the ovary and oviduct (Gast and Bead, 1990; Humphrey *et al.*, 1993). It is through these sites that SE is thought to enter egg contents (Okamura *et al.*, 2001; De Buck *et al.*, 2004). An additional route of contamination, pseudovertical transmission, occurs via fecal contamination on the egg shell. The observation that SE is introduced to laying hens through oral ingestion points to the environment as the critical source for the pathogen (van de Giessen *et al.*, 1994). Within the environment, insects (beetles and flies in particular) and rodents are considered significant sources (Henzler and Opits, 1992; Gray *et al.*, 1999; Schlosser *et al.*, 1999; Olsen and Hammack, 2000). Additional significant sources include the line of chickens used, management practices such as feed deprivation to induce molting (thus extending the productivity of the flock), and the particular pathogenic fitness of the SE itself (Holt and Porter, 1992; Beaumont *et al.*, 1994; Holt *et al.*, 1995; Protais *et al.*, 1996; Guard-Petter, 1998, 2001).

20.7 Pathogen monitoring strategies

The use of risk assessment for application to food safety is a relatively recent occurrence that originated from the creation of the World Trade Organization

(WTO) and the Sanitary and Phyto-Sanitary agreement (Wooldridge *et al.*, 1996). The Codex Alimentarius Commission (CAC) is a body, established in 1963 by the Food and Agricultural Organization of the United Nations (FAO) and the WHO. Guidelines set forth by the CAC are recognized by the WTO as international reference points for the resolution of disputes concerning food safety and consumer protection (Lammerding and Fazil, 2000). CAC guidelines are based on four components: hazard identification, exposure assessment, hazard characterization, and risk characterization. It is upon this framework that the Hazard Analysis Critical Control Point (HACCP) system was based. The HACCP concept was first developed by the United States National Aeronautical and Space Administration (NASA) in conjunction with the Pillsbury Corporation to ensure the safety of food for astronauts while in space. The HACCP system is a systematic approach for the identification, evaluation, and control of food safety hazards; it is based on the development and application of the following seven principles: (1) conduct a hazard analysis; (2) determine the critical control points (CCPs); (3) establish critical limits; (4) establish monitoring procedures; (5) establish corrective actions; (6) establish verification procedures; and (7) establish record-keeping and documentation procedures (http://www.fda.gov/Food/FoodSafety/HazardAnalysisCriticalControlPointsHACCP/default.htm).

Since FSIS mandated HACCP in 1997, many of the nation's poultry processing facilities implemented some or all of the HACCP principles into their operations. Initially, after adoption, the prevalence of *Salmonella* spp. on processed poultry was reduced to less than 15% (FSIS, 1999). However, the long-term results of the implementation of the HACCP system have been mixed. Overall, there has been a decrease in *Salmonella* spp. infections in humans. However, when the clinical cases were analyzed by serotype, serotype Typhimurium decreased significantly, while serotype Enteritidis increased (Vugia *et al.*, 2006). This observation was supported by a survey of *Salmonella* spp. contamination in broiler chickens conducted from 2000 to 2005 (Altekruse *et al.*, 2006). The authors reported an increase in positive tests for the Enteritidis serotype in broiler chickens (Altekruse *et al.*, 2006; Chui *et al.*, 2009). The authors of the study noted that the purpose of the FSIS HACCP *Salmonella* program is to assess performance of individual establishments rather than to estimate the national prevalence of poultry contamination (Altekruse *et al.*, 2006).

Currently, the United States Department of Agriculture (USDA) is pursuing a 'farm to fork' approach to food safety by taking steps to improve the safety of meat and poultry at each stage in the food production, processing, distribution, and marketing chain. Additionally, other pathogens, such as *Campylobacter* spp., are to be included. It is expected that difficulties will arise with the application of these principles to all zoonotic bacterial pathogens associated with poultry.

20.8 Improving pathogen control

As mentioned previously, the intensive nature of poultry meat and table egg production coupled with the integrated systems makes combating zoonotic bacterial pathogens difficult. Add to this the consumer demand for high-quality products for the lowest price, and the delivery of a wholly safe product to the consumer becomes even more problematical. The elimination of all food-borne pathogens is impossible (Issacson *et al.*, 2005). It is likely that combining efforts to reduce pathogens during production coupled with efforts at reduction during processing will ultimately minimize the number and frequency of pathogens in food thereby reducing the risk of consumer exposure. The additional strategy of consumer education should also contribute to the reduction of human illness. An immense number of strategies aimed at reducing or eliminating pathogens from poultry meat and table eggs have been investigated. These strategies are being implemented at all stages of production, processing, and post-processing levels (Lin, 2009). Because of the large number of interventions, a comprehensive description is impractical thus a list of several of the most described interventions is presented in Table 20.1.

While many of these strategies were demonstrated effective in the laboratory or when applied in limited field trials, implementation in extensive trials or commercial operations has proven problematic. It is absolutely clear that effective reduction under commercial conditions requires further research, especially during production. Additional data on horizontal transmission during commercial poultry production are essential to increase our knowledge of critical sources for contamination. This knowledge becomes especially important with the proposed application of the HACCP system for *Campylobacter* spp in poultry.

Again, the role of the consumer cannot be ignored in reducing the burden of food-borne illness associated with poultry. Management and implementation of intervention strategies during production and processing increase costs, which are likely passed on to the consumer. Proper food handling, the avoidance of cross-contamination, as well as proper cooking and storage by the consumer should also be considered in the contribution to the reduction and possible elimination of risk. Lastly, the consumer needs more effective education when it comes to 'alternative' interventions such as addition of genetically modified organisms and irradiation. Public resistance remains a significant barrier regarding the safety of irradiation. A recent study suggested that 33% of consumers still will not purchase irradiated foods, especially since critics argue that irradiated food is not natural, destroys vitamins, and is applied to 'dirty' food (Nayga *et al.*, 2005; Brewer and Rojas, 2008).

20.9 Antimicrobial resistance

The use of antimicrobial agents is essential in agricultural production to maintain both animal health and welfare. However, the intensive nature of agricultural production coupled with consumer demand for inexpensive commodities has pressed producers into using these drugs to promote animal growth rather than just treat illnesses (therapeutic) (Castanon, 2007). The growing observation of increased resistance to several antimicrobial agents is rapidly becoming a global

Table 20.1 Intervention/Mitigation strategies

Location	Intervention	Reference
Preharvest/Production		
Hatchery		
	environmental biosecurity	
	air	
	ventilation, filtration, UV treatment, biocidal agents	Dowd and Maier, 1999
	water	
	filtration, sedimentation, chemical disinfection, waste water disposal, flood/runoff control	Dowd *et al.*, 2004; Rangarajan *et al.*, 2000; Redman *et al.*, 2001
	insect control, pest control	
	physical structure hygiene	
	vaccines	Okamura *et al.*, 2007
	egg disinfection	Mitchell *et al.* 2002, Davies and Breslin, 2003a; Davies and Breslin, 2003b; Rodriguez-Romo and Yousef, 2005; Cox and Pavic, 2007
	UV light	
	ozone	
	chemicals	
	electrostatic chargings	
	pulsed light	
	gas plasma	
	refrigeration	USDA, 1998; USFDA, 2004
	pasteurization	USFDA ,2004
Production		
	environmental biosecurity	Shreeve *et al.*, 2002; Havelaar *et al.*, 2007
	domestic and wild animal control	Kwan *et al.*, 2008; Colles *et al.*, 2009
	all in – all out strategy	Hald *et al.*, 2001
	disinfectant foot dips	Anonymous 2005,
	competitive exclusion (addition of non-pathogenic bacterial culture originally derived from animal of interest)	Snoeyenbos, *et al.*, 1978; Nurmi *et al.*, 1992; Schoeni and Wong, 1994; Wagner 2006; Cox and Pavic, 2009

	probiotics generally defined microorganisms not necessarily of animal origin often lactic acid bacterian or yeasts)	Ziprin and Deloach, 1993; Schrezenmeir and de Vrese, 2001; Ehrmann *et al.*, 2002; Mountzouris *et al.*, 2009 Schrezenmeir and De Vrese, 2001; Biggs and Parsons, 2007; Geier *et al.*, 2009
	prebiotics (organic compounds unavailable to, or undigestible by, the host animal)	Wyszyńska *et al.*, 2004; de Zoete *et al*, 2007; Cox and Pavic, 2009
	vaccines	Goode *et al.*, 2003; Toro *et al.*, 2005; Atterbury *et al.*, 2007;
	bacteriophage	Sillankorya *et al.*, 2009
	phage lytic enzymes	Seal, personal communication 2010
	antimicrobial proteins (bacteriocins)	Joerger 2003; Stern *et al.*, 2006; Cole *et al.*, 2006;
	litter treatment	Cox and Pavic, 2009
	feed treatment	Van Immerseel *et al*, 2006; Cox and Pavic, 2009
Processing		
	transportation crate disinfection	McCrea and Macklin, 2006; Peyrat *et al.*, 2008
	scalding	Avens *et al.*, 2002; Cason and Hinton, 2006; Cason *et al.*, 2007
	immersion Chilling	
	air chilling	Huezo *et al.*, 2007; Cox and Pavic, 2009
	Chemical disinfectants (chlorine, chlorine dioxide, acidified sodium chlorite, peroxyacetic acid or trisodium phosphate	Hansen and Larson, 2007; James *et al.*, 2007 Kim and Day, 2006; Hugas and Tsigarida, 2008 Stern *et al.*, 2003; Sanberg *et al.*, 2005; Wagenaar *et al.*, 2006
	freezing	
	irradiation (gamma)	Tauxe 2001; Hugas and Tsigarida, 2008
Packing and RTE		
	active packing (modified atmosphere packing, antimicrobial packing)	Angelidis and Koutsoumanis, 2006; Audenaert *et al.*, 2010
	bacteriophage	Angelidis and Koutsoumanis, 2006
	high hydrostatic pressure (HPP)	Akhtar *et al.*, 2009
	high electric field pulse (PEF)	Dunn, 1996

concern. The treatment of food-producing animals with antimicrobial agents, also important in human medicine, presents a public health risk. The transfer of resistant zoonotic pathogens or resistance genes from animals to humans via consumption of contaminated food is a great concern. Additionally, the development of cross-resistance, such as the case with fluoroquinolones, could ultimately compromise the effectiveness of all fluoroquinolone drugs (Tollefson *et al.*, 1999). The public health impact of resistance includes treatment failures and increased health care costs; newer and more expensive antibiotics will be needed to treat infections (American Society for Microbiology Public and Scientific Affairs Board. Report of the ASM Task Force on Antibiotic Resistance, Washington, D.C., March 16, 1995). On January 1, 2006, the European Union banned the feeding of all antibiotics and related drugs to livestock, including poultry, for growth promotion purposes. With respect to the United States, a bill to ban the use of seven classes of medically important antibiotics in livestock and poultry, 'The Preservation of Antibiotics for Medical Treatment Act (PAMTA)', was introduced on March 17, 2009. It is expected that debate over this bill will be contentious at best.

Currently, the United States has established several public and private surveillance systems to track the emergence of antibiotic resistance. Public agencies include the National Antimicrobial Resistance Monitoring System (NARMS) and the Collaboration on Animal Health and Food Safety Epidemiology (CAHFSE). NARMS is a multi-agency program that monitors retail meats for resistance (FDA), resistant food-borne pathogens in humans (CDC), and resistant bacteria in animals on farms and animal products in slaughter and processing facilities (USDA). CAHFSE is a program conducted by USDA that collects information on disease status, associated antibiotic use, and resistance of bacteria in farm animals. While these surveillance systems have been invaluable in providing data on the trends and emergence of antibiotic resistances, it is imperative that alternative intervention strategies for the reduction of zoonotic pathogens be developed.

20.10 Future trends

To date, an enormous amount of research on the reduction of zoonotic bacteria in and on poultry and egg products has been completed; however, pathogens such as *Campylobacter* spp. and *Salmonella* spp. continue to remain a public health concern. Furthermore, the rise of antibiotic resistance in these pathogens, and possible transfer to humans, adds to these concerns. Several intervention strategies, applied during the poultry processing stage, have proven effective in reducing the levels of incoming pathogens. However, a greater concentration on the reduction of pathogens during production is needed. There exists a gap in knowledge regarding the basic mechanisms employed by these pathogens for environmental survival, colonization of the bird, and horizontal transfer. As new scientific technologies, such as pyroseqencing, proteomics, nucleic acid microarrays, and phenotype microarrays, are developed, they should be employed to assist in gaining knowledge of these basic processes. There is also an additional need to develop better detection and recovery methods to assist in future, comprehensive epidemiologic investigations.

It is expected that the simultaneous application of several intervention strategies during production, processing, and at the consumer and retail level will result in the reduction of these pathogens and thus reduce the risk of consumer illness.

20.11 Sources of further information and advice

There are several public and private agencies that provide scientific and policy updates on zoonotic pathogens in agriculture. These organizations include the Food and Agriculture Organization of the United Nations (FAO), the WHO, the European Food Safety Authority, the US Food and Drug Administration (US-FDA), the United States Department of Agriculture (USDA), and the CDC. Additionally, there are several texts on zoonotic pathogens, Foodborne Pathogens: Andrews and Bäumler (2005) and Juneja and Sofos (2010). An excellent text that specifically focuses on the poultry industry is given in Curtis (2005) and Kelly (2005).

20.12 References

AKHTAR S, SAREDES-SABJA D, TORRES J A and SARKER M R (2009), 'Strategy to inactivate *Clostridium perfringens* spores in meat products', *Food Microbiology*, **26**(3), 272–7.

ALEKSIC S, HEINZERLING F and BOCKEMUHL J (1996), 'Human infection caused by Salmonellae of subspecies II to VI in Germany, 1977–1992', *Zentralblatt fuer Bakteriologie*, **283**(3), 391–8.

ALLEN V M, BURTON C H, WILKINSON D J, WHYTE R T, HARRIS J A, HOWELL M, ET AL. (2008), 'Evaluation of the performance of different cleaning treatments in reducing microbial contamination of poultry transport crates', *British Poultry Science*, **49**(3), 233–40.

ALLOS B M (1998), '*Campylobacter jejuni* infection as a cause of the Guillain–Barré syndrome', *Infectious Disease Clinics of North America*, **12**(1), 173–84.

ALLOS B M (2001), '*Campylobacter jejuni* infections: Update on emerging issues and trends', *Clinical Infectious Diseases*, **32**(8), 1201–6.

ALLOS B M and BLASER M J (1995), '*Campylobacter jejuni* and the expanding spectrum of related infections', *Clinical Infectious Diseases*, **20**(5), 1092–9.

ALTEKRUSE S F, BAUER N, CHANLONGBUTRA A, DESAGUN R, NAUGLE A, SCHLOSSER W, UMHOLTZ R and WHITE P (2006), '*Salmonella enteritidis* in broiler chickens, United States, 2000–2005', *Emerging Infectious Diseases*, **12**(12), 1848–52.

ALTEKRUSE S F, HUNT J M, TOLLEFSON L K and MADDEN J M (1994), 'Food and animal sources of human *Campylobacter jejuni* infection', *Journal of the American Veterinary Medical Association*, **204**(1), 57–61.

ANDREWS H L and BÄUMLER A J (2005), '*Salmonella* species', in Fratamico, P., Bhunia, A. and Smith J.L. (Eds): *Foodborne Pathogens: Microbiology and Molecular Biology*, 1st ed., Caister Academic Press, Norfolk, UK, pp. 327–39.

ANGELIDIS A S and KOUTSOUMANIS K (2006), 'Prevalence and concentration of *Listeriamonocytogenes* in sliced ready-to-eat meat products in the Hellenic retail market', *Journal of Food Protection*, **69**(4), 938–42.

ANGULO F J and SWERDLOW D L (1999), 'Epidemiology of human *Salmonella enterica* serovar Enteritidis infections in the United States', in Saeed, A.M., Gast, R.K., Potter, M.E. and Wall, P.G., *Salmonella Enteric Serovar Enteritidis in Humans and Animals*, Iowa State University Press, Ames, Iowa, USA, pp. 33–42.

ANGULO F L, VOETSCH A C, VUGIA D, HADLER J L, FARLEY M, ET AL. (1998), 'Determining the burden of human illness from foodborne diseases – CDC's Emerging Infectious Disease Program Foodborne Diseases Active Surveillance Network (FoodNet)', *Veterinary Clinics of North America Food Animal Practice*, **14**, 165–72.

ANONYMOUS (2005), Advisory Committee on the Microbiological Safety of Food (ACMSF) Second report on *Campylobacter*, Her Majesty's Stationery Office (HMSO), London.

ANONYMOUS (2007), 'The community summary report on trends and sources of zoonoses, zoonotic agents, antimicrobial resistance and foodborne outbreaks in the European Union in 2006', *European Food Safety Authority Journal*, **130**, 118–45.

AOAC (1990), *Official Methods of Analysis of the AOAC (Association of Official Analytical Chemists). Vols. I and II*, Association of Official Analytical Chemists, Arlington, Virginia, USA.

APHA (1985), *Standard Methods for the Examination of Dairy Products*, American Public Health Association, Washington, DC, USA.

ARMSTRONG R W and FUNG P C (1993), 'Brainstem encephalitis (Rhombencephalitis) due to *Listeria monocytogenes*: Case report and review', *Clinical Infectious Diseases*, **16**(5), 689–702.

ASPINALL S T, WAREING D R, HAYWARD P G and HUTCHINSON D N (1996), 'A comparison of a new *Campylobacter* selective medium (CAT) with membrane filtration for the isolation of thermophilic Campylobacters including *Campylobacter upsaliensis*', *Journal of Applied Bacteriology*, **80**(6), 645–50.

ATTERBURY R J, VAN BERGEN M A, ORTIZ F, LOVELL M A, HARRIS J A, DE BOER A, ET AL. (2007), 'Bacteriophage therapy to reduce *Salmonella* colonization of broiler chickens', *Applied and Environmental Microbiology*, **73**(14), 4543–9.

AUDENAERT K, D'HAENE K, MESSENS K, RUYSSEN T, VANDAMME P and HUYS G (2010), 'Diversity of lactic acid bacteria from modified atmosphere packaged sliced cooked meat products at sell-by date assessed by PCR-denaturing gradient gel electrophoresis', *Food Microbiology*, **27**(1), 12–18.

AVENS J S, ALBRIGHT S N, MORTON A S, PREWITT B E, KENDALL P A and SOFOS J N (2002), 'Destruction of microorganisms on chicken carcasses by steam and boiling water immersion', *Food Control*, **13**, 445–50.

BAILEY J S (1988), 'Integrated colonization control of *Salmonella* in poultry', *Poultry Science*, **67**(6), 928–32.

BAILEY J S, STERN N J, FEDORKA-CRAY P, CRAVEN S E and COX N A (1999), 'A multi-state epidemiological investigation of sources and movement of *Salmonella* through integrated poultry operations', *Proceedings of the 103rd Annual Meeting of the United States Animal Health Association, San Diego, California, USA*, October 7–14: 471–81.

BEAUMONT C, PROTAIS J, COLIN P, GUILLOT J F, BELLATIF F, MOULINE C, ET AL. (1994), 'Comparison of resistance of different poultry lines to intramuscular or oral inoculation by *Salmonella enteritidis*', *Veterinary Research*, **25**, 412.

BELL C and KYRIAKIDES A (2005), *Listeria: A practical approach to the organism and its control in foods*, Blackwell Publishing, Oxford.

BERCHIERI A, JR, WIGLEY P, PAGE K, MURPHY C K and BARROW P A (2001), 'Further studies on vertical transmission and persistence of *Salmonella enterica* serovar Enteritidis phage type 4 in chickens', *Avian Pathology*, **30**(4), 297–310.

BERG D E, AKOPYONTS N S and KERSULYTE D (1994), 'Fingerprinting microbial genomes using the RAPD and AP-PCR method', *Methods in Molecular and Cell Biology*, **5**, 13–24.

BIGGS P and PARSONS C M (2007), 'The effects of several oligosaccharides on true amino acid digestibility and true metabolizable energy in cecectomized and conventional roosters', *Poultry Science*, **86**(6), 1161–5.

BLASER M J and RELLER L B (1981), '*Campylobacter* enteritis', *New England Journal of Medicine*, **305**(24), 1444–52.

BLATTNER F R, PLUNKETT G, 3RD, BLOCH C A, PERNA N T, BURLAND V, RILEY M, ET AL. (1997), 'The complete genome sequence of *Escherichia coli* K-12', *Science*, **277**(5331), 1453–62.

BLEARS M J, DE GRANDIS S A, LEE H and TREVORS J T (1998), 'Amplified fragment length polymorphism (AFLP): A review of the procedure and its applications', *Journal of Industrial Microbiology and Biotechnology*, **21**(3), 99–114.

BOKKENHEUSER V D and SUTTER V L (1981), '*Campylobacter* infections', in Balows, A. and Hausler, W.J., *Bacterial, Mycotic and Parasitic Infections*, 6th ed., American Public Health Association, Washington, D.C., USA, pp. 301–10.

BOLTON F J, COATES D, HINCHLIFFE P M and ROBERTSON L (1983), 'Comparison of selective media for isolation of *Campylobacter jejuni/coli*', *Journal of Clinical Pathology*, **36**(1), 78–83.

BOURKE B, CHAN V L and SHERMAN P (1998), '*Campylobacter upsaliensis*: Waiting in the wings', *Clinical Microbiology Reviews*, **11**(3), 440–9.

BOURKE B, SHERMAN P M, WOODWARD D, LIOR H and CHAN V L (1996), 'Pulsed-field gel electrophoresis indicates genotypic heterogeneity among *Campylobacter upsaliensis* strains', *FEMS Microbiology Letters*, **143**, 57–61.

BRENNER F W, VILLAR R G, ANGULO F J, TAUXE R and SWAMINATHAN B (2000), '*Salmonella* nomenclature', *Journal of Clinical Microbiology*, **38**(7), 2465–7.

BREWER M S and ROJAS M (2008), 'Consumer attitudes toward issues in food safety', *Journal of Food Safety*, **28**, 1–22.

BRUCE J (1996), 'Automated system rapidly identifies and characterizes microorganisms in food', *Food Technology*, **50**, 77–81.

BRYAN F L and DOYLE M P (1995), 'Health risks and consequences of *Salmonella* and *Campylobacter jejuni* in raw poultry', *Journal of Food Protection*, **58**, 326–44.

BRYAN F L and MCKINLEY T W (1974), 'Prevention of foodborne illness by time-temperature control of thawing, cooking, chilling, and reheating turkeys in school lunch kitchens', *Journal of Milk and Food Technology*, **37**, 420–9.

BRYNESTAD S and GRANUM P E (2002), '*Clostridium perfringens* and foodborne infections', *International Journal of Food Microbiology*, **74**(3), 195–202.

BURKHOLDER K M, THOMPSON K L, EINSTEIN M E, APPLEGATE T J and PATTERSON J A (2008), 'Influence of stressors on normal intestinal microbiota, intestinal morphology, and susceptibility to *Salmonella* Enteritidis colonization in broilers', *Poultry Science*, **87**(9), 1734–41.

BUTZLER J P and OOSTEROM J (1991), '*Campylobacter*: Pathogenicity and significance in foods', *International Journal of Food Microbiology*, **12**(1), 1–8.

BUTZLER J P and SKIRROW M B (1979), '*Campylobacter* enteritis', *Clinical Gastroenterology*, **8**(3), 737–65.

BUZBY J C and ROBERTS T (1997), 'Economic costs and trade impacts of microbial foodborne illness', *World Health Statistics Quarterly*, **50**(1–2), 57–66.

BYRD J A, DELOACH J R, CORRIER D E, NISBET D J and STANKER L H (1999), 'Evaluation of *Salmonella* serotype distributions from commercial broiler hatcheries and grower houses', *Avian Diseases*, **43**(1), 39–47.

BYRNE C, DOHERTY D, MOONEY A, BYRNE M, WOODWARD D, JOHNSON W, ET AL. (2001), 'Basis of the superiority of cefoperazone amphotericin teicoplanin for isolating *Campylobacter upsaliensis* from stools', *Journal of Clinical Microbiology*, **39**(7), 2713–16.

CALDWELL M B, GUERRY P, LEE E C, BURANS J P and WALKER R I (1985), 'Reversible expression of flagella in *Campylobacter jejuni*', *Infection and Immunity*, **50**(3), 941–3.

CARRAMIÑANA J J, YANGÜELA J, BLANCO D, ROTA C, AGUSTIN A I and HERRERA A (1997), '*Salmonella* incidence and distribution of serotypes throughout processing in a Spanish poultry slaughterhouse', *Journal of Food Protection*, **60**, 1312–17.

CASON J A and HINTON A, JR. (2006), 'Coliforms, *Escherichiacoli, Campylobacter*, and *Salmonella* in a counterflow poultry scalder with a dip tank', *International Journal of Poultry Science*, **5**, 846–9.

CASON J A, BUHR R J, RICHARDSON L J and COX N A (2007), 'Internal and external carriage of inoculated *Salmonella* in broiler chickens', *International Journal of Poultry Science*, **6**, 952–4.

CASTANON J I R (2007), 'History of the use of antibiotics as growth promoters in European poultry feeds', *Poultry Science*, **86**, 2466–72.

CEELEN L M, DECOSTERE A, CHIERS, K, DUCATELLE, R, MAES, D and HAESEBROUCK, F (2007), 'Pathogenesis of *Helicobacter pullorum* infections in broilers', *International Journal of Food Microbiology*, **116**(2), 207–13.

CFSAN (2003), 'Interpretive summary: Quantitative assessment of the relative risk to public health from foodborne *Listeria monocytogenes* among selected categories of ready-to-eat foods', available from: http://www.foodsafety.gov/~dms/lmr2-su.html

CHOWDHURY M N (1984), '*Campylobacter jejuni* enteritis; a review', *Tropical and Geographical Medicine*, **36**(3), 215–22.

CHU G, VOLLRATH D and DAVIS R W (1986), 'Separation of large DNA molecules by contour-clamped homogeneous electric fields', *Science*, **234**(4783), 1582–5.

CHUI K K, WEBB P, RUSSELL R M and NAUMOVA E N (2009), 'Geographic variations and temporal trends of *Salmonella*-associated hospitalization in the U.S. Elderly, 1991–2004: A time series analysis of the impact of HACCP regulation', *BMC Public Health*, **9**, 447.

COLE K, FARNELL M B, DONOGHUE A M, STERN N J, SVETOCH E A, ERUSLANOV B N, ET AL. (2006), 'Bacteriocins reduce *Campylobacter* colonization and alter gut morphology in turkey poults', *Poultry Science*, **85**(9), 1570–5.

COLLES F M, MCCARTHY N D, HOWE J C, DEVEREUX C L, GOSLER A G and MAIDEN M C (2009), 'Dynamics of *Campylobacter* colonization of a natural host, Sturnus vulgaris (European starling)', *Environmental Microbiology*, **11**(1), 258–67.

COOKSEY K (2005), 'Effectiveness of antimicrobial food packaging materials', *Food Additives and Contaminants*, **22**(10), 980–87.

CORNILLOT E, SAINT-JOANIS B, DAUBE G, KATAYAMA S, GRANUM P E, CANARD B, ET AL. (1995), 'The enterotoxin gene (*cpe*) of *Clostridium perfringens* can be chromosomal or plasmid-borne', *Molecular Microbiology*, **15**(4), 639–47.

CORRY J E, ALLEN V M, HUDSON W R, BRESLIN M F and DAVIES R H (2002), 'Sources of *Salmonella* on broiler carcasses during transportation and processing: Modes of contamination and methods of control', *Journal of Applied Microbiology*, **92**(3), 424–32.

COUFAL C D, CHAVEZ C, KNAPE K D and CAREY J B (2003), 'Evaluation of a method of ultraviolet light sanitation of broiler hatching eggs', *Poultry Science*, **82**(5), 754–9.

COX N A, BAILEY J S, MAULDIN J M and BLANKENSHIP L C (1990), 'Presence and impact of *Salmonella* contamination in commercial broiler hatcheries', *Poultry Science*, **69**(9), 1606–9.

COX J M and PAVIC A (2010), 'Advances in enteropathogen control in poultry production', *Journal of Applied Microbiology*, **108**, 745–55.

COX N A, RICHARDSON L J, BUHR R J, MUSGROVE M T, BERRANG M E and BRIGHT W (2007), 'Bactericidal effect of several chemicals on hatching eggs inoculated with *Salmonella* serovar Typhimurium', *Journal of Applied Poultry Research*, **16**, 623–7.

CRAVEN S E, STERN N J, BAILEY J S and COX N A (2001), 'Incidence of *Clostridium perfringens* in broiler chickens and their environment during production and processing', *Avian Diseases*, **45**(4), 887–96.

CURTIS P A (2005), 'HACCP in poultry processing', in Mead, G.C., *Food Safety Control in the Poultry Industry*, Cambridge,Woodhead Publishing Ltd, 380–92.

DAVIES R and BRESLIN M (2003a), 'Observations on *Salmonella* contamination of commercial laying farms before and after cleaning and disinfection', *Veterinary Record*, **152**(10), 283–7.

DAVIES R H and BRESLIN M (2003b), 'Investigations into possible alternative decontamination methods for *Salmonella* Enteritidis on the surface of table eggs', *Journal of Veterinary Medicine B Infectious Disease and Veterinary Public Health*, **50**(1), 38–41.

DAVIES R, BRESLIN M, CORRY J E, HUDSON W and ALLEN V M (2001), 'Observations on the distribution and control of *Salmonella* species in two integrated broiler companies', *Veterinary Record*, **149**(8), 227–32.

DE BOER E, TILBURG J J, WOODWARD D L, LIOR H and JOHNSON W M (1996), 'A selective medium for the isolation of *Arcobacter* from meats', *Letters in Applied Microbiology*, **23**(1), 64–6.

DE BUCK J, PASMANS F, VAN IMMERSEEL F, HAESEBROUCK F and DUCATELLE R (2004), 'Tubular glands of the isthmus are the predominant colonization site of *Salmonella* Enteritidis in the upper oviduct of laying hens', *Poultry Science*, **83**(3), 352–8.

DE JONG A E, VERHOEFF-BAKKENES L, NAUTA M J and DE JONGE R (2008), 'Cross-contamination in the kitchen: Effect of hygiene measures', *Journal of Applied Microbiology*, **105**(2), 615–24.

DE ZOETE M R, VAN PUTTEN J P and WAGENAAR J A (2007), 'Vaccination of chickens against *Campylobacter*', *Vaccine*, **25**(30), 5548–57.

DEN BAKKER H C, DIDELOT X, FORTES E D, NIGHTINGALE K K and WIEDMANN M (2008), 'Lineage specific recombination rates and microevolution in *Listeria monocytogenes*', *BMC Evolutionary Biology*, **8**, 277.

DOWD S E and MAIER M W (1999), 'Aeromicrobiology', in Mair, R.M., Pepper, I.L. and Gerba, C.P., *Environmental Microbiology*, Academic Press, San Diego, CA, pp. 91–122.

DOWD S E, THURSTON-ENRIQUEZ J A and BRASHEARS M (2004), 'Environmental Reservoirs and Transmission of Foodborne Pathogens', in Beier, R.C., Pillai, S.D., Phillips, T.D. and Ziprin, R.L. (eds), *Preharvest and Postharvest Food Safety*, Blackwell Publishing, Oxford, UK, pp. 161–200.

DUIM B, WASSENAAR T M, RIGTER A and WAGENAAR J (1999), 'High-resolution genotyping of *Campylobacter* strains isolated from poultry and humans with amplified fragment length polymorphism fingerprinting', *Applied and Environmental Microbiology*, **65**(6), 2369–75.

DUNN J (1996), 'Pulsed light and pulsed electric field for foods and eggs', *Poultry Science*, **75**(9), 1133–6.

DURAND A M, GIESECKE W H, BARNARD M L, VAN DER WALT M L and STEYN H C (1990), '*Salmonella* isolated from feeds and feed ingredients during the period 1982–1988: Animal and public health implications', *Onderstepoort Journal of Veterinary Research*, **57**(3), 175–81.

EDWARDS P R and EWING W H (1972), *Identification of Enterobacteriaceae*, 3rd ed., Burgess Publishing, Minneapolis, MN, USA.

EHRMANN M A, KURZAK P, BAUER J and VOGEL R F (2002), 'Characterization of *Lactobacilli* towards their use as probiotic adjuncts in poultry', *Journal of Applied Microbiology*, **92**(5), 966–75.

EKDAHL K, NORMANN B and ANDERSSON Y (2005), 'Could flies explain the elusive epidemiology of campylobacteriosis?', *BMC Infectious Disease*, **5**(1), 11.

FAO (2007a), 'Core production data', available from: http://faostat.fao.org/site/340/default.aspx [Accessed January 10 2010].

FAO (2007b), 'Prodstat: Livestock (primary and processed)', available from: http://faostat.fao.org/site/569/DesktopDefault.aspx?PageID=569ChickenMeat [Accessed January 10 2010].

FDA (1992), *Bacteriological Analytical Manual*, 6th ed., Association of Official Analytical Chemists, United States Food and Drug Administration, Arlington, VA, USA.

FIPULA D (2007), 'Antibody engineering and modification technologies', *Biomolecular Engineering*, **24**, 201–5.

FOX J G (1997), 'The expanding genus of *Helicobacter*: Pathogenic and zoonotic potential', *Seminars in Gastrointestinal Disease*, **8**, 124–41.

FRENZEN P, RIGGS T, BUZBY J, BREUER T, ROBERTS T, VOETSCH D, REDDY S, ET AL. (1999), '*Salmonella* cost estimate update using Foodnet data', *Food Review*, **22**, 10–15.

FRIEDMAN C R, HOEKSTRA R M, SAMUEL M, MARCUS R, BENDER J, SHIFERAW B, ET AL. (2004), 'Risk factors for sporadic *Campylobacter* infection in the United States: A case-control study in FoodNet sites', *Clinical Infectious Diseases*, **38**(3), S285–S296.

FSIS (1999), Second progress report on *Salmonella* testing for raw meat and poultry products.

GARBER L, SMELTZER M, FEDORKA-CRAY P, LADELY S and FERRIS K (2003), '*Salmonella enterica* serotype Enteritidis in table egg layer house environments and in mice in US Layer houses and associated risk factors', *Avian Diseases*, **47**(1), 134–42.

GAST R K and BEARD C W (1990), 'Isolation of *Salmonella* Enteritidis from internal organs of experimentally infected hens', *Avian Diseases*, **34**(4), 991–3.

GEIER M S, TOROK V A, ALLISON G E, OPHEL-KELLER K and HUGHES R J (2009), 'Indigestible carbohydrates alter the intestinal microbiota but do not influence the performance of broiler chickens', *Journal of Applied Microbiology*, **106**(5), 1540–8.

GENIGEORGIS C A, HASSUNEY M and COLLINS P (1986), '*Campylobacter jejuni* infection on poultry farms and its effect on poultry meat contamination during slaughtering', *Journal of Food Protection*, **49**, 895–903.

GEORNARAS I, DE JESUS E, VAN ZYL E and VON HOLY A (1997), 'Bacterial populations of different sample types from carcasses in the dirty area of a South African poultry abattoir', *Journal of Food Protection*, **60**, 551–4.

GIANNELLA R A (1996), 'Salmonella', in Baron, S. et al., *Medical Microbiology*, 4th ed., University of Texas Medical Branch, Galveston, TX, USA. Available from: http://www.ncbi.nlm.nih.gov/bookshelf/br.fcgi?book—med&part=A1221.

GODSCHALK P C, HEIKEMA A P, GILBERT M, KOMAGAMINE T, ANG C W, GLERUM J, ET AL. (2004), 'The crucial role of *Campylobacter jejuni* genes in anti-ganglioside antibody induction in Guillain–Barré syndrome', *Journal of Clinical Investigations*, **114**(11), 1659–65.

GOLDSCHMIDT M C (2006), 'The use of biosensor and microarray techniques in the rapid detection and identification of Salmonellae', *Journal of Association of Analytical Communities International*, **89**(2), 530–7.

GONZALEZ I, GRANT K A, RICHARDSON P T, PARK S F and COLLINS M D (1997), 'Specific identification of the enteropathogens *Campylobacter jejuni* and *Campylobacter coli* by using a PCR test based on the *ceuE* gene encoding a putative virulence determinant', *Journal of Clinical Microbiology*, **35**(3), 759–63.

GOODE D, ALLEN V M and BARROW P A (2003), 'Reduction of experimental *Salmonella* and *Campylobacter* contamination of chicken skin by application of lytic bacteriophages', *Applied and Environmental Microbiology*, **69**(8), 5032–6.

GOOSSENS H and BUTZLER J-P (1992), 'Isolation and identification of *Campylobacter* spp.', in Nachamkin, I., Blaser, M.J. and Tompkins, L.S., *Campylobacter jejuni: Current Status and Future Trends*, 1st ed., American Society for Microbiology, Washington, D.C., USA, pp. 93–109.

GRAJEWSKI B A J, KUSEK W and GELFAND H M (1985), 'Development of a bacteriophage typing scheme for *Campylobacter jejuni* and *Campylobacter coli*', *Journal of Clinical Microbiology*, **22**, 13–18.

GRAY J P, MADDOX C W, TOBIN P C, GUMMO J D and PITTS C W (1999), 'Reservoir competence of *Carcinopspumilio* for *Salmonella* Enteritidis (Eubacteriales: Enterobacteriaceae)', *Journal of Medical Entomology*, **36**(6), 888–91.

GRAY M J, ZADOKS R N, FORTES E D, DOGAN B, CAI S, CHEN Y, ET AL. (2004), '*Listeria monocytogenes* isolates from foods and humans form distinct but overlapping populations', *Applied and Environmental Microbiology*, **70**(10), 5833–41.

GRIFFITHS P L and PARK R W A (1990), 'Campylobacters associated with human diarrheal disease', *Journal of Applied Bacteriology*, **69**(3), 281–301.

GRIMONT F and GRIMONT P A D (1986), 'Ribosomal ribonucleic acid gene restriction patterns as potential taxonomic tools', *Annales de l'Institut Pasteur Microbiology*, 137B, 165–75.

GUARD-PETTER J (1998), 'Variants of smooth *Salmonella enterica* serovar Enteritidis that grow to higher cell density than the wild type are more virulent', *Applied and Environmental Microbiology*, **64**(6), 2166–72.

GUARD-PETTER J (2001), 'The chicken, the egg and *Salmonella* Enteritidis', *Environmental Microbiology*, **3**(7), 421–30.

HALD B, RATTENBORG E and MADSEN M (2001), 'Role of batch depletion of broiler houses on the occurrence of *Campylobacter* spp. in chicken flocks', *Letters in Applied Microbiology*, **32**(4), 253–6.

HANSEN D and LARSEN B S (2007), 'Reduction of *Campylobacter* on chicken carcasses by sonosteam treatment', *Proceedings of European Congress of Chemical Engineering (ECCE-6): Copenhagen*, September 16–20: 81.

HANSSON I, EDEROTH M, ANDERSSON L, VAGSHOLM I, AND ENGVALL E O (2005), 'Transmission of *Campylobacter* spp. to chickens during transport to slaughter', *Journal of Applied Microbiology*, **99**(5), 1149–57.

HAVELAAR A H, MANGEN M J, DE KOEIJER A A, BOGAARDT M J, EVERS E G, JACOBS-REITSMA W F, ET AL. (2007), 'Effectiveness and efficiency of controlling *Campylobacter* on broiler chicken meat', *Risk Analysis*, **27**, 831–44.

HELMS M, SIMONSEN J, OLSEN K E and MØLBAK K (2005), 'Adverse health events associated with antimicrobial drug resistance in *Campylobacter* species: A registry-based cohort study', *Journal of Infectious Disease*, **191**, 1050–5.

HENSEL A and NEUBAUER H (2002), 'Human pathogens associated with on-farm practices – implications for control and surveillance strategies', in Smulders, F.J.M. and Collins, J.D., *Food Safety Assurance and Veterinary Public Health, vol. 1. Food Safety Assurance in the Pre-harvest Phase 1*, Wageningen Academic Publishers, Wageningen, Netherlands, pp. 125–39.

HENZLER D J and OPITZ H M (1992), 'The role of mice in the epizootiology of *Salmonella* Enteritidis infection on chicken layer farms', *Avian Diseases*, **36**, 625–31.

HEYNDRICKX M, HERMAN L, VLAES L, BUTZLER J P, WILDEMAUWE C, GODARD C, ET AL. (2007), 'Multiple typing for the epidemiological study of the contamination of broilers with *Salmonella* from the hatchery to the slaughterhouse', *Journal of Food Protection*, **70**, 323–34.

HEYNDRICKX M, VANDEKERCHOVE D, HERMAN L, ROLLIER I, GRIJSPEERDT K and ZUTTER L (2002), 'Routes for *Salmonella* contamination of poultry meat: Epidemiological study from hatchery to slaughterhouse', *Epidemiology and Infection*, **129**, 253–65.

HIETT K L, STERN N J, FEDORKA-CRAY P, COX N A, MUSGROVE M T and LADELY S (2002), 'Molecular subtype analyses of *Campylobacter* spp. from Arkansas and California poultry operations', *Applied and Environmental Microbiology*, **68**(12), 6220–36.

HO H T, LIPMAN L J and GAASTRA W (2006), '*Arcobacter*, what is known and unknown about a potential foodborne zoonotic agent!', *Veterinary Microbiology*, **115**(1–3), 1–13.

HOLLAND S, ALFONSO E, GELENDER H, HEIDEMANN D, MENDELSOHN A, ULLMAN S and MILLER D (1987), 'Corneal ulcer due to *Listeria monocytogenes*', *Cornea*, **6**(2), 144–6.

HOLT P S, MACRI N P and PORTER R E, JR. (1995), 'Microbiological analysis of the early *Salmonella* Enteritidis infection in molted and unmolted hens', *Avian Diseases*, **39**(1), 55–63.

HOLT P S, AND PORTER R E, JR. (1992), 'Microbiological and histopathological effects of an induced-molt fasting procedure on a *Salmonellaenteritidis* infection in chickens', *Avian Diseases*, **36**(3), 610–18.

HOOD A M, PEARSON A D and SHAHAMAT M (1988), 'The extent of surface contamination of retailed chickens with *Campylobacter jejuni* serogroups', *Epidemiology and Infection*, **100**(1), 17–25.

HUEZO R, NORTHCUTT J K, SMITH D P, FLETCHER D L and INGRAM K D (2007), 'Effect of dry air or immersion chilling on recovery of bacteria from broiler carcasses', *Journal of Food Protection*, **70**, 1829–34.

HUGAS M and TSIGARIDA E (2008), 'Pros and cons of carcass decontamination: The role of the European Food Safety Authority', *Meat Science*, **78**(1–2), 43–52.

HULTON C S J, HIGGINS C F and SHARP P M (1991), 'Eric sequences – a novel family of repetitive elements in the genomes of *Escherichia coli, Salmonella* Typhimurium and other enterobacteria', *Molecular Microbiology*, **5**(4), 825–34.

HUMPHREY T J, BASKERVILLE A, WHITEHEAD A, ROWE B and HENLEY A (1993), 'Influence of feeding patterns on the artificial infection of laying hens with *Salmonella* Enteritidis phage type 4', *Veterinary Record*, **132**(16), 407–9.

HUMPHREY T, O'BRIEN S and MADSEN M (2007), '*Campylobacters* as zoonotic pathogens: A food production perspective', *International Journal of Food Microbiology*, **117**(3), 237–57.

ISAACSON R E, TORRENCE M and BUCKLEY M R (2005), 'Preharvest food safety and security', *Report from the American Academy of Microbiology*, available from: http://www.asm.org/?option=com_content&view=article&id=33269&Itemid=417 [Accessed on January **11**, 2010].

JAMES C, JAMES S J, HANNAY N, PURNELL G, BARBEDO-PINTO C, YAMAN H, ET AL. (2007), 'Decontamination of poultry carcasses using steam or hot water in combination with rapid cooling, chilling or freezing of carcass surfaces', *International Journal of Food Microbiology*, **114**(2), 195–203.

JAY J M (1992), *Modern food microbiology*, Van Nostrand Reinhold, New York, USA.

JEFFERS G T, BRUCE J L, MCDONOUGH P L, SCARLETT J, BOOR K J and WIEDMANN M (2001), 'Comparative genetic characterization of *Listeria monocytogenes* isolates from human and animal listeriosis cases', *Microbiology*, **147**, 1095–104.

JOERGER R D (2003), 'Alternatives to antibiotics: Bacteriocins, antimicrobial peptides and bacteriophages', *Poultry Science*, **82**(4), 640–7.

JUNEJA V K, SNYDER O P and CYGNAROWICZ-PROVOST M (1994), 'Influence of cooling rate on outgrowth of *Clostridium perfringens* spores and cooked ground beef', *Journal of Food Protection*, **57**, 1063–7.

JUNEJA V K, WHITING R C, MARKS H M and SNYDER O P (1999), 'Predictive model for growth of *Clostridium perfringens* at temperatures applicable to cooling of cooked meat', *Food Microbiology*, **16**(4), 335–49.

JUNEJA V K and SOFOS J N (2010), 'Preface', in Juneja, V.K. and Sofos, J.N., *Pathogens and Toxins in Foods*, ASM Press,Washington, DC, USA, pp. xi–xii.

KABEYA H, MARUYAMA S, MORITA Y, OHSUGA T, OZAWA S, KOBAYASHI Y, ET AL. (2004), 'Prevalence of *Arcobacter* species in retail meats and antimicrobial susceptibility of the isolates in Japan', *International Journal of Food Microbiology*, **90**(3), 303–8.

KAIJSER B (1988), '*Campylobacterjejuni/coli*', *AOMIS*, **96**(4), 283–8.

KASLOW R A, SULLIVAN-BOLYAI J Z, HAFKIN B, SCHONBERGER L B, KRAUS L, MOORE M J, ET AL. (1984), 'HLA antigens in Guillain–Barré Syndrome', *Neurology*, **34**(2), 240–2.

KATHARIOU S (2002), '*Listeriamonocytogenes* virulence and pathogenicity, a food safety perspective', *Journal of Food Protection*, **65**(11), 1811–29.

KAZWALA R R, COLLINS J D, HANNAN J, CRINION R A P and OMAHONY H (1990), 'Factors responsible for the introduction and spread of *Campylobacter jejuni* infection in commercial poultry production', *Veterinary Record*, **126**(13), 305–6.

KEAT A (1983), 'Reiter's Syndrome and reactive arthritis in perspective', *New England Journal of Medicine*, **309**(26), 1606–15.

KELLY L (2005), 'Microbial risk assessment in poultry production and processing', in Mead, G.C., *Food Safety Control in the Poultry Industry*, Woodhead Publishing Ltd, Cambridge, UK, pp. 255–72.

KETLEY J M (1995), 'Virulence of *Campylobacter* species – a molecular-genetic approach', *Journal of Medical Microbiology*, **42**(5), 312–27.

KETLEY J M (1997), 'Pathogenesis of enteric infection by *Campylobacter*', *Microbiology*, **143**, 5–21.

KIM D and DAY D F (2006), 'A biocidal combination capable of sanitizing raw chicken skin', *Food Control*, **18**, 1272–6.

KWAN P S, BARRIGAS M, BOLTON F J, FRENCH N P, GOWLAND P, KEMP R, ET AL. (2008), 'Molecular epidemiology of *Campylobacter jejuni* populations in dairy cattle, wildlife, and the environment in a farmland area', *Applied and Environmental Microbiology*, **74**(16), 5130–8.

LAMMERDING A M and FAZIL A (2000), 'Hazard identification and exposure assessment for microbial food safety risk assessment', *International Journal of Food Microbiology*, **58**(3), 147–57.

LASTOVICA A, LE ROUX E, WARREN R and KLUMP H (1993), 'Clinical isolates of *Campylobacter mucosalis*', *Journal of Clinical Microbiology*, **31**(10), 2835–6.

LEVIN B R, LIPSITCH M and BONHOEFFER S (1999), 'Evolution and disease – population biology, evolution, and infectious disease: Convergence and synthesis', *Science*, **283**(5403), 806–9.

LIN J (2009), 'Novel approaches for *Campylobacter* control in poultry', *Foodborne Pathogens and Disease*, **6**(7), 755–65.

LINDBLOM G B, SJOGREN E and KAIJSER B (1986), 'Natural *Campylobacter* colonization in chickens raised under different environmental conditions', *Journal of Hygiene (London)*, **96**(3), 385–91.

LIPMAN L, HO H and GAASTRA W (2008), 'The presence of *Arcobacter* species in breeding hens and eggs from these hens', *Poultry Science*, **87**(11), 2404–7.

LIPUMA J J (1998), 'Molecular tools for epidemiologic study of infectious diseases', *Pediatric Infectious Disease Journal*, **17**(8), 667–75.

LUBER P (2009), 'Cross-contamination versus undercooking of poultry meat or eggs – which risks need to be managed first?', *International Journal of Food Microbiology*, **134**(1–2), 21–8.

LUPSKI J R and WEINSTOCK G M (1992), 'Short, interspersed repetitive DNA sequences in prokaryotic genomes', *Journal of Bacteriology*, **174**(14), 4525–9.

MAIDEN M C J, BYGRAVES J A, FEIL E, MORELLI G, RUSSELL J E, URWIN R, ET AL. (1998), 'Multilocus sequence typing: A portable approach to the identification of clones within populations of pathogenic microorganisms', *Proceedings of the National Academy of Sciences of the United States of America*, **95**(6), 3140–5.

MARIN C, HERNANDIZ A and LAINEZ M (2009), 'Biofilm development capacity of *Salmonella* strains isolated in poultry risk factors and their resistance against disinfectants', *Poultry Science*, **88**(2), 424–31.

MARTIN B O, HUMBERT M, CAMARA E, GUENZI J, WALKER T, MITCHELL P, ET AL. (1992), 'A highly conserved repeated DNA element located in the chromosome of *Streptococcus pneumoniae*', *Nucleic Acids Research*, **20**, 3479–83.

MASLOW J N, MULLIGAN M E and ARBEIT R D (1993), 'Molecular epidemiology – application of contemporary techniques to the typing of microorganisms', *Clinical Infectious Diseases*, **17**(2), 153–64.

MBATA T I (2005), 'Poultry meat pathogens and its control', *Internet Journal of Food Safety*, V, 20–8.

MCCREA B A and MACKLIN K S (2006), 'Effect of different cleaning regimens on recovery of *Clostridiumperfringens* on poultry live haul containers', *Poultry Science*, **85**(5), 909–13.

MEAD P S, SLUTSKER L, DIETZ V, MCCAIG L F, BRESEE J S, SHAPIRO C, ET AL. (1999), 'Food-related illness and death in the United States', *Emerging Infectious Diseases*, **5**(5), 607–25.

MEER R R, SONGER J G and PARK D L (1997), 'Human disease associated with *Clostridium perfringens* enterotoxin', *Reviews in Environmental Contamination and Toxicology*, **150**, 75–94.

MENG J and DOYLE M P (1997), 'Emerging issues in microbiological food safety', *Annual Review of Nutrition*, **17**, 255–75.
MEUNIER J R and GRIMONT P A D (1993), 'Factors affecting reproducibility of random amplified polymorphic DNA-fingerprinting', *Research in Microbiology*, **144**(5), 373–9.
MICHELI M R and BOVA R (1997), *Fingerprinting Methods Based on Arbitrarily Primed PCR*, Springer-Verag, New York, USA.
MILKMAN R (1973), 'Electrophoretic variation in *Escherichia coli* from natural sources', *Science*, **182**, 1024–6.
MITCHELL B W, BUHR R J, BERRANG M E, BAILEY J S and COX N A (2002), 'Reducing airborne pathogens, dust and *Salmonella* transmission in experimental hatching cabinets using an electrostatic space charge system', *Poultry Science*, **81**(1), 49–55.
MIWA N, NISHINA T, KUBO S and ATSUMI M (1997), 'Most probable number method combined with nested polymerase chain reaction for detection and enumeration of enterotoxigenic *Clostridium perfringens* in intestinal contents of cattle, pig and chicken', *Journal of Veterinary Medical Science*, **59**(2), 89–92.
MOUNTZOURIS K C, BALASKAS C, XANTHAKOS I, TZIVINIKOU A and FEGEROS K (2009), 'Effects of a multi-species probiotic on biomarkers of competitive exclusion efficacy in broilers challenged with *Salmonella* Enteritidis', *British Poultry Science*, **50**(4), 467–78.
MULDER R W A W (1995), 'Impact of transport and related stresses on the incidence and extent of human pathogens in pigmeat and poultry', *Journal of Food Safety*, **15**(3), 239–46.
MULLIS K B and FALOONA F A (1987), 'Specific synthesis of DNA in vitro via a polymerase-catalyzed chain reaction', *Methods in Enzymology*, **155**, 335–50.
MURRAY P R, ROSENTHAL K S and PFALLER M A (2009), *Medical Microbiology*, 6th ed., Mosby Elsevier, St. Louis, MO, USA.
MURRAY E G D, WEBB R E and SWANN M B R (1926), 'A disease of rabbits characterized by a large mononuclear leucocytosis, caused by a hitherto undescribed *Bacillus* bacterium *monocytogenes* (n. sp.)', *Journal of Pathological Bacteriology*, **29**, 407–439.
NACHAMKIN I, ALLOS B M and HO T (1998), '*Campylobacter* species and Guillain-Barré Syndrome', *Clinical Mircobiology Reviews*, **11**(3), 555–67.
NACHAMKIN I, LIU J, LI M, UNG H, MORAN A P, PRENDERGAST M M and SHEIKH K (2002), '*Campylobacterjejuni* from patients with Guillain– Barré syndrome preferentially expresses a GD(1a)-like epitope', *Infection and Immunity*, **70**(9), 5299–303.
NAYGA R M, AIEW W and NICHOLS J P (2005), 'Information effects on consumers' willingness to purchase irradiated food products', *Review of Agricultural Economics*, **27**(1), 37–48.
NEWELL D G, SHREEVE J E, TOSZEGHY M, DOMINGUE G, BULL S, HUMPHREY T and MEAD G (2001), 'Changes in the carriage of *Campylobacter* strains by poultry carcasses during processing in abattoirs', *Applied and Environmental Microbiology*, **67**(6), 2636–40.
NIGHTINGALE K K, WINDHAM K and WIEDMANN M (2005), 'Evolution and molecular phylogeny of *Listeria monocytogenes* isolated from human and animal listeriosis cases and foods', *Journal of Bacteriology*, **187**(16), 5537–51.
NOVAK J S, PECK M W, JUNEJA V K and JOHNSON E A (2005), '*Clostridium botulinum* and *Clostridium perfringens*', in Fratamico, P.M., Bhunia, A.K. and Smith, J.L., *Foodborne Pathogens, Microbiology and Molecular Biology*, Caister Academic Press, Wymondham, pp. 383–407.
NURMI E, NUOTIO L and SCHNEITZ C (1992), 'The competitive-exclusion concept – development and future', *International Journal of Food Microbiology*, **15**(3–4), 237–40.
OKAMURA M, KAMIJIMA Y, MIYAMOTO T, TANI H, SASAI K and BABA E (2001), 'Differences among six *Salmonella* serovars in abilities to colonize reproductive organs and to contaminate eggs in laying lens', *Avian Diseases*, **45**(1), 61–69.
OKAMURA M, TACHIZAKI H, KUBO T, KIKUCHI S, SUZUKI A, TAKEHARA K and NAKAMURA M (2007), 'Comparative evaluation of a bivalent killed *Salmonella* vaccine to prevent egg contamination with *Salmonella enterica* serovars Enteritidis, Typhimurium, and

Gallinarum biovar Pullorum, using 4 different challenge models', *Vaccine*, **25**(25), 4837–44.

OLSEN A R and HAMMACK T S (2000), 'Isolation of *Salmonella* spp. from the housefly, *Muscadomestica* L., and the dump fly, *Hydrotaeaaenescens* (Wiedemann) (Diptera: Muscidae), at caged-layer houses', *Journal of Food Protection*, **63**(7), 958–60.

ON S L (2001), 'Taxonomy of *Campylobacter*, *Arcobacter*, *Helicobacter* and related bacteria: Current Status, Future Prospects and Immediate Concerns', *Symposium series (Society for Applied Microbiology)*, **30**, 1S–15S.

OOSTEROM J, NOTERMANS S, KARMAN H and ENGELS G B (1983), 'Origin and prevalence of *Campylobacter jejuni* in poultry-processing', *Journal of Food Protection*, **46**(4), 339–44.

PAOLI G C, BHUNIA A K and BAYLES D O (2005), '*Listeria monocytogenes*', in Fratamico, P.M., Bhunia, A.K. and Smith, J.L., *Foodborne Pathogens: Microbiology and Molecular Biology*, 1st ed., Caister Academic Press, Norfolk, UK, pp. 295–340.

PEARSON A D, GREENWOOD M, HEALING T D, ROLLINS D, SHAHAMAT M, DONALDSON J, ET AL. (1993), 'Colonization of broiler-chickens by waterborne *Campylobacter jejuni*', *Applied and Environmental Microbiology*, **59**(4), 987–96.

PEARSON A D, GREENWOOD M H, FELTHAM R K A, HEALING T D, DONALDSON J, JONES D M, ET AL. (1996), 'Microbial ecology of *Campylobacter jejuni* in a United Kingdom chicken supply chain: Intermittent common source, vertical transmission, and amplification by flock propagation', *Applied and Environmental Microbiology*, **62**(12), 4614–20.

PENNER J L (1988), 'The genus *Campylobacter*: A decade of progress', *Clinical Microbiology Reviews*, **1**(2), 157–72.

PERNA N T, PLUNKETT G, BURLAND V, MAU B, GLASNER J D, ROSE D J, ET AL. (2001), 'Genome sequence of enterohaemorrhagic *Escherichia coli* O157:H7', *Nature*, **409**(6819), 529–33.

PETERSON M C (1994), 'Clinical aspects of *Campylobacter jejuni* infections in adults', *Western Journal of Medicine*, **161**(2), 148–52.

PETIT L, GIBERT M and POPOFF M R (1999), '*Clostridium perfringens*: Toxinotype and genotype', *Trends in Microbiology*, **7**(3), 104–10.

PEYRAT M B, SOUMET C, MARIS P and SANDERS P (2008), 'Recovery of *Campylobacter jejuni* from surfaces of poultry slaughterhouses after cleaning and disinfection procedures: Analysis of a potential source of carcass contamination', *International Journal of Food Microbiology*, **124**(2), 188–94.

PIRIE J H H (1940), '*Listeria* – change of name for a genus of bacteria', *Nature*, **145**, 264–264.

POPOFF M Y and LE MINOR L (1997), 'Antigenic formulas of the *Salmonella* serovars', *WHO Collaborating Centre for Reference and Research on Salmonella*, 7th revision, Pasteur Institute, Paris, France.

POSFAY-BARBE K M and WALD E R (2009), 'Listeriosis', *Seminars in Fetal & Neonatal Medicine*, **14**(4), 228–33.

PROTAIS J, COLIN P, BEAUMONT C, GUILLOT J F, LANTIER F, PARDON P and BENNEJEAN G (1996), 'Line differences in resistance to *Salmonella* Enteritidis PT4 infection', *British Poultry Science*, **37**(2), 329–39.

QUINONES B, PARKER C T, JANDA J M JR, MILLER W G and MANDRELL R E (2007), 'Detection and genotyping of *Arcobacter* and *Campylobacter* isolates from retail chicken samples by use of DNA oligonucleotide arrays', *Applied and Environmental Microbiology*, **73**(11), 3645–55.

RANGARAJAN A E, BIHN A, GRAVANI R B, SCOTT D L and PRITTS M P (2000), *Food safety begins on the farm: A growers guide to good agricultural practices (GAP) publications*, College of Agricultural and Life Sciences, Cornell University, Ithaca, NY, USA.

RANTA J and MAIJALA R (2002), 'A probabilistic transmission model of *Salmonella* in the primary broiler production chain', *Risk Analysis*, **22**(1), 47–58.

REDMAN J A, GRANT S B, OLSON T M and ESTES M K (2001), 'Pathogen filtration, heterogeneity, and the potable reuse of wastewater', *Environmental Science & Technology*, **35**(9), 1798–805.

REEVES M W, EVINS G M, HEIBA A A, PLIKAYTIS B D and FARMER J J (1989), 'Clonal nature of *Salmonella* Typhi and its genetic relatedness to other salmonellae as shown by multilocus enzyme electrophoresis, and proposal of *Salmonella bongori* comb.nov', *Journal of Clinical Microbiology*, **27**(2), 313–20.

RILEY L W (2004), 'Principles and approaches', in, *Molecular Epidemiology of Infectious Diseases: Principles and Practices*, 1st ed., ASM Press, Washington, DC, USA, pp. 1–28.

RODRIGUEZ-ROMO L A and YOUSEF A E (2005), 'Inactivation of *Salmonella enterica* serovar enteritidis on shell eggs by ozone and UV radiation', *Journal of Food Protection*, **68**, 711–17.

ROWAN N J and ANDERSON J G (1998), 'Effects of above-optimum growth temperature and cell morphology on thermotolerance of *Listeria monocytogenes* cells suspended in bovine milk', *Applied and Environmental Microbiology*, **64**(6), 2065–71.

SAHIN O, ZHANG Q, MEITZLER J C, HARR B S, MORISHITA T Y and MOHAN R (2001), 'Prevalence, antigenic specificity, and bactericidal activity of poultry anti-*Campylobacter* maternal antibodies', *Applied and Environmental Microbiology*, **67**(9), 3951–7.

SANDBERG M, HOFSHAGEN M, OSTENSVIK O, SKJERVE E and INNOCENT G (2005), 'Survival of *Campylobacter* on frozen broiler carcasses as a function of time', *Journal of Food Protection*, **68**(8), 1600–5.

SARKER M R, SINGH U and MCCLANE B A (2000), 'An update on *Clostridium perfringens* enterotoxin', *Journal of Natural Toxins*, **9**(3), 251–66.

SARLIN L L, BARNHART E T, CALDWELL D J, MOORE R W, BYRD J A, CALDWELL D Y, ET AL. (1998), 'Evaluation of alternative sampling methods for *Salmonella* critical control point determination at broiler processing', *Poultry Science*, **77**(8), 1253–7.

SCANES C G (2007), 'The global importance of poultry', *Poultry Science*, **86**(6), 1057–8.

SCHLOSSER W D, HENZLER D J, MASON J, CRADLE D, SHIPMAN L, TROCK S, ET AL. (1999), 'The *Salmonella enterica* serovar enteritidis pilot project', in Saeed A.M., Gast, R.K., Potter, M.E. and Wall P.G., *Salmonella enterica serovar Enteritidis in Humans and Animals*, Iowa State University Press, Ames, Iowa, USA, pp. 353–66.

SCHLUNDT J, TOYOFUKU H, JANSEN J and HERBST S A (2004), 'Emerging food-borne zoonoses', *Reviews in Science and Technology*, **23**(2), 513–33.

SCHLUTER H, BEYER C, BEYER W, HAGELSCHUER I, GEUE L and HAGELSCHUER P (1994), 'Epidemiologic studies of *Salmonella* infections in poultry flocks', *Tierarztliche Umschau*, **49**(7), 400–10.

SCHOENI J L and WONG A C (1994), 'Inhibition of *Campylobacter jejuni* colonization in chicks by defined competitive exclusion bacteria', *Applied and Environmental Microbiology*, **60**(4), 1191–7.

SCHREZENMEIR J and DE VRIESE M (2001), 'Probiotics, prebiotics, and synbiotics – approaching a definition', *American Journal of Clinical Nutrition*, **73**, 361S–364S.

SELANDER R K, CAUGANT D A, OCHMAN H, MUSSER J M, GILMOUR M N and WHITTAM T S (1986), 'Methods of multilocus enzyme electrophoresis for bacterial population-genetics and systematics', *Applied and Environmental Microbiology*, **51**(5), 873–84.

SHREEVE J E, TOSZEGHY M, RIDLEY A and NEWELL D G (2002), 'The carry-over of *Campylobacter* isolates between sequential poultry flocks', *Avian Diseases*, **46**(2), 378–85.

SILLANKORVA S, PLETENEVA E, SHABUROVA O, SANTOS S, CARVALHO C, AZEREDO J and KRYLOV V (2010), '*Salmonella* Enteritidis bacteriophage candidates for phage therapy of poultry', *Journal of Applied Microbiology*, **108**, 1175–86.

SKIRROW M B (1991), 'Epidemiology of *Campylobacter* enteritis', *International Journal of Food Microbiology*, **12**(1), 9–16.

SKOVGAARD N (2007), 'New trends in emerging pathogens', *International Journal of Food Microbiology*, **120**, 217–24.

SLADER J, DOMINGUE G, JORGENSEN F, MCALPINE K, OWEN R J, BOLTON F J and HUMPHREY T J (2002), 'Impact of transport crate reuse and of catching and processing on *Campylobacter* and *Salmonella* contamination of broiler chickens', *Applied and Environmental Microbiology*, **68**(2), 713–19.

SLUTSKER L, ALTEKRUSE S F and SWERDLOW D L (1998), 'Foodborne diseases – emerging pathogen and trends', *Infectious Disease Clinics of North America*, **12**(1), 199–216.

SNOEYENBOS G H, WEINACK O M and SMYSER C F (1978), 'Protecting chicks and poults from *Salmonellae* by oral-administration of "normal" gut microflora', *Avian Diseases*, **22**(2), 273–87.

SONGER J G (1996), 'Clostridial enteric diseases of domestic animals', *Clinical Microbiology Reviews*, **9**(2), 216 –34.

STERN M J, AMES G F L, SMITH N H, ROBINSON E C and HIGGINS C F (1984), 'Repetitive extragenic palindromic sequences – a major component of the bacterial genome', *Cell*, **37**(3), 1015–26.

STERN N J, CLAVERO M R S, BAILEY J S, COX N A and ROBACH M C (1995), '*Campylobacter* spp. in broilers on the farm and after transport', *Poultry Science*, **74**(6), 937–41.

STERN N J, HIETT K L, ALFREDSSON G A, KRISTINSSON K G, REIERSEN J, HARDARDOTTIR H, ET AL. (2003), '*Campylobacter* spp. in Icelandic poultry operations and human disease', *Epidemiology and Infection*, **130**(1), 23–32.

STERN N J, SVETOCH E A, ERUSLANOV B V, PERELYGIN V V, MITSEVICH E V, MITSEVICH I P, ET AL. (2005), 'Isolation of a *Lactobacillus salivarius* strain and purification of its bacteriocin, which is inhibitory to *Campylobacter jejuni* in the chicken gastrointestinal system', *Antimicrobial Agents and Chemotherapy*, **50**, 3111–16.

STERN N J, SVETOCH E A, ERUSLANOV B V, PERELYGIN V V, MITSEVICH E V, MITSEVICH I P, ET AL. (2006), 'Isolation of a *Lactobacillus salivarius* strain and purification of its bacteriocin, which is inhibitory to *Campylobacter jejuni* in the chicken gastrointestinal system', *Antimicrobial Agents and Chemotherapy*, **50**, 3111–16.

STULL T L, LIPUMA J J and EDLIND T D (1988), 'A broad-spectrum probe for molecular epidemiology of bacteria: ribosomal RNA', *Journal of Infectious Diseases*, **157**(2), 280–6.

SWAMINATHAN B (2001), '*Listeria monocytogenes*', in Doyle M.P., Beuchat, L.R. and Montville, T.J. (eds), *Food Microbiology, Fundamentals and Frontiers*, ASM Press, Washington, DC, USA, pp. 383–409.

SWAMINATHAN B and MATAR G M (1993), 'Molecular typing methods', in Persing D.H., Smith T.F., Tenover, F.C. and White, T.J. (eds), *Diagnostic Molecular Microbiology: Priciples and Applications*, ASM Press, Washington, DC, USA, pp. 26–50.

SWAMINATHAN B, BARRETT T J, HUNTER S B, TAUXE R V and FORCE C P T (2001), 'PulseNet: The molecular subtyping network for foodborne bacterial disease surveillance, United States', *Emerging Infectious Diseases*, **7**(3), 382–9.

TAKAHASHI M, KOGA M, YOKOYAMA K and YUKI N (2005), 'Epidemiology of *Campylobacter jejuni* islotated from patients with Guillain-Barré and Fisher syndromes in Japan', *Journal of Clinical Microbiology*, **43**(1), 335–9.

TAUXE R V (1992), 'Epidemiology of *Campylobacter jejuni* infections in the United States and other industrialized nations', in Nachamkin, I., Tompkins, S. and Blaser, M., *Campylobacter jejuni: Current Status and Future Trends*, ASM Press, Washington, DC, USA, 9–19.

TAUXE R V (2001), 'Food safety and irradiation: Protecting the public from foodborne infections', *Emerging Infectious Diseases*, **7**(3), 516–21.

TAUXE R V (2002), 'Emerging foodborne pathogens', *International Journal of Food Microbiology*, **78**(1–2), 31–41.

TAYLOR D E, EATON M, YAN W and CHANG N (1992), 'Genome maps of *Campylobacter jejuni* and *Campylobacter coli*', *Journal of Bacteriology*, **174**(7), 2332–7.

TENOVER F C, ARBEIT R, ARCHER G, BIDDLE J, BYRNE S, GOERING R, ET AL. (1994), 'Comparison of traditional and molecular methods of typing isolates of *Staphylococcus aureus*', *Journal of Clinical Microbiology*, **32**(2), 407–15.

TENOVER F C, ARBEIT R D and GOERING R V (1997), 'How to select and interpret molecular strain typing methods for epidemiological studies of bacterial infections: A review for healthcare epidemiologists', *Infection Control and Hospital Epidemiology*, **18**(6), 426–39.

TODD E C, GREIG J D, BARTLESON C A and MICHAELS B S (2007), 'Outbreaks where food workers have been implicated in the spread of foodborne disease. Part 3. Factors contributing to outbreaks and description of outbreak categories', *Journal of Food Protection*, **70**(9), 2199–217.

TOLLEFSON L, FEDORKA-CRAY P J and ANGULO F J (1999), 'Public health aspects of antibiotic resistance monitoring in the USA', *Acta Veterinaria Scandinavica Supplementum*, **92**, 67–75.

TORNIEPORTH N G, JOHN J, SALGADO K, DEJESUS P, LATHAM E, MELO M C N, GUNZBURG S T and RILEY L W (1995), 'Differentiation of pathogenic *Escherichia coli* strains in Brazilian children by PCR', *Journal of Clinical Microbiology*, **33**(5), 1371–4.

TORO H, PRICE S B, MCKEE S, HOERR F J, KREHLING J, PERDUE M, ET AL. (2005), 'Use of bacteriophages in combination with competitive exclusion to reduce *Salmonella* from infected chickens', *Avian Diseases*, **49**(1), 118–24.

TSCHIRDEWAHN B, NOTERMANS S, WERNARS K and UNTERMANN F (1991), 'The presence of enterotoxigenic *Clostridium perfringens* strains in faeces of various animals', *International Journal of Food Microbiology*, **14**(2), 175–8.

TUERK C and GOLD L (1990), 'Systematic evolution of ligands by exponential enrichment – RNA ligands to bacteriophage-T4 DNA-polymerase', *Science*, **249**(4968), 505–10.

TURNBULL P C B and SNOEYENBOS G H (1974), 'Experimental salmonellosis in the chicken. 1. Fate and host response in alimentary canal, liver and spleen', *Avian Diseases*, **18**, 153–77.

TYLER K D, WANG G, TYLER S D and JOHNSON W M (1997), 'Factors affecting reliability and reproducibility of amplification-based DNA fingerprinting of representative bacterial pathogens', *Journal of Clinical Microbiology*, **35**(2), 339–46.

USDA (1998), 'Refrigeration and labeling requirements of shell eggs: Final rule', *Federal Register*, **63**, 45663–75.

USDA (2003), 'Control of *Listeria monocytogenes* in ready-to-eat meat and poultry products', *Federal Register*, **68**(109), 34208–54.

USDHHA-FDA-CFSAN (1999), 'Evaluation of risks related to microbiological contamination of ready-to-eat food by food preparation workers and the effectiveness of interventions to minimize those risks', available from: http://www.cfsan.fda.gov/~ear/rterisk.html

USFDA (2004), 'Prevention of *Salmonella* Enteritidis in shell eggs during production: Proposed rule', *Federal Register*, **69**, 56823–906.

VAN BELKUM A (1999), 'Short sequence repeats in microbial pathogenesis and evolution', *Cellular and Molecular Life Sciences*, **56**(9–10), 729–34.

VAN DE GIESSEN A W, AMENT A J H J AND NOTERMANS S H W (1994), 'Intervention strategies for *Salmonella* Enteritidis in poultry flocks: A basic approach', *International Journal of Food Microbiology*, **21**, 145–54.

VAN DE GIESSEN A W, VAN LEEUWEN W J and VAN PELT W (1999), '*Salmonella enterica* serovar Enteritidis in The Netherlands: Epidemiology, prevention, and control', in Saeed, A.M., Gast, R.K., Potter, M.E. and Wall, P.G. (eds), *Salmonella enterica serovar Enteritidis in humans and animals*, Iowa State University Press, Ames, Iowa, USA, pp. 71–80.

VAN IMMERSEEL F, MEULEMANS G, DE BUCK J, PASMANS F, CELGE P, BOTTREAU E, ET AL. (2004), 'Bacteria host interactions of *Salmonella* paratyphi B dt+ in poultry', *Epidemiology and Infection*, **132**, 239–43.

VANDAMME P, FALSEN E, ROSSAU R, HOSTE B, SEGERS P, TYTGAT R, ET AL. (1991), 'Revision of *Campylobacter*, *Helicobacter*, and *Wolinella* taxonomy – emendation of generic descriptions and proposal of *Arcobacter* gen. nov', *International Journal of Systematic Bacteriology*, **41**(1), 88–103.

VANDAMME P, HARRINGTON C S, JALAVA K and ON S L W (2000), 'Misidentifying Helicobacters: The *Helicobacter cinaedi* example', *Journal of Clinical Microbiology*, **38**(6), 2261–6.

VANDERZANT C and SPLITTSTOESSER D (1992), *Compendium of Methods for the Microbiological Examination of Foods*, American Public Health Association, Washington, DC, USA.

VAZQUEZ-BOLAND J A, KUHN M, BERCHE P, CHAKRABORTY T, DOMINGUEZ-BERNAL G, GOEBEL W, ET AL. (2001), '*Listeria* pathogenesis and molecular virulence determinants', *Clinical Microbiology Reviews*, **14**(3), 584–640.

VERSALOVIC J, KOEUTH T and LUPSKI J R (1991), 'Distribution of repetitive DNA sequences in eubacteria and application to fingerprinting of bacterial genomes', *Nucleic Acids Research*, **19**(24), 6823–31.

VOETSCH A C, VAN GILDER T J, ANGULO F J, FARLEY M M, SHALLOW S, MARCUS R, ET AL. (2004), 'FoodNet estimate of the burden of illness caused by nontyphoidal *Salmonella* infections in the United States', *Clinical Infectious Diseases*, **38**, S127–S134.

VUGIA D, CRONQUIST A, HADLER J, TOBIN-D'ANGELO M, BLYTHE D, SMITH K, ET AL. (2006), 'Preliminary FoodNet data on the incidence of infection with pathogens transmitted commonly through food – 10 states, United States, 2005 (reprinted from mmwr, vol 55, pg 392–395, 2006)', *Jama-Journal of the American Medical Association*, **295**(19), 2241–3.

WAGENAAR J A, MEVIUS D J and HAVELAAR A H (2006), '*Campylobacter* in primary animal production and control strategies to reduce the burden of human campylobacteriosis', *Revue Scientifique Et Technique-Office International Des Epizooties*, **25**(2), 581–94.

WAGNER R D (2006), 'Efficacy and food safety considerations of poultry competitive exclusion products', *Molecular Nutrition & Food Research*, **50**(11), 1061–71.

WALKER R I, CALDWELL M B, LEE E C, GUERRY P, TRUST T J and RUIZPALACIOS G M (1986), 'Pathophysiology of *Campylobacter* enteritis', *Microbiological Reviews*, **50**(1), 81–94.

WALL P G and WARD L R (1999), 'Epidemiology of *Salmonella enterica* serovar enteritidis Phage Type 4 in England and Wales', in Saeed, A.M., Gast, R.K., Potter, M.E. and Wall, P.G. (eds), *Salmonella enterica serovar Enteritidis in Humans and Animals*, Iowa State University Press, Ames, Iowa, USA, pp. 19–26.

WARD L R, DE SA J D H and ROWE B (1987), 'A phage typing scheme for *Salmonella* Enteritidis', *Epidemiology and Infection*, **99**, 291–4.

WARRELL D A, COX T M and FIRTH J D (2003), *Oxford Textbook of Medicine*, 4th ed., Oxford University Press, Oxford.

WELSH J and MCCLELLAND M (1991), 'Genomic fingerprinting using arbitrarily primed PCR and a matrix of pairwise combinations of primers', *Nucleic Acids Research*, **19**(19), 5275–9.

WENZEL R and HERRMANN R (1988), 'Repetitive DNA-sequences in *Mycoplasma pneumoniae*', *Nucleic Acids Research*, **16**(17), 8337–50.

WHITELOCK-JONES L, CARSWELL J and RASMUSSEN K C (1989), '*Listeria pneumonia*. A case report', *South Africal Medical Journal*, **75**(4), 188–9.

WHYTE P, MC GILL K, COLLINS J D and GORMLEY E (2002), 'The prevalence and PCR detection of *Salmonella* contamination in raw poultry', *Veterinary Microbiology*, **89**, 53–60.

WILLIAMS J G K, KUBELIK A R, LIVAK K J, RAFALSKI J A and TINGEY S V (1990), 'DNA polymorphisms amplified by arbitrary primers are useful as genetic markers', *Nucleic Acids Research*, **18**(22), 6531–5.

WOOLDRIDGE M, CLIFTON-HADLEY R and RICHARDS M (1996), 'I don't want to be told what to do by a mathematical formula overcoming adverse perceptions of risk analysis', *Proceedings of the Society of Veterinary Epidemiology and Preventive Medicine, Reading UK*, 36–47.

WYSZYNSKA A, RACZKO A, LIS M and JAGUSZTYN-KRYNICKA E K (2004), 'Oral immunization of chickens with avirulent *Salmonella* vaccine strain carrying *C. jejuni* 72dz/92 *cjaA* gene elicits specific humoral immune response associated with protection against challenge with wild-type *Campylobacter*', *Vaccine*, **22**(11–12), 1379–89.

YERAMIAN E and BUC H (1999), 'Tandem repeats in complete bacterial genome sequences: Sequence and structural analyses for comparative studies', *Research in Microbiology*, **150**(9–10), 745–54.

ZIPRIN R L and DELOACH J R (1993), 'Comparison of probiotics maintained by *in vivo* passage through laying hens and broilers', *Poultry Science*, **72**(4), 628–35.

21

Tracing zoonotic pathogens in dairy production

J. S. Van Kessel, M. Santin-Duran and J. S. Karns, US Department of Agriculture, USA and Y. Schukken, Cornell University, USA

Abstract: Dairy farming has become a highly productive system producing ample amounts of high-quality milk and meat from fewer cows on less land on fewer, but larger, farms. Despite this consolidation and modernization, zoonotic pathogenic bacteria and protozoans remain problems on the modern dairy farm. Although pasteurization has greatly reduced illness due to contaminated dairy products, post-processing contamination and an apparent increase in the consumption of raw milk, raw milk products and meat from dairy cows continue to result in outbreaks of gastrointestinal illness. Methods used for pathogen detection, identification, subtyping and characterization methods have shown the relationships between pathogens from cow feces and the surrounding environment and those contaminating milk and meat; however, control of these pathogens on the farm remains difficult.

Key words: dairy, zoonotic pathogens, *E. coli*, *Salmonella*, *Cryptosporidium*, *Listeria monocytogenes*.

21.1 Introduction

The presence of biological components in milk that can cause foodborne illness has been recognized for nearly 200 years and commercial pasteurization of milk was instituted around the turn of the twentieth century to help prevent such illness. Pasteurization destroys both spoilage organisms and zoonotic pathogens and is therefore effective in increasing the safety and shelf life of milk and milk products. Currently the vast majority of milk consumed in developed countries is pasteurized although there is a growing interest in consuming raw (unpasteurized) milk. Raw milk advocates often claim that pasteurization destroys many of the beneficial properties of milk and that consumption of the raw product is a healthier alternative. In the USA this is a controversial issue. The Food and Drug

Administration (FDA) bans the interstate shipment of raw milk, and approximately half of the states ban the sale of raw milk due to the risk of foodborne illness associated with the consumption of raw milk and raw milk products.

21.2 Foodborne pathogens in dairy production chains and their significance for public health

Milking systems vary substantially but most modern systems fall into the two broad categories: pipeline and parlor systems. Pipeline systems are used in stanchion barns where the cows are milked in their individual stall and the milking personnel move from stall to stall with the milking cluster that is attached to the udder. In parlor systems all cows in the herd are rotated through a group of milking stalls, the number of which varies from 2 to 50. In both systems milk is transported under vacuum through stainless steel pipes from the udder to a milk receiver and then pumped into a refrigerated tank that is generally referred to as a bulk tank. The milk is stored at approximately 4°C in the tank for up to two days until it is transported to a processing plant.

Zoonotic pathogens may contaminate bulk milk if milk is obtained from cows with udder infections (mastitis) or through fecal contamination of milk. Bulk tank milk contamination occurs by direct secretion of bacteria from mastitic udders. Mastitis is an inflammation in the mammary gland that occurs when bacteria invade the gland, multiply and release bacteria or bacterial products such as toxins. When a cow is identified as having mastitis, the affected teats are generally treated with antibiotics and the milk is withheld from the bulk tank until the infection is cleared and no residues of treatment are left in the milk. However, frequently mastitis is subclinical and therefore may go undetected. In these cases, the infecting organism is secreted with the milk and contaminates the milk in the bulk tank. Although some contamination occurs from mastitis milk, the primary source of zoonotic pathogen contamination in the bulk tank is fecal matter that enters the milking system during the milking process. Bovine feces contain high numbers of microorganisms ($>10^{10}$ cfu per g) and so even when a very small amount of feces gets into the milk a significant number of bacteria, protozoans, fungi and viruses can enter the bulk tank.

Most dairy producers follow a strict milking regime that is aimed at protecting the mammary glands from contamination and at minimizing transfer of pathogens between herd mates. The hygiene protocols are designed to protect the cow from infection and the milk and milk system from contamination. The goal is to minimize contamination of the bulk milk with feces and other debris. In general, the milking procedures often include a pre-milking step where the teats are treated with a bactericidal liquid and dried with a towel. After the milking is complete, the teats are again treated with a bactericidal, post-milking dip. Many aspects of the milking system are designed to protect the teats from physical damage and environmental contamination. Additionally, all lines and equipment that come in contact with the milk are cleaned and sanitized following standardized protocols to ensure that there is no carry over and growth of the microorganisms in the milk.

Despite extensive efforts to control and minimize the entry of pathogens into the bulk tank, contamination cannot be completely avoided. Most of the pathogens of concern are bacteria but protozoan contamination has also been demonstrated. The major mastitis-causing bacteria include *Streptococcus* spp., *Staphylococcus* spp., *Mycoplasma* spp., enterococci and coliforms (*Escherichia coli*, *Klebsiella* spp., *Enterobacter*) (Jayarao, 2001; Jayarao *et al.*, 2004). Fortunately, most of the mastitis pathogens are not zoonotic and are not typically considered to be important foodborne pathogens. However, some instances of mastitis associated with infection by *Listeria monocytogenes* and *Salmonella* spp. have been reported (Pearson and Marth, 1990; Jensen *et al.*, 1996; Wesley, 1999; Radke *et al.*, 2002). Although *E. coli* have been frequently isolated from mastitic udders (Burvenich *et al.*, 2003), reports of mastitis caused by enteropathogenic *E. coli* (EPEC) are exceptional. This may, in part, be due to lack of testing for strains with the pathogenic attributes or to the low percentage of the *E. coli* population represented by EPEC. Screening of mastitis-causing *E. coli* strains detected very few, if any, Shiga toxin (*stx*)-positive strains (Cullor, 1997; Murinda *et al.*, 2002b). Pathogens that enter the bulk tank via contamination during the milking process can include any organism that is being shed in the feces by the cows and also any other environmental organism that is picked up by the cows and tracked into the milking parlor. The major zoonotic pathogens that have been identified in the feces of dairy cattle include the bacteria *Salmonella*, enteropathogenic *E. coli*, *L. monocytogenes* and *Campylobacter*, as well as the protozoan, *Cryptosporidium*. In an ongoing intensive longitudinal study on three well-managed dairy farms in the Northeastern USA, major bacterial pathogens were identified on all of the farms in the study (Pradhan *et al.*, 2009). This study demonstrated the value of long-term longitudinal sampling as pathogens were often found to be shed in cow feces sporadically. Cross-sectional sampling would have resulted in detection of only a limited number of foodborne pathogens on these farms; the longitudinal repeated sampling of the cows and the farm environment identified all of the pathogens of interest at some point in time on each of the premises.

When pathogens are present, either in infected animals or in the dairy environment, there are several routes for transmission to humans. People who have direct contact with the animals are at obvious risk of exposure, primarily through unintentional ingestion of feces. Farm workers and service providers have the most frequent direct animal contact but the general public can be exposed at petting zoos, farm tours and other open events such as state fairs. Additionally people can become infected by consuming raw milk or raw milk products. Even if the initial pathogen load in the raw milk is low, temperature abuse of the product can lead to increased concentrations of pathogens and higher potential for human illness. There is an increased risk for post-processing contamination of pasteurized milk when the raw milk is contaminated; dairy processing plants follow strict hygiene protocols but there have been documented cases where a breakdown within the system caused contaminated product to leave the plant.

An often underestimated risk of human exposure to pathogens associated with dairy production is through ground beef and whole cuts from culled animals. Each year about 35% of the cows on a dairy farm are culled and since there are approximately nine million dairy cows in the USA, cull dairy animals are a significant part of the beef market. Estimates suggest that cull dairy animals account for approximately 17% of the ground beef sold in the USA and more than half of most carcasses are used for higher value cuts such as steaks and roasts. Cull animals that are harboring zoonotic pathogens or come from infected premises can therefore pose a risk for infected product entering the food supply (Troutt and Osburn, 1997; Troutt *et al.*, 2001).

21.2.1 Pathogens

Dairy farms have been identified as reservoirs for many zoonotic organisms that have been implicated as foodborne pathogens. Some of the pathogens such as *Salmonella* spp. have a variable impact on the animals; they may cause clinical or subclinical disease but there are many instances where very little impact on animal health or productivity was observed. Other zoonotic pathogens (such as Shiga toxin-producing *E. coli* and *Campylobacter*) consistently have no apparent affect on the animals. When infected animals are asymptomatic, dairy farmers will most likely not be aware that the animals and premises are harboring zoonotic pathogen. This adds burdensome complexity to the development of process controls that are designed to eradicate or minimize the presence of microbial pathogens in the dairy farm environment.

Salmonella spp.

Bacteria of the genus *Salmonella* are a major cause of human foodborne illness. Salmonellosis is generally self-limiting but can lead to serious or fatal complications in people who have other compromising health issues. Once ingested, the incubation period for *Salmonella* infection is 12–72 hours and symptoms generally include fever, diarrhea and abdominal cramps. Because most cases do not require treatment, salmonellosis is generally underreported, and so the frequency is likely substantially larger than the approximately 40 000 cases reported annually in the USA. The CDC estimates that the actual number of illnesses due to *Salmonella* is more than one million per year (Mead *et al.*, 1999).

Numerous outbreaks of disease due to *Salmonella* have been associated with milk and dairy product consumption. *Salmonella* outbreaks of varying size have been attributed to raw milk but post-pasteurization contamination has also been identified. If the milk that enters the processing facility is contaminated, there is an increased risk for contamination of product, particularly when there is a breakdown in sanitation protocols. An outbreak in 2000 due to multi-drug-resistant *Salmonella enterica* serotype Typhimurium in Pennsylvania and New Jersey was linked to pasteurized milk that was contaminated because of sub-standard environmental conditions in the processing plant (Olsen *et al.*, 2004). *Salmonella* Typhimurium isolates were obtained from 93 persons, and 44 isolates were compared by pulsed-field gel electrophoresis (PFGE) analysis. Based on PFGE analysis, 38 of these isolates were defined as outbreak-related strains, indicating that they came from a common source. The PFGE patterns of these outbreak strains did not match any of the *S. enterica* serotype Typhimurium PFGE patterns in the CDC's PulseNet database at that time. Pulsed-field gel electrophoresis patterns of isolates obtained from plant personnel (one) and dairy cows (two) obtained at the same time period were identified as outbreak-related strains indicating that the pathogen likely originated at the dairy farms.

Various surveys have indicated a significant presence of *Salmonella* in dairy production systems. In a study of dairy farms in England and Wales, Davison *et al.* (2005) sampled farms up to four times between October 1999 and February 2001. The number of *Salmonella*-positive farms ranged from 18.7% to 24.7% over the four sampling periods and, on average, a positive cow was identified on a farm 19.1% of the time. Through the course of the study, *Salmonella* was isolated from 4.7% of the nearly 20 000 fecal samples tested. National, single time point surveys of the US dairy herd have isolated *Salmonella* from 27% to 31% of the herds (USDA, 1998, 2003). Other surveys such as the one reported by Callaway *et al.* (2005) suggest an even greater prevalence. In a survey of dairy farms in four states, *Salmonella* was isolated from nearly 10% of fecal samples and 9 of 16 herds had at least one cow shedding *Salmonella*.

Given the observed presence of *Salmonella* in up to one-third of dairy herds, significant contamination of bulk milk is to be expected despite rigorous hygiene efforts during the milking process. Several regional surveys in the USA and Canada have indicated that between 0.2% and 9% of raw milk samples were contaminated with *Salmonella* (Hassan *et al.*, 2000; Jayarao *et al.*, 2001; Murinda *et al.*, 2002a; Oliver *et al.*, 2005; Rohrbach *et al.*, 1992; Steele *et al.*, 1997). The National Animal Health Monitoring System (NAHMS) Unit of the Animal and Plant Health Inspection Service (APHIS) conducted national surveys on the health and health management of dairy cattle in the USA in 1996, 2002 and 2007. Samples of bulk tank milk were collected in the 2002 and 2007 surveys to determine the incidence of raw milk contamination. Salmonellae were isolated from 2.6% of the bulk milk samples in 2002 (Van Kessel *et al.*, 2004) but PCR analysis indicated that up to 11.8% of the samples actually were contaminated (Karns *et al.*, 2005). In-line milk filter samples were collected in addition to bulk milk samples in the NAHMS 2007 survey. In this study, real time PCR analysis indicated that 11% of bulk milk samples and 25% of milk filters were positive for the presence of *Salmonella* (USDA, 2009). Hence, using a single point in time cross-sectional sampling scheme, *Salmonella* was isolated from either the milk and/or the milk filter from 36% of the dairy operations in the United States.

Most of the dairy-associated salmonellae are in the group *Salmonella enterica* subsp. *enterica*. This is a very diverse group with more than 2500 identified serotypes; however 100 of the serotypes from this group account for 98% of all isolates associated with human illness in the USA (CDC, 2008a). Many *Salmonella* serotypes have been identified in surveys of dairy cattle, dairy cattle facilities and in raw milk. Frequently, multiple serotypes are identified in a single dairy farm (Callaway *et al.*, 2005; Edrington, Schultz, *et al.*, 2004; Fitzgerald *et al.*, 2003; Hume *et al.*, 2004; Peek *et al.*, 2004). Although some of the dairy related serotypes are not commonly associated with human illness, more than half of the 'top 20' human clinical serotypes have been isolated from dairy operations (CDC, 2008a).

Listeria monocytogenes

In the USA approximately 2500 people are diagnosed annually with listeriosis, a foodborne disease caused by *L. monocytogenes*. Listeriosis is different from salmonellosis in that it is associated with much higher hospitalization and mortality rates (approximately 92% and 20%, respectively). Immunocompromised individuals are at an elevated risk for infection and pregnant women who become infected are at high risk for fetal loss. As with other foodborne illnesses, there are likely a substantial number of unreported cases, particularly those who have experienced a non-invasive *L. monocytogenes* infection that is self-limiting and primarily characterized by diarrhea. Dairy farms have been implicated as major reservoirs for human pathogenic *L. monocytogenes* strains. *Listeria* are also prevalent in the dairy environment and are commonly isolated from soil, decaying organic matter, feces and poorly fermented silage (Roberts and Weidmann, 2003; Weis and Seeliger, 1975; Wesley, 1999). *Listeria monocytogenes* is the only zoonotic *Listeria* species and is implicated in most cattle and human cases of listeriosis. Although *L. ivanovii* causes abortions in cattle and sheep, this species rarely infects humans.

Ready-to-eat (RTE) meats and dairy products are the foods most often implicated in listeriosis outbreaks (Farber and Peterkin, 1991; Meng and Doyle, 1997; Ryser, 2001). Although RTE meats are extensively tested for *L. monocytogenes* contamination, it is very difficult to completely eliminate this bacterium from the processing plants. These products undergo a killing step in the production process and are typically not cooked prior to consumption. Therefore, if post-processing contamination occurs there is an increased risk for human illness. Additionally, *Listeria* are psychrophilic so growth can occur between processing and consumption, even under refrigerated conditions.

As with *Salmonella*, *L. monocytogenes* has been implicated in foodborne disease associated with the consumption of raw milk as well as post-process contaminated milk. A *L. monocytogenes* outbreak in Massachusetts in 2008 was linked to pasteurized milk that had been contaminated after it was processed. This outbreak led to the deaths of three people (CDC, 2008c). Additionally Hispanic-style cheeses such as queso fresco have been identified as the source of *L. monocytogenes* in several outbreaks (CDC, 2001). These cheeses are frequently made from raw milk under uncontrolled conditions. If the raw milk source is contaminated with *L. monocytogenes*, the processing conditions are conducive for bacterial growth to levels that pose a risk for consumers of the product.

There are 13 known serotypes of *L. monocytogenes* (Farber and Peterkin, 1991) and various molecular typing methods have distributed these serotypes within three major lineages: Lineage 1 (serotypes 1/2b, 3b, 4b), Lineage 2 (1/2a,

3a, 1/2c, 3c) and Lineage 3 (4a, 4c and some 4b). Serotypes 1/2a, 1/2b and 4b are responsible for 98% of human listeriosis cases. Lineage 3 isolates are infrequently isolated from food and humans and are generally associated with animal listeriosis.

Clinical listeriosis is observed in cows but *L. monocytogenes* has also been frequently isolated from the feces of asymptomatic cows. Therefore, undetected fecal contamination can lead to *L. monocytogenes*-contaminated bulk milk. Surveys of raw, bulk tank milk have detected *L. monocytogenes* in 1–13% of bulk tank samples (Hassan *et al*., 2000; Jayarao *et al*., 2001; Mohammed *et al*., 2009; Oliver *et al*., 2005; Rohrbach *et al*., 1992; Steele *et al*., 1997). In the 2007 NAHMS Dairy survey, where both bulk milk and in-line milk filters were tested, *L. monocytogenes* was isolated from 7% of the randomly selected farm operations that were tested. In a more recent survey of 50 dairy cattle operations in New York state, Mohammed *et al*. (2009) detected *L. monocytogenes* in at least one cow from each farm using an enrichment followed by a PCR-based diagnostic test. In this New York study, prevalence in bulk tank milk and in-line milk filters was 16% and 45%, respectively. Half of the herds chosen for the study had previously been identified as harboring *L. monocytogenes*, and this along with PCR detection versus isolation of the organism may account for the high observed prevalence. In the longitudinal study of dairy herds mentioned above, *L. monocytogenes* was isolated regularly from fecal samples in one of the three herds with a prevalence ranging from 0% to 24.3% (Pradhan *et al*., 2009). Isolation of *L. monocytogenes* was rare in the other two herds.

There is significant serotype overlap between strains that are isolated from raw milk or dairy cattle and human clinical isolates. There appears to be substantial genotypic diversity within serotype isolate groups (Borucki *et al*., 2004, 2005; Nightingale *et al*., 2004) that allows for more precise source tracking in the case of human clinical disease epidemics. Borucki *et al*. (2004) used PFGE to identify common strains of *L. monocytogenes* among dairy farm associated isolates and isolates from human epidemic and sporadic cases. Thus, dairy farms appear to be a significant reservoir for human outbreak strains of *L. monocytogenes* (Borucki *et al*., 2004; Nightingale *et al*., 2004).

Enterohemorrhagic Escherichia coli

Escherichia coli are common intestinal inhabitants in most mammals, including humans, and play an important role in a healthy digestive system. However, a subset of this very diverse species is pathogenic to humans and is implicated in neonatal meningitis, urinary tract infections and gastrointestinal diseases (infantile and sporadic diarrhea, traveler's diarrhea). Enterohemorrhagic *E. coli* (EHEC), a subset of the enteropathogenic *E. coli*, are associated with foodborne disease in developed countries and have the potential to cause hemolytic uremic syndrome, a serious and often fatal illness. Human infection with EHEC has been associated with consumption of contaminated foods or water and close contact with contaminated farm animals (Bach *et al*., 2002). The O157 serotype is the most common EHEC associated with foodborne disease, but other serotypes such as O111, O103 and O26 are also significant causes of human disease. Isolation of these latter serotypes is more challenging and there appears to be geographical differences with respect to predominant serotypes (O'Brien and Kaper, 1998).

The primary reservoir for *E. coli* O157:H7 is considered to be cattle and, although beef cattle are most often implicated, dairy cattle are also known carriers of this pathogen. As mentioned earlier, cattle that are infected with salmonellae or *L. monocytogenes* can exhibit variable levels of clinical illness even though they are frequently asymptomatic carriers. However, except for young calves, cattle are generally asymptomatic carriers of *E. coli* O157:H7 or other Shiga toxin-producing *E. coli* (STEC). Estimates of *E. coli* O157:H7 prevalence in dairy cattle are variable but generally low and range from less than 0.2% to as high as 6% (Byrne *et al.*, 2003; Edrington, Hume, *et al.*, 2004; Faith *et al.*, 1996; Hancock *et al.*, 1998, Wells *et al.*, 1991). Herd level prevalence has been estimated to be as high as nearly 50% but is also variable (Edrington, Hume, *et al.*, 2004; Faith *et al.*, 1996; Wells *et al.*, 1991; Zhao *et al.*, 1995).

This variability in animal and herd prevalence is likely caused by multiple factors. The evolution of detection methods since the first reported outbreaks in the 1980s has improved the sensitivity of detection. This is particularly important for this pathogen since the bacterial load in an infected animal can be quite low. Additionally cattle are intermittent shedders and many of the prevalence estimates are based on single time point samplings. Shedding also appears to be seasonal, with higher prevalence rates detected in summer months versus winter months. The reason for this latter phenomenon has not been elucidated. Edrington *et al.* (2006) observed that higher shedding in summer months cannot be attributed to temperature alone. If warmer ambient temperatures result in greater carriage of *E. coli* O157:H7, one would expect higher observed prevalences in the southern states versus the northern US states and Canada. In fact, *E. coli* O157:H7 prevalence is higher in the northern climates. Edrington *et al.* (2006) have suggested changing length in day-light and subsequent physiologic changes in animals as the explanation for seasonal shedding of this pathogen.

Based on a few studies, STEC (also known as verotoxigenic *E. coli* [VTEC]) contamination of raw milk is, in general, relatively low. In 1997, Steele *et al.* (1997) detected VTEC in 0.9% of raw milk collected from more than 1700 bulk tanks and Murinda *et al.* (2002b) detected similar low prevalences in a survey of Tennessee dairy farms. Jayarao and Henning (2001) detected STEC in 3.8% of samples collected from 131 herds in South Dakota and Minnesota. However, *E. coli* O157:H7 was not isolated in any of these studies. In a national survey of US dairy herds, Karns *et al.* (2007) detected the *eaeA* gene the gamma allele of the translocated intimin receptor (γ-tir) found in EHEC strains of *E. coli* O157:H7, and one or both of the Shiga toxin genes (stx_1 and stx_2) in 4.2% of 859 raw milk samples. This combination of genes is indicative of the presence of O157:H7. However, further testing for the presence of the *fliC*, *rfbE* and *hlyA* genes found in EHEC O157:H7 showed that *E. coli* O157:H7 was a likely contaminant of only 0.2% (2) of the samples and isolation of *E. coli* O157:H7 was only achieved from one sample. This and other work shows that, although STEC is sometimes a contaminant of raw milk, the level of contamination of the raw milk supply is low.

Despite the low observed levels of bulk milk contamination, outbreaks of foodborne illness associated with EHEC-contaminated raw milk or milk products have been observed (CDC, 2007, 2008b). However, milk constitutes only a small portion of the potential dairy production-related outbreaks. Dairy animals may contribute pathogenic *E. coli* to the food supply through multiple routes. Ground beef and produce are the two most often implicated food groups in outbreaks of *E. coli* O157:H7. As was mentioned above, culled dairy animals contribute a significant proportion of the ground beef produced in the USA and consequently culled dairy cows entering the food chain represent a foodborne disease risk. Additionally the proximity of dairy production to produce production in some areas of the country also represents a risk that is often difficult to quantify. The potential avenues of cross contamination between the two production systems include contamination of the water supply (i.e. irrigation water), movement of feral animals and birds and even transport via aerosols when, for example, manure is spread on crop land. Finally dairy animals are often participants in agricultural fairs, another known source of public exposure.

Campylobacter spp.

Campylobacter spp. are a major cause of foodborne illness worldwide and are the most common pathogens associated with diarrheal illness in many developed countries including the USA, the UK and New Zealand. Mead *et al.* (1999) estimated that there are at least two million cases of campylobacteriosis annually in the USA. As with salmonellosis, campylobacteriosis in humans is characterized by fever, diarrhea and abdominal cramps. Generally the disease is self-limiting but Guillain-Barré syndrome and Reiter's syndrome are serious sequelae that occur following an estimated 0.1% and 1% of *Campylobacter* infections, respectively. Both of these syndromes have the potential for serious lifetime consequences and even death.

Although human *Campylobacter* infection is most widely associated with the handling and consumption of poultry meat, campylobacteriosis has also been linked to the consumption of contaminated milk or milk products. Wood *et al.* (1992) identified 20 campylobacteriosis outbreaks associated with the consumption of raw milk during youth activities in 11 US states from 1981 to 1988. Campylobacteriosis occurred in 45% (458/1013) of persons who drank the raw milk associated with these activities. The CDC identified an outbreak of *Campylobacter* infection in Wisconsin in 2001 that was associated with drinking raw milk obtained through a cow share program (2002). A total of 75 ill persons were associated with the outbreak. A smaller outbreak (nine reported illnesses) was associated with a cow share program in California in 2008. Cow share programs are operated in some states as a means of legally distributing raw milk from the farm to consumers. As indicated above with other zoonotic pathogens, outbreaks have also been documented with pasteurized milk. Fahey *et al.* (1995) described an outbreak of campylobacteriosis in the UK that affected more than 100 people. The outbreak was associated with the consumption of milk that had been inadequately pasteurized prior to distribution and sale.

Although five *Campylobacter* species have been identified to have caused human gastrointestinal illness, *Campylobacter jejuni* is the most common disease causing species. *Campylobacter coli* is the second most common species isolated from humans. These two species have also been identified in dairy cattle with *C. jejuni* being the predominant species. In a survey of feces from healthy dairy cattle on 96 dairy operations in the USA (Englen *et al.*, 2007), *Campylobacter* spp. were isolated from 51.2% of the 1435 cows sampled. At least one positive sample was identified on 97.9% of the dairy operations. In smaller surveys of US dairy operations, Sato *et al.* (2004) and Harvey *et al.* (2004) found much lower prevalences (2.9–29.1%) of *Campylobacter* in the feces of dairy cattle. Some of the observed variation in prevalence may be explained by the apparent seasonality of campylobacter shedding in cattle (Stanley and Jones, 2003; Sato *et al.*, 2004). As with the bacterial pathogens mentioned above, asymptomatic shedding of *Campylobacter* in dairy cows is not uncommon and therefore the farmer is often unaware that the animals are infected.

Cryptosporidium spp.

Cryptosporidium spp. is an important protozoan parasite that causes a diarrheal disease in humans and animals worldwide. Cryptosporidiosis is transmitted via a fecal-oral route by *Cryptosporidium* oocysts. Oocysts are very resistant to environmental challenges and can survive in the environment for long periods of time. *Cryptosporidium* spp. have a very low infectious dose (10–20 oocysts). Infection with *Cryptosporidium* oocysts can be acquired through person-to-person or animal-person transmission and through food or water. There are several documented outbreaks attributed to drinking water, the largest reported in Milwaukee, Wisconsin, in 1993. In this outbreak over 400 000 individuals became ill and 100 people died (MacKenzie *et al.*, 1994). *Cryptosporidium* has also been linked to a number of foodborne outbreaks involving other vehicles of infection including unpasteurized milk, apple cider and salads (Djuretic *et al.*, 1997; Millard *et al.*, 1994; Quiroz *et al.*, 2000). Many studies have suggested that drinking unpasteurized milk is a risk factor for cryptosporidiosis and three known outbreaks involving milk were consequences of drinking unpasteurized milk (Harper *et al.*, 2002), or drinking milk from a facility where milk was improperly pasteurized (Gelletlie *et al.*, 1997).

New molecular diagnostic tools have been developed to detect and differentiate *Cryptosporidium* at the species/genotype and subtype levels (Xiao and Ryan, 2004, 2008; Xiao, 2010). Small subunit rRNA-based tools are most commonly used for genotyping *Cryptosporidium*. The 60-kDa glycoprotein (GP60) is the most polymorphic marker identified so far in the *Cryptosporidium* genome, and DNA sequence analysis of the GP60 is widely used to subtype *C. parvum* in order to distinguish isolates for source tracking. The GP60 target is similar to a microsatellite sequence by having tandem repeats of the serine-coding trinucleotide TCA, TCG and TCT at the 5′ end of the gene. However, in addition to variation in the number of trinucleotide repeats, there are extensive sequence differences in the nonrepeat regions, which categorize *C. parvum* into several subtype families (alleles). Members of different subtype families differ from each other extensively in the primary sequence. Within each subtype family, subtypes differ from each other mostly in the number of nucleotide repeats (TCA, TCG and TCT). Subtypes of *C. parvum* and *C. hominis* at the GP60 locus are named based on their subtype family designation and the number of each type of trinucleotide repeats (Sulaiman *et al.*, 2005). For each subtype, the name starts with the subtype family designation followed by the number of TCA (represented by the letter A), TCG (represented by the letter G) and TCT (represented by the letter T) repeats found. Within the genus *Cryptosporidium* there is extensive genetic variation. In addition to the 20 named species of *Cryptosporidium*, over 40 genotypes have been described and new genotypes are continually being discovered (Fayer *et al.*, 2008; Xiao and Fayer, 2008; Fayer and Santín, 2009). It is worth noting that some *Cryptosporidium* species and genotypes have been found only in particular animal species while others have a broader host range.

Cryptosporidiosis is a very common infection in cattle worldwide (Santín and Trout, 2008). For dairy cattle, *Cryptosporidium* has become a concern not only because of the direct economic losses associated with lost production caused by the infection, but also from a public health perspective because of the potential for environmental contamination with *Cryptosporidium* oocysts. Four species are responsible for most cattle infections (*C. parvum*, *C. bovis*, *C. andersoni* and *C. ryanae*) (Lindsay *et al.*, 2000; Santín *et al.*, 2004, 2008; Feng *et al.*, 2007; Fayer *et al.*, 2008) and it is important to determine the species of *Cryptosporidium* in infected cattle because *C. parvum* is the only one of these that is a serious pathogen for humans. *Cryptosporidium parvum* is known to infect humans worldwide and is recognized as the major zoonotic *Cryptosporidium* species. Infection with *C. parvum* is most frequently observed in calves and often characterized by profuse watery diarrhea with acute onset. *Cryptosporidium andersoni* has been reported in humans but only sporadically (Leoni *et al.*, 2006). In cattle an age associated occurrence of different *Cryptosporidium* species has been documented with *C. parvum*, the most prevalent species in pre-weaned calves, *C. bovis* and *C. ryanae* in post-weaned calves, and *C. andersoni* in heifers and milking cows (Santín *et al.*, 2004, 2008; Langkjaer *et al.*, 2007; Feng *et al.*, 2007).

Multiple studies of cattle have shown prevalences of *Cryptosporidium* spp. ranging from 0% to 100% (Santín and Trout, 2008). There is a close association between prevalence of infection and age of the animals with the highest prevalence being in younger animals (Langkjaer *et al.*, 2007; Santín *et al.*, 2008). Cryptosporidiosis in calves is generally established during the first two weeks of life (Wade *et al.*, 2000; Santín *et al.*, 2008). A longitudinal study in which fecal samples were collected from 30 calves from birth to two years of age showed that all calves became infected with *C. parvum* at some time in the first two years (cumulative prevalence of 100%), whereas *C. bovis* and *C. ryanae* had cumulative prevalences of 80% and 60%, respectively, and *C. andersoni* was observed in only one animal (cumulative prevalence of 3.3%).

On commercial dairy farms fecal contamination from cattle of all ages is present in fields, pens, water supplies, buildings, tools and on animals themselves.

Animal handlers can also mechanically transport infective oocysts in feces on clothing and shoes. An animal infected with *C. parvum* could be excreting up to 10^7 oocysts per gram of feces (Fayer *et al.*, 1998), so a high degree of contamination can occur in the farm; this obviously represents a hazard to human health when such contamination occurs with zoonotic species. Proximity of dairy farms to agricultural fields may allow introduction of oocysts via contamination of irrigation water. There is a need to reduce the burden of contamination implementing good manure management practices. Direct access of cattle to water sources should be controlled and adequate treatment of feces for pathogenic inactivation should be conducted before disposal to reduce the transport of oocysts onto products that are consumed raw.

21.3 Tracking the sources, reservoirs and potential amplification steps of human pathogens in dairy production

21.3.1 Sources

Dairy farms are diverse, complex and relatively open environments. Although producers generally understand the need for biosecurity (Hoe and Ruegg, 2006), controlling exposure of the animals from outside sources is complex. There are many avenues by which dairy animals can be exposed to pathogens, some of which are easier than others to block. Bringing animals onto the premises from other farms is generally recognized as a risk factor for exposing the dairy herd to multiple disease vectors, including zoonotic pathogens. However, purchasing replacement animals is still a standard practice on many farms to replace low-producing, sick or injured animals or to increase herd size (Hoe and Ruegg, 2006). Contract heifer raising or rearing of heifers at another location is also a common practice and involves moving young calves off the farm for a period of approximately 12–18 months. The risk of introducing pathogens back into the parent herd upon re-entry of the animals increases substantially when heifers from multiple herds are co-mingled at the heifer-raising facility (Adhikari *et al.*, 2009; Hegde *et al.*, 2005). Human movement also presents risks for introducing pathogens onto the dairy farm. As with any farm system, there are numerous farm service people who necessarily come to the farm for deliveries, consultation or to work directly with the animals. Most of these people move from farm to farm and routinely follow appropriate footwear and equipment sanitation procedures; however, this traffic still poses a high risk for transport of pathogens between premises.

Other less manageable vehicles for potential pathogen entry to the farm are rodents, flies, birds and wildlife. Kirk *et al.* (2002) captured seven bird species from nine dairy farms in California and tested their gastrointestinal tracts for the presence of *Salmonella*. *Salmonella* was isolated from 22 of 892 (2.5%) captured birds and the within dairy prevalence of *Salmonella*-positive birds ranged from 0.7% to 16.7%. In a separate study, *Salmonella* was isolated from cloacal swabs collected from pigeons trapped on dairy farms in Colorado (Pedersen *et al.*, 2006). *Salmonella* was isolated from 9 of 106 (8.5%) samples. Pigeons were also trapped in urban locations but none of these pigeons (171) tested positive for *Salmonella*. In the same study, presumptive *E. coli* isolates from 406 samples from pigeons all tested negative for Shiga toxins. However, virulence genes (*eae*, *CNF-1*, *K1* and *hlyA*) were detected in isolates from 7.9% of the pigeons. Several studies have successfully isolated STEC from pigeons (Morabito *et al.*, 2001; Schmidt *et al.*, 2000) and seagulls (Makino *et al.*, 2000) so it does appear that birds can harbor both STEC and *Salmonella*. Birds can also be infected with *Cryptosporidium* and *C. parvum* was identified in feces from Canada geese (Zhou *et al.*, 2004). Also, *C. parvum* oocysts retained their viability and infectivity following passage through Canada geese (Graczyk *et al.*, 1998). Given the proximity of farms in farm-intense areas, it is highly likely that birds have a role in the dissemination of pathogens within the farming community. Similarly, *E. coli* O157 strains have been isolated from stable flies captured on dairy farms (Hancock *et al.*, 1998; Heuvelink *et al.*, 1998; Shere *et al.*, 1998) from houseflies on a cattle farm (Alam and Zurek, 2004), and from flies collected at agricultural fairs (Keen *et al.*, 2006). Flies that were collected from various livestock and wildlife facilities in Georgia were also found to be contaminated with *Cryptosporidium* (Conn *et al.*, 2007). Clearly, flies can act as mechanical vectors and contribute to the contamination, dissemination and persistence of pathogens in dairy cattle and their environment.

Dairy cattle diets are made up of ingredients that are grown on the farmland surrounding the animal housing facilities and also purchased from more geographically distant locations. Ingestion of contaminated feeds can lead to contamination in the herd. For example, in the USA corn and hay silage are very common ingredients in dairy cow diets and *Listeria* spp. have frequently been isolated from poorly fermented silages (i.e. pH > 4). In a study of 2000 dairy herds in France, Sanaa *et al.* (1993) determined that feeding improperly fermented silages was associated significantly with *L. monocytogenes* contamination of bulk milk. In this French study, hygiene in the milking process and cleanliness in the animal exercise area were also significant factors in bulk milk contamination. Presumably because the animal feces were infected, increased hygiene minimized the impact of fecal contamination of the bulk tank milk. Contamination of processed animal feeds has also been identified as a source of animal infection. The potential for contamination at the feed mill exists through a variety of means, such as contaminated ingredients and the presence of birds and rodents.

In several surveys conducted to specifically determine the risk factors for contamination of farms or bulk milk, herd size was implicated as being a risk factor with the risk of infection increasing with increased herd size (Antognoli *et al.*, 2009; Kabagambe *et al.*, 2000; Vaessen *et al.*, 1998; Warnick *et al.*, 2001). Additional risk factors for *Salmonella* contamination included components of the feeding regime (Kabagambe *et al.*, 2000; Vaessen *et al.*, 1998), the presence of liver fluke infections (Vaessen *et al.*, 1998), the presence of rodents or geese on the premise (Warnick *et al.*, 2001) and the use of a flush water system for manure removal (Kabagambe *et al.*, 2000). Antognoli *et al.* (2009) concluded that geographical location and herd size affected the risk factors for *L. monocytogenes* contamination of bulk milk. As was pointed out by Warnick *et al.* (2001), dairy farms are very diverse in size and management and there are multiple factors that are responsible for animal infection; therefore, it is difficult to pinpoint all the specific factors that present the greatest risks for infection.

21.3.2 Reservoirs

Once established, eliminating bacterial or protozoan pathogens from a dairy farm can be onerous. Besides the animal reservoirs (cows, birds, rodents), even on well managed, clean facilities, there are many moist 'nooks and crannies' where microbes will not only survive but potentially multiply. These reservoirs provide the potential for maintaining the on-farm pathogen presence and continued re-infection of animals. The combination of survival in the environment and asymptomatic infection in the animals often leads to a long-term contamination of the animals and premises.

As was discussed earlier in this chapter, dairy cattle can harbor many zoonotic bacteria and protozoan species without impact on their health or production. Therefore the cows and/or calves are frequently infected unbeknownst to the producer. For example, an endemic *Salmonella* infection was tracked in a dairy herd in Pennsylvania for more than four years (Pradhan *et al.*, 2009; Van Kessel, Karns, *et al.*, 2007; Van Kessel *et al.*, 2008). In the spring of 2004 this herd experienced a brief clinical outbreak of *S. enterica* ser. Typhimurium var. Copenhagen followed by a few clinical cases associated with *S. enterica* ser. Kentucky in the summer of 2004. Both of these outbreaks were short-lived and neither of these serotypes were isolated after less than two months of the initial outbreak. During the serotype Kentucky outbreak, *S. enterica* ser. Cerro was isolated from several environmental samples, marking the beginning of a sustained, subclinical outbreak of serotype Cerro. This approximately 100 cow herd was sampled every six to eight weeks over the next four years and herd prevalence ranged from 8% to 88% during this time (Fig. 21.1). After the cows had been shedding serotype Cerro for approximately 16 months, a gradual shift from serotype Cerro to serotype Kentucky was observed. As with the serotype Cerro infection, animals that were infected with serotype Kentucky were asymptomatic. Although serotype Cerro has caused disease in humans, it is an uncommon human isolate. Serotype Kentucky is more frequently associated with human infections than Cerro, but Kentucky is also a relatively rare human serotype. Although asymptomatic carriage did not apparently impact animal production in this herd (unpublished data), it still represents a significant risk for public safety. As was mentioned above, the results of many single time point surveys have shown that asymptomatic shedding of salmonellae, *L. monocytogenes*, or pathogenic *E. coli* is frequent on dairy farms and this scenario of long-term shedding, population shifts and contaminated environments is likely to be found on many farms.

When animal infection is on-going it is very difficult to clear the environment of the pathogen. Cows excrete more than 60 kg of feces per day and therefore an

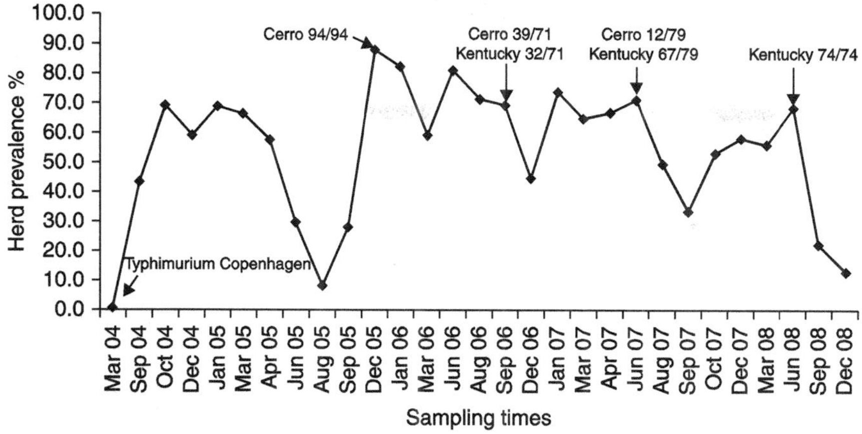

Fig. 21.1 Fecal shedding prevalence in a dairy herd with an endemic *Salmonella* infection. Arrows indicate the predominant serotypes at that time point.

infected cow is a continuous source of environmental contamination. Fecal contamination of water troughs and feed bunks can be minimized but cannot be avoided and therefore cycling of the pathogens between the animals and their near environment is essentially continuous and difficult to break.

Recent studies on a dairy farm in New York state identified the presence of a *L. monocytogenes*-containing biofilm as the source of continuous contamination of the bulk tank milk (Latorre *et al.*, 2009). The plastic milk meters used in the parlor for recording the milk weights from each animal were shown to contain biofilms. It is likely that biofilms also formed on sections of the stainless steel that make up most of the milk transport system. In this particular farm, it appeared that the specified protocols for cleaning and sanitizing of the milking equipment were not being followed. Although *L. monocytogenes* was identified as the problem in this farm, many bacterial species can form biofilms and so the problem could also occur with other bacteria such as EHEC.

21.3.3 Pathogen survival

Manure from the dairy animals can be handled in a variety of ways but in most cases it ultimately ends up on the land surrounding the animal housing premises. Little is known about how long manure-borne pathogens can survive once they are excreted and spread into the environment. Survival time can be influenced by many environmental factors such as temperature, moisture, pH, soil type and tillage practices for manure that has been spread on cropland. Based on a study of survival in cowpats deposited in sunny and shaded areas, it was demonstrated that cowpats may be a significant source of *E. coli* for at least 30 days (Van Kessel, Pachepsky, *et al.*, 2007). By simulating manure treatment and rainfall on soil cores, Gagliardi and Karns (2000) showed that *E. coli* O157:H7 can leach through the soil and survive for more than two months. An initial phase of bacterial growth was observed in both of these studies so it appears that the level of environmental contamination is sometimes amplified before it begins to subside. *Cryptosporidium* oocysts can also survive in the environment for long periods of time. In a study in which *C. parvum* oocysts were incubated between 5°C to 15°C, some oocysts remained infectious for six months (Fayer *et al.*, 1998).

21.4 Pathogen monitoring strategies

Preventative screening for food-associated pathogens is not usually conducted on dairy farms and pathogens are generally only detected when one or more animals exhibit clinical signs of infection. Therefore asymptomatic infections in the cows or replacement animals are only detected when there is a co-infection with a disease-causing organism. Zoonotic bacteria such as EHEC and *L. monocytogenes* are rarely detected unless a survey of some sort is conducted. Salmonellae are detected more frequently because salmonellosis is not uncommon in cattle and is often included in diagnostic testing of sick animals or in necropsy of dead animals.

Routine monitoring of bulk milk has been in place for decades for the majority of commercial milk produced in the USA. Milk is routinely monitored for bacterial load (standard plate counts, preliminary incubation count), and for somatic cell counts. Somatic cell counts are an indication of udder health and the presence of subclinical mastitis. Bulk milk or milk from individual cows is also frequently tested for microbial groups (coliforms, coagulase negative *Staphylococci*, *Streptococci*, Gram-negative non-coliforms) or specific bacterial species (*Staphylococcus aureus*, *Streptococcus agalactiae*, *Mycoplasma*) (Jayarao *et al.*, 2001, 2004). This screening is conducted for animal health purposes, to detect or identify mastitis-causing organisms and not for detection of food-associated, zoonotic pathogens. Raw milk sales are legal at either the farm or retail level in more than half of the states in the USA and additional screening for pathogens may be conducted on milk that is destined to be consumed without processing. Although some level of testing may be required by individual states, testing is often infrequent. For example, in the state of Pennsylvania, raw milk sales permit holders are only required to test for food-associated pathogens (*Salmonella*, *L. monocytogenes*, *Campylobacter*, and *E. coli* O157:H7) biannually while there is a monthly requirement in the state of New York. As has been discussed, the microbial ecology of dairy farms is dynamic and pathogen contamination is likely to go undetected with infrequent testing. In the case of Pennsylvania, when a foodborne outbreak occurs and the farm is identified as the source or when contamination is identified on routine testing, consecutive tests with negative results are then typically required before sales are resumed.

There is a growing interest in the consumption of raw milk in the USA. In order to reduce the risk of these consumers being exposed to pathogens in raw milk, the raw milk should be tested more frequently. In addition, testing of the milk filters rather than only the milk has the potential to increase the sensitivity of detection and further reduce the risks of human exposure.

21.5 Improving pathogen control

Improving on-farm zoonotic pathogen control is for most dairy farms not a specific strategy. As outlined above, most zoonotic pathogens are not important causes of disease or production loss in the dairy cows or replacement animals. Therefore, few specific programs to reduce zoonotic pathogen load on the farm are currently in place, with the potential exception of *Salmonella* prevention programs. However, more general good management practices may also lead to a reduced pathogen load on the farm and subsequently in food products coming from the farm.

Salmonella prevention programs are typically based on reducing the risk of entry of *Salmonella* pathogens onto the farm through biosecurity and reducing the risk of transmission of the infection on the farm through good management practices in the calf rearing facilities. This is especially important because once established in the milking population, carriage of zoonotic pathogens can be difficult to control. In some specific situations such as *Salmonella* Dublin infections, adult cows are tested to identify carriers, and control of macro-parasites (liver fluke) is practiced to reduce the incidence of carrier animals. Farms may institute *Salmonella* vaccination programs, but the efficacy of the *Salmonella* vaccines, either specific *Salmonella* spp. vaccines or general gram-negative core antigen vaccines, to reduce on-farm transmission of *Salmonella* spp. remains uncertain (Heider *et al.*, 2008; Hermesch *et al.*, 2008; House *et al.*, 2001). Manure treatments such as anaerobic digestion or composting have been shown to reduce the levels of pathogens and may help break cycles of re-infection or infection of incoming animals but they require strict process controls to assure effectiveness. Research has shown that adjusting the pH of manures or adding carbonates and/or ammonia to maintain high concentrations of carbonate anion or free ammonia can dramatically reduce the population of *E. coli* and *Salmonella* but the effects of these treatments on the cattle or on downstream use of the manure as fertilizer are not completely known (Arthurs *et al.*, 2001; Park and Diez-Gonzalez, 2003). Several studies have suggested that dietary factors such as grains, hay or by-products such as cotton seed or distillers grains may affect pathogen carriage in mature cows but little is known about how dietary changes designed to reduce pathogen loads affect milk production (Callaway *et al.*, 2009; Diez-Gonzalez *et al.*, 1998; Jacob *et al.*, 2008; Loneragan and Brashears, 2005; Tkalcic *et al.*, 2000).

General programs to stimulate the use of best management practices exist in several states. In New York the New York Cattle Health Assurance Program (NYSCHAP; http://www.nyschap.vet.cornell.edu) is promoting the use of best management practices for a variety of different disease control programs. For example, the milk quality and mastitis module of the program promotes standardized milk procedures that include adequate cleaning of the teat and udder before milking, wearing of gloves and clean farm worker clothing while harvesting milk in the milking parlor. Although such procedures are primarily aimed at reducing mastitis, the 'side effect' is also a lower zoonotic pathogen load in milk. As mentioned above, parlor hygiene has been identified as an important factor in bulk milk pathogen load. Similarly, in the calf health programs the use of adequate and timely colostrum is being promoted. Although this is primarily to affect calf health and the reduction of infection with *Mycobacterium avium* subspp. *paratuberculosis*, the generic side effect is that calves are more healthy and less susceptible to infections with other organisms such as *Salmonella* spp. and *Cryptosporidium*.

More specific programs to reduce the zoonotic pathogen load in bulk milk or beef are being promoted for raw milk producers or dairy farmers selling beef and dairy products directly to consumers. Weekly monitoring of bulk milk for zoonotic pathogens leads firstly to a much better awareness of the potential risks and secondly to the need for control programs once zoonotic pathogens have been identified in the bulk tank milk. In some cases molecular methods might reduce the time it takes to assess the contamination of milk and manure and allow for increased rates of sampling, however, such methods require specialized expertise and equipment and would incur significant extra costs.

21.6 Future trends

Food safety is of increasing concern to consumers, including those of dairy and beef products. This concern will be translated into a demand of consumers, retailers and processors for the producer of the primary product to implement more verifiable production processes that reduce the risk of foodborne pathogens in the raw product. To be able to meet such demands, dairy farmers will be more interested in implementing effective preventative programs to reduce the zoonotic pathogen load in products that leave the farm. Such programs will likely include a monitoring component, a risk-assessment component and verifiable implementation of the relevant best management procedures. Our current insights into preventative practices that reduce zoonotic bacterial load of milk and dairy beef are limited to common sense hygiene practices, ration changes to reduce the STEC shedding and isolation of known shedders of zoonotic pathogens. In our experience, dairy farmers are committed to providing quality milk to consumers but their ability to adopt new management practices is strongly affected by economics. Further insight into specific practices to reduce the load on the dairy farm of *Salmonella* spp., STEC, *L. monocytogenes*, *Campylobacter* and *Cryptosporidium* spp. and their impacts upon production will be essential to provide consumer of dairy products and dairy beef with more assurances of the safety of their food.

21.7 References

ADHIKARI B, BESSER T E, GAY J M, FOX L K, DAVIS M A, COBBOLD R N, ET AL. (2009), 'The role of animal movement, including off-farm rearing of heifers, in the interherd transmission of multidrug-resistant *Salmonella*', *J Dairy Sci*, **92**, 4229–38.

ALAM M J and ZUREK L (2004), 'Association of *Escherichia coli* O157:H7 with houseflies on a cattle farm', *Appl Environ Microbiol*, **70**, 7578–80.

ANTOGNOLI M C, LOMBARD J E, WAGNER B A, MCCLUSKEY B J, VAN KESSEL J S and KARNS J S (2009), 'Risk factors associated with the presence of viable *Listeria monocytogenes* in bulk tank milk from US dairies', *Zoonoses Public Health*, **56**, 77–83.

ARTHURS C E, JARVIS G N and RUSSELL J B (2001), 'The effect of various carbonate sources on the survival of *Escherichia coli* in dairy cattle manure', *Curr Microbiol*, **43**, 220–24.

BACH S J, MCALLISTER T A, VEIRA D M, GANNON V P J and HOLLEY R A (2002), 'Transmission and control of *Escherichia coli* O157:H7 – a review', *Can J Anim Sci*, **82**, 475–90.

BORUCKI M K, GAY C C, REYNOLDS J, MCELWAIN K L, KIM S H, CALL D R, ET AL. (2005), 'Genetic diversity of *Listeria monocytogenes* strains from a high-prevalence dairy farm', *Appl Environ Microbiol*, **71**, 5893–9.

BORUCKI M K, REYNOLDS J, GAY C C, MCELWAIN K L, KIM S H, KNOWLES D P and HU J (2004), 'Dairy farm reservoir of *Listeria monocytogenes* sporadic and epidemic strains', *J Food Pro*, **67**, 2496–9.

BURVENICH C, VAN MERRIS V, MEHRZAD J, EZ-FRAILE A and DUCHATEAU L (2003), 'Severity of *E. coli* mastitis is mainly determined by cow factors', *Vet Res*, **34**, 521–64.

BYRNE C M, EROL I, CALL J E, KASPAR C W, BUEGE D R, HIEMKE C J, ET AL. (2003), 'Characterization of *Escherichia coli* O157:H7 from downer and healthy dairy cattle in the upper Midwest region of the United States', *Appl Environ Microbiol*, **69**, 4683–8.

CALLAWAY T R, CARR M A, EDRINGTON T S, ANDERSON R C and NISBET D J (2009), 'Diet, *Escherichia coli* O157:H7, and Cattle: a review after 10 years', *Curr Issues Mol Biol*, **11**, 67–80.

CALLAWAY T R, KEEN J E, EDRINGTON T S, BAUMGARD L H, SPICER L, FONDA E S, ET AL. (2005), 'Fecal prevalence and diversity of *Salmonella* species in lactating dairy cattle in four states', *J Dairy Sci*, **88**, 3603–8.

CDC (2001), 'Outbreak of listeriosis associated with homemade mexican style cheese – North Carolina, October–January 2001', *Morb Mortal Wkly Rep*, **50**, 560–2.

CDC (2002), 'Outbreak of *Campylobacter jejuni* infections associated with drinking unpasteurized milk procured through a cow-leasing program – Wisconsin, 2001', *Morb Mortal Wkly Rep*, **51**, 548–9.

CDC (2007), '*Escherichia coli* O157:H7 infection associated with drinking raw milk – Washington and Oregon, November–December 2005', *Morb Mortal Wkly Rep*, **56**, 165–7.

CDC (2008a), *Salmonella surveillance: annual summary, 2006*. Department of Health and Human Services, CDC, Atlanta, GA.

CDC (2008b), '*escherichia coli* 0157:h7 infections in children associated with raw milk and raw colostrum from cows – California, 2006', *Morb Mortal Wkly Rep*, **57**, 625–8.

CDC (2008c), 'outbreak of *Listeria monocytogenes* infections associated with pasteurized milk from a local dairy – Massachusetts, 2007', *Morb Mortal Wkly Rep*, **57**, 1097–100.

CONN D B, WEAVER J, TAMANG L and GRACZYK T K (2007), 'Synanthropic flies as vectors of *Cryptosporidium* and *Giardia* among livestock and wildlife in a multispecies agricultural complex', *Vector-Borne Zoonotic*, **7**, 643–51.

CULLOR J S (1997), 'Mastitis and dairy environment pathogens of public health concern', *National Mastitis Council Annual Meeting Proceedings*, Tulare, California, February: pp. 20–3.

DAVISON H C, SMITH R P, PASCOE S J, SAYERS A R, DAVIES R H, WEAVER J P, ET AL. (2005), 'Prevalence, incidence and geographical distribution of serovars of *Salmonella* on dairy farms in England and Wales', *Vet Rec*, **157**, 703–11.

DIEZ-GONZALEZ F, CALLAWAY T R, KIZOULIS M G and RUSSELL J B (1998), 'Grain feeding and the dissemination of acid-resistant *Escherichia coli* from cattle', *Science*, **281**, 1666–8.

DJURETIC T, WALL P G and NICHOLS G (1997), 'General outbreaks of infectious intestinal disease associated with milk and dairy products in England and Wales: 1992 to 1996', *Commun Dis Rep CDR Rev*, **7**, R41–5.

EDRINGTON T S, CALLAWAY T R, IVES S E, ENGLER M J, LOOPER M L, ANDERSON R C, ET AL. (2006), 'Seasonal shedding of *Escherichia coli* O157:H7 in ruminants: a new hypothesis', *Foodborne Pathog Dis*, **3**, 413–21.

EDRINGTON T S, HUME M E, LOOPER M L, SCHULTZ C L, FITZGERALD A C, CALLAWAY T R, ET AL. (2004), 'Variation in the faecal shedding of *Salmonella* and *E. coli* O157:H7 in lactating dairy cattle and examination of *Salmonella* genotypes using pulsed-field gel electrophoresis', *Lett Appl Microbiol*, **38**, 366–72.

EDRINGTON T S, SCHULTZ C L, BISCHOFF K M, CALLAWAY T R, LOOPER M L, GENOVESE K J, ET AL. (2004), 'Antimicrobial resistance and serotype prevalence of *Salmonella* isolated from dairy cattle in the southwestern United States', *Microb Drug Resist*, **10**, 51–6.

ENGLEN M D, HILL A E, DARGATZ D A, LADELY S R and FEDORKA-CRAY P J (2007), 'Prevalence and antimicrobial resistance of *Campylobacter* in US dairy cattle', *J Appl Microbiol*, **102**, 1570–7.

FAHEY T, MORGAN D, GUNNEBURG C, ADAK G K, MAJID F and KACZMARSKI E (1995), 'An outbreak of *Campylobacter jejuni* enteritis associated with failed milk pasteurisation', *J Infect*, **31**, 137–43.

FAITH N G, SHERE J A, BROSCH R, ARNOLD K W, ANSAY S E, LEE M-S, ET AL. (1996), 'Prevalence and clonal nature of *Escherichia coli* O157:H7 on dairy farms in Wisconsin', *Appl Environ Microbiol*, **62**, 1519–25.

FARBER J M and PETERKIN P I (1991), '*Listeria monocytogenes*, a food-borne pathogen', *Microbiol Rev*, **55**, 476–511.

FAYER R, GASBARRE L, PASQUALI P, CANALS A, ALMERIA S and ZARLENGA D (1998), '*Cryptosporidium parvum* infection in bovine neonates: dynamic clinical, parasitic and immunologic patterns', *Int J Parasitol*, **28**, 49–56.

FAYER R and SANTÍN, M (2009), '*Cryptosporidium* xiaoi n. sp. (Apicomplexa: Cryptosporidiidae) in sheep (*Ovis aries*)', *Vet Parasitol*, **164**, 192–200.

FAYER R, SANTÍN M and TROUT J M (2008), '*Cryptosporidium ryanae* n. sp. (Apicomplexa: Cryptosporidiidae) in cattle (*Bos taurus*)', *Vet Parasitol*, **156**, 191–8.

FENG Y, ORTEGA Y, HE G, DAS P, XU M, ZHANG X, ET AL. (2007), 'Wide geographic distribution of *Cryptosporidium bovis* and the deer-like genotype in bovines', *Vet Parasitol*, **144**, 1–9.

FITZGERALD A C, EDRINGTON T S, LOOPER M L, CALLAWAY T R, GENOVESE K J, BISCHOFF K M, ET AL. (2003), 'Antimicrobial susceptibility and factors affecting the shedding of *E. coli* O157:H7 and *Salmonella* in dairy cattle', *Lett Appl Microbiol*, **37**, 392–8.

GAGLIARDI J V and KARNS J S (2000), 'Leaching of *Escherichia coli* O157:H7 in diverse soils under various agricultural management practices', *Appl Environ Microbiol*, **66**, 877–83.

GELLETLIE R, STUART J, SOLTANPOOR N, ARMSTRONG R and NICHOLS G (1997), 'Cryptosporidiosis associated with school milk', *Lancet*, **350**, 1005–6.

GRACZYK T K, FAYER R, TROUT J M, LEWIS E J, FARLEY C A, SULAIMAN I, ET AL. (1998), '*Giardia* sp. cysts and infectious *Cryptosporidium parvum* oocysts in the feces of migratory Canada geese (*Branta canadensis*)', *Appl Environ Microbiol*, **64**, 2736–8.

HANCOCK D D, BESSER T E, RICE D H, EBEL E D, HERRIOTT D E and CARPENTER L V (1998), 'Multiple sources of *Escherichia coli* O157 in feedlots and dairy farms in the northwestern USA', *Prev Vet Med*, **35**, 11–19.

HARPER C M, COWELL N A, ADAMS B C, LANGLEY A J and WOHLSEN T D (2002), 'Outbreak of *Cryptosporidium* linked to drinking unpasteurised milk', *Commun Dis Intell*, **26**, 449–50.

HARVEY R B, DROLESKEY R E, SHEFFIELD C L, EDRINGTON T S, CALLAWAY T R, ANDERSON R C, ET AL. (2004), '*Campylobacter* prevalence in lactating dairy cows in the United States', *J Food Prot*, **67**, 1476–9.

HASSAN L, MOHAMMED H O, MCDONOUGH P L and GONZALEZ R N (2000), 'A cross-sectional study on the prevalence of *Listeria monocytogenes* and *Salmonella* in New York dairy herds', *J Dairy Sci*, **83**, 2441–7.

HEGDE N V, COOK M L, WOLFGANG D R, LOVE B C, MADDOX C C and JAYARAO B M (2005), 'Dissemination of *Salmonella enterica* subsp. *enterica* serovar Typhimurium var. Copenhagen clonal types through a contract heifer-raising operation', *J Clin Microbiol*, **43**, 4208–11.

HEIDER L C, MEIRING R W, HOET A E, GEBREYES W A, FUNK J A and WITTUM T E (2008), 'Evaluation of vaccination with a commercial subunit vaccine on shedding of *Salmonella enterica* in subclinically infected dairy cows', *J Am Vet Med Assoc* **233**, 466–9.

HERMESCH D R, THOMSON D U, LONERAGAN G H, RENTER D R and WHITE B J (2008), 'Effects of a commercially available vaccine against *Salmonella enterica* serotype Newport on milk production, somatic cell count, and shedding of *Salmonella* organisms in female dairy cattle with no clinical signs of salmonellosis', *Am J Vet Res*, **69**, 1229–34.

HEUVELINK A E, VAN DEN BIGGELAAR F L, ZWARTKRUIS-NAHUIS J, HERBES R G, HUYBEN R, NAGELKERKE N, ET AL. (1998), 'Occurrence of verocytotoxin-producing *Escherichia coli* O157 on Dutch dairy farms', *J Clin Microbiol*, **36**, 3480–7.

HOE F G and RUEGG P L (2006), 'Opinions and practices of Wisconsin dairy producers about biosecurity and animal well-being', *J Dairy Sci*, **89**, 2297–308.

HOUSE J K, ONTIVEROS M M, BLACKMER N M, DUEGER E L, FITCHHORN J B, MCARTHUR G R, ET AL. (2001), 'Evaluation of an autogenous *Salmonella* bacterin and a modified live *Salmonella* serotype Choleraesuis vaccine on a commercial dairy farm', *Am J Vet Res*, **62**, 1897–902.

HUME M E, EDRINGTON T S, LOOPER M L, CALLAWAY T R, GENOVESE K J and NISBET D J (2004), '*Salmonella* genotype diversity in nonlactating and lactating dairy cows', *J Food Prot*, **67**, 2280–3.

JACOB M E, FOX J T, DROUILLARD J S, RENTER D G and NAGARAJA T G (2008), 'Effects of dried distillers' grain on fecal prevalence and growth of *Escherichia coli* O157 in batch culture fermentations from cattle', *Appl Environ Microbiol*, **74**, 38–43.

JAYARAO B M and HENNING D R (2001), 'Prevalence of foodborne pathogens in bulk tank milk', *J Dairy Sci*, **84**, 2157–62.

JAYARAO B M, PILLAI S R, SAWANT A A, WOLFGANG D R and HEGDE N V (2004), 'Guidelines for monitoring bulk tank milk somatic cell and bacterial counts', *J Dairy Sci*, **87**, 3561–73.

JAYARAO B M, PILLAI S R, WOLFGANG D R, GRISWOLD D R and HUTCHINSON L J (2001), 'Herd level information and bulk tank milk analysis: tools for improving milk quality and herd udder health', *Bovine Pr*, **35**, 23–35.

JENSEN N E, AARESTRUP F M, JENSEN J and WEGENER H C (1996), '*Listeria monocytogenes* in bovine mastitis: possible implication for human health', *Int J Food Microbiol*, **32**, 209–16.

KABAGAMBE E K, WELLS S J, GARBER L P, SALMAN M D, WAGNER B and FEDORKA-CRAY P J (2000), 'Risk factors for fecal shedding of *Salmonella* in 91 US dairy herds in 1996', *Prev Vet Med*, **43**, 177–94.

KARNS J S, VAN KESSEL J S, MCCLUSKEY B J and PERDUE M L (2005), 'Prevalence of *Salmonella enterica* in bulk tank milk from US dairies as determined by polymerase chain reaction', *J Dairy Sci*, **88**, 3475–9.

KARNS J S, VAN KESSEL J S, MCCLUSKY B J and PERDUE M L (2007), 'Incidence of *Escherichia coli* O157:H7 and *E. coli* virulence factors in US bulk tank milk as determined by polymerase chain reaction', *J Dairy Sci*, **90**, 3212–19.

KEEN J E, WITTUM T E, DUNN J R, BONO J L and DURSO L M (2006), 'Shiga-toxigenic *Escherichia coli* O157 in agricultural fair livestock, United States', *Emerg Infect Dis*, **12**, 780–6.

KIRK J H, HOLMBERG C A and JEFFREY J S (2002), 'Prevalence of *Salmonella* spp. in selected birds captured on California dairies', *J Am Vet Med Assoc*, **220**, 359–62.

LANGKJAER R B, VIGRE H, ENEMARK, H L and MADDOX-HYTTEL C (2007), 'Molecular and phylogenetic characterization of *Cryptosporidium* and *Giardia* from pigs and cattle in Demark', *Parasitology*, **134**, 339–50.

LATORRE A A, VAN KESSEL J A, KARNS J S, ZURAKOWSKI M J, PRADHAN A K, ZADOKS R N, ET AL. (2009), 'Molecular ecology of *Listeria monocytogenes*: evidence for a reservoir in milking equipment on a dairy farm', *Appl Environ Microbiol*, **75**, 1315–23.

LEONI F, AMAR C, NICHOLS G, PEDRAZA-DÍAZ S and MCLAUCHLIN J (2006), 'Genetic analysis of *Cryptosporidium* from 2414 humans with diarrhoea in England between 1985 and 2000', *J Med Microbiol*, **55**, 703–7.

LINDSAY D S, UPTON S J, OWENS D S, MORGAN U M, MEAD J R and BLAGBURN B L (2000), '*Cryptosporidium andersoni* n. sp. (Apicomplexa: Cryptosporiidae) from cattle, *Bos taurus*', *J Eukaryot Microbiol*, **47**, 91–5.

LONERAGAN G H and BRASHEARS M M (2005), 'Pre-harvest interventions to reduce carriage of *E. coli* O157 by harvest-ready feedlot cattle', *Meat Science*, **71**, 72–8.

MACKENZIE W R, HOXIE N J, PROCTOR M E, GRADUS M S, BLAIR K A, PETERSON D E, ET AL. (1994), 'A massive outbreak in Milwaukee of *Cryptosporidium* infection transmitted through the public water supply', *N Engl J Med*, **331**, 161–7.

MAKINO S, KOBORI H, ASAKURA H, WATARAI M, SHIRAHATA T, IKEDA T, ET AL. (2000), 'Detection and characterization of Shiga toxin-producing *Escherichia coli* from seagulls', *Epidemiol Infect*, **125**, 55–61.

MEAD P S, SLUTSKER L, DIETZ V, MCCAIG L F, BRESEE J S, SHAPIRO C, ET AL. (1999), 'Food-related illness and death in the United States', *Emerg Infect Dis*, **5**, 607–25.

MENG J and DOYLE M P (1997), 'Emerging issues in microbiological food safety', *Annu Rev Nutr*, **17**, 255–75.

MILLARD P S, GENSHEIMER K F, ADDISS D G, SOSIN D M, BECKETT G A, HOUCK-JANKOSKI A, ET AL. (1994), 'An outbreak of cryptosporidiosis from fresh-pressed apple cider', *J Am Med Assoc* **272**, 1592–6.

MOHAMMED H O, STIPETIC K, MCDONOUGH P L, GONZALEZ R N, NYDAM D V and ATWILL E R (2009), 'Identification of potential on-farm sources of *Listeria monocytogenes* in herds of dairy cattle', *Am J Vet Res*, **70**, 383–8.

MORABITO S, DELL'OMO G, AGRIMI U, SCHMIDT H, KARCH H, CHEASTY T, ET AL. (2001), 'Detection and characterization of Shiga toxin-producing *Escherichia coli* in feral pigeons', *Vet Microbiol*, **82**, 275–83.

MURINDA S E, NGUYEN L T, IVEY S J, GILLESPIE B E, ALMEIDA R A, DRAUGHON F A, ET AL. (2002a), 'Molecular characterization of *Salmonella* spp. isolated from bulk tank milk and cull dairy cow fecal samples', *J Food Prot*, **65**, 1100–5.

MURINDA S E, NGUYEN L T, IVEY S J, GILLESPIE B E, ALMEIDA R A, DRAUGHON F A, ET AL. (2002b), 'Prevalence and molecular characterization of *Escherichia coli* O157:H7 in bulk tank milk and fecal samples from cull cows: a 12-month survey of dairy farms in east Tennessee', *J Food Prot*, **65**, 752–9.

NIGHTINGALE K K, SCHUKKEN Y H, NIGHTINGALE C R, FORTES E D, HO A J, HER Z, ET AL. (2004), 'Ecology and transmission of *Listeria monocytogenes* infecting ruminants and in the farm environment', *Appl Environ Microbiol*, **70**, 4458–67.

O'BRIEN A D and KAPER J B (1998), 'Shiga toxin-producing *Escherichia coli*: Yesterday, today, and tomorrow', in Kaper, J.B. and O'Brien, A.D. (Eds): *Escherichia coli O157:H7 and Other Shiga Toxin-Producing E. coli Strains*, ASM Press, Washington, DC, pp. 1–11.

OLIVER S P, JAYARAO B M and ALMEIDA R A (2005), 'Foodborne pathogens in milk and the dairy farm environment: food safety and public health implications', *Foodborne Pathog Dis*, **2**, 115–29.

OLSEN S J, YING M, DAVIS M F, DEASY M, HOLLAND B, IAMPIETRO L, ET AL. (2004), 'Multidrug-resistant *Salmonella* Typhimurium infection from milk contaminated after pasteurization', *Emerg Infect Dis*, **10**, 932–5.

PARK G W and DIEZ-GONZALEZ F (2003), 'Utilization of carbonate and ammonia-based treatments to eliminate *Escherichia coli* O157:H7 and *Salmonella* Typhimurium DT104 from cattle manure', *J Appl Microbiol*, **94**, 675–85.

PEARSON L J and MARTH E H (1990), '*Listeria monocytogenes* – threat to a safe food supply: a review', *J Dairy Sci*, **73**, 912–28.

PEDERSEN K, CLARK L, ANDELT W F and SALMAN M D (2006), 'Prevalence of Shiga toxin-producing *Escherichia coli* and *Salmonella enterica* in rock pigeons captured in Fort Collins, Colorado', *J Wildl Dis*, **42**, 46–55.

PEEK S E, HARTMANN F A, THOMAS C B and NORDLUND K V (2004), 'Isolation of *Salmonella* spp from the environment of dairies without any history of clinical salmonellosis', *J Am Vet Med Assoc*, **225**, 574–7.

PRADHAN A K, VAN KESSEL J S, KARNS J S, WOLFGANG D R, HOVINGH E, NELEN K A, ET AL. (2009), 'Dynamics of endemic infectious diseases of animal and human importance on three dairy herds in the northeastern United States', *J Dairy Sci*, **92**, 1811–25.

QUIROZ E S, BERN C, MACARTHUR J R, XIAO L, FLETCHER M, ARROWOOD M J, ET AL. (2000), 'An outbreak of cryptosporidiosis linked to a foodhandler', *J Infect Dis*, **181**, 695–700.

RADKE B R, MCFALL M and RADOSTITS S M (2002), '*Salmonella* Muenster infection in a dairy herd', *Can Vet J*, **43**, 443–53.

ROBERTS A J and WIEDMANN M (2003), 'Pathogen, host and environmental factors contributing to the pathogenesis of listeriosis', *Cell Mol Life Sci*, **60**, 904–18.

ROHRBACH B W, DRAUGHON F A, DAVIDSON P M and OLIVER S P (1992), 'Prevalence of *Listeria monocytogenes, Campylobacter jejuni, Yersinia enterocolitica*, and *Salmonella* in bulk tank milk: risk factors and risk of human exposure', *J Food Prot*, **55**, 93–7.

RYSER E T (2001), 'Public Health Concerns', in Marth, E.H. and Steele, J.L. (Eds): *Applied Dairy Microbiology*, Marcel Dekker Inc., New York, pp. 397–545.

SANAA M, POUTREL B, MENARD J L and SERIEYS F (1993), 'Risk factors associated with contamination of raw milk by *Listeria monocytogenes* in dairy farms', *J Dairy Sci*, **76**, 2891–8.

SANTÍN M and TROUT J M (2008), 'Livestock', in Fayer, R. and Xiao, L. (Eds): *Cryptosporidium and Cryptosporidiosis*, 2nd ed., CRC Press, Boca Raton, FL, pp. 451–83.

SANTÍN M, TROUT J M and FAYER R (2008), 'A longitudinal study of cryptosporidiosis in dairy cattle from birth to 2 years of age', *Vet Parasitol*, **155**, 15–23.

SANTÍN M, TROUT J M, XIAO L, ZHOU L, GREINER E and FAYER R (2004), 'Prevalence and age-related variation of *Cryptosporidium* species and genotypes in dairy calves', *Vet Parasitol*, **122**, 103–17.

SATO K, BARTLETT P C, KANEENE J B and DOWNES F P (2004), 'Comparison of prevalence and antimicrobial susceptibilities of *Campylobacter* spp. isolates from organic and conventional dairy herds in Wisconsin', *Appl Environ Microbiol*, **70**, 1442–7.

SCHMIDT H, SCHEEF J, MORABITO S, CAPRIOLI A, WIELER L H and KARCH H (2000), 'A new Shiga toxin 2 variant (Stx2f) from *Escherichia coli* isolated from pigeons', *Appl Environ Microbiol*, **66**, 1205–8.

SHERE J A, BARTLETT K J and KASPAR C W (1998), 'Longitudinal study of *Escherichia coli* O157:H7 dissemination on four dairy farms in Wisconsin', *Appl and Environ Microbiol*, **64**, 1390–9.

STANLEY K and JONES K (2003), 'Cattle and sheep farms as reservoirs of *Campylobacter*', *J Appl Microbiol*, **94**(Suppl), 104S–13S.

STEELE M L, MCNAB W B, POPPE C, GRIFFITHS M W, CHEN S, DEGRANDIS S A, ET AL. (1997), 'Survey of Ontario bulk tank raw milk for food-borne pathogens', *J Food Prot*, **60**, 1341–46.

SULAIMAN I M, HIRA P R, ZHOU L, AL-ALI F M, AL-SHELAHI F A, SHWEIKI H M, ET AL. (2005), 'Unique endemicity of cryptosporidiosis in children in Kuwait', *J Clin Microbiol*, **43**, 2805–9.

TKALCIC S, BROWN C A, HARMON B G, JAIN A V, MUELLER E P, PARKS A, ET AL. (2000), 'Effects of diet on rumen proliferation and fecal shedding of *Escherichia coli* O157:H7 in calves', *J Food Prot*, **63**, 1630–6.

TROUTT H F, GALLAND J C, OSBURN B I, BREWER R L, BRAUN R K, SCHMITZ J A, ET AL. (2001), 'Prevalence of *Salmonella* spp. in cull (market) dairy cows at slaughter', *J Am Vet Med Assoc*, **219**, 1212–15.

TROUTT H F and OSBURN B I (1997), 'Meat from dairy cows: possible microbiological hazards and risks', *Rev Sci Tech*, **16**, 405–14.

USDA (1998), *E. coli O157 and Salmonella: Status on US Dairy Operations*, #N286.598, USDA:APHIS:VS, CEAH, National Animal Health Monitoring, Fort Collins, CO.

USDA (2003), *Salmonella and Listeria in Bulk Tank on US Dairies*, #N402.1203, USDA:APHIS:VS:CEAH, Fort Collins, CO.

USDA (2009), *Prevalence of Salmonella and Listeria in Bulk Tank Milk and In-line Filters on US Dairies*, #N528.0709, USDA:APHIS:VS:CEAH, Fort Collins, CO.

VAESSEN M A, VELING J, FRANKENA K, GRAAT E A and KLUNDER T (1998), 'Risk factors for *Salmonella* dublin infection on dairy farms', *Vet Quart*, **20**, 97–9.

VAN KESSEL J S, KARNS J S, GORSKI L, MCCLUSKEY B J and PERDUE M L (2004), 'Prevalence of Salmonellae, *Listeria monocytogenes*, and fecal coliforms in bulk tank milk on US dairies', *J Dairy Sci*, **87**, 2822–30.

VAN KESSEL J S, KARNS J S, WOLFGANG D R, HOVINGH E, JAYARAO B M, VAN TASSELL C P, ET AL. (2008), 'Environmental sampling to predict fecal prevalence of *Salmonella* in an intensively monitored dairy herd', *J Food Prot*, **71**, 1967–73.

VAN KESSEL J S, KARNS J S, WOLFGANG D R, HOVINGH E and SCHUKKEN Y H (2007), 'Longitudinal study of a clonal, subclinical outbreak of *Salmonella enterica* subsp. *enterica* Serovar Cerro in a US dairy herd', *Foodborne Pathog Dis*, **4**, 449–61.

VAN KESSEL J S, PACHEPSKY Y A, SHELTON D R and KARNS J S (2007), 'Survival of *Escherichia coli* in cowpats in pasture and in laboratory conditions', *J Appl Microbiol*, **103**, 1122–7.

WADE S E, MOHAMMED, H O and SCHAAF SL (2000), 'Prevalence of *Giardia* sp. *Cryptosporidium parvum* and *Cryptosporidium andersoni* in 109 dairy herds in five counties of southeastern New York', *Vet Parasitol*, **93**, 1–11.

WARNICK L D, CROFTON L M, PELZER K D and HAWKINS M J (2001), 'Risk factors for clinical salmonellosis in Virginia, USA cattle herds', *Prev Vet Med*, **49**, 259–75.

WEIS J and SEELIGER H P (1975), 'Incidence of *Listeria monocytogenes* in nature', *Appl Microbiol*, **30**, 29–32.

WELLS J G, SHIPMAN L D, GREENE K D, SOWERS E G, GREEN J H, CAMERON D N, ET AL. (1991), 'Isolation of *Escherichia coli* serotype O157:H7 and other Shiga-like toxin-producing *E. coli* from dairy cattle', *J Clin Microbiol*, **29**, 985–9.

WESLEY I V (1999), 'Listeriosis in animals', in Ryser, E.T. and Marth, E.H. (Eds): *Listeria, Listeriosis, and Food Safety*, Marcel Dekker Inc., New York, pp. 39–73.

WOOD R C, MACDONALD K L and OSTERHOLM M T (1992), '*Campylobacter enteritis* outbreaks associated with drinking raw milk during youth activities: a 10-year review of outbreaks in the United States', *J Am Med Assoc*, **268**, 3228–30.

XIAO L (2010), 'Molecular epidemiology of cryptosporidiosis: an update', *Exp Parasitol*, **124**, 80–9.

XIAO L and FAYER R (2008), 'Molecular characterization of species and genotypes of *Cryptosporidium* and *Giardia* and assessment of zoonotic transmission', *Int J Parasitol*, **38**, 1239–55.

XIAO L and RYAN U M (2004), 'Cryptosporidiosis: an update in molecular epidemiology', *Curr Opin Infect Dis*, **17**, 483–90.

XIAO L and RYAN U M (2008), 'Molecular epidemiology', in Fayer, R. and Xiao, L. (Eds): *Cryptosporidium and Cryptosporidiosis*, 2nd ed., CRC Press, Boca Raton, FL, pp. 119–71.

ZHAO T, DOYLE M P, SHERE J and GARBER L (1995), 'Prevalence of enterohemorrhagic *Escherichia coli* O157:H7 in a survey of dairy herds', *Appl Environ Microbiol*, **61**, 1290–3.

ZHOU L, KASSA H, TISCHLER M L and XIAO L (2004), 'Host-adapted *Cryptosporidium* spp. in Canada geese (*Branta canadensis*)', *Appl Environ Microbiol*, **70**, 4211–15.

22

Tracing pathogens in molluscan shellfish production chains

R. J. Lee and R. E. Rangdale, Centre for Environment, Fisheries and Aquaculture Science, UK

Abstract: Bivalves are recognised as high-risk products as they are exposed to several sources of pathogens in the harvesting area. Contamination and multiplication of bacterial pathogens may also occur after harvest. Regulatory controls have concentrated on the use of bacterial indicators to reflect on the risk of faecal contamination. This is unsatisfactory for vibrios and does not fully address contamination by enteric viruses. Molecular-based methods can now be used to detect such pathogens and molecular typing methods may be used to track pathogens following illness outbreaks and contamination events. Their potential use by regulatory bodies and industry is under consideration.

Key words: bivalve molluscs, pathogens, risk assessment, typing.

22.1 Introduction

Since the early 1970s, the global consumption of shellfish has increased considerably with a concomitant rise in the reports of outbreaks of infection (Potasman *et al*., 2002). Bivalve shellfish present a particular risk to the consumer as during the process of filter feeding, pathogens in the environment can be concentrated in the digestive tract. This risk is compounded by traditional practices of eating shellfish either raw or lightly cooked.

A number of aspects need to be considered by the shellfish industry in site selection, especially for aquaculture purposes. These include biological, environmental, legal and economic factors including the likely extent of microbiological contamination and the frequency of biotoxin events (Laing and Spencer, 2000). Additional advice on the selection of a site in relation to microbiological contamination is given in Lee *et al*. (2000).

In most countries, controls on the sanitary quality of molluscan shellfish are based on monitoring of faecal contamination at the harvesting area and to an extent end-product testing for compliance with specified microbiological criteria. It is acknowledged that this type of approach does not give absolute assurance of quality and safety, and a much higher degree of assurance can be provided by a preventative approach based on the application of the Hazard Analysis Critical Control Point (HACCP) principles at all steps in the food supply and processing system (Huss *et al.*, 2003). The ability to track pathogens throughout the production chain can assist in the development of HACCP plans, and also enable evaluation of the public health and economic impacts of monitoring, production and post-harvest strategies.

To date, relatively little systematic tracking of pathogens throughout the entire molluscan production chain has been undertaken. This chapter sets out to examine current practices with respect to the production of bivalve molluscs, identifies existing molecular-based methodologies for pathogen tracking and considers their future use in this sector.

22.2 Overview of shellfish production chains

Production chains for bivalve molluscan shellfish vary markedly depending on whether the shellfish are harvested from the wild or grown in aquaculture, on the species involved, on the local seabed and sea conditions and on local practices.

The basic steps are shown in Fig. 22.1. There may be a number of variations on this basic flow. Further details on the basic components are provided below. These describe general industry practices and do not necessarily provide full details of legislative requirements or recommended good practice. It should also be borne in mind that practices undertaken to meet legal or market requirements in developed countries may differ markedly from those undertaken for local sale in developing countries and that environmental conditions, including ambient temperature, will also vary greatly.

22.2.1 Harvest

Shellfish may be located in the water column (in bags/pouches on trestles, or suspended on lines or in nets), on rocks or structures (mussels), on the bottom (scallops and oysters) or in the bottom sediment (cockles, clams, razor clams). They may be in the intertidal zone, and thus exposed for part of the tidal cycle, or further out and thus continually submerged. As appropriate to the location and species, harvest may be undertaken by boat, from land or by diver. These factors will all affect the extent of physical damage and shock suffered by the shellfish and the amount of detritus and dirt associated with the harvested product. The location and method of harvest will also affect the practicality of post-harvest temperature controls and, in combination with the distance from the receiving centre, the nature and time of transport.

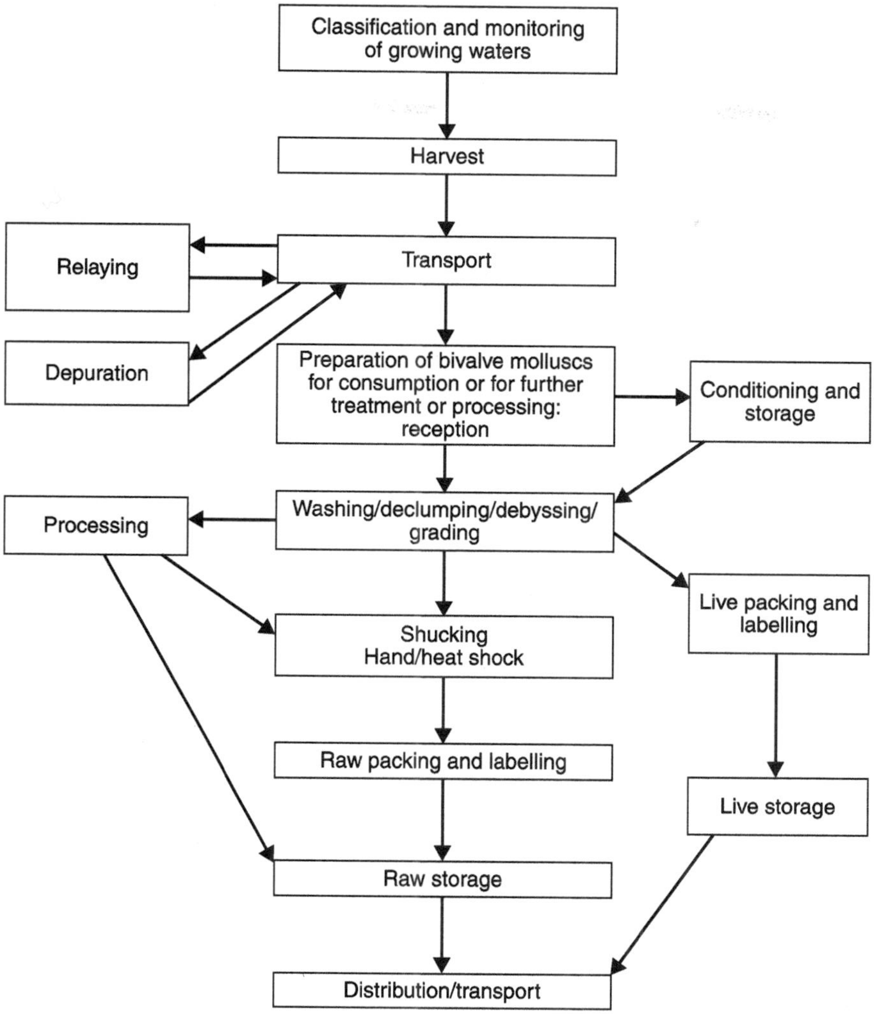

Fig. 22.1 Basic steps in bivalve mollusc production chains. Adapted from Lee *et al.*, 2008.

There has been significant interest in some parts of the world in locating shellfish farms significant distances offshore in order to overcome either problems caused by near-shore contamination from human or animal sources or to overcome planning/seabed ownership problems. Offshore farms require more robust equipment in order to withstand the rougher environment and obviously add to difficulties of access and transport.

In some operations, shellfish may be moved from the main harvest area to 'wet storage' areas near the receiving centre (especially if both are operated by the same company). This has the practical advantage of providing convenient raw material for the centre which may be less affected by tides and weather. However, such practices may have other implications with regard to hygiene controls.

22.2.2 Transport

Transport may occur at a number of steps in the production chain including those shown in Fig. 22.1. It may be by boat, lorry or even aircraft, or a combination of these. The stage at which the transport is undertaken will largely dictate whether it is of bulk shellfish (e.g. post-harvest) or packed (e.g. in net bags or on covered trays) and the state of cleanliness of the product. Generally, if the shellfish stay dry, and therefore shut, during transport, they will not become further contaminated. However, contamination of the outside of the shells may be taken up at a later stage if they are subsequently immersed in seawater. There are also some species that do not seal properly out of the water (e.g. scallops, cockles and razor clams). Control of temperature during transport is necessary to control the growth of bacterial, but not viral, pathogens – the actual use of this control varies greatly.

22.2.3 Receipt and storage at centre

Practices for receipt and storage vary greatly – the shellfish may be unloaded directly onto the ground outside and left exposed in an uncontrolled environment, raised off the ground and/or covered to protect from contamination and exposure, or may be placed inside for additional protection, sometimes under temperature control.

22.2.4 Washing, removal of broken shell and other debris (debyssing for mussels)

This is an essential operation prior to either depuration or packaging. For mussels, the byssal threads need to be removed in order to allow any clumps of shellfish to be separated. Any water used at this stage needs to be free from contaminants, both microbiological and chemical, in order to prevent them from being taken up by the shellfish if they become immersed.

Shellfish may also be subject to conditioning prior to packing. In EU hygiene legislation, conditioning is defined as: '. . . the storage of live bivalve molluscs coming from class A production areas, purification centres or dispatch centres in tanks or any other installation containing clean seawater, or in natural sites, to remove sand, mud or slime, to preserve or to improve organoleptic qualities and to ensure that they are in a good state of vitality before wrapping or packaging' (European Communities, 2004). Unless adequately controlled, this process may introduce additional contamination into the shellfish.

22.2.5 Post-harvest treatment (where necessary)

Post-harvest treatments may be dictated by legislation or may be applied by the industry for preservation or marketing purposes. These treatments include purification (depuration), relaying, cooking, pasteurisation, freezing and high-pressure treatment. The first three are more traditional while the latter three have largely been introduced in order to reduce the level of pathogenic vibrios. Purification is widely practised and involves placing the shellfish in tanks of clean seawater under conditions which optimise their activity (Lee *et al.*, 2008). This results in purging of contamination from the digestive tract. The treatments vary markedly in their effectiveness against different pathogens.

22.2.6 Washing, grading and packing

Washing and grading are usually undertaken prior to packing for species that are sold live, after relevant post-harvest treatment such as purification or relaying.

22.2.7 Post-packing storage

This storage is usually undertaken under protected temperature-controlled conditions. It is also necessary to ensure that the storage is undertaken under hygienic conditions to avoid recontamination.

22.2.8 Wholesale, catering and retail sale

Transport to other premises and storage and display need to be undertaken under hygienic conditions and temperature control in order to avoid recontamination of the product. Display systems involving either the immersion of shellfish or sprinkling with water need to be carefully maintained and any water used needs to be free from contaminants. Owing to the possibility of recontamination, and cross-contamination between batches, these systems may not be allowed or may be subject to strict controls.

Cooking in restaurants (and the home) may not raise the temperature of the shellfish high enough to inactivate pathogenic organisms. The most heat-resistant pathogen often involved in shellfish-associated outbreaks is the hepatitis A virus (Lees, 2000) although even bacterial pathogens, such as *Salmonella* spp., may survive cooking if this is not undertaken properly (Greenwood *et al.*, 1998).

22.3 Foodborne pathogens in shellfish

22.3.1 Pathogens associated with faecal contamination

As described earlier, bivalve shellfish are often grown in shallow, estuarine or coastal areas close to population centres and human activity. Such areas may be exposed to human faecal contamination through wastewater treatment outflows, combined sewer overflows, etc., or to zoonotic pathogens through run-off from agricultural practices contiguous with harvesting areas. Bivalve shellfish accumulate and concentrate bacterial and viral pathogens present in the surrounding water through the process of filter feeding. As they are often consumed either raw or lightly cooked any pathogens present are not inactivated by heat. Bivalve shellfish may therefore present a risk to the consumer and monitoring strategies to mitigate this risk are generally undertaken. A number of faecal-borne bacterial pathogens have been implicated in shellfish-associated illness (e.g. *Vibrio cholerae* O1, non-O1 *V. cholerae, Salmonella* spp., *Listeria monocytogenes*, *Campylobacter* species). It is however widely accepted that the incidence of bacterial infections has reduced in recent times and that the most common bivalve shellfish vectored illnesses are acute viral gastroenteritis caused by human noroviruses (NoVs) and infectious hepatitis caused by the hepatitis A virus.

Human noroviruses are single stranded, non-segmented RNA members of the Calicividae family. Most human infections are caused by two genogroups (I and II). Noroviruses are recognised as the most common cause of non-bacterial intestinal disease in the developed world. Norovirus illness exhibits a strong seasonality with the majority of outbreaks occurring during the winter months. Clinical infection results in relatively mild gastroenteritis, including diarrhoea and vomiting, with an onset of 24–28 hours, and a usual duration of approximately two days. During the 1980s and early 1990s a linkage between the consumption of sewage-contaminated shellfish and norovirus infection was identified using electron microscopy and serodiagnostics. Recent advances in molecular biology have enabled the confirmation of this association through detection of identical sequences of the norovirus genome in faecal specimens from patients and bivalve shellfish.

Another significant viral pathogen associated with the consumption of sewage-contaminated bivalve shellfish is hepatitis A virus. Hepatitis A is a spherical, non-enveloped enterovirus within the Picornaviridae family. It is structurally similar to norovirus with a RNA genome surrounded by a small protein capsid. Hepatitis A causes an acute infectious liver disease in adults, which can result in a severe debilitating illness, and occasionally a fatal fulminant hepatitis. Definitive linkage between food vector and disease transmission is often problematic without comprehensive epidemiological investigation due to the extended three- to six-week incubation period.

22.3.2 Naturally occurring pathogens of shellfish

A number of organisms potentially pathogenic to humans occur naturally in marine and estuarine environments. Among these are members of the genera *Shewanella*, *Photobacterium* and *Vibrio*. In terms of prevalence of infection and severity, the marine vibrios, notably *Vibrio parahaemolyticus* and *Vibrio vulnificus* are the most significant. Both are Gram-negative bacteria with an absolute requirement for Na^+ for growth. As with sewage-borne pathogens, bivalve shellfish can become contaminated with marine vibrios during filter feeding or through recontamination during processing.

V. parahaemolyticus can frequently be isolated from temperate estuarine and marine environments. It typically causes gastroenteritis following consumption of raw or undercooked seafoods, particularly oysters. Illness is characterised by watery or bloody diarrhoea, vomiting, abdominal cramps, headaches, fever and nausea (Joseph *et al.*, 1982) and rarely wound infections (Johnson *et al.*, 1984). Generally, infections have been restricted to regions where ambient seawater temperatures are in excess of 20°C such as the Gulf Coast of the USA and Southeast Asia. In such regions, it is the primary cause of seafood-associated gastroenteritis (Baker-Austin *et al.*, 2010b). Pathogenicity in humans is linked to the possession of thermostable direct haemolysins (TDH) and thermostable direct haemolysin-related haemolysins (TRH). Only a small proportion (2–3%) of the *V. parahaemolyticus* in the environment will possess these pathogenicity traits that enable them to cause disease in humans. As a result, the presence of total *V. parahaemolyticus* in seafoods may not lead directly to the risk of illness to the consumer (Anon, 2001).

V. vulnificus can be isolated under a wide range of salinity and temperature conditions from bivalve shellfish and fin fishes (Aznar *et al.*, 1994). It can cause disease in humans through invasion of pre-existing wounds or through consumption of raw shellfish (Oliver and Kaper, 1997). Most infections are reported during the warmer summer months when seawater temperatures exceed 20°C (Rippey, 1994) and are usually associated with the consumption of raw oysters. Men with elevated serum iron levels resulting from underlying medical conditions, primarily liver cirrhosis, are considered the highest risk group (Oliver and Kaper, 1997). *V. vulnificus* carries one of the highest mortality rates of any bacterial pathogen with mortality rate of more than 50% (Baker-Austin *et al.*, 2010a). To date, three biotypes of *V. vulnificus* have been reported. Biotype 1 causes most of the infections reported in humans, whereas biotype 2 is responsible for an infection in eels. Biotype 3 is reported to be recombinant clone of two *V. vulnificus* populations (Bisharat *et al.*, 2005). Recently, molecular typing has indicated the presence of virulent (clinical) and avirulent (environmental) genotypes within biotype 1 (Aznar *et al.*, 1994). As with *V. parahaemolyticus* only a minority of environmentally isolated *V. vulnificus* are considered pathogenic (Vickery *et al.*, 2007) thus isolation of undifferentiated genotypes may not truly reflect risk.

22.4 Typing methods for tracking pathogens in shellfish production chains

The basis for all molecular tracking methods is that microorganisms within an infectious chain are phylogenetically or clonally related. There are a wide variety of methodologies now available that have been used in suspected shellfish-related outbreak investigations, identification of origins and vehicles of infection in bivalve molluscan production. In some cases, these techniques have also contributed to epidemiological surveillance and have been used to evaluate control measures.

Molecular typing methods fall into four general categories depending upon the approach: plasmid analysis, hybridisation, nucleic acid amplification using the polymerase chain reaction (PCR) with or without sequencing (e.g. multilocus sequence typing) and macrorestriction of DNA followed by pulsed-field gel electrophoresis (PFGE) fragment resolution (Towner and Cockayne, 1993).

All approaches have been used to various extents in studies to elucidate the origin, and transmission patterns of shellfish pathogens, and to establish epidemiological links between infectious agents, patients and food vectors. Examples of their use are discussed hereafter.

22.5 Pathogen monitoring strategies

As noted earlier, illness associated with the consumption of bivalve molluscs is usually caused by contamination arising in the harvesting area, or by post-harvest immersion in contaminated seawater. There are therefore potentially two separate points where pathogen monitoring may be undertaken. One is in the harvesting area itself, and the other is in the production chain. The latter may be subdivided into the production chain in the strict sense and the second at retail. A more specialised application is in the investigation of outbreaks and this may involve monitoring at all levels. The occurrence of known contamination events in the harvesting area, such as caused by a spill of untreated sewage, will usually be assessed by appropriate monitoring in the harvesting area. However, supplementary testing of shellfish already in the production chain may be undertaken to inform management decisions.

Traditionally, monitoring for pathogens has tended to be limited and concentrated on the end-product. *Salmonella* is the only pathogen that has been included in formal monitoring requirements, usually in the production chain and/ or at end-product level. This may be a consequence of the fact that shellfish hygiene control systems were developed to address the typhoid outbreaks of the nineteenth and early twentieth centuries. However, the more limited incidence and severity of infections with non-typhoidal *Salmonella* spp. means that the role of such monitoring needs re-evaluation. Indeed, this was partly addressed in a report to the Food and Agriculture Office (FAO) of the United Nations undertaken in support of a Codex Committee on Fish and Fishery Products review of the recommended sampling plan for *Salmonella* in bivalves (Lee, 2009).

Harvesting area and production chain/end-product monitoring are considered and described separately hereafter.

22.5.1 Monitoring in harvesting areas

The bulk of pathogens causing shellfish-associated illness arise from faecal contamination of the harvesting areas, with the contamination being associated with human or animal sources. In general, these pathogens have not been monitored directly in the harvesting areas, but the risk of contamination with such pathogens has been evaluated using time series monitoring for faecal indicator bacteria, including *E. coli*. In general, this is a successful strategy and a number of investigations have shown that the incidence of pathogens increases as the evaluated risk increases (Lee and Younger, 2003; Guilfoyle *et al.*, 2007). In the UK, the proportion of samples positive for *Salmonella* was shown to increase from approximately 1% for samples taken from class A production areas to approximately 9% in areas prohibited for harvesting on the basis of *E. coli* monitoring. An international report has shown that the prevalence of *Salmonella* in harvesting areas varied from 0% to 34.2%, with differences being seen between geographical locations, species, years and seasons (Lee, 2009).

Some national authorities have supplemented the classification of harvesting areas based on indicator bacteria with tests for *Salmonella*. These have sometimes been undertaken as frequently as the indicator(s) and sometimes on a proportion of samples. This has tended to cause problems as a proportion of samples will tend to be positive even from areas where the product can, in theory, go for direct human consumption without further treatment (e.g. EU class A; US approved). There then needs to be a decision as to whether the area needs to be closed on the basis of the presence of the pathogen even though the indicator concentrations are within limits. The risk is difficult to evaluate as tests for *Salmonella* are usually simply presence/absence in 25 g with no estimation of concentration. There is some limited evidence that the risk of salmonellosis will increase with an increase in faecal indicator concentration – this may be due to the concentration of *Salmonella* being higher in such circumstances.

Norovirus remains an important cause of acute gastroenteritis following consumption of faecally polluted shellfish. While strategies for monitoring faecal contamination of harvesting areas based on indicator organisms have reduced the risk to the consumer, every year outbreaks of norovirus-related illness are reported (Lees, 2000). The detection of norovirus in bivalve shellfish is reliant upon the use of real-time reverse transcriptase PCR (Le Guyader *et al.*, 2000). This relatively complex molecular technology has until recently been restricted to research activities in a few specialist laboratories, and direct monitoring of norovirus has not been undertaken in the context of official control. The broad range of circulating strains of norovirus, necessitating specific primer and probe sets for identification, has further complicated its detection in foodstuffs. More recently, the identification of conserved regions of the norovirus genome and advances in PCR techniques enabling quantitation of genome fragments has enabled the development of broadly reactive assays which are more amenable to standardization and consequently more routine use.

In the EU, the potential for monitoring for norovirus in bivalve molluscs is foreshadowed in EU Regulation once methods are sufficiently developed (European Communities, 2004). The European Committee for Standardization (CEN) has established a working group comprising of European food virologists to address the development of a reference method for norovirus. This group has elaborated a quantitative method for the determination of norovirus (and hepatitis A) in foodstuffs, including bivalve shellfish. The method has not yet been formally validated or published; however, if it proves effective the previous technical restrictions on direct detection of norovirus in bivalves will no longer apply in the EU. Elsewhere in the world (e.g. Hong Kong, Korea and New Zealand) similar approaches are being used for testing both indigenously produced and imported shellfish. However, as yet no consistent global approach has been developed and no internationally recognised (ISO or equivalent) testing methods are currently available. The application of virus standards in harvesting area monitoring (or for end-products) may have significant consequences both in terms of public health and trade and application for official control will require careful consideration. Currently, limited norovirus testing of harvesting areas is undertaken by some authorities in association with suspected or proven food-poisoning outbreaks or after major sewage contamination events. Interpretation of the results of such testing is complicated by the absence of information on the background presence and levels of norovirus in such areas.

Globally, *Vibrio* spp. are the most important cause of shellfish-borne bacterial illness (Panicker *et al.*, 2004), and in the EU they are responsible for the largest numbers of rejections and detentions of imported seafood (Ababouch *et al.*, 2005). Despite this official monitoring of shellfish in harvesting areas is not widespread, testing for vibrios tends to be restricted to imported shellfish, *ad hoc* surveillance, outbreak investigations and in some places end-product testing.

There is a lack of systematic data on the *Vibrio* content of bivalve molluscs produced from the EU. Primarily, this is due to the reportedly low incidence of illness (Wagley *et al.*, 2008) and recommendations that existing internationally recognised methods were not sufficient to support specific standards for *V. parahaemolyticus* and *V. vulnificus* in raw and undercooked seafood (Anon, 2001). Currently, in the EU there is no legal basis for *Vibrio* testing, this coupled with the sporadic nature of seafood-associated *Vibrio* infections in the EU has resulted in an absence of systematic official monitoring in harvesting areas.

The situation is somewhat different in the USA where a substantial number of *V. parahaemolyticus* and *V. vulnificus* illnesses associated with bivalve molluscs (particularly oysters originating from the Gulf Coast states) occur annually particularly during the warmer summer months (DePaola *et al.*, 2003). This has prompted the Interstate Shellfish Sanitation Conference (ISSC), the body responsible for shellfish sanitation in the USA to develop risk-based strategies for assessing shellfish harvest areas (USFDA, 1986; Zimmerman *et al.*, 2007). Essentially these require shellfish producing states to conduct an annual risk evaluation that includes: the number of *Vibrio* cases linked to shellfish consumption; total numbers of *Vibrio* cells in oyster matrices (and pathogenic strains in the case of *V. parahaemolyticus* where data are available); measurements of water, air temperature, salinity, harvesting techniques and use of post-harvest processing. If risks are considered significant a plan identifying control measures, such as post-harvest treatment, site closures, restriction on time from harvesting to refrigeration and controls to ensure rapid cooling, is required.

22.5.2 Monitoring in the production chain post-harvest

Monitoring post-harvest will usually be undertaken by the food business operator (FBO) in support of their HACCP. However, some level of monitoring may be dictated by the formal control system, e.g. under the European Microbiological Criteria Regulations (European Communities, 2005). Some monitoring may be undertaken by the authorities post-harvest as a part of their audit on processing and retail operations. Published data on the presence of pathogens through the market chain and to retail level are extremely limited – the data that are available indicate that the isolation rate for *Salmonella* from samples taken at final sale in temperate northern countries is approximately 1% (Lee, 2009).

The monitoring undertaken by the FBO will depend on the type of processing, if any, undertaken prior to packing. For depurated shellfish, it will normally be undertaken on receipt and on the final product. The frequency of monitoring of the received shellfish should depend on the following factors:

- Are the shellfish sourced from a small number of readily identifiable areas?
- Is the level of contamination in each harvesting area relatively stable?

If the answer to both of these answers is 'Yes', then the frequency of monitoring at receipt would be low. If the answer to either is 'No', then the frequency should be relatively high. The frequency should be increased if a new source is being used, or if there is reason to suspect that contamination may be present. In the latter case, a judgement will need to be made as to whether the appropriate action should be taken to avoid the source rather than to increase monitoring (the decision may be made by the authorities). In general, for assessment of pathogens of faecal origin, the judgement is made on a combination of classification status and/or ongoing harvesting area faecal indicator monitoring results. Some FBOs are undertaking testing of stock from potential new areas for norovirus prior to deciding on the use of such sources. The results of such testing are therefore already influencing trade.

The frequency of monitoring post-processing will depend on 'the quality of the source material' and 'the consistency of the process' and may need to be changed if either of these aspects change.

The frequency of monitoring both received shellfish and final product should take into account both the results of historical monitoring and other information that informs the assessment of risk (such as an association of the source area with illness).

As well as frequency of monitoring, with regard to the number of batches to be tested, there is the consideration of the likely variability of the pathogen within an individual batch. This will depend on a number of factors:

- the size of the batch;
- the variability of the original contamination in the source material;
- the effect of post-harvest treatment on variability.

It should therefore not be assumed that a single sample from a batch will give adequate information on the presence, or concentration, of a pathogen throughout the batch. Although this is a fundamental concept in all food microbiology, it is accentuated with bivalve molluscs due to the discrete nature of the individual animals, and the effect of activity of individuals on the uptake and/or retention of pathogens.

In addition to the normal production chain monitoring, additional testing for faecal indicators and pathogens may be undertaken at import. In this regard, Hong Kong specifies the absence of norovirus in imported batches of shellfish.

22.6 Pathogen typing strategies

Pathogen typing as a component of harvesting area monitoring has generally been restricted to *ad hoc* surveillance activities or used for epidemiological purposes where human illness associated with shellfish consumption is suspected. In this context, however, molecular typing methodologies have been used extensively in food poisoning outbreaks and contamination events, and have been successfully applied to detect and monitor both bacterial and viral pathogens from molluscan shellfish sources. In addition, a limited number of studies have described typing methodologies to monitor infectious agents throughout the bivalve mollusc production chain. In one such study, Martinez-Urtaza and Liebana (2005) used PFGE to study the persistence and dissemination of the uncommon zoonotic pathogen, *Salmonella* Senftenberg in Spanish mussel processing factories. The authors identified a persistent pulse type in one facility during five-year study period and suggested that brine used in the processing lines was a potential source of contamination.

Leal *et al.* (2008) examined the virulence factors and the genetic relationship of serogroup O3 strains of *V. parahaemolyticus* implicated in outbreaks of diarrhoea in the northeast region of Brazil. The authors utilised multiplex PCR for the detection of the *tl*, *tdh* and *trh* genes, and by random-amplified polymorphic DNA (RAPD) analysis. Using this approach, the pandemic clone of *V. parahaemolyticus* (serovar 03:K6) was identified as the causal agent of disease. Sarkar *et al.* (2003) isolated *V. parahaemolyticus* in 26% of sewage samples in Calcutta, India. Five of these eight *V. parahaemolyticus* strains identified were *tdh* positive, and were subsequently compared by RAPD-PCR and PFGE with strains selected from a culture collection to determine their genetic relatedness. The results indicated that the strains isolated from sewage and clinical strains were serologically similar but genetically diverse and that they were not part of the pandemic clone. Khan *et al.* (2002) used enterobacterial repetitive intergenic consensus PCR (ERIC-PCR) to type strains of *V. parahaemolyticus* from gastroenteritis outbreaks in the USA (Texas, New York and Pacific Northwest) in 1997–1998. The Texas and New York outbreak isolates had a specific 850 bp DNA fragment that was absent in Pacific Northwest isolates, and which was later utilised as a fragment for testing geographically specific pathogenic *V. parahaemolyticus* strains. Such findings highlight the potential importance of utilising molecular typing methods to develop additional molecular targets in monitoring studies.

Unlike *V. parahaemolyticus*, the epidemiology of *V. vulnificus* is typically restricted to individual infections rather than large-scale food-related outbreaks, and this is almost certainly related to complex host-specific factors. Irrespective, several reports have successfully utilised molecular typing methods for pathogenic monitoring purposes. Rosche *et al.* (2005) utilised RAPD analysis of clinically derived and environmentally isolated *V. vulnificus* strains. The authors identified a DNA fragment termed the virulence correlated gene (*vcg*) variations in the *vcg* was associated with two distinct genotypes of the bacteria; one pathogenic (clinical) and one non-pathogenic (environmental). Clinical or C type has been correlated with 90% of isolates from human cases whereas the environmental (E type) variant has been identified in 87% of strains isolated from shellfish (Warner and Oliver, 2008). Recently, Gonzalez-Escalona *et al.* (2007) used a molecular typing methodology to target variability in the 16S-23S rRNA intergenic spacer region as a means of predicting potential virulence. The results indicated that clinical strains group very tightly, and could be readily distinguished from non-pathogenic isolates. Variability in the small subunit 16S rRNA gene using typing methods has also been used in other pathogen monitoring studies for *V. vulnificus*, with varying success (Aznar *et al.*, 1994; Nilsson *et al.*, 2003; Vickery *et al.*, 2007).

Following a large outbreak of hepatitis A virus (HAV) infection in the USA in 2005, the local authorities carried out investigations in order to characterise the pathogen, identify the potential source and the route of transmission and implement appropriate control measures. Molecular epidemiological methods utilising HAV RNA sequencing analysis identified shellfish consumption as the predominant risk factor. This was the first time identical hepatitis A virus sequences were obtained from both the implicated food product and case patients (Bialek *et al.*, 2007).

Similarly, molecular genotyping methods have been used successfully to detect and type human norovirus. Fukuda *et al.* (2006) used molecular typing methods to determine the molecular diversity of norovirus in numerous outbreaks in Japan. Based on the genotyping of noroviruses, the involvement of multiple genotypes was found in shellfish-related outbreaks, while outbreaks related to person-to-person contact and consumption of foods contaminated by food handlers exhibited only one genotype. Le Guyader *et al.* (2006) reported the use of a Pan-European and multi-laboratory effort utilising molecular typing and sequencing to identify an international norovirus outbreak linked to oyster consumption. In this study, the authors assimilated both clinical, epidemiological and molecular data that identified a single foodstuff (oysters) as the source of the outbreak. Such studies highlight the importance of molecular tools for pathogen monitoring in food-related outbreaks.

22.7 Improving pathogen control

Reducing the exposure of consumers to levels of pathogens that may cause infection is primarily the role of HACCP. However, where current approaches do not satisfactorily control the possibility of pathogen being present in the final product, e.g. for norovirus, there is the potential for instituting appropriate monitoring of the pathogens or some other assessment of the potential risk, or a combination of the two.

22.7.1 Risk assessment for pathogens

Specific risk profiles and/or risk assessments have been undertaken for a number of shellfish-associated pathogens. These are listed in Table 22.1. The risk assessment procedures were undertaken by FAO/WHO and followed the recommended procedures for full risk assessments (CAC, 1999). These require

Table 22.1 Risk profiles and risk assessments undertaken for shellfish-associated pathogens

Pathogen(s)	Profile or assessment	Seafood type	Authority	Reference
Cryptosporidium spp	Profile	Shellfish	New Zealand Food Safety Authority	http://www.nzfsa.govt.nz/science/risk-profiles/
Norwalk-like virus	Profile	Mollusca (raw)	New Zealand Food Safety Authority	http://www.nzfsa.govt.nz/science/risk-profiles/
Vibrio parahaemolyticus	Profile	Seafood	New Zealand Food Safety Authority	http://www.nzfsa.govt.nz/science/risk-profiles/
Clostridium botulinum	Profile	Ready-to-eat smoked seafood in sealed packaging	New Zealand Food Safety Authority	http://www.nzfsa.govt.nz/science/risk-profiles/
Vibrio cholerae	Assessment	Shrimp	FAO/WHO	http://www.fao.org/ag/agn/agns/jemra_riskassessment_vibrio_en.asp
Vibrio vulnificus	Assessment	Raw oysters	FAO/WHO	http://www.fao.org/ag/agn/agns/jemra_riskassessment_vibrio_en.asp
Vibrio parahaemolyticus	Assessment	Oysters	FAO/WHO	http://www.fao.org/ag/agn/agns/jemra_riskassessment_vibrio_en.asp
Vibrio parahaemolyticus	Assessment	Bloody clams	FAO/WHO	http://www.fao.org/ag/agn/agns/jemra_riskassessment_vibrio_en.asp
Norovirus	Assessment	Food (including bivalve molluscs)	FAO/WHO	http://www.fao.org/ag/agn/agns/jemra_riskassessment_viruses_en.asp
Listeria monocytogenes	Assessment	Ready-to-eat foods	FAO/WHO	http://www.fao.org/ag/agn/agns/jemra_riskassessment_listeria_en.asp

detailed information and are beyond the scope of local authorities or single food business operations. Risk profiles are intended to be preliminary statements of current knowledge, undertaken by risk managers, in order to determine whether a full risk assessment is necessary. Those listed in Table 22.1 are relatively comprehensive and tend to differ little from a formal risk assessment, except that they do not include quantitative modelling.

There are a number of simpler forms of risk assessment. For example, sanitary surveys, as used in both the USA and EU shellfish hygiene programmes, will provide an estimate of the faecal contamination of an area or source material (CRL, 2007; USFDA, 2008). Even simpler forms are risk matrices, where risk factors are given relative, but arbitrary values and a total or sum obtained which allows some comparison of risk. The approach may be applied to both harvesting areas and processes and an example of each is given in Tables 22.2 and 22.3, respectively. The final value obtained from the process can be either used for simple comparative purposes or ranges can be established from experience which then determines whether any management action is necessary.

22.7.2 Combination of risk assessment and monitoring

Combining both risk assessment and the monitoring approaches described in Section 22.5 will potentially provide the best approach to overall assessment of risk that may potentially, or actually, relate to products from the specific harvesting areas or production plants. There are three steps in this interaction:

1 The risk assessment will inform the production of a monitoring plan, informing which pathogens should be included in such monitoring, which species should be monitored and the location of monitoring points and timing of monitoring.
2 The results of the monitoring will supplement the contents of the risk assessment to identify the actual extent of contamination in the area or plant. The monitoring will not replace the initial risk assessment as it is likely that it will not be comprehensive enough to assess the risk of all products for all potential pathogens.
3 The outcome of the monitoring may result in the risk assessment being reviewed if the results are markedly at odds with those that would have been expected from the assessment. Results from faecal indicator monitoring have shown that unforeseen sources of faecal contamination may impact on an area.

22.7.3 Risk management based on both risk assessment and pathogen monitoring

FAO and WHO have undertaken development of risk management practices based on risk assessment (FAO/WHO, 2006). However, the risk assessments were those listed in Table 22.1 and, as identified previously, beyond local resourcing.

Table 22.2 Risk matrix for the evaluation of harvesting areas for enteric pathogens from human sources

Factor	Score			Distance from shellfishery (km)	Total (score ÷ distance)
	Low (0)	Medium (2)	High (4)		
Species	Non-oysters	–	Oysters	N/A	
Population	<1000	1000–5000	>5000		
Presence of continuous community sewage discharges	None	–	One or more		
Presence of combined sewer overflows	None	–	One or more		
Presence of private sewage discharges	None	One or more within 500 m of shellfishery but not within 100 m	One or more within 100 m of shellfishery	N/A	
Presence of boating activity	None or little	Moored boats or passing boat traffic	Marina, harbour or large number of live-aboard boats		
Classification status	A	B	C		
Unusual *E. coli* results over last 3 years	*All results* Class A ≤230 Class B ≤4600 Class C ≤10 000	*Each result* Class A >230 Class B >4600 Class C >10 000	*Each result* Class A>1000 Class B>10 000 Class C>18 000		
Confirmed outbreaks	None	–	One or more in three years		
Unconfirmed outbreaks	None	Between one and three in three years	More than three in three years		
Pathogen detected	No	In ≤25% of samples	In >25% of samples		
			Grand total		

Note: Other risk factors will need to be considered enteric pathogens of animal, or human and animal, origin

Table 22.3 Risk matrix for the evaluation of the processing of bivalve molluscs

Factor	Score			Total
	Low (0)	Medium (2)	High (4)	
Species	Non-oysters	–	Oysters	
Confidence in source of shellfish	High	Medium	Low	
Risk status of harvesting area	Low	Medium	High	
Microbiological testing undertaken of product from new sources	Yes	–	No	
Processing method validated as significantly reducing pathogen	Yes	–	No	
			Grand total	

The simpler risk assessment procedures outlined above may allow the results of monitoring to be set in context and to help assign a level of risk to the results of the pathogen monitoring. In turn, the monitoring results may identify whether a pathogen that may have been identified as potentially present by the assessment is actually present in the area or batch. Care needs to be taken in the latter case to ensure that account is taken of all potential pathogens that may be present and that monitoring is undertaken at the right locations and at frequent enough intervals (see Section 22.5). In the case of pathogens derived from faecal contamination, quantitative results from faecal indicator monitoring (faecal indicator bacteria and/or F-specific bacteriophages) may additionally inform the decision-making process in cases when pathogen monitoring is undertaken on a presence/absence or semi-quantitative basis.

Appropriate risk management action may involve one or more of the following actions: closure of the harvesting area, recall of already harvested and distributed product, greater than usual level of post-harvest treatment (in the broad sense). The action may be instigated on the basis of one of the following:

- the risk assessment indicating that the pathogen presence/level may be unacceptable;
- pathogen monitoring showing the unacceptable presence/level of a pathogen;
- both assessment and monitoring showing an elevated risk;
- one of the above plus elevated faecal indicators (where appropriate).

The combination and strength of evidence (and thus level of risk) may determine the choice or risk management action.

22.8 Future trends

To date, outside of specific outbreak investigations, molecular tracking approaches have not been widely used throughout bivalve molluscan production chains. However, over recent years techniques in molecular biology have improved substantially and in many cases the potential now exists for their broader application. This is particularly true for bacterial pathogens where typing strategies, like PFGE and MLST, used in existing shellfish-related epidemiological investigations could be used in studies to delineate the origin of pathogens, identify reservoirs and to track pathogens throughout production. Currently, however, most molecular tracking techniques still lack sufficient discriminatory capability, are technically complex and/or require non-portable laboratory-based equipment. Consequently, their use is still restricted to well-equipped specialist microbiology reference laboratories. New simplified tracking methods with greater discriminatory power without the requirement for complex equipment would greatly enhance their use within the shellfish industries and regulatory agencies. Ultimately rapid detection and differentiation of foodborne pathogens will help to target practices that are higher risk and assist in establishing robust risk-based preventative strategies. Tracking pathogens throughout the production chain will provide a greater understanding of the critical processes within the production chain and may therefore confer future advantages in terms of public health improvements by strengthening the preventative approach to managing food safety.

22.9 Sources of further information and advice

The following websites contain information related to the safety and microbiological monitoring of bivalve molluscs:

European Union Reference Laboratory for monitoring bacteriological and viral contamination of bivalve molluscs: www.crlcefas.org
European Commission – Food Safety: http://ec.europa.eu/food/index_en.htm
FAO Fisheries and Aquaculture Department: http://www.fao.org/fi/
FAO Agriculture and Consumer Protection Department: http://www.fao.org/ag/agn/
UC Davis Seafood Network Information Centre: http://seafood.ucdavis.edu/
WHO Food Safety: http://www.who.int/foodsafety/en/
National Food Safety/Standards Agencies will provide information specific to individual countries.

The website below contains information on an international initiative relating to typing pathogens associated with foodborne outbreaks of illness:

PulseNet: http://www.pulsenetinternational.org/

22.10 References

ABABOUCH L, GANDINI G and RYDER J (2005), *Causes of Detentions and Rejections in International Fish Trade*, FAO Fisheries Technical Paper-473.

ANON (2001), *Opinion of the Scientific Committee on Veterinary Measures relating to public health on Vibrio vulnificus and Vibrio parahaemolyticus*, European Commission, Geneva, Switzerland.

AZNAR R, LUDWIG W, AMANN R I and SCHLEIFER K H (1994), 'Sequence determination of rRNA genes of pathogenic *Vibrio* species and whole-cell identification of *Vibrio vulnificus* with rRNA-targeted oligonucleotide probes', *Int. J. Syst. Bacteriol.*, **44**, 330–7.

BAKER-AUSTIN C, GORE A, OLIVER J, RANGDALE R, MCARTHUR J V and LEES D N (2010a), 'Rapid in situ detection of virulent *vibrio vulnificus* strains in shellfish matrices using real-time PCR', *Environ. Microbiol. Reports*, **2**(1), 76–80.

BAKER-AUSTIN C, STOCKLEY L, RANGDALE R and MARTINEZ-URTAZA J (2010b), 'Environmental occurrence and clinical impact of *vibrio vulnificus* and *vibrio parahaemolyticus*: a European perspective', *Environ. Microbiol. Reports*, **2**(1), 7–18.

BIALEK S R, GEORGE P A, XIA G-L, GLATZER M B, MOTES M L, VEAZEY J E, HAMMOND R M, JONES T, SHIEH Y C, WAMNES J, VAUGHAN G, KHUDYAKOV Y and FIORE A E (2007), 'Use of molecular epidemiology to confirm a multistate outbreak of hepatitis A caused by consumption of oysters', *Clin. Infect. Dis.*, **44**, 838–40.

BISHARAT N, COHEN D I, HARDING R M, FALUSH D, CROOK D W and PETO T (2005), 'Hybrid *Vibrio vulnificus*', *Emerg. Infect. Dis.*, **11**, 30–5.

CAC (1999), *Principles and Guidelines for the Conduct of Microbiological Risk Assessment*, Codex Alimentarius Commission, Joint FAO/WHO Food Standards Programme, CAC/GL 30-1999.

CRL (2007), *Microbiological Monitoring of Bivalve Mollusc Harvesting Areas*, Guide to good practice, Technical application, EU working group on the microbiological monitoring of bivalve mollusc harvesting areas, issue 3; February 2007. Available online at: www.crlcefas.org

DEPAOLA A, ULASZEK J, KAYSNER C A, TENGE B J, NORDSTROM J L, WELLS J, PUHR N and GENDEL S M (2003), 'Molecular, serological, and virulence characteristics of *Vibrio parahaemolyticus* isolated from environmental, food, and clinical sources in North America and Asia', *Appl. Environ. Microbiol.*, **69**, 3999–4005.

EUROPEAN COMMUNITIES (2004), 'Regulation (EC) No. 854/2004 of the European Parliament and of the Council of 29 April 2004 laying down specific rules for the organization of official controls on products of animal origin intended for human consumption', *Off. J. Eur. Communities*, L.226, 25.6.04: 83–127.

EUROPEAN COMMUNITIES (2005), 'Commission Regulation (EC) No. 2073/2005 on microbiological criteria for foodstuffs', *Off. J. Eur. Communities*, L.338, 22.12.05: 1–26.

FAO/WHO (2006), *The Use of Microbiological Risk Assessment Outputs to Develop Practical Risk Management Strategies: Metrics to improve food safety*, Report, 3–7 April 2006, Kiel, Germany.

FUKUDA S, TAKAO S, KUWAYAMA M and MIYAZAKI K (2006), 'Molecular epidemiology of gastroenteritis outbreaks due to Norovirus', *Hiroshima J. Vet. Med.*, **21**, 64–8.

GONZALEZ-ESCALONA N, JAYKUS L A and DEPAOLA A (2007), 'Typing of *Vibrio vulnificus* strains by variability in their 16S-23S rRNA intergenic spacer regions', *Foodborne Pathog. Dis.*, **4**, 327–37.

GREENWOOD M, WINNARD G and BAGOT B (1998), 'An outbreak of *Salmonella* Enteritidis phage type 19 infection associated with cockles', *Commun. Dis. Public Health*, **1**, 35–7.

GUILFOYLE F, KEAVENEY S, FLANNERY J and DORÉ B (2007), 'REDRISK: reduction of the virus risk in shellfish harvesting areas', *Proceedings of the 7th Irish Shellfish Safety Workshop*, Marine Environment and Health Series No. 27, Marine Institute, Galway, pp. 20–7. Available online at: http://www.marine.ie/home/publicationsdata/publications/MEHS.htm

HUSS H H, ABABOUCH L and GRAM L (2003), *Assessment and Management of Seafood Safety and Quality*, FAO Technical paper 444, Food and Agriculture Organisation of the United Nations, Rome.

JOHNSON D E, WEINBERG L, CIARKOWSKI J, WEST P and COLWELL R R (1984), 'Wound infection caused by Kanagawa-negative *Vibrio parahaemolyticus*', *J. Clin. Microbiol.*, **20**, 811–12.

JOSEPH S W, COLWELL R R and KAPER J B (1982), '*Vibrio parahaemolyticus* and related halophilic *Vibrios*', *Crit. Rev. Microbiol.*, **10**, 77–124.

KHAN A A, MCCARTHY S, WANG, R-F and CERNIGLIA C E (2002), 'Characterization of United States outbreak isolates of *Vibrio parahaemolyticus* using enterobacterial repetitive intergenic consensus (ERIC) PCR and development of a rapid PCR method for detection of O3:K6 isolates', *FEMS Microbiol. Lett.*, **206**, 209–14.

LAING I and SPENCER B E (2000), *Bivalve Cultivation: Criteria for Selecting a Site*, Science Series: Technical Report No. 136, CEFAS, Lowestoft.

LEAL N C, DA SILVA S C, CAVALCANTI V O, DE A FIGUEIROA Â C T, NUNES V V F, MIRALLES I S and HOFER E (2008), '*Vibrio parahaemolyticus* serovar O3:K6 gastroenteritis in northeast Brazil', *J. Appl. Microbiol.*, **105**, 691–7.

LEE R J (2009), *Salmonella in Bivalve Molluscs: Sampling plan Considerations*, Report to the Food and Agriculture Office of the United Nations, FAO, Rome.

LEE R J, LOVATELLI A and ABABOUCH L (2008), *Bivalve Depuration: Fundamental and Practical Aspects*, FAO Fisheries Technical Paper. No. 511, FAO, Rome.

LEE R J and YOUNGER A D (2003), 'Determination of the relationship between faecal indicator organisms and the presence of human pathogenic micro-organisms in shellfish', in Villalba A, Reguera B, Romalde, J.L. and Beiras, R. (Eds): *Molluscan Shellfish Safety*, Consellería de Pesca e Asuntos Marítimos da Xunta de Galicia and Intergovernmental Oceanographic Commission of UNESCO, Santiago de Compostela, Spain, pp. 247–52.

LEE R J, YOUNGER A D and LEES D N (2000), 'The impact of microbiological contamination on the development of shellfish farming in England and Wales', *Shellfish News*, **44**, 14–6.

LE GUYADER F, HAUGARREAU L, MIOSSEC L, DUBOIS E and POMMEPUY M (2000), 'Three-year study to assess human enteric viruses in shellfish', *Appl. Environ. Microbiol.* **66**, 3241–4248.

LE GUYADER F S, BON F, DEMEDICI D, PARNAUDEAU S, BERTONE A, CRUDELI S, DOYLE A, ZIDANE M, SUFFREDINI E, KOHLI E, MADDALO F, MONINI M, GALLAY A, POMMEPUY M, POTHIER P and RUGGERI F M (2006), 'Detection of multiple noroviruses associated with an international gastroenteritis outbreak linked to oyster consumption', *J. Clin. Microbiol.*, **44**, 3878–82.

LEES D (2000), 'Viruses and bivalve shellfish', *Int. J. Food Microbiol.*, **59**, 81–116.

MARTINEZ-URTAZA J and LIEBANA E (2005), 'Use of pulsed-field gel electrophoresis to characterize the genetic diversity and clonal persistence of *Salmonella senftenberg* in mussel processing facilities', *Int. J. Food Microbiol.*, **105**, 153–63.

NILSSON W B, PARANJPYE R N, DEPAOLA A and STROM M S (2003), 'Sequence polymorphism of the 16S rRNA gene of *Vibrio vulnificus* is a possible indicator of strain virulence', *J. Clin. Microbiol.*, **41**, 442–6.

OLIVER J D and KAPER J B (1997), '*Vibrio* species', in Doyle M.P., Beuchat, L.R. and Montville, T.J. (Eds): *Food Microbiology – Fundamentals and Frontiers*, ASM Press, Washington, DC, pp. 228–64.

PANICKER G, CALL D R, KRUG M J and BEJ A K (2004), 'Detection of pathogenic *Vibrio* spp. in shellfish by using multiplex PCR and DNA microarrays', *Appl. Environ. Microbiol.*, **70**, 7436–444.

POTASMAN I, PAZ A and ODEH M (2002), 'Infectious outbreaks associated with bivalve shellfish consumption: a worldwide perspective', *Clin. Infect. Dis.*, **35**, 921–8.

RIPPEY S R (1994), 'Infectious diseases associated with molluscan shellfish consumption', *Clin. Microbiol. Rev.*, **7**, 419–25.

ROSCHE T M, YANO Y and OLIVER J D (2005), 'A rapid and simple PCR analysis indicates there are two subgroups of *Vibrio vulnificus* which correlate with clinical or environmental isolation', *Microbiol. Immunol.*, **49**, 381–9.

SARKAR B, CHOWDHURY N R, NAIR G B, NISHIBUCHI M, YAMASAKI S, TAKEDA Y, GUPTA S K, BHATTACHARYA S K and RAMAMURTHY T (2003), 'Molecular characterization of *Vibrio parahaemolyticus* of similar serovars isolated from sewage and clinical cases of diarrhoea in Calcutta, India', *World J. Microbiol. Biotech.*, **19**, 771–6.

TOWNER K J and COCKAYNE A (1993), *Molecular Methods for Microbial Identification and Typing*, Chapman & Hall, London.

US FOOD AND DRUG ADMINISTRATION (USFDA) (2008), *National Shellfish Sanitation Program (NSSP): Guide for the Control of Molluscan Shellfish, 2007 Revision*. Available online at: http://www.fda.gov/Food/FoodSafety/Product-SpecificInformation/Seafood/FederalStatePrograms/NationalShellfishSanitationProgram/ucm046353.htm

US FOOD AND DRUG ADMINISTRATION (USFDA) CENTER FOR FOOD SAFETY AND APPLIED NUTRITION (1986), 'Sanitation of the harvesting and processing of shellfish', *National Shellfish Sanitation Program Manual of Operations*, Part II, US Food and Drug Administration, Washington, DC.

VICKERY M C, NILSSON W B, STROM M S, NORDSTROM J L and DEPAOLA A (2007), 'A real-time PCR assay for the rapid determination of 16S rRNA genotype in *Vibrio vulnificus*', *J. Microbiol. Methods*, **68**, 376–84.

WAGLEY S, KOOFHETHILE K, WING J B and RANGDALE R (2008), 'Comparison of *V. parahaemolyticus* isolated from seafoods and cases of gastrointestinal disease in the UK', *Int. J. Environ. Health Res.*, **18**, 283–93.

WARNER E B and OLIVER J D (2008), 'Population structure of two genotypes of *Vibrio vulnificus* in oysters (*Crassostrea virginica*) and sea water', *Appl. Environ. Microbiol.*, **74**, 80–5.

ZIMMERMAN A M, DEPAOLA A, BOWERS J C, KRANTZ J A, NORDSTROM J L, JOHNSON C N and GRIMES D J (2007), 'Variability of total and pathogenic *Vibrio parahaemolyticus* densities in northern Gulf of Mexico water and oyster', *Appl. Environ. Microbiol.*, **73**, 7589–96.

23

Tracing pathogens in fruit and vegetable production chains

R. E. Mandrell, US Department of Agriculture, USA

Abstract: The health benefits of fresh fruits and vegetables and year-round availability of many produce commodities has led to increased consumption in the United States and other western countries. This has, however, been correlated with a rise in the number of foodborne outbreaks. Although there are logical sources of pathogens, definitive explanations of how contamination occurs have been lacking. Biotic or abiotic processes linking primary pathogen reservoirs to fields are dynamic and difficult to identify. Tracking reservoirs and movement of pathogens in the environment require intensive sampling and accurate fingerprinting/genotyping. Fresh produce outbreaks associated with enteric pathogens, their sources and fitness in the environment will be presented along with methods for tracking pathogens and outcomes from recent investigations. Next generation DNA sequencing and genomics methods being developed will enhance microbial source tracking studies in complex produce production environments.

Key words: microbial source tracking, lettuce, spinach, leafy greens, pre-harvest, wildlife, genotyping, serovars, reservoirs, rainfall, climate, prevalence.

23.1 Introduction

It's my belief that marshes were there once, and the roots of centuries rotted into the soil and made it black and fertilized it. And when you turn it up, a little greasy clay mixes and holds it together. That's from Gonzales north to the river mouth. Off to the sides, around Salinas and Blanco and Castroville and Moss Landing, the marshes are still there. And when one day those marshes are drained off, that will be the richest of all land in this red world.

Samuel Hamilton in John Steinbeck's *East of Eden*, 1952

This comment from a character in John Steinbeck's novel *East of Eden* was prescient of the future of this farming region in the Salinas Valley on the central coast of California in the United States. This unique region known as the 'Salad Bowl of America' has optimal climate and soil conditions for producing a large portion of the US supply of leafy vegetables (approximately 70%) and other fresh produce commodities. Of course, other regions of the United States have favorable, but different, growing conditions conducive to growing other major commodities, such as fresh tomatoes grown in the southeastern US locations, and a variety of fruits and vegetables produced in the United States and other parts of the world for consumption locally or for export (Beuchat, 2002; Calvin *et al.*, 2009). Increased interest in the nutritional and health benefits of fresh produce has led to development of new varieties of domestic and imported produce and new processing and packaging strategies to meet the increasing demand (Clemens, 2004).

Although the overall incidence of illness linked to contaminated produce is quite low relative to the total number of produce consumptions (e.g. bagged spinach = probably millions of consumptions/illness, on average) (Calvin *et al.*, 2009), there has been an increase in the number of outbreaks in the last two decades (Sivapalasingam *et al.*, 2004) (Table 23.1). Some recent major outbreaks have been multi-state and multi-country in size, and the pathogens associated with them appear to be hypervirulent (Manning *et al.*, 2008).

Definitive information about pathogen fate and transport processes relevant to produce contamination at any stage of the production chain has been lacking. Perhaps the most notable example of a recurring problem with foodborne illness associated with fresh produce is the 20 or more leafy greens-associated outbreaks occurring between 1995 and 2008 linked to *Escherichia coli* O157:H7 contamination (Table 23.1). The fact that at least ten of the outbreaks were traced to produce items that were grown in or near the Salinas Valley of California indicates that the implicated produce was contaminated before harvest. Similarly, recent *Salmonella enterica* outbreaks in Europe associated with lettuce, and multiple outbreaks in North America associated with tomatoes, cantaloupes and raw almonds, also fit epidemiology suggesting pre-harvest contamination of produce. Intensive investigations of some of these outbreaks have yielded information consistent with field/pre-harvest contamination (CALFERT, 2007a, b, 2008; Hedberg *et al.*, 1999; Gupta *et al.*, 2007; Greene *et al.*, 2008; Castillo *et al.*, 2004).

The potential sources of pre- and post-harvest contamination, such as livestock/wild animal/human feces, irrigation or processing water, cross-contamination in transport/distribution/processing, seem logical, but the animal vectors and watersheds that link primary environmental pathogen reservoirs to fields are dynamic and transient by nature and not easy to identify. The importance of minimizing pre-harvest contamination emphasizes the need for comprehensive investigations and studies of the fresh produce production 'farm-to-fork' continuum to identify risk factors associated with contamination, including the sources, transport and ecology of enteric pathogens that have been linked commonly to outbreaks.

Water, flooding, wind, roaming animals and possibly, some farming practices (e.g. vehicles, fertilization, spraying, harvesting), are candidate sources of produce contamination. Surveys of some of these logical sources of pre-harvest

Table 23.1 Summary of selected outbreaks suspected of being associated with pre-harvest contamination of fresh produce by enteric human pathogens[a]

Pathogen	Month-Year	Location[b]	No. ill	Known or suspected vehicle	Source region[c]	Reference
E. coli O157:H7	Jul-95	MT	74	Lettuce, Romaine	MT, WA	(Ackers *et al*., 1998)
E. coli O157:H7	Sep-95	ME	30	Lettuce, Iceberg	Unknown	(CDC, 1995)
E. coli O157:H7	Sep-95	ID	20	Lettuce, Romaine	Unknown	(CSPI, 2008)
E. coli O157:H7	Oct-95	OH	11	Lettuce	Unknown	(CDC, 1995)
E. coli O157:H7	May-96	IL, CT	61	Lettuce, Mesclun mix	CA (SV)	(Hilborn *et al*., 1999)
E. coli O157:H7	Jun-96	NY	7	Lettuce, Mesclun	Unknown	(CDC, 1996)
E. coli O157:H7	May-98	CA	2	Lettuce, salad	Unknown	(CDC, 1998)
E. coli O157:H7	Sep-98	MD	4	Lettuce	Unknown	(CDC, 1998)
E. coli O157:H7	Feb-99	NE	65	Lettuce, salad	Unknown	(CDC, 1999)
E. coli O157:H7	Sep-99	Sweden	11	Lettuce	Europe (Central)	(Welinder-Olsson *et al*., 2004)
E. coli O157:H7	Sep-99	CA	8	Lettuce, Romaine	CA (SV)	(CDC, 1999)
E. coli O157:H7	Sep-99	WA	6	Lettuce, Romaine	CA (SV)	(CDC, 1999)
E. coli O157:H7	Oct-99	OH, IN	47	Lettuce, salad	Unknown	(CDC, 1999)
E. coli O157:H7	Oct-99	OR	3	Lettuce, Romaine hearts	CA (SV)	(CDC, 1999)
E. coli O157:H7	Oct-99	PA	41	Lettuce, Romaine	CA (SV)	(CDC, 1999)
E. coli O157:H7	Jul-02	WA	29	Lettuce, Romaine	CA (SV)	(CDC, 2002)
E. coli O157:H7	Nov-02	IL, WI, MN, SD, UT	24	Lettuce	CA (SjoV)	(CDC, 2002)
E. coli O157:H7	Sep-03	CA	57	Lettuce, Iceberg/Romaine	CA (SV)	(CDHS, 2004a)
E. coli O157:H7	Sep-03	ND	5	Lettuce, mixed with Romaine	Unknown	(CDC, 2003)
E. coli O157:H7	Oct-03	CA	16	Spinach	CA (SV)	(CDHS, 2004b)
E. coli O157:H7	Nov-04	NJ	6	Lettuce	CA (SV)	(CDC, 2004)

E. coli O157:H7	Sep-05	MN	11	Romaine, also vegetables	CA (SV)	(MDPH, 2005)
E. coli O157:H7	Aug/Sep-05	Sweden	135	Lettuce, Iceberg	Sweden	(Soderstrom *et al.*, 2008)
E. coli O157:H7	Aug/Sep-06	Multi (26 states)	>200	Spinach, baby, bagged	CA (SjuV)	(CALFERT, 2007b, c)
E. coli O157:H7	Nov-06	NJ, NY, PA, DE	71	Lettuce, Iceberg	CA (CentV)	(CALFERT, 2007a)
E. coli O157:H7	Nov/Dec-06	MN, IA, WI	81	Lettuce, Iceberg	CA (CentV)	(CALFERT, 2008)
E. coli O157:H7	Sep/Oct-07	Netherlands, Iceland	50	Lettuce, shredded	Unknown	(Friesema *et al.*, 2008)
E. coli O157:H7	May-08	WA	10	Lettuce, Romaine	CA (SV)	(WDOH, 2008)
E. coli O157:H7	Sep-08	MI, IL and Ontario	>40	Lettuce, Iceberg	CA (?)	(Marler-Clark, 2009)
E. coli O145	Mar/Apr-2010	MI,NY,OH,TN	>30	Lettuce, Romaine Yuma	AZ	(CDC, 2010)
S. Senftenberg	Jan/Jun-07	UK	55	Basil	Israel	(Pezzoli *et al.*, 2008)
S. Thompson	Mar-99	CA	>41	Cilantro	Mexico, sus.	(Campbell *et al.*, 2001)
S. Saphra	Feb/May-97	Multi	24	Cantaloupe	Mexico	(Mohle-Boetani *et al.*, 1999)
S. Poona	Spring-00-02[d]	Multi and Canada	58	Cantaloupe	Mexico	(MMWR, 2002)
S. Saintpaul	Oct-06	Australia	36	Cantaloupe	Australia	(Munnoch *et al.*, 2009)
S. Litchfield	Jan/Mar-08	Multi and Canada	51	Cantaloupe	Honduras	(CDC, 2008a)
S. Newport	May/Jun-01	UK	19	Vegetables, bagged	Italy, Spain	(Sagoo *et al.*, 2003)
S. Typhimurium, DT204b	Aug/Sep-00	Iceland, other EU countries	392	Lettuce, shredded	Unknown (imported)	(Crook *et al.*, 2003)
S. Newport	Sep/Oct-04	N. Ireland, UK	>130	Lettuce, Iceberg	Unknown	(Irvine *et al.*, 2009)
S. Typhimurium	May-05	Finland	60	Lettuce, Iceberg	Spain	(Takkinen *et al.*, 2005)
S. Typhimurium	May/Jul-05	Multi	152	Oranges, juice	FL?	(Jain *et al.*, 2009)
S. Thompson	Oct/Dec-04	Multi in Europe	21	Rucola (arugula)	Italy	(Nygard *et al.*, 2008)
S. Javiana	Jun/Aug-90	IL, MI, MN, WI	176	Tomatoes	SC	(Hedberg *et al.*, 1999)

(Continued)

Table 23.1 Continued

Pathogen	Month-Year	Location[b]	No. ill	Known or suspected vehicle	Source region[c]	Reference
S. Montevideo	Jun/Aug-93	IL, MI, MN, WI	100	Tomatoes	SC	(Hedberg *et al.*, 1999)
S. Baildon	Dec98–Jan99	Multi	86	Tomatoes	FL	(Cummings *et al.*, 2001)
S. Javiana	Jun/Jul-02	FL	141	Tomatoes, pre-diced	?	(Srikantiah, 2002; Gupta *et al.*, 2007)
S. Newport	Sep/Oct-02	Multi	510	Tomatoes	VA	(Greene *et al.*, 2008)
S. Braenderup	Jul-04	Multi	125	Tomatoes	FL	(Gupta *et al.*, 2007)
S. Javiana and other serovars	Jul-04	Multi	429	Tomatoes, pre-sliced	?	(Gupta *et al.*, 2007)
S. Newport	Jul/Nov-05	Multi	72	Tomatoes	VA	(MMWR, 2007a; Greene *et al.*, 2008)
S. Braenderup	Nov/Dec-05	Multi	82	Tomatoes, pre-diced	FL	(MMWR, 2007a)
S. Newport	Jul/Nov-06	Multi	115	Tomatoes	?	(MMWR, 2007a)
S. Typhimurium	Sep/Oct-06	Multi	190	Tomatoes	OH	(MMWR, 2007a)
S. Enteritidis	Oct00–Jul01	Multi and Canada	168	Almonds, raw	CA	(Isaacs *et al.*, 2005)
S. Enteritidis	Sep03–Apr04	Multi and Canada	29	Almonds, raw	CA	(MMWR, 2004)
S. Enteritidis	Dec05–Aug06	Sweden	15	Almonds, raw	CA	(Ledet Muller *et al.*, 2007)
S. Litchfield	Oct06–Jan07	Australia	26	Papaya	Australia	(Gibbs *et al.*, 2009)
S. Saintpaul	Apr/Jul-08	Multi	>1200	Peppers[e]	Mexico (sus.)	(CDC, 2008b)
Shigella sonnei[f]	Jul/Aug-98	Multi and Canada	486	Parsley	Mexico	(Naimi *et al.*, 2003)
Shigella flexneri	May-01	NY	886	Tomatoes	FL	(Reller *et al.*, 2006)
Shigella sonnei	Aug-04	Multi	116	Carrots	CA?	(Gaynor *et al.*, 2008)

Shigella sonnei	Apr/Jun-09	Denmark, Norway	>30	Sugar peas	Kenya (sus.)	(Muller *et al.*, 2009; Heier *et al.*, 2009)
Shigella dysenteriae	May/Jun-09	Sweden	47	Sugar snap peas	Kenya (sus.)	(Lofdahl *et al.*, 2009)
Yersinia pseudotuberculosis	Oct-98	Finland	47	Lettuce, Iceberg	Finland	(Nuorti *et al.*, 2004)
Yersinia pseudotuberculosis	May-03	Finland	111	Carrots, grated	Finland	(Jalava *et al.*, 2006)
Yersinia pseudotuberculosis	Mar-04	Finland	53	Carrots	Finland	(Kangas *et al.*, 2008)
Yersinia pseudotuberculosis	Aug/Sep-06	Finland	>400	Carrots	Finland	(Rimhanen-Finne *et al.*, 2008)

[a] Outbreaks included have been selected based on location or suspected pre-harvest contamination. Outbreaks associated with almonds have been included because of recurrent outbreaks suspected of being linked to a common location.

[b] US states are designated by the two-letter abbreviation; Multi, multiple states involved.

[c] SV, Salinas Valley, CA; SJoV, San Joaquin Valley, CA; SJuV, San Juan Valley, CA; CentV, Central Valley, CA. Some location information was provided by California Dept. of Public Health (personal communication). Unknown = traceback not done or incomplete. Sus. = suspected.

[d] Represents three outbreaks (2000, 2001, 2002); the 2000 and 2002 outbreaks were caused by the same strain.

[e] Cases occurred in 43 states, Washington, DC and Canada; jalapeño and/or Serrano peppers grown in Mexico were suspected as the cause of a majority of cases.

[f] Parsley served in 8 restaurants. Also enterotoxigenic *E. coli* (ETEC) cases (no. = 76) in Minnesota associated with parsley served in two restaurants.

contamination such as livestock manure, wild animal feces, and surface and irrigation water have reported a wide range of percentage incidence of pathogenic *E. coli, Salmonella* species, and other foodborne pathogens. Tracking these potential sources and movement of pathogens requires intensive studies of complex environments using modern molecular methods for genotyping (fingerprinting) pathogens or surrogates of fecal contamination. Although numerous approaches for tracking pathogens and other fecal contaminants have been developed, it remains a relatively labour-intensive and somewhat inaccurate process. Modern molecular and genomics methods, however, have stimulated development of improved methods for isolating and/or detecting organisms and differentiating them at high enough resolution to monitor more effectively the prevalence and movement of pathogens. This review will not provide a comprehensive comparison of all strain discrimination methods for microbial source tracking (MST). Numerous recent reviews are available with details on advantages and disadvantages of various strain discrimination methods such as digestion of pathogen genomic DNA and running on gels (PFGE (pulsed-field gel electrophoresis), RFLP ribotyping, AFLP), indexing gene presence/absence (microarrays), DNA sequencing of conserved and/or virulence genes (MLST/MVLST), PCR-based genetic assays (RAPD, REP-PCR) and analysis of variable regions of the genome (VNTR, multilocus variable number tandem repeat analysis (MLVA)) (Simpson *et al.*, 2002; Scott *et al.*, 2002; Meays *et al.*, 2004; Whittam and Bergholz, 2007; Harwood, 2007; Foley *et al.*, 2009). Details of these and other methods will be presented also in other sections of this book. This chapter will review information related to the microbial food safety of fresh produce including selected relevant outbreaks, incidence of pathogens in the produce production environment, and some recent strategies and methods of tracking pathogens to determine reservoirs and movement of pathogens in the environment.

23.2 Summary of major outbreaks linked to pre-harvest contamination of produce

Increased production and consumption of fresh produce has been associated with an increase in the number of outbreaks of foodborne illness (CSPI, 2007; Sewell and Farber, 2001; Sivapalasingam *et al.*, 2004; Mandrell, 2009); the types of produce and pathogens associated most frequently with these outbreaks have been documented in previous review articles about this subject (Sivapalasingam *et al.*, 2003; Sewell and Farber, 2001; Nguyen-the and Carlin, 1994; Beuchat, 1996; Seymour and Appleton, 2001; Harris *et al.*, 2003; Mandrell and Brandl, 2004; Beuchat, 2006; Johnston *et al.*, 2006b; Heaton and Jones, 2008). Selected outbreaks linked to probable pre-harvest contamination are cited in Table 23.1 to document temporal and geographic factors and the commodities implicated; additional details for some of these outbreaks are available in a previous review of this subject (see Mandrell, 2009). In addition to these notable large outbreaks, numerous small outbreaks have occurred probably, but have not been detected or reported due to limited epidemiological information.

In recent reviews of outbreaks associated specifically with fresh produce, the US Centers for Disease Control and Prevention (CDC), analyzing data from the CDC Foodborne Outbreak Surveillance System for 1973–1997, identified 190 outbreaks associated with produce, 16 058 illnesses, 598 hospitalizations and eight deaths (Sivapalasingam *et al.*, 2003). CDC analyzed illnesses associated specifically with leafy greens between 1973 and 2006 and identified 502 outbreaks, >18 000 illnesses, and 15 deaths; 30 of the outbreaks were caused by *E. coli* O157:H7, 35 by *Salmonella*, and 196 by Norovirus (Herman *et al.*, 2008). Comparison of these two studies indicates that produce-associated outbreaks increased from 0.7% of all foodborne outbreaks in the 1970s to 6% in the 1990s, and increased further to the present (Herman *et al.*, 2008; Lynch *et al.*, 2009). The bacterial, viral and protozoal pathogens associated with fresh produce outbreaks (number of outbreaks) in the United States between 1973 and 1997 include: *Salmonella* (30 outbreaks), *E. coli* O157:H7 (13), non-O157 *E. coli* (2), *Shigella* (10), *Campylobacter* (4), *Bacillus cereus* (1), *Yersinia enterocolitica* (1), *Staphylococcus aureus* (1), Hepatitis A (12), Norovirus (9), *Cyclospora cayetanensis* (8), *Giardia lamblia* (5), and *Cryptosporidium parvum* (3); an additional 87 outbreaks were documented without any etiology identified (Sivapalasingam *et al.*, 2003). Multiple outbreaks of *S. enterica* and *E. coli* O157:H7 illness have been associated with fresh sprouts (e.g. alfalfa, mung bean, radish) grown from contaminated seed (Michino *et al.*, 1999; Breuer *et al.*, 2001; Mahon *et al.*, 1997; Proctor *et al.*, 2001; Mohle-Boetani *et al.*, 2008) harvested in different parts of the world (e.g. United States, Australia, China) under less than ideal agricultural conditions. These seed-related outbreaks emphasize the importance of the pre-harvest environment at every step of the production cycle, including seed and transplant production, harvesting, and the fields prior to and following harvest (water, fertilizers, crop debris, human and animal visits). Although there is no evidence that contaminated seeds have been implicated in the non-sprout outbreaks noted (Table 23.1), seeds should be appreciated as an early pre-harvest control point in fresh produce production and worthy of monitoring for contamination.

As noted previously, pre-harvest contamination has been suspected in numerous outbreaks associated with leafy vegetables (lettuce and spinach), tomatoes, cantaloupes, parsley, and probably, other commodities (e.g. jalapeño peppers, April–July, 2008). Of the more than 20 foodborne outbreaks since 1995 linked to contamination of US-grown leafy vegetables by *E. coli* O157:H7, at least 600 reported illnesses and five deaths occurred. Since 2000, at least 12 outbreaks have been linked to *Salmonella*-contaminated tomatoes (>1600 cases), and at least four outbreaks have been linked to *Salmonella*-contaminated cantaloupes from three different countries (166 cases) (Table 23.1). Limited details available from an investigation of a major outbreak of *Salmonella enterica* ser. Saintpaul associated with jalapeño and/or Serrano peppers grown in Mexico (distributed by a company in Texas) indicated that contamination occurred on the farm (CDC, 2008b). Several outbreaks suspected of being associated with pre-harvest contamination of tomatoes, lettuce and carrots by *Shigella* and *Yersinia* species have also been reported (Table 23.1). The questions remaining after investigations of these outbreaks concluded emphasize the importance of improving tracking sources and genotypes of pathogens related to pre-harvest contamination and their persistence/fitness in diverse geographic environments and ecosystems (Table 23.1; e.g. leafy vegetables – Western US/Sweden/Italy; tomatoes – Eastern US; cantaloupe – Mexico/Australia/Honduras; *Yersinia* – Finland).

The epidemiological studies of fresh produce outbreaks noted above conclude often without definitive evidence of the source of contamination. A series of outbreaks and traceback investigations linked to leafy greens in the United States stimulated public health agencies to determine the possible sources of contamination. Investigations of 12 leafy greens outbreaks associated with *E. coli* O157:H7 determined that the commodity was grown on farms in the Salinas Valley, a region located on the Central Coast of California and the major supplier of fresh produce to the US market (Table 23.1; see references for additional details). In addition, a major *E. coli* O157:H7 outbreak occurred in the summer of 2006 involving baby spinach grown in a valley adjacent to the Salinas Valley (CALFERT, 2007b; Cooley *et al.*, 2007). Similarly, recurrent outbreaks associated with tomatoes were suspected of being grown on farms in Virginia and Florida, and multiple outbreaks with cantaloupes on farms in Mexico (Table 23.1). Pathogen strains were isolated in some investigations from watersheds in the vicinity of implicated fields, and for the first time in recent outbreak investigations, *E. coli* O157:H7 and *Salmonella* strains indistinguishable from the clinical outbreak strains were isolated from environmental samples (CALFERT, 2007b, c; Cooley *et al.*, 2007; CALFERT, 2008; Greene *et al.*, 2008).

It has become evident that accurate information about the fate and transport processes relevant to contamination, and the fitness of pathogens near, on or in produce plants in the field, is critical for developing strategies for minimizing pre-harvest contamination of produce. Thus, information provided by these investigations assists in identifying common risk factors and differences between growing regions and commodities.

23.3 Incidence of human pathogens on fresh produce

Produce items are rarely contaminated with significant levels of enteric pathogens. An assessment of the number of consumptions of bagged baby spinach and number of estimated illnesses in the year of the 2006 outbreak associated with *E. coli* O157:H7 (approximately 1000) would equate roughly to >4 million consumptions/illness (Calvin, 2007; Calvin *et al.*, 2009). Nevertheless, even a very low contamination rate on raw produce is unacceptable considering the low infectious dose for pathogens like *E. coli* O157:H7 (approximately 30–100 cells (see Teunis *et al.*, 2004)).

Numerous surveys of fresh produce for incidence of foodborne pathogens have been reported, providing data relevant also for assessing the survival and fitness of pathogens in agricultural environments such as manure, water and soil. Beuchat published in 1996 one of the first and best reviews of reported incidence of common foodborne pathogens on ready-to-eat vegetables, and the potential sources of the pathogens and mechanisms of contamination (Beuchat, 1996). The incidence, growth and survival of foodborne pathogens in fresh and processed produce have also been reported in comprehensive reviews by Nguyen-the and Carlin (Nguyen-the and Carlin, 2000), Harris *et al.* (see Tables I-1 to I-7 in (Harris *et al.*, 2003)), and other recent reviews (Johnston *et al.*, 2006b; Beuchat, 2006; Mandrell and Brandl, 2004; Crepet *et al.*, 2007; Heaton and Jones, 2008; Diez-Gonzalez and Mukherjee, 2009). Although distinctions between pre- and post-harvest contamination are unclear, these reviews provide useful summaries of the different methods for isolating and detecting pathogens (e.g. *Salmonella, Listeria, Yersinia, Campylobacter* species, *E. coli* O157:H7, and generic *E. coli*) from multiple types of produce items grown in different regions of the world and information about incidence, fitness and epidemiology related to microbial food safety of produce.

The incidence of pathogens reported in studies of produce is often between 0% and <10% of all samples tested, with an incidence of >20% reported occasionally (Nguyen-the and Carlin, 1994; Harris *et al.*, 2003; Mandrell and Brandl, 2004). The occasional study reporting concentrations of pathogen indicate low levels in most studies, even for generic *E. coli*, a measure of potential fecal contamination. For example, the percentages of positives out of 774 total samples tested for *Salmonella* on leafy vegetables or salad in eight separate studies were 0, 0, 0.6, 0.9, 3.5, 6.3, 7.1, and 68% (Harris *et al.*, 2003). Of 1356 pre- and post-processed spinach samples tested in a separate study, 0.4% and 0.7% were positive for *Salmonella* and *L. monocytogenes*, respectively (Ilic *et al.*, 2008). In contrast, all 214 samples of lettuce or salad mix tested for *E. coli* O157:H7 in large UK and US studies were negative (Harris *et al.*, 2003). Of >3800 ready-to-eat salad vegetables from retail markets sold in the United Kingdom, only 0.2% were positive for *Salmonella*; an additional 0.5% were considered of poor quality due to contamination with *E. coli* or *L. monocytogenes* at >100 CFU per g of product (Sagoo *et al.*, 2003). A recent review of 165 prevalence studies of *L. monocytogenes* in fresh vegetables reported a 3% overall incidence (754/25 078 positive) (Crepet *et al.*, 2007). This high incidence is noteworthy considering the high mortality of the at-risk population, but also reflects the higher incidence of this potential pathogen in the environment compared to pathogenic *E. coli* and *Salmonella* and other pathogens. Additional recent surveys were reported for 'minimally processed' vegetables in Brazil (2.2% positive for *Salmonella*) (Froder *et al.*, 2007), fresh vegetable samples in South Africa (2.2% positive for *E. coli* O157:H7; 'spinach' as high as 1 600 000 CFU /g) (Abong'o *et al.*, 2008), fresh produce in Ontario Canada (0.17% positive for *Salmonella*) (Arthur *et al.*, 2007a), and fresh produce samples in Mexico (5.7% positive for *Salmonella*) (Quiroz-Santiago *et al.*, 2009). These results reflect the tremendous diversity of produce quality depending upon spatial and temporal factors, and possibly methodological factors.

Multiple outbreaks of *Salmonella* illness associated with tomatoes have occurred recently, but surveys of tomatoes for the incidence of pathogens have been limited. Of 123 samples of domestic (US) tomatoes tested by US FDA-CFSAN starting in May, 2001, none were positive for *Salmonella* or *E. coli* O157:H7 (FDA-CFSAN, 2001b); also, 0/20 imported tomato samples collected starting in March 1999 were negative for both pathogens (FDA-CFSAN, 2001a). However, 11 of 151 imported and 4 of 115 domestic cantaloupe samples in the same surveys were positive for *Salmonella* or *Shigella*. These results appear consistent with the multiple outbreaks occurring in 1997, 2000, 2001, 2002 and 2008 due to *Salmonella*-contaminated cantaloupe imported from Mexico or Honduras (Table 23.1). Surveys of cantaloupe and environmental samples from farms and packing plants in South Texas and Mexico identified *Salmonella* in 5/950 and 1/300 cantaloupes, respectively, and a higher incidence of *Salmonella* in water sources, tanks, and in the field (9.2–22.2%); generic *E. coli* was isolated also at significant levels from cantaloupes (3.9%, 25.7%) and irrigation water (22.8% and 31.1%), respectively (Castillo *et al.*, 2004). These results emphasize the role of water (well, river, aquifer, canal, and dam) as a significant potential source of both pre- and post-harvest contamination.

Outbreaks associated with fresh produce or juices contaminated with viruses or protozoa also occur, and pre-harvest contamination is a probable explanation. Hepatitis A virus (Niu *et al.*, 1992; Wheeler *et al.*, 2005; MMWR, 2003) and norovirus (see Figure 7, 'Produce-Pathogen Combinations' in CSPI, 2009) are the most common viruses associated with outbreaks linked to produce. The source of the contamination is suspected to be due most often to humans preparing the food in restaurants or institutions, rather than contamination in the field. However, the large number of produce-related norovirus outbreaks (CSPI, 2009; Ethelberg *et al.*, 2010), the obvious presence of norovirus in marine waters or watersheds (Tian *et al.*, 2007; Noda *et al.*, 2008), the ability of human noroviruses to replicate in cattle and pigs (Koopmans, 2008), and the fitness of norovirus on leafy vegetables (Wei *et al.*, 2010; Mara and Sleigh, 2010) suggest more research on noroviruses in the produce production environment is warranted. A recent cluster of at least 11 outbreaks in Denmark and Norway associated with norovirus on lettuce (possibly co-contaminated with enterotoxigenic *E. coli*), and suspected of being contaminated in fields in France, underscores why pre-harvest produce sources of norovirus should be considered more seriously (Ethelberg *et al.*, 2010).

Previous reviews note the role of important protozoa associated with foodborne illness (*Cryptosporidium*, *Cyclospora* and *Giardia*) and list selected outbreaks linked to fresh produce likely resulting from pre-harvest contamination or poor quality water (Rose and Slifko, 1999; Duffy and Moriarty, 2003). *Giardia* and *Cryptosporidium* are frequent colonizers of cattle (Olson *et al.*, 2004). Protozoa species that infect humans have been detected in water (Chaidez *et al.*, 2005; Phillip *et al.*, 2008; Mota *et al.*, 2009; Smith and Nichols, 2009) and in a variety of mammalian wildlife species (Appelbee *et al.*, 2005), thus illustrating the importance of MST to identify new environmental reservoirs of protozoa pathogens.

The fitness of a pathogen in the environment is related probably to the likelihood of survival on produce. Long-term persistence of foodborne pathogens in the environment is exemplified by a strain of *S. enterica* ser. Enteritidis implicated in at least one major outbreak, and possibly a minor outbreak, associated with raw almonds in 2000–01 (Isaacs *et al.*, 2005) and 2005–06 (Ledet Muller *et al.*, 2007), respectively. The serotype Enteritidis outbreak strain, subtyped as phage type 30, was isolated from a suspect orchard at multiple times over at least a five-year period, and with increasing frequency in samples collected during and following harvests (August–December) and following rain events (Uesugi *et al.*, 2007). *Salmonella* strains isolated during the study were all phage type (PT) 30, and indistinguishable from the clinical outbreak strains (or one band difference) by two-enzyme PFGE analysis (Uesugi *et al.*, 2007). Additional PT 30 strains isolated subsequently from orchards miles away, and an additional 2003–04 outbreak associated with serotype Enteritidis contaminated raw almonds, but of different PT (9c) and PFGE profile (MMWR, 2004), indicated a complex epidemiology and ecology, and the benefit of source tracking even with low resolution fingerprinting methods. Preliminary analysis by microarray gene indexing and MLVA of these PT30 and 9c strains representing a six-year period and from different environmental and clinical sources, revealed the PT30 and 9c strains were highly related phylogenetically. However, microarray gene expression analysis also revealed differences between the environmental and clinical strains correlating with biochemical and fitness differences (C. Parker, 2010). This type of long-term environmental MST study emphasizes the importance of high-resolution genotyping methods for characterizing pathogens for accurate matching of strains for identification of point sources, reservoirs and virulence.

Persistence (fitness) of a pathogen in an agricultural environment raises questions relevant to other produce-related outbreaks. Is contamination periodic and cumulative or is it due to a major isolated contamination event? Do persistent strains reflect selection and evolution of special fitness characteristics in a specific environment (e.g. almonds, leafy vegetables, tomatoes, cantaloupes, peppers, etc.)? Is pathogen incidence/concentration dynamic? Does pathogen fitness in a harsh environment (dry, high UV, low nutrients) with subsequent resuscitation/amplification (rain/moisture, low UV) correlate with virulence? Are some wildlife species (e.g. mammalian, avian, and amphibian) high shedders of pathogens? These and other questions are difficult to answer, but point to important areas for further research.

23.4 Incidence of generic *E. coli* on produce

What is the general quality of fresh produce? Produce-associated outbreaks (Table 23.1) have prompted multiple surveys to determine the incidence of pathogens or generic *E. coli* (as evidence of fecal contamination) on fresh produce. These data are beneficial as baseline indicators of the general microbiological quality of different types of produce grown in different regions conventionally or organically, and tested at different stages of the pre- and post-harvest cycle. Table 23.2 lists some of the selected recent studies of the incidence of generic *E. coli* in produce at different stages in the post-harvest cycle, including on the farm and in the distribution system.

The data summarized in Table 23.2 indicate major differences in *E. coli* incidence depending upon the size, time and location of the study, and possibly, differences in the sensitivity of methods. Generic *E. coli* incidence ranged from 2% to 36.5%. Details regarding differences in incidence of *E. coli* associated with commodity (parsley and cilantro are highest) (USDA-AMS-MDP, 2009), organic versus conventional (Mukherjee *et al.*, 2004; Arthur *et al.*, 2007a; Valentin-Bon *et al.*, 2008), and country/ecosystem are available in the references cited. In other studies, the approximate concentration of *E. coli* rather than incidence was determined. For example, a survey of produce items (e.g. arugula, cantaloupe, cilantro, parsley, spinach) collected between November 2000 and May 2002 from 13 farms in the southeastern United States revealed *E. coli* levels ranging from 0.7 to 1.5 log CFU/g for field or packing shed produce (Johnston *et al.*, 2005); a similar survey comparing produce grown in the southern United States and Mexico measured *E. coli* ranging between 0.7–1.9 and 0.7–4.0 log CFU/g for Mexican and southeastern US produce, respectively (Johnston *et al.*, 2006a). *E. coli* counts on positive produce samples in studies listed in Table 23.2 were reported as an average of >1200 MPN/g (Mukherjee *et al.*, 2004), >5 to 290 CFU/g for leaf lettuce, <5 to 7600 and <5 to 16 000 CFU/g for cilantro and parsley, respectively (Arthur *et al.*, 2007a), and 3 to 9.2 MPN/g for bagged cut spinach and lettuce mixes (conventional and organic) (Valentin-Bon *et al.*, 2008). A study initiated by the USDA Agricultural Marketing Service in 2002 and coordinated with state and other federal agencies to survey the microbial quality of fresh produce items available at terminal markets and wholesale distribution centres involved analysis of approximately 65 000 samples, both domestic and imported (e.g. cantaloupe, leaf and romaine lettuce, tomatoes, green onions, alfalfa sprouts, peppers) for generic *E. coli, E. coli* 'with pathogenic potential' (including *E. coli* O157:H7) and *Salmonella* (Table 23.2 (see USDA-AMS-MDP, 2009)). Although very low levels of generic *E. coli* were detected on >25% of samples, only 1.8% and 0.06% of the samples had >10 MPN/g and >100 MPN/g detected, respectively (USDA-AMS-MDP, 2009). More importantly, *E. coli* with pathogenic potential based on PCR results for various virulence factors, including shigatoxin 1 and 2 (Stx 1 and 2), ranged from 0.1% to 0.4% of all samples tested each year. Cantaloupe (26–32%), leaf/romaine lettuce (25–44%), cilantro (66–71%) and parsley (72%) were the commodities most frequently positive for generic *E. coli* ((USDA-AMS-MDP, 2009); data not shown). A recent study of >1300 samples of baby and savoy spinach yielded an overall generic *E. coli* incidence of 8.9%, but without a significant difference in the incidence and amount of *E. coli* pre- and post-processing with water and sanitizers (Ilic *et al.*, 2008). Thus, generic *E. coli* is frequently present on some produce types, but at significant concentrations only infrequently.

Although generic *E. coli* is evidence of fecal contamination, this finding does not correlate significantly with evidence of pathogen contamination. Thus, *E. coli* incidence is simply an indicator of potential contamination, and a potential risk

Table 23.2 Selected studies of the incidence of generic *E. coli* on selected fresh produce items

Region	Produce type	Period	% incidence[a]	Comments	References
US-MN	Multi	May–Sep, 2002	4.3	Certified organic	(Mukherjee *et al.*, 2004)
US-MN	Multi	May–Sep, 2002	11.4	Non-certified organic	(Mukherjee *et al.*, 2004)
US-MN	Multi	May–Sep, 2002	1.6	Conventional	(Mukherjee *et al.*, 2004)
Spain	Lettuce	~1999	25.7	University restaurants	(Soriano *et al.*, 2000)
US-DC	Multi	~2000	2	Retail and farmer's markets	(Thunberg *et al.*, 2002)
Canada-Ontario	Multi	Aug–Sep, 2004	5.3	Retail distribution and farmer's markets	(Arthur *et al.*, 2007a)
Spain-Catalonia	Multi	2005–06	14.8	Retail	(Abadias *et al.*, 2008)
US-DC area	Lettuce, spinach	2007	16	Organic and conventional	(Valentin-Bon *et al.*, 2008)
US and imported	Multi	2002	7.4	Distribution centre	(USDA-AMS-MDP, 2009)
US and imported	Multi	2003	6.7	Distribution centre	(USDA-AMS-MDP, 2009)
US and imported	Multi	2004[b]	28.8	Distribution centre	(USDA-AMS-MDP, 2009)
US and imported	Multi	2005	36.5	Distribution centre	(USDA-AMS-MDP, 2009)
US and imported	Multi	2006	20.5	Distribution centre	(USDA-AMS-MDP, 2009)
US and imported	Multi	2007[b]	12.0	Distribution centre	(USDA-AMS-MDP, 2009)
US and imported	Multi	2008	5.1	Distribution centre	(USDA-AMS-MDP, 2009)
US and Canada	Spinach	Jan06–Mar07	8.9	Pre- and post-processing	(Ilic *et al.*, 2008)

a Average incidence for multiple produce items (Multi).
b Generic *E. coli* method was modified in 2004 and again in 2007.

factor for further post-harvest contamination, cross-contamination during washing, or amplification of bacteria (pathogen) during transport and storage. Tracking *E. coli* incidence can serve as one measure of fecal microbial flora during the produce production and processing cycle, and for assessing the potential for pathogenic strains, if they were to be present, to survive under the same produce processing conditions.

Evidence of fecal contamination as high as 50–70% on some produce items does not correlate necessarily to a higher incidence of illness. The large number of consumptions of ready-to-eat produce (and tree nuts), which equates to many billions/year, results in an extremely high number of total consumptions per case (millions/case). Unidentified sporadic cases would lower the consumption/case number, but the unidentified to identified case ratio is uncertain (a number of 10 to >40 has been proposed). Regardless, the rare convergence-of-events resulting in a major outbreak are worthy of continued efforts to understand the reservoirs and movement of pathogens in the produce production environment and the conditions that may enhance their survival or growth. Assessment of potential foodborne pathogens rather than surrogates (e.g. generic *E. coli*, *Enterobacter*, fecal coliforms) requires sophisticated and expensive molecular and microbiological techniques, but the outcomes are more relevant to identification of point sources and risk factors.

23.5 Animal sources of enteric foodborne pathogens relevant to produce contamination

Identifying the primary and secondary reservoirs of enteric pathogens is a critical part of MST. Food animals have been linked to many outbreaks associated with produce, meat, milk and other food products. Numerous studies have documented the incidence in cattle (Elder *et al.*, 2000; Hussein and Bollinger, 2005; Fegan *et al.*, 2005; Low *et al.*, 2005; Dargatz *et al.*, 2003), swine (Chapman *et al.*, 1997; Jay *et al.*, 2007), sheep (Ogden *et al.*, 2005), poultry (Chapman *et al.*, 1997; Rose *et al.*, 2002; Foley *et al.*, 2008; McCrea *et al.*, 2006), and multiple species of wild animals (Ejidokun *et al.*, 2006; Hernandez *et al.*, 2003; Kirk *et al.*, 2002; Sargeant *et al.*, 1999; Pritchard *et al.*, 2001; Wetzel and LeJeune, 2006) of *E. coli* O157:H7, *S. enterica*, and *C. jejuni* (Miller and Mandrell, 2006). Selected studies documenting the incidence of pathogenic *E. coli*, *Salmonella* and *C. jejuni* in animals are summarized in Table 23.3. Again, details regarding the methods, periods, locations and samples studied can be obtained from the original papers cited.

Cattle are major primary carriers of enteric pathogens associated with foodborne outbreaks, including those associated with produce. However, other domestic animals and wildlife are common or intermittent carriers of these pathogens (Tables 23.1 and 23.3). The concentrations of pathogens in wildlife samples have not been well documented compared to livestock. The total quantity of viable pathogen cells shed by animals (i.e. concentration of pathogen x volume of feces x fitness), domestic or wild, is relevant epidemiologically for identifying

Table 23.3 Selected studies of the incidence of *E. coli* O157, *S. enterica* and *C. jejuni* in livestock and wild animal feces[a]

Animal	Pathogen	Reference
Cattle (mostly beef, some dairy)	*E. coli* O157	Review in (Hussein and Bollinger, 2005)
		(Matthews *et al.*, 2006)
		(Doane *et al.*, 2007)
		(Arthur *et al.*, 2007b)
		(Tuyet *et al.*, 2006)
	Non-O157 *E. coli*	Review in (Hussein and Bollinger, 2005)
	S. enterica	(Dargatz *et al.*, 2003)
		(Barkocy-Gallagher *et al.*, 2003)
		(Fegan *et al.*, 2004)
		(Milnes *et al.*, 2008)
		(Pangloli *et al.*, 2008)
Sheep	*E. coli* O157	(Ogden *et al.*, 2005)
		(Chapman *et al.*, 1997)
		(Milnes *et al.*, 2008)
		(Keen *et al.*, 2006)
		(Oporto *et al.*, 2008)
	S. enterica	(Branham *et al.*, 2005)
		(Milnes *et al.*, 2008)
Pigs	*E. coli* O157	(Chapman *et al.*, 1997)
		(Feder *et al.*, 2003)
		(Nakazawa and Akiba, 1999)
		(Cooper *et al.*, 2007)
		(Milnes *et al.*, 2008; Keen *et al.*, 2006)
		(Doane *et al.*, 2007)
		(Jay *et al.*, 2007)
	S. enterica	(Milnes *et al.*, 2008)
Chickens and turkeys	*E. coli* O157:H7	(Doane *et al.*, 2007)
		(Cooper *et al.*, 2007)
	S. enterica	(Bailey *et al.*, 2001)
		(Kinde *et al.*, 2004)
		(Rasschaert *et al.*, 2007)
Goats	*E. coli* O157	(Cooper *et al.*, 2007)
	S. enterica	(Branham *et al.*, 2005)
Deer	*E. coli* O157	(Keene *et al.*, 1997)
		(Sargeant *et al.*, 1999)
		(Fischer *et al.*, 2001)
		(Renter *et al.*, 2001)
		(Dunn *et al.*, 2004)
	S. enterica	(Branham *et al.*, 2005)
		(Renter *et al.*, 2006)
Rabbits[b], ducks, fish, rats	*E. coli* O157:H7 or *E. coli* O157	(Pritchard *et al.*, 2001; Leclercq and Mahillon, 2003)
		(Leclercq and Mahillon, 2003)
		(Tuyet *et al.*, 2006)
		(Cizek *et al.*, 1999)
		(Garcia and Fox, 2003)[a]
Wild tortoises/turtles/ water frogs, wild birds	*S. enterica*	(Hidalgo-Vila *et al.*, 2007)
		(Kirk *et al.*, 2002)

(Continued)

Table 23.3 Continued

Animal	Pathogen	Reference
		(Hughes *et al.*, 2008) (Fenlon, 1981) (CDC, 2009)
Cattle, poultry, geese, ducks, pigs, sheep	*C. jejuni/C. coli*	Review of >20 studies (Miller and Mandrell, 2006)

a Incidence, location, period of sampling and other details are presented in the cited articles and summarized in Mandrell, 2009.
b Non-O157 EHEC in rabbits.

potential sources of pathogen and relevant risk factors (Chase-Topping *et al.*, 2007). Some pathogen strains may be predominant in specific environments due to high shedding animals, population sizes, and high fitness characteristics of strains. Tracking specific pathogen strains would be beneficial for identifying dissemination from point sources to produce fields.

The data listed in Table 23.3 reflect the dynamic nature of pathogen incidence associated with differences in hosts, locations, periods of study and methods. Incidence ranges from 0.2% to 28% (Hussein and Bollinger, 2005), 0.1% to 62% (Duffy, 2003), 0.7% to 7.3%, and 0.3% to 8.9% (Table 23.3). *E. coli* O157:H7 was reported to persist up to 24 months at individual cattle feedlots or ranches, and some strains were isolated from farms separated up to 50 km apart (Rice *et al.*, 1999; LeJeune *et al.*, 2004; Wetzel and LeJeune, 2006). However, the epidemiological importance of the density of livestock to human illness with *E. coli* O157:H7 and other STEC in the same vicinity has been reported (Michel *et al.*, 1999). These data indicate the importance of cattle as a primary reservoir of pathogenic *E. coli*, the fitness of some strains in the ranch environment, the potential for transport from ranches as point sources to other locations, and the potential association of cattle density with sporadic illness with pathogenic *E. coli*. Similarly, investigations of recent outbreaks associated with leafy vegetables have indicated that beef and dairy cattle in the vicinity may be linked to contamination either directly or indirectly (Jay *et al.*, 2007; CALFERT, 2007b, 2008).

Salmonella incidence in livestock and wild animals also has been reported. *S. enterica* incidence in cow fecal samples in Australia, US and UK studies ranged from 1.4% to 9% (Table 23.3). A recent large study of 7680 animal and environmental samples from a single US dairy reported 13–72% positive cattle samples (depending upon period of testing), and >50% of air, soil, water, insect and bird feces samples yielded *S. enterica* (Pangloli *et al.*, 2008). Similarly, high incidences of *S. enterica* in pigs were reported in a UK study (23.4%), in poultry flocks (10.5–13%) in US and Belgium studies, and in poultry production environmental samples (12–51%) in a US study (Table 23.3). *S. enterica* has been isolated from other farm-related animals, deer (one to seven per cent in two studies), wild birds (up to 3%), rabbits, rats, fish and amphibians (see citations in Table 23.3 for more details). The persistence of select strains of *Salmonella* is evidenced by a report of isolation of a multidrug-resistant *S.* Newport strain for months on two farms (Cobbold *et al.*, 2006), and, as noted above, from an almond orchard soil periodically for at least five years (Uesugi *et al.*, 2007).

C. jejuni outbreak illnesses associated with fresh produce are less frequent and lower profile than *E. coli* and *Salmonella* outbreak illnesses (Mandrell and Brandl, 2004). This is in contrast to an overall higher incidence of *C. jejuni* and possibly *C. coli* in cattle, poultry, other farm and wild animals compared to other pathogens (Miller and Mandrell, 2006). Conversely, recent studies of *C. jejuni* incidence on fruits and vegetables have reported extremely low detection (Sagoo *et al.*, 2001; Thunberg *et al.*, 2002; Moore *et al.*, 2002; Sagoo *et al.*, 2003), suggesting that *C. jejuni* may be less fit on plants (Brandl *et al.*, 2004) compared to *E. coli* O157 and *Salmonella.* In contrast, high levels of *C. jejuni* sporadic illnesses compared to other enteric pathogens (MMWR, 2005b, 2007b) suggest surveillance to identify food sources associated with *C. jejuni* illness should be continued.

Although the majority of cows positive for *E. coli* O157:H7 in a herd have <100 CFU of per gram of feces, animals shedding 1000 up to 1 000 000 CFU/g of feces have been detected and termed high-level shedders or 'super shedders' (Low *et al.*, 2005; Chase-Topping *et al.*, 2007). Therefore, animals shedding high levels of pathogenic strains are extremely relevant epidemiologically, because they probably are responsible for the predominant strains in the environment. If 'predominant strains' are also virulent species, then they are candidates for causing infectious disease, including foodborne outbreaks (Matthews *et al.*, 2006). Models of shedding, fitness and virulence are consistent with the '80/20 rule' predicting that 80% of the transmission of an infectious agent results from 20% of the most infectious (or highest shedding) members of the population (Matthews *et al.*, 2006) and identifies a risk factor that should be examined further.

The fitness and/or virulence of a pathogen in water, soil and on field crops is also important epidemiologically in outbreaks. It is noteworthy that *E. coli* O157:H7 strains linked to four outbreaks associated with bagged leafy vegetables in 2005 and 2006 appear to be part of a phylogenetically distinct group ('clade 8'); indeed, these clade 8 strains were associated with outbreaks with higher than normal levels of hemolytic uremic syndrome or hospitalization (Manning *et al.*, 2008). Therefore, high shedders, increased strain fitness, and/or evolution/selection of virulent strains are important factors in the epidemiology of outbreaks.

The results summarized in Table 23.3 emphasize that there are many potential livestock and wildlife sources of pathogens and modes of transport of pathogens relevant to contamination. Small mammals (e.g. squirrels, mice, raccoons), large mammals (feral swine, deer, and elk) and birds illustrate the diversity of population sizes, barriers (fencing height, depth, and gage) and habitat that are issues in considering interventions to control exposure of wildlife to fields. Resident wildlife species cannot be contained generally so co-mingling of livestock and wildlife is inevitable, thus leading to cross-transmission of pathogens and dissemination to new locations. Measuring the concentration and total amount of pathogen disseminated is relevant to identifying potential risks in a produce production region, but definitive data about the major sources of pathogen and their movement in environments relevant to produce production have been lacking. Estimates of the dose of *E. coli* O157:H7 capable of causing illness in a population exposed to contaminated food ranges from 4 to <40 cfu/g of food (Strachan *et al.*, 2001; Teunis *et al.*, 2004), thus emphasizing the risks associated with even minor dissemination of pathogens near produce.

Tracking the sources and incidence of pathogens in any complex environment is difficult. The environment reflects dynamic processes involving many potential factors relevant to understanding an outbreak including, for example, changing point sources of pathogens, pathogen fitness in the animal, food, grass or water, animal intrusions and comingling, immunity and fecal shedding. Weather conditions (rain/humidity, temperature, UV, wind), plant disease, agricultural practices (fertilization, foliar applications), and antimicrobials also could influence pathogen survival in a produce-related ecosystem and, ultimately, the size of an outbreak.

23.6 Pathogens in municipal and agricultural watersheds

Pathogens shed onto range soil, in feedlots or other habitats, can be dispersed and disseminated further by runoff into watersheds. Selected recent studies of municipal or agricultural watersheds indicate the incidence of *E. coli* O157:H7 in watersheds is low generally (<2%) compared to *Salmonella.* Selected surveys of diverse watersheds in the United States (Higgins *et al.*, 2005; Cooley *et al.*, 2007; Meinersmann *et al.*, 2008; Fincher *et al.*, 2009), Canada (Johnson *et al.*, 2003; Gannon *et al.*, 2004; Walters *et al.*, 2007), the United Kingdom (Ihekweazu *et al.*, 2006), France (Baudart *et al.*, 2000), and Central Africa (Tuyet *et al.*, 2006) report incidences of *Salmonella* as high as 75% in a watershed in Georgia (US) in 2005 (Meinersmann *et al.*, 2008), an overall incidence of *E. coli* O157:H7 of 6.5% over a 19-month study (2005–06) in a central California coast watershed near leafy vegetable production (Cooley *et al.*, 2007), and a reported incidence of *E. coli* O157:H7 of 59% for a one-year study of a watershed in central Indiana (Fincher *et al.*, 2009). The California study was initiated as a result of three separate outbreaks of *E. coli* O157:H7 linked to leafy vegetables grown in the Salinas Valley region of California (Table 23.1), and possibly linked to a single farm (Cooley *et al.*, 2007). Monthly water sampling from >20 sites, within approximately 30 km of one another revealed: (i) increased frequency of *E. coli* O157:H7 isolation at sites nearby grazing cattle in elevated regions of the watershed, (ii) increased incidence of pathogen correlated with heavy water flow, and (iii) phylogenetically related strains isolated on the same day up to 30 km apart, or from the same sites months apart (Cooley *et al.*, 2007). Additional details of this MST study are presented below.

23.7 Fitness of human pathogens in the environment

Outbreak investigations of farms and ranches near leafy vegetable production on the central coast region of California have identified new sources of *E. coli*

O157:H7 and other enteric pathogens and hypotheses of fate and transport (Cooley *et al.*, 2007; Jay *et al.*, 2007). The results suggest that some pathogen strains may have evolved fitness to persist in different environments and become predominate. The incidence levels in livestock and wildlife (Table 23.3) indicate environments where evolution and selection of pathogen fitness characteristics occur. It is unclear whether pathogen fitness in animal reservoirs relates in any manner to fitness in other environments, most importantly, the human host. However, evolution of pathogen virulence associated with changes in food production (animal and plant) and processing is worth further study.

Outbreaks associated with fresh produce have stimulated abundant research to address the fitness and biology of enteric pathogens related to specific produce commodities. Table 23.4 provides a summary of selected studies that assessed the

Table 23.4 Selected studies of the survival/fitness of enteric pathogens in environmental samples or microcosms

Pathogen	Environment	Maximum survival (days)	Reference
E. coli O157:H7	Water, 8°C Water, 25°C	>91 <84	(Wang and Doyle, 1998)
E. coli O157	Water trough, sediment	245	(LeJeune *et al.*, 2001)
E. coli O157:H7	Water, <15°C Water+feces, <15°C	14 24	(McGee *et al.*, 2002)
E. coli O157	Water, biofilms	>30	(Cooper *et al.*, 2007)
E. coli O157:H7	Water: lake, river, drinking trough microcosms	6 to >60 Lake > river	(Avery *et al.*, 2008)
E. coli O157	Soil	105	(Ogden *et al.*, 2002)
E. coli O157	Soil, manure-amended (child illness)	69	(Mukherjee *et al.*, 2006)
E. coli O157:H7	Soil, manure-amended	>35	(Williams *et al.*, 2007)
E. coli O157:H7 (Stx-neg)	Soil, 36 types	54–105	(Franz *et al.*, 2008)
E. coli O157:H7	Soil, cover crops	40–96	(Gagliardi and Karns, 2002)
E. coli O157:H7	Manure, cow Manure, sheep	47 >600	(Kudva *et al.*, 1998)
E. coli O157 (Stx-neg)	Feces, cow Water	97 109	(Scott *et al.*, 2006)
E. coli O157:H7	Feces, cow, turned Feces, cow, unturned	42 90	(Fremaux *et al.*, 2007)
E. coli O157:H7	Manure, cow Manure, slurry	21 35	(Himathongkham *et al.*, 1999)
E. coli O157:H7 (attenuated, Stx-neg)	Soil, manure-amended Lettuce Parsley Onions Carrots	154–217 77 177 74 168	(Islam *et al.*, 2004, Islam *et al.*, 2005)
E. coli O157:H7 (Stx-neg)	Lettuce and spinach	<7	(Hutchison *et al.*, 2008)

(*Continued*)

Table 23.4 Continued

Pathogen	Environment	Maximum survival (days)	Reference
S. enterica	Water, river	>45	(Santo Domingo *et al.*, 2000)
S. enterica	Soil, chicken farm	240	(Davies and Breslin, 2003)
S. enterica	Soil	>120	(Holley *et al.*, 2006)
S. Newport	Soil, manure amended Manure, cow	107–332 49–184	(You *et al.*, 2006)
S. Enteritidis	Soil, almond orchard	>1500	(Uesugi *et al.*, 2007)
S. enterica	Soil, tomato crop debris (microcosm)[a]	56	(Barak and Liang, 2008)
S. Enteritidis	Lettuce and spinach	>14 to <21	(Hutchison *et al.*, 2008)

a Some soils include crop debris from tomato plants infected with the pathogen *Xanthomanas campestris* and colonized with *S. enterica.*

fitness of *E. coli* O157:H7 and *Salmonella* in environments relevant to fresh produce contamination. Each study involves different locations and experimental conditions, but the results reported, generally, are consistent. *E. coli* O157 was reported to survive in some studies for weeks or months depending upon the water source (river, lake, or trough), addition of manure or soil, water temperature or animal feces (Table 23.4). Attenuated *E. coli* O157 and fully pathogenic *Salmonella* survived on plants or plant debris for days or weeks under field conditions (Islam *et al.*, 2004, 2005; Hutchison *et al.*, 2008) or in microcosms mimicking field conditions (Barak and Liang, 2008; Hutchison *et al.*, 2008).

These results provide a limited 'snapshot' of pathogen fitness under the selected test or environmental conditions. However, two particular studies are worth noting. A survey of a family garden subsequent to a child's *E. coli* O157:H7 illness after playing in the manure fertilized garden revealed that the same strain was detectable in soil samples from the garden for >69 days (Mukherjee *et al.*, 2006). Similarly, *S. enterica* ser. Enteritidis phage type 30 strains associated with at least one outbreak linked to raw almonds, and possibly a second (Table 23.1), were isolated over at least a five-year period from soil drag swab samples obtained in an orchard linked to the outbreak (Table 23.4). These strains are relevant for MST, but they are useful also for comparative genomics of clinical versus environmental strains to identify clues to whether environmental fitness correlates with virulence (Manning *et al.*, 2008).

Manure-amended soil, plants and plant debris appear to be beneficial to the survival of *E. coli* O157:H7 and *Salmonella* (Table 23.4). Ruminant-digested grasses and feeds and crop debris have nutrients supporting survival and possibly growth of enteric pathogens under the appropriate environmental conditions, including temperature, moisture and atmosphere (see Brandl, 2006 for review), but differences in the decline of *E. coli* due to UV, humidity and soil types have been reported (Van Kessel *et al.*, 2007; Lang and Smith, 2007). *E. coli* O157, *S. enterica*, and generic *E. coli* as fecal indicator bacteria, survive for months and possibly years, under realistic environmental conditions; under optimal conditions, they may grow 1–3 logs (Table 23.4).

Crop debris may be conducive to amplification of pathogens if present. In a laboratory study, tomato seeds planted in soil with *Salmonella*-contaminated tomato crop debris resulted in plants with *Salmonella* on the roots and leaves (Barak and Liang, 2008). Tomato plants infected with *Xanthomonas campestris* pv. Vesicatoria, a plant pathogen, also supported *Salmonella* growth indicating the potential importance of debris, plant disease and fallow periods in the pre-harvest produce production cycle (Barak and Liang, 2008). Plant lesions created mechanically (cut edges) or biologically (senescence, plant pathogens) produce enhanced conditions for enteric pathogens to grow. Laboratory studies of cut lettuce and lettuce with soft rot caused by *Erwinia chrysanthemi* supported significant increases in *E. coli* O157:H7 within four hours after inoculation (Brandl, 2008). Thus, intact or rotted crop debris may enhance the conditions for even better survival or growth of human pathogens (Barak and Liang, 2008; Brandl, 2006; Brandl and Amundson, 2008; Brandl, 2008).

Animals are obvious potential point sources for pathogens (Table 23.3) and movement of livestock, wildlife intrusion, flooding, dust, manure/compost fertilizers and farm vehicles could transport and/or disseminate pathogens. High-shedding animals, manure, crop and/or ground cover debris, and produce plant seedlings, leaves and roots, are candidate sites for possibly limited growth of the pathogen. Therefore, the survival of pathogenic *E. coli* and *Salmonella* in manure, soil and water (Table 23.4) are relevant to hypotheses about how pre-harvest contamination occurs. Improvements in isolating and genotyping pathogens from the produce production ecosystem will enhance tracking the sources and movement of pathogens and identify risk factors contributing to outbreaks.

23.8 Fecal indicators of contamination in watersheds

Major produce-associated outbreaks in the last decade have confirmed that pre-harvest contamination occurs (Table 23.1). Although surveys of fresh produce at different stages of the production chain indicate that bacterial pathogens are at extremely low incidence generally (Beuchat, 1996; Harris *et al.*, 2003; Nguyen-the and Carlin, 1994, 2000), fecal indicator bacteria (*E. coli*) appear to increase in prevalence during transport and distribution (Table 23.2) to wholesale and retail markets (Valentin-Bon *et al.*, 2008), although, again, this does not correlate well with pathogen incidence. Pre-harvest contamination can occur through a variety of sources. Water (irrigation, flooding), intrusion by animals either directly

(Table 23.3; wildlife, domestic, humans) or indirectly (fertilizer, compost), and dust are potential mechanisms of pre-harvest contamination. Water quality is a primary factor in production of safe fresh produce; irrigation water can be from wells, holding ponds or surface water depending upon the type of produce and location. Surface water could be a source of pathogens affecting aquifer re-charging, exposure of animals to colonization and/or transport to produce fields by irrigation. However, the majority of leafy vegetable production on farms on the central California coast involves irrigation with well water of high quality relative to surface water. Indeed, irrigation of leafy vegetables associated with recent outbreaks was by well water (CALFERT, 2007b, 2008). It is noteworthy also that US winter produce production occurs mainly in the Imperial Valley of California and Yuma region of Arizona, where irrigation water is sourced often from surface water. In contrast, outbreaks associated with produce from these locations had been rare (Table 23.1) (Campbell, 2005), until a recent major outbreak associated with Romaine lettuce contaminated with *E. coli* 0145 (CDC, 2010). These paradoxical and now emerging trends emphasize the need for more environmental studies in regions associated, and also not associated, with outbreaks using improved methods of MST.

Watersheds are impaired by the presence of fecal bacteria from livestock, wildlife and humans. The geography and ecology within and surrounding the watershed, including the density of animals, hydrology, elevation/runoff, meteorological conditions (e.g. rainfall and temperature), pathogen fitness (Table 23.4), water composition (salinity, nutrients) and vegetation could be factors in the amount of contamination. Waterborne disease outbreaks in the United States (1948–1994) and Canada (1975–2001) occur more frequently following heavy rain events (Curriero *et al.*, 2001; Thomas *et al.*, 2006). Also, heavy rainfall correlates with increased illness associated with enteric virus and bacteria contamination of molluscan shellfish in different parts of the world (Lee and Morgan, 2003; Martinez-Urtaza *et al.*, 2004; Riou *et al.*, 2007). Details related to tracing pathogens in molluscan shellfish production chains are presented in Chapter 22 of this book. Although no definitive links between heavy rain events and human illness associated with produce have been reported, flood contamination of fields or irrigation water sources intended for growing produce is a potential risk factor for illness (CDHS, 2005).

Hydrological processes are relevant to transport of pathogens in the environment, including fecal disintegration and dispersion, resuscitation of pathogens in arid environments, trapping of pathogens in wetlands, concentration of pathogens on sediment particles, exposure of wildlife to pathogens and other transport processes (Ferguson *et al.*, 2003). Similarly, the soil and sediment particles present in flowing or static water bodies can interact and bind with microorganisms by mechanisms that are not well defined, and likely vary depending upon variations in soil, fecal and water composition, weather and other factors (Gagliardi and Karns, 2000; Brookes *et al.*, 2004; Ferguson *et al.*, 2003). Transport of pathogens in dust, on harvest equipment, in manure/compost and pesticide and herbicide sprays diluted with surface water, also should be considered.

23.9 Survival of human pathogens on pre-harvest plants

Outbreaks associated with produce suspected of being contaminated in the field confirm that enteric bacteria are capable of attaching and surviving on plants (Tables 23.1 and 23.4). Field studies with attenuated strains of *E. coli* O157:H7 and other pathogens on plants confirm that they can survive for weeks and months (Tables 23.2 and 23.4). Laboratory studies indicate that *E. coli* O157:H7 and *Salmonella* applied to a variety of plant roots, leaves and seeds can attach tenaciously (resisting sanitization), and survive, but also in some instances grow when conditions are ideal for a pathogen (warm temperature, high humidity, adequate nutrients) (Brandl, 2006). Sophisticated fluorescence microscopy experiments have revealed specific locations on leaves and roots where sub-cuticular cells, root hairs or breaks in the tissue (e.g. lateral root formation) provide sites and nutrients adequate for harbouring opportunistic pathogen cells. Aggregation of enteric pathogen cells with one another and with plant epiphytic or plant pathogen microflora, suggest active and complex interactions may occur on plants in the field, resulting possibly in interactions/contamination very difficult to remove by normal washing or sanitizing methods (Brandl, 2006). In addition, there appears to be emerging support for the hypothesis that some human pathogen cells on plants may become internalized through different routes of entry on roots, shoots and flowers (Guo *et al.*, 2001; Solomon *et al.*, 2002; Warriner *et al.*, 2003; Dong *et al.*, 2003; Franz *et al.*, 2007; Doyle and Erickson, 2008; Schikora *et al.*, 2008; Kroupitski *et al.*, 2009), although the results of some laboratory studies indicate internalization is not simple to demonstrate (Zhang *et al.*, 2009; Sharma *et al.*, 2009).

Plants may even respond to the presence of human pathogens. In model plant systems (*Arabidopsis thaliana* mutants and gene expression arrays) genes and gene pathways are up-regulated similar to plant resistance responses to plant pathogens (Dong *et al.*, 2003; Thilmony *et al.*, 2006; Schikora *et al.*, 2008), suggesting that some human pathogens have the capability to be endopathogenic for some plants. If endopathogenesis and/or internalization were to occur in the field (leafy greens, tomatoes, melons, etc.), it would raise concerns regarding post-harvest treatments for decontamination.

Reviews of different mechanisms that plant epiphytes and pathogens and human enteric pathogens use to attach to plants (Mandrell *et al.*, 2006; Solomon *et al.*, 2006), an excellent review of the general biology, ecology and fitness characteristics of human enteric pathogens on plants (Brandl, 2006) and additional reports of this subject (Aruscavage *et al.*, 2006; Whipps *et al.*, 2008; Heaton and Jones, 2008; Doyle and Erickson, 2008) have been published previously. Further details about the molecular interactions that can occur between bacterial human pathogens (e.g. flagellin, fimbriae, pili, curli, outer membrane proteins) and plants (generally undefined) and the microbial ecology on plants that may enhance or control pathogen survival are provided in these reviews.

23.10 Hydrology and microorganisms

The dynamic incidence of pathogens in livestock, wildlife (Table 23.3) and watersheds, the environmental fitness characteristics of foodborne pathogens (Table 23.4), and recurring outbreaks of foodborne illness associated with ready-to-eat produce (Table 23.1), are consistent with low level incidence of generic *E. coli* on fresh produce obtained from distribution centres and retail markets (Table 23.2). Post-harvest contamination could occur in a variety of ways and exacerbate what might have been initially minimal contamination.

Dispersion and dissemination of microbes in water and the use of microbes as tracers of water movement are relevant to understand dissemination of enteric pathogens in water. Heavy rainfall will disperse pathogens rapidly in fecal matter into surface and ground water (Ferguson *et al.*, 2003). Early tracing experiments decades ago with bacteria, yeast and virus are informative of dilution and dissemination of microorganisms in water and relevant to current questions about pathogens. *S. marcescens*, used as a tracer because it forms distinctive red colonies on culture media, was dosed into a river at a single point (approximately 10^{14} cells) and dissemination and movement was monitored at a point downstream (Wimpenny *et al.*, 1972). The bacteria moved at approximately 2.5 km/h over 2.9 km between dosing and detection points at a maximum of 500 cells/ml, corresponding to $>1.7 \times 10^8$-fold dilution of the bacteria during transport (Wimpenny *et al.*, 1972). To achieve a comparable amount of *E. coli* O157:H7 from a 'high-shedder' cattle feces (e.g. 10^6 cells/g), for example, would require >200 000 kg of feces. In a recent study of an elevated watershed, *E. coli* O157:H7 strains were tracked from a point source (small corral with a few head of cattle) into a small stream (Cooley *et al.*, 2007). Indistinguishable or related pathogen strains identified by MLVA genotyping were isolated at the point source and up to 135 m downstream (3 m lower altitude) from the point source, although water flow was low prior to and at the time of sampling (Cooley *et al.*, 2007). Studies done in the early 1980s that traced movement of cultured *E. coli* added to different pore-groundwater (water in spaces and fractures in rock and sediment) sites, at amounts ranging between 10^7 to 10^{12} cells, measured effective flow velocities ranging from 0.03 to 7 m/h depending upon the porosity and permeability of the rock/sediment/soil system (Tables 17 and 18 in (Käss, 1998)). The maximum ranges of detection and times of last detection for the added *E. coli* were between 12 to 180 m and 37 to 257 days, respectively (Käss, 1998). These data are relevant to contamination of aquifers and irrigation well water in regions with high livestock density and emphasize the critical importance of geohydrology to produce production. MST is beneficial for effective monitoring of specific pathogens and strains in similar geohydrologic ecosystems.

Isolation and/or detection of pathogens in water at distant sites from a suspected point source might involve transport mechanisms different than those reflected by laboratory-cultured microorganisms in tracer studies. Thus, water is both an efficient transporter and diluter of microorganisms, depending upon the state of the microorganism in water (e.g. aggregated; bound to particles, other microorganisms, vegetation or algae; in protozoa vacuoles). However, high-shedding cattle and water are not the only factors explaining contamination events, but rather, are two risk factors among others (e.g. reservoirs, amplification, dust, insects, etc.) that MST methods can identify.

Climate might play a role in some outbreaks. It is noteworthy that during the 2006 outbreak of *E. coli* O157:H7 associated with bagged baby spinach unusually high daily temperatures occurred at the time of planting (July 22–25, 2006: maximum daily 100–110°F (37.7–43.3°C); average daily 77–85°F (25–29.4°C)), and approximately five to six days prior to harvest (CALFERT, 2007b, c). This unusual condition stimulates questions regarding when contamination occurred in the crop cycle and whether high temperatures may have enhanced survival or growth of pathogen in the pre-harvest environment. For example, *E. coli* O157 has been shown to survive and increase in number with increasing temperature (10–30°C) in natural freshwater microcosms containing low concentrations of organic carbon (Vital *et al.*, 2008) and on plants at temperatures >20°C (Brandl, 2006). The direct correlation between pathogen growth and increasing water temperature is consistent with enteric bacteria that have evolved to grow optimally at body temperatures.

23.11 Microbial source tracking

Efforts to identify the sources of fecal pollution have evolved since the early epidemiological detective work in 1854 by John Snow, who identified a well at '40 Broad Street' in London as the source of a local concentration of cases of cholera (Snow, 1855). Advances in bacteriology and public health led to the identification of coliforms, *Enterococci and E. coli* as common bacteria in human and animal feces and, eventually, to methods to characterize them for purposes of identifying and tracking fecal contamination in food, water and the environment. Current MST methods involve identifying phenotypic and genetic characteristics of fecal indicator microorganisms to match with similar characteristics (fingerprints) of microorganisms isolated from an environment of interest. The two major approaches to MST are classified as library-dependent or -independent (Stewart *et al.*, 2007) and require databases containing characteristics of a significant number of strains isolated from known sources, such as animal feces (library-dependent); or animal- or animal GI microbiome-specific characteristics (e.g. mitochondrial DNA, specific DNA sequence) for matching with environmental strains or extracted sample DNA, thus establishing a link to an animal species as responsible for contamination (library-independent). The different phenotypic and genotypic methods developed, and their advantages and disadvantages, have been documented in other reviews (Simpson *et al.*, 2002; Meays *et al.*, 2004; Field and Samadpour, 2007; Harwood, 2007; Santo Domingo and Sadowsky, 2007; Whittam and Bergholz, 2007; Foley *et al.*, 2009), and in chapters in this book. In this section, I will focus on MST of enteric pathogens relevant to the microbial safety of fresh produce and provide details of a recent outbreak investigation that assist in illustrating the value of MST in identifying potential risk factors.

Pathogens exist in produce production environments (Tables 23.1, 23.3 and 23.4), usually at low concentrations and at a limited number of point sources, so any success in MST of pathogens in these environments, including identification of point sources, can be a major benefit to public health. MST of pathogens requires sensitive and efficient methods, intensive sampling, and successful isolation of pathogens or fecal indicator microorganisms for further characterization (Meays *et al.*, 2004; Field and Samadpour, 2007). MST of fecal indicator microorganisms rather than specific pathogens is informative when the goal is related generally to detecting the source(s) of fecal pollution and modes of transport. However, in recent produce outbreak or environmental investigations, it has been more informative to focus on specific pathogens implicated in outbreaks or recalls. Recent surveys of fresh produce at various stages of the distribution chain have revealed that the incidence of fecal indicator bacteria, for example generic *E. coli*, on produce or other environmental samples is significantly more common than pathogen (Table 23.2), thus confounding accurate correlations between indicators and pathogens. Nevertheless, as noted above, library-dependent and -independent MST also can be informative of animal sources of contamination, if accurate signatures (linked to animal/host) are detectable and/or a robust library of signatures/genotypes/profiles relevant to an area of study has been developed (Stewart *et al.*, 2007).

Pulsed-field gel electrophoresis (PFGE) remains the gold standard method in the public health sector for fingerprinting foodborne pathogens, primarily, because CDC's PulseNet database has stored PFGE profiles submitted by public health labs representing tens of thousands of sporadic and outbreak strains for comparison (Swaminathan *et al.*, 2001). Thus, PulseNet USA, PulseNet Europe and similar databases continue to serve the public health community in identifying outbreaks. Recently, sequenced-based typing methods, such as MLVA, Multilocus Sequence Typing (MLST), and single nucleotide polymorphism (SNP) microarrays, are gaining acceptance due to ease of use, speed and high-resolution data for comparisons. Indeed, MLVA, an effective method for genotyping *E. coli* O157:H7 (Lindstedt *et al.*, 2003; Noller *et al.*, 2003; Keys *et al.*, 2005; Hyytia-Trees *et al.*, 2006), is used routinely by the CDC to differentiate outbreak and sporadic isolates with indistinguishable PFGE profiles and related temporally. Similarly, MLVA methods developed for *S.* Typhimurium (Heck, 2009; Larsson *et al.*, 2009), *S.* Enteritidis (Boxrud *et al.*, 2007; Cho *et al.*, 2007) and *S.* Typhi (Octavia and Lan, 2009) are beneficial in identifying outbreak and sporadic isolates; similar MLVA methods are being used by CDC in outbreak investigations of *S.* Typhimurium and *S.* Enteritidis.

A major advantage of MLVA for environmental MST is the simplicity and speed possible using the capability of capillary DNA sequencers and fragment size analysis, and multiplexing of primers with different fluorophores for multiple MLVA loci. This facilitates genotyping of multiple isolates simultaneously and, therefore, is extremely beneficial in outbreak and environmental investigations for finding specific strains ('needle in the haystack'), phylogenetic relatedness among strains, and for monitoring movement of pathogens in a produce production ecosystem. In contrast, chromosome changes as a result of indels, phage, plasmids, and recombination events can affect PFGE profiles (Yoshii *et al.*, 2009), thus confounding assessment of strain relatedness. For example, in environmental survey studies, our laboratory has identified isolates highly related by 11-loci MLVA, but distinguishable and 'unrelated' by PFGE (as a result probably of an indel event). Conversely, we have isolated environmental strains indistinguishable by *Xba*I and *Bln*I PFGE analysis to clinical outbreak strains submitted to PulseNet, but that by MLVA are unrelated (unpublished data); these results occurring more frequently are consistent with outbreak/sporadic isolate differences identified by CDC during outbreak surveillance (Hyytia-Trees *et al.*, 2006). Also, in ongoing intensive surveys of a major produce production environment in California, my colleagues and I have identified multiple sets of environmental strains indistinguishable by both PFGE and MLVA to human clinical strains linked to small outbreaks with no known source of contamination. In the absence of convincing epidemiology and, possibly, higher resolution genotyping and MST, these 'matching fingerprints' are simply intriguing observations. However, suspicions of a regional connection between 'matched' agricultural environment and clinical isolates emphasize the need for high-resolution genotyping to enhance MST for robust epidemiological traceback investigations.

23.12 Microbial source tracking in recent produce outbreak investigations

MLVA has been very effective in recent environmental studies tracking *E. coli* O157:H7 strains in produce production environments, watersheds and cattle feedlots (Cooley *et al.*, 2007; Murphy *et al.*, 2008). One of the most intensive investigations to date involving MST of pathogen associated with fresh produce contamination was that in 2006 of farms and ranches linked by lot number to a multi-state outbreak associated with contamination of bagged baby spinach by *E. coli* O157:H7. Four farms and nearby ranches were visited and >1000 samples of soil, produce, wildlife, livestock, water and other relevant samples were tested, *E. coli* O157:H7 strains were isolated, then fingerprinted by PFGE and/or MLVA for comparison to human strains (CALFERT, 2007b, c). An intriguing finding in this investigation was the isolation of multiple strains of *E. coli* O157:H7 from a high percentage of cattle on one ranch and multiple feral swine trapped in the vicinity of the cattle and a suspected spinach field. More than 50% of the O157 isolates from positive feral swine and cattle on the nearby ranch, plus isolates from river and dirt samples collected within a mile of the field, were indistinguishable from the human outbreak strains (Jay *et al.*, 2007; Cooley *et al.*, 2007).

The overall incidence of *E. coli* O157:H7 in the cattle tested (34%) and at least one species of roaming wild animals (15%) suggested a 'pathogen hot zone' existed in the vicinity of produce fields. This is similar to transport of *E. coli* O157:H7 between dairy farms by wild birds (Wetzel and LeJeune, 2006). Co-mingling of wild animals (mammalian and avian) underscores the importance of initiating other surveys of wildlife species in food production environments to determine whether they are potential sources of pathogens. Nevertheless, even for this major and intensively investigated outbreak, it remains unclear how pathogen was transported to produce in the field. Although roaming O157-positive feral swine were considered logical sources of transmission, it was reported that no tracks in the suspect field were evident.

Multiple *E. coli* O157 strains isolated from samples collected from a small stream and a dry dirt sample located approximately a mile from the suspect spinach field point to water and dust as other possible transport scenarios in this outbreak event. Shallow aquifers re-charged with contaminated water and/or direct contamination through cracks in wells could be major risk factors (Käss, 1998). However, large volumes of well water tested in the 2006 spinach outbreak investigation were negative for the pathogen (CALFERT, 2007b). High winds transporting contaminated dust or aerosols from a 'hot zone' onto fresh produce is a potential risk factor also (Rosas *et al.*, 1997; Baertsch *et al.*, 2007). Although it is logical that enteric pathogens could be spread by rain, wind and insect dispersal similar to mechanisms of dispersal reported for plant epiphytic bacteria (Hirano and Upper, 2000), no evidence exists that this has been a significant factor in any outbreak. Nevertheless, regional differences in the surrounding ecosystem and changing microclimates involving temperature, relative humidity, wind, ultraviolet radiation, and other factors may play a part in some outbreaks.

A few additional recent outbreaks investigated less intensively than the 2006 baby spinach outbreak, but resulting in isolation of the pathogen species linked to the outbreak from environmental samples in some instances, and identification of additional pre-harvest contamination risk factors are: (i) *E. coli* O157:H7-contaminated romaine lettuce grown in fields flooded previous to planting by overflowing creek/ditch water from heavy winter rains (CDHS, 2005), (ii) *E. coli* O157:H7-contaminated commercial iceberg lettuce contaminated probably by dairy wastewater and irrigation water blended for discharge, but having inadequate backflow protection (CALFERT, 2008), (iii) multiple *Salmonella* outbreaks associated with tomatoes suspected of being contaminated in the field by wildlife, possibly amphibians, reptiles or birds exposed to contaminated surface water (MMWR, 2005a, 2007a; Greene *et al.*, 2008), and (iv) *Salmonella* outbreaks associated with cantaloupes contaminated probably by contaminated wash water (Castillo *et al.*, 2004). The pathogen species and/or serovar have been isolated from a suspected source farm environment in other outbreaks, but primary livestock or wildlife sources were not identified. For example, pathogen strains were isolated from environmental samples collected on farms/orchards growing papaya (Gibbs *et al.*, 2009), almonds (Uesugi *et al.*, 2007), carrots (Jalava *et al.*, 2006), and iceberg lettuce (Nuorti *et al.*, 2004) associated with outbreaks, but animal or other sources were not identified.

23.13 Next generation microbial source tracking

The rapid development of next generation DNA sequencing technologies (e.g. Roche 454, Illumina Genome Analyzer, ABI Solid) and third-generation technologies (e.g. Pacific Biosciences, US; Helicos Biosciences, US; Complete Genomics, US; Nanopore, UK) for human genome analyses are providing also major capabilities for sequencing microorganism genomes (100–1000 times smaller). Proof-of-principle high-resolution SNP arrays constructed from analysis of multiple genomes of a pathogen species or subtype (*S. enterica*, STECs, *E. coli* O157:H7) have revealed the advantages of comparing larger portions of strain genomes for identifying phylogenetic relationships (Zhang *et al.*, 2006; Octavia and Lan, 2007; Leopold *et al.*, 2009). However, this approach relies still on the DNA sequences of a limited number of strains sequenced.

23.14 Conclusions

The increased incidence of produce-related outbreaks tracked to specific regions, and *E. coli* O157:H7 outbreaks in particular, has stimulated questions about what might have changed over the last decade to explain this increase. Is it related to growing (fertilization, water, shallow tilling, seeds, cultivars), production practices (cutting, transport, bagging, atmosphere), pathogen changes (increased fitness in animals, water), livestock (transport, incidence of pathogens), or better detection (methods, public health system, media)? MST the incidence and genetic relatedness of pathogens in the produce production ecosystem will assist in identifying point sources of enteric pathogens and their movement that result in contamination events that affect public health.

Animals on or near fresh produce fields are potential risk factors and worthy of attempts to prevent transfer of their fecal material through intrusion of animals, runoff from adjacent ranches, flooding, or manure/compost amendments inadequately cycled or treated. Lacking convincing evidence of pathogen carriage by a suspect animal species, however, is problematic for making informed decisions about mitigation approaches (predation, fencing, monitoring/testing). Indeed, lack of definitive proof of the sources of pathogen contamination creates significant conflicts between conservationists, environmentalists, and growers on one side, versus those in the produce industry responsible for addressing pre-harvest produce food safety issues. Creation of vegetative zones for filtering run-off from fields and for wildlife habitat contradicts food safety goals of removing habitat to prevent attracting wildlife possibly colonized with pathogens (Berreti and Stuart, 2008). Some compromise between these competing interests will be necessary for sustaining the desirable produce locations, while still maintaining the quality and safety of produce.

As noted above, major outbreaks may occur as a result of a convergence of multiple, but infrequent events, implying that each event alone may be insufficient. Pathogen incidence and virulence in a pre-harvest food production environment might be associated with corresponding changes in the ecology, hydrology, meteorology, and agricultural practices in an environment. Considering the impossibility of controlling certain aspects of the ready-to-eat produce production environment, it is logical to assume that additional outbreaks will occur. Intensive practices leading to exposure of pathogens to complex environments, or significant replication of microorganisms, will increase the rates of new mutations and 'fitness' in environment(s) where mutations are beneficial (LeClerc *et al.*, 1996; Leopold *et al.*, 2009). Modern molecular biology techniques (genomics) are facilitating the fingerprinting of outbreak-related pathogen strains for purposes of high resolution tracking of the possible sources of contamination in pre-harvest environments. Also, comparative genomics of these data are revealing insights about pathogen evolution and emergence of virulence-related factors that raise questions about whether produce outbreak-related pathogens are more virulent and have special fitness characteristics (Zhang *et al.*, 2006; Manning *et al.*, 2008). The rapid changes possible in bacterial genomes by mutations, phage insertions and deletions, and recombination, as examples, predict the emergence from food production environments of organisms with selected fitness characteristics that reflect the environment. Intense farming through multiple cycles of soil preparation, fertilization, irrigation and planting of different crop varieties may play a role in evolution of pathogen fitness. If some of these fitness characteristics are virulence traits in humans (i.e. pathogens) also, then the pathogens will be identified through our response to human illness.

Considering the known potential risk factors in the pre-harvest environment documented above, some approaches for preventing contamination of food can be offered. Common sense approaches include maintaining water quality and minimizing exposure of fields to wild animals and surface water (flooding), and dust from agricultural activity. Other less obvious approaches requiring more resources are identifying high-shedding livestock or wildlife, treatment of livestock with effective vaccines or other antimicrobials, checking and maintaining feed quality, observing field conditions (wildlife intrusions), re-directing or destroying suspect produce, and control of wild animal habitat. Post-harvest approaches involve sample testing (test and hold), clean water, novel sanitizers (chemical or biological), and irradiation, to name a few.

MST of pathogens in produce production environments can provide valuable information regarding sources, movement and persistence. Next generation DNA sequencing methods will change significantly MST strategies as bioinformatics analysis of the immense amount of new sequence data for enteric pathogens of interest becomes available. Comparative genomics of produce outbreak-related pathogens (Table 23.1) will assist in identification of regions in multiple genomes (e.g. variable number tandem repeats, common loci, virulence factor loci, SNPs, phage) advantageous for developing higher resolution methods critical for accurate phylogeny of strains lacking strong epidemiological associations (e.g. spatial and temporal), virulence typing, and creation of DNA databases for simple and accurate comparisons.

The 'Holy Grail' of pathogen genotyping for MST will be complete genome sequencing (i.e. at least high quality draft sequence) in reasonable time and cost of all bacterial pathogen strains included in a study. There are ongoing projects at major sequencing centres and other laboratories involving sequencing hundreds to thousands of pathogen strains both for discovery and development of much more robust detection and genotyping methods. Cost decreases and bioinformatics capability increases are needed for full genome sequencing of pathogen strains routinely. However, full genome analysis of multiple strains will enable high-resolution MST, including MST at much wider spatial and temporal distances than available currently, and will reveal the evolutionary history and fitness of pathogens in diverse environments. Full genome phylogenetic comparisons of strains leading to matches among strains isolated years, and perhaps thousands of miles, apart will be possible and will enable tracking of sources and movement of pathogens not possible previously.

Finally, modern MST molecular methods emerging will assist in environmental investigations of future outbreaks by providing tools for rapid, sensitive and high-resolution analysis of pathogens in complex ecosystems involving plants, livestock, wildlife, soil, water and dynamic climate conditions. MST strategies will improve significantly, but the pace will be dictated by certain realities of microbiology and database management, including the need for intensive sampling for identifying accurate trends, slow improvements in culture-based methods, transitions to new DNA sequence or SNP databases, compatibility of methods between labs and agencies, and acceptance of new approaches. Nevertheless, MST will be a critical component in understanding better the environmental aspects associated with pre-harvest contamination of fresh produce ('hot zones', point sources, risk factors) and, hopefully, will provide guidance on interventions that can minimize contamination.

23.15 Acknowledgements

The author thanks representatives of the USDA Agricultural Marketing Service for providing data collected for the 'Microbial Data Program', M. Jay-Russell for source information regarding *E. coli* O157:H7 leafy vegetable outbreaks, and to his colleagues and collaborators in National Research Initiative Competitive Grant nos. 2006–55212–16927 and 2007–35212–18239 from the USDA National Institute of Food and Agriculture.

23.16 References

ABADIAS M, USALL J, ANGUERA M, SOLSONA C and VINAS I (2008), 'Microbiological quality of fresh, minimally-processed fruit and vegetables, and sprouts from retail establishments', *Int J Food Microbiol*, **123**, 121–9.

ABONG'O B O, MOMBA M N B and MWAMBAKANA J N (2008), 'Prevalence and antimicrobial susceptibility of *Escherichia coli* O157:H7 in vegetables sold in the Amathole District, Eastern Cape Province of South Africa', *J Food Prot*, **71**, 816–19.

ACKERS M L, MAHON B E, LEAHY E, GOODE B, DAMROW T, HAYES P S, BIBB W F, RICE D H, BARRETT T J, HUTWAGNER L, GRIFFIN P M and SLUTSKER L (1998), 'An outbreak of *Escherichia coli* O157:H7 infections associated with leaf lettuce consumption', *J Infect Dis*, **177**, 1588–93.

APPELBEE A J, THOMPSON R C and OLSON M E (2005), '*Giardia* and *Cryptosporidium* in mammalian wildlife – current status and future needs', *Trends Parasitol*, **21**, 370–6.

ARTHUR L, JONES S, FABRI M and ODUMERU J (2007a), 'Microbial survey of selected Ontario-grown fresh fruits and vegetables', *J Food Prot*, **70**, 2864–7.

ARTHUR T M, BOSILEVAC J M, NOU X, SHACKELFORD S D, WHEELER T L and KOOHMARAIE M (2007b), 'Comparison of the molecular genotypes of *Escherichia coli* O157:H7 from the hides of beef cattle in different regions of North America', *J Food Prot*, **70**, 1622–6.

ARUSCAVAGE D, LEE K, MILLER S and LEJEUNE J T (2006), 'Interactions affecting the proliferation and control of human pathogens on edible plants', *J Food Sci*, **71**, R89–99.

AVERY L M, WILLIAMS A P, KILLHAM K and JONES D L (2008), 'Survival of *Escherichia coli* O157:H7 in waters from lakes, rivers, puddles and animal-drinking troughs', *Sci Total Environ*, **389**, 378–85.

BAERTSCH C, PAEZ-RUBIO T, VIAU E and PECCIA J (2007), 'Source tracking aerosols released from land-applied class B biosolids during high-wind events', *Appl Environ Microbiol*, **73**, 4522–31.

BAILEY J S, STERN N J, FEDORKA-CRAY P, CRAVEN S E, COX N A, COSBY D E, LADELY S and MUSGROVE M T (2001), 'Sources and movement of *Salmonella* through integrated poultry operations: a multistate epidemiological investigation', *J Food Prot*, **64**, 1690–7.

BARAK J D and LIANG A S (2008), 'Role of soil, crop debris, and a plant pathogen in *Salmonella enterica* contamination of tomato plants', *PLoS ONE*, **3**, e1657.

BARKOCY-GALLAGHER G A, ARTHUR T M, RIVERA-BETANCOURT M, NOU X, SHACKELFORD S D, WHEELER T L and KOOHMARAIE M (2003), 'Seasonal prevalence of Shiga toxin-producing *Escherichia coli*, including O157:H7 and non-O157 serotypes, and *Salmonella* in commercial beef processing plants', *J Food Prot*, **66**, 1978–86.

BAUDART J, LEMARCHAND K, BRISABOIS A and LEBARON P (2000), 'Diversity of *Salmonella* strains isolated from the aquatic environment as determined by serotyping and amplification of the ribosomal DNA spacer regions', *Appl Environ Microbiol*, **66**, 1544–52.

BERRETI M and STUART D (2008), 'Food safety and environmental quality impose conflicting demands on Central Coast growers', *California Agriculture*, **62**, 68–73.

BEUCHAT L R (1996), 'Pathogenic microorganisms associated with fresh produce', *J Food Prot*, **59**, 204–16.

BEUCHAT L R (2002), 'Ecological factors influencing survival and growth of human pathogens on raw fruits and vegetables', *Microbes Infect*, **4**, 413–23.

BEUCHAT L R (2006), 'Vectors and conditions for preharvest contamination of fruits and vegetables with pathogens capable of causing enteric diseases', *Bri Food J*, **108**, 38–53.

BOXRUD D, PEDERSON-GULRUD K, WOTTON J, MEDUS C, LYSZKOWICZ E, BESSER J and BARTKUS J M (2007), 'Comparison of multiple-locus variable-number tandem repeat analysis, pulsed-field gel electrophoresis, and phage typing for subtype analysis of *Salmonella enterica* serotype Enteritidis', *J Clin Microbiol*, **45**, 536–43.

BRANDL M T (2006), 'Fitness of human enteric pathogens on plants and implications for food safety', *Annu Rev Phytopathol*, **44**, 367–92.

BRANDL M T (2008), 'Plant lesions promote the rapid multiplication of *Escherichia coli* O157:H7 on postharvest lettuce', *Appl Environ Microbiol*, **74**, 5285–9.

BRANDL M T and AMUNDSON R (2008), 'Leaf age as a risk factor in the contamination of lettuce with *Escherichia coli* O157:H7 and *Salmonella enterica*', *Appl Environ Microbiol*, **74**, 2298–306.

BRANDL M T, HAXO A F, BATES A H and MANDRELL R E (2004), 'Comparison of survival of *Campylobacter jejuni* in the phyllosphere with that in the rhizosphere of spinach and radish plants', *Appl Environ Microbiol*, **70**, 1182–9.

BRANHAM L A, CARR M A, SCOTT C B and CALLAWAY T R (2005), '*E. coli* O157 and *Salmonella* spp. in white-tailed deer and livestock', *Curr Issues Intest Microbiol*, **6**, 25–9.

BREUER T, BENKEL D H, SHAPIRO R L, HALL W N, WINNETT M M, LINN M J, NEIMANN J, BARRETT T J, DIETRICH S, DOWNES F P, TONEY D M, PEARSON J L, ROLKA H, SLUTSKER L and GRIFFIN P M (2001), 'A multistate outbreak of *Escherichia coli* O157:H7 infections linked to alfalfa sprouts grown from contaminated seeds', *Emerg Infect Dis*, **7**, 977–82.

BROOKES J D, ANTENUCCI J, HIPSEY M, BURCH M D, ASHBOLT N J and FERGUSON C (2004), 'Fate and transport of pathogens in lakes and reservoirs', *Environ Int*, **30**, 741–59.

CALFERT (2007a), Environmental investigation of *Escherichia coli* O157:H7 outbreak associated with Taco Bell restaurants in Northeastern United States. California Food Emergency Response Team, Sacramento, CA. http://www.dhs.ca.gov/fdb/local/PDF/Taco_Bell_final_report_redacted_11_19_2007.pdf.

CALFERT (2007b), Investigation of an *Escherichia coli* O157:H7 outbreak associated with Dole pre-packaged spinach. California Food Emergency Response Team, Sacramento, CA. http://www.dhs.ca.gov/ps/fdb/HTML/Food/EnvInvRpt.htm.

CALFERT (2007c), Investigation of an *Escherichia coli* O157:H7 outbreak associated with Dole pre-packaged spinach. Attachment 10: Environmental samples from farms and watersheds. California Food Emergency Response Team. Sacramento, CA. http://www.dhs.ca.gov/ps/fdb/HTML/Food/EnvInvRpt.htm.

CALFERT (2008), Investigation of the Taco John's *Escherichia coli* O157:H7 outbreak associated with iceberg lettuce. California Food Emergency Response Team, Sacramento, CA. http://www.dhs.ca.gov/fdb/HTML/Food/EnvInvRpt.htm.

CALVIN L (2007), *Outbreak linked to spinach forces reassessment of food safety practices* http://www.ers.usda.gov/AmberWaves/June07/Features/Spinach.htm.

CALVIN L, JENSEN H and LIANG J (2009), 'The economics food safety: The 2006 foodborne illness outbreak linked to spinach', In Fan X., Niemira B.A., Doona C.J., Feeherry, F.E. and Gravani, R.B. (eds.) *Microbial Safety of Fresh Produce*, Ames, IA, IFT Press, Wiley-Blackwell.

CAMPBELL J V, MOHLE-BOETANI J, REPORTER R, ABBOTT S, FARRAR J, BRANDL M, MANDRELL R and WERNER S B (2001), 'An outbreak of *Salmonella* serotype Thompson associated with fresh cilantro', *J Infect Dis*, **183**: 984–7.

CAMPBELL K (2005), March 27, 2008. *Imperial Valley farmers face critical shortage of workers*. http://www.cfbf.com/agalert/AgAlertStory.cfm?ID=466&ck=E836D813FD184325132FCA8EDCDFB40E.

CASTILLO A, MERCADO I, LUCIA L M, MARTINEZ-RUIZ Y, PONCE DE LEON J, MURANO E A and ACUFF G R (2004), '*Salmonella* contamination during production of cantaloupe: a binational study', *J Food Prot*, **67**, 713–20.

CDC (1995), *Outbreak Surveillance Data, Annual Listing of Foodborne Disease Outbreaks, 1995 Foodborne Disease Outbreak Line Listing*. http://www.cdc.gov/foodborneoutbreaks/outbreak_data.htm.

CDC (1996), *Outbreak Surveillance Data, Annual Listing of Foodborne Disease Outbreaks, 1996 Foodborne Disease Outbreak Line Listing*. http://www.cdc.gov/foodborneoutbreaks/outbreak_data.htm.

CDC (1998), *Outbreak Surveillance Data, Annual Listing of Foodborne Disease Outbreaks, 1998 Foodborne Disease Outbreak Line Listing*. http://www.cdc.gov/foodborneoutbreaks/outbreak_data.htm.

CDC (1999), *Outbreak Surveillance Data, Annual Listing of Foodborne Disease Outbreaks, 1999 Foodborne Disease Outbreak Line Listing*. http://www.cdc.gov/foodborneoutbreaks/outbreak_data.htm.

CDC (2002), *Outbreak Surveillance Data, Annual Listing of Foodborne Disease Outbreaks, 2002 Foodborne Disease Outbreak Line Listing*. http://www.cdc.gov/foodborneoutbreaks/outbreak_data.htm.

CDC (2003), *Outbreak Surveillance Data, Annual Listing of Foodborne Disease Outbreaks, 2003 Foodborne Disease Outbreak Line Listing*. http://www.cdc.gov/foodborneoutbreaks/outbreak_data.htm.

CDC (2004), *Outbreak Surveillance Data, Annual Listing of Foodborne Disease Outbreaks, 2004 Foodborne Disease Outbreak Line Listing*. http://www.cdc.gov/foodborneoutbreaks/outbreak_data.htm.
CDC (2008a), *Investigation of outbreak of infections caused by Salmonella Litchfield.* http://www.cdc.gov/salmonella/litchfield/.
CDC (2008b), *Investigation of outbreak of infections caused by Salmonella Saintpaul.* http://www.cdc.gov/salmonella/saintpaul/.
CDC (2009), *Outbreak of human Salmonella Typhimurium infections associated with contact with water frogs*. http://www.cdc.gov/salmonella/typh1209/index.html.
CDC (2010), Multistate outbreak of human *E. coli* O145 infections linked to shredded romaine lettuce from a single processing facility. Centers for Disease Control and Prevention. http://www.cdc.gov/print.do? url=http://www.cdc.gov/ecoli/2010/ecoli_O145/index.html. [Online.]
CDHS (2004a), Investigation of *E. coli* O157:H7 Illnesses in San Diego and Orange Counties California Department of Health Services, Sacramento, CA. http://www.dhs.ca.gov/fdb/HTML/Food/EnvInvRpt.htm.
CDHS (2004b), Investigation of *E. coli* O157:H7 Outbreak in San Mateo Retirement Facility. California Department of Health Services, Sacramento, CA. http://www.dhs.ca.gov/fdb/HTML/Food/EnvInvRpt.htm.
CDHS (2005), Addendum Report to Investigation of Pre-washed Mixed Bagged Salad following an Outbreak of *Escherichia coli* O157:H7 in San Diego and Orange County. California Department of Health Services, Sacramento, CA. http://www.dhs.ca.gov/fdb/HTML/Food/EnvInvRpt.htm.
CHAIDEZ C, SOTO M, GORTARES P and MENA K (2005), 'Occurrence of Cryptosporidium and Giardia in irrigation water and its impact on the fresh produce industry', *Int J Environ Health Res*, **15**, 339–45.
CHAPMAN P A, SIDDONS C A, GERDAN MALO A T and HARKIN M A (1997), 'A 1-year study of *Escherichia coli* O157 in cattle, sheep, pigs and poultry', *Epidemiol Infect*, **119**, 245–50.
CHASE-TOPPING M E, MCKENDRICK I J, PEARCE M C, MACDONALD P, MATTHEWS L, HALLIDAY J, ALLISON L, FENLON D, LOW J C, GUNN G and WOOLHOUSE M E (2007), 'Risk factors for the presence of high-level shedders of *Escherichia coli* O157 on Scottish farms', *J Clin Microbiol*, **45**, 1594–603.
CHO S, BOXRUD D J, BARTKUS J M, WHITTAM T S and SAEED M (2007), 'Multiple-locus variable-number tandem repeat analysis of *Salmonella* Enteritidis isolates from human and non-human sources using a single multiplex PCR', *FEMS Microbiol Lett*, **275**, 16–23.
CIZEK A, ALEXA P, LITERAK I, HAMRIK J, NOVAK P and SMOLA J (1999), 'Shiga toxin-producing *Escherichia coli* O157 in feedlot cattle and Norwegian rats from a large-scale farm', *Lett Appl Microbiol*, **28**, 435–9.
CLEMENS R (2004), Feb. 14, 2007 2004. *The expanding U.S. market for fresh produce, review paper*. http://www.agmrc.org/NR/rdonlyres/EEDD83B9-E4FE-433B-A818-327B6B5E221A/0/expandingusmarketproduce.pdf.
COBBOLD R N, RICE D H, DAVIS M A, BESSER T E and HANCOCK D D (2006), 'Long-term persistence of multi-drug-resistant *Salmonella enterica* serovar Newport in two dairy herds', *J Am Vet Med Assoc*, **228**, 585–91.
COOLEY M, CARYCHAO D, CRAWFORD-MIKSZA L, JAY M T, MYERS C, ROSE C, KEYS C, FARRAR J and MANDRELL R E (2007), 'Incidence and tracking of *Escherichia coli* O157:H7 in a major produce production region in California', *PLoS ONE*, **2**, e1159.
COOPER I R, TAYLOR H D and HANLON G W (2007), 'Virulence traits associated with verocytotoxigenic *Escherichia coli* O157 recovered from freshwater biofilms', *J Appl Microbiol*, **102**, 1293–9.
CREPET A, ALBERT I, DERVIN C and CARLIN F (2007), 'Estimation of microbial contamination of food from prevalence and concentration data: application to *Listeria monocytogenes* in fresh vegetables', *Appl Environ Microbiol*, **73**, 250–8.

CROOK P D, AGUILERA J F, THRELFALL E J, O'BRIEN S J, SIGMUNDSDOTTIR G, WILSON D, FISHER I S, AMMON A, BRIEM H, COWDEN J M, LOCKING M E, TSCHAPE H, VAN PELT W, WARD L R and WIDDOWSON M A (2003), 'A European outbreak of *Salmonella enterica* serotype Typhimurium definitive phage type 204b in 2000', *Clin Microbiol Infect*, **9**, 839–45.

CSPI (2007), *Outbreak Alert: Closing the gaps in our federal food safety net.* http://www.cspinet.org/foodsafety/outbreak_alert.pdf.

CSPI (2008), *Outbreak Alert! Database, 1990–2005.* http://www.cspinet.org/foodsafety/outbreak/pathogen.php.

CSPI (2009), Outbreak Alert! Analyzing foodborne outbreaks, 1998–2007. Washington, DC. http://cspinet.org/new/pdf/outbreakalertreport09.pdf.

CUMMINGS K, BARRETT E, MOHLE-BOETANI J C, BROOKS J T, FARRAR J, HUNT T, FIORE A, KOMATSU K, WERNER S B and SLUTSKER L (2001), 'A multistate outbreak of *Salmonella enterica* serotype Baildon associated with domestic raw tomatoes', *Emerg Infect Dis*, **7**, 1046–8.

CURRIERO F C, PATZ J A, ROSE J B and LELE S (2001), 'The association between extreme precipitation and waterborne disease outbreaks in the United States, 1948–1994', *Am J Public Health*, **91**, 1194–9.

DARGATZ D A, FEDORKA-CRAY P J, LADELY S R, KOPRAL C A, FERRIS K E and HEADRICK M L (2003), 'Prevalence and antimicrobial susceptibility of *Salmonella* spp. isolates from US cattle in feedlots in 1999 and 2000', *J Appl Microbiol*, **95**, 753–61.

DAVIES R H and BRESLIN M (2003), 'Persistence of *Salmonella* Enteritidis phage type 4 in the environment and arthropod vectors on an empty free-range chicken farm', *Environ Microbiol*, **5**, 79–84.

DIEZ-GONZALEZ F and MUKHERJEE A (2009), *Produce safety in organic vs. conventional crops*. In Fan X., Niemira B., Doona C.J., Feeherry, F. and Gravani, R.B. (eds.) *Microbial Safety of Fresh Produce*. Oxford, UK, IFT/Blackwell Publishing, 83–99.

DOANE C A, PANGLOLI P, RICHARDS H A, MOUNT J R, GOLDEN D A and DRAUGHON F A (2007), 'Occurrence of *Escherichia coli* O157:H7 in diverse farm environments', *J Food Prot*, **70**, 6–10.

DONG Y, INIGUEZ A L, TRIPLETT E W and AHMER B M M (2003), 'Kinetics and strain specificity of rhizosphere and endophytic colonization by enteric bacteria on seedlings of *Medicago sativa* and *Medicago truncatula*', *Applied and Environmental Microbiology*, **69**, 1783–90.

DOYLE M P and ERICKSON M C (2008), 'Summer meeting 2007 – the problems with fresh produce: an overview', *J Appl Microbiol*, **105**, 317–30.

DUFFY G (2003), 'Verocytoxigenic *Escherichia coli* in animal faeces, manures and slurries', *J Appl Microbiol*, **94**(Suppl), 94S–103S.

DUFFY G and MORIARTY E M (2003), '*Cryptosporidium* and its potential as a food-borne pathogen', *Anim Health Res Rev*, **4**, 95–107.

DUNN J R, KEEN J E, MORELAND D and THOMPSON R A (2004), 'Prevalence of *Escherichia coli* O157:H7 in white-tailed deer from Louisiana', *Journal of Wildlife Diseases*, **40**, 361–5.

EJIDOKUN O O, WALSH A, BARNETT J, HOPE Y, ELLIS, S, SHARP M W, PAIBA G A, LOGAN M, WILLSHAW G A and CHEASTY T (2006), 'Human Vero cytotoxigenic *Escherichia coli* (VTEC) O157 infection linked to birds', *Epidemiol Infect*, **134**, 421–3.

ELDER R O, KEEN J E, SIRAGUSA G R, BARKOCY-GALLAGHER G A, KOOHMARAIE M and LAEGREID W W (2000), 'Correlation of enterohemorrhagic *Escherichia coli* O157 prevalence in feces, hides, and carcasses of beef cattle during processing [see comments]', *Proc Natl Acad Sci USA*, **97**, 2999–3003.

ETHELBERG S, LISBY M, BÖTTIGER B, SCHULTZ A C, VILLIF A, JENSEN T, OLSEN K E, SCHEUTZ F, KJELSØ C and MÜLLER L (2010), 'Outbreaks of gastroenteritis linked to lettuce, Denmark, January 2010. *Eurosurveillance*', **15**, pii=19484.

FDA-CFSAN (2001a). FDA survey of imported fresh produce, FY 1999 field assignment. http://www.cfsan.fda.gov/~dms/prodsur6.html.

FDA-CFSAN (2001b). Survey of domestic fresh produce: interim results. http://vm.cfsan.fda.gov/~dms/prodsur9.html.

FEDER I, WALLACE F M, GRAY J T, FRATAMICO P, FEDORKA-CRAY P J, PEARCE R A, CALL J E, PERRINE R and LUCHANSKY J B (2003), 'Isolation of *Escherichia coli* O157:H7 from intact colon fecal samples of swine', *Emerg Infect Dis*, **9**, 380–3.

FEGAN N, VANDERLINDE P, HIGGS G and DESMARCHELIER P (2004), 'Quantification and prevalence of *Salmonella* in beef cattle presenting at slaughter', *J Appl Microbiol*, **97**, 892–8.

FEGAN N, VANDERLINDE P, HIGGS G and DESMARCHELIER P (2005), 'A study of the prevalence and enumeration of *Salmonella enterica* in cattle and on carcasses during processing', *J Food Prot*, **68**, 1147–53.

FENLON D R (1981), 'Seagulls (*Larus* spp.) as vectors of salmonellae: an investigation into the range of serotypes and numbers of salmonellae in gull faeces', *J Hyg (Lond)*, **86**, 195–202.

FERGUSON C, DE RODA HUSMAN A M, ALTAVILLA N, DEERE D and ASHBOLT N (2003), 'Fate and transport of surface water pathogens in watersheds', *Critical Reviews in Environmental Science and Technology*, **33**, 299–361.

FIELD K G and SAMADPOUR M (2007), 'Fecal source tracking, the indicator paradigm, and managing water quality', *Water Res*, **41**, 3517–38.

FINCHER L M, PARKER C D and CHAURET C P (2009), 'Occurrence and antibiotic resistance of *Escherichia coli* O157:H7 in a watershed in north-central Indiana', *J Environ Qual*, **38**, 997–1004.

FISCHER J R, ZHAO T, DOYLE M P, GOLDBERG M R, BROWN C A, SEWELL C T, KAVANAUGH D M and BAUMAN C D (2001), 'Experimental and field studies of *Escherichia coli* O157:H7 in white-tailed deer', *Appl Environ Microbiol*, **67**, 1218–24.

FOLEY S L, LYNNE A M and NAYAK R (2008), 'Salmonella challenges: prevalence in swine and poultry and potential pathogenicity of such isolates', *J Anim Sci*, **86**, E149–62.

FOLEY S L, LYNNE A M and NAYAK R (2009), 'Molecular typing methodologies for microbial source tracking and epidemiological investigations of Gram-negative bacterial foodborne pathogens', *Infection, Genetics and Evolution*, **9**, 430–40.

FRANZ E, SEMENOV A V, TERMORSHUIZEN A J, DE VOS O J, BOKHORST J G and VAN BRUGGEN A H (2008), 'Manure-amended soil characteristics affecting the survival of *E. coli* O157:H7 in 36 Dutch soils', *Environ Microbiol*, **10**, 313–27.

FRANZ E, VISSER A A, VAN DIEPENINGEN A D, KLERKS M M, TERMORSHUIZEN A J and VAN BRUGGEN A H (2007), 'Quantification of contamination of lettuce by GFP-expressing *Escherichia coli* O157:H7 and *Salmonella enterica* serovar Typhimurium', *Food Microbiol*, **24**, 106–12.

FREMAUX B, DELIGNETTE-MULLER M L, PRIGENT-COMBARET C, GLEIZAL A and VERNOZY-ROZAND C (2007), 'Growth and survival of non-O157:H7 Shiga-toxin-producing *Escherichia coli* in cow manure', *J Appl Microbiol*, **102**, 89–99.

FRIESEMA I, SIGMUNDSDOTTIR G, VAN DER ZWALUW K, HEUVELINK A B, DE JAGER C, RUMP B, BRIEM H, HARDARDOTTIR H, ATLADOTTIR A, GUDMUNDSDOTTIR E and VAN PELT W (2008), 'An international outbreak of Shiga toxin-producing *Escherichia coli* O157 infection due to lettuce, September–October 2007', *Euro Surveill*, **13**, pii=19065.

FRODER H, MARTINS C G, DE SOUZA K L, LANDGRAF M, FRANCO B D and DESTRO M T (2007), 'Minimally processed vegetable salads: microbial quality evaluation', *J Food Prot*, **70**, 1277–80.

GAGLIARDI J V and KARNS J S (2000), 'Leaching of *Escherichia coli* O157:H7 in diverse soils under various agricultural management practices', *Appl Environ Microbiol*, **66**, 877–83.

GAGLIARDI J V and KARNS J S (2002), 'Persistence of *Escherichia coli* O157:H7 in soil and on plant roots', *Environ Microbiol*, **4**, 89–96.

GANNON V P, GRAHAM T A, READ S, ZIEBELL K, MUCKLE A, MORI J, THOMAS J, SELINGER B, TOWNSHEND I and BYRNE J (2004), 'Bacterial pathogens in rural water supplies in Southern Alberta, Canada', *J Toxicol Environ Health A*, **67**, 1643–53.

GARCIA A and FOX J G (2003), 'The rabbit as a new reservoir host of enterohemorrhagic *Escherichia coli*', *Emerg Infect Dis*, **9**, 1592–7.

GAYNOR K, PARK S Y, KANENAKA R, COLINDRES R, MINTZ E, RAM P K, KITSUTANI P, NAKATA M, WEDEL S, BOXRUD D, JENNINGS D, YOSHIDA H, TOSAKA N, HE H, CHING-LEE M and EFFLER P V (2008), 'International foodborne outbreak of *Shigella sonnei* infection in airline passengers', *Epidemiol Infect*, **137**, 335–41.

GIBBS R, PINGAULT N, MAZZUCCHELLI T, O'REILLY L, MACKENZIE B, GREEN J, MOGYOROSY R, STAFFORD R, BELL R, HILEY L, FULLERTON K and VAN BUYNDER P (2009), 'An outbreak of *Salmonella enterica* serotype Litchfield infection in Australia linked to consumption of contaminated papaya', *J Food Prot*, **72**, 1094–8.

GREENE S K, DALY E R, TALBOT E A, DEMMA L J, HOLZBAUER S, PATEL N J, HILL T A, WALDERHAUG M O, HOEKSTRA R M, LYNCH M F and PAINTER J A (2008), 'Recurrent multistate outbreak of *Salmonella* Newport associated with tomatoes from contaminated fields, 2005', *Epidemiol Infect*, **136**, 157–65.

GUO X, CHEN J, BRACKETT R E and BEUCHAT L R (2001), 'Survival of salmonellae on and in tomato plants from the time of inoculation at flowering and early stages of fruit development through fruit ripening', *Appl Environ Microbiol*, **67**, 4760–4.

GUPTA S K, NALLUSWAMI K, SNIDER C, PERCH M, BALASEGARAM M, BURMEISTER D, LOCKETT J, SANDT C, HOEKSTRA R M and MONTGOMERY S (2007), 'Outbreak of *Salmonella* Braenderup infections associated with Roma tomatoes, northeastern United States, 2004: a useful method for subtyping exposures in field investigations', *Epidemiol Infect*, **135**, 1165–73.

HARRIS L J, FARBER J N, BEUCHAT L R, PARISH M E, SUSLOW T V, GARRET E H and BUSTA F F (2003), Outbreaks associated with fresh produce: incidence, growth, and survival of pathogens in fresh and fresh-cut produce. *Comprehensive Reviews in Food Science and Food Safety*, Wiley Interscience, Malden, MA, pp. 78–141.

HARWOOD V (2007), *Assumptions and limitations associated with microbial source tracking*. In Santo Domingo, J. and Sadowsky, M. (eds.) *Microbial Source Tracking*. Washington, DC, ASM Press, Washington, DC, pp. 33–64.

HEATON J C and JONES K (2008), 'Microbial contamination of fruit and vegetables and the behaviour of enteropathogens in the phyllosphere: a review', *J Appl Microbiol*, **104**, 613–26.

HECK M (2009), 'Multilocus variable number of tandem repeats analysis (MLVA) – a reliable tool for rapid investigation of *Salmonella* Typhimurium outbreaks', *Euro Surveill*, **14**, 1.

HEDBERG C W, ANGULO F J, WHITE K E, LANGKOP C W, SCHELL W L, STOBIERSKI M G, SCHUCHAT A, BESSER J M, DIETRICH S, HELSEL L, GRIFFIN P M, MCFARLAND J W and OSTERHOLM M T (1999), 'Outbreaks of salmonellosis associated with eating uncooked tomatoes: implications for public health. The Investigation Team', *Epidemiol Infect*, **122**, 385–93.

HEIER B T, NYGARD K, KAPPERUD G, LINDSTEDT B A, JOHANNESSEN G S and BLEKKAN H (2009), *Shigella sonnei* infections in Norway associated with sugar peas, May–June 2009', *Euro Surveill*, **14**, 1–3.

HERMAN K M, AYERS T L and LYNCH M (2008), *Foodborne disease outbreaks associated with leafy greens, 1973–2006. International Conference on Emerging Infectious Diseases*. Atlanta, GA, Center for Disease Control and Prevention.

HERNANDEZ J, BONNEDAHL J, WALDENSTROM J, PALMGREN H and OLSEN B (2003), '*Salmonella* in birds migrating through Sweden', *Emerg Infect Dis*, **9**, 753–5.

HIDALGO-VILA J, DIAZ-PANIAGUA C, DE FRUTOS-ESCOBAR C, JIMENEZ-MARTINEZ C and PEREZ-SANTIGOSA N (2007), 'Salmonella in free living terrestrial and aquatic turtles', *Vet Microbiol*, **119**, 311–15.

HIGGINS J A, BELT K T, KARNS J S, RUSSELL-ANELLI J and SHELTON D R (2005), '*tir*- and *stx*-positive *Escherichia coli* in stream waters in a metropolitan area', *Appl Environ Microbiol*, **71**, 2511–19.

HILBORN E D, MERMIN J H, MSHAR P A, HADLER J L, VOETSCH A, WOJTKUNSKI C, SWARTZ M, MSHAR R, LAMBERT-FAIR M A, FARRAR J A, GLYNN M K and SLUTSKER L (1999), 'A multistate outbreak of *Escherichia coli* O157:H7 infections associated with consumption of mesclun lettuce', *Arch Intern Med*, **159**, 1758–64.

HIMATHONGKHAM S, BAHARI S, RIEMANN H and CLIVER D (1999), 'Survival of *Escherichia coli* O157:H7 and *Salmonella* Typhimurium in cow manure and cow manure slurry', *FEMS Microbiol Lett*, **178**, 251–7.

HIRANO S S and UPPER C D (2000), 'Bacteria in the leaf ecosystem with emphasis on *Pseudomonas syringae*-a pathogen, ice nucleus, and epiphyte', *Microbiol Mol Biol Rev*, **64**, 624–53.

HOLLEY R A, ARRUS K M, OMINSKI K H, TENUTA M and BLANK G (2006), '*Salmonella* survival in manure-treated soils during simulated seasonal temperature exposure', *J Environ Qual*, **35**, 1170–80.

HUGHES L A, SHOPLAND S, WIGLEY P, BRADON H, LEATHERBARROW A H, WILLIAMS N J, BENNETT M, DE PINNA E, LAWSON B, CUNNINGHAM A A and CHANTREY J (2008), 'Characterisation of *Salmonella* enterica serotype Typhimurium isolates from wild birds in northern England from 2005 to 2006', *BMC Vet Res*, **4**, 4.

HUSSEIN H S and BOLLINGER L M (2005), 'Prevalence of Shiga toxin-producing *Escherichia coli* in beef cattle', *J Food Prot*, **68**, 2224–41.

HUTCHISON M L, AVERY S M and MONAGHAN J M (2008), 'The airborne distribution of zoonotic agents from livestock waste spreading and microbiological risk to fresh produce from contaminated irrigation sources', *J Appl Microbiol*, **105**, 848–57.

HYYTIA-TREES E, SMOLE S C, FIELDS P A, SWAMINATHAN B and RIBOT E M (2006), 'Second-generation subtyping: a proposed PulseNet protocol for multiple-locus variable-number tandem repeat analysis of Shiga toxin-producing *Escherichia coli* O157 (STEC O157)', *Foodborne Pathog Dis*, **3**, 118–31.

IHEKWEAZU C, BARLOW M, ROBERTS S, CHRISTENSEN H, GUTTRIDGE B, LEWIS D and PAYNTER S (2006), 'Outbreak of *E. coli* O157 infection in the south west of the UK: risks from streams crossing seaside beaches', *Euro Surveill*, **11**, 128–30.

ILIC S, ODOMERU J and LEJEUNE J T (2008), 'Coliforms and prevalence of *Escherichia coli* and foodborne pathogens on minimally processed spinach in two packing plants', *J Food Prot*, **71**, 2398–403.

IRVINE W N, GILLESPIE I A, SMYTH F B, ROONEY P J, MCCLENAGHAN A, DEVINE M J and TOHANI V K (2009), 'Investigation of an outbreak of *Salmonella enterica* serovar Newport infection', *Epidemiol Infect*, **137**, 1449–56.

ISAACS S, ARAMINI J, CIEBIN B, FARRAR J A, AHMED R, MIDDLETON D, CHANDRAN A U, HARRIS L J, HOWES M, CHAN E, PICHETTE A S, CAMPBELL K, GUPTA A, LIOR L Y, PEARCE M, CLARK C, RODGERS F, JAMIESON F, BROPHY I and ELLIS A (2005), 'An international outbreak of salmonellosis associated with raw almonds contaminated with a rare phage type of *Salmonella* Enteritidis', *J Food Prot*, **68**, 191–8.

ISLAM M, DOYLE M P, PHATAK S C, MILLNER P and JIANG X (2004), 'Persistence of enterohemorrhagic *Escherichia coli* O157:H7 in soil and on leaf lettuce and parsley grown in fields treated with contaminated manure composts or irrigation water', *J Food Prot*, **67**, 1365–70.

ISLAM M, DOYLE M P, PHATAK S C, MILLNER P and JIANG X (2005), 'Survival of *Escherichia coli* O157:H7 in soil and on carrots and onions grown in fields treated with contaminated manure composts or irrigation water', *Food Microbiology*, **22**, 63–70.

JAIN S, BIDOL S A, AUSTIN J L, BERL E, ELSON F, LEMAILE-WILLIAMS M, DEASY M, MOLL M E, REA V, VOJDANI J D, YU P A, HOEKSTRA R M, BRADEN C R and LYNCH M F

(2009), 'Multistate Outbreak of *Salmonella* Typhimurium and Saintpaul infections associated with unpasteurized orange juice-United States, 2005', *Clin Infect Dis*, **48**, 1065–71.

JALAVA K, HAKKINEN M, VALKONEN M, NAKARI U M, PALO T, HALLANVUO S, OLLGREN J, SIITONEN A and NUORTI J P (2006), 'An outbreak of gastrointestinal illness and erythema nodosum from grated carrots contaminated with *Yersinia pseudotuberculosis*', *J Infect Dis*, **194**, 1209–16.

JAY M T, COOLEY M, CARYCHAO D, WISCOMB G W, SWEITZER R A, CRAWFORD-MIKSZA L, FARRAR J A, LAU D K, O'CONNELL J, MILLINGTON A, ASMUNDSON R V, ATWILL E R and MANDRELL R E (2007), '*Escherichia coli* O157:H7 in feral swine near spinach fields and cattle, central California coast', *Emerg Infect Dis*, **13**, 1908–11.

JOHNSON J Y, THOMAS J E, GRAHAM T A, TOWNSHEND I, BYRNE J, SELINGER L B and GANNON V P (2003), 'Prevalence of *Escherichia coli* O157:H7 and *Salmonella* spp. in surface waters of southern Alberta and its relation to manure sources', *Can J Microbiol*, **49**, 326–35.

JOHNSTON L M, JAYKUS L A, MOLL D, ANCISO J, MORA B and MOE C L (2006a), 'A field study of the microbiological quality of fresh produce of domestic and Mexican origin', *Int J Food Microbiol*, **112**, 83–95.

JOHNSTON L M, JAYKUS L A, MOLL D, MARTINEZ M C, ANCISO J, MORA B and MOE C L (2005), 'A field study of the microbiological quality of fresh produce', *J Food Prot*, **68**, 1840–7.

JOHNSTON L M, MOE C L, MOLL D and JAYKUS L (2006b), 'The epidemiology of produce-associated outbreaks of foodborne disease', in James, J (ed.) *Microbial Hazard Identification in Fresh Fruit and Vegetables*, Hoboken, NJ, John Wiley and Sons, pp. 37–72.

KANGAS S, TAKKINEN J, HAKKINEN M, NAKARI U M, JOHANSSON T, HENTTONEN H, VIRTALUOTO L, SIITONEN A, OLLGREN J and KUUSI M (2008), '*Yersinia pseudotuberculosis* O:1 traced to raw carrots, Finland', *Emerg Infect Dis*, **14**, 1959–61.

KÄSS, W (1998), *Tracing Technique in Geohydrology*. Rotterdam-NE/Brookfield-VT, A.A. Balkema, p. 581.

KEEN J E, WITTUM T E, DUNN J R, BONO J L and DURSO L M (2006), 'Shiga-toxigenic *Escherichia coli* O157 in agricultural fair livestock, United States', *Emerg Infect Dis*, **12**, 780–6.

KEENE W E, SAZIE E, KOK J, RICE D H, HANCOCK D D, BALAN V K, ZHAO T and DOYLE M P (1997), 'An outbreak of *Escherichia coli* O157:H7 infections traced to jerky made from deer meat', *Jama*, **277**, 1229–31.

KEYS C, KEMPER S and KEIM P (2005), 'Highly diverse variable number tandem repeat loci in the *E. coli* O157:H7 and O55:H7 genomes for high-resolution molecular typing', *J Appl Microbiol*, **98**, 928–40.

KINDE H, CASTELLAN D M, KASS P H, ARDANS A, CUTLER G, BREITMEYER R E, BELL D D, ERNST R A, KERR D C, LITTLE H E, WILLOUGHBY D, RIEMANN H P, SNOWDON J A and KUNEY D R (2004), 'The occurrence and distribution of *Salmonella* Enteritidis and other serovars on California egg laying premises: a comparison of two sampling methods and two culturing techniques', *Avian Dis*, **48**, 590–4.

KIRK J H, HOLMBERG C A and JEFFREY J S (2002), 'Prevalence of *Salmonella* spp in selected birds captured on California dairies', *J Am Vet Med Assoc*, **220**, 359–62.

KOOPMANS M (2008), 'Progress in understanding norovirus epidemiology', *Curr Opin Infect Dis*, **21**, 544–52.

KROUPITSKI Y, GOLBERG D, BELAUSOV E, PINTO R, SWARTZBERG D, GRANOT D and SELA S (2009), 'Internalization of *Salmonella enterica* in leaves is induced by light and involves chemotaxis and penetration through open stomata', *Appl Environ Microbiol*, **75**, 6076–86.

KUDVA I T, BLANCH K and HOVDE C J (1998), 'Analysis of *Escherichia coli* O157:H7 survival in ovine or bovine manure and manure slurry', *Appl Environ Microbiol*, **64**, 3166–74.

LANG N L and SMITH S R (2007), 'Influence of soil type, moisture content and biosolids application on the fate of *Escherichia coli* in agricultural soil under controlled laboratory conditions', *J Appl Microbiol*, **103**, 2122–31.

LARSSON J T, TORPDAHL M, PETERSEN R F, SORENSEN G, LINDSTEDT B A and NIELSEN E M (2009), 'Development of a new nomenclature for *Salmonella* typhimurium multilocus variable number of tandem repeats analysis (MLVA)', *Euro Surveill*, **14**, 1–5.

LECLERC J E, LI B, PAYNE W L and CEBULA T A (1996), 'High mutation frequencies among *Escherichia coli* and *Salmonella* pathogens', *Science*, **274**, 1208–11.

LECLERCQ A and MAHILLON J (2003), 'Farmed rabbits and ducks as vectors for VTEC O157:H7', *Vet Rec*, **152**, 723–4.

LEDET MULLER L, HJERTQVIST M, PAYNE L, PETTERSSON H, OLSSON A, PLYM FORSHELL L and ANDERSSON Y (2007), 'Cluster of *Salmonella* Enteritidis in Sweden 2005–2006 – suspected source: almonds', *Euro Surveill*, **12**, E9–10.

LEE R J and MORGAN O C (2003), 'Environmental factors influencing the microbiological contamination of commercially harvested shellfish', *Water Sci Technol*, **47**, 65–70.

LEJEUNE J T, BESSER T E and HANCOCK D D (2001), 'Cattle water troughs as reservoirs of *Escherichia coli* O157', *Appl Environ Microbiol*, **67**, 3053–7.

LEJEUNE J T, BESSER T E, RICE D H, BERG J L, STILBORN R P and HANCOCK D D (2004), 'Longitudinal study of fecal shedding of *Escherichia coli* O157:H7 in feedlot cattle: predominance and persistence of specific clonal types despite massive cattle population turnover', *Appl Environ Microbiol*, **70**, 377–84.

LEOPOLD S R, MAGRINI V, HOLT N J, SHAIKH N, MARDIS E R, CAGNO J, OGURA Y, IGUCHI A, HAYASHI T, MELLMANN A, KARCH H, BESSER T E, SAWYER S A, WHITTAM T S and TARR P I (2009), 'A precise reconstruction of the emergence and constrained radiations of *Escherichia coli* O157 portrayed by backbone concatenomic analysis', *Proc Natl Acad Sci USA*, **106**, 8713–18.

LINDSTEDT B A, HEIR E, GJERNES E, VARDUND T and KAPPERUD G (2003), 'DNA fingerprinting of shiga-toxin producing *Escherichia coli* O157 based on multiple-locus variable-number tandem-repeats analysis (MLVA)', *Ann Clin Microbiol Antimicrob*, **2**, 12.

LOFDAHL M, IVARSSON S, ANDERSSON S, LANGMARK J and PLYM-FORSHELL L (2009), 'An outbreak of *Shigella dysenteriae* in Sweden, May-June 2009, with sugar snaps as the suspected source', *Euro Surveill*, **14**, 1–3.

LOW J C, MCKENDRICK I J, MCKECHNIE C, FENLON D, NAYLOR S W, CURRIE C, SMITH D G, ALLISON L and GALLY D L (2005), 'Rectal carriage of enterohemorrhagic *Escherichia coli* O157 in slaughtered cattle', *Appl Environ Microbiol*, **71**, 93–7.

LYNCH M F, TAUXE R V and HEDBERG C W (2009), 'The growing burden of foodborne outbreaks due to contaminated fresh produce: risks and opportunities', *Epidemiol Infect*, **137**, 307–15.

MAHON B E, PONKA A, HALL W N, KOMATSU K, DIETRICH S E, SIITONEN A, CAGE G, HAYES P S, LAMBERT-FAIR M A, BEAN N H, GRIFFIN P M and SLUTSKER L (1997), 'An international outbreak of *Salmonella* infections caused by alfalfa sprouts grown from contaminated seeds', *J Infect Dis*, **175**, 876–82.

MANDRELL R E (2009), 'Enteric human pathogens associated with fresh produce: sources, transport and ecology', in Fan X., Niemira B., Doona C.J., Feeherry, F. and Gravani, R.B. (eds.) *Microbial Safety of Fresh Produce*, Ox.ford, UK, IFT/Blackwell Publishing, pp. 3–42.

MANDRELL R E and BRANDL M T (2004), '*Campylobacter* species and fresh produce: outbreaks, incidence and biology', in Beier R.C., Pillai S.D., Phillips T.D. and Ziprin R.L. (eds.) *Preharvest and Postharvest Food Safety: Contemporary Issues and Future Directions*, Ames, IA, IFT Press and Blackwell Publishing, pp. 59–72.

MANDRELL R E, GORSKI L and BRANDL M T (2006), 'Attachment of microorganisms to fresh produce', in Sapers G.M., Gorny, J.R. and Yousef, A.E. (eds.) *Microbiology of Fruits and Vegetables*, Boca Raton, FL, CRC, Taylor & Francis, pp. 33–74.

MANNING S D, MOTIWALA A S, SPRINGMAN A C, QI W, LACHER D W, OUELLETTE L M, MLADONICKY J M, SOMSEL P, RUDRIK J T, DIETRICH S E, ZHANG W, SWAMINATHAN B,

ALLAND D and WHITTAM T S (2008), 'Variation in virulence among clades of *Escherichia coli* O157:H7 associated with disease outbreaks', *Proc Natl Acad Sci USA*, **105**, 4868–73.

MARA D and SLEIGH A (2010), 'Estimation of norovirus infection risks to consumers of wastewater-irrigated food crops eaten raw', *J Water Health*, **8**, 39–43.

MARLER-CLARK (2009), *September 2008 Aunt Mid's lettuce E. coli O157:H7 outbreak linked to Santa Barbara Farms*. http://www.marlerblog.com/2009/06/articles/legal-cases/september-2008-aunt-mids-lettuce-e-coli-o157h7-outbreak-linked-to-santa-barbara-farms/.

MARTINEZ-URTAZA J, SACO M, DE NOVOA J, PEREZ-PINEIRO P, PEITEADO J, LOZANO-LEON A and GARCIA-MARTIN O (2004), 'Influence of environmental factors and human activity on the presence of *Salmonella* serovars in a marine environment', *Appl Environ Microbiol*, **70**, 2089–97.

MATTHEWS L, LOW J C, GALLY D L, PEARCE M C, MELLOR D J, HEESTERBEEK J A, CHASE-TOPPING M, NAYLOR S W, SHAW D J, REID S W, GUNN G J and WOOLHOUSE M E (2006), 'Heterogeneous shedding of *Escherichia coli* O157 in cattle and its implications for control', *Proc Natl Acad Sci USA*, **103**, 547–52.

MCCREA B A, TONOOKA K H, VANWORTH C, BOGGS C L, ATWILL E R and SCHRADER J S (2006), 'Prevalence of *Campylobacter* and *Salmonella* species on farm, after transport, and at processing in specialty market poultry', *Poult Sci*, **85**, 136–43.

MCGEE P, BOLTON D J, SHERIDAN J J, EARLEY B, KELLY G and LEONARD N (2002), 'Survival of *Escherichia coli* O157:H7 in farm water: its role as a vector in the transmission of the organism within herds', *J Appl Microbiol*, **93**, 706–13.

MDPH (2005), *Health officials investigate E. coli O157:H7 cases related to Dole prepackaged lettuce mixes sold at Rainbow Foods*. http://www.health.state.mn.us/news/pressrel/ecoli093005.html.

MEAYS C L, BROERSMA K, NORDIN R and MAZUMDER A (2004), 'Source tracking fecal bacteria in water: A critical review of current methods', *Journal of Environmental Management*, **73**, 71–9.

MEINERSMANN R J, BERRANG M E, JACKSON C R, FEDORKA-CRAY P, LADELY S, LITTLE E, FRYE J G and MATTSSON B (2008), '*Salmonella, Campylobacter* and *Enterococcus* spp.: their antimicrobial resistance profiles and their spatial relationships in a synoptic study of the Upper Oconee River basin', *Microb Ecol*, **55**, 444–52.

MICHEL P, WILSON J B, MARTIN S W, CLARKE R C, MCEWEN S A and GYLES C L (1999), 'Temporal and geographical distributions of reported cases of *Escherichia coli* O157:H7 infection in Ontario', *Epidemiol Infect*, **122**, 193–200.

MICHINO H, ARAKI K, MINAMI S, TAKAYA S, SAKAI N, MIYAZAKI M, ONO A and YANAGAWA H (1999), 'Massive outbreak of *Escherichia coli* O157:H7 infection in school children in Sakai City, Japan, associated with consumption of white radish sprouts', *Am J Epidemiol*, **150**, 787–96.

MILLER W G and MANDRELL R E (2006), 'Campylobacter in the food and water supply: prevalence, outbreaks, isolation, and detection', in Ketley, J. and Konkel, M.E. (eds.) *Campylobacter jejuni: New Perspectives in Molecular and Cellular Biology*, Norfolk, UK, Horizon Scientific Press, pp. 101–63.

MILNES A S, STEWART I, CLIFTON-HADLEY F A, DAVIES R H, NEWELL D G, SAYERS A R, CHEASTY T, CASSAR C, RIDLEY A, COOK A J, EVANS S J, TEALE C J, SMITH R P, MCNALLY A, TOSZEGHY M, FUTTER R, KAY A, and PAIBA G A (2008), 'Intestinal carriage of verocytotoxigenic *Escherichia coli* O157, *Salmonella*, thermophilic *Campylobacter* and *Yersinia enterocolitica*, in cattle, sheep and pigs at slaughter in Great Britain during 2003', *Epidemiol Infect*, **136**, 739–51.

MMWR (2002), 'Multistate outbreaks of *Salmonella* serotype Poona infections associated with eating cantaloupe from Mexico – United States and Canada, 2000–2002', *MMWR Morb Mortal Wkly Rep*, **51**, 1044–7.

MMWR (2003), 'Hepatitis A outbreak associated with green onions at a restaurant – Monaca, Pennsylvania, 2003', *MMWR Morb Mortal Wkly Rep*, **52**, 1155–7.

MMWR (2004), 'Outbreak of *Salmonella* serotype Enteritidis infections associated with raw almonds – United States and Canada, 2003–2004', *MMWR Morb Mortal Wkly Rep*, **53**, 484–7.

MMWR (2005a), 'Outbreaks of *Salmonella* infections associated with eating Roma tomatoes – United States and Canada, 2004', *MMWR Morb Mortal Wkly Rep*, **54**, 325–8.

MMWR (2005b), 'Preliminary foodnet data on the incidence of infection with pathogens transmitted commonly through food – 10 sites, United States, 2004', *MMWR Morb Mortal Wkly Rep*, **54**, 352–6.

MMWR (2007a), 'Multistate outbreaks of *Salmonella* infections associated with raw tomatoes eaten in restaurants – United States, 2005–2006', *MMWR Morb Mortal Wkly Rep*, **56**, 909–11.

MMWR (2007b), 'Preliminary FoodNet data on the incidence of infection with pathogens transmitted commonly through food – 10 states, 2006', *mmwr Morb Mortal Wkly Rep*, **56**, 336–9.

MOHLE-BOETANI J C, FARRAR J, BRADLEY P, BARAK J D, MILLER M, MANDRELL R, MEAD P, KEENE W E, CUMMINGS K, ABBOTT S and WERNER S B (2008), '*Salmonella* infections associated with mung bean sprouts: epidemiological and environmental investigations', *Epidemiol Infect*, **137**, 357–66.

MOHLE-BOETANI J C, REPORTER R, WERNER S B, ABBOTT S, FARRAR J, WATERMAN S H and VUGIA D J (1999), 'An outbreak of *Salmonella* serogroup Saphra due to cantaloupes from Mexico', *J Infect Dis*, **180**, 1361–4.

MOORE J E, WILSON T S, WAREING D R, HUMPHREY T J and MURPHY P G (2002), 'Prevalence of thermophilic *Campylobacter* spp. in ready-to-eat foods and raw poultry in Northern Ireland', *J Food Prot*, **65**, 1326–8.

MOTA A, MENA K D, SOTO-BELTRAN M, TARWATER P M and CHAIDEZ C (2009), 'Risk assessment of *Cryptosporidium* and *Giardia* in water irrigating fresh produce in Mexico', *J Food Prot*, **72**, 2184–8.

MUKHERJEE A, CHO S, SCHEFTEL J, JAWAHIR S, SMITH, K. AND DIEZ-GONZALEZ, F (2006), 'Soil survival of *Escherichia coli* O157:H7 acquired by a child from garden soil recently fertilized with cattle manure', *J Appl Microbiol*, **101**, 429–36.

MUKHERJEE A, SPEH D, DYCK E and DIEZ-GONZALEZ F (2004), 'Preharvest evaluation of coliforms, *Escherichia coli, Salmonella*, and *Escherichia coli* O157:H7 in organic and conventional produce grown by Minnesota farmers', *J Food Prot*, **67**, 894–900.

MULLER L, JENSEN T, PETERSEN R F, MOLBAK K and ETHELBERG S (2009), 'Imported fresh sugar peas as suspected source of an outbreak of *Shigella sonnei* in Denmark, April–May 2009', *Euro Surveill*, **14**, 1–3.

MUNNOCH S A, WARD K, SHERIDAN S, FITZSIMMONS G J, SHADBOLT C T, PIISPANEN J P, WANG Q, WARD T J, WORGAN T L, OXENFORD C, MUSTO J A, MCANULTY J and DURRHEIM D N (2009), 'A multi-state outbreak of *Salmonella* Saintpaul in Australia associated with cantaloupe consumption', *Epidemiol Infect*, **137**, 367–74.

MURPHY M, MINIHAN D, BUCKLEY J F, O'MAHONY M, WHYTE P and FANNING S (2008), 'Multiple-locus variable number of tandem repeat analysis (MLVA) of Irish verocytotoxigenic *Escherichia coli* O157 from feedlot cattle: uncovering strain dissemination routes', *BMC Vet Res*, **4**, 2.

NAIMI T S, WICKLUND J H, OLSEN S J, KRAUSE G, WELLS J G, BARTKUS J M, BOXRUD D J, SULLIVAN M, KASSENBORG H, BESSER J M, MINTZ E D, OSTERHOLM M T and HEDBERG C W (2003), 'Concurrent outbreaks of *Shigella sonnei* and enterotoxigenic *Escherichia coli* infections associated with parsley: implications for surveillance and control of foodborne illness', *J Food Prot*, **66**, 535–41.

NAKAZAWA M and AKIBA M (1999), 'Swine as a potential reservoir of shiga toxin-producing *Escherichia coli* O157:H7 in Japan', *Emerg Infect Dis*, **5**, 833–4.

NGUYEN-THE C and CARLIN F (1994), 'The microbiology of minimally processed fresh fruits and vegetables', *Crit Rev Food Sci Nutr*, **34**, 371–401.

NGUYEN-THE C and CARLIN F (2000), 'Fresh and processed vegetables', in Lund B, Baird-Parker, T. and Gould, G. (eds.) *The Microbiological Safety and Quality of Food*, Gaithersburg, MD, Aspen, pp. 620–84.

NIU M T, POLISH L B, ROBERTSON B H, KHANNA B K, WOODRUFF B A, SHAPIRO C N, MILLER M A, SMITH J D, GEDROSE J K, ALTER M J, ET AL. (1992), 'Multistate outbreak of hepatitis A associated with frozen strawberries', *J Infect Dis*, **166**, 518–24.

NODA M, FUKUDA S and NISHIO O (2008), 'Statistical analysis of attack rate in norovirus foodborne outbreaks', *Int J Food Microbiol*, **122**, 216–20.

NOLLER A C, MCELLISTREM M C, PACHECO A G, BOXRUD D J and HARRISON L H (2003), 'Multilocus variable-number tandem repeat analysis distinguishes outbreak and sporadic *Escherichia coli* O157:H7 isolates', *J Clin Microbiol*, **41**, 5389–97.

NUORTI J P, NISKANEN T, HALLANVUO S, MIKKOLA J, KELA E, HATAKKA M, FREDRIKSSON-AHOMAA M, LYYTIKAINEN O, SIITONEN A, KORKEALA H and RUUTU P (2004), 'A widespread outbreak of *Yersinia pseudotuberculosis* O:3 infection from iceberg lettuce', *J Infect Dis*, **189**, 766–74.

NYGARD K, LASSEN J, VOLD L, ANDERSSON Y, FISHER I, LOFDAHL S, THRELFALL J, LUZZI I, PETERS T, HAMPTON M, TORPDAHL M, KAPPERUD G and AAVITSLAND P (2008), 'Outbreak of *Salmonella* Thompson infections linked to imported rucola lettuce', *Foodborne Pathog Dis*, **5**, 165–73.

OCTAVIA S and LAN R (2007), 'Single-nucleotide-polymorphism typing and genetic relationships of *Salmonella enterica* serovar Typhi isolates', *J Clin Microbiol*, **45**, 3795–801.

OCTAVIA S and LAN R (2009), 'Multiple-locus variable-number tandem-repeat analysis of *Salmonella enterica* serovar Typhi', *J Clin Microbiol*, **47**, 2369–76.

OGDEN I D, HEPBURN N F, MACRAE M, STRACHAN N J, FENLON D R, RUSBRIDGE S M and PENNINGTON T H (2002), 'Long-term survival of *Escherichia coli* O157 on pasture following an outbreak associated with sheep at a scout camp', *Lett Appl Microbiol*, **34**, 100–4.

OGDEN I D, MACRAE M and STRACHAN N J (2005), 'Concentration and prevalence of *Escherichia coli* O157 in sheep faeces at pasture in Scotland', *J Appl Microbiol*, **98**, 646–51.

OLSON M E, O'HANDLEY R M, RALSTON B J, MCALLISTER T A and THOMPSON R C (2004), 'Update on *Cryptosporidium* and *Giardia* infections in cattle', *Trends Parasitol*, **20**, 185–91.

OPORTO B, ESTEBAN J I, ADURIZ G, JUSTE R A and HURTADO A (2008), '*Escherichia coli* O157:H7 and non-O157 Shiga toxin-producing *E. coli* in healthy cattle, sheep and swine herds in Northern Spain', *Zoonoses Public Health*, **55**, 73–81.

PANGLOLI P, DJE Y, AHMED O, DOANE C A, OLIVER S P and DRAUGHON F A (2008), 'Seasonal incidence and molecular characterization of *Salmonella* from dairy cows, calves, and farm environment', *Foodborne Pathog Dis*, **5**, 87–96.

PARKER CT, HUYNH S, QUINONES B, HARRIS LJ, MANDRELL RE (2010), 'Comparison of genotypes of *Salmonella enterica* serovar Enteritidis phage type 30 and 9c strains isolated during three outbreaks associated with raw almonds', *Appl Environ Microbiol*, **76**, 3723–31.

PEZZOLI L, ELSON R, LITTLE C L, YIP H, FISHER I, YISHAI R, ANIS E, VALINSKY L, BIGGERSTAFF M, PATEL N, MATHER H, BROWN D J, COIA J E, VAN PELT W, NIELSEN E M, ETHELBERG S, DE PINNA E, HAMPTON M D, PETERS T and THRELFALL J (2008), 'Packed with Salmonella-investigation of an international outbreak of *Salmonella* Senftenberg infection linked to contamination of prepacked basil in 2007', *Foodborne Pathog Dis*, **5**, 661–8.

PHILLIP D A, RAWLINS S C, BABOOLAL S, GOSEIN R, GODDARD C, LEGALL G and CHINCHAMEE A (2008), 'Relative importance of the various environmental sources of *Cryptosporidium* oocysts in three watersheds', *J Water Health*, **6**, 23–34.

PRITCHARD G C, WILLIAMSON S, CARSON T, BAILEY J R, WARNER L, WILLSHAW G and CHEASTY T (2001), 'Wild rabbits – a novel vector for verocytotoxigenic *Escherichia coli* O157', *Vet Rec*, **149**, 567.

PROCTOR M E, HAMACHER M, TORTORELLO M L, ARCHER J R and DAVIS J P (2001), 'Multistate outbreak of *Salmonella* serovar Muenchen infections associated with alfalfa sprouts grown from seeds pretreated with calcium hypochlorite', *J Clin Microbiol*, **39**, 3461–5.

QUIROZ-SANTIAGO C, RODAS-SUAREZ O R, CARLOS R V, FERNANDEZ F J, QUINONES-RAMIREZ, E I and VAZQUEZ-SALINAS C (2009), 'Prevalence of *Salmonella* in vegetables from Mexico', *J Food Prot*, **72**, 1279–82.

RASSCHAERT G, HOUF K, VAN HENDE J and DE ZUTTER, L (2007), 'Investigation of the concurrent colonization with *Campylobacter* and *Salmonella* in poultry flocks and assessment of the sampling site for status determination at slaughter', *Vet Microbiol*, **123**, 104–9.

RELLER M E, NELSON J M, MOLBAK K, ACKMAN D M, SCHOONMAKER-BOPP D J, ROOT T P and MINTZ E D (2006), 'A large, multiple-restaurant outbreak of infection with *Shigella flexneri* serotype 2a traced to tomatoes', *Clin Infect Dis*, **42**, 163–9.

RENTER D G, GNAD D P, SARGEANT J M and HYGNSTROM S E (2006), 'Prevalence and serovars of *Salmonella* in the feces of free-ranging white-tailed deer (*Odocoileus virginianus*) in Nebraska', *J Wildl Dis*, **42**, 699–703.

RENTER D G, SARGEANT J M, HYGNSTORM S E, HOFFMAN J D and GILLESPIE J R (2001), '*Escherichia coli* O157:H7 in free-ranging deer in Nebraska', *J Wildl Dis*, **37**, 755–60.

RICE D H, MCMENAMIN K M, PRITCHETT L C, HANCOCK D D and BESSER T E (1999), 'Genetic subtyping of *Escherichia coli* O157 isolates from 41 Pacific Northwest USA cattle farms', *Epidemiol Infect*, **122**, 479–84.

RIMHANEN-FINNE R, NISKANEN T, HALLANVUO S, MAKARY P, HAUKKA K, PAJUNEN S, SIITONEN A, RISTOLAINEN R, POYRY H, OLLGREN J and KUUSI M (2008), '*Yersinia pseudotuberculosis* causing a large outbreak associated with carrots in Finland, 2006', *Epidemiol Infect*, **137**, 342–7.

RIOU P, LE SAUX J C, DUMAS F, CAPRAIS M P, LE GUYADER S F and POMMEPUY M (2007), 'Microbial impact of small tributaries on water and shellfish quality in shallow coastal areas', *Water Res*, **41**, 2774–86.

ROSAS I, SALINAS E, YELA A, CALVA E, ESLAVA C and CRAVIOTO A (1997), '*Escherichia coli* in settled-dust and air samples collected in residential environments in Mexico City', *Appl Environ Microbiol*, **63**, 4093–5.

ROSE B E, HILL W E, UMHOLTZ R, RANSOM G M and JAMES W O (2002), 'Testing for *Salmonella* in raw meat and poultry products collected at federally inspected establishments in the United States, 1998 through 2000', *J Food Prot*, **65**, 937–47.

ROSE J B and SLIFKO T R (1999), '*Giardia*, *Cryptosporidium*, and *Cyclospora* and their impact on foods: a review', *J Food Prot*, **62**, 1059–70.

SAGOO S K, LITTLE C L and MITCHELL R T (2001), 'The microbiological examination of ready-to-eat organic vegetables from retail establishments in the United Kingdom', *Lett App Microbiol*, **33**, 434–9.

SAGOO S K, LITTLE C L, WARD L, GILLESPIE I A and MITCHELL R T (2003), 'Microbiological study of ready-to-eat salad vegetables from retail establishments uncovers a national outbreak of salmonellosis', *J Food Prot*, **66**, 403–9.

SANTO DOMINGO J and SADOWSKY M (2007), *Microbial Source Tracking*, Washington, DC, ASM Press.

SANTO DOMINGO J W, HARMON S and BENNETT J (2000), 'Survival of *Salmonella* species in river water', *Curr Microbiol*, **40**, 409–17.

SARGEANT J M, HAFER D J, GILLESPIE J R, OBERST R D and FLOOD S J (1999), 'Prevalence of *Escherichia coli* O157:H7 in white-tailed deer sharing rangeland with cattle', *J Am Vet Med Assoc*, **215**, 792–4.

SCHIKORA A, CARRERI A, CHARPENTIER E and HIRT H (2008), 'The dark side of the salad: *Salmonella typhimurium* overcomes the innate immune response of *Arabidopsis thaliana* and shows an endopathogenic lifestyle', *PLoS ONE*, **3**, e2279.

SCOTT L, MCGEE P, SHERIDAN J J, EARLEY B and LEONARD N (2006), 'A comparison of the survival in feces and water of *Escherichia coli* O157:H7 grown under laboratory conditions or obtained from cattle feces', *J Food Prot*, **69**, 6–11.

SCOTT T M, ROSE J B, JENKINS T M, FARRAH S R and LUKASIK J (2002), 'Microbial source tracking: current methodology and future directions', *Appl Environ Microbiol*, **68**, 5796–803.

SEWELL A M and FARBER J M (2001), 'Foodborne outbreaks in Canada linked to produce', *J Food Prot*, **64**, 1863–77.

SEYMOUR I J and APPLETON H (2001), 'Foodborne viruses and fresh produce', *J Appl Microbiol*, **91**, 759–73.

SHARMA M, INGRAM D T, PATEL J R, MILLNER P D, WANG X, HULL A E and DONNENBERG M S (2009), 'A novel approach to investigate the uptake and internalization of *Escherichia coli* O157:H7 in spinach cultivated in soil and hydroponic medium', *J Food Prot*, **72**, 1513–20.

SIMPSON J M, SANTO DOMINGO J W and REASONER D J (2002), 'Microbial source tracking: state of the science', *Environ Sci Technol*, **36**, 5279–88.

SIVAPALASINGAM S, BARRETT E, KIMURA A, VAN DUYNE S, DE WITT W, YING M, FRISCH A, PHAN Q, GOULD E, SHILLAM P, REDDY V, COOPER T, HOEKSTRA M, HIGGINS C, SANDERS J P, TAUXE R V and SLUTSKER L (2003), 'A multistate outbreak of *Salmonella enterica* serotype Newport infection linked to mango consumption: impact of water-dip disinfestation technology', *Clin Infect Dis*, **37**, 1585–90.

SIVAPALASINGAM S, FRIEDMAN C R, COHEN L and TAUXE R V (2004), 'Fresh produce: a growing cause of outbreaks of foodborne illness in the United States, 1973 through 1997', *J Food Prot*, **67**, 2342–53.

SMITH H V and NICHOLS R A (2009), 'Cryptosporidium: Detection in water and food', *Exp Parasitol*, **124**, 61–79.

SNOW J (1855), *On the Mode of Communication of Cholera*, London, UK, John Churchill.

SODERSTROM A, OSTERBERG P, LINDQVIST A, JONSSON B, LINDBERG A, BLIDE ULANDER S, WELINDER-OLSSON C, LOFDAHL S, KAIJSER B, DE JONG B, KUHLMANN-BERENZON S, BOQVIST S, ERIKSSON E, SZANTO E, ANDERSSON S, ALLESTAM G, HEDENSTROM I, LEDET MULLER L and ANDERSSON Y (2008), 'A large *Escherichia coli* O157 outbreak in Sweden associated with locally produced lettuce', *Foodborne Pathogens and Disease*, **5**, 339–49.

SOLOMON E B, BRANDL M T and MANDRELL R E (2006), 'Biology of foodborne pathogens on produce', in Matthews, K.R. (ed.) *Emerging Issues in Food Safety: Microbiology of Fresh Produce*, Washington, DC, ASM Press, pp. 55–83.

SOLOMON E B, YARON S and MATTHEWS K R (2002), 'Transmission of *Escherichia coli* O157:H7 from contaminated manure and irrigation water to lettuce plant tissue and its subsequent internalization', *Appl Environ Microbiol*, **68**, 397–400.

SORIANO J M, RICO H, MOLTO J C and MANES J (2000), 'Assessment of the microbiological quality and wash treatments of lettuce served in University restaurants', *Int J Food Microbiol*, **58**, 123–8.

SRIKANTIAH P (2002), Outbreak of *Salmonella* Javiana infections among participants of the 2002 US Transplant Games – Orlando, Florida, June 2002. Centers for Disease Control and Prevention. Atlanta, GA.

STEWART J, SANTO DOMINGO J and WADE T (2007), 'Fecal pollution, public health, and microbial source tracking', in Santo Domingo, J. and Sadowsky, M. (eds.) *Microbial Source Tracking*, Washington, DC, ASM Press, pp. 1–32.

STRACHAN N J, FENLON D R and OGDEN I D (2001), 'Modelling the vector pathway and infection of humans in an environmental outbreak of *Escherichia coli* O157', *FEMS Microbiol Lett*, **203**, 69–73.

SWAMINATHAN B, BARRETT T J, HUNTER S B and TAUXE R V (2001), 'PulseNet: the molecular subtyping network for foodborne bacterial disease surveillance, United States', *Emerg Infect Dis*, **7**, 382–9.

TAKKINEN J, NAKARI U M, JOHANSSON T, NISKANEN T, SIITONEN A and KUUSI M (2005), 'A nationwide outbreak of multiresistant *Salmonella* Typhimurium in Finland due to contaminated lettuce from Spain, May 2005', *Euro Surveill*, **10**, E050630 1.

TEUNIS P, TAKUMI K and SHINAGAWA K (2004), 'Dose response for infection by *Escherichia coli* O157:H7 from outbreak data', *Risk Anal*, **24**, 401–7.

THILMONY R, UNDERWOOD W and HE S Y (2006), 'Genome-wide transcriptional analysis of the *Arabidopsis thaliana* interaction with the plant pathogen *Pseudomonas syringae* pv. tomato DC3000 and the human pathogen *Escherichia coli* O157:H7', *Plant J*, **46**, 34–53.

THOMAS K M, CHARRON D F, WALTNER-TOEWS D, SCHUSTER C, MAAROUF A R and HOLT J D (2006), 'A role of high impact weather events in waterborne disease outbreaks in Canada, 1975–2001', *Int J Environ Health Res*, **16**, 167–80.

THUNBERG R L, TRAN T T, BENNETT R W, MATTHEWS R N and BELAY N (2002), 'Microbial evaluation of selected fresh produce obtained at retail markets', *Journal of Food Protection*, **65**, 677–82.

TIAN P, ENGELBREKTSON A L, JIANG X, ZHONG W and MANDRELL R E (2007), 'Norovirus recognizes histo-blood group antigens on gastrointestinal cells of clams, mussels, and oysters: a possible mechanism of bioaccumulation', *J Food Prot*, **70**, 2140–7.

TUYET D T, YASSIBANDA S, NGUYEN THI P L, KOYENEDE M R, GOUALI M, BEKONDI C, MAZZI J and GERMANI Y (2006), 'Enteropathogenic *Escherichia coli* O157 in Bangui and N'Goila, Central African Republic: A brief report', *Am J Trop Med Hyg*, **75**, 513–15.

UESUGI A R, DANYLUK M D, MANDRELL R E and HARRIS L J (2007), 'Isolation of *Salmonella* Enteritidis phage type 30 from a single almond orchard over a 5-year period', *J Food Prot*, **70**, 1784–9.

USDA-AMS-MDP (2009), *U.S. Department of Agriculture, Agricultural Marketing Service, Microbial Data Program. Data summaries, 2002–2008.* http://www.ams.usda.gov/science/mpo/Mdp.htm. [Online.]

VALENTIN-BON I, JACOBSON A, MONDAY S R and FENG P C (2008), 'Microbiological quality of bagged cut spinach and lettuce mixes', *Appl Environ Microbiol*, **74**, 1240–2.

VAN KESSEL J S, PACHEPSKY Y A, SHELTON D R and KARNS J S (2007), 'Survival of *Escherichia coli* in cowpats in pasture and in laboratory conditions', *J Appl Microbiol*, **103**, 1122–7.

VITAL M, HAMMES F and EGLI T (2008), '*Escherichia coli* O157 can grow in natural freshwater at low carbon concentrations', *Environ Microbiol*, **10**, 2387–96.

WALTERS S P, GANNON V P and FIELD K G (2007), 'Detection of Bacteroidales fecal indicators and the zoonotic pathogens *E. coli* O157:H7, *Salmonella*, and *Campylobacter* in river water', *Environ Sci Technol*, **41**, 1856–62.

WANG G and DOYLE M P (1998), 'Survival of enterohemorrhagic *Escherichia coli* O157:H7 in water', *J Food Prot*, **61**, 662–7.

WARRINER K, IBRAHIM F, DICKINSON M, WRIGHT C and WAITES W M (2003), 'Internalization of human pathogens within growing salad vegetables', *Biotechnol Genet Eng Rev*, **20**, 117–34.

WDOH (2008), *E. coli infections detected in two Puget Sound counties.* http://www.doh.wa.gov/Publicat/2008_news/08-092.htm.

WEI J, JIN Y, SIMS T and KNIEL K E (2010), 'Manure- and biosolids-resident murine norovirus 1 attachment to and internalization by Romaine lettuce', *Appl Environ Microbiol*, **76**, 578–83.

WELINDER-OLSSON C, STENQVIST K, BADENFORS M, BRANDBERG A, FLOREN K, HOLM M, HOLMBERG L, KJELLIN E, MARILD S, STUDAHL A and KAIJSER B (2004), 'EHEC outbreak among staff at a children's hospital – use of PCR for verocytotoxin detection and PFGE for epidemiological investigation', *Epidemiol Infect*, **132**, 43–9.

WETZEL A N and LEJEUNE J T (2006), 'Clonal dissemination of *Escherichia coli* O157:H7 subtypes among dairy farms in northeast Ohio', *Appl Environ Microbiol*, **72**, 2621–6.

WHEELER C, VOGT T M, ARMSTRONG G L, VAUGHAN G, WELTMAN A, NAINAN O V, DATO V, XIA G, WALLER K, AMON J, LEE T M, HIGHBAUGH-BATTLE A, HEMBREE C, EVENSON S, RUTA M A, WILLIAMS I T, FIORE A E and BELL B P (2005), 'An outbreak of hepatitis A associated with green onions', *N Engl J Med*, **353**, 890–7.

WHIPPS J, HAND P, PINK D and BENDING G (2008), 'Human pathogens and the phyllosphere', in Laskin A.I., Sariaslani, S. and Gadd, G.M. (eds.) *Adv Appl Microbiol*, Amsterdam, Elsevier, pp. 183–221.

WHITTAM T and BERGHOLZ T (2007), 'Molecular subtyping, source tracking, and food safety', in Santo Domingo, J. and Sadowsky, M. (eds.) *Microbial Source Tracking*, Washington, DC, American Society of Microbiology, pp. 93–136.

WILLIAMS A P, AVERY L M, KILLHAM K and JONES D L (2007), 'Survival of *Escherichia coli* O157:H7 in the rhizosphere of maize grown in waste-amended soil', *J Appl Microbiol*, **102**, 319–26.

WIMPENNY, J W T, COTTON N and STATHAM M (1972), 'Microbes as tracers of water movement', *Water Res.*, **6**, 731–73.

YOSHII N, OGURA Y, HAYASHI T, AJIRO T, SAMESHIMA T, NAKAZAWA M, KUSUMOTO M, IWATA T and AKIBA M (2009), 'Spontaneous chromosomal deletions in enterohemorrhagic *Escherichia coli* O157:H7 through passage in cattle led to changes in pulsed-field gel electrophoresis profiles', *Appl Environ Microbiol*, **75**, 5719–26.

YOU Y, RANKIN S C, ACETO H W, BENSON C E, TOTH J D and DOU Z (2006), 'Survival of *Salmonella enterica* serovar Newport in manure and manure-amended soils', *Appl Environ Microbiol*, **72**, 5777–83.

ZHANG G, MA L, BEUCHAT L R, ERICKSON M C, PHELAN V H and DOYLE M P (2009), 'Heat and drought stress during growth of lettuce (Lactuca sativa L.) does not promote internalization of *Escherichia coli* O157:H7', *J Food Prot*, **72**, 2471–5.

ZHANG W, QI W, ALBERT T J, MOTIWALA A S, ALLAND D, HYYTIA-TREES E K, RIBOT E M, FIELDS P I, WHITTAM T S and SWAMINATHAN B (2006), 'Probing genomic diversity and evolution of *Escherichia coli* O157 by single nucleotide polymorphisms', *Genome Res*, **16**, 757–67.

Index

FSC
www.fsc.org
MIX
Papier | Fördert gute Waldnutzung
FSC® C083411

Kolibri 360 GmbH
Ferdinand-Jühlke-Straße 7
99095 Erfurt, Deutschland
produktsicherheit@kolibri360.de